MODERN PROBLEMS IN CONDENSED MATTER SCIENCES

*Oh, how many of them there
are in the fields!
But each flowers in its
own way—
In this is the highest achievement
of a flower!*

*Matsuo Bashó
1644–1694*

PREFACE TO THE SERIES

"Modern Problems in Condensed Matter Sciences" is a series of contributed volumes and monographs on condensed matter science that is published by North-Holland Publishing Company. This vast area of physics is developing rapidly at the present time, and the numerous fundamental results in it define to a significant degree the face of contemporary science. This being so, it is clear that the most important results and directions for future developments can only be covered by an international group of authors working in cooperation.

Both Soviet and Western scholars are taking part in the series, and each contributed volume has, correspondingly, two editors. Furthermore, it is intended that the volumes in the series will be published subsequently in Russian by the publishing house "Nauka".

The idea for the series and for its present structure was born during discussions that took place in the USSR and the USA between the former President of North-Holland Publishing Company, Drs. W.H. Wimmers, and the General Editors.

The establishment of this series of books, which should become a distinguished encyclopedia of condensed matter science, is not the only important outcome of these discussions. A significant development is also the emergence of a rather interesting and fruitful form of collaboration among scholars from different countries. We are deeply convinced that such international collaboration in the spheres of science and art, as well as other socially useful spheres of human activity, will assist in the establishment of a climate of confidence and peace.

The General Editors of the Series,

V.M. Agranovich   A.A. Maradudin

PREFACE

The purpose of this volume is to provide a snapshot of current problems associated with excitations in the molecular condensed phase. Molecular systems can be chosen to display a wide range of level structures and interlevel dynamical processes. They can therefore manifest a remarkably varied collection of optical properties, that expose information on the structure of the molecules and of the materials they compose. It is this great variety that brings life and excitement to the field.

This volume is largely concerned with electronic and vibrational excitations having a distinctly molecular identification so that the spectroscopy has a molecular character. These molecular levels are subject to perturbations resulting from interactions amongst the molecules or between the molecules and their environment. Such properties set this class of materials apart from wider band molecular systems and conducting polymers. The subject of this compilation is therefore more concerned with fundamental processes and novel properties, than with new molecular materials. Nevertheless, the subject matter deals with the behavior of many different classes of materials, including molecular crystals, chemically and isotopically mixed crystals, and glasses.

The authors were not asked to write comprehensive reviews of their subjects in the expectation that the set of articles would then capture the individuality of the approaches. We tried to provide a balance between theoretical and experimental research currently underway in this field.

In one sense this volume represents our view of the state of the art of condensed phase spectroscopy. The field originated from the early explorations of molecular luminescence and electronic states, especially in glasses, followed by the postwar decades of experimental and theoretical study of insulating molecular crystals. But new technology has originated new directions for this field. Many of the contributions to the present volume—and indeed a large proportion of current researches in spectroscopy—are dependent on laser methods. Experiments based on lasers have resulted in measurements being made of condensed phase properties that were hitherto unknown. These include homogeneous lineshapes, inelastic interactions, coherent energy transport, quasi-particle scattering and new optical effects. The measurements have in turn stimulated new theoretical investigation especially in quantum statistical mechanics. The continual interchange between experiment and theory is at the heart of the vitality of this field of research.

Spectroscopy of condensed phases seemed such a natural choice for an international volume of this nature. Soviet and Western scientists have each made key contributions to the development of our present understanding of the structure and dynamics of the many types of excitations that occur in molecular solids and other condensed phases. We owe special thanks to the contributors to this volume for the careful preparation of their manuscripts and thoughtful maintenance of schedules.

V.M. Agranovich   R.M. Hochstrasser
Moscow, USSR   Philadelphia, PA, USA

CONTENTS

CHAPTER 1

Theories of Energy Transport

ROBERT SILBEY

*Department of Chemistry
and Center for Materials Science and Engineering
Massachusetts Institute of Technology
Cambridge, MA 02139
U.S.A.*

*Spectroscopy and Excitation Dynamics
of Condensed Molecular Systems
Edited by
V.M. Agranovich and R.M. Hochstrasser*

Contents

1. Introduction

The present chapter deals with the theories of exciton transport in molecular crystals. (For reviews see Powell and Soos 1975, Silbey 1976, Zwemer and Harris 1978.) The basic quantity which we want to compute is the exciton diffusion coefficient, D, which describes the rate of relaxation of a nonequilibrium distribution of excitons towards the uniform distribution found at infinite time. We define

$$D = \frac{1}{2d} \lim_{t \to \infty} \frac{\langle R^2(t) \rangle}{t} = \frac{1}{2d} \lim_{t \to \infty} \frac{d\langle R^2(t) \rangle}{dt},$$

where d is the dimensionality of the system (taken as isotropic for simplicity), and $\langle R^2(t) \rangle$ is the mean square displacement of an exciton at time t from its position at time zero. Other definitions of D, for example using the Kubo formalism (Kubo 1957, 1958), are equivalent to this.

Before we consider the technical aspects of the calculation, we should remark that transport is diffusive only if there is some mechanism for scattering an exciton from one state to another. In the absence of such a mechanism, the exciton moves as a perfect wave through the crystal; this is sometimes called "coherent" transport, but as this term has other connotations, we shall avoid using it here. In this limit, the mean square displacement never becomes proportional to time, and no diffusion coefficient can be defined. When there is some mechanism for scattering, repeated scattering events cause the exciton distribution to lose memory of its initial state and to attain equilibrium in a diffusive way. Scattering occurs most commonly from phonons and from impurities or other crystal imperfections.

Two limiting cases of transport are often distinguished. In band transport, the exciton states are labelled by a wave vector k, and phonons or impurities scatter the exciton from one k state to another. The transport is then governed by the scattering times of the k states. In hopping transport, the exciton is almost localized in states labelled by a site index n, and phonons and imperfections scatter these states. These two limits and the transition between them have been studied for a number of years, and a great deal is known about them. Calculations of transport may be performed in terms of wave vector or site states, irrespective of which limit applies. Unfortunately, neither choice is ideal: the wave vector states are most suitable

3

for taking account of translational symmetry in a crystal, but the site states are most suitable for calculating D.

In most calculations of the diffusion coefficient, a model (i.e. hopping or band transport) is assumed and the parameters of the model calculated. For example, in band transport, D is assumed to be the average of the square of the velocity in a k state multiplied by the relaxation time in that state. The velocities are found from the exciton band structure, the relaxation time by a scattering calculation. In hopping transport, it is assumed that the exciton states are localized and a second-order perturbation theory calculation is made of the rate of transfer (via the Fermi Golden Rule) from one molecule to another. It is not necessary, however, to assume a particular model before doing the calculation if we are willing to deal with a bit more complexity. In the general case, a single equation can reproduce the various transport limits. This is obtained at the cost of deriving the equation of motion of the density matrix of the system. This operator contains all the necessary information about the system, and has a correspondingly complicated structure. However, for various model Hamiltonians, we can derive all the necessary equations and solve them, at least approximately.

There are a number of parameters of importance in this calculation: the exciton bandwidth (B), the phonon frequency (ω), phonon bandwidth (Δ), the strength of the phonon–exciton scattering or coupling, the strength of the phonon impurity scattering and the temperature. If the exciton bandwidth is large compared to the exciton–phonon coupling, then the band model will be appropriate; if the exciton bandwidth is small compared to this coupling, then the exciton states will be more localized, and the hopping model will be appropriate. Of course, the phonon–exciton scattering is temperature dependent, and this dependence may be complex. For molecular crystals, the exciton bandwidths vary from a few cm^{-1} (triplet states) to a few thousand cm^{-1} (dipole allowed singlet states). The phonon frequencies vary from 10–40 cm^{-1} for librational modes, a few hundred cm^{-1} for low frequency intramolecular modes, up to 2500 cm^{-1} for high frequency intramolecular modes, and (typically) $\approx 50 \, cm^{-1}$ for the Debye frequency. The phonon bandwidths are usually small ($\leq 20 \, cm^{-1}$) except for the acoustic and intermolecular optical modes. The exciton–phonon coupling strengths vary also: we expect this coupling to be most important for the low and mid frequency modes, since the high frequency modes are not thermally excited. The low lying optical modes are usually thought to be most strongly coupled to the exciton, and in addition are substantially thermally excited below room temperature. The impurity–exciton coupling depends on impurity concentration and is thus most in the control of the experimenter.

From these brief comments, it is clear that a completely general theory of transport in such systems will be impossibly complex. In this chapter, we will discuss phonon–exciton scattering (section 2) for various models of the coupling to different kinds of modes, and exciton–impurity coupling (section 3) concentrating on the limit of very strong scattering.

2. Transport in pure crystals

2.1. General remarks

Theories and models of exciton transport in pure molecular crystals are so numerous that it is hard to appreciate the relationships and differences between them. In part, this variety and complexity is essential to molecular crystals. As we stated above, transport is governed by a large number of parameters. Some of these parameters may be quite anisotropic and no one of them is necessarily much bigger or smaller than the others. As a result, there is a whole range of different regimes defined by different orderings of the parameters. These regimes correspond not only to different mechanisms of transport, such as hopping or band motion, but also to different physical processes determining these mechanisms.

The usual transport theories have been constructed specifically for one or two regimes (Trlifaj 1956, Agranovich and Konobeev 1959, 1963, 1968, Davydov 1971). They have often assumed a particular physical picture of transport, such as whether or not localized states are important, using methods and approximations appropriate to that picture (but not always explicitly stated). However, it is possible to compare the various theories as special cases of a more general theory.

In this section, we present a more general theory of electronic transport in perfect molecular crystals. Inevitably, such a general theory cannot be evaluated exactly; we do not derive a single algebraic expression for the diffusion coefficient from which the results for all regimes follow on simplification. Instead, we show how different regimes permit or require different ways of evaluating the quantities in the general theory.

The theory we describe is necessarily mathematical, but we concentrate on results rather than derivations and on broad principles rather than matters of detail. The different regimes appear initially as mathematical consequences of inequalities among the parameters. We then show how the regimes may be interpreted physically. Much of the mathematics can be found in Munn and Silbey (1980). In the present article, a localized state at position R_n in the crystal is labelled n, a delocalized state is labelled k. In general n, R_n and k are vectors with d components for a d-dimensional crystal; the reader should note that we do not use vector notation for these quantities.

The diagonal matrix elements of the density operator in *any* choice of basis states give the probabilities that these states are occupied. For the diffusion coefficient, we require the $P_n(t)$, which are diagonal elements of the exciton density matrix σ in the site representation. These follow from the density matrix of the coupled exciton–phonon system integrated over all phonon variables:

$$P_n(t) = \sigma_{nn}(t) = \mathrm{Tr}_L[\hat{\rho}(t)]_{nn}, \tag{1}$$

where Tr_L is a trace over phonon lattice states. The density matrix $\hat{\rho}(t)$ for the coupled system obeys the Liouville equation

$$\dot{\hat{\rho}}(t) = -\mathrm{i}[H, \hat{\rho}(t)] \equiv \mathrm{i}\hat{\mathscr{L}}\hat{\rho}(t), \tag{2}$$

where $\hat{\mathscr{L}}$ is the Liouville operator and the superposed dot denotes a time derivative.

Using mathematical procedures derived from irreversible statistical mechanics (Zwanzig 1960), one can derive an exact equation for the reduced exciton density operator, $\hat{\sigma}(t)$, or rather its matrix elements in the wave vector representation. The equation assumes only that the exciton and phonon variables were uncoupled at time zero, i.e. that the initial phonon density matrix was thermal. The result is (Grover and Silbey 1971)

$$\dot{\sigma}_{kk'}(t) = -\mathrm{i}(\epsilon_k - \epsilon_{k'})\sigma_{kk'}(t) - \int_0^t \mathrm{d}\tau \sum_q K_{kk';qs}(t - \tau)\sigma_{qs}(\tau), \tag{3}$$

because of conservation of wave vector $s = k' - k + q$; here ϵ_k is the energy of the exciton state k. Because this equation is exact, the formula for the kernel K is impossible to evaluate except in the most exceptional (and least interesting) cases, and will not be given here.

However, in an approximation equivalent to second-order time-dependent perturbation theory in the exciton–phonon coupling $\hat{V}$, eq. (3) becomes (Munn and Silbey 1980)

$$\dot{\sigma}_{kk'}(t) = -\mathrm{i}(E_k - E_{k'})\sigma_{kk'}(t) - \Gamma_{kk'}\sigma_{kk'}(t) + \sum_{q \neq k} W_{kk';qs}(t)\sigma_{qs}(t). \tag{4}$$

The Hamiltonian of the system has been taken to be $H = H_0 + V$, where

$$H_0 = H_{ph}^0 + \sum_k \epsilon_k a_k^+ a_k, \tag{5}$$

$$V = \sum_{k,k'} V_{kk'} a_k^+ a_{k'}, \tag{6}$$

$$\epsilon_k = \sum_n J_{nm}\, \mathrm{e}^{\mathrm{i}k \cdot (R_n - R_m)}, \tag{7}$$

and the phonon operators, $V_{kk'}$, are zero when averaged over the canonical phonon ensemble (assumed to be the initial phonon ensemble):

$$\langle V_{kk'} \rangle = 0. \tag{8}$$

Here, the energies E_k differ from the ϵ_k by terms up to second order in the coupling. Because the results are restricted to second order, it is necessary that the coupling V should be small, and this may require the Hamiltonian to be transformed in a suitable way. Equation (4) is also restricted to a limit where initial transient behaviour has been removed, so that it may *not* be applicable at short times. For the diffusion coefficient it is the long time behaviour which is required, and eq. (4) does give this behaviour correctly.

The quantities $W_{kk';qs}$ are given by

$$W_{kk';qs} = \frac{1}{2}\int_{-\infty}^{+\infty} dt \ [\langle V_{sk'}V_{kq}(\tau) \rangle \, e^{i\Omega_{sk'}\tau} + \langle V_{sk'}(\tau)V_{kq} \rangle \, e^{i\Omega_{kq}\tau}], \tag{9}$$

where the angle brackets denote an average over phonon states and

$$\Omega_{kq} = E_k - E_q. \tag{10}$$

The remaining quantities in eq. (4) are the $\Gamma_{kk'}$, which satisfy

$$\Gamma_{kk'} = \tfrac{1}{2}\Gamma_{kk} + \tfrac{1}{2}\Gamma_{k'k'} + \Gamma_{kk'}^{PD}, \tag{11}$$

where $\Gamma_{kk'}^{PD}$ is called the pure dephasing rate and is given by

$$\Gamma_{kk'}^{PD} = \frac{1}{2}\int_{-\infty}^{+\infty} d\tau \ \langle [V_{kk}(\tau) - V_{k'k'}(\tau)][V_{kk} - V_{k'k'}] \rangle. \tag{12}$$

The nature of the diagonal elements Γ_{kk} can be seen from eq. (4) which, for the diagonal elements, reduces to

$$\dot{\sigma}_{kk}(t) = -\Gamma_{kk}\sigma_{kk}(t) + \sum_{q \neq k} W_{kk;qq}\sigma_{qq}(t). \tag{13}$$

This contains only the probabilities of finding the exciton in the states k, and is therefore a Pauli master equation. Equation (13) shows that Γ_{kk} represents the rate of scattering out of exciton state k, i.e. the inverse lifetime of the state, while $W_{kk;qq}$ represents the rate of scattering from exciton state q into exciton state k. From these considerations, or by summing eq. (13) over k, it follows that

$$\Gamma_{kk} = \sum_{q \neq k} W_{qq;kk}. \tag{14}$$

The form (13) for W guarantees that

$$W_{qq;kk} = W_{kk;qq} \, e^{-\beta(E_q - E_k)}, \tag{15}$$

irrespective of $\hat{V}$, as required by the principle of detailed balance. (For a discussion of these points, see Oppenheim et al. 1979.) This ensures that

$$\sigma_{kk}(\infty) \equiv \sigma_{kk}^{eq} = e^{-\beta E_k} \bigg/ \sum_q e^{-\beta E_q}, \qquad (16)$$

with off-diagonal elements zero, as expected from the Boltzmann distribution.

Equation (11) shows that the rate at which off-diagonal elements $\sigma_{kk'}(t)$ decay is related to the decay rates of the occupation probabilities of the individual states k and k'. This simple result, and the simple form of eq. (4), where the coefficients Γ and W are independent of time, are valid only in the k representation at long time because only in that representation is the equilibrium exciton density matrix diagonal. If one works in the site representation, eq. (16) shows that the $\sigma_{nm}(\infty)$ are complicated functions which are in general nonzero for $n \neq m$. Transformation of eq. (4) to this representation shows that the time derivatives of the diagonal elements σ_{nn} are in general coupled to the off-diagonal elements σ_{nm}, so that a master equation like eq. (13) cannot be written. However, provided the density matrix σ was diagonal at time zero (for example, with a single exciton at the origin), the off-diagonal elements at time t can be related to the diagonal elements at all previous times. This leads to a generalized master equation (GME) of the form (Kenkre 1975, Kenkre and Knox 1973, 1974, 1976)

$$\dot{\sigma}_{nn}(t) = \int_0^t \sum_m G_{nm}(t - \tau)\sigma_{mm}(\tau)\,d\tau, \qquad (17)$$

where the kernels G are complicated oscillatory functions. An equation of this form can also describe initial transient behaviour in either representation. In order that diffusion may be described by the GME, the kernels G_{nm} must decay fast enough so that $\langle R^2(t) \rangle$ becomes proportional to t (as $t \to \infty$). Some simplified models may not have this property, and thus do not predict diffusive behaviour. For decay to be fast enough, there must be a dissipative mechanism such as exciton–phonon coupling which introduces exponential decays into the G_{nm}, or into the correlation functions in eq. (9). Recently, the exact kernels for the Haken–Strobl model (Haken and Strobl 1973) have been calculated (Kenkre 1978, Kuhne and Reineker 1978); in this work it is clear how the exciton–phonon coupling introduces exponential decays into the memory.

The diffusion coefficient can be written

$$D = \frac{1}{2d} \lim_{t \to \infty} \sum_n R_n^2 \dot{\sigma}_{nn}. \qquad (18)$$

To evaluate D we either need to substitute the complicated expression (17) in this equation, or else to transform to the wave vector representation, with the result

$$D = \frac{1}{2d} \lim_{t \to \infty} \sum_k \nabla_K^2 \dot{\sigma}_{k,k+K}(t) \Big|_{K=0}. \tag{19}$$

In the latter case, the insertion of eq. (4) for $\sigma_{kk'}(t)$ in the limit $t \to \infty$ gives an equation which can be somewhat simplified by neglect of small terms. The result is still rather complicated, and so for present purposes we shall use an approximate formula which gives a good qualitative description of the behaviour of D in various limits. We take (Munn and Silbey 1980, see also Yarkony and Silbey 1977)

$$D = \sum_k \sigma_{kk}^{eq} (v_k^2 / \Gamma_{kk} + \gamma_{kk}), \tag{20}$$

where $v_k = \nabla_k E_k$ is the exciton velocity in state k and the form of γ_{kk} is given by

$$\gamma_{kk} = \left((\tfrac{1}{2} \nabla_k^2 - \nabla_K^2) N^{-1} \sum_q W_{q,q+K;k,k+K} \right)_{K=0}. \tag{21}$$

The first term in eq. (20) looks like a band theory result (recall that $1/\Gamma_{kk}$ is the exciton scattering time in state k). It can thus be related to the phenomenological theories of exciton diffusion, which often use scattering times or scattering lengths to calculate mean free paths and diffusion coefficients. The second term in eq. (20), on the other hand, looks like a hopping result. We shall see that these terms can indeed be interpreted in this way. For large $\Gamma_{kk'}$, small v_k or both, the second term dominates, while for large v_k (wide exciton bands) the first term dominates. In general, though, both terms contribute and the transport cannot be described as either pure hopping or pure band motion.

In the case of a general exciton–phonon coupling term,

$$V = \frac{1}{\sqrt{N}} \sum_{k,q} V_{kq} a_{k+q}^{+} a_k, \tag{22}$$

where V_{kq} is a phonon operator of unspecified form. Then

$$\Gamma_{kk} = \frac{1}{N} \sum_{q \neq k} \frac{1}{2} \int_{-\infty}^{+\infty} d\tau \, e^{i\Omega_{kq}\tau} \langle V_{kq} V_{qk}(\tau) \rangle. \tag{23}$$

To get an idea of the structure of these terms, let us assume that the phonon correlation function decays on a time scale α^{-1}, where α will be proportional to the phonon bandwidth. To avoid mathematical complication, we will assume that this decay is Gaussian in time. That is, we take

$$\langle V_{kq} V_{qk}(\tau) \rangle = V^2 \, e^{-\alpha^2 \tau^2 / 4}, \tag{24}$$

and we assume that the exciton band has a Gaussian density of states:

$$N_{exc}(E) = \frac{1}{\sqrt{\pi}B}\, e^{-E^2/B^2}, \tag{25}$$

where B is the exciton bandwidth. Then we find

$$\Gamma_{kk} = \frac{V^2}{(B^2+\alpha^2)^{1/2}} \exp[-E_k^2/(\alpha^2+B^2)]. \tag{26}$$

Thus the scattering rate of an exciton is largest in the center of the band and decreases towards the top and the bottom. This is merely a consequence of the fact that there are more processes allowable in the center in this simple picture (where temperature and phonon populations have been neglected). The most important result for our purposes is that Γ_{kk} is inversely proportional to $(B^2+\alpha^2)^{1/2}$, where α is (proportional to) the phonon bandwidth. This shows that in the limit of a narrow phonon band $\Gamma_{kk} \sim B^{-1}$, while in the limit of a narrow exciton band $\Gamma_{kk} \sim 1/\alpha$. This corresponds qualitatively to the slow phonon and slow exciton limits of Munn and Siebrand (1970). When the exciton bandwidth B is large, it plays the role of the continuum for relaxation, while when the phonon bandwidth α is large, it plays this role. Since Γ_{kk} also is proportional to the width of the kth state (for $k = 0$, it is the optical linewidth), we see that $V/(B^2+\alpha^2)^{1/2}$ is the narrowing factor for the perturbation V. In the limit of *no* motion at all, the linewidth Γ_{kk} would be proportional to V, while the motion narrows this to $V^2/(B^2+\alpha^2)^{1/2}$ (Toyozawa 1976).

2.2. Haken–Strobl model

A very useful model was introduced by Haken and Strobl (1973) and extended by many others (Haken and Reineker 1972, Schwarzer and Haken 1972, 1973, 1974). The basic assumptions are that

$$\langle V_{nn}(\tau)V_{nn}\rangle = \gamma_0\delta(\tau), \tag{27}$$

$$\langle V_{nm}(\tau)V_{mn}\rangle = \langle V_{nm}(\tau)V_{nm}\rangle = \gamma(R_n - R_m)\delta(\tau), \tag{28}$$

and all other correlation functions are zero. In addition, it was assumed that the V's were Gaussian, so that only the second cumulant was nonzero. This, added to the delta function correlation in time, makes the model exactly soluble. Of course, the form of γ_0 and $\gamma(R)$ have to be found from a microscopic side calculation; however the dynamics can be solved with γ_0 and $\gamma(R)$ as is. Using the above correlation functions, we find ($s = k' - k + q$)

$$\langle V_{sk'}V_{sq}(\tau)\rangle = \gamma_0\delta(\tau) + \delta(\tau)\sum_{R_n}\gamma(R_n)[\cos(k - k')\cdot R_n + \cos(s + k)\cdot R_n],$$
$$\tag{2.29}$$

where the sum is over all R_n in the lattice except $R_n = 0$. Using this form we find

$$\Gamma_{kk} = \gamma_0 + \sum_{R_n} \gamma(R_n), \tag{30}$$

$$\gamma_{kk} = \sum_{R_n} R_n^2 \gamma(R_n), \tag{31}$$

$$D = \left\langle \frac{v_k^2}{\gamma_0 + \Sigma_{R_n} \gamma(R_n)} \right\rangle + \sum_n R_n^2 \gamma(R_n). \tag{32}$$

The simplest model is a one-dimensional crystal, in which only nearest neighbor J_{nm} and $\gamma(R_n)$ are taken to be nonzero. Then, (a = nearest neighbor distance)

$$\frac{D}{a^2} = \left\langle \frac{4J^2 \sin^2 k}{\gamma_0 + 2\gamma_1} \right\rangle + 2\gamma_1, \tag{33}$$

and at high temperatures ($k_B T \gg 4J$)

$$\frac{D}{a^2} = \frac{2J^2}{\gamma_0 + 2\gamma_1} + 2\gamma_1. \tag{34}$$

From the form of the V_{nn} and V_{nm} operators, we can find the temperature dependence of the parameters γ_0 and γ_1. We assume that V_{nn}, which represents the energy at a site, is a sum of terms linear and quadratic in phonon variables:

$$V_{nn} = \sum_i g_i q_i \omega_i + \sum_i f_i \omega_i^2 (q_i^2 - \langle q_i^2 \rangle), \tag{35}$$

where the sum is over local phonon variables, q_i, with frequency ω_i and coupling constants g_i and f_i. Since the Haken–Strobl model is only valid at high T (see below) we may calculate the averages classically and find (note that we *assume* the delta function correlation in time is valid):

$$\gamma_0 = \sum_i g_i k_B^2 T + \frac{1}{2} \sum f_i^2 (k_B T)^2 = \alpha_0 (k_B T) + \alpha_1 (k_B T)^2. \tag{36}$$

The constants α_0 and α_1 can be fit from experiment. The parameter γ_1 can be found from the assumed form of the term $V_{n,n+1}$. For singlet states which are dipolar coupled, the first term is

$$V_{n,n+1} = \frac{J_0}{a} \frac{1}{\sqrt{N}} \sum_q Q_q e^{iq \cdot a}, \tag{37}$$

leading to γ_1 proportional to $k_B T$. For triplet states, which are exchange coupled, we take, following Haken and Reinecker (1972)

$$V_{n,n+1} = J_0 (1 - e^{-\lambda |x|}), \tag{38}$$

where x represents the fluctuating part of the distance between the molecules at n and $n + 1$. We then find $[y = (k_B T)^{1/2}/2\omega]$

$$\gamma_1 \propto T \qquad\qquad\qquad \text{at small } y, \qquad\qquad (39)$$

$$\gamma_1 \propto \left(\text{const.} - \frac{3}{2\sqrt{\pi}}\frac{1}{y}\right) \quad \text{at large } y. \qquad\qquad (40)$$

These results indicate the usefulness of the Haken–Strobl model for
discussing exciton transport: it is an exactly soluble model which indicates
the correct physical behavior in many limits. Recall that γ_0 represents the
site diagonal (local) fluctuations, and note that it appears only in the
denominator of a term in D. This shows that the local fluctuations always
impede transport. Note however that γ_1, the nonlocal fluctuations, appears
in two places: once in the same denominator as γ_0 and once in a numerator
as a hopping term. This indicates that nonlocal fluctuations can impede
transport (in the band term) and can induce transport (in the hopping term).
These results are in accord with our physical intuition in that nonlocal
lattice fluctuations can increase the hopping integral while also destroying
the phase relationships needed for band transport.

Unfortunately, the Haken–Strobl model cannot be universally applied.
The reasons for this are: (i) because the assumption is made that the
temporal behavior of the correlation functions is a delta function, the
model cannot be used at low temperatures, where these functions must
decay more slowly; (ii) in addition, this temporal behavior is tantamount to
assuming that the phonon fluctuations are very fast compared to exciton
motion, so $B \ll \Delta$ (the phonon bandwidth) (Toyozawa 1976); (iii) because of
the assumption that $\langle V_{nn}(\tau) V_{nn}(0) \rangle = \langle V_{nn}(0) V_{nn}(\tau) \rangle$, detailed balance is lost
and the equilibrium values of *all* exciton state populations are equal,
suggesting that $k_B T \gg B$ for this model. Attempts to address (ii) have been
made by Sumi (1977) and Blumen and Silbey (1978), by taking
$\langle V_{nn}(t) V_{nn}(0) \rangle = A\, e^{-\Gamma t}$. This allows a richer dynamics, but renders the
model only approximately soluble. The finiteness of Γ shows the depen-
dence on both the exciton bandwidth B and the phonon bandwidth Δ in the
line shape and diffusion coefficient. Wertheimer and Silbey (1980) have
recently examined (i) and (iii) for a simple two-molecule system. In order
to examine the dynamics in detail, however, more microscopic approaches
are necessary, to which we now turn.

2.3. Dynamical models

We turn now to microscopic phonon models, in order to obtain a better
understanding of the various coupling mechanisms and their effect on the
transport. In general we expand V_{kq} in terms of phonon operators keeping

linear and quadratic terms only:

$$V_{kq} = g(k; k - q)\omega_{k-q}(b_{k-q} + b^+_{-k+q})$$

$$+ \sum_{q_1} g(k; q; q_1)(b_{q_1} + b^+_{-q_1})(b_{k-q-q_1} + b^+_{-k+q+q_1}). \tag{41}$$

The forms are dictated by the translational invariance of the lattice.

In the following pages, we treat the linear phonon term in the V_{kq} in a variety of cases (weak and strong coupling to both optical and acoustic modes) in the limit of local coupling, i.e. that $g(k; k - q) = g(k - q)$ only. This form arises when the exciton–phonon coupling term is diagonal in the exciton site representation. We then treat the quadratic term in the same limit. The forms of the transport coefficient in more general models are more complicated; we concentrate on local coupling which already shows a large range of behavior.

If the electron–phonon coupling is sufficiently weak, the first term in eq. (41) can be treated as a perturbation and the second neglected. Then in eq. (6) the operator V_{kq} is given by

$$\hat{V}_{kq} = g_{k-q}\omega_{k-q}(b_{k-q} + b^+_{-k+q}). \tag{42}$$

The required correlation functions are of the form

$$\langle V_{k+K,q+K} V_{qk}(\tau) \rangle = |g_{k-q}|^2 \omega^2_{k-q}[(n_{k-q} + 1) e^{-i\omega_{k-q}\tau} + n_{k-q} e^{i\omega_{k-q}\tau}], \tag{43}$$

where we have used the results $g_{-q} = g^*_q$ and $\omega_{-q} = \omega_q$. In eq. (43) n_{k-q} is the thermal equilibrium number of phonons in mode $k - q$, given by

$$n_{k-q} = (e^{\beta\omega_{k-q}} - 1)^{-1} \quad \begin{matrix} \sim kT/\omega_{k-q} & \text{at high } T, \\ \sim e^{-\beta\omega_{k-q}} & \text{at low } T, \end{matrix} \tag{44}$$

where $\beta = 1/k_B T$, with k_B the Boltzmann constant. Substitution of eq. (43) in eq. (9) yields integrals leading to δ functions:

$$W_{q,q+K;k,k+K} = \pi|g_{k-q}|^2 \omega^2_{k-q}$$

$$\times \{(n_{k-q} + 1)[\delta(E_{k+K} - E_{q+K} - \omega_{k-q}) + \delta(E_k - E_q - \omega_{k-q})]$$

$$+ n_{k-q}[\delta(E_{k+K} - E_{q+K} + \omega_{k-q}) + \delta(E_k - E_q + \omega_{k-q})]\}. \tag{45}$$

The scattering rates are then obtained as (Agranovich and Konobeev 1959, 1963, 1968, Munn and Silbey 1980):

$$\Gamma_{kk} = 2\pi N^{-1} \sum_q |g_q|^2 \omega^2_q \{(n_q + 1)\delta(E_{k-q} - E_k + \omega_q) + n_q\delta(E_{k-q} - E_k - \omega_q)\}. \tag{46}$$

At this level of approximation, there is no scattering unless the excitation bandwidth B is wider than some phonon energy so that single-phonon processes are energetically feasible. The hopping rates are obtained by substituting eq. (45) in eq. (21) and changing the summation variable to $k - q$:

$$\gamma_{kk} = \left(\frac{1}{2}\frac{d^2}{dk^2} - \frac{d^2}{dK^2}\right)\pi N^{-1}\sum_q |g_q|^2 \omega_q^2$$
$$\times \{(n_q + 1)[\delta(E_{k+K} - E_{k-q+K} - \omega_q) + \delta(E_k - E_{k-q} - \omega_q)]$$
$$+ n_q[\delta(E_{k+K} - E_{k-q+K} + \omega_q) + \delta(E_k - E_{k-q} + \omega_q)]\}_{K=0}. \qquad (47)$$

In the limit $K = 0$, both terms in a given square bracket have the same derivative with respect to k, which equals the derivative of the first term with respect to K. Hence γ_{kk} is zero; with weak electron–phonon coupling there is no hopping, as expected, and the diffusion coefficient is given by the standard band expression involving the scattering rates Γ_{kk}.

2.3.1. Optical phonons

For a narrow optical phonon band we can set $g_q = g$, $\omega_q = \omega$, and $n_q = n$ for all q. The scattering rates are then

$$\Gamma_{kk}^{op} = 2\pi g^2 \omega^2[(n + 1)N_{ex}(E_k - \omega) + nN_{ex}(E_k + \omega)], \qquad (48)$$

where $N_{ex}(E)$ is the excitation density of states. In general the resulting diffusion coefficient is difficult to obtain except by numerical means. However, for wide parabolic excitation bands such that $B \gg \omega$, we can set

$$\Gamma_{kk}^{op} \approx 2\pi g^2 \omega^2(2n + 1)N_{ex}(E_k), \qquad (49)$$

$$v_k^2 = 2E_k/m^*, \qquad (50)$$

where m^* is the effective mass. Conventional procedures then lead to

$$D/a^2 = [2\pi k_B T/(m^*)^5]^{1/2}[g^2\omega^2(2n + 1)]^{-1}, \qquad (51)$$

where a is the intermolecular distance. Since $m^* \sim 1/B$, then $D \sim B^{5/2}/\omega^2$, and $D \sim T^{1/2}$ at low T and $\sim T^{-1/2}$ at high T.

2.3.2. Acoustic phonons

We assume that acoustic phonons can be adequately described by a Debye spectrum, cut-off frequency ω_D, and that the electron–phonon coupling is given by the deformation potential approximation,

$$g_q = (A/\omega_q)^{1/2}. \qquad (52)$$

The scattering rate is then

$$\Gamma_{kk}^{ac} = 2\pi AN^{-1}\sum_q \omega_q[(n_q + 1)\delta(E_{k-q} - E_k + \omega_q) + n_q\delta(E_{k-q} - E_k - \omega_q)]. \qquad (53)$$

Further formal results can be obtained by defining the joint density of states at total wave vector k by

$$\rho_k(E, \omega) = N^{-1} \sum_q \delta(\omega - \omega_q)\delta(E - E_{k-q}), \tag{54}$$

which satisfies

$$N^{-1} \sum_k \rho_k(E, \omega) = N_{ph}(\omega)N_{ex}(E), \tag{55}$$

where $N_{ph}(\omega)$ is the phonon density of states. Using eq. (54) in eq. (53) we obtain

$$\Gamma_{kk}^{ac} = 2\pi A \int dE \int d\omega\, \rho_k(E, \omega)\omega$$
$$\times \{[n(\omega) + 1]\delta(E - E_k + \omega) + n(\omega)\delta(E - E_k - \omega)\}. \tag{56}$$

In some limits, results can be obtained more directly. In the conventional semiconductor limit $B \gg k_B T \gg \omega_D$, we have $n_q \approx k_B T/\omega_q \gg 1$, and $|E_{k-q} - E_k| \gg \omega_q$ except for a few sets of wave vectors. These results yield

$$\Gamma_{kk}^{ac} = 4\pi A k_B T N_{ex}(E_k), \tag{57}$$

$$D/a^2 = [\pi/k_B T(m^*)^5]^{1/2}/A, \tag{58}$$

so that $D \sim B^{5/2}/A \sim B^{3/2}$ if we assume A is proportional to B. Equation (58) gives a $T^{-1/2}$ temperature dependence for the diffusion constant.

In molecular crystals one may also require the limit $k_B T \gg \omega_D \gg B$. Then, in eq. (58) nonzero contributions arise only for frequencies $\omega_q \lesssim B$. Taking on average $|E_{k-q} - E_k| \approx \frac{1}{2}B$ yields for all k,

$$\Gamma^{ac} \approx 4\pi A k_B T N_{ph}(\tfrac{1}{2}B), \tag{59}$$

where for a Debye spectrum $N_{ph}(\omega) = 3\omega^2/\omega_D^3$. Taking similarly $\langle v_k^2 \rangle \approx (\frac{1}{2}Ba)^2$, we obtain

$$D/a^2 = \omega_D^3/12\pi A k_B T, \tag{60}$$

so that $D \sim \omega_D^3/A \sim \omega_D^3/B$ (with the previous assumption $A \sim B$) and D varies as T^{-1}. The narrow excitation band in this limit greatly reduces the number of allowed one-phonon scattering processes, so that the diffusion coefficient may be large, as the ratio ω_D/B indicates.

Finally, in the limit $k_B T \gg \omega_D$, the scattering rates are given by eq. (56) but the thermal averages in D have to be taken over a narrow excitation band. We take $v_k^2 \sim (\frac{1}{2}Ba)^2$ and $N_{ex}(E_k) \sim 1/B$, obtaining

$$D/a^2 = B^3/(16\pi A k_B T), \tag{61}$$

so that D varies as T^{-1}. This result accords with the narrow excitation band treatments of Glarum (1963) and Friedman (1965) in the study of electron transport.

2.3.3. *Transformed coupling*

The transformation of the Hamiltonian (5), (6) which yields a weak residual excitation–phonon coupling even when the g_q are large, has been discussed several times (Grover and Silbey 1971, Merrifield 1964, Cho and Toyozawa 1970, Fischer and Rice 1970). It produces a uniform shift in the excitation energy levels and a displacement in the equilibrium position of the phonons corresponding to the formation of a polaron. Since the transfer interactions J_{nm} compete with this tendency to form a localized state, the optimum transformation should be determined variationally (Yarkony and Silbey 1976). However, for present purposes we use the full clothing transformation which is exact for $J = 0$ and yields the correct untransformed results for large J and weak coupling. The results are qualitatively similar to those which would be obtained with the full variational transformation, but are simplified by the absence of the temperature-dependent variational parameters.

After the transformation, the excitation part of the Hamiltonian (5), (6) is

$$H_{ex} = \sum_k \left(\epsilon - N^{-1}\sum_q \omega_q|g_q|^2 + \sum_n \tilde{J}_h\, e^{ik\cdot R_h}\right) a_k^+ a_k, \tag{62}$$

where R_h is a lattice vector and

$$\tilde{J}_h \equiv J_{n+h,n}\langle \theta_{n+h}^+\theta_n\rangle, \tag{63}$$

$$\theta_n \equiv \exp\left(N^{1/2}\sum_q (g_q^n)^*(b_q^+ - b_{-q})\right), \tag{64}$$

$$\langle\theta_{n+h}^+\theta_n\rangle = \exp\left(-N^{-1}\sum_q (2n_q+1)|g_q|^2(1-\cos q\cdot R_h)\right). \tag{65}$$

The phonon part is

$$H_{ph} = \sum_q \omega_q(b_q^+ b_q + \tfrac{1}{2}). \tag{66}$$

The residual coupling is described by the operators

$$V_{kq} = N^{-1/2}\sum_{n,h} J_h\, e^{ik\cdot R_h} e^{i(k-q)\cdot R_n}(\theta_{n+h}^+\theta_n - \langle\theta_{n+h}^+\theta_n\rangle). \tag{67}$$

The excitation part and the coupling are temperature dependent through the thermal averages in eqs. (63) and (67); this partition of the Hamiltonian ensures the correct thermal equilibrium behavior.

To simplify the treatment of the correlation functions, we assume that the phonons belong to a narrow optical band and ignore any anisotropy.

We neglect any terms which are exactly zero for zero phonon bandwidth (Grover and Silbey 1971, Munn 1973, 1974) and retain only terms corresponding to sites h which are nearest neighbors of the origin. Then we obtain, after making the approximations that the phonon bandwidth $\Delta \ll k_B T$, $g_q = g$, $\omega_q = \omega$, for all q,

$$\Gamma_{kk} \approx \frac{2\pi^{1/2}[I_0(y) - 1]z\tilde{J}^2}{(\tilde{B}^2 + \Delta^2)^{1/2}} \exp[-E_k^2/(\tilde{B}^2 + \Delta^2)],$$

$$\gamma_{kk} \approx a^2\Gamma_{kk},$$

(68)

where z is the number of nearest neighbor molecules, each with transfer integral $\tilde{J}$, $\tilde{B} = 2z\tilde{J}$, $I_0(y)$ is the zeroth-order modified Bessel function and $y = 4g^2[n(n+1)]^{1/2}$.

With these results, Munn and Silbey (1980) computed the diffusion constant for this model for various coupling strengths (g^2), exciton bandwidths and temperatures. For very large temperature, $\tilde{B} < \Delta$ and $g^2 \gg 1$, the result agrees with Holstein (1959) and Gosar (1971), leading to activated hopping. For other limits (e.g. $\tilde{B} > \Delta$, $g^2 \approx 1$, etc.) the results differ from that result, and a more diverse temperature dependence can be seen. The reader is referred to Munn and Silbey (1980) for further discussion.

2.3.4. Quadratic coupling to optical phonons

Munn and Siebrand (1970) and later Munn and Silbey (1978) have examined the effect of local quadratic coupling to optical phonons on the diffusion constant. The latter authors used a transformation similar to that used in the linear coupling problem (see previous subsection) and computed the diffusion constant as a function of temperature. The most important effect of quadratic phonon coupling is to change the frequency of the phonons (this should be most important for intramolecular and librational modes). The ratio of the new frequency to the old is given by

$$\frac{\omega'}{\omega} = e^{-4\eta} = \left(1 + \frac{4g}{\omega}\right)^{1/2},$$

(69)

where g is the coupling constant in eq. (41) assumed here to be independent of q' and q. Since frequency changes upon excitation are relatively small, g and η are small numbers. Using the formulae given above, these authors found (Munn and Silbey 1978):

$$D = \frac{J^2}{\Delta} e^{-4\eta^2} e^{-4y} \left(y^2 + \frac{\Delta^2}{4y\omega}\right),$$

(70)

where

$$y = 2\eta^2 \bar{n}(\bar{n} + 1),$$

and Δ is the optical phonon bandwidth with average frequency ω. As

$T \to 0$, n and y go to zero, D gets very large, because the scattering decreases. As T increases, D has a minimum near $y = \frac{1}{2}(\Delta/\omega)^{2/3}$ and then increases slowly until $y \approx \frac{1}{2}$ where it begins to fall to zero as $T \to \infty$. These results are for $B < \Delta$, so that they are valid only for triplet excitons. It is clear that these results are very similar to those for linear coupling in the same limit (see previous subsection). One difference is that a term resembling the Haken–Strobl γ_0 appears in the present results and it does not in the case of transformed linear coupling. This is due, in the latter case, to the assumption that $B < \Delta$, which means no one-phonon process can give rise to a scattering within the exciton band; in the present case, even though $B < \Delta$, a two-phonon process *can* give rise to such scattering.

A treatment of quadratic coupling to acoustic phonons has been given by Pesz and Luty (1977), who find that, although this coupling decreases the diffusion constant, the temperature dependence is rather weak. Thus, adding weak quadratic coupling seems only to change the results slightly from those of linear coupling.

A final point should be noted here. Strong quadratic phonon effects can have the drastic effect of producing bound states, that is, an exciton and a local phonon bound to each other. This can be seen most simply by examining the case of no exciton coupling, $J_{nm} = 0$, and Einstein phonons, $\omega_q = \omega$. Then the one-phonon, one-exciton levels have energies $\epsilon + \omega$ (ϵ being the electronic excitation energy and ω the phonon frequency if the phonon is not on the same site) and $\epsilon + \omega'$ (when the two excitations are on the same site). The exciton and phonon interactions broaden these into *two* bands if $\omega' - \omega$ is large enough. One band is made up of scattering states, and the other of bound states. The latter should profoundly affect the transport; no satisfactory treatment of this has been given.

2.4. Concluding remarks

In the last few subsections, we have assumed that when the phonon–exciton coupling is weak we use simple perturbation theory, but if the coupling is strong, we first transform the Hamiltonian to a new form (given in subsection 2.3.3) in which the major part of the coupling has been treated exactly. The criterion of weak and strong means, of course, the relative size of the coupling to the exciton bandwidth. The effective exciton–phonon coupling is temperature dependent, increasing as T increases. The exciton bandwidth also depends on T, decreasing as T increases (because of the T-dependent Debye–Waller factor in $\tilde{J}$). Thus we may have weak coupling for small T and strong coupling at high T. In order to treat this systematically, Yarkony and Silbey (1977) introduced a variational transformation which equals the one used in subsection 2.3.3 at high T, but is essentially the unit operator (i.e. no transformation) at low T.

The variational parameter is then chosen to minimize the free energy of the exciton at all T; this parameter is then T dependent. The results show the behavior outlined above, of a gradual change from band to hopping behavior as T increases, for most cases. However, if the exciton band-width is very large and the exciton–phonon coupling is also very large, a new possibility emerges: the transport will change *abruptly* from band to hopping at some temperature. The reason for this is the coexistence of band and localized states in the same crystal, which was first predicted by Toyozawa (1976).

There are undoubtedly more possibilities in the transport behavior when the anisotropies in the exciton bands are taken into account. There has not yet been a careful study of these effects.

3. *Exciton–impurity scattering*

In this section, we will consider the effect of randomly placed impurities on the motion of an exciton. In subsection 3.1 we consider the weak scattering of an exciton by a low concentration of impurities. In subsection 3.2, we consider the opposite extreme of a localized exciton scattered strongly by a high concentration of impurities.

3.1. *Weak impurity scattering*

If the scattering by impurities (Agranovich and Konobeev 1959, 1963, 1968) is weak, then we may calculate Γ_{kk} for an exciton band state by second-order perturbation theory. We find from the Fermi Golden rule,

$$\Gamma_{kk} \propto \sum_{k' \neq k} |V_{kk'}|^2 \delta(E_k - E_{k'}), \qquad (71)$$

where $V_{kk'}$ is the matrix element of the impurity potential between the band states k and k'. We assume for simplicity that the impurities are at a set of sites, $\{n\}$, randomly placed in the lattice and that at each site n the perturbation is local, i.e. V_{nn}; then

$$V_{kk'} = \frac{1}{N} \sum_n e^{i(k-k') \cdot R_n} V_{nn}, \qquad (72)$$

and

$$\Gamma_{kk} = \frac{1}{N^2} \sum_n \sum_m \sum_{k'} \delta(E_k - E_{k'}) e^{i(k-k') \cdot (R_n - R_m)} V_{nn} V_{mm}. \qquad (73)$$

We now average this quantity over all possible configurations of impurities, and assume as a consequence that only the terms $n = m$ are nonzero. Thus,

$$\langle \Gamma_{kk} \rangle = \frac{1}{N} \sum_n \langle V_n^2 \rangle \frac{1}{N} \sum_{k'} \delta(E_k - E_{k'}) = c \langle V^2 \rangle N(E_k), \tag{74}$$

where c is the fraction of sites occupied by impurities and $N(E_k)$ is the exciton density of states at E_k. The diffusion constant is then given by

$$D = \sum_k \frac{v_k^2}{\langle \Gamma_{kk} \rangle} \sigma_{kk}^{eq}, \tag{75}$$

and so is proportional to c^{-1}. Using this formula, we can define a mean free path and a cross section for impurity scattering. Both of these will depend on the form taken for V_{nn}, of course.

   In the case that both phonon and impurity scattering are present, and the impurity scattering is weak, the total scattering rate Γ_{kk} will be the sum of the Γ_{kk} for phonons and the Γ_{kk} for impurity scattering.

3.2. Strong scattering limit: dispersive transport

We shall now discuss another case of current interest: Consider a large number of molecules, each capable of being excited, *randomly* placed on a lattice. For a given concentration or density of molecules on the lattice, there are an extremely large number of possible configurations or arrangements of the molecules on the lattice. We will only consider excitation transfer on these molecules: the other lattice sites are empty or occupied by molecules incapable of being excited. Now the impurity concentration is so high that the k states of subsection 3.1 are no longer a valid description. This is the case of strong impurity scattering. In the limit of strong scattering, the excitations are largely localized, and the probabilities of finding the excitation at a site n, $p_n(t)$ (in a given configuration), obey the Pauli master equation:

$$\frac{d}{dt} p_n(t) = \sum_{m \neq n} W_{nm} p_m(t) - \sum_{m \neq n} W_{mn} p_n(t). \tag{76}$$

The sum is over the occupied lattice sites and we will assume $W_{nm} = W_{mn}$ (equal energy sites). Then the W_{nm} are functions only of the distance $R_n - R_m$; however, the distribution of distances is different in each configuration. Because of this, the equation for the *average* probabilities,

$$\langle p_n(t) \rangle = \frac{1}{\mathcal{N}} \sum_{\text{config.}} p_n(t), \tag{77}$$

where $\mathcal{N}$ is the number of configurations, has been shown to be the Generalized Master Equation (GME) (see for example Haan and Zwanzig

1978):

$$\frac{\mathrm{d}\langle p_n(t)\rangle}{\mathrm{d}t} = \sum_{m \neq n} \int_0^t \mathrm{d}\tau \, [M_{nm}(t - \tau)\langle p_m(\tau)\rangle - M_{mn}(t - \tau)\langle p_n(\tau)\rangle]. \tag{78}$$

In this equation, because of the average over all configurations, the transition rates have a *memory* and the sum is over *all* lattice sites. Since on the average every site is the same, this equation is for average probabilities on a translationally invariant lattice. Because of this

$$M_{nm}(t) = M(R_n - R_m; t), \tag{79}$$

and the GME can be solved by Fourier and Laplace transforming. Define (N is the number of lattice sites)

$$\langle \tilde{p}_K(z)\rangle = \int_0^\infty \mathrm{e}^{-zt} \frac{1}{N} \sum_n \mathrm{e}^{iK \cdot R_n} \langle p_n(t)\rangle \, \mathrm{d}t, \tag{80}$$

$$\tilde{M}(K, z) = \int_0^\infty \mathrm{e}^{-zt} \frac{1}{N} \sum_n \mathrm{e}^{iK \cdot (R_n - R_m)} M(R_n - R_m; t) \, \mathrm{d}t, \tag{81}$$

then

$$\langle \tilde{p}_k(z)\rangle = \frac{1}{N} [z - \tilde{M}(k, z) + \tilde{M}(0, z)]^{-1}, \tag{82}$$

where we have assumed $p_n(0) = \delta_{n0}$ for all configurations. By inverting the Fourier transform and calculating the Laplace transform of the square displacement, we find

$$\langle \tilde{R}^2(z)\rangle \equiv \frac{1}{N} \sum_n \langle \tilde{p}_n(z)\rangle R_n^2$$

$$= \frac{1}{z^2} \{\nabla_K^2 \tilde{M}(K, z)\}_{K=0}. \tag{83}$$

The time dependence of $\langle R^2(t)\rangle$ can be found by inverting this Laplace transform. If $\tilde{M}(K, z)$ were independent of z, then $\langle R^2(t)\rangle$ would be proportional to t, and thus indicate a simple diffusive process. However, in general $\tilde{M}(K, z)$ will be dependent on z, thus $\langle R^2(t)\rangle$ will be a rather complicated function of time. Haan and Zwanzig (1978) have considered this problem (for a glass, not a lattice, but the results are similar) and have shown by an elegant scaling argument that, if the W_{mn} are given by the Förster theory:

$$W_{mn} = \tau^{-1}(R_0/R_{mn})^6, \tag{84}$$

then

$$\langle R^2(t)\rangle \propto t^{1/3} f(ct^{1/2}), \tag{85}$$

where f is some unknown function and c is the density of molecules on which the excitation can travel. Thus at short time, when the mean square displacement is linear in c, $\langle R^2(t) \rangle \sim t^{5/6}$, and so the classical diffusion equation is not valid. At long times, we should have $\langle R^2(t) \rangle$ proportional to t, so that $\langle R^2(t) \rangle \sim t c^{4/3}$, and the diffusion constant is proportional to $c^{4/3}$ as in the Förster theory of transfer on a translationally invariant lattice. We see therefore that the diffusion coefficient is a function of time. Because of the complexity of the averaging process, Haan and Zwanzig (1978) only gave the first few terms in the expansion of $\langle R^2(t) \rangle$ in the density. However, it is clear from (85) that the high density limit is the same as the long time limit (and the low density and short time limits are the same). Thus a calculation of the long time diffusion coefficient requires a rather complex calculation. In a later development, Gochanour et al. (1979) have given a self-consistent diagrammatic calculation of $\tilde{M}(K, z)$. The calculation is complex and only low-order numerical results were possible. This tour de force calculation, however, shows quite nicely the time dependence of the diffusion coefficient.

Another approach to this problem is known as the continuous time random walk model (CTRW) of Montroll and Weiss (1965), Scher and Montroll (1973) and Scher and Lax (1973). The basic equations of this model are

$$\langle p_n(t) \rangle = \delta_{n0}\phi(t) + \int_0^t d\tau \sum_{m \neq n} \psi_{nm}(t - \tau)\langle p_m(\tau) \rangle, \tag{86}$$

where $\phi(t)$ is the probability of the excitation initially at site 0 to remain at that site *without* leaving, and $\psi_{nm}(t - \tau)$ is the probability distribution of the excitation going from m to n after waiting a time $t - \tau$ on m, so that

$$\phi(t) = 1 - \sum_{m \neq n} \int_0^t d\tau \, \psi_{nm}(\tau), \tag{87}$$

and again because of the configurational averaging $\psi_{nm}(\tau)$ is a function of $R_n - R_m$. The equation has been shown to be identical to the GME (Klafter and Silbey 1980a,b, Kenkre et al. 1973), using the definition

$$\psi_{nm}(t) \equiv \int_0^t d\tau \, M_{nm}(t - \tau)\phi(\tau). \tag{88}$$

The calculation based on the CTRW are just as complex and difficult as those based on the GME, so nothing is gained by the exact correspondence. In other words, an *exact* calculation of $\langle R^2(t) \rangle$ in either formalism requires a solution to a rather difficult statistical problem; however, approximations are sometimes easier and more useful for one equation than for the other. As an illustration, Godzik and Jortner (1979, 1980) have used the CTRW to compute $\langle R^2(t) \rangle$ for the Förster transfer problem in a second-order approximation, and found close agreement with the calculation of Gochanour et al. (1979) based on the GME.

The standard approximation is to take $\phi(t)$, the probability of the excitation never having left the initial site, as

$$\phi(t) = \left\langle \exp\left(-\sum_{n \neq 0} W_{0n} t\right)\right\rangle = \int_0^\infty e^{-\omega t} \rho(\omega)\, d\omega, \tag{89}$$

where the angular brackets represent an average over all configurations or over all sets of transition rates W_{0n} consistent with the density. The last form is obtained by assuming that there is a distribution $\rho(\omega)$ of leaving rates ω, due to the variety of possible configurations. It is just this form, introduced by Scher and Lax (1973), which was used by Godzik and Jortner (1979, 1980) in their calculation. The second form of this approximation can be very useful when it is noticed that the long time behavior of $\langle R^2(t)\rangle$ is governed by the long time behavior of $\phi(t)$; thus this behavior is governed by the small z dependence of $\tilde{\phi}(z)$:

$$\tilde{\phi}(z) = \int_0^\infty \frac{\rho(\omega)}{z + \omega}\, d\omega. \tag{90}$$

It has been shown (Klafter and Silbey 1980a,b) that the long time diffusion constant is proportional to $[\tilde{\phi}(z = 0)]^{-1}$ or

$$D(t \to \infty) \sim [\tilde{\phi}(z \to 0)]^{-1} = \langle 1/\omega \rangle^{-1} = \omega_{\text{eff}}. \tag{91}$$

This shows that at long times, the excitation has visited so many sites that its mean square displacement is governed by an effective jump rate.

This short discussion indicates that the calculation of the time-dependent diffusion constant even in the simplest and least strongly fluctuating system (randomly distributed, equal energy sites) is a difficult task. The introduction of unequal energies (that is, a distribution of site energies) makes the system more complex; very few calculations have been made on such systems.

There is a further possibility: Suppose $\tilde{\phi}(z \to 0)$ does not exist. For example, if $\rho(\omega) \sim \omega^{-\alpha}$, $0 \leq \alpha < 1$, then $\langle 1/\omega \rangle = \infty$. This seemingly pathological case has been the focus of a number of investigations (Alexander and Bernasconi 1979, Bernasconi et al. 1978, Alexander 1981) because it may be experimentally realizable. Consider an excitation which travels on a one-dimensional path (for example, the triplet excitations in dibromonaphthalene crystals). If the transfer rate, $W(R)$, is very short ranged,

$$W(R) \sim e^{-\gamma R}, \tag{92}$$

then, because the probability distribution of *nearest* neighbors in one dimension is e^{-cR} (where c is the density of molecules on which the excitation travels) we find the distribution of leaving rates to be

$$\rho(\omega) \sim \omega^{(c-\gamma)/\gamma}. \tag{93}$$

Since $c < 1$ and usually $\gamma > 1$ (all in units of the inverse lattice constant) this is of the form for which $\langle 1/\omega \rangle = \infty$. We then predict that excitation transfer in this system will be nonclassical, in that even at long times it will not be diffusive. A similar calculation of the rate of trapping at a low energy sink shows that this too will be nonclassical (Klafter and Silbey 1980c, 1981). However, it should be noted that every triplet state has a small component of singlet state mixed in, and this component will produce a small addition to the transfer rate which has a more *gentle* dependence on R than exponential. This means the small ω behavior (large R behavior) may be dominated by another functional form than that given above. When this is translated into the time domain, we predict that the trapping rate and diffusion should be nonclassical until a time at which the small nonexponential part of ω comes into play; for times after this we predict a cross-over to classical diffusive and trapping behavior. Since this time will depend on the density (c), time-resolved, density-dependent experiments should be able to see this effect.

Another example in which $\langle 1/\omega \rangle$ does not exist comes about if the excitation is traveling from site to site via an activated process where the activation energy, Δ, is a random variable with distribution function $g(\Delta)$:

$$\omega = \omega_0 \, e^{-\Delta/k_B T}, \tag{94}$$

$$g(\Delta) = g_0 \, e^{-\gamma \Delta}. \tag{95}$$

Such a distribution of activation barriers could come about from a frozen-in configuration at some temperature $T_0 = (k_B \gamma)^{-1}$ (Higashi and Kastner 1979, Klafter and Silbey 1980d, Bernasconi et al. 1979). The distribution in ω becomes:

$$\rho(\omega) \sim \omega^{(\gamma k_B T - 1)} \sim \omega^{(T/T_0 - 1)}. \tag{96}$$

Thus for $T < T_0$, the diffusion coefficient will never become time independent, and nonclassical diffusion occurs. Note that as T is increased to T_0, a cross-over to classical diffusion takes place.

This short discussion of the strong scattering limit does not exhaust the possibilities. This is an area of intense activity at the present time.

4. Conclusions

In this chapter, we reviewed the theories of exciton transport, concentrating on phonon and impurity scattering of excitons. The theories are complex, due to the variety of possible orderings of the relevant parameters. In general, the qualitative results of the theories are in accord with our intuitive ideas; however, as the discussion of the strong impurity scattering case shows, there are still surprises in store for us. In addition, there are still a number of outstanding unsolved problems.

Acknowledgements

I would like to thank my colleagues, A. Blumen, M. Grover, J. Klafter, R.W. Munn, and D. Yarkony, who have contributed greatly to many of the ideas in this chapter. In particular, most of the work reported in section 2 was done with R.W. Munn.

References

Agranovich, V.M. and Y.V. Konobeev, 1959, Opt. Spectrosc. **6**, 241.
Agranovich, V.M. and Y.V. Konobeev, 1963, Sov. Phys. Solid State **5**, 999.
Agranovich, V.M. and Y.V. Konobeev, 1968, Phys. Stat. Sol. **27**, 435.
Alexander, S., 1981, Phys. Rev. **B23**, 2951.
Alexander, S. and J. Bernasconi, 1979, J. Phys. **C12**, L1.
Bernasconi, J., S. Alexander and R. Orbach, 1978, Phys. Rev. Lett. **41**, 185.
Bernasconi, J., H. Beyeler, S. Strasler and S. Alexander, 1979, Phys. Rev. Lett. **42**, 819.
Blumen, A. and R. Silbey, 1978, J. Chem. Phys. **69**, 3589.
Cho, K. and Y. Toyozawa, 1970, J. Phys. Soc. Jpn. **30**, 1555.
Davydov, A.S., 1971, Theory of Molecular Excitons (Plenum Press, New York).
Fischer, S. and S.A Rice, 1970, J. Chem. Phys. **52**, 2091.
Friedman, L., 1965, Phys. Rev. **140**, A1649.
Glarum, S. 1963, J. Phys. Chem. Sol. **24**, 1577.
Gochanour, C., H.C. Andersen and M. Fayer, 1979, J. Chem. Phys. **70**, 4254.
Godzik, K. and J. Jortner, 1979a, Chem. Phys. **38**, 227.
Godzik, K. and J. Jortner, 1979b, Chem. Phys. Lett. **63**, 428.
Godzik, K. and J. Jortner, 1980, J. Chem. Phys. **72**, 4471.
Gosar, P., 1971, Phys. Rev. **B3**, 1991.
Grover, M. and R. Silbey, 1971, J. Chem. Phys. **54**, 4843.
Haan, S. and R. Zwanzig, 1978, J. Chem. Phys. **68**, 1879.
Haken, H. and P. Reineker, 1972, Z. Phys. **249**, 253; 1972, **250**, 300.
Haken, H. and G. Strobl, 1973, Z. Phys. **262**, 185.
Higashi, G. and M. Kastner, 1979, J. Phys. **C12**, L821.
Holstein, T., 1959, Ann. Phys. NY **8**, 343.
Kenkre, V., 1975a, Phys. Rev. **B11**, 1741.
Kenkre, V., 1975b, Phys. Rev. **B12**, 2150.
Kenkre, V., 1978, Phys. Lett. **65A**, 391.
Kenkre, V. and R. Knox, 1973, Phys. Rev. **B9**, 5279.
Kenkre, V. and R. Knox, 1974, Phys. Rev. Lett. **33**, 803.
Kenkre, V. and R. Knox, 1976, J. Lumin. **12/13**, 187.
Kenkre, V., E. Montroll and M. Schlesinger, 1973, J. Stat. Phys. **9**, 45.
Klafter, J. and R. Silbey, 1980a, Phys. Rev. Lett. **44**, 55.
Klafter, J. and R. Silbey, 1980b, J. Chem. Phys. **72**, 843.
Klafter, J. and R. Silbey, 1980c, J. Chem. Phys. **72**, 849.
Klafter, J. and R. Silbey, 1980d, Surf. Sci. **92**, 393.
Klafter, J. and R. Silbey, 1981, J. Chem. Phys., **74**, 3510.

Kubo, R., 1957, J. Phys. Soc. Jpn. **12**, 570.
Kubo, R., 1958, Lec. Theor. Phys. Boulder **1**, 120.
Kuhne, R., and P. Reineker, 1978, Phys. Lett. **69A**, 133.
Merrifield, R., 1964, J. Chem. Phys. **40**, 445.
Montroll, E. and G. Weiss, 1965, J. Math. Phys. NY **6**, 167.
Munn, R.W., 1973, J. Chem. Phys. **58**, 3230.
Munn, R.W., 1974, Chem. Phys. **6**, 469.
Munn, R.W. and W. Siebrand, 1970, J. Chem. Phys. **52**, 47.
Munn, R.W. and R. Silbey, 1978, J. Chem. Phys. **68**, 2439.
Munn, R.W. and R. Silbey, 1980a, Mol. Cryst. Liq. Cryst. **57**, 131.
Munn, R.W. and R. Silbey, 1980b, J. Chem. Phys. **72**, 2763.
Oppenheim, I., K. Shuler and G. Weiss, 1979, Stochastic Processes in Chemical Physics: The Master Equation (MIT Press, Cambridge, MA, USA).
Pesz, K. and T. Luty, 1977, Phys. Stat. Sol. (b) **83**, 565.
Powell, R.C. and Z.G. Soos, 1975, J. Lumin. **11**, 1.
Silbey, R., 1976, Ann. Rev. Phys. Chem. **27**, 203.
Scher, H. and M. Lax, 1973a, Phys. Rev. **B7**, 4491.
Scher, H. and M. Lax, 1973b, Phys. Rev. **B7**, 4502.
Scher H. and E. Montroll, 1973, Phys. Rev. **B12**, 2455.
Schwarzer, E. and H. Haken, 1972, Phys. Lett. **42A**, 317.
Schwarzer, E. and H. Haken, 1973, Opt. Commun. **9**, 64.
Schwarzer, E. and H. Haken, 1974, Chem. Phys. Lett. **27**, 41.
Sumi, H., 1977, J. Chem. Phys. **67**, 2943.
Toyozawa, Y., 1976, J. Lumin. **12/13**, 13, and refs. therein.
Trlifaj, M., 1956, J. Czech. Phys. **6**, 533.
Wertheimer, R. and R. Silbey, 1980, Chem. Phys. Lett. **75**, 243.
Yarkony, D. and R. Silbey, 1976, J. Chem. Phys. **65**, 1042.
Yarkony, D. and R. Silbey, 1977, J. Chem. Phys. **67**, 5818.
Zwanzig, R.W., 1960, Lectures in Theoretical Physics, Boulder, vol. III (Interscience, New York).
Zwemer, D. and C.B. Harris, 1978, Ann. Rev. Phys. Chem. **29**, 473.

CHAPTER 2

Excitons and Polarons in Organic Weak Charge Transfer Crystals

DIETRICH HAARER* and MICHAEL R. PHILPOTT

IBM Research Laboratory
San Jose, CA 95193
U.S.A.

*Present and permanent address: Lehrstuhl Experimentalphysik, Universität Bayreuth, Universitätsstrasse 30, Postfach 3008, 8580 Bayreuth, FRG.

Spectroscopy and Excitation Dynamics
of Condensed Molecular Systems
Edited by
V.M. Agranovich and R.M. Hochstrasser

Contents

1. Introduction

In this chapter we will review some recent developments in the understanding of the optical and photoconductive properties of organic solids composed of donor and acceptor molecules alternating in quasi-one-dimensional arrays. These solids are also referred to as mixed stack electron donor–acceptor (EDA) crystals. The origins of this subject go back a long way. It has been known for at least a hundred years that mixing solutions of organic substances can result in the formation of "molecular complexes" or "addition products" (Werner 1909, Pfeiffer 1927). These are detectable in some cases by the precipitation of crystalline compounds that are dissociable into the original components, whereas in other cases the formation of complexes is discernible only by the colors they impart to the solutions. It is now known that the colors are due to the appearance of a long wavelength absorption band, often in the red or yellow, due to the formation of a molecular complex with an electronic transition in this region. The class of complexes formed when one species is an electron donor and the second an electron acceptor is of particular interest. The first quantum-mechanical explanation of the color changes was given by Mulliken (1950, 1951) whose theory of charge transfer interactions between donor and acceptor molecules served to stimulate a great deal of experimental and theoretical work in this area.

The field of optical spectroscopy of solid crystalline charge transfer (CT) complexes has been actively investigated since the 1950's. It can be divided into two categories. The first category, consisting of molecular crystals with neutral ground states, which have interesting excitonic and photo-conductive CT properties, is discussed in detail in this review. In the course of this chapter we will propose experiments that should provide deeper physical insight in the optical and electronic processes of these very interesting materials. This category is also referred to as weak CT solids in the literature. The second category consists of salts with ionic ground states. These materials show one-dimensional, metallic-like behavior (Keller 1977, Mayerle 1980). They are not discussed further except to show distinctions with the first category.

Donor molecules are characterized by a low ionization potential I_D. In molecular orbital language their highest occupied molecular orbital (HOMO) is not too far from the single ionization continuum of the

29

molecule (generally $I_D < 7\,\mathrm{eV}$). Acceptor molecules are characterized by a large electron affinity E_A (generally $E_A > 2\,\mathrm{eV}$) which is a measure of the energy which can be gained by adding an extra electron to the molecule and thus creating a radical anion with a half-filled lowest unoccupied molecular orbital (LUMO). According to the molecular orbital theory, the first excited electronic state of an EDA complex is obtained by transferring an electron from the HOMO of the donor to the LUMO of the acceptor. This state is illustrated schematically in fig. 1, which also shows the relative positions of donor and acceptor levels, again in a schematic way. The energy of a photon that can cause the transfer of an electron from the donor (D) to the acceptor (A) in the donor–acceptor (DA) pair is simply

$$h\nu_{CT} = I_D - E_A - \Delta_C + V_B, \tag{1}$$

where Δ_C is the Coulombic energy (including polarization energy) gained in the approach of the two ions to the typical donor–acceptor distance (3.5 Å) found from X-ray structure determinations, and V_B is a term arising from overlap of the electronic wavefunctions of donor and acceptor. The quantity V_B occupies an important position in the simple theory since it is responsible for the stabilization of the ground electronic state of the complex. Transitions of the isolated donor and acceptor molecules, denoted sche-

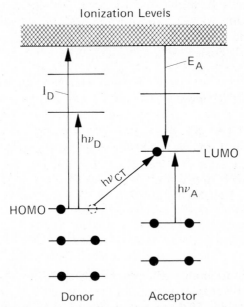

Fig. 1. Schematic energy level diagram for transitions in a donor–acceptor pair. I_D is the ionization energy of the donor and E_A the electron affinity of the acceptor. $h\nu_A$ and $h\nu_D$ are transitions of isolated acceptor and donor. $h\nu_{CT}$ is the energy of the charge transfer transition.

matically by the vertical arrows labelled $h\nu_D$ and $h\nu_A$ in fig. 1, lie higher in energy than the charge transfer (CT) transition $h\nu_{CT}$. In the early theoretical papers of Mulliken (1950, 1951) the valence bond picture was used extensively. The ground electronic state of a DA pair is described in this picture by the wavefunction

$$\psi_G(AD) = a\psi_1(A, D) + b\psi_2(A^-, D^+), \tag{2}$$

where $|a| \approx 1$ and b is small. The corresponding excited state wavefunction is given by

$$\psi_E(AD) = a^*\psi_2(A^-, D^+) + b^*\psi_1(A, D), \tag{3}$$

where a^* and b^* are the complex conjugates of a and b, respectively. Note that both ground and excited state wavefunctions are linear combinations of a polar configuration ψ_2 describing the ion pair A^-D^+ and the nonionic molecular configuration ψ_1 describing the AD pair of neutral (i.e., not ionized) molecules. This empirical approach, which appears to describe many systems quite well, rests on the assumption that the interactions between A and D are so weak compared to bonding interactions within the D and A molecules that the electron transfer does not result in any large rearrangement of the intramolecular electronic structure of the two subunits as would occur if a strong covalent bond were formed. For a detailed discussion of the various aspects of the valence bond picture we refer the reader to some of the comprehensive review articles and books that have appeared over the years (McGlynn 1958, Briegleb 1961, Foster 1969, Yarwood 1973).

2. Structure of CT crystals

At distances of 3.4 Å and less the electron clouds of planar aromatic molecules in face-to-face orientation overlap and electron exchange interactions can become large. A measure of the interaction between a donor and an acceptor due to actual transfer of a whole electron is the integral for overlap between the HOMO of the donor and the LUMO of the acceptor. This integral is

$$S_{DA} = \int \psi_D^* \psi_A \, d\tau. \tag{4}$$

There has been considerable discussion in the literature concerning the idea that donor and acceptor molecules adopt a configuration to maximize the overlap integral S_{DA}. If there is a driving force that tends to maximize S_{DA} for the range of interplanar separations found in crystalline EDA complexes then the manifold of geometries available to the complex would be severely limited. However, since the actual equilibrium geometry is determined by the interplay of several interactions (i.e., Coulombic, Pauli exchange, polarization, ...) of comparable magnitude, it is not clear that this idea has general validity or that the equilibrium geometry of donor–acceptor pairs can be related in detail to that of stacked arrays (Mayoh and Prout 1972). The structures of crystalline EDA complexes are determined not only by electronic overlap, but also, as in other molecular crystals, by steric interactions and Van der Waals interactions (Kitaigorodskii 1978). Given a donor and acceptor it is not yet possible to make an a priori prediction of the crystal structure with any reliability. However, in the past this has not prevented many authors from postulating a maximum overlap rule and interpreting the geometry and interactions of donor–acceptor complexes on this basis. Later we will discuss one case in this fashion, in order to illustrate the general line of reasoning in such arguments. Fortunately the difficulty with overlap does not pose a major obstacle in the field of optical spectroscopy because the crystal structures of complexes of planar donor and acceptor molecules are commonly either of the segregated stack or the mixed stack type. Figure 2 shows the chief difference

(a) Segregated Stacks

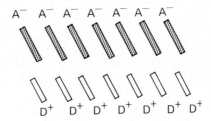

(b) Mixed Stack

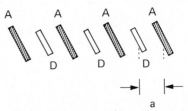

Fig. 2. (a) Typical structure of an ion radical salt as, for example, TTF–TCNQ. (b) Mixed stack CT solid, as for example, anthracene–PMDA. Most members of this last family have little charge transfer in the ground state.

schematically. In segregated stacks there are separate columns of donors and acceptors, whereas in the mixed stack systems donors and acceptors alternate along the same stack.

All the segregated stack organic solids have ionic ground electronic states $[0.7 < b < 1$, eq. (2)] and some show appreciable electrical conductivity ($\sigma > 1\,\Omega^{-1}\,cm^{-1}$ at room temperature). The enormous interest in these materials was sparked by the suggestion that it might be possible to synthesize high temperature superconductors (Little 1964) and the sub-

Anthracene-PMDA

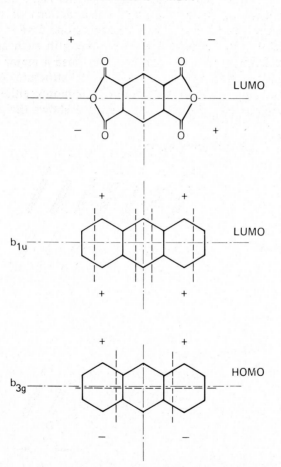

Fig. 3. Symmetry of the various highest occupied (HOMO) and lowest unoccupied (LUMO) molecular orbitals of PMDA (top) and anthracene (middle and bottom). The dashed lines are nodal planes.

sequent discovery of anomalously high conduction in TTF–TCNQ (tetra-thiafulvalene tetracyano-*p*-quinodimethane) at 60 K (Coleman et al. 1973). However, they represent only a small fraction of all the EDA complex solids known, most of which have mixed stack crystal structures. Considerable effort has been devoted to finding conducting CT salts that do not belong to the TTF–TCNQ family (Torrance et al. 1979). In passing we note that low temperature superconductivity has recently been demonstrated for one organic solid of the segregated stack variety (Jerome et al. 1980).

In contrast to the segregated stack materials almost all, but not quite all, mixed stack solids have neutral ground states. To illustrate further the use of the overlap integral in discussing the properties and the influence of molecular symmetry on the overlap we consider the crystalline complex formed from anthracene as donor and pyromellitic dianhydride (PMDA) as acceptor. The complex crystallizes with a 1:1 stoichiometry, the crystal space group is triclinic P$\bar{1}$, and there is one donor–acceptor pair per unit cell (Boeyens and Herbstein 1965, Robertson and Stezowski 1978, Mayerle 1978). Figure 3 shows both molecules as well as the symmetry of the pertinent orbital wavefunctions, such as the acceptor LUMO (top) and both the LUMO and HOMO of the donor molecule. (The dashed lines are nodal planes at which the sign of the wavefunction changes.) In fig. 4, the overlap of wavefunctions resulting from different configurations is depicted

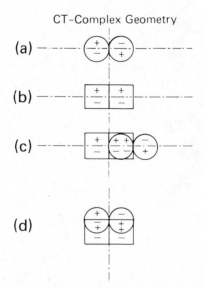

Fig. 4. Schematic representation of: (a) the symmetry of the LUMO of PMDA, (b) the symmetry of the HOMO of anthracene, (c) and (d) the various different overlap configurations (see text) of anthracene and PMDA.

schematically. The acceptor is represented by the two circles and the donor is represented by the rectangle. If one would put the two molecules straight on top of each other (not shown in the figure), the resulting overlap integral S_{DA} would be zero because the various contributions cancel. A positive overlap results, however, if one slides the two molecules partly along their long axis (x-axis) (see fig. 4c). Sliding along the short axis (y-axis) does not result in a positive overlap (see fig. 4d). The projection of the two

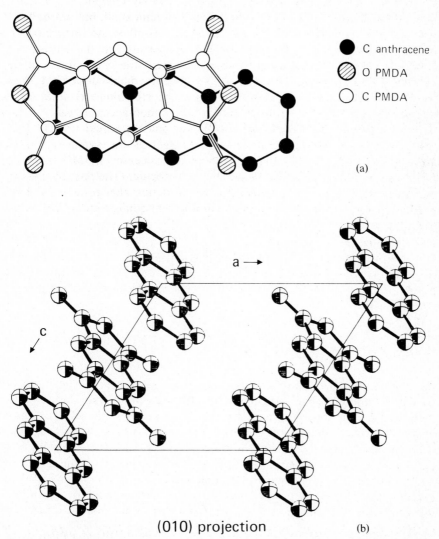

● C anthracene

◐ O PMDA

○ C PMDA

(a)

a ⟶

c

(010) projection   (b)

Fig. 5. (a) Projection of anthracene and PMDA molecule in the CT single crystal. (b) Projection onto the (010) plane showing the alternation of anthracene and PMDA molecules.

molecules as it exists in the crystal structure (Mayerle 1978) is shown in fig. 5a and a projection on the crystallographic plane (010) showing the alternation of donors and acceptors is shown in fig. 5b. The slippage along the molecular x-axis that could have been postulated on the basis of qualitative overlap considerations is clearly visible in both views. It should be reemphasized again that molecular overlap considerations are at best a guideline which allow one to rationalize some of the observed packing structures of CT solids. The real crystal structure is a result of competition between overlap optimization and steric restraints which have to be fulfilled to allow for space filling molecular packing. When external parameters like pressure or temperature are varied, neutral-to-ionic phase transitions are possible (Torrance et al. 1980), and a ground electronic state of the mixed stack consisting of alternating ions D^+ and A^- is created. The existence of these transitions graphically illustrates the dangers existing in predictions based on overlap alone.

In the following we will discuss optical and photoconduction data which will provide experimental evidence for the delocalized nature of excited states in CT solids. For simplicity we will not attempt to give an exhaustive historical review of the entire field but will instead concentrate on some of the more recent data on those molecular donor–acceptor systems that have been prepared in ultrapure crystalline form. PMDA (pyromellitic dianhydride) is one of the few acceptor molecules which can be purified by zone-refining (Haarer and Karl 1973); it will serve as acceptor molecule for most of the experimental data presented. The majority of the commonly used donor molecules such as naphthalene, anthracene, phenanthrene and pyrene can be zone-refined without major technical difficulties (Pfann 1966).

3. Excitons in CT crystals

In solution the formation of CT complexes gives rise to new optical transitions in the visible in regions where the separated donor and acceptor molecules do not absorb light. It was discovered early in the study of these charge transfer transitions that their energy varied in a systematic manner for a series of donors coupled to a given acceptor and vice versa. This is of course very strong evidence for the intermolecular nature of the transition and by inference the existence of a complex. For a donor–acceptor pair in solution the energy $h\nu_{CT}$ given by eq. (1) is that needed to transfer an electron from the donor to the acceptor a few angstroms away. In the excited electronic state generated by this electron transfer process the separation of charge may be viewed as almost complete, the donor being positively charged and the acceptor negatively charged.

Let us consider an array of molecules as typically found in organic molecular crystals. In organic solids composed of only one chemical species, the excited electronic state can be thought of as a linear combination of highly localized excitations of electrons residing almost exclusively on the molecular unit. The theory of Frenkel excitons (Frenkel 1931, Davydov 1962, Craig and Walmsley 1968, Philpott 1973) in molecular crystals describes how this localized state can propagate through the crystal in a coherent or incoherent fashion, depending upon the relative magnitudes of the exciton transfer integral

$$\langle \phi_n^r \phi_m^0 | V_{nm} | \phi_n^0 \phi_m^r \rangle \tag{5}$$

compared to the exciton–lattice phonon and exciton–molecular vibration coupling parameters. The superscripts r and 0 denote the excited and the ground state, respectively, and the subscripts n and m denote the sites occupied by the molecules. The characteristic of Frenkel excitons is that the orbital from which the electron is promoted resides on the same site (i.e. molecule) as the excited state orbital. It makes little difference whether we consider the optical excitations in atomic crystals as originally conceived by Frenkel (1931), or in molecular crystals where both orbitals are confined to the same nuclear skeleton of a particular molecule. Put another way, we can say that the electron and its hole are on the same site. The hole is just a way of describing the charge distribution of the orbital vacated by the promoted electron.

The lowest excited states of CT solids are excitons of quite a different character. The electron and hole are delocalized over two sites with the electron residing mostly on the acceptor and the hole on the donor. In other words, the electron spends most of its time on acceptors and the hole on donors. Thus the excited states resemble ion pair states. In principle coherent motion of a pair is possible. In inorganic systems with large dielectric constants (e.g. ZnO, CdS) the hole and electron may be separated far enough for the excited states to resemble hydrogenic or positronium systems, i.e. the electron and hole orbit each other. Excitons of this type are called Wannier excitons (Wannier 1973). Such isotropic motions are not possible in the highly anisotropic materials considered here. However, in some ways the creation of ion-pair excited states in crystalline EDA complexes can be thought of as a step towards the realization of more delocalized excited states of the Wannier type (Philpott 1977).

The extreme dipolar nature of the CT exciton implies the existence of strong intermolecular Coulomb forces that can create pronounced geometry change in the excited state. These geometry changes can lead to a self-trapping of the excitonic states. Self-trapping phenomena of a related kind are known to exist for excess electrons or holes in insulators. In ionic crystals these excess charges interact strongly with the lattice and are known as polarons (Fröhlich 1963, and references therein). It will be shown below that polaron effects (Holstein 1959, Foerster 1965, Toyozawa 1970) can reduce the measured exciton bandwidth by several orders of magnitude.

The unperturbed ground state of the donor–acceptor stack ψ_0 is an antisymmetrized product of neutral ground state donor and acceptor molecule wavefunctions. A manifold of excited electronic states is formed by the promotion of an electron from any donor molecule n to any acceptor molecule m. Each CT configuration corresponds to an ordered pair $l = (n, m)$. For a chain on which the electron and hole are always on adjacent molecules, we use the alternative representation $n\sigma$, where $\sigma = \pm 1$ is the vector from n to m. The localized charge transfer states are denoted by $\psi(n, \sigma)$. The sign of σ tells the direction of charge transfer and hence the direction of the dipole of the localized excited state. Promotion of an electron beyond the range of the nearest neighbor, $|\sigma| > 1$, is not considered energetically important for the lowest CT exciton band in crystals with small dielectric constants.

The transfer of CT excitons along the chain is governed by matrix elements of the crystal Hamiltonian $\mathcal{H}$. For example,

$$V(n, \sigma, 0; 0, n', \sigma') = \langle \psi(n, \sigma) | \mathcal{H} | \psi(n', \sigma') \rangle \tag{6}$$

transfers an excitation from the pair $n\sigma$ to $n'\sigma'$. No change in internal polarity occurs if $\sigma = \sigma'$. Thus

$$M_1 = \langle \psi(n, \sigma) | \mathcal{H} | \psi(n + 1, \sigma) \rangle \tag{7}$$

is independent of σ, the direction of electron transfer along the chain. Examples of exciton transfer with a change in polarity are

$$M_h = \langle \psi(n, +1) | \mathcal{H} | \psi(n + 1, -1) \rangle, \tag{8}$$

$$M_e = \langle \psi(n, +1) | \mathcal{H} | \psi(n, -1) \rangle, \tag{9}$$

$$M_{2h} = \langle \psi(n, +1) | \mathcal{H} | \psi(n + 2, -1) \rangle, \tag{10}$$

$$M_{2e} = \langle \psi(n, +1) | \mathcal{H} | \psi(n - 1, -1) \rangle. \tag{11}$$

The types of exciton transfer with the resulting polarity change are shown schematically in fig. 6. The $|I\rangle$ represents the initial state (left-hand side of the matrix element) and $|II\rangle$ the final (right-hand side of the matrix element) state. The matrix elements M_h and M_e correspond to the flipping of polarity by means of hole and electron transfer through an intermediate acceptor and donor molecule, respectively. The two M_2 matrix elements describe the coherent exciton transfer with polarity reversal over a distance of three intermolecular separations.

At this point we should also mention that in one-component quasi-one-dimensional organic crystals like 9,10-dichloroanthracene and

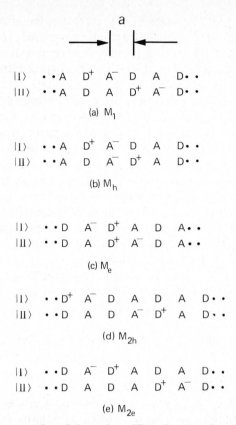

Fig. 6. Schematic representation of the various CT exciton transfer matrix elements along a mixed stack of donor and acceptor molecules. $|I\rangle$ and $|II\rangle$ represent the site configuration before and after transfer of the exciton. Adjacent donors and acceptors are separated by a distance a. The matrix elements displayed are: (a) transfer without polarity flip, (b) and (c) hole and electron transfer with polarity flip, (d) and (e) exciton transfer over a distance $3a$ with polarity flip.

9-cyanoanthracene (Syassen and Philpott 1977, Macfarlane and Philpott 1976) and organic crystals such as pyrene with close pairs of molecules another type of excited state exists, called excimers (Birks 1975), that, like CT exciton states, lie below the molecular exciton states. These excimer states are not as polar as charge transfer excitons since identical moieties (i.e., the donor and acceptor molecules are identical) are involved. Detailed electronic structure calculations of excimers show that a high degree of charge separation occurs so that these states represent a situation intermediate between the molecular exciton with no charge transfer and the CT exciton with complete charge transfer between adjacent molecules.

4. Polarons in CT crystals

The polaron concept was introduced to describe excess electrons in ionic materials like the alkali halides (see, e.g., Fröhlich 1963). The mathematical treatment of polaron properties relies heavily on the use of isotropic continuum electrostatics, which is reasonable for alkali halides considering their large dielectric constants and high crystal symmetry. Even though these materials have little resemblance to the quasi-one-dimensional CT crystals of interest in this chapter, some of the concepts and mathematical formalism used to describe polarons are general enough to carry over to highly anisotropic materials like the anthracene–PMDA crystal. By a polaron we mean an excess electron or an excess hole (i.e. a ground state molecular orbital not occupied by an electron) that interacts strongly with the phonons to form a new kind of quasi-particle. The new particle can be visualized as a charge dressed in a phonon cloud, moving through the crystal. These excess charges may be generated by a variety of means, absorption of light to photoconduction states, injection of electrons or holes from electrodes, dissociation of excitons at imperfections or phase boundaries. We will not elaborate further. In weak CT crystals the electron polaron resides almost exclusively on the acceptor and as a first approximation it occupies the LUMO of the neutral acceptor. The hole polaron sits on the donor sites and as a first approximation may be thought of as a half-filled HOMO of the neutral molecule. Motion of the polaron will depend very strongly on the interaction with the lattice vibrations including the intramolecular modes of the donors and acceptors. The motion can be coherent, giving rise to band states, or incoherent with hopping motion between lattice sites. In both cases sensitivity to temperature is expected. The impurity, dislocation and defect concentrations of real samples will obviously pose problems when interpreting experimental data to extract information concerning phenomena intrinsic to the quasi-one-dimensional system.

5. Theories of CT exciton transitions

The simplest model used to interpret the optical spectra of CT crystals assumes that the process of light absorption is similar to that of color centers in alkali halide crystals. This picture is basically a local one, ignoring the existence of dispersion in the CT exciton band. Its usefulness stems from the strong resemblance between color center spectra (Fowler 1968), which typically have a zero phonon line and an accompanying phonon sideband, and the low temperature reflection and luminescence spectra of anthracene–PMDA. We will refer to this phenomenological model as the local state model. More complex models that include dispersion of CT excitons have been proposed to explain field induced CT transitions and Franck–Condon effects. These will be described briefly since they provide additional physical insights and, in principle, a mathematical foundation for further analysis of CT exciton spectra.

5.1. Local model

The optical absorption spectrum of anthracene–PMDA at low temperatures appears to consist of a zero phonon line accompanied by a phonon sideband. The higher vibronic transitions have a similar appearance. In the local model it is assumed that each intramolecular vibration excited by the electron transfer acts independently as a separate excited state which couples to a set of lattice modes. The exciton bandwidth is assumed to be negligible compared to the lattice phonon and intramolecular phonon energies, i.e., the exciton energy does not depend on wavevector. Absorption of the photon is conceived to be a local process accompanied by the transfer of an electron from a donor to the nearest acceptor. At this point some conceptual difficulties are encountered because in an infinite array the electron can go to the acceptor on either side of the donor, or two donors can simultaneously give half of an electron to an acceptor that they sandwich. In other words, it is not completely established how a theory relevant to a single pair DA, or even a triple like ADA or DAD, applies to an extended array ... ADADADAD.... This is an embedding problem.

In the local model it is assumed that the lattice phonons are dispersionless with energy $\hbar\omega_p$, and are coupled linearly by a configurational coordinate Q to the electronic transition. The potential energy surfaces in the direction of the configuration coordinate Q are shown schematically in fig. 7 for the 0–0 vibronic transition. In real crystals several vibronic transitions are observed in both absorption and fluorescence so that a more complicated picture is needed with manifolds of nested potential surfaces representing the intramolecular vibrational modes of the ground and excited electronic state (fig. 8). At absolute zero the energies and intensities of transitions from the ground level are

$$E_n^{(v)} = \hbar\omega_{CT} + \hbar\omega_v + n\hbar\omega_p, \tag{12}$$

and

$$W_n^{(v)} = S_v^n\, e^{-S_v}/n!, \tag{13}$$

respectively (Fowler 1968). Note that for a given intramolecular mode v the intensities $W_n^{(v)}$ sum to unity. In eqs. (12) and (13) $\hbar\omega_{CT}$ is the energy of the electronic transition, $\hbar\omega_v$ is the energy of the intramolecular vibration, and $\hbar\omega_p$ is the energy of the dispersionless lattice phonon. At $T = 0\,K$ the

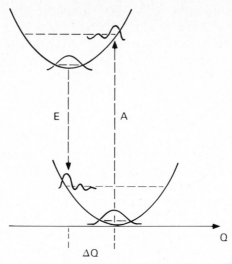

Fig. 7. Schematic diagram of the simple two-surface local model for CT transitions showing the origin of the Stokes shift of a CT complex as the consequence of an excited state relaxation according to the Franck–Condon scheme.

fractional intensity in the zero phonon line (ZPL) is $\exp(-S)$, and so the ratio of intensity in the ZPL to the rest of the band provides a simple method, at least in principle, of determining the magnitude of the exciton–phonon coupling constant S [not to be confused with overlap S_{DA} defined by eq. (4)]. If S is smaller than unity then exciton–phonon coupling is weak and the zero phonon line appears prominently. For very large values of S ($S > 10$) one has a strong exciton–phonon coupling regime and the zero phonon line is too weak to be detected. In this limit the phonon sideband appars to be Gaussian with a full width at half maximum (FWHM) given by

$$FWHM = 2.35 \times \hbar\omega_p \times \sqrt{S}. \tag{14}$$

The difference in energy between $\hbar(\omega_{CT} + \omega_v)$ and peak in the sideband is called the deformation or displacement energy,

$$E_u = \hbar\omega_p \times S. \tag{15}$$

In the configuration coordinate model an explicit formula for S can be derived:

$$S = K^2/2\hbar\omega_n^3 m. \tag{16}$$

Here K is the linear coupling constant, ω_n the oscillator frequency and m is the oscillator mass.

This simple picture can be elaborated by introducing terms for quadratic

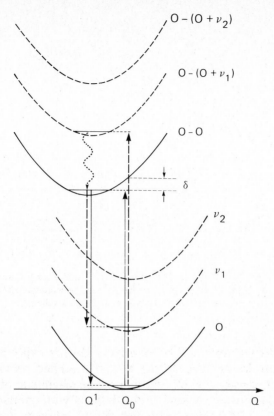

Figure 8. Franck–Condon scheme for transitions in CT complexes with several intramolecular vibrational modes. The intramolecular vibrational mode is ν_1. The binding energy in the excited state is δ.

lattice phonon coupling and the interaction between different vibrational states v. By using a ball and spring model with Coulombic interactions it is also possible to derive explicit expressions for the coupling constant S and the contraction of the lattice around the donor site (Haarer 1977).

At finite temperature the intensity ratio of the zero phonon line I_{ZPL} and the total transition I_{tot} (zero phonon line plus phonon sideband) is given by

$$I_{ZPL}/I_{tot} = \exp\{-S[1 + 6.6(T/\theta_D)^2]\}, \tag{17}$$

where θ_D is the Debye temperature of the crystal. Using the measured temperature dependent intensity of the zero phonon line and the Franck–Condon maximum of the luminescence band, one can get a complete set of model parameters for the CT transition in anthracene–PMDA (Haarer 1974, 1977). These parameters are listed in table 1. Using the local model

Table 1
Anthracene–PMDA parameters.

Debye temp. (K)	ω_n (cm^{-1}) oscillator frequency	S (expt) Huang–Rhys factor	S (theory) Huang–Rhys factor	δ (cm^{-1}) binding energy
100	62 ± 5	5.95 ± 0.3	5.0	372 ± 10

Haarer derived a value for S of about 6 which would place anthracene–PMDA in the class of intermediate phonon coupled systems with barely measurable zero phonon lines. According to this estimate anthracene–PMDA belongs to the class of materials which should show self-trapping of the CT exciton states at liquid helium temperatures. However, it must be pointed out that these estimates of the model parameters were derived from luminescence data and do not agree completely with those derived from reflection data (Brillante and Philpott 1980).

The origin of this discrepancy may be the simplistic one-coordinate model used in the interpretation of the data. The essential features of the model which is based on the Franck–Condon picture are depicted in fig. 8. The generalized coordinate Q is assumed to couple to the electronic transition to phonons in both the absorption and the emission. The various parabolic energy surfaces correspond to the ground state (0) and intramolecular vibrational levels of the ground state $v_1, v_2, \ldots$. The corresponding excited state surfaces are denoted by 0–0, 0–(0 + v_1), etc. The binding energy, namely the energy difference between the excited state with the Franck–Condon maximum and the phononless level, is given by δ. Figure 8 is based on the following simplifying assumptions:

(i) Linear electron–phonon coupling, i.e., the curvature of the parabola in the ground and excited states is identical. This assumption implies that there is no difference in force constants for the ground and excited states. This assumption is certainly a gross simplification. The introduction of quadratic electron–phonon coupling might explain some of the experimental discrepancies. A different curvature of the energy parabolae in the excited state would lead to different S values for the absorption and emission process as is observed experimentally.

(ii) The assumption that the ground state vibrational energy parabolae $v_1, v_2, \ldots$ are parallel to the vibrationless states 0 and 0–0 is also a simplifying approximation. This latter assumption, however, is expected to hold quite well.

(iii) The model does not take into account any distortion of the energy surfaces that can occur during the relaxation of the system from 0–0 + δ to 0–0.

Further discussion of the above features and their incorporation into the exciton picture goes well beyond the scope of this chapter and would, in some aspects, even go beyond the presently available theories of excitons in weak CT crystals.

5.2. Band theory with phonons

The Hamiltonian of the one-dimensional chain (Agranovich and Zakhidov 1977, Brillante and Philpott 1980) can be written

$$\mathscr{H} = \mathscr{H}_{ex} + \mathscr{H}_{phonon} + \mathscr{H}_{int}. \tag{18}$$

The simplest part of $\mathscr{H}$ is the phonon part, namely

$$\mathscr{H}_{phonon} = \sum_q \hbar\omega_q(a_q^+ a_q + \tfrac{1}{2}), \tag{19}$$

where q is the wavevector and a (a^+) is the phonon annihilation (creation) operator.

As was described earlier in section 3, in the site representation a CT exciton basis state corresponds to a positively charged donor (D) molecular ion on site n and a negatively charged acceptor (A) molecular ion on site m. Each CT configuration corresponds to an ordered pair $l = (n, m)$. For a chain on which the electron and hole are always on adjacent molecules, we use the alternative representation $l = n\sigma$, where $\sigma = \pm 1$ is the vector from n to m.

If we restrict exciton transfer to nearest neighbors then only the terms M_e and M_h are nonzero (these electron and hole transfer interactions are defined by eqs. (8) and (9) and are shown schematically in fig. 6) and the Hamiltonian $\mathscr{H}_{ex}$ in the site representation is given by

$$\mathscr{H}_{ex} = \sum_{n,\sigma} \epsilon_\sigma b_{n\sigma}^+ b_{n\sigma} + \sum_n [M_e(b_{n,-1}^+ b_{n,1} + b_{n,1}^+ b_{n,-1})$$
$$+ M_h(b_{n-1,1}^+ b_{n+1,-1} + b_{n+1,-1}^+ b_{n-1,1})], \tag{20}$$

where b (b^+) are CT exciton annihilation (creation) operators, and ϵ_σ is the vertical excitation energy, corrected for the Coulombic and polarization interactions. The latter may also contain the effect of an externally applied static field E. Thus,

$$\epsilon_\sigma = I_D - E_A + \Delta_C + P - \boldsymbol{\mu}_\sigma \cdot \boldsymbol{E}, \tag{21}$$

where I_D is the ionization energy of the donor, E_A is the electron affinity of the acceptor, Δ_C is the Coulomb interaction between adjacent molecular ions (D^+A^-), P is the lattice polarization energy and $\boldsymbol{\mu}_\sigma$ the static dipole moment of the excited state D^+A^- corresponding to the pair (n, σ).

The interaction Hamiltonian is given by

$$\mathcal{H}_{\text{int}} = \sum_{n,\sigma} b^+_{n\sigma} b_{n\sigma} \sum_q \hbar\omega_q c_{n\sigma}(q)(a^+_q + a_{-q}), \tag{22}$$

where

$$c_{n\sigma}(q) = -N^{-1/2}\gamma_q \exp[-iq(n + \tfrac{1}{2})a], \tag{23}$$

and

$$\gamma_q = V'_C\, 2i \sin(\tfrac{1}{2}qa)/(2m\hbar N\omega_q^3)^{1/2} \tag{24}$$

is the coupling constant containing the derivative V'_C of the Coulombic interaction with respect to displacements along the chain. For simplicity we have assumed that the donor and acceptor molecules have the same masses. Note that $\mathcal{H}_{\text{int}}$ is almost the same as the conventional deformation interaction Hamiltonian of neutral exciton–phonon coupling theories (Davydov 1971, Fischer and Rice 1970, Grover and Silbey 1970).

The Hamiltonian $\mathcal{H}$, eq. (18), is partially diagonalized by a canonical transformation,

$$\tilde{\mathcal{H}} = e^Q \mathcal{H}\, e^{-Q}, \tag{25}$$

where

$$Q = \sum_{n,\sigma} b^+_{n\sigma} b_{n\sigma} \sum_q c_{n\sigma}(q)(a^+_q - a_{-q}). \tag{26}$$

This procedure eliminates the main effects of the strong interaction $\mathcal{H}_{\text{int}}$. The diagonalization process is carried one step further by transforming from local site operators to delocalized operators,

$$b_{k\sigma} = N^{-1/2} \sum_n b_{n\sigma} \exp(-ikna). \tag{27}$$

After we neglect certain small four-point operators, the crystal Hamiltonian (18) can be written

$$\tilde{\mathcal{H}} = \tilde{\mathcal{H}}_{\text{CT}} + \mathcal{H}_{\text{phonon}}, \tag{28}$$

where

$$\tilde{\mathcal{H}}_{\text{CT}} = \sum_k \sum_{\sigma,\sigma'} b_{k\sigma} b_{k\sigma'} \{\delta_{\sigma\sigma'}(\epsilon_\sigma - \Delta\epsilon_1)$$
$$+ (1 - \delta_{\sigma\sigma'})(M_e + M_h \exp[ik(\sigma' - \sigma)a]) \exp(-S)\}. \tag{29}$$

Here

$$\Delta\epsilon_1 = N^{-1} \sum_q \hbar\omega_q |\gamma_q|^2 \tag{30}$$

is the deformation or relaxation energy resulting from the distortion of the lattice around the CT exciton and e^{-S} is a generalized Franck–Condon factor that describes the compression of the CT exciton band. Notice that in this simple model S is the same for both the electron and hole transfer interactions. The explicit equation for S is

$$S = N^{-1} \sum_q |\gamma_q|^2 (1 - \cos qa). \tag{31}$$

In deriving eq. (29) we took the ensemble average corresponding to a temperature of absolute zero.

We note that apart from the Franck–Condon factor e^{-S} the Hamiltonian is the same as that considered in Haarer et al. (1975). It has energy eigenvalues

$$E_{k\xi} = (\epsilon - \Delta\epsilon_1) \pm [|V_{2k}|^2 + (\boldsymbol{\mu}_1 \cdot \boldsymbol{E})^2]^{1/2}, \tag{32}$$

where $\xi = \pm$, $\boldsymbol{\mu}_1 = \boldsymbol{\mu}_\sigma (\sigma = +1)$,

$$\epsilon = I_D - E_A + \Delta_C + P, \tag{33}$$

$$V_{2k} = e^{-S}(M_e\, e^{ika} + M_h\, e^{-ika})\, e^{ika}. \tag{34}$$

To determine the excitation spectrum of $\mathcal{H}$, eq. (28), one needs to calculate the imaginary part of the retarded Green's operator $G_R(k\xi, \omega)$ for the creation and annihilation operators of the completely diagonalized form of $\mathcal{H}_{CT}$. This is a more involved calculation that will not be pursued here. Similarities with the neutral exciton coupling problem are sufficiently great for us to anticipate that the results of this calculation in its simplest form will be similar to those obtained for the neutral case; see, for example, eq. (33) given by Fischer and Rice (1970) or eq. (50) given by Grover and Silbey (1970). The optical absorption spectrum consists of a sharp ZPL followed by a sequence of multiphonon transitions. However, there will be one important difference caused by the fact that in our problem the phonon belongs to the acoustic branch and is not a dispersionless optical phonon like the ones considered in the theories of neutral molecular excitons. The main consequence of dispersion will be a more rapid broadening of the optical absorption spectrum of the multiphonon peaks. Indeed, this offers a ready explanation of the width of the fundamental phonon mode of $30\ \text{cm}^{-1}$ observed in reflection spectra (Brillante and Philpott 1980) and of why the multiple phonon peaks are poorly resolved. In real crystals, factors contributing to the existence of a broadened one-phonon peak are (i) the peak in the density of states near the boundary of the first Brillouin zone, and (ii) the absence of perturbing low frequency optical modes (librations).

A more complete treatment of exciton–phonon coupling should include both the optical librational modes as well as the intramolecular modes of the donors and acceptors. The intramolecular vibrations are not likely to change either the overlap of donor and acceptor wavefunctions, or the distance between molecules. The appearance of particular intramolecular modes will be due to changes in bond order caused by the departure or arrival of the electron. Librations represent an intermediate case since they can change overlaps between adjacent donors and acceptors in the excited state and consequently may appear prominently in absorption spectra. However, their role in contributing to lattice distortions in which the ions move closer together would seem to be of secondary importance. Strictly speaking, this last statement is true only for "spherical" ions. For planar ions relaxation in the excited state could also involve slippage or rotation of neighbors past one another. We should not, therefore, take the simple models given too seriously.

6. *Optical experiments on the singlet states of CT crystals*

The intensities and therefore the oscillator strength of CT transitions in the mixed stack systems are observed to vary greatly (Philpott and Brillante 1979, Tanaka and Tanaka 1979). In phenanthrene–PMDA the first singlet transition is very weak, the crystals being pale yellow in color. Others are highly colored, anthracene–PMDA is deep red, anthracene–TCNQ is midnight blue. For the more intensely absorbing solids the optical properties cannot be completely understood without the use of polariton theory, i.e., the coupling of the exciton to the photon must be properly described. Polaritons are the coupled exciton–photon states and as such may be regarded as the electromagnetic normal modes of insulators. We commence this section with a brief review of the optical properties of anisotropic solids, using the language of polariton theory.

6.1. *Polaritons in CT crystals*

In this section, we briefly review the properties of excitons and polaritons in crystals of the orthorhombic class (Mills and Burstein 1974, Tosatti and Harbeke 1974, Koch et al. 1974). For simplicity, the plane of incidence of the external exciting light is assumed to contain two crystallographic axes. Let the crystal axes, a, b, c, be parallel to the laboratory system, x, y, z; then, in the laboratory system, the dielectric tensor is diagonal,

$$(\epsilon)_{\alpha\beta} = \epsilon_\alpha \delta_{\alpha\beta}, \tag{35}$$

where $\alpha = x, y, z$. If the main contributions to the dielectric come from single-particle exciton transitions, then a useful phenomenological expression for ϵ_α ($\alpha = x, y, z$) is

$$\epsilon_\alpha(\omega) = \epsilon_{\alpha\infty} + \sum_{j=1}^{N} \frac{\omega_{j\alpha}^2 f_{j\alpha}}{\omega_{j\alpha}^2 - \omega^2 - i\omega\gamma_{j\alpha}(\omega)}, \tag{36}$$

where $\gamma(\omega)$ is the phenomenological damping, which may be a function of frequency, and f is the oscillator strength of the exciton transition. It should be pointed out that f is not the true oscillator strength; this quantity is actually given by $F = (\omega_T/\omega_P)^2 f$, where ω_P is the so-called plasma frequency defined by $\omega_P^2 = 4\pi e^2/(m_e \nu_c)$ and ν_c is the volume of one unit cell.

If we designate by k the three-dimensional crystal wavevector, then Maxwell's equations for electromagnetic wave propagation within the bulk solid yield the following well-known equation for the electric field intensity:

$$\sum_\beta \left[\left(k^2 - \frac{\omega^2}{c^2} \epsilon_\alpha \right) \delta_{\alpha\beta} - k_\alpha k_\beta \right] E_\beta = 0. \tag{37}$$

The condition for a nontrivial solution of this last equation gives the famous Fresnel equation,

$$\det \left| \left(k^2 - \frac{\omega^2}{c^2} \epsilon_\alpha \right) \delta_{\alpha\beta} - k_\alpha k_\beta \right| = 0. \tag{38}$$

We take the plane of incidence external to the crystal to be the xz plane. Then all polaritons in the crystal excited by the external light wave will only propagate in the xz plane. In this case, the Fresnel equation factorizes into two separate equations which are referred to as the equations for ordinary and extraordinary polariton modes. The ordinary polaritons have E vectors parallel to the y axis and are therefore TE modes (transverse electric or s-polarized). The extraordinary polaritons, on the other hand, have E vectors parallel to the xz plane and are TM modes (transverse magnetic or p-polarized).

6.1.1. Ordinary exciton polaritons
The dispersion equation for the ordinary polariton is

$$k^2(\omega) = \frac{\omega^2}{c^2} \epsilon_y(\omega). \tag{39}$$

The transverse ordinary excitons have frequencies corresponding to the poles of $\epsilon_y(\omega)$, namely, the frequencies, $\omega_T = \omega_{jy}$ in eq. (36). The longitudinal ordinary excitons occur at the zeros of $\epsilon_y(\omega)$. In the case of an isolated exciton transition, with oscillator strength f, the longitudinal frequency occurs at

$$\omega_L = \omega_T(1 + f/\epsilon_{\infty y})^{1/2}. \tag{40}$$

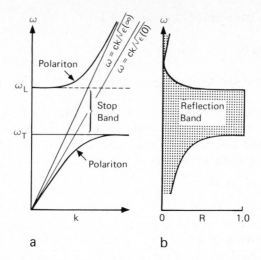

Fig. 9. Schematic diagram showing the interrelationships between the polariton stop band and the reflection band. In (a) the straight lines represent the asymptotes to the polariton dispersion curve for high and low frequency and at the transverse and longitudinal frequencies ω_T and ω_L.

Figure 9 shows schematically the relation between the polariton dispersion relation calculated from eq. (39) and the reflection band for normal incidence $\theta_0 = 0$ from eq. (39), assuming there to be a single undamped exciton transition at ω_T. Note that for this simple example the polariton has two branches. In coupling the exciton $\omega = \omega_T$ with the photon $\omega = ck/\sqrt{\epsilon(\infty)}$ an avoided crossing region, called the stop band, has been created. For an undamped exciton ($\gamma = 0$) the reflectivity is unity for the polariton stop band $\omega_T \leqslant \omega \leqslant \omega_L$.

6.1.2. Extraordinary exciton polaritons

The dispersion relation for the extraordinary exciton polariton can be reduced to the following equation:

$$c^2 k^2/\omega^2 = \epsilon_x \epsilon_z (\epsilon_z \cos^2\phi + \epsilon_x \sin^2\phi)^{-1}, \tag{41}$$

where ϕ is the angle between the wavevector k and the unit vector $\hat{z}$ parallel to the z axis. An explicit formula for $\cos\phi$ can be obtained from the equation

$$k \cos\phi = (\epsilon_x/\epsilon_z)^{1/2} \left(\frac{\omega^2}{c^2}\epsilon_z - k_x^2\right)^{1/2}, \tag{42}$$

where $k_x = k \sin\phi$ is fixed by Snell's law to be equal to the x component of the momentum of the photon incident on the crystal. If the outside medium is isotropic and has refractive index n_0, then

$$k_x = n_0 \frac{\omega}{c} \sin \theta_0, \tag{43}$$

where θ_0 is the angle of incidence of the photon striking the xy crystal surface.

The extraordinary exciton energies $\hbar\omega_e$ correspond to the poles in the right-hand side of eq. (41). For the special case of one isolated exciton transition in $\epsilon_x(\omega)$ and no transition at all in $\epsilon_z = \epsilon_{z\infty} = \epsilon_{x\infty} = \epsilon_\infty$ a constant, the extraordinary exciton frequency is given by

$$\omega_e = \omega_T[1 + (f \sin^2\phi/\epsilon_\infty)]^{1/2}, \tag{44}$$

where $\omega_T \leq \omega_e \leq \omega_L$. Equation (44) shows that the extraordinary exciton frequency exhibits orientational dispersion, i.e., the frequency depends on the angle ϕ between k and $\hat{z}$. Orientational dispersion has not yet been observed for the CT transition in anthracene–PMDA or for any other CT crystal. However, it most certainly exists and could probably be measured easily for the CT solids studied by Tanaka and Tanaka (1979), since these have very intense CT transitions.

6.1.3. Reflection of light from orthorhombic crystals

In the vicinity of an exciton transition the reflectivity of crystals can be large, making transmission spectroscopy a poor tool for monitoring exciton transitions. If the plane of incidence is the xz plane then for s-polarized (or TE) light, the crystal reflectivity is given by the squared magnitude of

$$r_s = -\frac{(\epsilon_y - \epsilon_0 \sin^2\theta_0)^{1/2} - \cos \theta_0 \sqrt{\epsilon_0}}{(\epsilon_y - \epsilon_0 \sin^2\theta_0)^{1/2} + \cos \theta_0 \sqrt{\epsilon_0}}, \tag{45}$$

where θ_0 is the external angle of incidence. For p-polarized (or TM) light, the crystal reflectivity is more complicated because of the anisotropic response of the crystal to external fields. The p-polarized reflectivity (Koch et al. 1974) is equal to the squared magnitude of

$$r_p = \frac{(\epsilon_x\epsilon_z/\epsilon_0)^{1/2} \cos \theta_0 - (\epsilon_z - \epsilon_0 \sin^2\theta_0)^{1/2}}{(\epsilon_x\epsilon_z/\epsilon_0)^{1/2} \cos \theta_0 + (\epsilon_z - \epsilon_0 \sin^2\theta_0)^{1/2}}. \tag{46}$$

6.2. The polar nature of the excited state

It is known from the early experiments on CT complexes that the low lying CT transitions show little or no structure compared to the well-resolved vibronic structure which is characteristic of electronic transitions of aromatic molecules in the gas phase, solution or in the crystalline state. A typical example is given by the absorption data of fig. 10a, which are taken

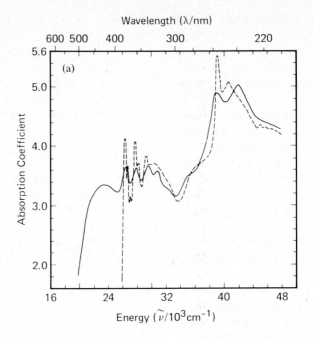

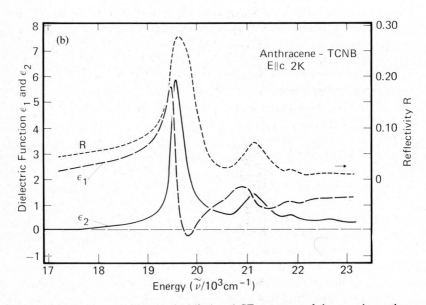

Fig. 10. (a) Molecular transition (dashed line) and CT spectrum of the complex anthracene–TCPA (solid line) (adapted from Czekalla 1959) in solution. (b) The reflection spectrum R and dielectric function $\epsilon_1 + i\epsilon_2$ of the crystalline CT complex anthracene–TCNB (tetracyanobenzene).

from a paper of Czekalla (1959). The dotted line shows a superposition of the molecular absorptions of the acceptor TCP (tetrachlorophthalic anhydride) and the donor anthracene in solid solution (propylether, methylcyclohexane). The data show well-resolved vibrational structures at frequencies corresponding to molecular transitions with linewidths which are typical for noncrystalline hosts ($\Delta \bar{\nu} \approx 400{-}800 \, cm^{-1}$). The most remarkable feature of fig. 10a is the lack of structure in the low lying CT band. This lack of structure was discussed by Czekalla et al. (1957) and was attributed to statistical fluctuations of the average complex distance. The authors argued that a variation of the equilibrium distance of 0.1 Å would change the CT transition energy by as much as $100 \, cm^{-1}$. This argument, which rationalizes a large inhomogeneous broadening, is applicable to CT transitions in glasses and in solution, but not to CT transitions in crystals.

In CT crystals there is a well-defined geometry and one expects that the inhomogeneous broadening due to fluctuations in the separation between donor and acceptor is reduced to a very small value. By analogy with results observed for transitions of comparable intensity in crystalline naphthalene etc., one expects linewidths of several wave numbers, and one would hence expect quite narrow absorption and emission lines. The first low temperature optical data on a crystalline CT complex anthracene–TNB (Lower et al. 1961, Hochstrasser et al. 1964) showed, however, that the absorption and emission bands were broad (hundreds of wave numbers). From these data it was evident that the CT solid is not adequately described by a rigid lattice model (without phonons) used to describe single-component crystals. The possibility of a localization of the CT exciton was described as a basis for a qualitative evaluation of the optical data (Lower et al. 1961, Hochstrasser et al. 1964). Subsequent studies of a large number of CT crystals have shown that with one exception all singlet transitions are broad with little if any resolvable structure. The exception anthracene–PMDA will be discussed in more detail below. An example of the general broad type spectrum obtained by reflection spectroscopy is shown in fig. 10b. Note that there is no vibronic structure in the imaginary part ϵ_2 of the dielectric function, so that the absorption in the crystal has the same general characteristic as the solution spectrum of fig. 10a.

As has been pointed out before, there are two common line broadening mechanisms which one has to consider: the static inhomogeneous broadening, which will dominate the linewidth in disordered media (glasses, polycrystalline materials), and the dynamic broadening, which is basically a consequence of the Franck–Condon principle and which was discussed qualitatively by Mulliken (1952). The mechanism of the dynamic broadening is depicted in fig. 7. If one assumes that Q is an intermolecular coordinate, such as, for instance, the donor–acceptor distance, then a large change in the excited state equilibrium configuration leads to a considerable Stokes shift between the absorption and emission bands as has been observed experimentally in numerous mixed stack CT crystals. The first organic CT material for which this Stokes shift was evaluated in a quantitative fashion by evaluating the first CT zero phonon spectra was anthracene–PMDA; its optical spectra will be discussed in the next section.

There is a third mechanism which can, in principle, give rise to large optical linewidths. If optical transitions are allowed throughout the whole Brillouin zone, then the ground and excited state dispersion contribute to the optical linewidth as depicted in fig. 11 (upper part). For this mechanism

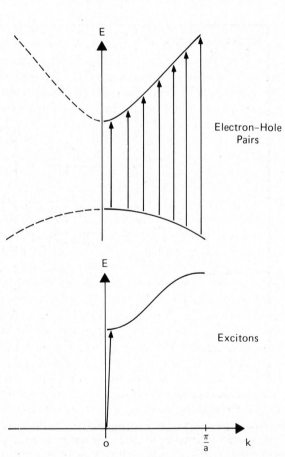

Fig. 11. Optical selection rules for electron–hole pair creation (top) and for the creation of an exciton with small wavevector ($k \approx 0$) at the Brillouin zone center (bottom).

to hold, one has to create a two-particle state like a free electron–hole pair (semiconductors) or an exciton–vibrational exciton state. In this case the selection rules can be fulfilled for a $k \approx 0$ phonon throughout the Brillouin zone. If, however, one concentrates on pure excitonic states, as will be done in subsequent sections, one can assume a strict $k = 0$ selection rule and thus neglect the k state broadening. This situation is depicted schematically in the lower half of fig. 11.

6.3. Singlet CT excitons in anthracene–PMDA

Figure 12 shows the reflection spectrum of the first singlet of anthracene–PMDA for incident light polarized parallel to the stack axis (Brillante and

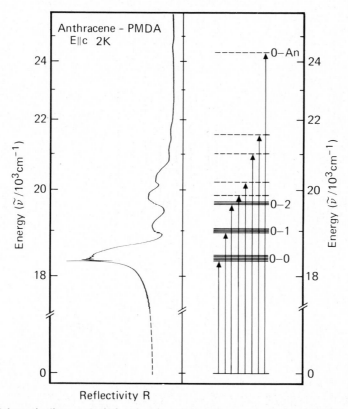

Fig. 12. Schematic diagram depicting the vibronic transitions of anthracene–PMDA from the ground state and their relation to structure in the 2 K reflection spectrum. The major band systems are classified 0–0, 0–1, 0–2, . . . , etc., corresponding to the intramolecular vibronic transitions of the excited electronic state. The transition 0–An is probably a neutral Frenkel exciton transition involving the donor molecule anthracene.

Philpott 1980, Brillante et al. 1978, Philpott and Brillante 1979, Philpott 1980). Also shown is a scheme for labelling the vibrational fine structure. Compare this spectrum with the one shown in fig. 10b, for anthracene–TCNB (tetracyanobenzene), which has no vibronic fine structure.

The first basic assumption is that each of the regions 0–0, 0–1, 0–2, etc. correspond to a vibronic CT transition in which there is simultaneously an electronic CT transition and an excitation of intramolecular vibration modes of either anthracene or PMDA or both. This assumption is supported by the observation that the zero lattice phonon line (ZLPL) $0-1_0$ coincides with $0-0_0 + 635 \text{ cm}^{-1}$, a known vibrational mode of PMDA. The 0–2 band is really composite since several excited state counterparts of the Raman modes of anthracene and PMDA are excited. Thus although region 0–2 appears at first glance as a set of lattice phonons with $0-2_0$ as origin, it really consists of many superimposed series based on several origins. In

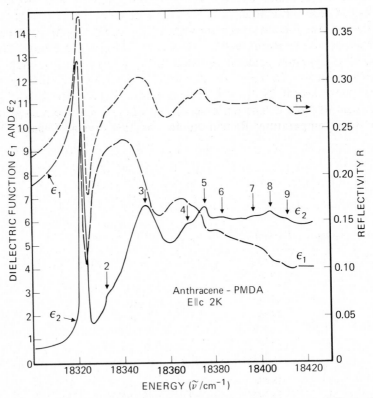

Fig. 13. Detail of the reflectivity R of anthracene–PMDA single crystal at 2 K, together with the real and imaginary parts of the dielectric function $\epsilon(\omega) = \epsilon_1(\omega) + i\epsilon_2(\omega)$ near the zero phonon line of the 0–0 region.

passing we note that the existence of multiple origins can also arise for structural reasons. For example in the weak CT crystal HMB–TCNE (hexamethylbenzene tetracyanoethylene) there are different CT origins because there are several geometrically different sets of complexes (Eckhardt and Hood 1980). Each of these is expected to have different potential energy surfaces. The second basic assumption made in the analysis is that the shape of the 0–0 and 0–1 regions and in part at least the 0–2 region is due to the participation of lattice phonons and librons in the CT process. In other words, there are vibronic transitions that cannot be assigned to intramolecular modes. The involvement of librons is made possible in solids consisting of large anisotropic molecules in face-to-face contact, since rocking and sliding motions can be as effective in achieving relaxation in the excited state as translational displacements.

The reflection spectrum of the zero phonon region of the anthracene–PMDA at 2 K is shown in fig. 13, along with the real and imaginary parts of the dielectric function, $\epsilon_1(\omega)$ and $\epsilon_2(\omega)$, obtained by Kramers–Kronig analysis of the reflectivity data (Brillante and Philpott 1980). The zero phonon line 0–0_0 at approximately 18 320 cm^{-1} is by far the sharpest and most intense feature in the reflection spectrum and relative to this line there are phonon peaks at 27, 47, 53, 62, 83, 111 and 142 cm^{-1} in the reflection spectrum. In the $\epsilon_2(\omega)$ "absorption" spectrum the corresponding phonon peaks are at 29, 47, 55, 62, 111 and 142 cm^{-1}. These results are summarized in table 2. The peaks at 47 and 62 cm^{-1} may be related to the known low temperature Raman modes of the crystal (Haarer 1977).

Table 2
Summary of structure in the $\epsilon_2(\omega)$ absorption spectrum for the 0–0 region of anthracene–PMDA.

Band	$\hbar\omega_T$ (cm^{-1})	Difference $\hbar\omega_T - \hbar\omega_{0-0_0}$	Comments
1	18 321	0	zero phonon line 0–0_0
2	18 334	≈ 13	weak shoulder
3	18 347	26	weak shoulder
4	18 350	29	max.
5	18 363	42	shoulder
6	18 369	47	shoulder
7	18 376	55	max. $2 \times \nu_4$
8	18 383	62	max.
9	18 398	≈ 77	weak shoulder
10	18 404	83	max. $3 \times \nu_4$
11	18 409	88	shoulder
12	18 434	111	max. $4 \times \nu_4$
13	18 463	142	max. $5 \times \nu_4$

However, the lowest peak at 29 cm^{-1} and its multiples at 55, 83, 111 and 142 cm^{-1} have no known counterparts in either the normal or pre-resonance Raman spectrum (Syme et al. 1979). Phonons of similar low frequency have also been detected in the 0–1 and 0–2 regions as will be described briefly later on. The zero phonon line 0–0$_0$ was first observed by Haarer (1977) in absorption measurement studies. Its existence allows quantitative estimates of the magnitude of exciton–phonon coupling constants to be made using various model theories, and is a clear indication of the polar nature of

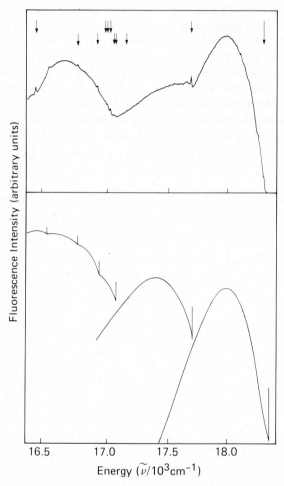

Fig. 14. 2 K fluorescence spectrum of anthracene–PMDA (top); symbolic deconvolution of the spectrum into electronic origin, vibrons and phonons (bottom). [This figure is adapted from Haarer (1977).]

the excited CT states. Although it is the reflection spectrum that shows the greatest detail, vibrational fine structure is also evident in the absorption and luminescence spectra of this crystal.

Figure 14 (upper part) gives a high resolution fluorescence spectrum of a single crystal of anthracene–PMDA at 2 K (Haarer 1974, 1977). The small, narrow peaks whose width is on the order of two wave numbers, are zero phonon transitions belonging to the electronic CT origin (long arrow) and to vibrational transitions of the donor anthracene (lower set of arrows) and the acceptor PMDA (upper set of arrows), respectively. A symbolic deconvolution of the entire spectrum into zero phonon peaks with adjacent broad phonon sidebands is given in the lower part of fig. 14.

The zero phonon origin of the emission spectrum of fig. 14 does not show up as a narrow peak, as one would expect from the simple con-volution consisting of purely electronic transitions, vibrational transitions and their corresponding phonon continua. It shows up as a little shoulder instead, which can be barely recognized in fig. 14 (upper part). The corresponding absorption spectrum, however (which is reproduced in fig. 15 together with the first part of the emission data), shows a very well-defined zero phonon origin at $18\,320\,cm^{-1}$. From these data one can conclude that the strong crystal reabsorption at the electronic origin is

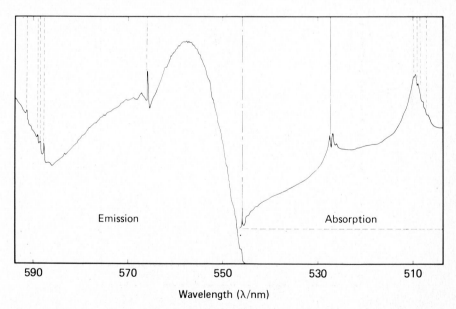

Fig. 15. 2 K absorption and fluorescence of anthracene–PMDA. The dashed lines show a vibronic mirror image situation with respect to the electronic origin (dash–dotted). [This figure is adapted from Haarer (1977).]

responsible for the suppression of the zero phonon peak in the emission spectra.

6.4. *Stark effect experiments*

A CT crystal, like anthracene–PMDA with one DA pair per unit cell and inversion symmetry along the stack axis, was depicted symbolically in fig. 2b. If one creates by light absorption a CT excited state, one has to take into account the symmetry of the crystal and therefore the excited state wavefunction must be an appropriate linear combination of D^+A^- pairs. There are two different manifolds of polar states, one made up from the site excitation functions $\ldots D^+A^- \ldots$ and the other from $\ldots A^-D^+ \ldots$. These are represented in fig. 6 by the kets $|I\rangle$ and $|II\rangle$. The crystal eigenstates are

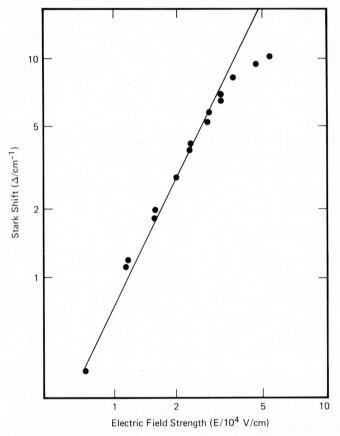

Fig. 16. Stark shift of the CT zero phonon line in anthracene–PMDA in the low and high field regimes. [This figure is adapted from Haarer et al. (1975).]

linear combinations like

$$\psi_{k\xi} = \sum_n [c_{k,+1}\psi_{n,+1} \exp(ikna) + c_{k,-1}\psi_{n,-1} \exp(-ikna)] \qquad (47)$$

($\xi = \pm 1$). The coefficients $c_{k,\pm 1}$ are obtained by diagonalizing the Hamiltonian [see eq. (29)]. The energies of the states at the zone center ($k = 0$) are

$$E_{k\pm} = \epsilon + V_1 \pm V_2, \qquad (48)$$

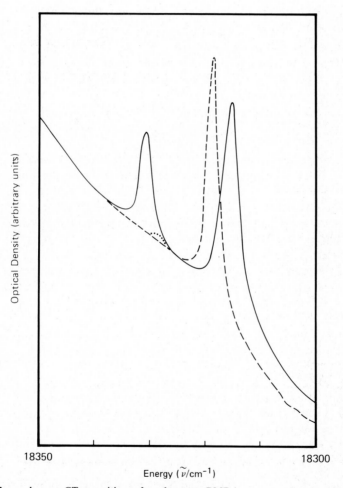

Fig. 17. Zero phonon CT transition of anthracene–PMDA at zero electric field strength (dashed) and at intermediate electric strength (solid line). [This figure is adapted from Haarer et al. (1975).]

Table 3.

Compound	$\Delta\mu$ (Debye)	$2V_{12} = 4M_1$ (cm^{-1})	Electronic bandwidth (cm^{-1})
Anthracene–PMDA	11 ± 1 (change in dipole moment)	10 ± 1 (Franck–Condon reduced bandwidth)	$\geqslant 1500$ cm^{-1} (not reduced by Franck–Condon factor)

where V_1 represents $2M_1$, a CT exciton migration matrix element as depicted in fig. 6a, and V_2 represents the sum of the two polarity flip matrix elements as depicted in figs. 6b and c (Haarer et al. 1978). An interesting aspect of the exciton migration in CT crystals is its polar nature which in principle can lead to unusual dispersion phenomena (Sakata and Nagakura 1970, Haarer et al. 1978).

The excited state energies in an electric field E are:

$$E_{k\pm} = (\epsilon - \Delta\epsilon_1) + V_{1k} \pm [|V_{2k}|^2 + (\boldsymbol{\mu}_1 \cdot \boldsymbol{E})^2]^{1/2}. \tag{49}$$

If the above quantum-mechanical picture of a delocalized excitonic excited state holds, one would expect a quadratic Stark effect at low electric fields with only one zero phonon transition being dipole allowed (ground state A_g to B_u); at high fields ($|\boldsymbol{\mu}_1 \cdot \boldsymbol{E}| > |V_{2k}|$) a linear Stark effect is expected with two zero phonon lines being dipole allowed. At this field strength the wavefunctions can be designated ψ_R and ψ_L. These symmetry adapted eigenfunctions describe physical states in which the charge is transferred either to the right (R) or left (L), respectively. Figure 16 shows the first experimental Stark data on a zero phonon line of a polar excited state. The figure shows a quadratic Stark effect at low electric fields (the straight line has a slope of two) and a bend-over to a linear effect at the highest experimental fields on the order of 10^5 V/cm (Haarer 1974, 1975). The quadratic shift and doubling of the zero phonon origin is shown in fig. 17. In the high field limit (not shown in the figure) both transitions have equal intensity. At this field strength the polar eigenstates ψ_L and ψ_R are the true eigenstates of the system. From the observed Stark shifts and from the zero field splitting, one gets the main parameters of the polar excited CT state. A compilation of the data is given in table 3.

6.5. Higher vibronic CT exciton transitions

Analysis of reflection spectra in the $0-n$ ($n = 1, 2, \ldots$) regions has revealed the existence of vibronic transitions that act as origins for series in lattice phonon vibrations (Brillante and Philpott 1980). The absorption spectra

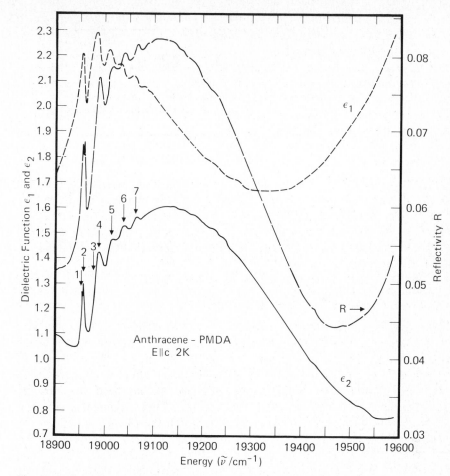

Fig. 18. Reflectivity R and the real and imaginary parts of the dielectric function $\epsilon(\omega) = \epsilon_1(\omega) + i\epsilon_2(\omega)$ for the 0–1 region of anthracene–PMDA crystal at 2 K.

derived by the Kramers–Kronig analysis show details parallel to those detected in the luminescence spectra. Figure 18 shows the 0–1 band. The origin $0–1_0$ is actually split, an effect also noticeable in its luminescence counterpart. In absorption this split may be due to two-particle bands superimposed on the allowed single-particle transition. The richness of these absorption and luminescence spectra promises to yield much detailed information in future investigations.

6.6. Summary of conclusions and suggestions for future experiments

From the results summarized in the last four subsections one can draw the following conclusions:

(i) The CT exciton has a very polar excited state with a comparatively large electron–hole separation (more than one lattice distance) but with zero electric dipole moment.

(ii) The CT exciton has a large electronic bandwidth due to the good molecular overlap between the tightly packed molecules in sandwich configuration (3.3–3.6 Å separation).

(iii) Exciton–phonon coupling plays an important role and reduces the excitonic bandwidth considerably. Whether or not the CT exciton is self-trapped at low temperatures is an open question.

A more detailed understanding of the anthracene–PMDA system will undoubtedly result from more refined experiments. We mention some of these here. Two-photon excitation profiles should reveal the presence of g-states, especially the forbidden $0–0_0^g$ component without the use of Stark spectroscopy. In addition progressions based on this dipole forbidden transition should be revealed. Raman spectroscopy in the pre-resonance and resonance region, in particular Raman excitation profiles for the librational modes, should reveal in more detail which nuclear motions are "turned-on" by the electron transfer. Fluorescence lifetime measurements on pure and isotopically labelled crystals should provide more information about energy conversion and the nature of trap states near the bottom of the CT exciton band. Brillouin scattering can be used to map out the polariton dispersion curve. The pressure dependence of the CT spectrum may reveal the physical parameters necessary to induce a neutral-to-ionic phase transition. We also anticipate that fundamentally important information will be obtained using modulation spectroscopies. Systematic piezomodulation of fluorescence and reflection spectra as have been reported recently (Eckhardt and Merski 1973, Merski and Eckhardt 1981) should contribute greatly to our understanding of the optical properties of this material. Electromodulation of the reflection spectra at 2 K should provide a wealth of detail concerning the degree of charge transfer especially for the congested higher vibronic transitions. This belief is encouraged by the report of Sebastian and Weiser (1979) for polymerized diacetylene crystals whose exciton transitions are believed to have a high percentage of CT character (Philpott 1977).

7. Triplet CT exciton states

7.1. Electronic structure of triplet CT states

In the previous sections singlet CT states were discussed. Singlet states are characterized by wavefunctions with antiparallel electron and hole spins, i.e. by spin $S = 0$. Singlet states correspond to allowed optical transitions and thus dominate the main features of the optical absorption and reflectivity spectra of CT crystals. It was also shown that singlet CT states are well described by polar excited states in which an almost complete electron transfer has occurred. Detailed experimental data on the crystal anthracene–PMDA were used to support the concept of a singlet CT exciton state.

The situation is more complicated for triplet states, spin forbidden in optical transitions, in which electron and hole spins are parallel and the total wavefunction is characterized by spin $S = 1$. For triplet CT states the two extreme situations that can occur are shown in fig. 19, which also shows the singlet manifold for reference. On the one hand, one can have "ionic" triplet states, whose wavefunction is dominated by polar contributions [$a^* \geqslant b^*$ in eq. (3)] (Hiyashi and Nayakura 1969, Briegleb and Wolf 1970). In these cases the CT phosphorescence spectra tend to be broad and structureless, quite similar to the majority of singlet CT spectra. On the other hand, there are examples where the phosphorescence spectra of CT crystals are sharp and structured (Haarer and Karl 1973, McGlynn et al. 1960) and correspond to triplet states "localized" on the donor (or acceptor) molecule. For these triplet states one can assume $a^* \ll b^*$ [see eq. (3)]. It is of interest to note that for the acceptor TCNB (tetracyanobenzene), a variation of the donor molecules from anthracene to pyrene to phenanthrene leads from the localized molecular triplet state limit to the triplet CT exciton limit (Möhwald and Sackmann 1974, Dalal et al. 1976).

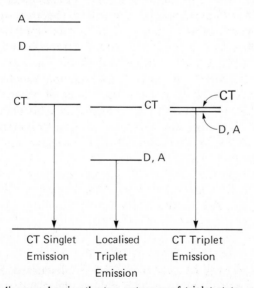

Fig. 19. Schematic diagram showing the two extremes of triplet states encountered in weak CT crystals. At left is the singlet manifold for reference.

This implies that the triplet states in weak CT crystals can have varying degree of charge transfer character depending on the separation of the donor (acceptor) triplet levels from the CT levels. Figure 19 shows schematically the energy level diagram for the singlet manifold with the localized singlet states of D and A well above the CT excited state. The position of the triplet CT level is always quite close to the position of the singlet CT level; it is slightly lower in energy due to the exchange interaction between the two electron spins which tends to minimize the Coulombic energy. The total singlet–triplet splitting, however, is small because of the large distance between the electron and the hole in a CT state.

If the triplet state of the donor or acceptor molecule is much lower than the CT level, the emitting triplet state will have a wavefunction that resembles the triplet state of the donor or acceptor molecule and hence be nonpolar in nature (see fig. 19). These are the cases where the CT phosphorescence spectra are sharp and structured (Haarer and Karl 1973, McGlynn et al. 1960). The wavefunction is primarily nonionic and corresponds to a local donor or acceptor wavefunction with $a^* \ll b^*$ in eq. (3). Anthracene–PMDA is a good example for this case because the triplet level of anthracene lies near $16\,000\ \text{cm}^{-1}$, well below the singlet CT level at $18\,300\ \text{cm}^{-1}$.

In the sequence anthracene–PMDA, naphthalene–PMDA, phenanthrene–PMDA, the triplet level of the donor molecule rises to the vicinity of the CT level (see fig. 19). By changing the donor we can vary the character of the wavefunction from a "localized" molecular state to a polar delocalized state with CT character [$a^* \geqslant b^*$ in eq. (3)]. Another experimentally well-investigated series is TCNB (tetracyanobenzene) with the donor molecules anthracene, pyrene and phenanthrene. This series also shows a graduation from localized molecular states to a delocalized CT triplet state in accordance with the aforegoing discussion (Möhwald and Sackmann 1974, Dalal et al. 1976).

CT systems with anthracene as donor molecule belong to the class of localized triplet states for all CT combinations which have been examined so far. Hence their lowest triplet states are only slightly perturbed by CT interactions (Haarer and Karl 1973) and therefore the phosphorescence spectra show mostly ZPL character. A quite interesting spectroscopic feature of localized CT states is the observed blue shift of the phosphorescence origin (McGlynn et al. 1960, Haarer et al. 1977). This blue shift was accounted for by the CT stabilization of the ground state. A theoretical calculation of the observed CT shift (Haarer et al. 1977) was possible and illustrates that the Mulliken theory (1950, 1951) can, often, yield useful semiquantitative results.

One CT system, naphthalene–TCPA (tetrachlorophthalic anhydride), shows that the details of the CT interactions are far from being understood. It is an example of a system in which small perturbations can lead to a large change in CT character (Haarer 1979). The naphthalene–TCPA crystal shows two completely different phosphorescence spectra stemming from two different trap states (see fig. 20). One spectrum, which dominates the phosphorescence of purified crystals (fig. 20, top spectrum), shows intermediate to strong exciton–phonon coupling. This strong coupling is indicative of a large CT in the excited state and hence the emission resembles the singlet spectra of anthracene–PMDA (see above). The un-purified naphthalene–TCPA crystals, however, show a ZPL-type phos-phorescence which is due to a different trap (Haarer 1979). This second

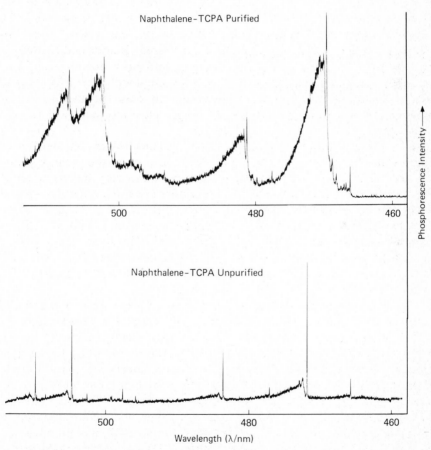

Fig. 20. Phosphorescence spectra of two different trap states in naphthalene–TCPA. The top spectrum shows intermediate exciton–phonon coupling; the bottom spectrum shows weak exciton–phonon coupling.

trap represents a weakly phonon coupled spectrum and hence a triplet state with small ionicity.

The large differences between the two trap spectra reflects the sensitivity of the CT wavefunction to geometry changes. It is expected that small changes of CT character lead to large changes of the electron–phonon coupling parameters. It is also expected that pressure changes and electric fields will strongly effect the triplet spectra of CT solids.

The above example demonstrates, that our present knowledge of CT complexes is frequently limited by the lack of detailed knowledge of donor–acceptor configurations and of overlap integrals [eq. (4)]. Some of these details are accessible by evaluating EPR spin density data, as will be described in the following section.

7.2. EPR and spin density experiments

Spin resonance data are extremely useful to map out spin densities in the excited triplet states of CT complexes. Therefore, they can be used to shed light on the charge distribution, i.e. the polarity of excited triplet states. This is symbolically shown in fig. 21 for a "molecular" triplet state (upper part of fig. 21) and for a CT triplet state (lower part); in the latter configuration both spins are localized in an ion pair configuration (the molecules are PDMA and anthracene). The spin Hamiltonian which describes the triplet CT state has two main dipolar interaction terms. If

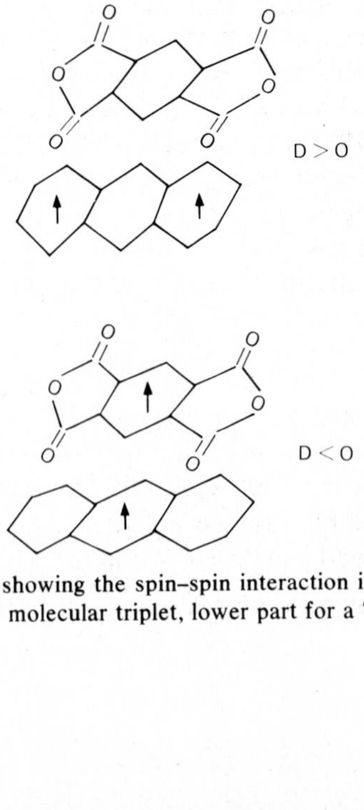

Fig. 21. Schematic diagram showing the spin–spin interaction in different geometries. Upper part for a "localized" molecular triplet, lower part for a "delocalized" CT triplet.

one neglects the nuclear Zeeman term it can be written as

$$H = \mu_B B \cdot g \cdot S + S \cdot F \cdot S + S \cdot A \cdot I. \tag{50}$$

μ_B is the Bohr magneton, g is the g-tensor and B is the external magnetic field. F and A are tensors which describe the electron spin–spin and the electron spin–nuclear spin interactions. In principle both the dipolar electron spin–spin interaction, which is given by the zero-field splitting tensor F, as well as the hyperfine tensor A reflect the excited state spin densities and charge densities if one neglects subtle spin polarization effects.

The F tensor arises from the dipolar interaction between the electron and hole spins. If the electron and hole wavefunction are localized on a "flat" molecule, both spins have a repulsive interaction, whose absolute value is given by their average distance. If, however, the two spins are in a sandwich-like CT configuration, they have an attractive dipolar interaction and in this case some of the tensor elements of F differ in sign.

As mentioned already, the electron–nuclear spin interaction is governed by the A tensor. Its diagonal part contains the probability distribution of the wavefunction at the various carbon atoms as measured through the dipolar proton–electron spin interaction. If one, for instance, reduces the electron density of a donor molecule by removing an electron from a π-orbital, one would expect a reduction of electron density at all the carbon atoms. Hence the dipolar electron–proton interaction is reduced and a smaller hyperfine splitting should be measured. Clearly the change in the splitting is a measure of the electron transfer to an acceptor molecule (Dalal et al. 1967).

In the following, we will briefly describe how the dipolar electron–electron spin interaction can be evaluated in a quantitative way. The zero field splitting (ZFS) problem has been discussed by several authors since 1967 (Möhwald and Sackmann 1974, Iwata et al. 1967, Beens et al. 1969, Keijzers and Haarer 1977). A full tensor evaluation (five independent tensor elements) was performed for phenanthrene–PMDA (Keijzers and Haarer 1977). In dealing with interactions of triplet states one must use two-electron wavefunctions.

If one assumes no charge transfer in the ground state one has the following wavefunction:

$$\psi(1, 2) = D(1)D(2), \tag{51}$$

where D is the HOMO of the donor molecule. For the excited state orbitals one has to consider a linear combination of LUMO's of the donor molecule (D^*) and of the acceptor molecule (A^*). If the overlap between D^* and A^* is neglected—as justified by a recent calculation (Haarer et al. 1977)—then the excited state two-electron wavefunction is

$$\psi^*(1, 2) = C_1|D(1)D^*(2)| + C_2|D(1)A^*(2)|, \quad C_1^2 + C_2^2 = 1, \tag{52}$$

where the normalization factor $2^{-1/2}$ is included in the Slater determinants. The above wavefunction corresponds, in a somewhat different notation, to the wavefunction proposed by Iwata et al. (1966). The first term in eq. (52) corresponds to a wavefunction representing the local donor triplet, the second term corresponds to an ionic wavefunction of the (A^-D^+) triplet state. The local PMDA triplet has been neglected since it is expected to lie considerably higher in energy and hence not to mix with the wavefunction of the lowest CT triplet state. (This assumption is certainly true for acceptor molecules like PMDA and for donor molecules like phenanthrene or naphthalene.)

In case one wants to consider a donor molecule which is surrounded by two acceptor molecules, the function A^* has to be replaced by $(A_I^* + A_{II}^*)/\sqrt{2}$, if again intermolecular overlap is neglected. An element of the experimental ZFS tensor $\mathbf{F}^{\text{exptl}}$ in the Hamiltonian (50) can be calculated as follows:

$$F_{ij} = \langle \psi^*(1,2)|H_{ij}^{\text{dip}}(1,2)|\psi^*(1,2)\rangle, \tag{53}$$

$$F_{ij}^{\text{exptl}} = C_1^2 F_{ij}^{\text{loc}} + C_2^2 F_{ij}^{\text{ion}}, \tag{54}$$

where the operator $H_{ij}^{\text{dip}}(1,2)$ represents the dipolar interaction of the two electron spins which make up the triplet state, $\mathbf{F}^{\text{loc}}$ is the ZFS tensor for a triplet state localized on the donor molecule, and $\mathbf{F}^{\text{ion}}$ is the tensor in case of complete charge transfer. Cross terms $\langle D(1)D^*(2)|H_{ij}^{\text{dip}}(1,2)|D(1)A^*(2)\rangle$ can be neglected because they are expected to be small compared with $\mathbf{F}^{\text{loc}}$ and $\mathbf{F}^{\text{ion}}$. So far the derivation of Hayashi et al. (1969) has been followed. In the following, however, a more detailed comparison between the calculated tensor elements and the experimental data will be made, which does not only focus on the principal values D and E of the dipolar interaction, but which also takes into account the different symmetry of the three tensors which appear in eq. (54). $\mathbf{F}^{\text{loc}}$ has the symmetry of the donor molecule, while $\mathbf{F}^{\text{ion}}$ reflects the symmetry of the entire CT complex. These symmetries are, in general, different. Therefore, the directions of the resulting principal axes will depend on the degree of charge transfer, and there is no reason why any one of the resulting axes should coincide with a symmetry axis of the donor molecule.

From the above considerations one can conclude that the complete ZFS tensor has to be measured, i.e. the principal values as well as the principal axes, in order to check the validity of eq. (54). Note that this equation is overdetermined since each tensor represents five parameters and since the whole equation has only one adjustable parameter (C_1). From eq. (54) one can also conclude that a large degree of CT can lead to a change of sign of "D" because of the different symmetry of the local triplet tensor and the ionic CT tensor.

The dipolar operator $H_{ij}^{dip}(1, 2)$ in eq. (53) is

$$H_{ij}^{dip}(1, 2) = \tfrac{1}{2}(g_e\mu_B)^2 \frac{3X_{12}^i X_{12}^j - R_{12}^2\delta_{ij}}{R_{12}^5},$$ (55)

where g_e is the free electron g value and X_{12}^i is the ith component of R_{12}. The ionic contribution to eq. (54) can be calculated with a point charge approximation, in which the spin density is assumed to be located on the nuclei. In this approximation F_{ij}^{ion} becomes

$$F_{ij}^{ion} = \tfrac{1}{2}(g_e\mu_B)^2 \sum_{p=1}^{N_D} \sum_{q=1}^{N_A} c_p^2 d_q^2 \frac{3X_{pq}^i X_{pq}^j - R_{pq}^2\delta_{ij}}{R_{pq}^5},$$ (56)

where c_p and d_q are coefficients in the donor HOMO and the acceptor LUMO, respectively, and R_{pq} connects atom p of the donor with atom q of the acceptor. The localized contribution (F^{loc}) cannot be calculated with a theoretical model as simple and straightforward as the point charge model. Therefore, it is simpler to take the experimentally determined donor (or acceptor) values (Brandon et al. 1954).

Using this approach the degree of charge transfer in the excited triplet state of phenanthrene–PMDA was determined to be $76 \pm 5\%$ (Keijzers and Haarer 1977). It is interesting to note that triplet wavefunctions of CT complexes can be analyzed in a more quantitative fashion than the corresponding singlet wavefunctions. This is due to the submolecular nature of the spin Hamiltonian and to the accuracy of EPR experiments. In principle, the hyperfine tensor A yields the same information. A complete evaluation of the proton hyperfine tensors, however, is, in most instances, more tedious than an evaluation of the ZFS tensor (for a more detailed comparison see Keijzers and Haarer 1977, Dalal et al. 1976).

8. Polarons in CT crystals

8.1. Relative position of the lowest conduction band and the CT exciton band

In the previous sections, it was shown that excited states in organic CT crystals are well described by wavefunctions which are delocalized over two or more molecules. This delocalization distinguishes CT crystals from other molecular crystals like anthracene and naphthalene, where the electron–hole pair is localized on one molecule. The larger extent of delocalization in CT materials puts them in a class between conventional molecular crystals, whose localized excited states are described by Frenkel excitons, and semiconductor crystals, whose delocalized excited states are described by Wannier excitons.

The comparison between CT states and semiconductor-like states is depicted symbolically in fig. 22. On the left-hand side it shows the optical absorption and the photoconductive response of an anthracene-like crystal (energy is plotted along the positive ordinate direction). As indicated in the figure, the strong singlet excitonic states with Frenkel character ($S_1, S_2, \ldots$) dominate the optical absorption spectra. It is interesting to note that the photoconductive response, whose onset is related to the energy of the conduction band (C.B.), is completely uncorrelated with the absorption spectrum. Both the optical transitions into the conduction band and the transitions into lower lying charge transfer states have very little absorption oscillator strength (Hanson 1973), and hence do not show up in the optical spectrum which is dominated by the strongly absorbing Frenkel states. Information on the delocalized band states can only be gained from photoconduction experiments capable of detecting a small number of free charge carriers in the presence of many excitonic states (Kepler 1960, LeBlanc 1960). Experiments of this kind based on the "time of flight" method, by observing the drift current of carriers, were first performed for anthracene and they established that the position of the conduction band is at approximately $4\,eV$, i.e. about $1\,eV$ above the lowest singlet exciton state (Castro and Hornig 1965, Chance and Braun 1976).

In CT crystals the energy level scheme changes in a way which makes the photoelectric response very different. This change is symbolized in the scheme on the right-hand side of fig. 22. Here the CT states have been

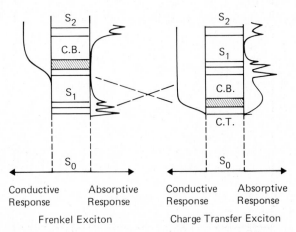

Fig. 22. Schematic diagram of the optical spectra and photoconduction excitation spectra of a typical organic molecular crystal (left) and of a typical CT solid (right). The CT solid shows similarities with direct band gap semiconductors. $S_0, S_1, S_2, \ldots$ signify the singlet ground and excited states, C.B. denotes the conduction band.

shifted to lower energy and are located below the molecular exciton states. At the same time, the CT transitions gain oscillator strength and become clearly visible in the optical spectrum. The numerical value for the oscillator strength is proportional to $(S_{DA})^2 z^2$, where S_{DA} is the overlap integral between the donor and acceptor orbitals involved [see eq. (4)], and z is the donor–acceptor distance at which S_{DA} has its maximum value (Mulliken 1950). As has been mentioned before, S_{DA} can be quite large for CT crystals due to the parallel, sandwich-like arrangement of donor and acceptor molecules which characterizes the mixed stack class of materials.

The excitation spectrum for photoconduction in CT crystals starts at much lower energies than in the single-component crystals, and its onset coincides with the onset of the optical absorption spectrum of the crystal, i.e. the CT exciton bands. This very important fact has been verified experimentally (Karl and Ziegler 1975). Figure 23 from Haarer and Möhwald (1977) shows the low energy tail of the optical absorption spectrum and of the photoaction spectrum of the CT crystal phenanthrene–PMDA. The onset of both bands practically coincides, a feature which resembles the situation encountered in the case of direct band gap semiconductors, where exciton transitions lie just below the conduction band edge. We

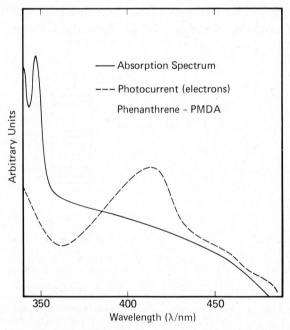

Fig. 23. Absorption spectrum and excitation spectrum of photocurrents as measured for phenanthrene–PMDA.

interpret the experimental results as follows. A photon which is being absorbed in the lowest CT band of the crystal creates a delocalized CT exciton (see the CT level shown in fig. 22). The optically excited CT exciton lies quite close to the lowest conduction band ($\Delta E < 0.5$ eV) and therefore the CT exciton can be thermally ionized before it decays to the ground state. The experimental evidence which supports this model is the fact that charge carrier production in phenanthrene–PMDA is thermally activated with an activation energy of about 0.5 ± 0.1 eV (Haarer and Möhwald 1977). This activation energy coincides roughly with the gap estimated to occur between the CT exciton band and the conduction band.

8.2. CT material with novel photoconductive properties

The close interrelation between polar excited states and electron–hole pair states in CT systems is the main reason for the important role which CT states play in the field of biology and in the area of technological applications of organic materials. It was a CT complex between the polymeric donor PVK (poly-*n*-vinylcarbazole) and the monomeric acceptor TNF (trinitrofluorenone) which constituted the first organic polymeric photoconductor and which was widely used in electrophotography (Schaffert 1971).

Since a detailed understanding of polymeric organic systems is quite involved and since the measured charge carrier mobilities in disordered media are dominated by complicated trapping phenomena (Gill 1972, Scher and Montroll 1975), we have chosen to discuss here some features of intrinsic photoconduction in CT solids using phenanthrene–PMDA as an example. This material is available as high purity single crystals. One aspect which seems to be unique for these systems is their one dimensionality, which is a consequence of their columnar structure (Herbstein 1971). Figure 24 shows the unit cell of phenanthrene–PMDA with two translationally inequivalent CT stacks; the stack axis makes an angle of 71° with the molecular planes; the intermolecular distance is about 3.5 Å (Evans and Robins 1977).

The first preliminary experiments on the photoconduction of crystalline CT complexes were performed in the mid-1960's by Tobin and Spitzer (1965) and Sharp (1967). These studies were not extensive enough to reveal details of the charge carrier transport mechanism at the molecular level. Only through a detailed knowledge of the electron–phonon coupling processes and of the mobility anisotropies can questions be answered as to whether the charge carriers in these materials can be described as small polarons (Fröhlich 1963, Holstein 1959). Detailed anisotropy data first became available for phenanthrene–PMDA (Möhwald et al. 1975) and anthracene–PMDA (Karl and Ziegler 1975) both available as single crystals of high purity. Both systems revealed

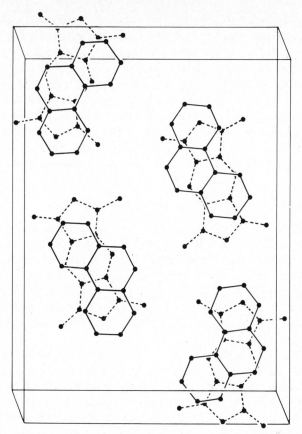

Fig. 24. Crystal structure of phenanthrene–PMDA. [This figure was taken from Keijzers and Haarer (1977), fig. 1.]

microscopic charge carrier mobilities with an anisotropy of about ten and with the axis of high mobility being parallel to the molecular stacking axis. A typical transient signal of a photocurrent experiment is shown in fig. 25, for electrons in phenanthrene–PMDA single crystals. Shown are the signals for perpendicular (a) and parallel directions (b) and for both directions on a logarithmic plot (c). The voltages applied across the 1.2 mm thick crystal were I: 1000 V, II: 2500 V, III: 1000 V, IV: 1500 V and V: 2500 V. In the absence of space charge effects the mobility μ can be obtained from the transit time τ (drop of the photocurrent, fig. 25), if one knows the crystal thickness d and the applied voltage V, by using the known relation

$$\mu = d^2/V\tau. \tag{57}$$

The measured temperature dependences of the (electron) mobilities are

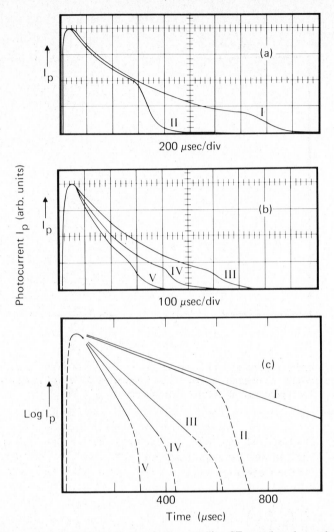

Fig. 25. Electron transient signals for the single crystalline CT complex phenanthrene–PMDA in directions perpendicular (a) and parallel (b) to the crystal stacking axis. Note that (c) is a logarithmic plot. The voltages applied across a 1.2 mm thick crystal are I: 1000 V, II: 2500 V, III: 1000 V, IV: 1500 V, V: 2500 V.

shown in fig. 26. The slopes of the various curves reflect activation energies on the order of 0.1 eV. The activated process can, in principle, be attributed to two different physical mechanisms. The first possibility is a trap limited mobility μ_t which follows the equation

$$\mu_0/\mu_t = 1 + (N_t/N_c) \exp(\Delta E/kT), \tag{58}$$

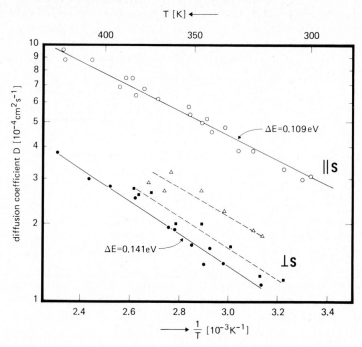

Fig. 26. Temperature dependence of the electron diffusion coefficients in phenanthrene–PMDA single crystals for orientations parallel and perpendicular to the stack axis S. [This figure was taken from Möhwald et al. (1975), fig. 2.]

where μ_0 is the intrinsic mobility, N_t is the density of traps and N_c is the density of states in the conduction band. The second possible transport mechanism can be described within the framework of the small polaron theory as

$$\mu_0 = \frac{ea^2}{kT^2}\frac{J^2}{\hbar}\left(\frac{\pi}{4kTE_a}\right)^{1/2}\exp(-E_a/kT), \tag{59}$$

where a is the lattice spacing, J is the electron transfer (hopping) integral and E_a is the activation energy corresponding to the polaron binding energy.

If one compares the two different processes, then the one described by eq. (58) can be ruled out because it would require an unreasonably high trap concentration of 10^{-2}, a number which seems far too high for the zone-refined crystals used for the experiments. This establishes that the transport is best described by the small polaron picture, in which the charge carrier mobility is limited by the polarization of the surrounding lattice.

Further support of the small polaron description of charge carriers in weak CT crystals comes from experiments on triplet excitons whose propagation is, like that of charge carriers, dominated by short range electron exchange interactions. EPR experiments on related systems (Keijzers and Haarer 1977, Möhwald and Sackmann 1973) yield activated diffusion coefficients with activation energies on the order of 0.1 eV. The only discrepancy between the two classes of experiments, however, seems to be the large anisotropy of the EPR experiments ($D_\parallel/D_\perp \geqslant 100$), which is not measured in the photoconduction data. Here the measured anisotropy of the mobilities is much smaller.

8.3. The anisotropy of the charge carrier transport

The transport properties of charge carriers and triplet excitons are governed by short range interactions (exchange integrals). It is known from EPR experiments (Keijzers and Haarer 1977, Möhwald and Sackmann 1973) that triplet excitons in CT crystals are very one dimensional with anisotropies of 100:1 or larger. Therefore it is rather surprising that the measured "macroscopic" anisotropies of the charge carrier mobility tensor are on the order of 10 or less.

Most published mobility data are based on drift experiments in which charge carriers have to overcome macroscopic distances on the order of millimeters. These experiments have the drawback that the low mobility direction $\mu_\perp$ tends to give unreliable results, because small crystal misalignments and chain imperfections along the "fast axis" lead to large experimental errors. More reliable mobility data can be gained from evaluating "microscopic" experiments such as spin relaxation or microscopic charge carrier trapping data. In the following the microscopic trapping phenomenon is discussed, which leads to the interesting conclusion, that the "intrinsic" mobility anisotropies are on the order of 100 or larger. The experiments thus suggest that CT crystals are indeed good model systems for studying one-dimensional transport phenomena.

It is known from statistical calculations that one-dimensional diffusion is a very inefficient process for sampling new lattice sites (Montroll and Weiss 1965). This effect is due to the fact in three dimensions almost every single hopping event leads to a new site (i.e. a site not visited before in the course of the random walk). In one dimension, however, the number of new sites does not increase linearly, but with the square root of the hopping events, since the charge carrier samples the same sites over and over again.

If the random walk of charge carriers in a CT solid is strictly one dimensional, then one can calculate that the expected charge carrier trapping time should become a function of the applied external electric field at very low field strengths. A simple model calculation shows that the critical field strength at which the characteristic trapping time becomes field dependent occurs at $E_{c,3} \approx 4 \times 10^6$ V/cm in three dimensions, and at $E_{c,1} \approx 1$ V/cm in one dimension (Haarer and Möhwald 1975). Note that the change in dimensionality gives rise to a difference of approximately six orders of magnitude. The critical field in three dimensions is so high that it lies above the breakdown field of most crystals and hence a field induced trapping experiment has never been successfully performed in three-dimensional systems. In a real one-dimensional system, the field is so low that most experiments should be dominated by the field induced trapping effect. Figures 25b and 25c show some charge carrier transients in phenanthrene–PMDA at fields above 2×10^3 V/cm. The figure shows that at these field strengths the trapping time constant becomes field dependent. The linear field dependence, which follows from a simple model calculation, is shown in fig. 27. As can be expected, the perpendicular direction does not show the one-dimensional trapping effect.

Phenanthrene–PMDA is, to our knowledge, the first one-dimensional photoconductive system to show field induced trapping effects. Even though the field at which the phenomenon occurs (2×10^3 V/cm) is higher

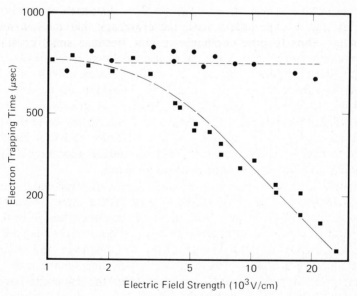

Fig. 27. Charge carrier trapping times of electrons in phenanthrene–PMDA, as a function of external electric field. The field was oriented perpendicular to the stack axis (squares) and parallel to the stack axis (circles). [This figure was taken from Haarer and Möhwald (1975), fig. 2.]

than the purely one-dimensional estimate of 1 V/cm, recent calculations have shown that it corresponds to a very large one-dimensional anisotropy of about $100:1$ ($\mu_{\parallel}/\mu_{\perp}$) (Scher et al. 1980, 1981). These calculations reconcile the EPR experiments on one-dimensional excitons with the microscopic photoconduction experiments.

9. Conclusions

In this chapter, we have discussed the photophysics of weak CT crystals using optical, EPR and photoconduction data and attempted to link it where possible to their one-dimensional structures. Two aspects dominate the results in both the spectroscopic experiments and the transport experiments:

(a) The polar nature of the excited CT state and the charge carrier leads to strong interactions with the lattice. In the spectroscopic experiments, the polar nature reveals itself in the phonon sideband structures of the observed spectra; in the photoconduction data the self-trapping phenomena limit the observed charge carrier mobilities.

(b) The one-dimensionality of CT crystals. This latter feature is a consequence in part of the space filling packing of large planar molecules by stacking. In quasi-one-dimensional systems CT excitons and polarons have properties not found in materials of higher effective dimensions. Consequently these materials show promise as model systems for the study of phenomena involving energy absorption and migration, and charge separation and migration.

Acknowledgment

We thank our friend and colleague H. Morawitz for a critical reading of the manuscript.

References

Agranovich, V.M. and A.A. Zakhidov, 1977, Chem. Phys. Lett. **50**, 278.
Beens, H., J. deJong and A. Weller, 1969, Coll. Amp. **15**, 289.
Brandon, W., R.E. Gerkin and C.A. Hutchison, Jr., 1964, J. Chem. Phys. **41**, 3717.
Brieglet, G. and D. Wolf, 1970, Z. Naturforsch. **A25**, 1925.
Brillante, A. and M.R. Philpott, 1980, J. Chem. Phys. **72**, 4019.
Brillante, A., M.R. Philpott and D. Haarer, 1978, Chem. Phys. Lett. **56**, 218.
Birks, J.B., 1975, Rept. Prog. Phys. **38**, 903.
Boeyens, J.C.A. and F.H. Herbstein, 1965, J. Phys. Chem. **69**, 2153.

Briegleb, G., 1961, Elektronen Donor–Acceptor-Komplexe (Springer, Berlin).

Castro, G. and J.F. Hornig, 1965, J. Chem. Phys. **42**, 1459.

Chance, K.R. and C.L. Braun, 1976, J. Chem. Phys. **64**, 3573.

Coleman, L.B., M.J. Cohen, D.J. Sandman, F.G. Yamagishi, A.F. Jarito and A.J. Heeger, 1973, Solid State Commun. **12**, 1125.

Craig, D.P. and S.H. Walmsley, 1968, Excitons in Molecular Crystals (Benjamin, New York).

Czekalla, J., 1959, Z. Elektrochem. **63**, 1157.

Czekalla, J., G. Briegleb, W. Herre and R. Gier, 1957, Z. Elektrochem. **61**, 537.

Dalal, N.S., D. Haarer, J. Bargon and H. Möhwald, 1976, Chem. Phys. Let. **40**, 326.

Davydov, A.S., 1962, Theory of Molecular Excitons (McGraw-Hill, New York/London).

Davydov, A.S., 1971, Theory of Molecular Excitons (Plenum Press, New York).

Eckhardt, C.J. and R.J. Hood, 1979, J. Am. Chem. Soc. **101**, 6170.

Eckhardt, C.J. and J. Merski, 1973, Surf. Sci. **37**, 937.

Evans, D.L. and W.T. Robins, 1977, Acta Cryst. **B33**, 2891.

Fischer, S. and S.A. Rice, 1970, J. Chem. Phys. **52**, 2089.

Foerster, Th., 1965, Modern Quantum Chemistry, vol. 3 (Academic Press, New York/London) pp. 93 ff.

Foster, R., 1969, Organic Charge-Transfer Complexes (Academic Press, London/New York).

Fowler, W.B., 1968, Physics of Color Centers (Academic Press, New York).

Frenkel, J., 1931, Phys. Rev. **37**, 17, 1276.

Fröhlich, H., 1963, in: Polarons and Excitons, eds. G.G. Kuyper and G.D. Whitfield (Plenum Press, New York).

Gill, W.D., 1972, J. Appl. Phys. **43**, 5033.

Grover, M.K. and R. Silbey, 1970, J. Chem. Phys. **52**, 2099.

Hanson, D.M., 1973, Crit. Rev. Solid State Sci. **3**, 243.

Haarer, D., 1974, Chem. Phys. Lett. **27**, 91.

Haarer, D., 1975, Chem. Phys. Lett. **37**, 192.

Haarer, D., 1977, J. Chem. Phys. **67**, 4076.

Haarer, D., 1979, J. Lumin. **18/19**, 435.

Haarer, D., 1980, Adv. Solid State Phys., Festkörperprobleme **20**, 341.

Haarer, D. and K. Karl, 1973, Chem. Phys. Lett. **21**, 49.

Haarer, D. and H. Möhwald, 1975, Phys. Rev. Lett. **34**, 1447.

Haarer, D. and H. Möhwald, 1977, unpublished results.

Haarer, D., M.R. Philpott and H. Morawitz, 1975, J. Chem. Phys. **63**, 5238.

Haarer, D., C.P. Keijzers and R. Silbey, 1977, J. Chem. Phys. **66**, 563.

Hayashi, H., and S. Nagakura, 1969, J. Chem. Phys. **50**, 993.

Hayashi, H., S. Iwata and S. Nagakura, 1969, J. Chem. Phys. **50**, 993.

Herbstein, F.F., 1971, in: Perspectives in Structural Chemistry, vol. 4, eds. J.D. Dunitz and J.A. Ibers (Wiley, New York).

Hochstrasser, R.M., S.K. Lower and C. Reed, 1964, J. Chem. Phys. **41**, 1072.

Holstein, T., 1959, Ann. Phys. **8**, 325.

Iwata, S., J. Tanaka and S. Nagakura, 1966, J. Am. Chem. Soc. **88**, 894.

Iwata, S., J. Tanaka and S. Nagakura, 1967, J. Chem. Phys. **47**, 2203.

Jerome, D., A. Mazaud, M. Ribault and K. Bechgaard, 1980, J. Physique Lett. **4**, L95.

Karl, N. and J. Ziegler, 1975, Chem. Phys. Lett. **32**, 438.

Keijzers, C.P. and D. Haarer, 1977, J. Chem. Phys. **67**, 925.

Keller, H.J., 1977, Chemistry and Physics of One-Dimensional Metals (Plenum Press, New York).

Kelper, K.G., 1969, Phys. Rev. **199**, 1226.

Kitaigorodskii, A.I., 1978, Chem. Soc. Rev. **7**, 133.

Koch, E.E., A. Otto and K. Kliewer, 1974, Chem. Phys. **3**, 362.

LeBlanc, O.H., 1960, J. Chem. Phys. **33**, 626.

Little, W., 1964, Phys. Rev. **A134**, 1416.

Lower, S.K., R.M. Hochstrasser and C. Reed, 1961, Mol. Phys. **4**, 161.

Macfarlane, R.M. and M.R. Philpott, 1976, Chem. Phys. Lett. **41**, 33.

Mayerle, J.J., 1978, unpublished measurements.

Mayerle, J.J., 1980, in: Mixed-Valence Compounds, ed. D.B. Brown (Reidel, New York) p. 451.

Mayoh, B. and C.K. Prout, 1972, J. Chem. Soc. **68**, 1072.

Merski, J. and C.J. Eckhardt, 1981, J. Chem. Phys., in press.

McGlynn, S.P., 1958, Chem. Rev. **58**, 1113.

McGlynn, S.P., J.D. Boggus and E. Elder, 1960, J. Chem. Phys. **32**, 357.

Mills, D.L. and E. Burstein, 1974, Rept. Prog. Phys. **37**, 817.

Möhwald, H. and E. Sackmann, 1973, Chem. Phys. Lett. **21**, 41.

Möhwald, H. and E. Sackmann, 1974, Z. Naturforsch. **A29**, 1216.

Möhwald, H., D. Haarer and G. Castro, 1975, Chem. Phys. Lett. **32**, 433.

Montroll, E.W. and G.H. Weiss, 1965, J. Math. Phys. **6**, 167.

Morawitz, H., 1981, Bull. Am. Phys. Soc. **26**, 310.

Mulliken, R.S., 1950, J. Am. Chem. Soc. **72**, 600.

Mulliken, R.S., 1951, J. Chem. Phys. **19**, 514.

Mulliken, R.S., 1952, J. Am. Chem. Soc. **74**, 811.

Pfann, W.G., 1966, Zone Melting, 2nd Ed. (Wiley, New York/London/Sydney).

Pfeiffer, P., 1927, Organische Molekulverbindungen, 2nd Ed. (Ferdinand Enke, Stuttgart).

Philpott, M.R., 1973, Adv. Chem. Phys. **23**, 227.

Philpott, M.R., 1977, Chem. Phys. Lett. **50**, 18.

Philpott, M.R. and A. Brillante, 1979, Mol. Cryst. Liq. Cryst. **50**, 163.

Philpott, M.R., 1980, Ann. Rev. Phys. Chem. **31**, 97.

Robertson, D.E. and J. Stezowski, 1978, Acta Cryst. **B34**, 3005.

Sakata, T. and S. Nagakura, 1970, Bull. Chem. Soc. Jpn. **43**, 1346.

Schaffert, R.M., 1971, IBM J. Res. Dev. **15**, 75.

Scher, H. and E.W. Montroll, 1975, Phys. Rev. **B12**, 2455.

Scher, H., S. Alexander and E.W. Montroll, 1980a, Bull. Am. Phys. Soc. **25**, 369.

Scher, H., S. Alexander and E.W. Montroll, 1980b, Proc. Nat. Acad. Sci. **77**, 3758.

Sebastian, L. and G. Weiser, 1979, Chem. Phys. Lett. **64**, 396.

Sharp, H., 1967, J. Phys. Chem. **71**, 2587.

Syassen, K. and M.R. Philpott, 1977, Chem. Phys. Lett. **50**, 14.

Syme, R.W., H. Morawitz and R.M. Macfarlane, 1979, Solid State Commun. **32**, 1059.

Tanaka, J. and M. Tanaka, 1979, Mol. Cryst. Liq. Cryst. **52**, 221–226.

Tobin, M.C. and D.P. Spitzer, 1965, J. Chem. Phys. **42**, 3652.

Torrance, J.B., J.J. Mayerle, V.Y. Lee and K. Bechgaard, 1979, J. Am. Chem. Soc. **101**, 4747.

Torrance, J.B., J.E. Vasquez, J.J. Mayerle and V.Y. Lee, 1981, Phys. Rev. Lett. **46**, 253.

Tosatti, E. and G. Harbeke, 1974, Nuovo Cimento **522**, 87.

Toyozawa, Y., 1970, J. Lumin. **12**, 632.

Wannier, G.H., 1937, Phys. Rev. **52**, 191.

Werner, A., 1909, Physikalische Berichte **42**, 4324.

Yarwood, J., 1973, Spectroscopy and Structure of Molecular Complexes (Plenum Press, New York).

CHAPTER 3

Biphonons and Fermi Resonance in Vibrational Spectra of Crystals

V.M. AGRANOVICH

Institute of Spectroscopy
USSR Academy of Sciences
142092 Troitsk, Moskovskaya Oblast
U.S.S.R.

Translated from the Russian by Nicholas Weinstein.

Spectroscopy and Excitation Dynamics
of Condensed Molecular Systems
Edited by
V.M. Agranovich and R.M. Hochstrasser

Contents

1. Introduction—Effects of strong anharmonicity in spectra of crystals

The application of lasers in optical experimental techniques has led to a rapid development of research into the properties of elementary excitations in condensed matter. In addition to the conventional methods of linear crystal optics, Raman scattering of light (RSL) has become one of the principal research methods, as have its various modifications, such as coherent active Raman spectroscopy (CARS) and others.

Up-to-date optical methods enable one to obtain sufficiently accurate and comprehensive information on various processes, in particular on those accompanied by the simultaneous production or annihilation of several quasi-particles (two in the simplest case). These processes are of special interest because in them a "residual" interaction between quasi-particles, which need not necessarily be weak in all cases, should be manifest to a greater or lesser degree.

For phonons such a "residual" interaction is anharmonicity, which is commonly ignored in calculating the frequencies and amplitudes of the normal vibrations of the crystal lattice. In this harmonic approximation, advanced at the very beginning of the development of present-day solid-state theory (Einstein 1906, 1911, Debye 1912), the excited states of the lattice are associated with sets of various numbers of phonons of one or another kind. The energy, for example, of the excited state of a lattice with two phonons, reckoned from the energy of its ground state, equals

$$E_{l_1 l_2}(\mathbf{k}_1, \mathbf{k}_2) = \hbar\Omega_{l_1}(\mathbf{k}_1) + \hbar\Omega_{l_2}(\mathbf{k}_2), \tag{1}$$

where l_1 and l_2 are the numbers of the branches of the phonon spectrum, and $\mathbf{k}_1$ and $\mathbf{k}_2$ are the wave vectors of the phonons. In contrast to a state with a single phonon, the state being considered is characterized by the values of two quasi-momenta, $\hbar\mathbf{k}_1$ and $\hbar\mathbf{k}_2$, and, consequently, is a two-particle state. The energy of the states of a lattice with a large number of phonons can be written in a similar manner. Since in such many-particle states the phonons $(\mathbf{k}_1 l_1)$, $(\mathbf{k}_2 l_2)$ etc. do not interact with one another when anharmonicity is neglected, the bandwidth of a many-particle state is found to be equal to the sum of the widths of the energy bands of the separate phonons.

Taking anharmonicity of the lattice vibrations into account leads to interaction of the phonons with one another. When this interaction turns out to be sufficiently strong, the formation of states of bound quasi-particles becomes possible along with the above-mentioned many-particle states. Such bound states are absent in the description of the crystal in the harmonic approximation. In such states the quasi-particles move about in the crystal as a whole and, therefore, like the separate phonons, they are characterized by a single value of the wave vector.

The bound state of two phonons is usually called a biphonon (Agranovich 1973). Quite a comprehensive theory of biphonons has been developed in the last ten years and, what is of prime significance, convincing evidence has been obtained of their existence in various kinds of crystals.

The study of bound states of optical phonons or, stated more generally, the study of the effects of strong anharmonicity in the vibrational spectra of crystals, has not only grown out of the development of modern solid-state theory. To no less extent, it was motivated by the requirements of experiment.

Biphonons, as well as other more complicated phonon complexes, should appear in the spectra of inelastically scattered neutrons. Nevertheless, up to the present, the most vital experimental data were obtained in analyzing the spectra of RSL by polaritons.

As is well known, the selection rules allow RSL by polaritons only in crystals without a centre of inversion. This is precisely the kind of crystal in which Fermi resonance with polaritons (to be discussed below) was found to be the physical phenomenon in which the special features of the biphonon spectrum were most evident.

Besides the region of basic (fundamental) frequencies of lattice vibrations, the polariton (light) branch in crystals also intersects the region of two-particle, three-particle, etc. states. Resonance with these states influences the dispersion law of the polariton and the result of this influence can expediently be investigated by the observation of the spectra of RSL by polaritons. What actually occurs here is a resonance, similar to the Fermi resonance in molecules, since one of the normal waves in the crystal (the polariton) resonates with states that are analogous to overtones or to combination tones of intramolecular vibrations.

The most useful experimental data were obtained in investigations of the Fermi resonance of the polariton with two-particle states. This led to the discovery of biphonons in many crystals. Before beginning a discussion of the results obtained in these investigations, we have several comments to make on the development of this research from the historical point of view.

It should be noted, to begin with, that biphonons are quite analogous to bound states of two magnons (Bethe 1931, Wortis 1963, Hanus 1962). As a matter of fact, states of this kind were investigated in crystalline hydrogen by van Kranendonk (1959, 1968) over twenty years ago, and soon afterward they were observed experimentally as well (Gush et al., 1960). Truly, in the papers by van Kranendonk (1959, 1968), only the bound states of two different quasi-particles were considered under the condition that the motion of one of them can be ignored in a first approximation (the van Kranendonk model, see subsection 2.2). This made the van Kranendonk model inapplicable for analysis of the biphonon spectrum in the frequency region of overtones, as well as combination tones, corresponding to phonons with comparable bandwidths. But his model was found to be suitable for analysis of the biphonon spectrum in the region of combination tone frequencies for many crystals [CO_2, NO_2, OCS, etc.; see the paper by Bogani (1978)], and also for analysis of the spectrum of vibronic states in molecular crystals [see the review by Sheka (1971)].

A generalization of biphonon theory, beyond the scope of the van Kranendonk model, was made by Agranovich (1970), Ruvalds and Zawadowski (1970a,b) and Maradudin (1971). Subsequently, the effect of biphonons on polariton dispersion in the spectral region of two-particle states was investigated in a number of papers (Agranovich 1973, Agranovich and Lalov 1971a,b, 1976a, Agranovich et al. 1979b), and the contribution of biphonons to the nonlinear polarizability of a crystal was discussed by Agranovich et al. (1971) and Efremov and Kaminskaya (1972). Problems of the theory of local and quasi-local biphonons in disordered media were discussed in a number of papers (Agranovich 1970, Agranovich et al. 1970, 1979a, Agranovich and Dubovskii 1980). The influence of anharmonicity in crystals on the spectra of inelastically scattered neutrons was considered by Krauzman et al. (1974), Prevot et al. (1977) and Agranovich and Lalov (1976b).

In the following discussion we shall again touch upon the results of the above-mentioned investigations to a greater or lesser extent. Here we only note that the investigations by Ruvalds and Zawadowski (1970a,b) were undertaken in connection with the attempts to interpret the second-order RSL spectra in diamond. The interest provoked by this crystal was due to the fact that as far back as 1946 Krishnan (1946) observed a sharp peak in the RSL spectrum of diamond at a frequency exceeding, as it was considered at that time, twice the maximum frequency of a single optical phonon ($\Omega_0 \approx 1332.5 \pm 0.5$ cm^{-1}) by about 1.9 ± 1.5 cm^{-1}. Since the nature of this peak remained unclear, Ruvalds and Zawadowski (1970a,b) advanced the hypothesis that the peak was due to the excitation of a biphonon. The mentioned phonon frequency Ω_0 corresponds to the value $k \approx 0$, optical phonons have negative effective mass at low k values and, to form biphonons with an energy exceeding that of two-particle states, repulsion is required and not attraction.

Subsequently, a peak in the RSL spectra, similar to the one observed by Krishnan, was not found in some crystals, such as silicon and germanium, with the same type of structure as diamond and with an even stronger anharmonicity than diamond. This incited Tubino and Birman (1975) to improve the accuracy of the calculations of the structure of the phonon bands in crystals with a diamond-type structure. It was shown as a result of comprehensive investigations that the dispersion curve of the above-mentioned high-frequency optical phonon in diamond has its highest maximum not at $k = 0$, but at $k \neq 0$. The result of these calculations indicates that the peak experimentally observed in the RSL spectra of diamond falls within the region of the two-phonon continuum. It cannot correspond to a biphonon and is most likely related to features of the density of two-particle (dissociated) states.

The bound state of two phonons for the overtone frequency region was evidently first identified by Ron and Hornig (1963). In this investigation they measured the absorption spectrum of the HCl crystal in the region of overtone frequencies of the fundamental vibration (i.e. at $\omega \approx 2\Omega$, where $\Omega = 2725 \text{ cm}^{-1}$). It was found that along with the wide absorption band, corresponding to the excitation of two free phonons (this bandwidth equals 2Δ, where $\Delta \approx 90 \text{ cm}^{-1}$ is the phonon bandwidth), there was also an absorption peak in the region of lower frequencies with the maximum at $\omega = 5313 \text{ cm}^{-1}$ and a half width approximately equal to 20 cm^{-1}.

Later, biphonons were detected in the region of overtone vibrations in many crystals.

A review of results obtained in recently conducted experimental investigations is given in section 8. In turning our attention to a systematic discussion of the theory of the effects resulting from strong anharmonicity for the overtone and combination tone frequency region of optical lattice vibrations, we contend that in the theory of biphonons or more complicated phonon complexes, anharmonicity cannot be taken into account within the framework of perturbation theory as is done, for instance, in the theory of heat conduction, thermal expansion, etc. (Leibfried and Ludwig 1961, Reisland 1973). For the optical branches of the spectrum in many crystals, the dimensionless parameter A/Δ, equal to the ratio of the anharmonicity constant A to the phonon bandwidth Δ, may be of the order of unity. It is exactly in these kinds of situations that anharmonicity is strong, leading (see below in section 2) to the development of qualitatively new features in phonon spectra.

It should be also borne in mind that, due to anharmonicity, "elementary" excitations are also possible in crystals, along with quantum objects formed of a small number of quasi-particles, such as biphonons. These "elementary" excitations correspond to the propagation in the crystal of purely classical nonlinear waves (of the soliton type), which are solutions of the corresponding nonlinear equations of motion. In quantum language it can be contended that the bound states of a large number of quasi-particles correspond to such excitations. An analysis of the experimental conditions under which the processes of RSL by these waves can be observed, for instance, is a matter for future investigations and could appreciably facilitate the development of research of these waves. The energy of nonlinear "elementary" excitations should be larger than that of a single quasi-particle (phonon or polariton). Hence, the investigation of Raman scattering by nonlinear waves, at any rate away from points of structural phase transitions and accompanied by the formation of a soft mode, would require the application of sufficiently powerful sources of light. But at high pumping levels the concentrations of various kinds of quasi-particles (for instance, phonons or polaritons) in crystals can drastically increase. In this connection, the question arises, also of exceptional interest, as to the conditions required for the existence of second sound or other excitations in a system of quasi-particles that are not in a state of thermodynamic equilibrium, and as to the feasibility of observing this second sound by optical methods. If, for example, we take polaritons, for which the velocity of second sound should be of the order of the velocity of light, the observation of second sound, as well as the observation of the polaritons themselves, would require, for instance, an investigation of RSL at small angles.

By mentioning here, along with biphonons to which this review is devoted, also classical nonlinear waves of the soliton type and of second sound in a system of quasi-particles, we only wished to draw attention to the extensive opportunities for and to possible trends in research and analysis of the effects of strong phonon anharmonicity. We also wished to indicate the place of the study of such extremely simple creations of strong anharmonicity as biphonons in this field of solid-state theory.

In the subsequent sections of this review we discuss the fundamentals of biphonon theory, consider the special features of Fermi resonance, including Fermi resonance with polaritons, and also analyze the data obtained in the study of infrared (IR) absorption and RSL spectra. It should be noted that a discussion of the effects of strong anharmonicity in RSL spectra has already been the subject of a review (Agranovich, 1973). Therefore, in the present contribution, attention is directed primarily to the results obtained in theory and experiment following the publication of the above-mentioned review, i.e. during the last ten years.

2. Biphonon theory

2.1. Biphonons in the overtone frequency region of an intramolecular vibrational spectrum

The conclusion that the anharmonicity of optical vibrations in crystals can be remarkably strong in the spectral region of overtone frequencies of intramolecular vibrations follows even from purely qualitative considerations. As a matter of fact, in isolated molecules the anharmonicity energy A for some intramolecular vibration usually amounts to 1–3% of the energy $\hbar\Omega$ of a quantum of the fundamental vibrations. Here the anharmonicity energy A is understood to be the quantity $A = (2\hbar\Omega - E_2)/2$, where E_2 is the energy of the excited state with quantum number $n = 2$. For $\hbar\Omega = 1000 \text{ cm}^{-1}$, for example, A is usually found to be approximately 10 to 30 cm^{-1}. At the same time, the energy of intermolecular interaction in crystals, determining the phonon energy bandwidth Δ for the indicated frequency region, can also have a value of the order of several scores of inverse centimetres, as follows, for instance, from measurements of second-order RSL spectra (Poulet and Mathieu 1970). This is precisely why the dimensionless ratio A/Δ, as previously mentioned in the Introduction, is not generally a small quantity. Therefore, when the phonon bandwidth is of the order of A, the optical vibrational spectrum in the region of overtone or combination tone frequencies may have an extremely complex structure.

To find a form of model Hamiltonian (Agranovich 1973, 1970) that would be sufficiently simple for analysis and, at the same time, would allow a discussion of the most interesting physical effects, we shall consider the problem of the occurrence of biphonons in more detail, using, as an example, a molecular crystal in the frequency region corresponding to the first overtone (i.e. at $\omega \approx 2\Omega$). If there are two quanta of molecular vibrations in the crystal, localized on different molecules, the energy of the crystal, intermolecular interaction being neglected, is $E = 2\hbar\Omega$. If both quanta are localized on a single molecule, $E = \hbar 2\Omega - 2A$ owing to the intramolecular anharmonicity. Hence, in the language of quasi-particles, we can contend that in the case being discussed the intramolecular anharmonicity ($A > 0$) leads to a reduction in the energy of the crystal as the particles approach one another and, consequently, corresponds to their attraction. But the localization of quasi-particles (intramolecular phonons) on a single molecule leads to an increase in the kinetic energy of their relative motion. Since, in order of magnitude, this energy is equal to the phonon energy bandwidth, the state with two phonons bound to each other (a biphonon) certainly occurs when $A \gg \Delta$. Then, besides the energy band of two-particle states, eq. (1), corresponding to the independent motion of two phonons, there are, in the molecular overtone frequency region of the crystal spectrum, biphonon states with a lower energy (for $A > 0$). The number of such states, even for $A > \Delta$, as will be shown below [see also Agranovich (1970)], depends upon the structure of the unit cell and is not equal, in general, to the number of molecules per unit cell, as is the case with the number of optical branches corresponding to the fundamental tone.

In the limiting case of weak intermolecular interaction ($\Delta \to 0$ and $|A| \gg$

Δ), biphonons have an extremely simple structure; they go over into the states of molecules excited to the second vibrational level. Here the spectrum of the crystal in the frequency region being considered consists of two lines, with the corresponding crystal energies $E = 2\hbar\Omega - 2A$ (both quanta "sit" on one molecule and the state of the crystal is N-fold degenerate, where N is the number of molecules in the crystal) and $E = 2\hbar\Omega$ (the quanta "sit" on different molecules and the state of the crystal is $[N(N-1)/2]$-fold degenerate).

If, on the contrary, anharmonicity is weak ($|A| \ll \Delta$), biphonons are not formed outside the band of two-particle states. But inside the band of two-particle states, as shown by Pitaevsky (1976), only weakly bound states of biphonons are formed [even for the smallest value of $|A|$; it is necessary, of course, that the value of the binding energy in the biphonon be greater than the width δ of the phonon level; regarding the feasibility of observing the states discussed by Pitaevsky (1976), see below].

A model Hamiltonian that describes the excitation spectrum of the crystal in the energy region $E = 2\hbar\Omega$ can be readily constructed on the basis of the qualitative considerations presented above. As a matter of fact, the Hamiltonian of the crystal, describing the effect of intermolecular interaction on the spectrum, for example, of nondegenerate molecular vibrations can be written in the harmonic approximation as follows:

$$\hat{H}_0 = \sum_n \hbar\Omega B_n^+ B_n + \sum_{n,m}' V_{nm} B_n^+ B_m, \tag{2}$$

where B_n^+ and B_n are the Bose creation and annihilation operators of a quantum of intramolecular vibrations with energy $\hbar\Omega$ in molecule n, and V_{nm} is the matrix element of the interaction of molecules n and m corresponding to the transfer of one quantum from molecule m to molecule n. If a unit cell of the crystal contains several molecules, then the subindex n is composite, $n \equiv (n, \alpha)$, where n is an integral-valued lattice vector, and α is the number of the molecule in the unit cell: $\alpha = 1, 2, \ldots, \sigma$.

To determine the energy of the phonons it is necessary to diagonalize Hamiltonian (2). In a crystal with a single molecule per unit cell, for instance, this leads to the relation

$$E \equiv E(k) = \hbar\Omega + V(k) \tag{3}$$

for an optical phonon with wave vector k, where $V(k)$, the so-called band addition, is given by the expression

$$V(k) = \sum_m V_{nm} \exp[ik(m-n)]. \tag{3a}$$

A degenerate intramolecular vibration in a crystal with even a single molecule per unit cell corresponds to a number of phonon bands equal to the degeneracy multiplicity ν. If, moreover, the number of molecules per unit cell $\sigma > 1$, then the number of phonon bands that are related to the given intramolecular vibration of frequency Ω becomes equal to $\sigma\nu$. Hence the possible energies of the phonon are determined by the quantities $E_l(k)$, where the subindex l assumes the values $l = 1, 2, \ldots, \sigma\nu$.

The aforesaid concerning the structure of the optical phonon spectrum is, of course, well known. We dwelt on this question in such detail because in the spectral region with $\omega \approx 2\Omega$ being discussed, the number of bands of two-particle states of the form of eq. (1) [equal to $\sigma\nu(\nu\sigma + 1)/2$] can be very large in the general case ($\nu \neq 1$ and $\sigma \neq 1$). Since the distance between the bands $E_l(k)$ is of the order of the bandwidth, the bands may overlap, forming a quite complex spectrum of two-particle states in the energy region $E = 2\hbar\Omega$. An experimental investigation of two-particle states therefore presents formidable difficulties, requiring the application of not only the most effective optical techniques, but also careful theoretical analysis, including group-theoretic analysis.

To take intramolecular anharmonicity into consideration, it is necessary to add the operator

$$\hat{H}_A = -A \sum_n (B_n^+)^2 B_n^2 \tag{4}$$

to the Hamiltonian (2). Hence, the total Hamiltonian is

$$\hat{H}' = \hat{H}_0 + \hat{H}_A. \tag{5}$$

Having been derived on the basis of purely qualitative considerations, Hamiltonian (5) naturally requires some substantiation. It can be shown [see Bogani (1978) and references therein] that when the cubic terms of the intramolecular anharmonicity are taken into account, as well as anharmonicity of the fourth order, we obtain a Hamiltonian of the form of eq. (5), provided that the natural frequencies of the intramolecular vibrations appearing in the Hamiltonian $\hat{H}_0$ are considered to have been found taking anharmonicity into account. This provides a correction of the order of $A/\hbar\Omega$.

We shall not present here the relations between the phenomenological quantity A and the anharmonicity constants appearing in the potential energy of the molecule. Such relations are not required here because these anharmonicity constants can be found only by making use of some model, whereas the quantity A can be determined directly from a comparison of the first and second order molecular spectra.

When we take many intramolecular vibrations into account, the Hamiltonian $\hat{H}'$ is of the following form:

$$\hat{H}' = \sum_{n,j} \hbar\Omega_j (B_n^j)^+ B_n^j + \sum_{\substack{n,m \\ j \leq j'}} V_{nm}^{jj'} (B_n^j)^+ B_m^{j'} - \sum_{n, j \leq j'} A(jj')(B_n^j)^+ (B_n^{j'})^+ B_n^j B_n^{j'}. \tag{5a}$$

It becomes necessary to make use of this Hamiltonian (see also subsection 2.2) when discussing the spectra of crystals containing molecules with degenerate or close frequencies. Assuming, however, that the frequency Ω of the intramolecular vibration is nondegenerate, we shall continue our discussion of expression (5).

An important feature of operator (5) is that it commutes with the operator of the total number of vibration quanta $\hat{N} = \Sigma_n B_n^+ B_n$. Hence, in the steady state of the crystal the number of such quanta is a conserved quantity. In particular, in a state with a single l quantum $|1\rangle_l$, the anharmonicity, eq. (4), is inessential:

$$\hat{H}_A|1\rangle_l = 0,$$

so that

$$\hat{H}'|1\rangle_l = \hat{H}_0|1\rangle_l = E_l(k)|1\rangle_l.$$

Operator (4) is chosen so as to provide the correct energy values for a crystal with two vibration quanta when intermolecular interaction is neglected. If, for instance, these quanta are located on different molecules, n and m, the corresponding wave function of the crystal is

$$\psi_{nm} = B_n^+ B_m^+|0\rangle,$$

where $|0\rangle$ is the ground state of the crystal. Since in this case $\hat{H}_A\psi_{nm} = 0$, we have $\hat{H}'\psi_{nm} = 2\hbar\Omega\psi_{nm}$.

If, however, both quanta are located on a single molecule, i.e. $n = m$, then, using the properties of a Bose operator, we find that $\hat{H}_A\psi_{nn} = -2A\psi_{nn}$. Hence $\hat{H}'\psi_{nn} = (2\hbar\Omega - 2A)\psi_{nn}$.

Thus, strictly speaking, operator (5) can be employed for investigating the states of a crystal with two vibration quanta only when the intramolecular anharmonicity of the form of eq. (4) dominates, and the part of the anharmonicity that is associated with the presence of intermolecular interaction can be neglected*. Since, by assumption, $A/\hbar\Omega \ll 1$ and $\Delta/\hbar\Omega \ll 1$, the total Hamiltonian (5) also neglects terms that do not conserve the number of quasi-particles (inessential corrections are introduced if they are taken into account).

We underline, however, that even in the limit of large A values, intermolecular anharmonicity may turn out to be important in calculating biphonon bandwidths. Here the Hamiltonian $\hat{H}'$ should include the term

$$\hat{H}_T = \frac{1}{2}\sum_{n,m}' W_{nm}(B_n^+)^2 B_m^2, \tag{6}$$

which leads to the transfer of two vibration quanta at once from molecule

---

*The opposite situation has been investigated by Lalov (1974).

m to molecule n and back again ($n \neq m$). This transfer $n \to m$ of two
vibration quanta is allowed, of course, when relation (5) is used as well.
But in this approximation, the corresponding matrix element is nonzero
only in the second order of perturbation theory with respect to intermolecular interaction V_{nm}. It is readily evident that this matrix element
equals V_{nm}^2/A, so that terms with W_{nm} may be omitted under the condition
that

$$V_{nm}^2/A \gg |W_{nm}|. \tag{7}$$

Even if this inequality is satisfied for small values of $|n - m|$, it may, in
general, be violated for large $|n - m|$ values because for dipole-active
overtones, for instance, $|W_{nm}| \sim |n - m|^{-3}$, whereas the quantity V_{nm}^2 can
decrease with increasing $|n - m|$ proportionally to $|n - m|^{-6}$ or more rapidly.
It should also be noted that for dipole-active overtones, it is important to
take operator (6) into account also because this corresponds to Coulomb
long-range interaction and in cubic crystals, for instance, leads to longitudinal–transverse splitting of the biphonon. Fortunately, the inclusion of
operator (6) in the total Hamiltonian [see also Lalov (1974) and Agranovich
et al. (1976)] only slightly modifies the calculation procedure, as will be
illustrated below.

When the anharmonicity is so large that the inverse inequality holds
instead of (7), then terms with W_{nm} precisely make the main contribution to
the energy bandwidth of the biphonon. Here the energy of the biphonon is

$$E_b(k) = 2\hbar\Omega - 2A + \sum_m W_{nm} \exp[ik(m - n)] + O(\Delta^2/A). \tag{7a}$$

Since this relation, as previously noted, is exact in the limit of large A
values, the quantities W_{nm}, appearing in the relation and determining the
matrix element for the transfer of two quanta from molecule n to molecule
m, should be found taking intramolecular anharmonicity into account. To
emphasize this, we shall write the corresponding matrix elements in the
form W_{nm}^A. In connection with the aforesaid, for further analysis of the
biphonon states we shall make use of a Hamiltonian of the form

$$\hat{H} = \hat{H}_0 + \hat{H}_A + \hat{H}_T, \tag{8}$$

which is more general than (5).

To obtain Hamiltonian (8) we proceeded from the model of a molecular
crystal. Actually, its range of application includes nonmolecular crystals as
well, provided we are concerned with optical phonons whose bandwidth is
much narrower than the phonon frequency. In these spectral regions the
vibrations of the atoms inside the unit cell are similar to intramolecular
vibrations in molecular crystals, since the comparatively narrow phonon
bandwidth is indicative of the weakness of the interaction between the
vibrations of atoms located in different unit cells.

The states of the crystal with two vibration quanta, of interest to us, can be written in the form

$$|2\rangle = \sum_{n,m} \psi(n, m) B_n^+ B_m^+ |0\rangle, \qquad \psi(n, m) = \psi(m, n), \tag{9}$$

where $\psi(n, m)$ has the meaning of the wave function of two phonons in the Schrödinger representation. This function should satisfy the Schrödinger equation

$$\hat{H}|2\rangle = E_2|2\rangle,$$

where E_2 is the required excitation energy of the crystal with two quanta. Since

$$\hat{H}_A|2\rangle = -A \sum_n (B_n^+)^2 B_n^2 \sum_{l,m} \psi(l, m) B_l^+ B_m^+ |0\rangle$$

$$= -2A \sum_{l,m} \psi(l, m) \delta_{lm} B_l^+ B_m^+ |0\rangle,$$

$$\hat{H}_0|2\rangle = 2\hbar\Omega|2\rangle + \sum_{n,m} \left(\sum_l [V_{nl}\psi(l, m) + V_{ml}\psi(l, n)] \right) B_n^+ B_m^+ |0\rangle,$$

$$\hat{H}_T|2\rangle = \frac{1}{2} \sum_{n,m} \sum_p (W_{np}^A + W_{mp}^A)\psi(p, p)\delta_{nm} B_n^+ B_m^+ |0\rangle,$$

we find that $\psi(n, m)$ satisfies the following system of equations:

$$(E_2 - 2\hbar\Omega)\psi(n, m) - \sum_l [V_{nl}\psi(l, m) + V_{ml}\psi(l, n)]$$

$$= \left(-2A\psi(n, n) + \sum_p{}' W_{np}^A \psi(p, p) \right) \delta_{nm}. \tag{10}$$

Before solving this system of equations we note that biphonon states, like other crystal states, transform according to the irreducible representations of the crystal space group. Therefore, the range of allowed values of its wave vector $\mathbf{K}$, as for a separate quasi-particle (a phonon, for instance), is determined by the first Brillouin zone. Consequently, the wave function of the biphonon, corresponding to the wave vector $\mathbf{K}$, can be searched for in the form

$$\psi(n, m) = \exp[\tfrac{1}{2}i\mathbf{K}(n + m)]\varphi_{\alpha\beta}(n - m), \tag{11}$$

where the function φ, determining the internal structure of the biphonon, satisfies, in accordance with eq. (9), the symmetry condition

$$\varphi_{\alpha\beta}(n - m) = \varphi_{\beta\alpha}(m - n). \tag{11a}$$

To find the biphonon energy levels, we introduce the two-particle Green's function in the zeroth-order approximation $G_E^0(n, m \mid l, l')$, satisfying the equation

$$(E - 2\hbar\Omega)G_E^0(n, m \mid l, l')$$

$$- \sum_p [V_{np}G_E^0(p, m \mid l, l') + V_{mp}G_E^0(p, n \mid l, l')] = \delta_{nl}\delta_{ml'}. \qquad (12)$$

As is known, the Green's function of an arbitrary self-adjoint operator $\hat{L}$ has the form

$$G_E(r \mid r') = \sum_\lambda \frac{\varphi_\lambda(r)\varphi_\lambda^*(r')}{E - E_\lambda}, \qquad (13)$$

where E_λ and φ_λ are its λth eigenvalue and eigenfunction. In our case, the role of this operator is played by the energy operator $\hat{H}_0$, whereas the states λ correspond to the set of states of a crystal with two phonons when anharmonicity is neglected. If μ is the number of the phonon band and k is the wave vector of the phonon, then the wave function of a state with two free phonons is determined by the equation

$$\Psi_{\mu k, \mu' k'}(n\alpha, m\beta) = \frac{a_\alpha^\mu(k)a_\beta^{\mu'}(k')}{N} \exp[i(kn + k'm)], \qquad (14)$$

i.e., it is equal to the product of the wave functions of the separate phonons $\psi_{\mu k}(n\alpha) = a_\alpha^\mu(k) \exp(ikn)/\sqrt{N}$, where N is the number of unit cells in the volume of the crystal. Making use of eq. (13), we now obtain

$$G_E^0(n, m \mid l, l') = \sum_{\mu,\mu',k,k'} \frac{\psi_{\mu k}(n)\psi_{\mu' k'}(m)\psi_{\mu k}^*(l)\psi_{\mu' k'}^*(l')}{E - \epsilon_\mu(k) - \epsilon_{\mu'}(k')}. \qquad (15)$$

Hence, as follows from eq. (10),

$$\psi(n, m) = \sum_l G_E^0(n, m \mid l, l) \left(-2A\psi(l, l) + \sum_p{}' W_{lp}^A\psi(p, p)\right). \qquad (16)$$

Assuming $n = m$ in this equation, and making use of relation (11), we obtain a system of σ equations for the quantities $\varphi_{\alpha\alpha}(0)$. This system is of the following form:

$$\varphi_{\alpha\alpha}(0) = \sum_\beta R_{\alpha\beta}(E, K)\varphi_{\beta\beta}(0), \qquad (17)$$

where

$$R_{\alpha\beta}(E, K) =$$

$$= \sum_l \left(-2AG_E^0(n\alpha, n\alpha \mid l\beta, l\beta) + \sum_{p,\gamma} G_E^0(n\alpha, n\alpha \mid p\gamma, p\gamma)W_{p\gamma,l\beta}^A\right)$$

$$\times \exp[iK(l - n)]. \qquad (18)$$

The energy values E of the states being investigated of crystals with two phonons are obtained from the condition that the determinant of the system of equations (17) equals zero, i.e. from the equation

$$|\delta_{\alpha\beta} - R_{\alpha\beta}(E, K)| = 0. \tag{19}$$

Crystals with a single molecule per unit cell are most readily analyzed. Here eq. (19) is of the form

$$1 = R(E, K). \tag{20}$$

Moreover, since in this case

$$G_E^0(n, n \mid l, l) = \frac{1}{N^2} \sum_{k,k'} \frac{e^{i(n-l)(k+k')}}{E - \epsilon(k) - \epsilon(k')},$$

where

$$\epsilon(k) = \hbar\Omega + \sum_m V_{nm} \exp[ik(m-n)] \equiv \hbar\Omega + V(k),$$

we find that eq. (20) can be written as follows:

$$1 = \frac{1}{N} \sum_q \frac{-2A + W^A(K)}{E - \epsilon(K/2 + q) - \epsilon(K/2 - q)}, \tag{20a}$$

where

$$W^A(K) = \sum_l W_{pl}^A \exp[iK(l - p)].$$

For $K = 0$ and $W^A = 0$, the form of eq. (20a) coincides with that of the equation for the energy of a quantum of local vibration in the vicinity of an isotopic impurity, and its analysis is actually known [see, e.g., Lifshits (1956)]. Here we only stress that for $|A| \gg |V(k)|$ the root of eq. (20a), lying outside the energy region of two-particle states $E(k, k') = 2\hbar\Omega + V(k) + V(k')$, always exists. To terms of the order of $|V(k)|^2/|A|$ it is given by the expression [cf. eq. (7a)]

$$E_2(K) = 2\hbar\Omega - 2A + W^A(K) - \frac{1}{2AN} \sum_q [V(\tfrac{1}{2}K + q) + V(\tfrac{1}{2}K - q)]^2.$$

If the anharmonicity is small, so that $|A| < |V(k)|$, eq. (20a) may not have a solution $E_2(k)$ lying outside the spectral region of two-particle states. In this case no biphonons are formed, but, by the effect of the anharmonicity, new maxima, not directly associated with van Hove points, can sometimes occur in the density of two-particle states. These maxima may appear, in particular, in RSL spectra and in the spectra of inelastically scattered neutrons. They are due to the possibility of the formation of quasi-steady ("resonance") quasi-bound states of two quasi-particles that are capable of dissociating, at the same energy, into two free phonons. Such maxima have not yet been identified experimentally. We shall dwell upon the ensuing special features in discussing the experimental data.

If a unit cell of the crystal contains several molecules, then, as has already been noted, the pattern of the spectrum of two-particle states becomes more complicated even when anharmonicity is ignored. As concerns the number of biphonon bands, it is equal, under conditions of strong anharmonicity ($|A| \gg |V_{nm}|$) for nondegenerate vibrational transitions, to the number σ of molecules in the unit cell.

As a matter of fact, as follows from eq. (15), the energy E of the biphonon in this case is

$$E - \epsilon_\mu(k) - \epsilon_{\mu'}(k') \approx E - 2\hbar\Omega,$$

so that

$$G_E^0(n, m \mid l, l') \approx \frac{1}{E - 2\hbar\Omega}\, \delta_{nl}\delta_{ml'}.$$

In this limit of large $|A|$ values the expression for $R_{\alpha\beta}$ [see eq. (18)] simplifies:

$$R_{\alpha\beta} = -\frac{2A\delta_{\alpha\beta}}{E - 2\hbar\Omega} + \frac{W_{\alpha\beta}^A(\mathbf{K})}{E - 2\hbar\Omega}.$$

Substituting this relation into eq. (17) we find that the quantities $\varphi_{\alpha\alpha}(0)$ satisfy a system of σ equations:

$$(E - 2\hbar\Omega + 2A)\varphi_{\alpha\alpha}(0) = \sum_\beta W_{\alpha\beta}^A(\mathbf{K})\varphi_{\beta\beta}(0),$$

similar to the system of equations for the Frenkel exciton in a crystal with σ molecules per unit cell. This proves the statement made above on the number of biphonon bands in the limiting case of large $|A|$ values.

If, however, anharmonicity is not too strong, then, as has been pointed out (Agranovich 1970), the number of biphonon bands may not be equal to σ (for $A = 0$ and $W^A = 0$ there are absolutely no such bands).

Light absorption by biphonons in anisotropic crystals can be strongly polarized, the corresponding polarization of the absorption lines being closely associated with the crystal symmetry. This should be kept in mind in discussing experimental investigations.

2.2. Biphonons in the combination tone frequency region of the spectrum—The van Kranendonk model

We shall now discuss the special features in the spectrum of excited states of a crystal in the combination tone frequency region of intramolecular vibrations. Assume, for instance, that we are concerned with the frequency region $\omega \approx \Omega_1 + \Omega_2$, where Ω_1 and Ω_2 are the frequencies of two nondegenerate intramolecular vibrations. Here the model Hamiltonian of the crystal, for the spectral region being considered, can be represented in the form [see also eq. (5a)]

$$\hat{H} = \hat{H}_0^{(1)} + \hat{H}_0^{(2)} - 2A \sum_n (B_n^{(1)})^+ (B_n^{(2)})^+ B_n^{(1)} B_n^{(2)}$$

$$+ \frac{1}{2} \sum_{n,m}' W_{n,m}^A (B_n^{(1)})^+ (B_n^{(2)})^+ B_m^{(1)} B_m^{(2)}, \tag{21}$$

where the operators $\hat{H}_0^{(1)}$ and $\hat{H}_0^{(2)}$ are expressed in terms of the creation and annihilation operators $(B_n^{(i)})^+$ and $B_n^{(i)}$ (where $i = 1, 2$) by a relation of the type of eq. (1).

The Hamiltonian (21) commutes with the operator

$$I_{nm} = (B_n^{(1)})^+ (B_m^{(2)})^+ B_n^{(2)} B_m^{(1)} + (B_n^{(2)})^+ (B_m^{(1)})^+ B_n^{(1)} B_m^{(2)},$$

which effects the interchange of the quanta $\hbar\Omega_1$ and $\hbar\Omega_2$, localized on molecules n and m. Consequently, the wave functions of a crystal with two quanta of different kinds can be either even or odd with respect to the interchange of the coordinates n and m. Since, in the case being considered,

$$|2\rangle = \sum_{n,m} \psi(n, m)(B_n^{(1)})^+ (B_m^{(2)})^+ |0\rangle, \tag{22}$$

the aforesaid means that there are even steady states, for which $\psi(n, m) = \psi(m, n)$, as well as odd ones, for which $\psi(n, m) = -\psi(m, n)$.

Since for odd states $\psi(n, n) = 0$, the Hamiltonian (21) cannot lead to the occurrence of odd biphonon states. This conclusion follows directly from the fact that the result of the action of the anharmonicity operator in Hamiltonian (21) on the odd wave functions of the biphonon is equal to zero. For example,

$$\hat{H}_A^{(2)}|2\rangle = -2A \sum_n \psi(n, n)(B_n^{(1)})^+ (B_n^{(2)})^+ |0\rangle = 0,$$

and in a similar way for the operator containing the quantity W^A.

If we include into the Hamiltonian (21) not only the intramolecular anharmonicity (proportional to A) and the anharmonicity due to the intermolecular interaction W^A, but also the part of the intermolecular anharmonicity that is of the form $\sum_{nm} L_{nm} \hat{I}_{nm}$, where L_{nm} is the matrix element of the operator of intermolecular interaction between molecules n and m, corresponding to an interchange of quanta, the odd biphonon states can also separate from the band of two-particle states. But since $|L_{nm}| \lesssim |V_{nm}^{(i)}|$, where $i = 1, 2$, the odd biphonon levels, in contrast to the even ones, must always be close to the band of two-particle states.

So far the odd biphonon states have not been discovered experimentally. In molecular crystals their contribution to the absorption spectrum or Raman scattering spectrum should be relatively small because in the odd states (22) there are no superposed configurations in which both quanta "sit" on one and the same molecule.

Transitions to these superposed configurations are exactly those that usually correspond to relatively high oscillator strengths, because the intramolecular constants of both mechanical and electrical anharmonicity exceed, as a rule, the corresponding constants of intermolecular anharmonicity. Later on we shall return again to these problems in finding the dielectric constant of a crystal for the spectral region of two-particle states. We shall now continue our discussion of the Hamiltonian (21).

If in expression (21) we put the quantity $V_{nm}^{(2)}$, for example, equal to zero, i.e., ignore the possibility of motion of one of the quanta, we arrive at the case that was investigated by van Kranendonk (1959). Assuming that the quantum $\hbar\Omega_2$ is localized on the molecule $n = 0$ and averaging the Hamiltonian (21) over the indicated state, we obtain a simpler expression for $\hat{H}$:

$$\hat{H} = \sum_n \hbar\Omega_1 (B_n^{(1)})^+ B_n^{(1)} + \sum_{n,m}{}' V_{nm}^{(1)} (B_n^{(1)})^+ B_m^{(1)} - 2A(B_0^{(1)})^+ B_0^{(1)} + \hbar\Omega_2,$$

which corresponds to the motion of the quantum $\hbar\Omega_1$ in a crystal with an "isotopic" substitutional impurity located at the lattice point $n = 0$ and with the shifted frequency of intramolecular vibration $\Omega_1' = \Omega_1 - 2A/\hbar$. Thus, in the van Kranendonk model, the energy calculation for the biphonon reduces to the well-known problem of calculating the energy of a local vibration in the region of an isotopic defect. In the given case this energy is determined by the equation

$$1 = -\frac{2A}{N} \sum_k{}' \frac{1}{E_2 - \hbar(\Omega_1 + \Omega_2) - V^{(1)}(k)}$$

$$\equiv -\frac{2A}{N} \int \frac{\rho(\epsilon)\, d\epsilon}{E_2 - \hbar\Omega_2 - \epsilon},$$

where $V^{(1)}(k)$ is determined by relation (3a) with $V_{nm} = V_{nm}^{(1)}$, $\rho(\epsilon)$ is the density of $B^{(1)}$-phonon states, and $\int \rho(\epsilon)\, d\epsilon = N$. It is clear that the van Kranendonk model, within the framework of which the bound states of two phonons were first discussed, is not suitable for analysis of biphonon spectra in the overtone frequency region or in the region of combination tones, corresponding to phonons with comparable bandwidths. But the application of this model was found to be highly successful in analysis of the infrared (IR) spectrum of crystalline hydrogen in the region of its rotation–vibration band [see the review of theoretical and experimental investigations compiled by van Kranendonk and Karl (1968)].

As a conclusion of this subsection we write down the equation which determines the energy of the biphonon when the motion of both quanta is taken into account. It can readily be seen that within the framework of the model (21) this equation is of the form

$$1 = \frac{1}{N} \sum_q \frac{-2A + W^A(\boldsymbol{K})}{E_2 - \hbar\Omega_1 - \hbar\Omega_2 - V^{(1)}(\boldsymbol{K}/2 + \boldsymbol{q}) - V^{(2)}(\boldsymbol{K}/2 - \boldsymbol{q})},$$

which is a simple generalization of eq. (20a).

3. Green's function method in biphonon theory—Fermi resonance in crystals

The relations obtained above for the energies of biphonons can also be derived by applying the Green's function method. This method is found to be extremely useful within the framework of the model being discussed, because all Green's functions required for calculating the dielectric constant of the crystal, its nonlinear polarizabilities, its density of states, its RSL cross section and other physical properties can be found exactly, without resorting to perturbation theory, notwithstanding the fact that anharmonicity is taken into account. We shall illustrate the aforesaid below, but shall employ a model that is more general than that previously used. To be quite exact, we shall find the Green's functions for the case when Fermi resonance is present in the crystal.

Upon Fermi resonance in an isolated molecule, the frequency of one of the molecular vibrations turns out to be close to the overtone frequency (or the combination tone frequency) of some other vibration. In the case, for example, of nondegenerate vibrations, resonance occurs between two excited states of the molecule. Owing to the anharmonicity of intramolecular vibrations, this leads to characteristic doublets of comparable intensity in the absorption spectra or the RSL spectra or (depending upon the symmetry of the molecule and the type of vibration) in both. If degenerate vibrations also participate in the Fermi resonance, the number of lines in these spectra can even be large (Herzberg 1945). In going over from an isolated molecule to a crystal, branches of optical phonons appear in the region of the fundamental vibrations of the molecule due to the translational symmetry and to the effect of the intermolecular interaction. In the region of overtone and combination tone frequencies, bands of many-particle states appear and, if anharmonicity is sufficiently strong, bands of states with quasi-particles bound to one another (for instance, biphonons) as well. Therefore, in general, a large number of excited states of the crystal resonate with one another in the Fermi resonance, substantially complicating the spectra obtained.

In order to analyze these spectra and, in particular, to investigate the effects of Fermi resonance on biphonon spectra, it is necessary to generalize the Hamiltonian (8) to some extent.

We shall assume that the conditions for Fermi resonance are satisfied in a free molecule, i.e., that there are two (for the sake of simplicity) nondegenerate vibrations with the frequencies Ω_1 and Ω_2, for which, for instance, $2\Omega_1 \approx \Omega_2$. In this case, when taking the intramolecular anharmonicity (with the constant Γ) into account, it is necessary to add to the Hamiltonian (8) the sum of two terms: $\hat{H}_0(C)$ and $\hat{H}_F(B, C)$, where

$$\hat{H}_0(C) = \sum_n \hbar \Omega_2 C_n^+ C_n + \sum_{n,m}' V_{nm}^{(2)} C_n^+ C_m, \tag{23}$$

$$\hat{H}_F(B, C) = \Gamma \sum_n [(B_n^+)^2 C_n + C_n^+ (B_n)^2], \tag{24}$$

so that the total Hamiltonian $\hat{H}$ assumes the form

$$\hat{H} = \hat{H}_0(B) + \hat{H}_0(C) + \hat{H}_A(B) + \hat{H}_T(B) + \hat{H}_F(B, C). \tag{25}$$

We shall assume for simplicity that there is a single molecule per unit cell of the crystal and proceed to a momentum representation for the operators B and C:

$$B_n = \frac{1}{\sqrt{N}} \sum_k B_k \, e^{ikn},$$

$$C_n = \frac{1}{\sqrt{N}} \sum_k C_k \, e^{ikn}.$$

In this representation, the Hamiltonian (25) is of the form

$$\hat{H} = \sum_k [\epsilon_1(k) B_k^+ B_k + \epsilon_2(k) C_k^+ C_k]$$
$$- \frac{1}{N} \sum_{k,k',q} \tilde{A}(k + k') B_k^+ B_{k'}^+ B_q B_{k+k'-q} + \frac{\Gamma}{\sqrt{N}} \sum_{k,k'} (B_k^+ B_{k'}^+ C_{k+k'} + \text{h.c.}), \tag{25a}$$

where

$$\tilde{A}(k) = A - \tfrac{1}{2} W^A(k).$$

If we introduce the operator

$$\hat{T}(k) = \frac{1}{\sqrt{N}} \sum_q B_{k/2-q} B_{k/2+q}, \tag{25b}$$

then the Hamiltonian (25a) can be written in the more concise form:

$$\hat{H} = \sum_k \{ \epsilon_1(k) B_k^+ B_k + \epsilon_2(k) C_k^+ C_k - \tilde{A}(k) T^+(k) T(k)$$
$$+ \Gamma [T^+(k) C_k + C_k^+ T(k)] \}. \tag{25c}$$

Assuming the temperature of the crystal to be zero, we shall find the retarded Green's function

$$G_k^{(1)}(t) = -i\theta(t)\langle 0|C_k(t)C_k^+(0)|0\rangle,$$

where $\theta(t) = 1$ for $t > 0$ and $\theta(t) = 0$ for $t < 0$. We find the equation for this function by making use of the equation of motion for the Heisenberg operator $\hat{A}$:

$$i\hbar \, d\hat{A}/dt = \hat{A}\hat{H} - \hat{H}\hat{A},$$

the relation $d\theta/dt = \delta(t)$ for the θ-function and the well-known commutation relations for the Bose operators B and C. Differentiating the expression for $G_k^{(1)}$ with respect to t we obtain

$$-i \, dG_k^{(1)}/dt = -\delta(t) - \theta(t)\langle 0|dC_k/dt, C_k^+(0)|0\rangle. \tag{26}$$

Since

$$i\hbar \, dC_k/dt = C_k\hat{H} - \hat{H}C_k = \epsilon_2(k)C_k + \Gamma T(k),$$

eq. (26) assumes the form

$$i\frac{dG_k^{(1)}}{dt} = \delta(t) + \frac{1}{\hbar}\epsilon_2(k)G_k^{(1)} + \frac{\Gamma}{\hbar}G_k^{(2)}, \tag{27}$$

where

$$G_k^{(2)}(t) = -i\theta(t)\langle 0|T(k, t)C_k^+(0)|0\rangle. \tag{28}$$

Next we introduce the Green's function

$$G_{k,q}^{(3)}(t) = -i\theta(t)\frac{1}{\sqrt{N}}\langle 0|B_{k/2+q}(t)B_{k/2-q}(t)C_k^+(0)|0\rangle, \tag{29}$$

such that obviously

$$\sum_q G_{k,q}^{(3)}(t) = G_k^{(2)}(t). \tag{30}$$

Differentiating relation (29) with respect to time, we obtain in a similar way

$$-i\frac{dG_{k,q}^{(3)}}{dt} = -\frac{1}{\hbar}[\epsilon_1(\tfrac{1}{2}k + q) + \epsilon_1(\tfrac{1}{2}k - q)]G_{k,q}^{(3)}$$

$$+ i\theta(t)\frac{\tilde{A}(k)}{\hbar\sqrt{N}}\langle 0|B_{k/2-q}(t)B_{k/2+q}(t)T^+(k, t)T(k, t)C_k^+(0)|0\rangle$$

$$- i\theta(t)\frac{\Gamma}{\hbar\sqrt{N}}\langle 0|B_{k/2-q}(t)B_{k/2+q}(t)T^+(k, t)C_k(t)C_k^+(0)|0\rangle. \tag{31}$$

This equation simplifies if we take into account that for arbitrary operators D_1 and D_2 [see the definition (25b)]

$$\langle 0|B_{k/2-q}(t)B_{k/2+q}(t)T^+(k,t)D_1D_2|0\rangle = \langle 0|D_2^+D_1^+T(k,t)B_{k/2+q}^+(t)B_{k/2-q}^+(t)|0\rangle$$

$$= \frac{2}{\sqrt{N}}\langle 0|D_2^+D_1^+|0\rangle = \frac{2}{\sqrt{N}}\langle 0|D_1D_2|0\rangle.$$

Hence, eq. (31) can also be written as follows:

$$-i\frac{dG_{k,q}^{(3)}}{dt} = -\frac{1}{\hbar}[\epsilon_1(\tfrac{1}{2}k+q)+\epsilon_1(\tfrac{1}{2}k-q)]G_{k,q}^{(3)} + \frac{2\tilde{A}(k)}{\hbar N}G_k^{(2)} - \frac{2\Gamma}{\hbar N}G_k^{(1)}. \qquad (32)$$

Turning now to the Fourier representation with respect to time, we find that

$$G_{k,q}^{(3)}(\omega) = -\frac{(2\tilde{A}/N)G_k^{(2)}(\omega)-(2\Gamma/N)G_k^{(1)}(\omega)}{\hbar\omega - \epsilon_1(\tfrac{1}{2}k+q)-\epsilon_1(\tfrac{1}{2}k-q)}, \qquad (32a)$$

so that [see eq. (30)]

$$G_k^{(2)}(\omega) = -[2\tilde{A}(k)G_k^{(2)}(\omega) - 2\Gamma G_k^{(1)}(\omega)]R(E,k), \qquad (33)$$

where

$$R(E,k) = \frac{1}{N}\sum_q \frac{1}{\hbar\omega - \epsilon_1(\tfrac{1}{2}k+q)-\epsilon_1(\tfrac{1}{2}k-q)}$$

$$= 2\int \frac{\rho_0(\epsilon,k)}{E-\epsilon}\,d\epsilon, \qquad (33a)$$

$$E = \hbar\omega,$$

$$\rho_0(\epsilon,k) = \frac{1}{2N}\sum_q \delta[\epsilon - \epsilon_1(\tfrac{1}{2}k+q)-\epsilon_1(\tfrac{1}{2}k-q)]$$

is the density of two-particle states with the total wave vector k.

In a similar manner, making use of eq. (27), we can obtain a second relation linking the functions $G_k^{(1)}(\omega)$ and $G_k^{(2)}(\omega)$:

$$G_k^{(1)}(\omega)[\hbar\omega - \epsilon_2(k)] - \Gamma G_k^{(2)}(\omega) = \hbar. \qquad (33b)$$

From relations (33) and (33b) we obtain the required Green's functions:

$$G_k^{(1)}(\omega) = \frac{\hbar[1+2\tilde{A}(k)R(E,k)]}{[E-\epsilon_2(k)]\Delta(E,k)}, \qquad (34)$$

where

$$\Delta(E,k) = 1 + 2\left(\tilde{A}(k) - \frac{\Gamma^2}{E-\epsilon_2(k)}\right)R(E,k), \qquad (34a)$$

$$G_k^{(2)}(\omega) = \frac{2\hbar\Gamma R(E,k)}{[E-\epsilon_2(k)]\Delta(E,k)}. \qquad (35)$$

The function $G_{k,q}^{(3)}(\omega)$ is also found to be completely determined, in accordance with eq. (32a).

Along with the functions $G_k^{(1)}$ and $G_k^{(2)}$, a number of other Green's functions must be known in order to calculate the dielectric constant of a crystal in the overtone frequency region, as well as the RSL cross section and the cross sections of nonlinear optical processes. Among others is required the two-particle Green's function $G_{k,q,q'}^{(4)}(t)$, which is determined by the relation

$$G_{k,q,q'}^{(4)}(t) = -i\theta(t)\langle 0|B_{k/2+q}(t)B_{k/2-q}(t)B_{k/2+q'}^+(0)B_{k/2-q'}^+(0)|0\rangle. \tag{36}$$

The calculation of this function is perfectly analogous to the foregoing so that only the final result will be given below. It can be shown that

$$G_{k,q,q'}^{(4)}(\omega) = -\frac{2}{\sqrt{N}} \frac{\tilde{A}(k)G_{k,q'}^{(5)}(\omega) - \Gamma G_{k,q'}^{(6)}(\omega)}{E - \epsilon_1(\tfrac{1}{2}k + q) - \epsilon_1(\tfrac{1}{2}k - q)}$$

$$+ \frac{\hbar(\delta_{q+q'} + \delta_{q-q'})}{E - \epsilon_1(\tfrac{1}{2}k + q) - \epsilon_1(\tfrac{1}{2}k - q)}, \tag{36a}$$

where $G_{k,q'}^{(5)}(\omega)$ is the Fourier component of the Green's function

$$G_{k,q'}^{(5)}(t) = -i\theta(t)\langle 0|T(k, t)B_{k/2-q'}^+(0)B_{k/2+q'}^+(0)|0\rangle,$$

and is determined by the relation

$$G_{k,q'}^{(5)}(\omega) = \frac{2\hbar}{\sqrt{N}\,[E - \epsilon_1(\tfrac{1}{2}k + q') - \epsilon_1(\tfrac{1}{2}k - q')]} \frac{1}{\Delta(E, k)}, \tag{37}$$

whereas

$$G_{k,q'}^{(6)}(\omega) = \frac{\Gamma}{E - \epsilon_2(k)} G_{k,q'}^{(5)}(\omega) \tag{38}$$

is the Fourier component of the Green's function

$$G_{k,q'}^{(6)}(t) = -i\theta(t)\langle 0|C_k(t)B_{k/2+q'}^+(0)B_{k/2-q'}^+(0)|0\rangle.$$

It follows from the expressions given above for the Green's functions $G^{(i)}(\omega)$, where $i = 1, 2, \ldots, 6$, that, when anharmonicity is taken into consideration in the system of phonons, characterized within the framework of the model being considered by the anharmonicity constants $\tilde{A}$ and Γ, this leads to the appearance, along with poles of the type of eq. (1), of a new type of poles for the Green's function. These poles are determined by the equation

$$\Delta(k, \omega) = 0. \tag{39}$$

This equation, a generalization of eq. (20a), enables one to calculate the energy of the biphonons, taking into account the Fermi resonance of the two-particle B-phonon states with the band of C phonons. A comparison of relations (39) and (20a) indicates that taking the Fermi resonance into account amounts to a renormalization of the anharmonicity constant:

$$\tilde{A}(k) \rightarrow \tilde{A}(k) - \frac{\Gamma^2}{E - \epsilon_2(k)}, \quad E = \hbar\omega.$$

The new anharmonicity "constant" becomes a function of the energy E and its effective magnitude in the energy region being considered is found to depend substantially on the position of the C-phonon energy with respect to the band of two-particle states. Hence, Fermi resonance, in general, strongly affects the conditions for the formation of biphonons and the positions of their levels. This equation, determining the energy $E = E' + i\gamma$ of the biphonon, can be rewritten, for convenience, in the form

$$\Phi_1(E) = \Phi_2(E), \tag{40}$$

where

$$\Phi_1(E) = -1 + \frac{\Gamma^2/\tilde{A}(k)}{\tilde{\epsilon}_2(k) - E}, \quad \tilde{\epsilon}_2 = \epsilon_2 + \Gamma^2/\tilde{A}, \tag{41}$$

$$\Phi_2(E) = 2\hat{A}(k) \int \frac{\rho_0(\epsilon, k)}{E - \epsilon} \, d\epsilon + 2\pi \, i\tilde{A}(k)\rho_0(E, k). \tag{42}$$

The function $\Phi_1(E)$ for the case $\tilde{A} > 0$ is given schematically in fig. 1 (it is taken into account here that $\Gamma^2/\tilde{A}\epsilon_2(k) \ll 1$). In the same figure a dashed line shows the value of $\Phi_1(E)$ for the case when the Fermi resonance is neglected (i.e. for $\Gamma = 0$; in this case $\Phi_1(E) = -1$).

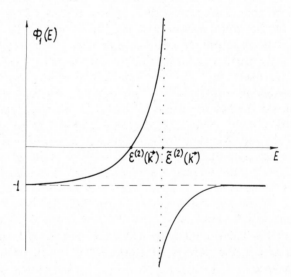

Fig. 1. The function $\Phi_1(E)$. $\epsilon_2(k)$ is the C-phonon energy and $\tilde{\epsilon}_2 = \epsilon_2 + \Gamma^2/A$.

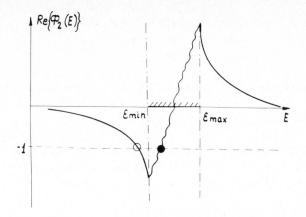

Fig. 2. The function $\text{Re}\{\Phi_2(E)\}$ is shown by the full line. The dashed line represents the function $\Phi_1(E) = -1$. The open circle represents a biphonon, the filled circle a quasi-biphonon.

Presented in fig. 2 for the same case of $\tilde{A} > 0$ is the relationship $\text{Re}\{\Phi_2(E)\}$ with damping ignored. Within the framework of the model being discussed, the damping of the biphonon states may be associated with dissociation into two free phonons. It is clear that such quasi-steady (resonance) states—quasi-biphonons—can have a physical meaning only in the region of small $\rho_0(E, k)$ values, where their width $\gamma \ll \text{Re}\{E\} = E'$.

The function $\text{Re}\{\Phi_2(E)\}$ is everywhere bounded in three-dimensional crystals. In particular, for states with small k values $\rho(\epsilon, k) \approx \rho(\epsilon, 0)$ and in the vicinity of the edge $\bar{\epsilon}$ (where $\bar{\epsilon} = \epsilon_{\min}$ or $\bar{\epsilon} = \epsilon_{\max}$) of the band of two-particle states $\rho_0(\epsilon, 0) = \rho_0 |\bar{\epsilon} - \epsilon|^{1/2}$. Therefore, as $E \to \bar{\epsilon}$, then outside the band of two-particle states the function $\Phi_2'(E) = \text{Re}\{\Phi_2(E)\}$, while remaining finite itself, only has an infinite derivative (see fig. 2). Also shown by a dashed line in fig. 2 is the function $\Phi_1 = -1$ for $\Gamma = 0$. It follows from this figure that when the value of $|\Phi_2(E)|$ for $E \to \epsilon_{\min}$ tends to a value less than unity (i.e. for $|\Phi_2(\epsilon_{\min})| < 1$), a bound state of two phonons—a biphonon—is formed. Along with the biphonon state, with energy $E_b < \epsilon_{\min}$ (open circle in fig. 2), a quasi-biphonon (solid circle in fig. 2) is also formed with energy E_{qb}, lying within the band of two-particle states. When anharmonicity is not too strong, so that the value of $|\Phi_2(\epsilon_{\min})|$ does not exceed unity too much, the quasi-biphonon falls in the region of low density of two-particle states, and its width is found to be small compared to the bandwidth. Then a distinct peak, not associated with van Hove points, is formed in the density of states found when anharmonicity is taken into account.

As a matter of fact, the density of states with the total wave vector

$q + q' = k$ is determined by the relation

$$\rho(E, k) = -\frac{1}{2\pi N} \sum_q (1 + \delta_{q0}) \operatorname{Im}\{G^{(4)}_{k,q,q}(E + i\gamma)\}, \quad \gamma \to +0, \tag{43}$$

while for noninteracting phonons (i.e. for $\tilde{A} = \Gamma = 0$), the relation is

$$\rho(E, k) \equiv \rho_0(E, k) = \frac{1}{2N} \sum_q (1 + \delta_{q0}) \delta[E - \epsilon_1(\tfrac{1}{2}k + q) - \epsilon_1(\tfrac{1}{2}k - q)].$$

If $\tilde{A} \neq 0$, but $\Gamma = 0$ (no Fermi resonance), the function $G^{(4)}_{k,q,q}(E)$, according to eq. (36), is determined by the relation

$$G^{(4)}_{k,q,q}(E) = [E - \epsilon_1(\tfrac{1}{2}k + q) - \epsilon_1(\tfrac{1}{2}k - q)]^{-1}$$
$$- \frac{4\tilde{A}(k)\Delta^{-1}(E, k)}{N[E - \epsilon_1(\tfrac{1}{2}k + q) - \epsilon_1(\tfrac{1}{2}k - q)]^2},$$

so that

$$\rho(E, k) = \rho_0(E, k) + \frac{16\tilde{A}^2(k)D(E, k)\rho_0(E, k)}{[1 + 2\tilde{A}(k)R'(E, k)]^2 + 16\pi^2\tilde{A}^2(k)\rho_0^2(E, k)}, \tag{44}$$

where

$$R'(E, k) = \operatorname{Re}\{R(E, k)\} = 2 \int \frac{\rho_0(\epsilon, k)}{E - \epsilon} \, d\epsilon,$$

$$D(E, k) = \int \frac{\rho_0(\epsilon, k)}{(E - \epsilon)^2} \, d\epsilon. \tag{44a}$$

In the range of energies E in which $\rho_0(E, k) = 0$, i.e. outside the band of two-particle states, relation (44) reduces to the following:

$$\rho(E, k) = 4D(E, k)|\tilde{A}(k)|\delta[1 + 2\tilde{A}(k)R(E, k)], \tag{44b}$$

or

$$\rho(E, k) = \delta(E - E_b),$$

where the energy E_b of the biphonon is the root of the equation

$$1 + 4\tilde{A}(k) \int \frac{\rho_0(\epsilon, k)}{E - \epsilon} \, d\epsilon = 0. \tag{45}$$

Within the band of two-particle states, in the range of energies $E \approx E_{qb}$, where E_{qb} is the root of the equation

$$1 + 4\tilde{A}(k) \int \frac{\rho_0(\epsilon, k)}{E - \epsilon} \, d\epsilon = 0,$$

relation (44) can be written in the form

$$\rho(E, k) = \rho_0(E, k) + \frac{1}{\pi} \frac{\gamma}{(E - E_{qb})^2 + \gamma^2}, \tag{46}$$

where $\gamma = \pi\rho_0(E_{qb}, k)/D(E_{qb}, k)$ is the half width of the quasi-biphonon level. From this expression for γ it follows that this half width can be sufficiently small only when the level of the quasi-biphon is within the region of low density $\rho_0(E, k)$ of the levels of two-particle states.

Next we turn to a discussion of the case with Fermi resonance ($\tilde{A} \neq 0$ and $\Gamma \neq 0$).

In this case the position and number of roots of eq. (40) depends substantially on the relation between the quantities $\tilde{A}$ and Γ, and on the position of the energy $\epsilon_2(k)$ with respect to the band of two-particle states. To illustrate the aforesaid, the curves $\Phi_1(E)$ and $\Phi_2(E)$ are shown in figs. 3, 4 and 5, and the roots of eq. (40) are indicated for three limiting situations.

Figure 3 corresponds to the case in which the energy $\epsilon_2(k)$ of the C phonon, as well as the energy $\tilde{\epsilon}_2(k) = \epsilon_2(k) + \Gamma^2/A(k)$, are located below the band of two-particle states and sufficiently distant from the bottom of the band, ϵ_{min}. In this case the number of solutions to eq. (40), lying outside the band of two-particle states, is equal to two. One of these (the lower one) is generically associated with the C-phonon state and goes over into this state as the energy of the C phonon moves away from the band of two-particle states.

In the case depicted in fig. 4, the energy $\tilde{\epsilon}_2(k)$ is within the band of two-particle states, and the quantity $\tilde{A} \approx \Gamma$. For the energy region $E < \epsilon_{min}$, eq. (40) has only a single solution, and no quasi-biphonon state is formed.

If the energy of the C phonon lies above the band of two-particle states

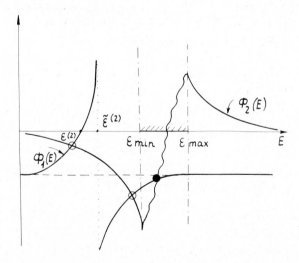

Fig. 3. The functions $\Phi_1(E)$ and $\Phi_2(E)$ in the presence of a Fermi resonance: ϵ_{min} and ϵ_{max} are the boundaries of the band of two-particle dissociated states. The open and filled circles represent the biphonon and quasi-biphonon, respectively. $\tilde{\epsilon}_2 < \epsilon_{min}$.

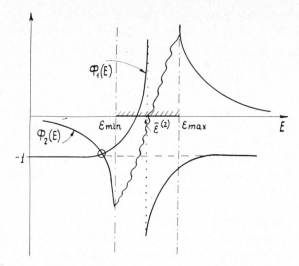

Fig. 4. Same as fig. 3, with $\epsilon_{min} < \tilde{\epsilon}_2 < \epsilon_{max}$.

and its bandwidth Δ is large compared to Γ, a situation is possible that corresponds to fig. 5. Here the solutions of eq. (40) lie on different sides of the band of two-particle states, and the formation of a quasi-biphonon state is possible.

Relations (36), (37) and (38) are required to find the density of states in the presence of Fermi resonance, and this can be done in a way similar to the case of $\Gamma = 0$. We shall not give the corresponding calculations here, but shall turn to a discussion of a more general situation arising in the case of Fermi resonance with a polariton.

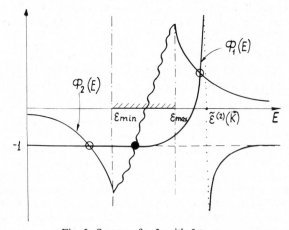

Fig. 5. Same as fig. 3, with $\tilde{\epsilon}_2 > \epsilon_{max}$.

4. Fermi resonance with polaritons

4.1. Microscopic theory

Let us consider the effects that arise when the branch of C phonons corresponds to dipole-active vibrations. In the region of small values of $k = 2\pi/\lambda$, where λ is the wavelength of light with frequency $\omega = \epsilon_2/\hbar$, C phonons of this kind strongly interact with the transverse photons. As a result, in the region of long wavelengths, new elementary excitations— polaritons*—are formed instead of C phonons and transverse photons. In their properties (spectrum and polarization) at low k values, these polaritons appreciably differ from both transverse photons in vacuum and phonons.

The spectrum of polaritons can be found by means of Maxwell's macroscopic equations, provided that the dielectric tensor of the medium (Agranovich and Ginzburg 1983) is assumed to be known. The corresponding results of analyses, concerned with the application of lattice vibration theory in the harmonic approximation, are also well known and have been expounded in many monographs and textbooks on solid-state theory. Without going into details, we emphasize here that always a gap appears in the polariton spectrum (here we ignore spatial dispersion) in the region of the fundamental dipole-active vibration (C phonon, exciton, etc.). The gap width is proportional to the oscillator strength of the corresponding resonance. If many phonon branches are taken into account, many gaps are formed; this is also well known, of course. At present, there is a sufficiently detailed theory for RSL by polaritons, taking many phonon bands into consideration. With this theory the RSL cross section can be calculated for various scattering angles provided that the dielectric tensor of the crystal is known, as well as the dependence of the polarizability of the crystal on the displacement of the lattice sites and the electric field generated by this displacement (Wright 1969).

It is an essential fact that the above-mentioned gaps in the polariton spectrum, as well as the corresponding interaction between the photon and phonon, are nonzero even within the framework of linear theory and, in

---

*The term "polariton" was proposed by Hopfield (1958). But in his paper he gave the name "polariton" to elementary excitations (phonons, excitons, etc.) capable, in the dipole approximation, of interacting with light. Being quite sure that it proves expedient to introduce new terms only to denote new phenomena, the author of the present chapter, also occupied in those years with the quantum theory of electromagnetic waves in the exciton region of the spectrum (Agranovich 1959), committed an error in his paper (Agranovich 1960), employing the term "polariton" to denote new elementary excitations that occur in the "mixing" of excitons and photons. It is precisely this definition of the term that has turned out to be generally accepted today.

general, do not require that anharmonicity be taken into account. Therefore, it makes sense to denote as polariton Fermi resonance only such
situations where vibrations of overtone or combination tone frequencies
resonate with the polariton. We now turn our attention to an analysis of
such rather complex situations, requiring that many-particle excited states
of the crystal be taken into consideration. Shown schematically in fig. 6 is a
typical polariton spectrum, as well as a band of two-particle states of B
phonons. If, under the effect of anharmonicity, biphonons with energy
$E = E_b$ are formed, these states also resonate with the polariton, influencing its spectrum.

Since RSL by polaritons is extremely intense for many crystals that lack
a centre of inversion, whereas second order RSL (i.e. RSL accompanied
by the simultaneous excitation of two quasi-particles) is relatively weak, as
a rule, the intersection of the biphonon levels and bands of two-particle
states with the polariton branch, shown in fig. 6 and resulting from
anharmonicity, leads to the partial transfer of RSL intensity from the
polariton to the biphonon and two-particle states. This can be taken as a
typical indication of the phenomenon. We note, however, that along with
this effect, which is of great significance in the experimental investigation
of the above-mentioned states, substantial changes occur in the spectra of
polaritons and two-particle states in the region of the intersection (see fig.
6). These changes may also appear in the RSL spectra. Since, as will be
shown in the following, the nature of these changes in a fundamental way
depends on whether or not biphonons are formed in the region of overtone
or combination tone frequencies, we can come to a conclusion on the
existence of bound states of phonons (biphonons) experimentally by in-

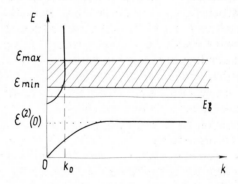

Fig. 6. Polariton dispersion in the Fermi resonance region, neglecting anharmonicity: $\epsilon_2(0)$ is
the energy of the fundamental vibration at $k = 0$; ϵ_{min} and ϵ_{max} are the minimum and maximum
energy values in the band of two-particle states.

vestigating the spectra of RSL by polaritons in a polariton Fermi resonance. Extensive experimental data are available at present in this field of investigation and they are reviewed below in section 8. We now return to the results of the preceding section and intend to show how Fermi resonance with polaritons can be investigated within the framework of microtheory [see also Agranovich and Lalov (1971a)].

To take the interaction between phonons and photons into consideration, it is necessary to add to the Hamiltonian (25), the Hamiltonian $\hat{H}_0(a)$ of the free field of transverse photons and the Hamiltonian $\hat{H}_{int}$ for the interaction of the field of transverse photons with phonons.

In taking anharmonicity into account, the Hamiltonian $\hat{H}_{int}$ should include not only terms that are quadratic with respect to the Bose operators a, B and C, but cubic terms as well. We shall discuss the structure of these terms below. Here we only point out that the linear transformation from the operators a, B and C to the polariton creation and annihilation operators, i.e. to the operators $\xi_\rho^+(k)$ and $\xi_\rho(k)$, where ρ is the number of the polariton branch and k is the polariton wave vector, diagonalizes the quadratic part of the total Hamiltonian. But such a diagonalization of the quadratic part of the total Hamiltonian is actually unnecessary for analyzing spectra in the Fermi resonance region. The transition from the operators a and C to the polaritons in the spectral region being considered is really essential because the structure and spectrum of photons and C phonons depend substantially on taking retardation into account. But as to the two-particle states of the crystal, corresponding to the excitation of two B phonons, taking retardation into consideration may prove to be significant only in exceptional cases. In the two-particle states, even at low values of the total momentum k (only these states resonate with a polariton), the principal contribution, if we are not concerned with the edges of the band of two-particle states, is made by the states $B_{k/2+q}^+ B_{k/2-q}^+ |0\rangle$ with high q values. Since it is inessential to take retardation into account for B phonons with high wave vector values, the meaning of the above statement on the degree of the influence exerted by retardation on the two-particle states of B phonons becomes clear.

The exceptional cases mentioned above may occur in crystals where small k values in the B-phonon band correspond to minimum or maximum energy of the B phonon, where the B phonon is dipole-active and the oscillator strength corresponding to it is sufficiently large. In these cases, the band edge of two-particle states, a minimum or maximum, respectively, is diffuse over a range of the order of the energy $\epsilon_{\parallel,\perp}$ of longitudinal–transverse splitting of the B phonon. Here it may become important to take retardation into account if the B-phonon bandwidth Δ is small compared to $\epsilon_{\parallel,\perp}$. But if the inequality $\Delta \gg \epsilon_{\parallel,\perp}$ is valid, as will be assumed below, it is inessential to take retardation into account for B phonons in investigating

Fermi resonance with a polariton. In view of the aforesaid, the quadratic part of the total Hamiltonian with respect to the Bose operators will be written in the form of the sum $\hat{H}_0(B) + \hat{H}_0(\xi)$, where

$$\hat{H}_0(\xi) = \sum_{\rho,k} \mathcal{E}_\rho(k)\xi_\rho^+(k)\xi(k),$$

and $\mathcal{E}_\rho(k) \equiv \hbar\omega_\rho(k)$ is the energy of the polariton ρk.

Let us now direct our attention to the Hamiltonian $\hat{H}_{int}$ and discuss the structure of its terms that are cubic in the quasi-particle creation and annihilation operators.

In the case being considered of a crystal with a single molecule per unit cell, these terms are evidently of the form:

$$\hat{H}_{int}^{(3)} = \sum_n \sum_{\rho,k} \left(\Gamma_\rho(k)(B_n^+)^2\xi_\rho(k) \frac{e^{ikn}}{\sqrt{N}} + \text{h.c.} \right). \tag{47}$$

To find the quantity $\Gamma_\rho(k)$ it is necessary to take into account the fact that the operators C_n are expressed in terms of the polariton creation and annihilation operators in the following way [see Agranovich (1959)]:

$$C_n = \frac{1}{\sqrt{N}} \sum_{\rho,k} e^{ikn}[u_\rho(k)\xi_\rho(k) + v_\rho^*(-k)\xi_\rho^+(-k)],$$

where u_ρ and v_ρ are certain factors. Hence, taking eq. (24) into account, we find that $\Gamma_\rho(k) = \Gamma u_\rho(k)$. Actually, however, along with the operator (24), $\hat{H}_{int}$ also includes an operator corresponding to direct interaction between the transverse photons and overtones. This operator,

$$\hat{H} = \sum_{n,k,j} [D(kj)(B_n^+)^2 a_{kj} + \text{h.c.}], \tag{48}$$

appears when terms proportional to $(B_n + B_n^+)^2$ in the dipole moment operator of the molecule are also taken into account, in addition to those linear in $(B_n + B_n^+)$. Thus, the operator being considered complies with the requirements made when the so-called electrooptic anharmonicity (Poulet and Mathieu 1970) is taken into account. Hence the constants $\Gamma_\rho(k)$, appearing in eq. (47), are determined by the total contribution of both the mechanical anharmonicity, eq. (24), and the electrooptic anharmonicity, eq. (48). Consequently, these constants depend, in general, on two independent phenomenological constants, Γ and D.

Now, when we compare the Hamiltonian (25a) with the Hamiltonian

$$\hat{H} = \hat{H}_0(B) + \hat{H}_0(\xi) + \hat{H}_{int}^{(3)} + \hat{H}^{(4)}(B),$$

where $\hat{H}^{(4)}(B)$ is the third term in eq. (25a), we come to the conclusion that the general structure of the Hamiltonian is conserved in going over to polaritons. This releases us from the obligation to repeat the calculations, so that we can proceed directly to the formulation of results.

We direct our attention, first of all, to an analysis of the dispersion law for polaritons in the region of the Fermi resonance. For this purpose, by analogy with eq. (34), we write the expression for the Fourier components of the Green's function

$$G_{k\rho}^{(1)}(t) = -i\theta(t)\langle 0|\xi_\rho(k, t)\xi_\rho^+(k, 0)|0\rangle.$$

Taking into account only the polariton branch ρ that intersects the region of two-particle states, we find that when we take anharmonicity into consideration

$$G_{k\rho}^{(1)}(E) = \frac{1 + 2\tilde{A}(k)R(E, k)}{\tilde{\Delta}(E, k)[E - \mathscr{E}_\rho(k)]}, \tag{49}$$

where

$$\tilde{\Delta}(E, k) = 1 + 2\left(\tilde{A}(k) - \frac{|\Gamma_\rho(k)|^2}{E - \mathscr{E}_\rho(k)}\right)R(E, k). \tag{50}$$

The poles of the Green's function (49) determine the dispersion of the polariton at the Fermi resonance. Here the energy of the polariton is determined from the condition $\tilde{\Delta}(E, k) = 0$, which can be conveniently written in the form

$$1 + 2\tilde{A}(k)R(E, k) = \frac{2|\Gamma_\rho(k)|^2}{E - \mathscr{E}_\rho(k)}R(E, k). \tag{51}$$

The left-hand side of eq. (51) may be equal to zero outside the spectrum of two-particle states if a biphonon level exists for $\Gamma_\rho(k) = 0$. If, in this case, the energy of the biphonon is $E_b(k)$, the left-hand side of eq. (51) can be written, for $E \approx E_b(k)$, in the form $\alpha^2(k)[E - E_b(k)]$, so that eq. (51) assumes the form

$$[E - E_b(k)][E - \mathscr{E}_\rho(k)] = \frac{|\Gamma_\rho(k)|^2}{\alpha^2(k)\tilde{A}(k)}. \tag{51a}$$

It follows from this relation that at a certain k_0, such that $E_b(k_0) = \mathscr{E}_\rho(k_0)$, i.e. at the point where the polariton intersects the biphonon level, a gap is formed in the polariton spectrum (see fig. 7) with the half width

$$\delta = |\Gamma_\rho(k_0)/\alpha(k_0)\sqrt{\tilde{A}(k_0)}|.$$

If, however, there is no biphonon, no gap appears in the polariton spectrum outside the region of two-particle states. Consequently, the experimental observation of such a gap is, at the same time, experimental proof of the existence of the biphonon (see section 8).

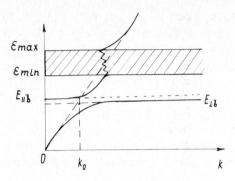

Fig. 7. Same as fig. 6, but with anharmonicity of vibrations taken into account.

4.2. Macroscopic theory—Transverse, longitudinal and surface biphonons

It was mentioned previously that the polariton spectrum can also be found within the framework of macroscopic electrodynamics, which requires that the dielectric tensor of the crystal be known. The results of a proper analysis, as could be expected, are equivalent to those obtained in micro-theory. Nevertheless, we shall make use of macrotheory in the following, in application to cubic crystals. We shall show how the existence of longitudinal and surface biphonons can be predicted within the framework of this approach [see also Agranovich (1973)].

We stress that the microscopic theory of Fermi resonance with polaritons, developed above, cannot be directly applied to cubic crystals, because triply degenerate states correspond to dipole-active transitions in such crystals [for the corresponding generalization of the theory, see Agranovich et al. (1976)].

If we do not take damping processes into consideration, the dielectric tensor reduces, as is well known, to the scalar $\epsilon(\omega)$ when spatial dispersion is neglected. In the region of the band of two-particle states, this scalar can be presented in the form:

$$\epsilon(\omega) = \epsilon_\infty - \frac{F_b \Omega_{\perp b}^2}{\omega^2 - \Omega_{\perp b}^2} - \int \frac{F(\omega')\omega'^2 \, d\omega'}{\omega^2 - \omega'^2}, \tag{52}$$

where ϵ_∞ is a quantity that is determined by the contribution of distant resonances and, in the spectral region being considered, can be assumed independent of ω, $\Omega_{\perp b} = E_b^\perp(0)/\hbar$, $E_b^\perp(k)$ is the energy of a transverse biphonon with wave vector k, F_b is a factor proportional to the biphonon oscillator strength, and $F(\omega')$ is a quantity proportional to the strength of an oscillator corresponding to the excitation of two free B phonons with total energy $\hbar\omega'$. The quantity $F(\omega')$ is also proportional to the density of energy levels with the total wave vector $k = 0$ in the band of two-particle states, so that $F(\omega') = 0$ if the frequency ω' is outside this band. The integral over frequencies in eq. (52) is taken in the sense of the principal value. It is also taken into consideration in this equation that resonances $\epsilon(\omega)$, corresponding to the frequencies of fundamental vibrations of the lattice (i.e. frequencies found in the harmonic approximation), do not fall within the frequency range being considered.

Polaritons are strictly transverse when spatial dispersion is neglected in cubic crystals, so that their dispersion law, i.e. the dependence of the frequency on the wave vector, can be determined from the relation

$$\epsilon(\omega) = k^2 c^2 / \omega^2. \tag{53}$$

If we let the velocity of light c approach infinity, i.e. neglect retardation, relation (53) assumes the form

$$\epsilon(\omega) = \infty. \tag{53a}$$

From this relation it follows that the frequencies of the transverse vibrations, found when only the instantaneous Coulomb interaction is taken into account, correspond to the resonances $\epsilon(\omega)$. Consequently, a resonance in eq. (52) at frequency $\Omega_{\perp b}$ corresponds to taking a resonance at the frequency of the transverse biphonon into account; this is indicated by the subscripts and superscripts.

Besides transverse polaritons, longitudinal waves may also exist in the spectral region being considered. The frequencies of these waves obey the equation

$$\epsilon(\omega) = 0. \tag{54}$$

These waves are not observed in the infrared absorption spectra, but, with proper selection of the polarization of the incident and scattered light, they can be observed in RSL spectra (see section 8).

To find the dispersion of polaritons it is necessary to determine, by means of eq. (53), the quantity ω as a function of k. The form of this function will be discussed below. First we shall consider the singularities of $\epsilon(\omega)$ that follow from eq. (52). We note, in the first place, that the quantity $\epsilon(\omega)$ assumes the value $+\infty$ or $-\infty$ depending on whether the frequency ω tends to $\Omega_{\perp b}$ from the right or from the left. But, if the frequency ω lies within the band of two-particle states, then, since the function $F(\omega')$ is finite, the quantity $\epsilon(\omega)$ is also finite and remains so even when the frequency ω approaches one of the boundaries of the band of two-particle states from outside. A singularity appears only in the derivative of $\epsilon(\omega)$, because in three-dimensional crystals the density of states $\rho(\omega')$ and, consequently, the quantity $F(\omega')$, are of the form $\rho(\omega') = \rho_0 |\hbar\omega' - \bar{\epsilon}|^{1/2}$ in the vicinity of any of the boundaries $\bar{\epsilon}$ of the band, as has been noted previously.

This is precisely why the function $\epsilon(\omega)$, though it remains finite, has an infinite derivative when the frequency ω approaches one of the boundaries of the band of two-particle states from outside.

In virtue of the aforesaid, in the frequency region being considered for $\Omega_{\perp b} < \epsilon_{min}$, the relation $\epsilon(\omega)$ can be schematically represented in the form shown in fig. 8. The fact that expression (52) has a term due to the presence of a biphonon leads to the appearance of a frequency range, near the frequency $\Omega_{\perp b}$, in which $\epsilon(\omega) < 0$, as can be seen in fig. 7. This indicates that in the given frequency range bulk electromagnetic waves, i.e. bulk polaritons, cannot exist [by virtue of eq. (53), they would require a negative value of k^2]. Hence, in full accordance with the results of microtheory, a gap in the bulk polariton spectrum corresponds to this frequency region.

It should also be noted that the vanishing of $\epsilon(\omega)$ at a certain frequency value $\omega \equiv \Omega_{\parallel b}$, where $\epsilon(\Omega_{\parallel b}) = 0$ and $\Omega_{\parallel b} > \Omega_{\perp b}$, points to the following fact. In the case being considered of a cubic crystal, a longitudinal biphonon with energy $\hbar\Omega_{\parallel b}$ is also formed, simultaneously with the transverse biphonon that leads to the appearance of a gap in the polariton spectrum. The frequency of this longitudinal biphonon in the region of small k values can depend only slightly on k (see fig. 7).

We point out that longitudinal–transverse splitting of the biphonon in the case being discussed does not comply with the well-known Lidden–Sax-

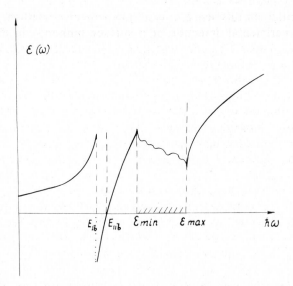

Fig. 8. Dependence of the dielectric constant on the frequency in the region of overtone frequencies: $E_{\parallel b}$ and $E_{\perp b}$ are the energies of the longitudinal and transverse biphonons, respectively; ϵ_{min} and ϵ_{max} are the minimum and maximum energy values in the band of two-particle states.

Teller formula, which is valid for the region of an isolated frequency of the fundamental vibration. Owing to the contribution of the integral the low-frequency value of $\epsilon(\omega)$ is determined by the relation

$$\epsilon(0) = \epsilon_\infty + F_b + \int F(\omega')\,d\omega',$$

so that the Lidden–Sax–Teller relation $\Omega_{\|b}^2 = \epsilon(0)\Omega_{\perp b}^2/\epsilon(\infty)$ is not, in general, valid for the frequencies $\Omega_{\|b}$ and $\Omega_{\perp b}$ [see also Agranovich et al. (1976)].

In the region of longitudinal–transverse splitting $\Omega_{\perp b} < \omega < \Omega_{\|b}$, where $\epsilon(\omega) < 0$, surface biphonons should also exist. At the boundary with the vacuum, their dispersion law satisfies the relation

$$k^2 = \frac{\omega^2}{c^2}\frac{\epsilon(\omega)}{\epsilon(\omega)-1}, \tag{55}$$

when retardation is taken into account. It follows from eq. (55) that as $c \to \infty$, i.e. when retardation is neglected, the frequency Ω_{sb} of a surface biphonon satisfies the relation

$$\epsilon(\Omega_{sb}) = -1.$$

Surface biphonons could be investigated, for example, by the attenuated total reflection (ATR) method. In contrast to RSL by polaritons, this method is effective, as is well known, both for crystals with and without inversion center. In this sense, it is a more universal method. At the same time, the experimental detection of a surface biphonon in the frequency range $\omega < \epsilon_{min}/\hbar$ would be circumstantial evidence of the existence of both a transverse and a longitudinal biphonon. Actually, by virtue of the fact that $\epsilon_\infty > 0$, the quantity $\epsilon(\omega)$, determined by eq. (52), can be negative in a certain frequency range $\omega < \epsilon_{min}/\hbar$ only if eq. (52) includes a resonance term with frequency $\Omega_{\perp b} < \epsilon_{min}/\hbar$.

In an entirely analogous way we can consider the situation when repulsion between phonons dominates in the spectral region of two-particle states and the biphonon level is found to be above the band of two-particle states ($\Omega_{\perp b} > \epsilon_{max}/\hbar$).

It should be underlined that the statement made above, based on the application of eq. (52) for $\epsilon(\omega)$, can be justified only for crystals in which the linewidth is small compared to the magnitude of longitudinal–transverse splitting of the biphonon. In the discussion of experimental data in section 8 we shall indicate crystals in which such a situation can really exist.

An analysis of the singularities of dispersion in the region of two-particle states may prove useful for anisotropic crystals as well. Here, in writing a phenomenological relation of the type of eq. (52) to determine the dependence of the dielectric tensor $\epsilon_{ij}(\omega)$ on ω, consideration must be given to the fact that dipole-active biphonons, like dipole-active phonons, at $k = 0$ transform according to the vector representations of the point group of the crystal. Hence, in uniaxial crystals, for example, dipole-active biphonons can be polarized either along (with the transition frequency $\Omega_{\|b}$) or across the optical axis (transition frequency $\Omega_{\perp b}$). Thus

$$\epsilon_{ij}(\omega) = \epsilon_{ij}^{\infty} - \frac{F_b^{\perp}}{\omega^2 - \Omega_{\|b}^2}\,\delta_{i3}\delta_{j3} - \frac{F_b^{\|}(\delta_{i1}\delta_{j1} + \delta_{i2}\delta_{j2})}{\omega^2 - \Omega_{\perp b}^2} - \int \frac{F_{ij}(\omega')\omega'^2\,d\omega'}{\omega^2 - \omega'^2}, \qquad (56)$$

where the tensor element F_{ij} equals zero if $i \neq j$, and $F_{11} = F_{22}$. In writing relation (56) we assumed that the optical axis of the crystal is along the z-axis, and that spatial dispersion and damping are ignored.

In conclusion we point out that in degenerate semiconductors Fermi resonance with plasmons (Agranovich and Mekhtiev 1971) is also possible along with Fermi resonance with phonons and polaritons.

In such crystals, the spectrum of long-wave longitudinal vibrations is not always restricted to plasmons and longitudinal phonons, even in the region of the fundamental frequencies. Whenever the frequency of a longitudinal optical phonon is found to be close to that of a plasmon, so-called plasmophonons appear, even in the harmonic approximation. These are elementary excitations that are a "mixture" of a plasmon and an optical phonon. The spectrum of plasmophonons has been measured in many semiconductors to great accuracy by the RSL method [see, e.g., Mooradian and McWhorter (1969)].

It is found that similar "mixing" effects should also occur, in general, when the frequency of the plasmon is close to that of the overtone or combination tone, i.e., when the conditions for Fermi resonance with a plasmon are satisfied. As distinct from the situation that has already been mentioned [and which was discussed by Mooradian and McWhorter (1969)], it is extremely important, in this case, to take anharmonicity into account. The resonance of a biphonon, if such occurs, leads to the formation of a gap in the plasmon spectrum. This effect can be most simply understood and described on the basis of purely phenomenological considerations.

As a matter of fact, when free electrons are taken into account, the following equation should be used to find the dielectric constant $\epsilon(\omega)$ instead of eq. (52):

$$\epsilon(\omega) = \epsilon_{\infty} - \frac{\Omega^2(k)}{\omega^2} - \frac{F_b\Omega_{\perp b}^2}{\omega^2 - \Omega_{\perp b}^2} - \int \frac{F(\omega')\omega'^2\,d\omega'}{\omega^2 - \omega'^2}, \qquad (52a)$$

where $\Omega = \Omega(k)$ is the frequency of a plasmon with wave vector k when anharmonicity is neglected, $\Omega(k) = \Omega(0) + \hbar k^2/2\mu$, and μ is the effective mass of the plasmon.

The use of eqs. (54) and (52a) enables one to determine the dispersion of longitudinal waves, plasmobiphonons, when the anharmonicity of lattice vibrations is taken into account. If the resonance frequency $\Omega_{\perp b} > \Omega(0)$, a gap is formed in the spectrum of longitudinal waves [for details, as well as the microscopic theory of this effect, see Agranovich and Mekhtiev (1971); we also refer to the work by Efremov and Kaminskaya (1972b), in which the intensity of RSL by plasmons and phonons was calculated under Fermi resonance conditions]. We point out that the use of eq. (53) enables one to determine the dispersion of polaritons in the region of the Fermi resonance between a plasmon and a biphonon.

5. *Dielectric constant of a crystal in the spectral region of two-particle phonon states*

A knowledge of the dielectric constant of a crystal for the frequency region being discussed is necessary, not only to find the dispersion of polaritons, but also for analyzing the structure of absorption spectra. In this connection (see the following section) and to illustrate the calculation procedure, we shall demonstrate how the dielectric tensor ϵ_{ij} can be found in the spectral region of two-particle states, within the framework of the same simplest model of a crystal as used previously in section 2.

According to the theory of linear response the transverse dielectric tensor of a crystal at temperature $T = 0$ is determined by the relation

$$\epsilon_{ij}^{\perp}(\omega) = \delta_{ij} + 4\pi\chi_{ij}^{\perp}(\omega), \tag{57}$$

where

$$\chi_{ij}^{\perp}(\omega) = -\frac{1}{\hbar V}[\Phi_{ij}(\omega + i\epsilon) + \Phi_{ij}(-\omega + i\epsilon)], \quad \epsilon \to +0, \tag{57a}$$

$$\Phi_{ij}(t) = -\tfrac{1}{2}i\theta(t)\langle 0|M_i(t)M_j(0) + M_j(t)M_i(0)|0\rangle, \tag{57b}$$

V is the volume of the crystal, M is its dipole moment,

$$M = \sum_n M_n, \tag{58}$$

with M_n the dipole moment of the molecule, where $n \equiv (n\alpha)$.

In determining the quantities $\Phi_{ij}(\omega)$ we shall make use of the Hamiltonian (25) in which, as assumed, Coulomb interaction has been completely taken into account. In this case, linear response theory determines only the so-called [see Agranovich and Ginzberg (1981)] transverse dielectric constant tensor $\epsilon_{ij}^{\perp}(\omega, k)$. This tensor relates the induction vector D to the transverse part of the macrofield E: $D_i(\omega, k) = \epsilon_{ij}^{\perp}(\omega, k)E_j^{\perp}(\omega, k)$. As is well known, the use of the tensor $\epsilon_{ij}^{\perp}(\omega, k)$ is sufficient for investigating the properties of normal waves in a medium.

Expanding the quantity M_n in a series in terms of the normal coordinates of the molecule and going over to the excitation creation and annihilation operators, we obtain up to second-order terms

$$M_{ni} = d_i^C(C_n + C_n^+) + d_i^{BB}(B_n + B_n^+)^2 + \cdots, \quad i = x, y, z. \tag{59}$$

It is precisely the terms that have been written out here that make, in the spectral region $\omega \approx 2\Omega_1 \approx \Omega_2$ being considered, the main contribution to the dipole moment of the molecule. Applying relations (58) and (24a) we find that

$$\frac{1}{\sqrt{N}} M_i = d_i^C(C_{k=0} + C_{k=0}^+) + d_i^{BB}[T(k = 0) + T^+(k = 0)] + \tilde{M}_i, \tag{60}$$

where the operator $\tilde{M}_i$ includes terms of the form of $B_q^+ B_q$, retaining the number of B phonons and making no contribution to the quantity $\Phi_{ij}(\omega)$. Making use of eq. (57b) we obtain

$$\frac{1}{N} \Phi_{ij}(t) = d_i^C d_j^C G_0^{(1)}(t) + \tfrac{1}{2}(d_i^{BB} d_j^C + d_j^{BB} d_i^C)[G_0^{(2)}(t) + G_0^{(7)}(t)]$$

$$+ d_i^{BB} d_j^{BB} G_0^{(8)}(t),$$

where

$$G_k^{(7)}(t) = \sum_q G_{k,q}^{(6)}(t) = -i\theta(t)\langle 0|C_k(t) T_k^+(0)|0\rangle,$$

$$G_k^{(8)}(t) = \sum_q G_{k,q}^{(5)}(t) = -i\theta(t)\langle 0|T_k(t) T_k^+(0)|0\rangle.$$

From relations (37) and (38) we obtain

$$G_k^{(8)}(\omega) = \frac{2\hbar R(E, k)}{\Delta(E, k)},$$

$$G_k^{(7)}(\omega) = \frac{\Gamma}{E - \epsilon_2(k)} G_k^{(8)}(\omega) = \frac{2\hbar \Gamma R(E, k)}{[E - \epsilon_2(k)]\Delta(E, k)} = G_k^{(2)}(\omega),$$

so that

$$\frac{1}{\hbar N} \Phi_{ij}(\omega) = \frac{d_i^C d_j^C}{\hbar\omega - \epsilon_2(0)} + \frac{2R(\omega, 0)}{\Delta(\omega, 0)}\left(\frac{\Gamma d_i^C}{\hbar\omega - \epsilon_2(0)} + d_i^{BB}\right)\left(\frac{\Gamma d_j^C}{\hbar\omega - \epsilon_2(0)} + d_j^{BB}\right). \tag{61}$$

Together with relations (57a) and (57), relation (61) completely determines the dielectric tensor of the crystal in the spectral region being discussed. Let us now turn to certain limiting cases. In the first place we point out that if anharmonicity is neglected ($\tilde{A} = 0$ and $\Gamma = 0$)

$$\frac{1}{\hbar N} \Phi_{ij}(\omega) = \frac{d_i^C d_j^C}{\hbar\omega - \epsilon_2(0)} + 2R(\omega, 0)d_i^{BB} d_j^{BB},$$

so that [see eq. (57a); in the following $\epsilon \to +0$]

$$\chi_{ij}^{\perp}(\omega) = -\frac{1}{v}\left[\frac{2\epsilon_2(0)d_i^C d_j^C}{(\hbar\omega)^2 - \epsilon_2^2(0) + i\epsilon} + 4d_i^{BB}d_j^{BB}\left(\int \frac{2x\rho_0(x)\,dx}{E^2 - x^2} - i\pi\rho_0(E)\right)\right],$$

where $v = V/N$, $E = \hbar\omega$.

If we ignore the effect of two-phonon transitions on the dispersion of the refractive index n, then in the region of the energies $E = \hbar\omega$, corresponding to the excitation of two-particle states, the absorption factor κ is found to be directly associated with the density of two-particle states. In fact, for light polarized, for instance, along the x-axis, which coincides, by definition, with the direction of the vector d^{BB}, the dielectric constant is

$$\epsilon_{11}^{\perp} = 1 + 4\pi\chi_{11}^{\perp} = (n + i\kappa)^2 \approx n^2 + 2in\kappa,$$

so that

$$\kappa(E) = \frac{8\pi^2}{vn_0}|d^{BB}|^2\rho_0(E),$$

where n_0 is the value of the refractive index if the contribution of the two-particle states is neglected.

Next we shall show that taking anharmonicity into account leads, in general, to the appearance of new maxima of $\kappa(E)$. We assume, for simplicity, that $\Gamma = 0$, but that $\tilde{A} \neq 0$. Then, for the region of two-particle states

$$\kappa(E) = -\frac{4\pi|d^{BB}|^2}{n_0 v}\,\mathrm{Im}\left\{\frac{R(\omega + i\epsilon, 0)}{1 + 2\tilde{A}(0)R(\omega + i\epsilon, 0)}\right\}$$

$$= \frac{|d^{BB}|^2}{n_0 v}\frac{8\pi^2\rho_0(E)}{[1 + 2\tilde{A}(0)R'(E, 0)]^2 + 16\pi^2|\tilde{A}(0)|^2\rho_0^2(E)}.$$

The analysis of this expression is similar to that for relation (44). Specifically, if eq. (45) is satisfied for a certain value of $E = E_b$ outside the region of two-particle states, then, for the mentioned region of the spectrum,

$$\kappa(E) = \frac{\pi^2|d^{BB}|^2}{2n_0 v|\tilde{A}(0)|^2 D(E_b, 0)}\,\delta(E - E_b). \tag{62}$$

In the vicinity of the quasi-biphonon energy, where $E \approx E_{qb}$ [see eq. (45a)],

$$\kappa(E) = \frac{\pi^2|d^{BB}|^2}{2n_0 v|\tilde{A}(0)|^2 D(E_{qb}, 0)}\frac{1}{\pi}\frac{\gamma}{(E - E_{qb})^2 + \gamma^2}. \tag{63}$$

A discussion of the more general situation with $\Gamma \neq 0$ is more cumbersome and is not given here. We only point out that the appearance of biphonon states also leads to new resonances of $\kappa(E)$ when Fermi resonance is taken into account. But the appearance of quasi-biphonon states substantially alters the density of states inside the band of two-particle states.

6. Bound states of quasi-particles (biphonons and biexcitons) in the nonlinear polarizabilities of a crystal—Gigantic nonlinear polarizabilities

An extensive range of nonlinear optical effects can be described within the framework of macroscopic electrodynamics, as is well known [see, e.g., Bloembergen (1965)], by applying the nonlinear relation between the induction vector D and the strength E of the macroscopic electric field. When the value of E in the light wave is small compared to the intra-atomic electric fields, this nonlinear relation can be written for plane waves, ignoring spatial dispersion, in the form of the expansion

$$D_i(r, t) = \epsilon_{ij}(\omega) \exp(-i\omega t)$$
$$+ 4\pi\chi_{ijl}(\omega, \omega')E_j(\omega)E_l(\omega') \exp[-i(\omega + \omega')t]$$
$$+ 4\pi\chi_{ijlm}(\omega, \omega', \omega'')E_j(\omega)E_l(\omega')E_m(\omega'') \exp[-i(\omega + \omega' + \omega'')t]$$
$$+ \cdots,$$

where χ_{ijl}, χ_{ijlm}, etc. are nonlinear polarizabilities, characterizing the nonlinear optical properties of the medium.

In the preceding section it was shown that the formation of bound states of phonons leads to the appearance of a new type of resonances of the dielectric tensor $\epsilon_{ij}(\omega)$. It is clear, of course [this fact was pointed out in Agranovich et al. (1971)], that the nonlinear polarizabilities should have analogous resonances, and that this also concerns, besides biphonons, other types of bound states of quasi-particles, such as biexcitons, electron–exciton complexes, etc.

An investigation of the contribution of the bound states of quasi-particles to the nonlinear polarizabilities is of interest for many reasons. The main ones are the opportunities for studying the properties of bound states, offered by the investigation of nonlinear optical processes, as well as the gigantic values of the nonlinear polarizabilities that can be reached, precisely as a result of the new type of resonances.

The very existence of nonlinear polarizabilities is due to the presence of some anharmonicity in the medium. Anharmonicity is usually regarded as a weak perturbation in calculating these polarizabilities. It is clear, however, that when anharmonicity leads to the formation of states of quasi-particles bound to each other, the polarizabilities, in the region of the resonances corresponding to these states, become nonanalytic functions of the anharmonicity constants. For this reason ordinary perturbation theory is found to be inapplicable in their calculation, and more general methods are required. In accordance with two papers by Agranovich et al. (1971) and Efremov and Kaminskaya (1972a), we intend to show how the method of many-time Green's functions can be employed for this case.

Let us first consider the dispersion of the nonlinear polarizability tensor of third order. The relation of this tensor to the triple-time correlation function is determined by the following expression (Fain and Khanin 1965) (here and in the following $\hbar = 1$):

$$\chi_{ijl}(\omega, \omega') = -\frac{\pi}{V} [\mathcal{K}_{ijl}(\omega + \omega', \omega') + \mathcal{K}_{ijl}(\omega + \omega', \omega)], \tag{64}$$

where

$$\mathcal{K}_{ijl}(\omega + \omega', \omega') = \frac{1}{2\pi} \int_{-\infty}^{+\infty} d\tau_1 \int_{-\infty}^{+\infty} d\tau_2 \exp\{i[(\omega + \omega')\tau_1 + \omega'\tau_2]\}$$

$$\times \theta(\tau_1)\theta(\tau_2)\langle 0|[[\hat{M}_i(\tau_1), \hat{M}_j(0)], \hat{M}_l(-\tau_2)]|0\rangle.$$

Here $[\hat{A}, \hat{B}]$ denotes the anticommutator, $[\hat{A}, \hat{B}] = \hat{A}\hat{B} - \hat{B}\hat{A}$.

Assuming that the Hamiltonian is of the form of eq. (25) and that in the dipole moment M of the crystal [see eq. (58)] the term $\sqrt{N}\, d_i^B(B_0 + B_0^+)$ is also included in the expansion (60), we reach the conclusion that in order to find the tensor χ_{ijl} it is necessary to know double-time Green's functions of the form

$$G^{(1)}(\tau_1, \tau_2) = \theta(\tau_1)\theta(\tau_2)\langle 0|[[B_0(\tau_1)B_0(0)], C_0^+(-\tau_2)]|0\rangle,$$

$$G^{(2)}(\tau_1, \tau_2) = \theta(\tau_1)\theta(\tau_2)\langle 0|[[B_0(\tau_1)C_0^+(0)], B_0(-\tau_2)]|0\rangle,$$

$$G^{(3)}(\tau_1, \tau_2) = \theta(\tau_1)\theta(\tau_2)\langle 0|[[B_0(\tau_1)B_0(0)], \hat{\Phi}(-\tau_2)]|0\rangle,$$

where

$$\hat{\Phi}(\tau_2) = \sum_k B_k^+(\tau_2)B_{-k}^+(\tau_2), \quad \text{etc.}$$

In applying the model Hamiltonian (25) all Green's functions can be exactly determined, although, admittedly, they correspond to the crystal temperature $T = 0$. Their calculation is extremely cumbersome, but is similar, to a considerable extent, to that set forth in section 3. Consequently, we shall not give here the relations obtained for them, but only the final expression for the tensor χ_{ijl}. We shall assume here that the frequency $\omega \approx \omega'$ is close to the limiting frequency of the B phonon, i.e., that $\hbar\omega \approx \hbar\omega' \approx \epsilon_1(0)$, whereas the sum $\hbar(\omega + \omega')$ is close to the energy of the biphonon. It is convenient, in this case, to separate out terms which have a resonance at the biphonon frequency, putting the remaining terms in $\tilde{\chi}_{ijl}$. In view of the aforesaid, the expression for χ_{ijl} can be presented as follows:

$$\chi_{ijl}(\omega, \omega') = \tilde{\chi}_{ijl} + \frac{\Gamma d_i^C d_j^B d_l^B/v}{[\hbar(\omega + \omega') - \epsilon_2(0)][\hbar\omega' - \epsilon_1(0)][\hbar\omega - \epsilon_1(0)]\Delta(\omega + \omega', 0)}$$

$$+ \frac{d_i^{BB} d_j^B d_l^B}{[\hbar\omega - \epsilon_1(0)][\hbar\omega' - \epsilon_1(0)]\Delta(\omega + \omega', 0)}, \tag{65}$$

where the quantity $\Delta(\omega, 0)$ is determined by relation (34a). It follows from expression (65) that along with the ordinary resonances at $\hbar\omega = \epsilon_1(0)$ or $\hbar(\omega' + \omega) = \epsilon_2(0)$, the nonlinear polarizability tensor has a resonance at the bound state of quasi-particles, i.e. at $\hbar(\omega + \omega') = E_b$, where E_b is the biphonon energy. If the binding energy of the biphonon substantially exceeds the line width, then at $\omega = \omega'$ (in this case the tensor χ_{ijl} corresponds to the generation of a second harmonic) the tensor $\chi_{ijl}(\omega, \omega)$ will manifest a resonance if the frequency ω lies within the transparency region.

Note that for $\hbar(\omega + \omega') = E_b$ it is necessary to take the damping of the biphonon into account in the expression for $\Delta(\omega + \omega', 0)$. In this case, along with the real part of the tensor $\chi_{ijl}(\omega, \omega)$, an imaginary part is also present. This corresponds, as is well known, to the occurrence of two-photon absorption that is accompanied, in the given case, by the excitation of a biphonon. As applied to excitons this question has been discussed by Hanamura (1973) within the framework of a somewhat different approach. We refer to it here [see also Flytzanis (1978)] because both the generation of a second harmonic and two-photon absorption are processes that are completely described by the nonlinear polarizability of the crystal found above with bound states taken into account. Actually, these processes can be investigated by a single method.

A relation analogous to eq. (65) also obtains for the nonlinear polarizability tensor χ_{ijlm}. In this case, for $\omega > 0$, $\omega' > 0$ and $\omega'' > 0$ as well,

$$\chi_{ijlm}(\omega, \omega', -\omega'') = \tilde{\chi}_{ijlm}$$

$$+ \frac{2A d_i^{BB} d_j^{BB} d_l^B d_m^B}{[\hbar\omega - \epsilon_1(0)][\hbar\omega' - \epsilon_1(0)][\hbar\omega'' - \epsilon_1(0)][\hbar\omega''' - \epsilon_1(0)]\Delta(\omega + \omega', 0)}. \tag{66}$$

This tensor corresponds to the four-photon process of scattering of light by light, for which the conservation law

$$\hbar\omega + \hbar\omega' = \hbar\omega'' + \hbar\omega'''$$

holds. If the energy $\hbar(\omega + \omega') \approx E_b$ and the frequencies ω, ω' and ω'' (and consequently ω''' as well) are close to the frequency $\epsilon_1(0)/\hbar$ of the B phonon, the nonlinear polarizability becomes very large (gigantic). With respect to excitons and biexcitons in a CuCl crystal, a gigantic nonlinear polarizability χ_{ijlm} was experimentally recorded by Grun et al. (1978). It follows from the aforesaid that the formation of such gigantic nonlinear polarizabilities in the region of biphonon or biexciton resonances, or the occurrence of intense two- or three-photon absorption is not a chance effect. Such gigantic resonance effects should be observed in all spectral regions in which dipole-active bound states of quasi-particles exist. In particular, such effects may exist in the infrared region of the spectra as well in all crystals in which dipole-active biphonons are detected.

7. Biphonons at finite temperatures

In the preceding it was assumed everywhere in discussing biphonon spectra that the crystal temperature was $T = 0$ and, consequently, that the crystal was in the ground state. This assumption does not impose any appreciable limitations if we are concerned with the frequency region corresponding to an overtone ($\omega \approx 2\Omega_1$) or combination tone ($\omega \approx \Omega_1 + \Omega_2$) of lattice vibrations for which $\hbar\Omega_1$, $\hbar\Omega_2 \gg kT$. If these inequalities do not hold and the temperature cannot be considered sufficiently low, two interesting effects are possible: temperature renormalization of the anharmonicity constants and formation of bound two-particle states (biphonons) in the region of difference tones of vibrations, i.e. at $\omega \approx \Omega_1 - \Omega_2$, where $\Omega_1 > \Omega_2$.

For the sake of brevity we shall not calculate here the two-particle Green's functions at finite temperatures in considering the above-mentioned effects; we shall only derive the biphonon dispersion equation [see also Maradudin (1971)].

It is sufficient for this purpose, within the framework of the diagrammatic technique at finite temperatures [for such techniques, see, e.g., Abrikosov et al. (1962)], to write the condition for occurrence of a pole in the full vertex part of the Dyson equation for the two-particle Green's function.

Assuming that the anharmonicity operator is of the form

$$\hat{H}_A = \frac{1}{4!} \sum_{k_1,k_2,k_3,k_4} \sum_{j_1,j_2,j_3,j_4} V(k_1j_1; k_2j_2; k_3j_3; k_4j_4) \, \varphi_{k_1}(j_1)\varphi_{k_2}(j_2)\varphi_{k_3}(j_3)\varphi_{k_4}(j_4), \quad (67)$$

where $\varphi_k(j) = (B_k^j)^+ + B_{-k}^j$, and j is the subscript or superscript of the phonon branch, we come to the conclusion that in the ladder approximation this vertex part can be brought into agreement with the sum of diagrams so that the condition for the appearance of a pole at the full vertex part, i.e. the condition

$$1 = \bigcirc\!\!\!\!\longrightarrow \quad ,$$

can be written in explicit form as follows [taking eq. (67) into account]:

$$1 = -\tfrac{1}{2} T \sum_{j_1 j_2} \sum_n \int \frac{dk}{(2\pi)^3} V(kj_1; q - k j_2; kj_1; q - k j_2)$$
$$\times D_{j_1}^{(0)}(k, \omega_n) D_{j_2}^{(0)}(q - k, \Omega - \omega_n),$$

where

$$D_j^{(0)}(k, \omega_n) = \frac{2\omega_j(k)}{\omega_j^2(k) - (i\omega_n)^2}.$$

After analytic continuation of the functions for $D_{j_1}^{(0)}$ and $D_{j_2}^{(0)}$ in the upper half Ω-plane and summation over n, we finally obtain the following equation for determining the biphonon energy $E = \hbar\Omega(q)$:

$$1 = \sum_{k, j_1, j_2} \frac{1}{\hbar} V(kj_1; q - k j_2; kj_1; q - k j_2)$$
$$\times \left([1 + n(\omega_{j_1}) + n(\omega_{j_2})] \frac{\omega_{j_1} + \omega_{j_2}}{(\omega_{j_1} + \omega_{j_2})^2 - \Omega^2} \right.$$
$$\left. + [n(\omega_{j_1}) - n(\omega_{j_2})] \frac{\omega_{j_2} - \omega_{j_1}}{(\omega_{j_2} - \omega_{j_1})^2 - \Omega^2} \right), \quad (68)$$

where $n(\omega)$ is the average (Planckian) number of phonons with energy $\hbar\omega$. At the temperature $T = 0$, the quantity $n(\omega) = 0$ and eq. (68) is converted, up to small (nonresonance) terms and in the frequency region $\Omega \approx \omega_1 + \omega_2$, into an equation of the form of (20b). But when $T \neq 0$ the most interesting feature of eq. (68) is the possibility of a root appearing in the region $\Omega = \omega_1 - \omega_2$, where $\omega_1 > \omega_2$. Equation (68) can be simplified for this frequency region by discarding small nonresonance terms. In this approximation we obtain, instead of eq. (68),

$$1 = -\frac{2}{N} \sum_q \frac{V_{12}(q, K - q; q, K - q)}{\hbar\omega_1(q) - \hbar\omega_2(K - q) - \hbar\Omega} [n(\omega_1) - n(\omega_2)], \quad (69)$$

and for $kT \gg \hbar\omega_1$ and $kT \gg \hbar\omega_2$, $n(\omega_1) = kT/\hbar\omega_1$ and $n(\omega_2) = kT/\hbar\omega_2$. Consequently, owing to the increase in the effective anharmonicity constant ($V_{eff} \sim TV$) at sufficiently high T values, eq. (69) always provides a solution $\hbar\Omega(K)$, lying outside the band of "difference" two-particle states

$$E_2^{(-)}(K, q) = \hbar\omega_1(q) - \hbar\omega_2(K - q).$$

This does not imply, however, that at high T values "difference" biphonons always exist. The point is that at high T values the width of the phonon levels also increases due to the possibility of phonon decay into two or more acoustic phonons. These widths increase with T proportionally to T^2 or even more rapidly. Hence, at sufficiently high T values the level of a "difference" biphonon should blur because of the aforesaid. This means that "difference" biphonons can be observed only in the region of not very high temperatures and at favourable relations, in this respect, between the anharmonicity constants V [see eq. (67)] and the constants that determine the decay rate of optical phonons. In any case, the possibility of the formation of "difference" biphonons exists and it should be allowed for in treating all the problems touched upon in the preceding sections [see also Lalov (1981)].

8. Experimental biphonon research

As has been pointed out previously, in many crystals biphonons were observed both in infrared absorption spectra and in RSL spectra.

In discussing infrared absorption spectra, we refer, first of all, to the papers of Dows and Schettino (1973) and of Schettino and Salvi (1975).

Dows and Schettino (1973) investigated the CO_2 crystal spectrum in the frequency region corresponding to the combination tone of the intramolecular vibrations ν_1 and ν_3 ($\nu_1 + \nu_3 \approx 3720\ \text{cm}^{-1}$).

Schettino and Salvi (1975) measured the infrared (IR) spectra of N_2O and OCS crystals. The CO_2 and N_2O molecules are linear, have no permanent dipole moments and form a simple cubic lattice upon crystallization. This lattice has four molecules per unit cell, which are oriented along the axes of a tetrahedron. The OCS molecule is also linear, but it forms a crystal of the trigonal system with one molecule per unit cell.

Since the CO_2 molecule is symmetrical, its stretching vibration ν_1 is IR inactive and has practically no dispersion. Hence, the van Kranendonk model can be employed to interpret the experimental results obtained in the region of the combination frequency $\nu_1 + \nu_3$. This has actually been taken into account by Bogani (1978).

Shown in fig. 9 is the transmission spectrum, measured by Dows and Schettino (1973), of a crystal of 1.8 μm thickness. Calculations carried out by Bogani (1978) indicate that the sharp absorption peak obtained in this case corresponds to the excitation of a biphonon.

Similar results were obtained (Schettino and Salvi 1975) for the N_2O crystal in the frequency region that corresponds to the combination frequency $\nu_2 + \nu_3$. The N_2O molecule is not symmetrical and therefore all of its three intramolecular vibrations, ν_1, ν_2 and ν_3, are IR active. Its flexural vibration ν_2 is doubly degenerate and, owing to the small value of the

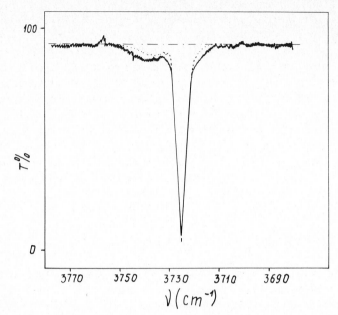

Fig. 9. The $\nu_1 + \nu_3$ band of CO_2. Observed and calculated transmission spectrum for a thickness of approximately 1.8 μm. The dotted line represents the calculated absorption spectrum (Bogani 1978).

dipole moment, the dispersion is weak throughout the Brillouin zone (less than 3 cm^{-1}). Hence, the van Kranendonk model can be applied for the frequency region $\nu_2 + \nu_3$, as well as for the region $\nu_1 + \nu_2$ [see Bogani (1978)]. A transmission spectrum, measured by Schettino and Salvi (1975), of a film of N_2O crystal, 18 μm thick, in the $\nu_2 + \nu_3$ band, is shown in fig. 10.

Extensive experimental data have been obtained in recent years in studying the effects of anharmonicity in crystals by the RSL method and, in particular, in the observation of RSL by polaritons.

As a result of this research, and in full agreement with the predictions of theory it was shown that:

(a) The formation of dipole-active biphonons leads to a discontinuity in the polariton branch outside the band of two-particle states of phonons, and that such a shape of the polariton branch cannot be explained on the basis of the theory in the harmonic approximation [see Mavrin and Sterin (1972), Winter and Claus (1972), Gorelik et al. (1975), Polivanov (1979a,b)].

(b) In cubic crystals, along with the transverse biphonons that lead to a discontinuity in the polariton branch [see (a)], longitudinal biphonons are also formed [see Gorelik et al. (1977)].

(c) In many crystals Fermi resonance of the phonon branches very

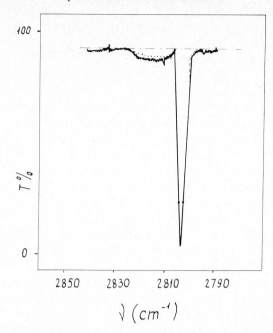

Fig. 10. Transmission spectrum of the $\nu_3 + \nu_3$ band of N_2O. The dotted line shows the calculated spectrum for a thickness of approximately 18 μm (Bogani 1978).

strongly influences the position and other characteristics of biphonons [see, e.g., Polivanov (1978)].

(d) In a number of crystals discontinuities of the polariton branch are observed inside the band of two-particle states as well [see, e.g., Aktsipetrov et al. (1977, 1978), Gorelik et al. (1978), Polivanov (1978); see also fig. 7].

The cited experiments stimulated a more comprehensive theoretical analysis of the dispersion of polaritons within the band of two-particle states (Agranovich and Lalov 1976a, Agranovich et al. 1979b). Since this problem is beyond the scope of the present review, we only point out that inside the band of two-particle states a channel opens up for polariton decay into two phonons. This process leads to the broadening of the polariton line [see, e.g., Polivanov (1979a,b)], and also to a change in the polariton dispersion law. The two most important effects in the latter case are: (a) interference of scattering by a polariton and two-particle states [this can lead to drops in intensity of the Fano antiresensive type; see Agranovich et al. (1979b)], and (b) the presence of singularities in the density of two-particle states (those, in particular, that correspond to quasi-biphonons).

Experimental investigations of the effects of strong anharmonicity in phonon and polariton spectra are reported in scores of papers. Though a discussion of them would be timely, we do not have the opportunity here for any detailed analysis. Nevertheless, in order to give some idea of the state-of-the-art concerning experimental research along these lines, as well as in the study of crystals, we shall make certain comments.

We note, first of all, that the first convincing proof of the presence of a gap in the polariton spectrum, appearing in the vicinity of the biphonon level, was obtained by Mavrin and Sterin (1972) and by Winter and Claus (1972) in investigating the lithium niobate crystal ($LiNbO_3$). In this crystal the dispersion branch of an ordinary polariton undergoes splitting (with a magnitude of approximately $5\,cm^{-1}$) at a frequency of $537\,cm^{-1}$. There are no fundamental vibrations of the lattice in this spectral region (the closest has a frequency of $582\,cm^{-1}$) and, consequently, the observed splitting is associated with the presence of a biphonon. This same kind of splitting was subsequently observed in many crystals, in particular in HIO_3 (Polivanov 1979a,b) and NH_4Cl (Gorelik et al. 1975, 1977, Polivanov 1978).

In the low-temperature phase IV, NH_4Cl crystals belong to the cubic class T_d and have one molecule per unit cell. Though these are not molecular crystals, the optical vibrations of the NH_4^+ ions, like the molecules in a molecular crystal, interact relatively weakly with one another. The vibrational spectra of this crystal have been investigated very comprehensively. This, precisely, is the crystal in which the longitudinal biphonon state was first identified and the magnitude of the longitudinal–transverse splitting (see fig. 11) was determined by Gorelik et al. (1975, 1977) from the RSL spectra for various light polarizations.

No less interesting results were obtained in previously mentioned papers (Polivanov 1979a,b). Polivanov investigated the spectra of RSL by polaritons in the HIO_3 crystal. This crystal is biaxial (symmetry point group 222). The phonon spectrum of this crystal has been much studied [see Krauzman et al. (1973)] and is usually subdivided into four groups [lattice vibrations (0 to $220\,cm^{-1}$), deformation vibrations of the IO_3 group (290 to $400\,cm^{-1}$), stretching vibrations of the IO_3 group (600 to $845\,cm^{-1}$), and vibrations of the OH group: out-of-plane (torsional) vibrations ($560\,cm^{-1}$), plane (deformation) vibrations ($1160\,cm^{-1}$) and stretching vibrations ($2940\,cm^{-1}$)]. Thus, for $\omega > 1160\,cm^{-1}$, only a single first-order line with frequency $\omega \approx 2940\,cm^{-1}$ can be observed in the scattering spectra. Nevertheless, a wide band can be observed in RSL spectra in the frequency region of approximately $2270\,cm^{-1}$ (scattering angle $\theta \approx 90°$). On the low-frequency edge of this band, i.e. at $\omega = 2270\,cm^{-1}$, there is a quite narrow and intense peak. It was pointed out by Polivanov (1979a) that scattering in the frequency region $2270\,cm^{-1} < \omega < 2940\,cm^{-1}$ corresponds to the excitation of two- and three-particle states. It was also postulated that the peak at

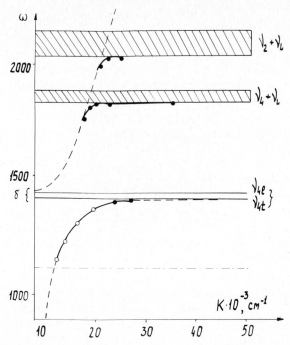

Fig. 11. Polariton dispersion in the NH$_4$Cl crystal. Longitudinal–transverse splitting of the biphonon is observed in the frequency region $\nu \approx 1460\ \text{cm}^{-1}$ (from Gorelik et al. 1977).

$\omega = 2270\ \text{cm}^{-1}$ indicates the existence of a biphonon that split off the overtone band of the fundamental vibration with frequency $\Omega = 1160\ \text{cm}^{-1}$. A second paper (Polivanov 1979b) confirmed this hypothesis. In the investigation of RSL by polaritons (scattering angle $\varphi = 0°$ to $4°$) it was shown (fig. 12) for the same crystal that the polariton branch has a discontinuity at frequency $\omega = 2270\ \text{cm}^{-1}$ and is greatly broadened in the band of two-particle (dissociated) states. Upon cooling the crystal from $T = 300\ \text{K}$ to $T = 80\ \text{K}$ a substantial decrease in the biphonon line width was observed (Polivanov 1979b), as well as the occurrence of polariton branch discontinuities within the band of two-particle states.

The broadening of the polariton branch in the region of dissociated states is due to the opening of a new decay channel, an effect mentioned above. As to the discontinuities of the polariton branch that occur in the same spectral region upon lowering of the temperature of the crystal, their cause is still not quite clear. Only in subsequent investigations can we hope to establish which of these discontinuities are due to critical points in the density of two-particle states, to the formation of quasi-biphonons, or to interference of the Fano antiresonance type [see Agranovich and Lalov (1976a), Agranovich et al. (1979b)].

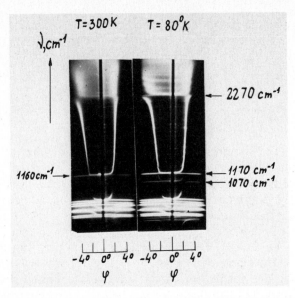

Fig. 12. Fragments of phonon spectra in scattering by polaritons in the HIO_3 crystal that were obtained by Polivanov (1979b) at $T = 300$ K and $T = 80$ K; φ is the scattering angle.

Experimental investigation of biphonons in disordered media has also been initiated in recent years by Belousov and Pogarev (1978). The interest shown in connection with this problem is due to the fact that the presence of impurities or lattice defects leads, in general, to the formation of local biphonons in the frequency region $\omega \approx 2\Omega_1$ or $\omega \approx \Omega_1 + \Omega_2$. It has been shown (Agranovich 1970) that the conditions for the occurrence of these states substantially differ from those required for the formation of local phonons ($\omega \approx \Omega_1$). It has been found that situations are even possible in which no local or quasi-local states are formed in the region of the fundamental tone ($\omega \approx \Omega_1$), notwithstanding the presence of impurities, and their splitting off the band of bulk states occurs only in the region of overtones or combination tones of vibrations. This, exactly, was the situation studied in the experimental research mentioned above (Belousov and Pogarev 1978). This work investigated first and second order RSL spectra in NH_4Cl crystals doped with the isotope ^{15}N. It has been shown that in such crystals doping with the isotope ^{15}N does not lead to the occurrence of a local or quasi-local state in the region of the fundamental vibration at the frequency $\Omega = 1400$ cm^{-1}. At the same time, in the overtone region of this vibration, i.e. at $\omega \approx 2800$ cm^{-1}, the introduction of this impurity leads to the appearance of a new and sharp maximum. The theoretical analysis of two-particle states, carried out recently by Agranovich et al. (1979a) based on the principle of the coherent potential method, resulted in a quantitative interpretation of the results obtained by Belousov and Pogarev (1978). In accordance with this interpretation, the new maximum observed in the RSL spectra of the NH_4Cl crystal upon introducing the isotopic impurity ^{15}N is due to the formation of a local quasi-biphonon.

9. Conclusion and prospects for further investigations

The possibility of the formation of biphonons and other larger phonon complexes substantially enriches the vibrational spectrum of many-particle states of crystals. Though many vital results have been obtained in this line of research, both on the experimental and theoretical sides, much still remains to be done. Among others, three-phonon and other, more complex, bound states of phonons deserve further study [their analysis has only been initiated by Lalov (1975), Krivenko et al. (1978), and Gotshev and Lalov (1979)]. However, even biphonons and their role in many optical processes have not yet been investigated to a sufficiently comprehensive degree. In connection with the aforesaid, we should like to call attention to the timeliness of calculations of biphonons in crystals of various structures for the region of degenerate vibrational transitions.

The investigations of biphonons in one- and two-dimensional crystals may also be of special interest. In such crystals, as well as in three-dimensional crystals in which certain phonons can be assumed to be quasi-one-dimensional or quasi-two-dimensional, the conditions for the formation of biphonons should be more favourable, all other things being equal, than in ordinary three-dimensional crystals.

Speaking of biphonons in one-dimensional crystals, we should also like to draw attention to their possible role in the transfer of energy along protein molecules (Green 1973). In such molecules there are vibrations with energy $\hbar\Omega \approx 0.2\,eV$. Hence, the formation of stable and mobile complexes, consisting of two or three vibration quanta, may prove to be an essential factor in the transfer of the energy $E \approx 0.5\,eV$ [see Green (1973)]. It is quite clear, of course, that such biphonon energy transfer over macroscopic distances, which are the only important ones in biology, can be considered possible only under the condition that the lifetime of the vibration quanta is sufficiently long. In this sense, the problem posed here has the same significance as in the assessment of the role of soliton transfer (Davydov and Kislukha 1973, 1976).

It has already been pointed out that the conditions for the formation of local or quasi-local biphonons in disordered crystals can differ substantially from the analogous conditions for the formation of local states in the region of the fundamental frequencies. In view of the development of experimental research of spectra of disordered crystals in the region of overtones and combination tones of the fundamental vibrations, what seems to be especially timely and of great theoretical interest is the further analysis of the above-mentioned conditions for the formation of local biphonons. In a more general sense, this means the analysis of the spectra of disordered crystals in the spectral region of many-particle states. In the theoretical papers cited above (Agranovich et al. 1979a, Agranovich and Dubovskii 1980) only the simplest version of the coherent potential method was applied. In these papers the method was employed to investigate the spectra of isotopically disordered crystals. Hence, the analysis of the spectra of mixed crystals, consisting of different molecules, is also timely. The coherent potential method (Elliot et al. 1974) was devised, as is well known, especially for calculating the spectra of one-particle states (electrons, phonons and excitons). In dealing with two-particle states, however, a number of theoretical problems arise when anharmonicity is taken into account [see also Agranovich and Dubovskii (1981)]. These problems require additional analysis and are associated primarily with the need to provide proper asymptotic behaviour of the method at low impurity concentrations (going over to the biphonon or quasi-biphonon equation).

Also worthy of further development is the theory of surface biphonons. The conditions required for the formation of these states are different from those for the formation of surface states for the spectral region of the fundamental vibrations. It was shown in one paper (Agranovich et al. 1970) on the example of the model of a bounded one-dimensional crystal, that situations may exist, in general, in which the surface state of the phonon is not formed and the spectrum of surface states begins only in the frequency region of the overtones or combination tones of the vibrations.

When the frequency of the surface biphonon lies within the band of the surface polariton, Fermi resonance occurs and the dispersion curve of the polariton is subject to a number of essential changes [gaps appear, etc. (Agranovich and Lalov 1976c)]. Consequently, experimental research of surface polariton dispersion under these conditions could yield, like similar investigations of bulk polaritons, a great deal of interesting information, not only about the surface biphonons themselves, but about the density of states of surface phonons and the magnitude of their anharmonicity constants as well.

In conclusion we stress the fact that biphonons and other effects of strong anharmonicity should appear, not only in the absorption and luminescence spectra of pure and impure crystals, in nonlinear processes, in RSL spectra and in the spectra of inelastically scattered neutrons, but, possibly, also in radiationless decay processes of electronically excited states of the crystal. An analysis of this problem, as well as of many other manifestations of states of phonons bound to one another, is of exceptional interest and will undoubtedly be the object of future research.

References

Abrikosov, A.A., L.P. Gor'kov and I.I. Dzaloshinskii, 1962, Methods of Quantum Field Theory in Statistical Physics (Fizmatgiz, Moscow) (in Russian).

Agranovich, V.M., 1959, Zh. Eksp. Teor. Fiz. **37**, 430.

Agranovich, V.M., 1960, Usp. Fiz. Nauk **71**, 141.

Agranovich, V.M., 1970, Fiz. Tverd. Tela **12**, 562 [1970, Sov. Phys. Solid State **12**, 430].

Agranovich, V.M., 1973, Supplement to the Russian edition (Mir Publishers, Moscow) of H. Poulet and J.-P. Mathieu, 1970, Spectres de vibration et symetrie des cristaux (Gordon and Breach, Paris).

Agranovich, V.M. and O.A. Dubovskii, 1980, Mol. Cryst. Liq. Cryst. **57**, 175.

Agranovich, V.M. and O.A. Dubovskii, 1981, Fiz. Tverd. Tela **23**, 2197.

Agranovich, V.M. and V.L. Ginzburg, 1983, Crystal Optics with Spatial Dispersion and the Theory of Excitons, 2nd Ed. (Springer. Berlin).

Agranovich, V.M. and I.I. Lalov, 1971a, Fiz. Tverd. Tela **13**, 1032 [1971, Sov. Phys. Solid State **13**, 859].

Agranovich, V.M. and I.I. Lalov, 1971b, Zh. Eksp. Teor. Fiz. **61**, 656 [1972, Sov. Phys. JETP **34**, 350].

Agranovich, V.M. and I.I. Lalov, 1976a, Solid State Commun. **19**, 503.

Agranovich, V.M. and I.I. Lalov, 1976b, Fiz. Tverd. Tela **18**, 1971.

Agranovich, V.M. and I.I. Lalov, 1976c, Opt. Commun. **16**, 239.

Agranovich, V.M. and M.A. Mekhtiev, 1971, Fiz. Tverd. Tela **13**, 2424.

Agranovich, V.M., Yu.E. Lozovik and M.A. Mekhtiev, 1970, Zh. Eksp. Teor. Fiz. **59**, 246.

Agranovich, V.M., N.A. Efremov and E.P. Kaminskaya, 1971, Opt. Commun. **3**, 387.

Agranovich, V.M., N.A. Efremov and I.K. Kobozev, 1976, Fiz. Tverd. Tela **18**, 3421.

Agranovich, V.M., O.A. Dubovskii and K.Ts. Stoichev, 1979a, Fiz. Tverd. Tela **21**, 3012.

Agranovich, V.M., E.P. Ivanova and I.I. Lalov, 1979b, Fiz. Tverd. Tela **21**, 1629.

Aktsipetrov, O.A., G.Kh. Kitaeva and A.N. Penin, 1977, Fiz. Tverd. Tela **19**, 1001.

Aktsipetrov, O.A., G.Kh. Kitaeva and A.N. Penin, 1978, Fiz. Tverd. Tela **20**, 402.

Belousov, M.V. and D.E. Pogarev, 1978, Zh. Eksp. Teor. Fiz. Pis'ma **28**, 692.

Bethe, H., 1931, Z. Phys. **71**, 205.

Bloembergen, N., 1965, Nonlinear Optics (Benjamin, New York).

Bogani, F., 1978, J. Phys. C: Solid State Phys. **11**, 1283, 1297.

Davydov, A.S. and N.I. Kislukha, 1973, Phys. Stat. Sol. (b) **59**, 465.

Davydov, A.S. and N.I. Kislukha, 1976, Zh. Eksp. Teor. Fiz. **71**, 293.

Debye, P., 1912, Ann. der Phys. **39**, 789.

Dows, D.A. and V. Schettino, 1973, J. Chem. Phys. **58**, 5009.

Efremov, N.A. and E.P. Kaminskaya, 1972a, Fiz. Tverd. Tela **14**, 1185.

Efremov, N.A. and E.P. Kaminskaya, 1972b, Fiz. Tverd. Tela **14**, 2105.

Einstein, A., 1906, Ann. der Phys. **22**, 1800.

Einstein, A., 1911, Ann. der Phys. **35**, 679.

Elliot, R.S., J.A. Krumhansl and P.L. Leath, 1974, Rev. Mod. Phys. **46**, 465.

Fain, V.M. and Ya.I. Khanin, 1965, Quantum Radiophysics (Sov. Radio Publishers, Moscow) (in Russian).

Flytzanis, C., 1978, in: Treatise in Quantum Electronics, eds. N. Rabin and C.L. Tang, vol. **1A** (Plenum Press, New York) p. 111.

Gorelik, V.S., G.G. Mitin and M.M. Sushinskii, 1975, Zh. Eksp. Teor. Fiz. **69**, 1823.

Gorelik, V.S., O.P. Maximov, G.G. Mitin and M.M. Sushinskii, 1977, Solid State Commun. **21**, 615.

Gorelik, V.S., G.G. Mitin and Yu.N. Polivanov, 1978, Kristallogr. **23**, 561.

Gotshev, I. and I.I. Lalov, 1979, Bulg. J. Phys. **6**, No. 4.

Green, D.E., 1973, Science **181**, 583.

Grun, J.B., Vu Duy Phach, A. Bivas and B. Honelage, 1978, Phys. Stat. Sol. (b) **86**, 159.

Gush, H.P., W.F.I. Hare, E.I. Allin and H.L. Welsh, 1960, Can. J. Phys. **38**, 176.

Hanamura, E., 1973, Solid State Commun. **12**, 951.

Hanus, I., 1962, Phys. Rev. Lett. **11**, 336.

Herzberg, G., 1945, Infrared and Raman Spectra of Polyatomic Molecules (Van Nostrand Co., New York).

Hopfield, J., 1958, Phys. Rev. **112**, 1555.

Krauzman, M., M. Le Postollec and J.-P. Mathieu, 1973, Phys. Stat. Sol. (b) **60**, 761.

Krauzman, M., R. Pick, H. Poulet, G. Hamel and B. Prevot, 1974, Phys. Rev. Lett. **33**, 528.

Krishnan, R.S., 1946, Proc. Ind. Acad. Sci. **24**, 25.

Krivenko, T.A., E.F. Sheka and E.I. Rashba, 1978, Mol. Cryst. Liq. Cryst. **47**, 119.

Lalov, I.I., 1974, Fiz. Tverd. Tela **16**, 2476.

Lalov, I.I., 1975, Phys. Stat. Sol. (b) **68**, 319, 681.

Lalov, I.I., 1981, Solid State Commun. **39**, 501.

Leibfried, G. and W. Ludwig, 1961, Theory of Anharmonic Effects in Crystals (Academic Press, New York).

Lifshits, I.M., 1956, Nuovo Cimento Suppl. **3** (X), 716.

Maradudin, A.A., 1971, in: Phonons, ed. M.A. Nusimovici (Flammarion Sciences, Paris) p. 427.

Mavrin, B.N. and Kh.E. Sterin, 1972, Zh. Eksp. Teor. Fiz. Pis'ma **16**, 265.

Mooradian, A. and A.L. McWhorter, 1969, Phys. Rev. **177**, 1231.

Pitaevsky, L.P., 1976, Zh. Eksp. Teor. Fiz. **70**, 738.

Polivanov, Yu.N., 1978, Usp. Fiz. Nauk **126**, 185.

Polivanov, Yu.N., 1979a, Zh. Eksp. Teor. Fiz. Pis'ma **30**, 415.

Polivanov, Yu.N., 1979b, Fiz. Tverd. Tela **21**, 1884.

Poulet, H. and J.-P. Mathieu, 1970, Spectres de vibration et symetrie des cristaux (Gordon and Breach, Paris).

Prevot, B., B. Hennion and B. Dorner, 1977, J. Phys. **C10**, 3999.

Reisland, J.A., 1973, The Physics of Phonons (Wiley, New York).

Ron, A. and D.F. Hornig, 1963, J. Chem. Phys. **39**, 1129.

Ruvalds, I. and A. Zawadowski, 1970a, Phys. Rev. **B2**, 1172.

Ruvalds, I. and A. Zawadowski, 1970b, Phys. Rev. Lett. **24**, 1111.

Schettino, V. and P.R. Salvi, 1975, Spectrochim. Acta **31A**, 399.

Sheka, E.F., 1971, Usp. Fiz. Nauk **104**, 593.

Tubino, R. and J.L. Birman, 1975, Phys. Rev. Lett. **35**, 670.

Van Kranendonk, J., 1959, J. Phys. **25**, 1080.

Van Kranendonk, J. and G. Karl, 1968, Rev. Mod. Phys. **40**, 451.

Winter, F.X. and R. Claus, 1972, Opt. Commun. **6**, 22.

Wortis, M., 1963, Phys. Rev. **132**, 85.

Wright, G.B., ed., 1969, Light Scattering Spectra of Solids (Plenum Press, New York).

CHAPTER 4

Energy Transport in Mixed Molecular Crystals

RAOUL KOPELMAN

Department of Chemistry
The University of Michigan
Ann Arbor, Michigan 48109
U.S.A.

Spectroscopy and Excitation Dynamics
of Condensed Molecular Systems
Edited by
V.M. Agranovich and R.M. Hochstrasser

Contents

1. Introduction: transfer, transport, percolation and clusters

Perfect crystals are characterized by their translational symmetry. The concept of translational symmetry is so powerful both in theory and in applications that both theorists and experimentalists are extremely reluctant to do without it. Thus we find approaches using effective Hamiltonians, effective Green's functions or "superlattices" applied to mixed crystals, all of which preserve the translational symmetry of the perfect crystal (Huber 1981, Gochanour et al. 1979, Haan and Zwanzig 1978, Klafter and Silbey 1980, Blumen et al. 1980, Godzik and Jortner 1980, Kopelman 1975). Many practical systems are described quite well by such approaches. An extreme example may be a "pure" anthracene crystal, which contains about 15% (fifteen mole percent) of the isotopic "impurity" molecules $^{13}C^{12}C_{13}H_{10}$, not to mention smaller amounts of isotopic and chemical impurities and defects. Nevertheless, the singlet electronic exciton states of this crystal are well described by the amalgamation limit (Kopelman 1975, Hoshen and Jortner 1971), i.e. preserving the translational symmetry, and giving results very similar to those derived from perfect crystal theory. Actually, practically all theoretical descriptions of normal ("neat") anthracene crystals quite successfully assumed the crystal to be an isotopically "pure" system.

In this chapter we purposely try to describe "idealized" mixed crystal systems where the translational disorder dominates the properties of the system. Thus the choice of both theory and experiments is biased by the above considerations. In such systems new concepts emerge which we believe to be of basic importance to the understanding of most real mixed crystals, whether natural or synthetic. Furthermore, we believe that the new qualitative characteristics of such systems may inspire new ideas, new designs and new materials for the transport of elementary excitations.

Our "idealized" systems are not only a manifestation of perfect order but also of "perfect randomness". To explore this limit fully, we emphasize heavily doped (highly concentrated) mixed crystals with excitations that exhibit short-range interactions and conform to a separated band limit (Kopelman 1975). The reasons for the above are: (i) Lightly doped systems can usually be described as slightly perturbed perfect crystals (Kopelman 1975). (ii) Long-range interactions tend to "average out" the local disorder. (iii) The separated band limit is the opposite extreme of the amalgamated band limit (Kopelman 1975), where again translational symmetry is applicable. Obviously, many real systems fall in-between the perfectly ordered and the perfectly random limits. Our task here is to emphasize the importance and beauty of the limit of "perfect randomness".

The cornerstone of the substitutionally random lattice is the cluster (the microdomain within which substitutional order is preserved). The average cluster size is the equivalent of a correlation length in one-dimensional systems, a "correlation area" in two-dimensional systems or a "correlation volume" in three-dimensional systems. The actual distributions of cluster sizes and shapes, and their dependence on order parameters are of much current interest (Stauffer 1979, Hoshen et al. 1979). They are believed to underly both the static and the dynamic physical properties of many interesting systems (Stauffer 1979, Hoshen et al. 1979).

We present an overview of some very recent developments: (i) lattice cluster theory, with its critical behaviour and universality; (ii) random and correlated walks on random lattices.

We furthermore present physical models of energy transport that incorporate the above and related mathematical approaches. Generalized diffusion and rate constants, the coherence and spread of the excitations and related concepts and approaches are also discussed, within the limitations posed by the extreme recency of these developments. Here the reader is also referred to chapter 1.

Experiments on energy transport are described only for substitutionally disordered crystals. We emphasize isotopically mixed crystals but also include chemical substitution. A critical comparison of experiments and theories is attempted. Following an interim balance sheet of the exciting new field of endeavor some suggestions for further theoretical and experimental investigations are made. Potential biological and technological applications are also briefly mentioned.

2. Localization and delocalization

The question of whether the excitonic states are extended or localized is of prime importance to the energy transport in a given system. Assuming that excitonic energy transport in a pure crystal is described by the band model (i.e. delocalized states) gives the system a "metallic" character in the sense of an analogy to electron conduction in metals. Certain alloys show an electrical metal-to-insulator transition as a function of the mole fraction of one of the alloy's components. Examples are certain bronzes (Mott and Davies 1978, p. 148) where the electronic conductivity drops sharply below a given concentration of an alkali metal (e.g. Na). The empirical facts have been interpreted in terms of (i) a Mott transition, (ii) an Anderson localization (Anderson–Mott transition) and (iii) classical percolation (Mott and Davies 1978, p. 148). We remind the reader that a Mott transition relates to sharp structural changes (phase transition), while both the Anderson localization and the classical percolation models depend on subtle changes in some parameter (Mott and Davies 1978, p. 148). In the Anderson model a disorder-caused energy spread approaches the magnitude of the bandwidth (of the ordered system). In the classical percolation picture a microscopically connected "metallic network" (on the atomic scale) falls apart below a certain point of the relevant parameter (metallic concentration or density). For interpretations of increasing complexity the reader is referred to the recent text of Mott and Davies (1978) and for further discussion of the theoretical models the reader is referred to the recent text by Ziman (1979) and the review by Thouless (1979). Indeed, electronic conduction in substitutionally and otherwise disordered materials is in the forefront of current condensed materials research.

An important question is: May an alloying of molecular crystals produce an exciton analog of the metal–insulator transition, viz. a transition from an exciton conducting band to some exciton semiconductivity i.e. thermally assisted hopping? We note that metallic conductivity decreases with temperature, while non-metallic conductivity increases with temperature and, due to its activation energy, is expected to go to zero at zero temperature. Before we can answer the above question we have to answer another question: is excitonic energy transport in pure molecular crystals "metallic"? In other words: is exciton energy transport in pure crystals in accordance with a simple band model? Only if the answer to the latter question is positive for at least one of the mixed crystal components ("parent") will it be reasonable to expect "metallic" behaviour for some concentration range of the randomly mixed crystals.

If the pure parent crystals of the mixed molecular crystals exhibit "nonmetallic" properties, e.g. exciton hopping, for the same thermodynamic parameters, then we expect all relevant exciton states to also be nonmetallic in the mixed crystals. While "hopping" or phonon-assisted transport may redelocalize the excitons, this should not be semantically confused with exciton delocalization (i.e. extended states, see below). We note that electron hopping has also been discussed within a percolation model (Pollak and Knotek 1979, Mott and Davies 1978) which should not be confused with the classical percolation theory. For instance, rigorous treatments of variable-range electron hopping have utilized the "most favoured path of sites" or "percolation channel" approach (Mott and Davies 1978, p. 35). This is a quantum-mechanical percolation approach dealing with hopping probabilities (or tunneling times) between pairs of sites. Questions of interest are: Can exciton hopping be described by a percolation approach? Can exciton tunneling be described by a percolation theory? These cardinal questions are dealt with in the present section and throughout the rest of this chapter.

Finally, we warn the reader that the mere term "delocalization" has different meanings for the chemist and the theoretical solid state physicist. In this section we adopt the physicist's semantics by which a delocalized state is extended to infinity (the theoretician's definition) or over macroscopic distances (the experimentalist's definition) with amplitudes that do not approach zero asymptotically. It is the localized state which has amplitudes approaching asymptotically to zero with increasing distance from the microscopic domain of localization.

2.1. Band model and Anderson–Mott transition

To illustrate this approach we use the Klafter–Jortner (KJ) model (1977, 1978) of an Anderson–Mott excitation mobility edge, which is based, in part, on the Lyo–Orbach approach (Lyo 1971, Orbach 1974). This model has the advantages of conceptual simplicity and consistency. However, with the passage of time it gained some complexity (Klafter and Jortner 1979, 1980a,b).

For a binary random molecular alloy (AB), KJ assume the separated band limit (Kopelman 1975), i.e. that the excitation is confined to the quasilattice of one of the components (say A). This leads to a guest (A component) exciton band in zeroth order, which for the limit $C_A = 1$ (mole fraction of A equals unity) becomes the exciton band for the parent A crystal. The exciton band and bandwidth are determined by short-ranged pairwise exciton interactions J_n, where n designates the relative coordinates of the interacting sites. Furthermore, both in the pure and the mixed crystals there is a disorder-induced local random strain (and/or inhomogeneity) energy W. The strain energy (W) does not depend on C_A. The latter assumption is somewhat plausible as this model was introduced for isotopically mixed molecular crystals, i.e. guest (A) and host (B) molecules that differ only by (symmetric) isotopic substitution. Now the basic and simple idea is that $\langle J_n \rangle$ decreases monotonically with C_A and that some critical concentration C_c exists, for which $J_c \equiv \langle J_n(C_c) \rangle$ meets the Anderson localization criterion (Klafter and Jortner 1979):

$$J_c \approx W/\kappa, \tag{1}$$

where κ is a numerical constant of order 10, related to the lattice connectivity $(\approx Z - 1$, where Z is the coordination number).

Below C_c all exciton states are localized ["Anderson localization" (Mott and Davies 1978)]. Above C_c there are extended (band) states, energetically flanked by localized states [Anderson–Mott model (Mott and Davies 1978)]. Thus the exciton transport becomes "metallic" above C_c but is "nonmetallic" below C_c. The critical exciton transport concentration C_c is thus an exciton mobility edge, in analogy to Mott's electron mobility edge (Mott and Davies 1978).

2.2. Band model and percolation transition

This model (Monberg and Kopelman 1980) again assumes a separated band limit (see below), resulting in the confinement of the excitation to the guest (A) quasilattice. Again the parent guest (A) crystal has an exciton band determined by the short-ranged pairwise exciton interactions J_n. For simplicity we now assume that $J_n = J$ when n is a nearest neighbor separation and $J_n = 0$, otherwise.

In the random binary mixed crystal there are clusters of guest (A) sites. Any two sites in the same cluster are connected either directly or via a chain of directly connected A sites (a direct connection between two sites exists only if they are nearest neighbors). The percolation theory (Stauffer 1979) tells us that for an infinite lattice there are either one or zero infinite A clusters. The rest of the clusters are finite. There is a critical concentration $C_A = C_c$ above which one infinite cluster exists and below which none exist. Extensive simulations for large (up to 10^9 sites) finite lattices (Hoshen et al. 1978, 1979) have again shown the existence of a critical concentration C_c above which one "maxicluster" exists (extending from one edge of the lattice to any other edge) and below which none exists. Finite clusters ("miniclusters") exist both above and below C_c. However, the cardinal point is that below C_c only finite clusters (miniclusters) exist.

For our model of only nearest neighbor exciton interactions (J) it is elementary that any exciton state is confined within a given cluster. Thus, in the regime of finite clusters only localized excitons exist. It is also quite plausible that an infinite cluster gives rise to extended exciton states, having assumed that the pure A crystal has only extended states. These extended states form a band (as the infinite cluster has an infinite number of associated extended states).

The conclusion is simple: below C_c there are only localized exciton states; above C_c there are both delocalized (extended) and localized states. We thus get a percolation transition from localized to extended (band) states at the critical percolation concentration. This is a transition from hopping ("nonmetallic") to band ("metallic") exciton transport as the concentration (C_A) crosses the critical concentration point (C_c).

We note that the value of C_c depends only on the lattice topology (Stauffer 1979). Values for C_c (the "site percolation concentration") for common lattices are available now with two to four digit precision (Hoshen et al. 1981, 1979). Furthermore, we note that one can easily relax the condition of nearest-neighbor-only interactions, as long as a sharp J_n cut-off occurs for some n. Qualitatively, the results are the same. Quantitatively, the value of C_c is smaller, as the effective cluster ("percolation") topology has been modified, increasing the lattice coordination number.

2.3. Hopping model and Anderson–Mott transition

We assume here that the parent A crystal ($C_A = 1$) has only localized exciton states, giving rise to a hopping ("nonmetallic") energy transport. The reason for this could be due to an Anderson localization [with eq. (1) applicable for $C_A = C_c = 1$] at $T = 0$, followed by phonon-assisted hopping at $T = T$. However, there may be other reasons for the localization and hopping, such as strong exciton–phonon interactions (small polaron model) or exciton–phonon scattering (Holstein et al. 1981).

Whatever the reason for exciton localization in the pure (parent) crystal, the same reasons should apply equally well or better for the mixed crystals (at the same temperature). If Anderson localization was the cause for hopping, then [see eq. (1)] J will decrease with dilution ($C_A < 1$) and W might increase with mixing (disorder). If exciton–phonon scattering is the reason for hopping, this scattering is likely to increase due to decreasing $\langle J \rangle$. We note that the phonons in isotopically mixed crystals are usually in the amalgamation limit (Prasad and Kopelman 1972) and are thus little affected by the mixing.

The above discussion leads us to the conclusion that an Anderson–Mott transition cannot account for an empirically observed exciton transport transition if the mechanism of transport consists of hopping (tunneling) both above and below the transition point (observed C_c).

2.4. Hopping model and percolation transition

Similarly to the approach of subsection 2.2, we now assume a cut-off in J_n for a given n (nearest neighbor or larger). This produces a cut-off in the hopping rate for a given hopping distance (better physical reasons for a cut-off in the hopping rate are given below). Only pairs of sites with distances smaller than or equal to the "cut-off" (n) are technically directly connected. This connectivity determines the cluster distribution for each C_A as well as C_c (the critical percolation transition concentration). Again, below C_c there are only miniclusters while above C_c there is a maxicluster (as well as miniclusters).

An exciton created on a minicluster could be confined there throughout the excitation lifetime. On the other hand, an exciton created on a maxicluster is likely to undergo significant transport (via hopping) within its lifetime. This again gives rise to a drastic change in the exciton transport ability at C_c. While the transport mechanism involves hopping both above and below C_c, the hopping below C_c may lead to endless hopping within a "cage" (the minicluster), while above C_c the exciton (on the maxicluster) is "free" to hop over large distances.

The hopping (or tunneling) percolation transition is defined by a time-dependent connectivity. Practically, the hopping time has to be shorter than the excitation lifetime. Thus it is the finite lifetime of the excitons which determines the cut-off rate and thus the cut-off distance and effective topology. Semantically, a "dynamic" cut-off gives rise to "dynamic" clusters and "dynamic percolation". This is in contrast to the "static" cut-off, "static" clusters and "static" percolation described in subsection 2.2 (which were all time independent).

The overall hopping time is actually limited by the timescale of a given experiment (for measuring the energy transport). Thus the dynamic percolation transition depends not only on the physical parameters but also on the time scale of the experiment! Therefore, in principle, it is easy to distinguish between a static (band model) percolation transition and a dynamic (hopping) percolation transition. On the other hand, we note that percolation transitions can arise for both extremes of exciton transport (band and hopping) and thus, by interpolation, also for intermediate transport cases (see below). Mathematically, it is the cluster structure of the random binary lattice that gives rise to some critical transition concentration C_c, no matter how short or long the effective interaction distance is, as long as a sharp cut-off does exist.

3. Cluster model, critical points, scaling and universality

Exciton transport in mixed molecular crystals can be monitored in a number of ways: (i) supertrapping (monitored by emission from an impurity acceptor), (ii) annihilation (exciton fusion monitored kinetically or via delayed fluorescence, (iii) spatial spread (monitored microscopically or by a diffraction grating method), (iv) surface sensors (adsorbed dye emission), (v) variations or combinations of the above mentioned methods (Francis and Kopelman 1981). We note here that the monitoring method cannot be totally divorced from the measured transport. (This is a good example of the inseparability of "apparatus" and "system".) We emphasize below a formalism based on supertrapping, as most of the experimental work is based on this method. Adaptation to other detection methods is straightforward but not necessarily trivial.

3.1. Cluster formalism of exciton transport

The basic assumptions are: (i) the exciton is confined within a cluster of size m (number of guest sites); (ii) the creation probability of an exciton inside a given cluster is proportional to the cluster size (m); (iii) the probability that a given guest cluster site is occupied by a supertrap is given by the fraction (S) of guest molecules that are supertraps (the fraction of donors is $1 - S$); (iv) the quantum-mechanical probability $\pi_m(t)$ that an exciton confined to a cluster of size m is supertrapped by an acceptor of unit cross section, within a time t, is independent of the cluster shape.

For a finite lattice with G guest sites and $S \ll 1$, the probability P of supertrapping an exciton ("percolation probability") is given by (Kopelman 1976)

$$P = G^{-1} \sum_m i_m m [1 - (1 - \pi_m m / G)^{\bar{\gamma} SG}], \tag{2}$$

where $i_m(C)$ is the number of clusters if size m at guest concentration C and $\bar{\gamma}$ is the supertrap efficiency ($\bar{\gamma} = 1$ for a supertrap with unit cross section). We note that P depends on the guest mole fraction (C), the time (t) and the relative supertrap concentration S. The C dependence arises via the cluster frequency i_m. We emphasize that the distribution of i_m contains all the pertinent information. Also note that:

$$1 = G^{-1} \sum_m i_m m.$$

The summation $\sum_m i_m m = G$ is the "first moment" related to the cluster distribution. The "second moment" gives the average cluster size:

$$I_{av} \equiv G^{-1} \sum_m i_m m^2. \tag{3}$$

For the average cluster size we have $I_{av} \to 1$ as $C \to 0$. I_{av} rises slowly above unity as C rises from zero. Only close to the critical concentration $(C \to C_c)$ does I_{av} rise catastrophically from order unity to order G. The functional dependence of I_{av} on C is not known (except for one dimension). However, an empirical (Monte Carlo based relationship) gives

$$I_{av} = \bar{K} |1 - C/C_c|^{-\gamma}, \quad C < C_c, \tag{4}$$

where $\bar{K}$ is a constant of order unity and γ is the percolation critical exponent which has been shown empirically to depend only on dimensionality, giving γ values of about 1.6 for three dimensions and 2.2 for two dimensions.

The quantities I_{av} and the critical exponent γ (not to be confused with the supertrapping efficiency) are very relevant to exciton transport. At low concentration (C), not only I_{av} but all m values (cluster sizes) are of order unity. Given a reasonable time (t), the exciton supertrapping probability approaches unity:

$$\pi_m \to 1, \quad C \ll C_c. \tag{5}$$

Equation (5) gives a major simplification (Kopelman 1976) for the "exciton percolation" (supertrapping) probability:

$$P = SI_{av}, \quad C \ll C_c. \tag{6}$$

This result is easily rationalized. If all guest sites (including both donor and acceptor) were evenly distributed through the host lattice (forming a guest superlattice), one would simply get $P = S$, as S is the acceptor (supertrap) fraction of the guest sites. However, due to the clusterization effect, one gets an amplification of the supertrapping because the cluster acts like an antenna. The average antenna size is simply given by I_{av}, which is also the average amplification factor. Combining eqs. (4) and (6) gives an important result:

$$P = \bar{K}S|1 - C/C_c|^{-\gamma}, \quad C \ll C_c, \tag{7}$$

which is our first example of a "critical exponent" behaviour for exciton transfer in doped binary crystals (see below).

3.2. Supertransfer limit: hopping and quasiband

The supertransfer limit is simply given by

$$\pi_m = 1, \quad t > t_{min}, \tag{8}$$

for all clusters m at all concentrations C, where the characteristic time t is longer than some minimal time t_{min}, below which eq. (8) is invalid. The supertransfer limit might apply for a "band" model, i.e., where the exciton is initially delocalized over the entire cluster (size m) but is quickly "sucked-in" by the supertrap. It might equally apply for a "hopping" model, i.e., where the hopping rate is fast enough so that within t_{min} the exciton had enough time to hop over the entire cluster. The latter case is the same as Huber's "rapid transfer" limit (Huber 1981). We note that, in contrast to eq. (5), eq. (8) applies even for "infinitely" large clusters m, i.e. above the critical percolation concentration (C_c). The condition of eq. (8) can be relaxed and "supertransfer" is still achieved if (Kopelman 1976):

$$\pi_m m \bar{\gamma} S \gg 1. \tag{9}$$

Combining eq. (8) with eq. (2) gives (Hoshen et al. 1978):

$$P = 1 - G^{-1} \sum_m i_m m \lambda^m, \quad \lambda \equiv 1 - S. \tag{10}$$

This interesting equation (see below) results in two important limiting equations and one important special case equation: (i) Not surprisingly one again obtains eq. (6) (Hoshen et al. 1978):

$$P(t_{min}) = SI_{av}, \quad C \ll C_c. \tag{6'}$$

(ii) In addition, one also obtains (Hoshen et al. 1978):

$$P(t_{min}) = \bar{P}_\infty, \quad C \gg C_c, \tag{11}$$

where $\bar{P}_\infty$ is the mathematical percolation probability, i.e. the probability that a given guest site belongs to the infinite cluster ("maxicluster"). Obviously, this equation gives $P = 1$ for $C = 1$ in the supertransfer limit. (iii) The important special equation is (Hoshen et al. 1978):

$$P(t_{min}) = S^{1/\delta}, \quad C = C_c, \tag{12}$$

where δ is another percolation critical exponent (Stauffer 1979), given in the critical region ($C = C_c$) by:

$$\delta = 1 + \gamma/\beta, \tag{13}$$

where γ is the critical exponent related to I_{av} [eq. (4)] and β is the critical exponent related (Stauffer 1979) to $\bar{P}_\infty$:

$$\bar{P}_\infty = \bar{k}|1 - C/C_c|^\beta, \quad C > C_c, \tag{14}$$

where $\bar{k}$ is a constant and, similar to γ, the exponent β is also believed (Stauffer 1979) to depend only on dimensionality. The values of β are about 0.41 for three dimensions and 0.14 for lattices in two dimensions. Thus δ also depends only on dimensionality. The value of δ is about 5 for lattices in three dimensions and 17 for two dimensions. We note that for S values of about 10^{-3} to 10^{-5} eq. (12) gives a P value of about 0.66 to about 0.5 for two dimensions (but 0.25 to 0.1 for three dimensions).

Combining eq. (11) with eq. (14) gives:

$$P(t_{min}) = \bar{k}|1 - C/C_c|^\beta, \quad C \gg C_c, \tag{15}$$

which is somewhat analogous to eq. (7). The latter can now be rewritten as

$$P(t_{min}) = \bar{K}|1 - C/C_c|^{-\gamma}, \quad C \ll C_c, \tag{7'}$$

We thus get "critical" behaviour in the supertransfer case for both limits ($C \gg C_c$ and $C \ll C_c$) and for the "special" case ($C = C_c$).

3.3. Analogy to magnetism, percolation exponents and scaling range

Noting that $S \ll 1$, we can rewrite eq. (10) as

$$P = 1 - G^{-1} \sum_m i_m m \lambda^m, \quad \lambda = e^{-S}. \tag{10'}$$

Now we notice that the supertrap fractional concentration S plays the role of a "notional" field; S is analogous to the magnetic field, provided that P is analogous to the magnetization (which can be viewed as a magnetization "probability"). Thus the appearance of the critical exponent δ [eq. (12)] is less surprising. The same is true for the appearance of the exponents γ and β, provided that we also note the well-known (Stauffer 1979) analogies between I_{av} and the magnetic susceptibility, between the concentration C and the temperature, and between the percolation probability $\bar{P}_\infty$ and the zero-field magnetization. However, we note that the critical exponents related to magnetism, like all known nonclassical critical exponents, are believed to be valid only over a narrow "scaling range", very close to the critical point (Stanley 1981).

The percolation exponents have been found to be valid over a "scaling range" (C/C_c range) that is extremely wide (Newhouse et al. unpublished). For instance, the critical exponent γ was found to change little from $C/C_c = 1$ to $C/C_c = 0.1$, for both two- and three-dimensional lattices. This justifies the derivation of eq. (7), where $C/C_c \ll 1$, and eq. (15), where $C/C_c \gg 1$. Thus one may preferably use the term "percolation exponents" rather than "critical exponents" to designate β and γ far away from the critical concentration (C_c). However, the exponent δ is still used [eq. (12)] just at the critical concentration.

3.4. Leaky cluster model and universality

The above discussion assumed a given cluster definition, i.e. bond cut-off, over the entire concentration range. We note that it is the bond cut-off which defines the effective topology of the lattice. An example is the square lattice, which gives rise to the "simple" square lattice, which gives rise to the "simple" square lattice site percolation problems ($C_c = 0.593$) when only nearest neighbor bonds are considered, but gives the "square-1,2" topology ($C_c = 0.407$) when next nearest neighbor bonds (along diagonals) are added. Assuming, for simplicity, a hopping exciton transfer we can get "dynamic percolation" (see above) under the following conditions:

$$t_n \ll \tau, \tag{16}$$

$$t_{nn} \gg \tau, \tag{17}$$

where t_n is a nearest neighbor hopping time, t_{nn} is a next nearest neighbor hopping time and τ is the time scale of the experiment (e.g. excitation lifetime). It is obvious that the excitation is totally confined within the clusters defined by n (nearest neighbor) type bonds. If a minicluster contains a supertrap, the exciton will most probably be supertrapped. If, in addition,

$$t_n \ll S\tau, \tag{18}$$

it is intuitively obvious that, even in an infinite cluster, the exciton will most probably be supertrapped, thus justifying the supertransfer limit [i.e. eqs. (8–11)].

We now partially relax the condition of eq. (17) replacing it by

$$t_{nn} \gg S\tau, \qquad t_{nn} \approx \tau. \tag{19}$$

This relaxation has little effect on the results at or above C_c, where the very large clusters dominate the transport behaviour, and where an occasional "leak" out of the maxicluster via a hop along a next-nearest bond (e.g. diagonal) will be compensated, statistically, by a similar hop into the maxicluster. However, for $C \ll C_c$, such occasional "leaks" will usually occur only out of miniclusters not containing a supertrap and into mini-clusters which might contain one. This will, on the average, increase the overall supertrapping probability P. A detailed time balance (Ahlgren 1979) results in an "amplification factor" Y for P [eqs. (6), (6')]:

$$Y = (\eta f \bar{\gamma} S)^{-1}, \quad C \ll C_c, \quad Y \gg 1, \tag{20}$$

where $\bar{\gamma}$ is a trapping probability, the factor f is defined by the ratio of the average hopping times,

$$f = \bar{t}_{nn} / \bar{t}_{n}, \tag{21}$$

and η is a redundancy factor related to the inverse of the probability that a next nearest neighbor hop results in a leak into a new cluster (not yet traversed by the exciton). Combining eq. (20) with eq. (6) gives:

$$P = (\eta f \bar{\gamma})^{-1} I_{av}, \tag{22}$$

which appears to be independent of S. However, I_{av} still depends on S implicitly (see below).

Even in the above leaky cluster model, S still defines the cluster topology (bond definition), including C_c, because of the pair of relation-ships given by eqs. (18) and (19). A drastic variation in S may thus change the cluster topology. For example, increasing S from say 10^{-6} to 10^{-3}, may result in the replacement of eq. (18) by:

$$t_{nn} \ll S\tau, \tag{23}$$

and the replacement of eq. (19) by:

$$t_{nnn} \gg S\tau, \qquad t_{nnn} \approx \tau, \tag{24}$$

where t_{nnn} refers to a hopping over a third nearest neighbor bond. Thus a drastic variation in S may result in a variation in the percolation (cluster) connectivity and thus in a drastic variation not only in P curves, but also in C_c. However, a universal description can still be achieved via the device of reduced concentration C/C_c. For instance, a curve of $\bar{P}_\infty$ vs C [cf. eq. (14)] or one of I_{av} vs C [cf. eq. (4)] does depend on the cluster connectivity (and specifically on C_c). Nevertheless, a curve of $\bar{P}_\infty$ vs C/C_c or of I_{av} vs C/C_c will not depend on the connectivity, provided that the basic underlying lattice is the same, and there is no change in dimensionality (Kopelman et al. 1979). Thus eq. (4) can be replaced by

$$I_{av}(C/C_c) = K|1 - C/C_c|^{-\gamma}, \quad C/C_c < 1, \tag{25}$$

where K as well as γ are independent of the bond connectivity, for a given underlying lattice and dimensionality; similarly, eq. (14) can be replaced by

$$\bar{P}_\infty(C/C_c) = k|1 - C/C_c|^\beta, \quad C/C_c > 1, \tag{26}$$

where k and β are independent of bond connectivity, as above. An apparently trivial, but important, transformation to reduced coordinates of eq. (22),

$$P(C/C_c) = (\eta f \bar{\gamma})^{-1} I_{av}(C/C_c), \quad C/C_c \ll 1, \tag{27}$$

now results in a universal relationship, truly independent of S. It is also independent of the physical parameters t_n, t_{nn} except their relative value (represented by f) and thus is also independent of time (e.g. τ) and temperature (see below). Utilizing eq. (25), eq. (27) can be written as

$$P(C/C_c) = K(\eta f \bar{\gamma})^{-1}|1 - C/C_c|^{-\gamma}, \quad C/C_c \ll 1, \tag{28}$$

where K is independent of C_c. Similarly, eq. (11) is still valid for the leaky minicluster model, as it only concerns the maxicluster. It can thus be transformed into:

$$P(C/C_c) = \bar{P}_\infty(C/C_c), \quad C/C_c \gg 1, \tag{29}$$

or its equivalent [cf. eq. (15)]:

$$P(C/C_c) = k|1 - C/C_c|^\beta, \quad C/C_c \gg 1, \tag{30}$$

where again k is independent of C_c, but only depends on the underlying lattice topology (and both k and β depend on the dimensionality). We thus obtained a pair of universal limiting equations [eqs. (28) and (30)]. We also note that the "special" equation [eq. (12)], can be rewritten as

$$P(C/C_c = 1) = S^{1/\delta}, \tag{31}$$

which is also independent of time (τ), hopping times and temperature (see below). In addition, its S dependence is very weak, especially for a two-dimensional system where $\delta = 17$. We may thus expect a nearly universal behaviour throughout the reduced concentration range, especially for two-dimensional systems, for the leaky dynamic cluster model, i.e. the most realistic cluster model that preserves the dimensionality of the exciton interactions (bonds).

4. Lattice walk, diffusion, percolation and kinetics

In this section we give a somewhat unconventional approach to the relation of random walk to "coherence", diffusion, percolation and kinetics. The definition of "coherence" is stretched somewhat. The applicabilities of "diffusion constants" and "rate constants" are tested. Percolation-limited kinetics is found to be a boundary case between homogeneous and heterogeneous kinetics.

4.1. Random walk on random binary lattices

Very elegant solutions exist for the problem of random walk on an ordered lattice (Montroll 1964, 1969). Some attempts have been made to apply analytical solutions to binary lattices (Lakatos–Lindenberg 1972). These have met with some success for very dilute concentrations, i.e. where the second component is a very "dilute guest". While further attempts may be on their way towards solving the problem of concentrated random binary lattices we do not believe that such attempts are likely to succeed in the near future. The reason for this belief has to do with the so far unsuccessful attempts to solve simple problems related to percolation, such as the critical site percolation concentration for a square lattice with nearest neighbor interactions only. Imagine a random walker limited to move on only one component of the binary random lattice and restricted to nearest neighbor hops. The number of sites visited as a function of the number of steps should obviously have a discontinuity at the critical concentration. Thus any successful analytical solution of the problem of the number of sites visited as a function of the number of steps, with concentration as a parameter, should give the critical concentration for the site percolation problem on a square lattice. This explains our pessimistic view stated above, which motivated us to perform Monte Carlo simulations for the purpose of learning about random walk properties on random binary lattices (Argyrakis and Kopelman 1980).

A little reflection on the part of the reader will convince him that simulations of random walk on binary random lattices will be very different above and below the critical percolation concentration. We have thus limited our simulations to the concentration regime above the critical percolation concentration. We also believe that, below the critical percolation concentration, it is more productive to use simple approximations (Kopelman 1976).

The results of simulations of random walk on binary random lattices may not be very surprising but we feel they are still important (figs. 1–3). For instance, both for three-dimensional, i.e. cubic, lattices and for two-dimensional, i.e. square, lattices, the efficiency of visitation goes down with

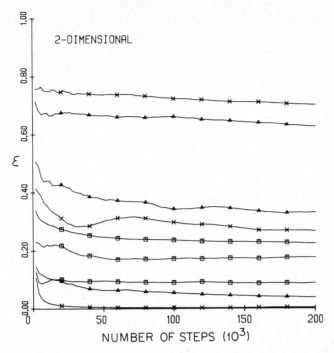

Fig. 1. Visitation efficiency versus the number of steps as a function of the coherency L and the guest concentration. We report here the cases of $L = 1$ ($\square$), $L = 10$ ($\blacktriangle$), and $L = 100$ ($\times$) (averages of Gaussian distributions with a standard deviation of $d = 0$, 3.0, 30.0, respectively). The several curves in each case correspond to different guest concentrations as follows: 1.0 (pure crystal), 0.85, and 0.70, respectively, from top to bottom. The lattice size is a square of 1022×1022, and only the four nearest neighbor interactions are considered here. All four carry equal probability. The cross-over concentration can easily be detected in each case (see text). The results are averages of several runs, typically 5–30 runs (depending on the fluctuations that each case had).

the concentration of the active site, i.e. the site on which the random walker is allowed to move. Clearly, this efficiency should approach zero at the critical percolation concentration. The usefulness and application of such data should become obvious from our discussion below.

4.2. Correlated walk on random binary lattices

By correlated walk we mean a walk in which directional memory is retained over a certain number L of steps. We call L the correlation or coherence parameter. The correlation parameter L itself may be distributed exponentially or gaussianly or otherwise. For computational ease we have generally preferred the gaussian distributions. This correlated walk problem is obviously of interest for both ordered and disordered lattices.

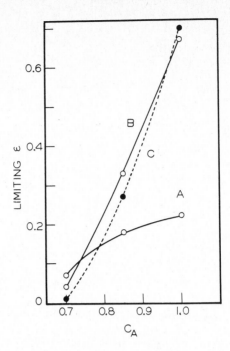

Fig. 2. The limiting visitation efficiency (from the calculations of the previous figure) versus guest concentration as a function of the coherence parameter L. The cases of $L = 1$ (curve A), 10 (curve B), and 100 (curve C) are reported here (with a Gaussian distribution having standard deviations of $d = 0$, 3.0, 30.0, respectively). Only the four nearest neighbors are considered here. See text for details.

For ordered lattices some analytical solutions have been derived, even though not rigorously (Kopelman and Argyrakis 1980).

For random binary lattices the correlated random walk gives interesting new results that differ qualitatively from the simple random walk results. For instance, there are some drastic changes with the decrease in the concentration of the active sites. For a concentration near unity the correlated random walk gives more efficient visitation than the simple random walk. However, at lower concentrations, which are still well above the critical concentration, this is no longer true. Examples (Argyrakis and Kopelman 1980) can be seen in figs. 1–3 in terms of visitation efficiency ϵ (number of distinct sites visited over number of steps).

4.3. Percolation, diffusion and rate constants

The relation between random walk (or Brownian motion) and diffusion is an old one (Einstein 1926). The relation of random walk and Brownian motion to kinetics and "rate constants" is also known (but may not be well

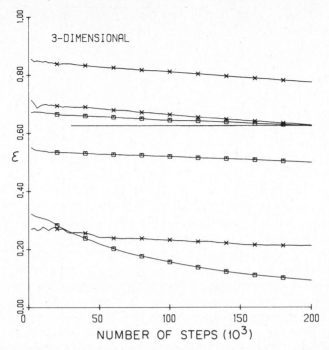

Fig. 3. Visitation efficiency versus the number of steps with the coherency L and the guest concentration as parameters. These simulations are for the three-dimensional case ($101 \times 101 \times 101$) with a total of six interactions, four in-plane and two out-of-plane. The coherency values shown are for $L = 1$ ($\square$) and $L = 10$ ($\times$) with standard deviation equal to 3.0. The guest concentrations reported here are 1.0, 0.75, and 0.50, from top to bottom. The results are averages of several runs, typically 3–30 runs (depending on the fluctuations that each case had). The solid line represents the prediction of the exact formalism for $L = 1$.

known). In a gaseous phase, for instance, the Brownian motion of particle A combined with the Brownian motion of particle B results in a constant probability of collision (per unit time). However, a random walker A on a fixed lattice, including fixed but randomly distributed particles B, results in a collision probability per unit time which is not constant with time. This results in a rate coefficient which is not a rate constant (Kopelman and Argyrakis 1980).

In a typical energy migration experiment, with supertrap monitoring, it has been shown (Kopelman and Argyrakis 1980) that the rate "constant" for supertrapping is given by

$$K(t) = (1 - P)^{-1}(\partial P / \partial t), \tag{32}$$

where P is the time-dependent supertrapping probability. The above can be expressed in simple cases (diffusion or percolation on connected lattices) as

$$K(t) \propto [\epsilon + t(\partial P/\partial t)], \tag{33}$$

where ϵ is the migration efficiency (number of distinct sites per total number of sites visited). This efficiency is a meaningful concept for either random or correlated walk on a connected lattice. It can easily be shown that only for special simple limits (Kopelman and Argyrakis 1980) will ϵ, and thus K, be a time-independent constant. One such case is the asymptotic behaviour ($t \to \infty$) of simple random walk on a one-component simple cubic lattice. Other cases involve certain types of correlated walks (Kopelman and Argyrakis 1980) on one-component, two- and three-dimensional lattices. On the other hand, K is time dependent even for the very simple cases of simple random walk on a one-component square lattice or linear chain, even for $t \to \infty$ (Kopelman and Argyrakis 1980).

When we deal with random percolation, which is a microscopic diffusion in a microscopically heterogeneous medium, the results are somewhat similar to those for diffusion. However, the quantitative deviations from oversimplified pictures are more severe, i.e. the deviation from the naive model of kinetic rate constants or Stern–Volmer kinetics is much more drastic. For instance, the so-called rate constant could approach zero asymptotically or even suddenly (below the critical concentration). Examples for this behaviour can be seen in Kopelman and Argyrakis (1980).

4.4. Dynamic percolation and kinetics

The topology or connectivity of a random binary lattice together with its concentration obviously defines the clusters which define the percolation, i.e. the "heterogeneous kinetics" of particles on this lattice. Even if this connectivity is fixed in time one gets a kinetic behaviour for a motion, migration or "percolation" of particles like excitons. However, we can also have a situation where the connectivity itself becomes a function of time (subsection 2.4). Thus the cluster structure becomes a function of time and thus the percolation dynamics is a function of time in more than one way. This leads to the concept of dynamic percolation which involves dynamic clusters, leakage from clusters, the formation of conglomerates, which are like clusters of clusters, and the effects of all these on the kinetics.

In subsection 2.4 we discussed the percolation transition in the context of a hopping model. What we had there was a cut-off in transfer rate or transfer probability which defines the topology of the lattice and thus the cluster distribution and percolation behaviour. Near the percolation transition (i.e. critical concentration), where the average cluster size is very large, an occasional leakage (i.e. transfer between nominally nonconnected A sites) will have little effect on the total energy transfer picture. However, well below the critical concentration, where the average cluster size is quite small, even one occasional leakage will effectively double the migration efficiency. If we have X such leakages, the migration efficiency will increase by a factor Y which is on the order of X. Then Y such leaky "clusters" (see subsection 3.4) define a weakly connected "conglomerate" within which energy migration is possible. The size of such a conglomerate, like the number of leaks, will rise monotonically with time, on the average.

In the above "leaky cluster model" the time not only defines the connectivity of the lattice but also, for a given connectivity, the amount of leakage. This has some interesting effects on the kinetics of such "dynamic percolation" (Francis and Kopelman 1981, Ahlgren and Kopelman unpublished). In addition, it also contributes to the universal "scaling" properties of such energy transport (see subsection 3.4).

The kinetics of exciton transport and exciton supertrapping (as well as exciton–exciton annihilation) on such binary and ternary alloys are essentially "percolation-limited kinetics" or "critical kinetics", as they exhibit a critical behaviour. One could even consider the critical transition as a transition from homogeneous kinetics (above C_c) to heterogeneous kinetics (below C_C). We return to this point below.

5. Exciton transport experiments

The term exciton transport is used by us for multiple-step exciton transfer, usually implying exciton migration over a domain that is orders of magnitude larger than a unit cell. While in certain situations energy transfer and energy transport experiments result in similar information (e.g. cluster structure), this is not always so. Two experimental methods are particuarly suitable for the monitoring of energy transport: (i) supertrapping, (ii) fusion. Usually the excitons and the exciton supertrap (acceptor) are present at low enough concentrations (10^{-2} to 10^{-7} or less) so as to require multiple energy transfer steps, i.e. energy transport, in order to register the "death" of a donor exciton (either by metamorphosis into an acceptor excitation or by exciton–exciton annihilation). We notice that the low exciton and supertrap concentrations are relative to the donor ("guest") concentration.

5.1. Transport dimensionality

Molecular crystals are seldom isotropic materials. They thus abound in low dimensionalities (two and even one) for the effective energy transport.

Even the typical three-dimensional case—orthorhombic benzene—is not an isotropic system but happens to have similar exciton interactions in all three dimensions. [This is not so for certain benzene vibrational excitons, where the interaction topology is effectively two-dimensional (LeSar and Kopelman 1978).] Colson and co-workers (Tudron and Colson 1976, Colson et al. 1977a,b) studied the energy transport for both the singlet and triplet excitons of benzene. They discovered (with the help of supertraps) critical concentrations (of C_6H_6 in C_6D_6) which are in good qualitative accord with a percolation model (Kopelman 1976). However, the same experiments are also in good accord with a macroscopic rate model (Blumen and Silbey 1979) on the one hand, and with a percolation-type simulation (Colson et al. 1977b) on the other hand.

Colson and Okumura (1980) have also studied a chemically mixed crystal, which exhibits an effective two-dimensional topology, in similarity to the much studied (Kopelman et al. 1975a,b, 1977a,b) isotopic mixed naphthalene system ($C_{10}H_8/C_{10}D_8$). The latter is discussed below in detail. Another effectively two-dimensional system is pyrazine which was studied by Zewail and co-workers (Smith et al. 1977, 1980) even though they mainly conducted energy transfer experiments. Strictly one-dimensional systems do not appear to exist in nature, but quasi-one-dimensional crystals (with respect to energy transport) have been popular systems for energy transport studies (Francis and Harris 1971, Breiland and Saylor 1980). These studies mainly involved the problem of exciton coherence (see below). While topological problems, such as clusterization, are easy to handle in strictly one-dimensional systems, this is not so for quasi-one-dimensional systems, where even a little "leakage" into other dimensions is followed by drastic qualitative changes in the exciton transport behaviour (Dlott et al. 1977). This is not true for two-dimensional systems. There a small leakage into the third dimension does not drastically change the basic aspects of the system (e.g. the critical concentration or the critical exponents, see below).

5.2. Steady-state experiments: naphthalene

The naphthalene system has been studied earlier and probably more thoroughly than any other system. It also reveals an interesting dichotomy between two- and three-dimensional behaviour. We thus use it as our main experimental example, with due apologies to the investigators of other systems.

5.2.1. Guest concentration effect
Phenomenologically, the most intriguing aspect of energy transport in mixed molecular crystals is the appearance of critical concentrations (guest

mole fractions). This was predicted on the basis of cluster percolation (Hong and Kopelman 1971) and indeed soon thereafter discovered (Ochs 1974, Kopelman et al. 1975a,b), but with some interesting twists.

Ochs (1974) expected to find the critical concentration for the singlet naphthalene exciton at about 30% (three-dimensional lattices) and that for the triplet at about 60% (square lattice). Instead, he discovered the triplet transition already at about 10% $C_{10}H_8$ (in $C_{10}D_8$), while not finding any sign of the singlet "transition" up to as high as 35%. We note that all samples were made of highly purified naphthalenes, from which the BMN (betamethylnaphthalene) was first carefully removed (via potassium fusion, repeated zone refining, etc.) and then added again in a measured amount (via "saturation"). The singlet transition (critical concentration) was eventually found (Monberg 1977) at about 50% [subsequently, in benzene, the singlet transition was indeed found at about 30%, but the triple transition was found at the low value of a few percent]. The particular shape of the curve of supertrapping probability (vs guest concentration) did agree quantitatively with a percolation model (Kopelman 1976) but not with effective medium models (Kopelman et al. 1975b). On the other hand, the critical concentration itself can be varied with the help of other parameters (see below), such as the supertrap (exciton) concentration, the temperature, the time scale of the experiment and the guest–host (molecular) energy separations (Kopelman et al. 1977a). More details are given below.

The energy transport "transitions" at certain critical concentrations resemble higher order phase transitions. The guest concentration plays the role of the temperature or density. Critical exponents have also been fitted phenomenologically (see below).

5.2.2. Supertrap concentration effect

In recent series of experiments, the supertrap concentration could be controlled well enough for a careful study of its effect on the energy transport. The quality control techniques are described elsewhere (Ahlgren 1979). In addition, it is the relative donor/acceptor ratio (guest/supertrap) S that is kept constant throughout a series of samples with different guest concentration (C).

Figure 4 (Ahlgren and Kopelman 1981) shows the effects of the supertrap concentration (S). An order of magnitude reduction in S causes a large shift (upwards) in the critical concentration, i.e. by about 100% in the triplet naphthalene case. On the other hand, for the singlet naphthalene system, the change in critical concentration is much milder (on the order of 10%). These changes go hand in hand with the temperature effects observed for these systems (see below). Similarly drastic effects are observed for exciton–exciton fusion experiments as a function of the exciton density (Kopelman et al. 1980).

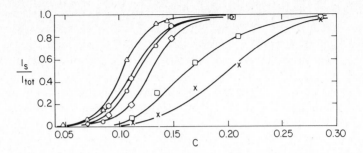

Fig. 4. Donor concentration dependence of the energy transport measure $I_s/I_{tot} = I_s/(I_s + I_d)$, where I_s is the acceptor ("supertrap") phosphorescence (0–0) and I_d is that of donor $C_{10}H_8$, for series A (X-trap, $S \approx 10^{-4}$; diamonds: 1.7 K, hexagons: 4.2 K), series B (BMN, $S = 10^{-3}$; circles: 1.7 K, triangles: 4.2 K) and series C (BMN, $S = 10^4$; crosses 1.7 K, squares: 4.2 K). The lines are visual guides. BMN: betamethylnaphthalene.

5.2.3. Temperature effects

Temperature effects of primary interest are those related to the energy migration throughout the donor ("guest", "impurity") quasilattice. Temperature effects of secondary and tertiary interest involve guest–host thermalization and supertrap–guest (acceptor-to-donor) thermalization. These secondary and tertiary effects are characterized by specific activation energies that are simply given by the appropriate energy denominators (guest–host energy separation and supertrap–guest energy separation). The secondary effect has been studied both for the naphthalene singlet and triplet excitons (Monberg and Kopelman 1980) and is weakly dependent on guest concentration. The tertiary effect has been studied, involuntarily, for both naphthalene and pyrazine, in situations where the acceptor is not a deep trap ("supertrap") but rather a shallow trap compared to kT, i.e. an X-trap (Monberg and Kopelman 1980) or a guest dimer (Smith et al. 1977, 1980).

The primary temperature effect may also be related, in part, to activation energies, e.g. cluster–cluster energy separations, as already shown by the elegant energy transfer study of Mauser et al. (1973) on the naphthalene singlet excitons. However, even where such activation energies are small (compared to kT), as in the naphthalene triplet excitons, the primary temperature effect is still significant and is undoubtedly related to exciton–phonon interactions.

Experimentally, the primary temperature effect is demonstrated for triplet naphthalene excitons in fig. 5 (Ahlgren and Kopelman 1980). The most important observation is that energy transport increases with temperature (in this T range) over the entire guest concentration range, and also over an order of magnitude variation in supertrap concentration. Quantitatively a factor of 2.5 or 3 increase in temperature (1.4 or 1.7 to 4.2 K) has

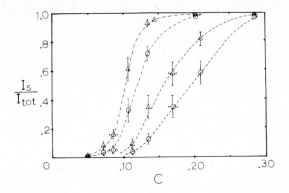

Fig. 5. Temperature dependence of critical concentration. The threshold for energy transport, C_c, decreases with increasing temperature from 1.7 K (circles) to 4.2 K (triangles). The dot–dashed curves are samples with $S = 10^3$ while the dashed curves are samples with $S = 10^4$. Note that $I_s/I_{tot} \equiv I_s/(I_s + I_d)$, where I_s is betamethylnaphthalene phosphorescence (0–0) and I_d is that of $C_{10}H_8$.

the same result as an order of magnitude increase in supertrap concentration (Fig. 4). Thus, there is a very significant downward shift in the critical concentration with increasing temperature. However, there is no "erosion" of the sharpness of the critical transition with temperature.

5.2.4. Universality and scaling behaviour

Empirically, the energy transport experiments can be expressed in terms of universal curves. It is only necessary to replace the guest concentration coordinate C by a reduced concentration coordinate $C_r \equiv C/C_c$, where C_c is the critical concentration. The critical concentration can either be defined by an arbitrary empirical criterion (like in a titration), say where the supertrapping probability (P) is 0.5, or it can be defined via a theoretical relationship [compare eqs. (12) and (31)]:

$$P(C_c) = S^{1/\delta}, \quad \sigma \equiv 1 + \gamma/\beta, \tag{34}$$

where γ and β are the critical exponents. In the latter case one has either to assume an effective dimensionality (to get β, γ and thus δ) or use an iterative procedure that defines the dimensionality (see below). Fortunately, the empirical and the theoretical criteria for C_c give practically indistinguishable results. An example, using the theoretical criterion, is given in fig. 6. It is based on six (fig. 4) different energy transport curves, where the difference is due to a different supertrap species, different supertrap concentration, different temperature or a combination of the above. This figure speaks for itself.

The traditional way of checking for scaling and critical exponents consists of appropriate log–log plots. Here one plots log P versus log$|C - C_c|$.

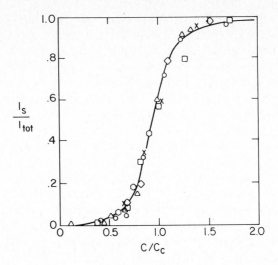

Fig. 6. Universal energy transport curve. The data points and designations are the same as in fig. 4. For each family of data points C_c was derived from fig. 4 via eq. (34), using $P = I_s/I_{tot}$.

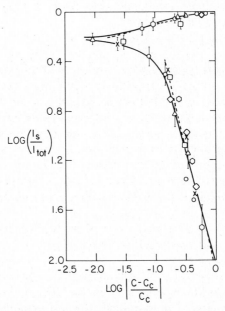

Fig. 7. "Scaled" energy transport curve. The data points and designations are the same as in figs. 4 and 6. Error bars were added to a few points to indicate experimental uncertainties. The dashed lines and least-squares fits to the experimental data, giving $\gamma = 2.1 \pm 0.2$ and $\beta = 0.13 \pm 0.05$.

Alternatively, one plots $\log P$ versus $\log|(C - C_c)/C_c| = \log|C/C_c - 1|$, using the universal plot. The two methods are equivalent. For the naphthalene triplet this results again (fig. 7) in a universal plot, as well as two asymptotic straight lines (at high and low C), giving "critical percolation exponents" (β and γ). The striking phenomena are: (i) the universality of the entire curve; (ii) the wide range over which the lines are straight as well as the fact that lines are still straight for values of C far above and far below C_c (e.g. smaller by an order of magnitude). Furthermore, the values of the critical exponents agree very well with theoretical (two-dimensional) values, i.e. $\beta = 0.13$ and $\gamma = 2.2$.

The far less complete data for naphthalene singlet excitons do appear to show a universal behaviour, as well as scaling (straight lines) for the log–log representation. However, the values of the slopes are different. This is discussed further below. We note that for other systems (benzene, pyrazine) there are not yet enough data, qualitatively and quantitatively, for a meaningful check of universality or scaling.

5.3. Time-resolved experiments: naphthalene

5.3.1. Supertransfer vs homogeneous kinetic regime

In a homogeneous kinetics scheme of exciton transfer from donor to acceptor one has the simple equation:

$$dN/dt = -kN - KN, \tag{35}$$

where N is the donor exciton population (naphthalene excitons) and K depends on B, the acceptor concentration (BMN). For time-independent rate constants K (k is always time independent) the total rate constant of donor excitation decay is $(k + K)$, which is obviously faster than the "natural" decay rate constant $k = \tau_0^{-1}$ (where τ_0 is the natural lifetime). Thus, the mere opening of a second decay channel speeds up the overall decay process.

Our steady state experiments (subsection 5.1) show for both singlets and triplets a monotonic increase of energy transport with guest concentration (C). Does it follow that K (or $k + K$) should also increase monotonically with C?

For the naphthalene triplet decay one indeed observes a monotonic increase in $(k + K)$ with guest concentration (C), as is shown in fig. 8. Thus our simple kinetic equation [eq. (35)] gives at least a qualitatively correct behaviour in this case. However, the naphthalene singlet exciton decay time (fig. 8) is nearly constant with guest concentration (Ahlgren et al. 1979). This behaviour appears to be paradoxically "nonkinetic". However, the correct answer is that only according to "homogeneous" kinetics do we have a paradox. Locally heterogeneous kinetics allows for this apparently bizzare observation. Specifically our model of "supertransfer" (Kopelman

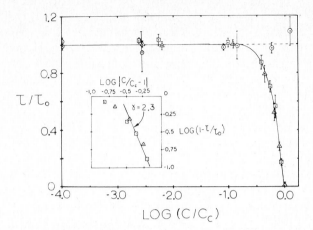

Fig. 8. Ratio of experimental $C_{10}H_8$ lifetime to lifetime at low concentration (τ_0) for singlet (circles, 2–4 K), and triplet (triangles: 4.2 K, squares: 1.7 K). Critical concentrations were determined from steady state experiments (fig. 5). Decay times for the non exponential triplet data were determined from $t \lesssim 1$ s, where the data were close to exponential. Insert: log–log plot shows scaling of triplet time-resolved data. The slope of 2.3 was added for comparison with the predicted critical exponent, $\gamma = 2.3$. A deviation from the slope of 2.3 is expected as C approaches C_c. The meaning of the triplet decay data is not clear, due to possible major contributions from triplet–triplet annihilation (Ahlgren 1979, Klymko and Kopelman 1981 and unpublished).

1976) predicts just this kind of behaviour. For instance, if the guest excitations are actually totally limited inside a guest cluster, then it either "finds" a supertrap in a very short time (say less than 1 ns) or it does not find it at all. Thus on a time scale of say 10 to 100 ns we only observe those naphthalene excitons which have no alternative decay channel (no super-trapping) available to them. Recent supertrap decay measurements lead to the same conclusion (Parson and Kopelman unpublished).

5.3.1.1. The limit of two exciton populations. The supertrap limit discussed above is most likely to occur in the "subcritical" region, that is in the concentration regime just below the critical concentration, where effectively the exciton population is confined within finite guest clusters. We thus have two exciton populations. (i) The guest excitons confined to finite guest clusters not containing a supertrap site. Such guest excitons have a decay that is characteristic of an isolated guest exciton, i.e. the "natural" guest exciton lifetime. (ii) The guest excitons confined to guest clusters which do contain a supertrap (or several). This population under-goes a metamorphosis into a "supertrapped" exciton population within a transient time much shorter than the experimental time scale. These supertrap excitons now have the "natural" lifetime of isolated supertrap

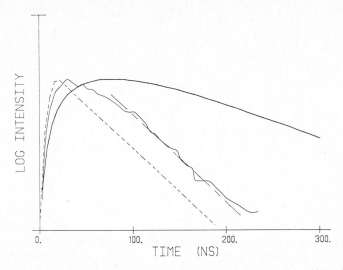

Fig. 9. Calculated and observed singlet exciton acceptor (BMN) decay for a crystal containing 0.42 donor ($C_{10}H_8$) and 0.58 host ($C_{10}D_8$). Light curve: experimental. Heavy curve: theoretical, based on homogeneous kinetics (coupled differential equations). Dashed curve: theoretical, based on the local (intracluster) supertransfer model. A copy of the lower portion of the last curve is superimposed (long dashes) on the experimental curve to show the agreement of the asymptotic slopes, assuming that the supertransfer model is justified only for $t \geq 100$ ns (after Parson and Kopelman 1982).

excitons. These two natural decays can be mathematically described as two uncoupled first order decay equations, in contrast to the familiar "mother–daughter" coupled differential equations that would be expected for homogeneous kinetics.

Figure 9 shows the experimental time evolution of supertrap singlet excitons in a ternary system (naphthalene guest, naphthalene-d_8 host). The difference between the predicted "homogeneous kinetics" behaviour and the predicted "two-population" behaviour is quite striking. The experiment follows the "two-population" (supertransfer) kinetics quite closely, under the particular conditions (subcritical guest concentration, relatively high supertrap concentration, specific low temperature).

5.3.2. Lifetime and critical exponents
For the simple kinetics discussed above it follows that the percolation (supertrapping) probability is

$$P = K/(K + k) = (\tau^{-1} - \tau_0^{-1})/\tau^{-1} = 1 - \tau/\tau_0, \tag{36}$$

where the first equality is always true for the steady state case. Now using eq. (8) one gets

$$(1 - \tau/\tau_0) \propto |1 - C/C_c|^{-\gamma}, \tag{37}$$

i.e. another scaling relationship.

In fig. 8 (insert) we see indeed a scaling (log–log) representation giving a slope γ of about 2.3, again in excellent agreement with two-dimensional energy transport as already discussed in subsection 5.2.4. We note, however, that these triplet experiments are complicated by exciton fusion (Klymko and Kopelman 1981), especially at early times, so that the above result may be fortuitous.

5.3.3. Nonexponential "homogeneous" decays

A gedanken experiment should convince one that for the limit of super-transfer (subsection 5.3.1) to be valid, a minimal acceptor concentration is necessary so that even excitons on very large clusters (including the supercluster) will be supertrapped within a very short time (i.e. below 1 ns). In addition, no significant cluster-to-cluster leakage is permitted to occur within the time scale of the experiment (i.e. about 10 to 100 ns).

Reducing the acceptor (supertrap) concentration has little effect on the time decay curves as long as the condition of supertransfer prevails. Well below the critical percolation concentration, i.e. for any exciton landing on a minicluster, the probability of supertrapping is reduced, on the average, but not the time it takes (i.e. it still takes less than 1 ns). However, above the percolation concentration the reduced acceptor density reduces the probability of supertrapping and also reduces the rate of supertrapping. Thus, for the percolating cluster, simple kinetic behaviour is again valid, qualitatively.

Because the supercluster is a connected quasilattice, its exciton kinetics is essentially a "homogeneous" kinetics. This does not necessarily mean an oversimplified, Stern–Volmer-type, kinetics (see subsection 4.3). As a matter of fact these decays are distinctly nonexponential, similar to some simulated curves (Argyrakis and Kopelman 1979, 1980). We note that our simulations have shown (Argyrakis and Kopelman 1979, 1980) that a random walk on a random binary lattice results in a more prolonged period of "nonexponentiality" than a random walk on a perfect lattice. Furthermore, a correlated random walk results in an even longer time regime of "nonexponentiality". We show below that a correlated random walk on a binary random lattice qualitatively accounts for the observed decay curves.

Triplet naphthalene exciton decay curves have also been observed as a function of guest conccencentration, supertrap concentration, temperature, etc. (see subsection 5.3.2). However, the nonexponential decay curves observed there (Ahlgren et al. 1979) are not as easily interpretable. One difficulty is duè to the spin–lattice relaxation among the three spin sublevels. Another difficulty is due to the exciton–exciton annihilation process, which proceeds simultaneously with the triplet exciton super-trapping by the acceptor (Kopelman et al. 1980, Argyrakis et al. 1980). Recent fusion decay experiments (Klymko and Kopelman, 1981) are also of much interest in this respect.

5.3.4. "Coherence" experiments

The above described time-resolved experiments on the singlet exciton migration of naphthalene have been fitted quantitatively with a percolation kinetics (see subsection 4.3), utilizing the correlated walk mode (subsection 4.2). The only freely adjustable parameter is the "coherency" (correlation parameter) L. While no fit was possible for a mean free path (in lattice units) of $L = 1$ (i.e. incoherent random walk), a reasonable fit is achieved for $L \geqslant 10^2$, i.e. for quasi-coherent exciton motion. We note that the value of the correlation parameter L is extrapolated to the pure, perfect naphthalene crystal at the experimental temperature (2 K). This value is also consistent with the results derived from steady state experiments performed on the very same samples (Argyrakis and Kopelman 1977, 1981).

6. Interpretation of transport experiments

This section, by nature, must be somewhat subjective, somewhat tentative and somewhat speculative. Again we limit ourselves to the experimental systems discussed above and thus, mostly, to the naphthalene singlet and triplet excitons.

6.1. Nature of transport

6.1.1. An Anderson–Mott transition?

The interpretation of the critical transition in terms of an Anderson–Mott transition has been suggested for the triplet excitons of naphthalene, as well as for other systems. We list here a number of objections to the interpretation of the critical transition as a direct manifestation of an Anderson–Mott transition.

(i) Above such a transition (or "mobility edge"), i.e. in the regime where a Bloch-type band does exist, we expect a negative temperature coefficient as is the case for typical "metallic" conduction. Furthermore, for excitons there is no analog of electron–electron (Coulomb) interactions, so the temperature-dependent part of the "resistance" should be solely due to phonon scattering. Experimentally, we observe (fig. 5) a positive temperature coefficient throughout the regime of the critical concentration for the triplet excitons (and the same is true for the singlet exciton). This appears to be inconsistent with the model of Anderson transition (Klafter and Jortner 1977, 1979).

(ii) Because of "thermally activated hopping" among localized states, it has been suggested (Klafter and Jortner 1977, 1979) that the sharp nature of the Anderson–Mott transition will "erode" with small temperature increases. No "erosion" is observed (for the triplet) in the temperature range of 1.5 to 4.2 K (Ahlgren and Kopelman 1979, 1980). However, the transition appears (fig. 5) to become sharper with temperature (in terms of an absolute concentration scale, but not in terms of a reduced concentration scale). There is certainly no "erosion" (broadening) of the transition with increasing temperature.

(ii) The critical concentration (C_c) shows (fig. 5) a significant temperature shift. Again, this is inconsistent with a mobility edge (Anderson–Mott transition) interpretation. Even if we postulate a "mobility edge" causing a critical concentration at zero temperature, there is no direct relationship between it and the observed critical concentrations at finite temperatures. We note that these critical concentrations increase with decreasing temperature, and an extrapolation to zero temperature, while dubious, may result in $C_c \rightarrow 1$. (It appears very reasonable to us that only a perfect and pure crystal will exhibit energy transport at zero temperature.)

(iv) The critical concentration shows a drastic shift with supertrap concentration (S), as can be seen from fig. 4. While a small shift can also be accounted for by an Anderson localization model (Klafter and Jortner, 1980a,b) we believe that the shift is much too drastic (Ahlgren and Kopelman 1980). This belief is based on quantitative calculations of cluster size and localization volume vs C/C_c (Kopelman, unpublished).

(v) A similar large shift in critical concentration is observed in very recent exciton annihilation experiments (Kopelman et al. 1980), where the exciton density effectively replaces the supertrap concentration. In these experiments there is no supertrap at all. The energy transport is monitored via delayed fluorescence, which gives the efficiency of triplet–triplet annihilation ("fusion"). The observed critical concentrations for this annihilation process depend drastically on the exciton density (which is simply monitored via phosphorescence). Again, it would be difficult to account for this shift via an Anderson model.

(vi) The time-resolved experiments show that the energy transport and the critical concentration depend drastically on the time scale of the experiment (Ahlgren 1979). The shorter the time, the higher the C_c. This is a typical manifestation of a kinetic process (see below) but is difficult to explain via an Anderson localization model. This effect is equivalent to the supertrap concentration effect (and probably also to the temperature shift effect).

(vii) The numerical values for the critical parameters for the Anderson localization model, B (bandwidth) and W (strain energy) are difficult to obtain. How B varies with guest concentration (C) has been under dispute resulting in discrepencies of many orders of magnitude (Kopelman and Monberg 1978a,b, 1980). We only note here that using simple averages for exchange integrals, e.g. $J(\langle n \rangle)$ or $\langle J_n \rangle$, gives unduly high weight to inter-actions that are responsible for the formations of small clusters (dimers, trimers) but should lead to a gross overestimation of B (it is obvious that using $\bar{J} = \langle J_n^{-1} \rangle^{-1}$ will give a much smaller value of B). Also only the upper limit for the strain energy is given by the inhomogeneous broadening (Kopelman and Monberg 1978a,b, 1980) and the variation of the latter with concentration is largely unknown. Thus any quantitative agreement be-tween the observed and predicted C_c values may be fortuitous.

In summary, the above listed objections to the Anderson localization model force us to look for alternatives. Some of these are discussed below.

6.1.2. Validity of kinetic and diffusion models

Qualitatively we can account for the critical energy transport by kinetic models. As mentioned above, the time dependence of the critical concen-tration is a kinetic manifestation. The more time there is available for exciton migration, the more time there is available for longer-range hops, the larger the allowed distances between donor (guest) sites, and thus the lower the critical guest concentration (the threshold for effective trans-port).

The supertrap concentration (S) effect is equivalent to the time effect. The higher the supertrap concentration, the fewer guest sites have to be sampled by the exciton during its search for the supertrap, the more time there is available for long-range hops, the larger the allowed distances between donor sites, and thus the lower the critical guest concentration.

The temperature effect can be interpreted in a similar fashion. Each hop requires, on the average, an activation process. The higher the temperature, the easier the activation, the shorter the hopping time, the more time there is available for long-range hops, and thus the lower the critical concen-tration.

Usually, the kinetic model for energy migration and the assumption of diffusion-limited kinetics go together. A simple but detailed quantitative model has been given by Blumen and Silbey (1979). It accounts well for many aspects of the experimental phenomena, including the steep rise of transport at some (critical) donor concentration (C_c).

Upon closer examination, the simple kinetic (Stern–Volmer) approach does not account for the experiments. Neither the singlet nor the triplet exciton decays are exponential (see subsection 5.3). For the triplet decay this may be due in part to slow spin–lattice relaxation (among spin sublevels). No such complication can account for the nonexponential decay of the singlet excitation (see subsection 5.3.3). Also, no ordinary kinetic "coupled equations" approach can account for the rise and fall times of the supertrap excitation (Argyrakis and Kopelman 1979). As mentioned (see section 4), it is possible to account for the excitation dynamics only by "coupled equations" where the rate constant (for supertrapping) is replaced by a time-dependent rate coefficient $K(t)$. Because of the persistence of the non-Stern–Volmer behaviour even at long times (Kopelman and Argyrakis 1980) it can be shown that no ordinary diffusion model can account for the results. Again the diffusion constant has to be replaced by a time-dependent diffusion coefficient $D(t)$, and this is true even for long times.

Very recent fusion (annihilation) experiments show (Klymko and Kopelman 1981) that under certain conditions the delayed fluorescence is not proportional to the square of the phosphorescence. This is in contradiction to any available kinetic model (Swenberg and Geacintov 1973). Any such model assumes that triplet–triplet annihilation must be a binary process. The critical exponent behaviour mentioned earlier (subsections 5.2.4 and 5.3.2) is also not derivable from any simple kinetics or diffusion model. The models of cluster kinetics or dynamic percolation are examined below.

6.1.3. Validity of cluster models
The cluster model ("leaky cluster model") (Ahlgren 1979, Kopelman unpublished) does account quantitatively for the critical energy transport, including the critical exponents, which are derived experimentally with no adjustable parameters (see subsection 5.2.4). With a single adjustable parameter this leaky cluster (dynamic percolation) model accounts for the whole range of the steady state experiments. Furthermore, the "adjustable" parameter result is physically very reasonable (Ahlgren 1979, Ahlgren and Kopelman unpublished). Another triplet exciton experiment that points out the importance of the cluster concept is the exciton fusion experiment (Klymko and Kopelman 1981), which reveals very useful fusion kinetics. While a homogeneous medium approach predicts for this binary reaction (binary exciton collision) a second-power dependence on triplet exciton density, experimentally one observes much higher powers (Klymko and Kopelman 1981), up to powers of twenty (Kopelman and Klymko unpublished); A "heterogeneous" kinetics, based on cluster-trapped excitons, is a reasonable interpretation of these observations. Further evidence for the importance of clusters, and for the percolative versus diffusive approach to exciton transport, has been derived recently from the singlet exciton time-resolved studies (Parson and Kopelman 1982). The nanosecond time evolution curves of the supertrap excitation cannot be related to the guest decay curves via a coupled equations, homogeneous

kinetics (diffusive transport). However, the simplest percolative model, based on the two-population idea (subsection 5.3.1.1), appears to work well, without any adjustable parameters.

6.2. Nature of clusters

6.2.1. Primary clusters

If we define clusters via nearest-neighbor-only interactions we get the common notion of clusters and cluster states (Kopelman 1975, 1976). Many of the simple examples, as well as the exciton literature interpretations, invoke only primary clusters (Hong and Kopelman 1971, Smith et al. 1980).

6.2.2. Static clusters

We define "static clusters" as clusters that are associated with static cluster states, i.e. cluster-wide delocalized exciton states. These static quantum states are usually monitored via optical absorption spectra and have been closely characterized for naphthalene and other simple molecular crystals (Kopelman 1975). Because we assume that there is real quantum-mechanical exciton delocalization throughout the cluster there can be no exciton hopping inside the cluster, only intercluster hopping. Often it is implied that primary clusters are static clusters and vice versa. We have good evidence (Kopelman 1975) that for singlet and triplet naphthalene excitons the primary clusters are indeed "static clusters". However, we also have evidence that secondary and tertiary clusters are static clusters for the singlet excitons (Kopelmann 1975) and this is most probably true for the triplet excitons (Port, private communication). However, we do not know yet the longest interaction range that defines static clusters for any given excitonic system.

6.2.3. Dynamic clusters

We define as dynamic clusters those clusters that are formed by n nearest neighbor distances, where n is the largest such distance (in a "taxicab geometry", Francis and Kopelman 1981) that guarantees highly efficient exciton hopping inside such a cluster. By "highly efficient" we mean very fast compared to the natural decay rate of the excitation (or a very short hopping time compared to the experimental time scale). Obviously, the definition of a dynamic cluster thus depends on the system, the excitation and/or the time scale of the experiment.

Inside the dynamic clusters static clusters are imbedded, among which the efficient hoppings occur. Usually inside the static cluster primary clusters are embedded (which possibly define the loci of the homogeneous annihilation process, Kopelman unpublished).

6.2.4. Conglomerates of leaky clusters

Still "inefficient" (slow) hopping is possible among the dynamic clusters. The dynamic clusters are thus "leaky clusters" (see above) and thus form conglomerates of leaky clusters. Such conglomerates play an important role in quantitative descriptions of the percolation-limited kinetics (Ahglren 1979, Francis and Kopelman 1981, Kopelman unpublished), as briefly described in section 5.

6.3. Excitation and trapping radius

In the case of heavily doped crystals it is important to realise that the effective static radius of a trap or excitation may be smaller than its "dynamic radius". The extent of an impurity excitation radius has been dealt with before (e.g. Kopelman 1975, Francis and Kopelman 1981). The effective radius of the steady state excitation is usually obtained via absorption spectra. The "dynamic" radius depends much more on the nature of the experiments but usually implies some kind of an energy funnel (Auweter 1978) that assures exciton trapping (and/or annihilation).

6.3.1. Static and dynamic excitation radius

The "amplification" of the radius due to dynamic effects is presumably due to the "funnel" (or "antenna") created around the excitation site. It is usually assumed (Swenberg and Geacintov 1973) that exciton–exciton annihilation occurs only at nearest neighbor distances, thus effectively limiting the interaction radius to the molecular site. However, this is not so in the case of "caging" of the excitation in binary lattices (Swenberg and Geacintov 1973). It is then the size of the cage that becomes the effective dynamic excitation radius. By our definition (subsection 6.2) this cage is the dynamic cluster. Very recent annihilation experiments (Kopelman et al. 1980, Klymko and Kopelman 1981) have corroborated and quantified this concept. The dynamic average cluster size is of prime importance and shows the expected critical behaviour. However, the annihilation process itself may occur on smaller subclusters, i.e. static clusters. We do not believe that annihilation happens only at nearest neighbor positions, but if it does, then the actual annihilation process is limited to the primary miniclusters (embedded in the static subclusters of the dynamic clusters).

6.3.2. Trapping radius and critical concentration

The static excitation radius on a trap (see above) may be effectively limited to the impurity site. Examples are the chemical traps: naphthalene in durene, in biphenyl and in hexamethylbenzene (Kopelman unpublished) as well as betamethylnaphthalene in naphthalene and anthracene in naphthalene (Kopelman et al. 1980). On the other hand, isotopic traps often have a much larger radius, which is directly related to the Rashba effect (Kopelman 1975, Rashba 1966).

The dynamic "radius" of a trap again depends on definition. The literature is rich in contradictions in this respect. For dilute anthracene in naphthalene the singlet excitation "funnel" has been assumed to be anywhere from about 5 to about 10^3 sites (Powell and Soos 1975, Auweter 1978, Argyrakis and Kopelman 1980). Recent experiments (Gentry and Kopelman unpublished) support the smallest value. The size of such a trapping funnel ("antenna") might of course be different in a mixed host crystal, i.e. anthracene in isotopically mixed naphthalene. It has been observed (Kopelman et al. 1980) that the relative dynamic radius of two traps, anthracene and betamethylnaphthalene, varies drastically with host concentration ($C_{10}H_8/C_{10}D_8$ ratio), and shows a critical behaviour as well as a critical temperature dependence behaviour (for the singlet exciton) about $C = 0.5$, which is the same concentration region where the critical transport behaviour has been observed (section 5) for similar samples under similar conditions (C, S and T).

6.3.3. Correlation length and localization volume

The extent of excitation localization (or delocalization) on a guest quasilattice has been dealt with using different semantics. Terms such as correlation or localization length (Klafter and Jortner 1979, 1980) deal with the one-dimensional aspects of a three- (or two-) dimensional problem. As there is no preferred direction to the exciton transport, it appears to us to be more natural for the problem to define a localization (or "correlation") volume (or "area") for excitations in these very anisotropic molecular crystals. For the dynamic (transport) problem considered by Klafter and Jortner (1979, 1980) this excitation volume is then the averaged lattice (or superlattice) analog of our concept of dynamic cluster or conglomerate (subsection 6.2). We believe that, quantitatively, the concept of average dynamic cluster size is probably the best current approximation of physical reality.

6.4. Coherence, correlation and transport

Exciton coherence in pure molecular crystals has been of interest since the 1960s (Agranovich and Konobeev 1963, Wolf and Port 1976, Silbey 1976, Harris and Zwemer 1978, Burland and Zewail 1979). For a pure and perfect crystal the only cause for exciton incoherence would be exciton–phonon scattering (and/or exciton–exciton scattering at high exciton densities, e.g. triplet–triplet fusion resulting in another triplet exciton). The mean free time between such scattering events can be related to a mean free path and thus to a coherence parameter l. The latter is a dimensionless mean free path obtained by dividing the mean free path by a lattice parameter (i.e. the nearest neighbor center-to-center distance). While the mean free time is a scalar, the mean free path or l may have vector properties in an anisotropic lattice. For instance, assuming nearest-neighbor-only interactions for naphthalene results in l being restricted to the ab-plane (the exciton band is likewise restricted). As the coherency (l) is reduced, there is a gradual transition from a band model to a hopping model (Silbey 1976).

The question arises whether the coherence parameter l is equivalent to the hopping correlation parameter L defined in section 4. If it is then one can generalize this coherence parameter from pure to mixed crystals while retaining its physical meaning as an expression of exciton–phonon scattering (exclusive of impurity or defect scattering). Furthermore, measurements on mixed crystals can then provide, in principle (and by extrapolation), a way of estimating coherence in pure crystals. Obviously, whether one measures exciton coherence or hopping correlation, the result is always related to exciton transport.

6.4.1. Correlated walk in mixed and pure crystals

Argyrakis and Kopelman (1980) have attempted to interpret old and new energy transport experiments in very dilute mixed crystals (Hammer and Wolf 1968, Ochs 1974, Auweter 1978, Auweter et al. 1979). Both steady state and time-resolved results could be interpreted semiquantitatively via the correlated hopping model (section 4) using a parameter L variation from 1 to 10^2 over the temperature 300 to 2 K. Argyrakis and Kopelman (1979, 1981) have used the same model for interpretation of similar experiments on heavily doped mixed crystals (1% to 50% $C_{10}D_8$ in $C_{10}H_8$), which gave similar results for the parameter L. We note that this parameter is assumed to be the same over the entire concentration domain (assuming invariant exciton–phonon scattering, see above), while the effects of impurity scattering are changing drastically. Thus one can get, by extrapolation, an L value for the pure perfect crystal at a given temperature. This value is of the order of 10^2 (lower limit?) for the singlet naphthalene excitons at liquid helium temperature, and is consistent with other literature estimates (Argyrakis and Kopelman 1979, 1981), Francis and Kopelman 1981).

6.4.2. Temperature effects and coherence

As stated above, the coherence parameter (L) decreases from low (2K) to high (300K) temperatures, as expected (Agranovich and Konobeev 1963) for pure crystals. This is obviously associated with higher energy transport at lower temperatures. However, for mixed or imperfect crystals this is not the case, due to the increased importance of defect (trap, impurity) scattering at lower temperatures. In the older experiments this was the explanation for the sudden reversal in energy transport efficiency at low temperatures (Hammer and Wolf 1968). This was indeed corroborated by the disappearance of such effects with purer and/or more perfect samples (Auweter 1978).

It should be noted that the correlated hopping model for mixed crystals gives essentially a cross-over behaviour from hopping to band-like behaviour at some cross-over impurity concentration, for a fixed temperature and L value (Argyrakis and Kopelman 1981). This exciton transport model thus automatically converts a negative temperature effect for a pure crystal to a positive temperature effect for a highly doped crystal. It does not, however, take into account the specific phonon assistance and relaxation effects related to the specific energy inhomogeneities in a disordered crystal (see below). However, it presents an interesting situation where a combination of impurity and phonon scattering might increase transport, relative to the same impurity scattering with no phonon scattering (Argyrakis and Kopelman 1981). The experimental results appear to be consistent with this picture (Argyrakis and Kopelman 1981) but further studies are necessary.

6.5. Phonons and transport

We give here an extremely short shrift to a very important effect which is at the forefront of research (e.g. Holstein et al. 1981; Monberg and Kopelman 1980).

6.5.1. Phonons as scatterers

The obvious role of phonons as exciton scatterers, resulting in a negative temperature effect on the energy transport, has been mentioned above in relation with band models and their role in Anderson–Mott transitions (subsections 6.1) and in causing exciton incoherence (subsection 6.4). This effect is analogous to the well-known role of phonon scattering in metallic electron conductivity (Mott and Davies 1978). However, we have also pointed out above how in disordered crystals, with specific impurity scattering effects, the phonon scattering can lead to a positive temperature effect (subsections 6.4 and 4.2). Obviously, the latter case falls within the realm of a "hopping" rather than a "band" model and thus should not change the standard notion that metallic-type transport should exhibit a negative temperature effect when the temperature effect expresses electron–phonon (rather than electron–electron) or exciton–phonon-type scattering.

It was mentioned that in isotopically mixed crystals (e.g. $C_{10}H_8/C_{10}D_8$) the phonons are in the amalgamation limit (Prasad and Kopelman 1972) and thus are changed very little throughout the concentration region. We also assume that the same is true for the exciton–phonon coupling, based on observations of the phonon side bands over the entire concentration range (Gentry and Kopelman unpublished). This justifies our above assumption of constant L over the whole concentration range (subsection 6.4).

6.5.2. Phonons as assistors

The role of phonons as assistors is well known for the case of electron hopping or electron promotion into a conducting band for semiconductors (Mott and Davies 1978). Similar effects are under much study for excitation transfer in ruby and doped glasses (Holstein et al. 1981). For excitons in mixed crystals there are two major kinds of phonon assistance, related to the two major kinds of energy barriers: guest–host energy barrier and guest–guest energy barrier. Occasionally there is a third barrier involved ("supertrap"–guest), but only where the energy acceptor is not a "real" supertrap, i.e., where this third barrier is not significantly large compared to the thermal (kT) energy. For examples of all these phonon assistance cases the reader is referred to the paper of Monberg and Kopelman (1980) and subsection 5.2.3.

6.5.3. Phonons as exciton cluster definers

In the dynamic cluster picture (subsection 6.2.3) the cluster is defined via the excitation transfer rate. Obviously this rate is affected by phonon assistance and phonon relaxation (for positive and negative energy mismatches). Thus the phonon processes (i.e. temperature) do define the extent of the dynamic clusters.

The various static clusters (subsection 6.2.2) have various eigenstates, with energy mismatches between the lowest eigenstates of two neighboring static clusters. The relation between such energy mismatches and the phonon relaxation and assistance processes determines whether two static clusters belong to the same dynamic cluster or not. Alternatively, they may determine the degree of "leakage" (subsection 6.2.4) among dynamic clusters and thus their degree of linkage into conglomerates as well as the resulting kinetics.

For a discussion of exciton–phonon coupling, effects of the phonon density of states, single vs multiple-phonon processes, and the relative importance of rotational vs translational and extended vs localized phonons, the reader is referred elsewhere (Holstein et al. 1981, Francis and Kopelman 1981).

7. Summary

7.1. Exciton transport in mixed crystals

The study of excitation transport in highly concentrated mixed crystals has been fruitful in a number of ways. It has brought to light phenomena not suspected before, mainly those related to critical phenomena. Indeed, it has provided one of the most elegant connections between apparently complex physical phenomena and basically simple mathematical models: cluster statistics and percolation theory. These models are of much current interest because they are suspected to underlie some of the most important approaches of statistical thermodynamics. Also, the critical regime of the transport and the range of validity of critical exponents are wider for "critical energy transport" than for any other critical physical phenomenon. Experimentally, the use of mixed crystals opened up a new variety of studies, as the control of the concentration parameters admits some flexibility in the choices for other parameters such as time and temperature. It also offers an opportunity to study a simplified case of "heterogeneous" kinetics.

7.2. Exciton transport in pure crystals

There are still a few mysteries considering energy transport in pure crystals. One of them is the theoretical question of "coherence". To what extent does "coherence" represent wave-like motion of some exciton packet? How much is the wave spread (delocalized) at any given instant? What are the roles of exciton–phonon and exciton–defect scattering? Related practical questions are: Can a thin slice of a pure, perfect crystal be used as an excitation anode or antenna? What are the nature and roles of surfaces, interfaces and defects in such pure crystals? How well can they be controlled? Hopefully such questions can be answered in part by studying mixed crystal systems (see above) in combination with proper theoretical extrapolations.

7.3. Exciton transport in disordered systems

Energy transport in highly disordered systems (molecular glasses, polymers, liquids and interfaces) may turn out to be basically predictable from our model systems of purely translationally (substitutionally) disordered lattices. The latter problems have been expressed in terms of mathematical simulations and simple analytical formulae. This has been the leitmotiv of this chapter. There are potential practical uses for such disordered materials, e.g. solar energy materials and surface materials for biologically active materials. Furthermore, there are a number of biological systems, such as photosynthetic antenna, which may be better understood as disordered molecular aggregates, rather than as "quasiordered" (Kopelman 1976). Biomimetic materials and systems may also be developed, based on the theoretical understanding gained from simple model systems such as the isotopically mixed crystals described in this chapter.

7.4. Epilogue

Energy transport in concentrated mixed crystals is drastically different from that in neat and lightly doped crystals. There is a strong microscopic heterogeneity (due to clusters) which dominates the energy transport. Similar effects are expected for many other systems: surfaces, liquid crystals, amorphous materials. Among the latter will be both synthetic polymeric (and multipolymeric) materials, as well as natural biopolymers and other biological aggregates where the clustering of the energy transporting chemical groups is either very likely or even well known (e.g. photosynthetic units). For all these materials and for their various applications (solar energy devices, radiation damage, heterogeneous photochemistry and excitochemistry) the mixed crystal studies may serve as the starting point.

The mixed crystal studies themselves may be much improved via selective (cluster) excitation, via more sophisticated pulse and pulse sequence studies, as well as by studying magnetic field effects (e.g. on annihilation) and electric field effects (e.g. field-induced heterogeneity).

Will magnetic or electric field effects exhibit "critical" behaviour with critical exponents? Could they be related to electrical excitations in biological materials such as nerves and brain? Mixed liquid crystal studies may be relevant to basic biological problems. Mixed crystal studies at extremely low temperatures may be relevant to basic physical problems (e.g. Anderson transition), and mixed crystal surface studies may be relevant to basic problems of surface science and chemical kinetics.

Acknowledgement

This research was supported by NSF Grant No. DMR 800679 and NIH Grant No. R 01 NS 08116-14.

References

Agranovich, V.M. and Y.V. Konobeev, 1963, Solid State **5**, 999.
Ahlgren, D.C., 1979, Ph.D. Thesis, Univ. of Michigan (Ann Arbor, MI).
Ahlgren, D.C. and R. Kopelman, 1979, J. Chem. Phys. **70**, 3133.
Ahlgren, D.C. and R. Kopelman, 1980, J. Chem. Phys. **73**, 1005.
Ahlgren, D.C. and R. Kopelman, 1981, Chem. Phys. Lett. **77**, 136.
Ahlgren, D.C. and R. Kopelman, unpublished.
Ahlgren, D.C., E.M. Monberg and R. Kopelman, 1979, Chem. Phys. Lett. **64**, 122.
Argyrakis, P., 1978, Ph.D. Thesis Univ. of Michigan (Ann Arbor, MI).
Argyrakis, P. and R. Kopelman, 1977, J. Chem. Phys. **66**, 3301.
Argyrakis, P. and R. Kopelman, 1979, Chem. Phys. Lett. **61**, 187.
Argyrakis, P. and R. Kopelman, 1980, Phys. Rev. **B22**, 1830.

Argyrakis, P. and R. Kopelman, 1981, Chem. Phys. **57**, 29.

Argyrakis, P., J. Hoshen and R. Kopelman, 1980, High Density Excitation Calculations in Molecular Solids: A Monte Carlo Study, in: Fast Reactions in Energetic Systems, eds. C. Capellos and R.F. Walker (Reidel, Dordrecht, Netherlands) p. 685.

Auweter, H., 1978, Doctoral Dissertation, Univ. of Stuttgart.

Auweter, H., A. Braun, U. Mayer and D. Schmid, 1979, Z. Naturforsch. **34a**, 761.

Blumen, A. and R. Silbey, 1979, J. Chem. Phys. **70**, 3707.

Blumen, A., J. Klafter and R. Silbey, 1980, J. Chem. Phys. **72**, 5320.

Breiland, W.G. and M.C. Saylor, 1980, J. Chem. Phys. **72**, 6485.

Burland, D.M. and A.H. Zewail, 1979, Advances in Chemical Physics, Vol. **40** eds. I. Prigogine and S.A. Rice (Wiley, New York).

Colson, S.D. and M. Okumura, 1980, Mol. Cryst. Liq. Cryst. **57**, 255 (Broude Memorial Issue, ed. D.M. Hanson).

Colson, S.D., R.E. Turner and V. Vaida, 1977a, J. Chem. Phys. **66**, 2187.

Colson, S.D., S.M. George, T. Keyes and V. Vaida, 1977b, J. Chem. Phys. **67**, 4941.

Dlott, D.D., M.D. Fayer and R.D. Wieting, 1977, J. Chem. Phys. **67**, 3808.

Einstein, A., 1926, Investigations on the Theory of the Brownian Movement (Dutton, New York).

Francis, A.H. and C.B. Harris, 1971, Chem. Phys. Lett. **9**, 188.

Francis, A.H. and R. Kopelman, 1981, Excitation Dynamics in Molecular Solids, in: Topics in Applied Physics, Vol. 49, eds. W.M. Yen and P.M. Selzer (Springer, Berlin) p. 241.

Gentry, S.T. and R. Kopelman, unpublished.

Gentry, S.T., R.P. Parson and R. Kopelman, unpublished.

Godzik, K. and J. Jortner, 1980, J. Chem. Phys. **72**, 4471.

Gochanour, G.R., H.C. Anderson and M.D. Fayer, 1979, J. Chem. Phys. **70**, 4254.

Haan, S.W. and R. Zwanzig, 1978, J. Chem. Phys. **68**, 1879.

Hammer, A. and H.C. Wolf, 1968, Mol. Cryst. **4**, 191.

Harris, C.B. and D.A. Zwemer, 1978, Ann. Rev. Phys. Chem. **29**, 473.

Holstein, T., S.K. Lyo and R. Orbach, 1981, Excitation Transfer in Disordered Systems, in: Topics in Applied Physics, Vol. 49, eds. W.M. Yen and P.M. Selzer (Springer, Berlin) ch. 2, p. 39.

Hong, H.-K. and R. Kopelman, 1971, J. Chem. Phys. **55**, 5380.

Hoshen, J. and J. Jortner, 1972, J. Chem. Phys. **56**, 933.

Hoshen, J., R. Kopelman and E.M. Monberg, 1978, J. Stat. Phys. **19**, 219.

Hoshen, J., P. Klymko and R. Kopelman, 1979, J. Stat. Phys. **21**, 583.

Huber, D.L., 1981, Dynamics of Incoherent Transfer, in: Topics of Applied Physics, vol. 49, eds. W.M. Yen and P.M. Selzer (Springer, Berlin) p. 83.

Klafter, J. and J. Jortner, 1977, Chem. Phys. Lett. **49**, 410.

Klafter, J. and J. Jortner, 1978, Chem. Phys. Lett. **60**, 5.

Klafter, J. and J. Jortner, 1979, J. Chem. Phys. **71**, 1961.

Klafter, J. and J. Jortner, 1980a, J. Chem. Phys. **73**, 1004.

Klafter, J. and J. Jortner, 1980b, J. Non-Cryst. Solids **35/36**, 147.

Klafter, J. and R. Silbey, 1980, J. Chem. Phys. **72**, 843.

Klymko, P.W. and R. Kopelman, 1981, J. Lumin. in press.

Kopelman, R. 1975, Excitons in Pure and Mixed Crystals, in: Excited States, vol. 2, ed. E.C. Lim (Academic Press, New York) ch. 2.

Kopelman, R., 1976, Exciton Percolation in Molecular Alloys and Aggregates, in: Topics in Applied Physics, vol. 15: Radiationless Processes in Molecules and Condensed Phases, ed. F.K. Fong (Springer, Berlin) ch. 5.

Kopelman, R., unpublished.

Kopelman, R. and P. Argyrakis, 1980, J. Chem. Phys. **72**, 3053.

Kopelman, R., E.M. Monberg, F.W. Ochs and P.N. Prasad, 1975a, J. Chem. Phys. **62**, 292.

Kopelman, R., E.M. Monberg, F.W. Ochs and P.N. Prasad, 1975b, Phys. Rev. Lett. **34**, 1506.

Kopelman, R., E.M. Monberg and F.W. Ochs, 1977a, Chem. Phys. **19**, 413.

Kopelman, R., E.M. Monberg and F.W. Ochs, 1977b, Chem. Phys. **21**, 373.

Kopelman, R., E.M. Monberg, J.S. Newhouse and F.W. Ochs, 1979, J. Lumin. **18/19**, 41.

Kopelman, R., D.C. Ahlgren, P. Argyrakis, S. Gentry, D. Hooper, J. Hoshen, P. Klymko and J.S. Newhouse, 1980, Critical Exciton Transport and Excitation Radius, in: 9th Molecular Crystal Symp., Mittelberg-Kleinwalsertal.

Lakatos-Lindenberg, K., R.P. Hemenger and R.M. Pearlstein, 1972, J. Chem. Phys. **56**, 4852.

Lyo, S.K., 1971, Phys. Rev. **B3**, 3331.

Mauser, K.E., H. Port and H.C. Wolf, 1973, Chem. Phys. **1**, 74.

Monberg, E.M., 1977, Ph.D. Thesis, Univ. of Michigan (Ann Arbor, MI).

Monberg, E.M. and R. Kopelman, 1978a, Chem. Phys. Lett. **58**, 492.

Monberg, E.M. and R. Kopelman 1978b, Chem. Phys. Lett. **58**, 497.

Monberg, E.M. and R. Kopelman, 1980, Mol. Cryst. Liq. Cryst. **57**, 271 (Broude Memorial Issue, ed. D.M. Hanson).

Montroll, E.W., 1964, Proc. Symp. Appl. Math. **XVI**, 193.

Montroll, E.W., 1969, J. Math. Phys. (N.Y.) **10**, 753.

Mott, N.F. and E.A. Davies, 1978, Electronic Processes in Non-Crystalline Materials, 2nd Ed. (Oxford Univ. Press, New York).

Newhouse, J.S., J. Hoshen and R. Kopelman, unpublished.

Orbach, R., 1974, Phys. Lett. **48A**, 417.

Ochs, F.W., 1974, Ph.D. Thesis, Univ. of Michigan (Ann Arbor, MI).

Parson, R. and R. Kopelman, 1982, Chem. Phys. Lett. **87**, 528.

Pollak, M. and M.L. Knotek, 1979, J. Non-Cryst. Solids **32**, 141.

Powell, R.C. and Z.G. Soos, 1975, J. Lumin. **11**, 1.

Prasad, P.N. and R. Kopelman, 1972, J. Chem. Phys. **57**, 863.

Rashba, E.I., 1966, Zh. Eskp. Teor. Fiz. **50**, 1164.

Silbey, R., 1976, Ann. Rev. Phys. Chem. **27**, 203.

Smith, D.D., R.D. Mead and A.H. Zewail, 1977, Chem. Phys. Lett. **50**, 358.

Smith, D.D., D.P. Millar and A.H. Zewail, 1980, J. Chem. Phys. **72**, 1187.

Stanley, H.E., 1981, Phase Transitions, 2nd Ed., in press.

Stauffer, D., 1979, Phys. Rept. **54**, 1.

Swenberg, C.E. and N.F. Geacintov, Exciton Interactions in Organic Solids, in: Organic Molecular Photophysics, vol. 1, ed. J.B. Birks (Wiley, New York, 1973) ch. 10, p. 489.

Thouless, D.J., 1979, in: Ill-Condensed Matter, Les Houches 1978, vol. XXXI, eds. R. Balian, R. Maynard and G. Thoulouse (North-Holland, Amsterdam) ch. 1.

Tudron, F.B. and S.D. Colson, 1976, J. Chem. Phys. **65**, 4084.

Wolf, H.C. and H. Port, 1976, J. Lumin. **12/13**, 33, Proc. XIIth European Congress on Molecular Spectroscopy (Elsevier, Amsterdam, 1975).

Ziman, J.M., 1979, Models of Disorder: the Theoretical Physics of Homogeneously Disordered Systems (Cambridge Univ. Press, Cambridge, England/New York).

CHAPTER 5

Exciton Coherence

M.D. FAYER

Department of Chemistry
Stanford University
Stanford, CA 94305
U.S.A.

Spectroscopy and Excitation Dynamics
of Condensed Molecular Systems
Edited by
V.M. Agranovich and R.M. Hochstrasser

Contents

1. Introduction

Excited states of pure molecular crystals (Davydov 1971), i.e. excitons, have a variety of physical characteristics which arise from intermolecular interactions among the molecules comprising the solid. Perhaps the feature which has been most intriguing to researchers in the field is the spatial mobility of excited states of pure molecular crystals. Exciton transport has been demonstrated in a variety of experiments. For example, the front surface of a crystal is illuminated with a wave length of light that is strongly absorbed. The crystal has a Beer's length that is very short compared to the crystal thickness so excitons are only produced very close to the front surface. If the excitation is produced using a pulsed source, at short times certain wavelengths of fluorescence are observed emerging from the front crystal surface but not from the back crystal surface. At later times fluorescence is observed from the back crystal surface. This experiment (Simpson 1956, Gallus and Wolf 1966) and a number of variations of this experiment are interpreted in terms of exciton mobility. Excitations migrate through the crystal, leaving the front surface region. Some of these excitons move to the vicinity of the back surface. Fluorescence originating from the vicinity of the back surface can escape the crystal from the back surface without reabsorption. The radiationless transport of excitations through the crystal is an essential feature of the interpretation of these experiments.

Another example demonstrating the existence of exciton transport involves the trapping of excitations by very low concentration impurities in nearly pure molecular crystals (Powell 1971, 1973). An impurity with excited state energy below the pure crystal excited states, i.e. exciton band states, will behave as a potential well. An excitation in the vicinity of the impurity can fall into the potential well giving up excess energy to the mechanical degrees of freedom of the lattice, i.e. the phonon states of the lattice. The excitation is localized or "trapped" on the impurity and is no longer mobile. Experimentally, this phenomenon can be monitored by pulsed excitation of the crystal with wave lengths chosen to produce excitons. At early times, optical emission occurs primarily from the crystal exciton states. However, as time progresses trap emission increases at the expense of the exciton emission. The interpretation of this type of experiment is based upon the radiationless transport of excitons to the

187

vicinity of the impurity followed by trapping and subsequent trap optical emission. Here, as in the example cited above, exciton mobility is the key to understanding the system dynamics.

Given the clear experimental demonstrations of exciton mobility by its influence on a variety of experimental observables, the important physical problems are to understand the microscopic mechanisms and the factors influencing exciton transport. In certain types of systems in which excited state transport occurs, such as in room temperature dilute dye solutions, the process is "incoherent." By incoherent it is meant that the transport process and observables relating to it can be described strictly in terms of probabilities using, for example, the master equation (Gochanour et al. 1979) or continuous time random walk formalisms (Godzik and Jortner 1980). In situations where these formalisms are applicable it is unnecessary to consider probability amplitudes, phase relationships between wave functions or interference effects, i.e. the transport problem can be handled completely using classical concepts.

In molecular crystals the situation can be quite different. The translational symmetry of the lattice gives the system a periodic potential. A common example of a system with a periodic potential is the benzene molecule's π electron system. The periodic potential around the ring yields π electron eigenfunctions which are superpositions of the p_z atomic orbitals. The six molecular orbitals are six different linear combinations of atomic orbitals with well-defined phase relationships. In the same manner, the excited state eigenfunctions of a crystal having N molecules are N linear ·combinations of the molecular excited state eigenfunctions with well-defined phase relationships. Thus for each molecular excited state, the molecular crystal will have N corresponding excited states. Intermolecular interactions lift the degeneracy of these N states and a band of closely spaced states results. This is referred to as an exciton band. As in the benzene molecule, where the π electrons are delocalized over the entire molecule, in a molecular crystal the excited state eigenfunctions, i.e. the exciton band states, are delocalized over the entire crystal.

The nature of the exciton band states with well-defined phase relationships between molecular functions immediately suggests that this problem is different from the incoherent example cited above in which phase relationships between excitations on different molecules were unimportant. In situations in which the exciton transport process and observables relating to it require consideration of probability amplitudes, phase relationships and interference effects, the exciton is "coherent". The exciton can be described as a quasi-particle and transport as wave-like in a manner that closely resembles the description of a photon wave packet. Exciton coherence is not an all or nothing proposition. It is believed that for various crystals under a variety of physical conditions a range of possibilities exist.

Thus a system can be in the coherent limit in which coherence effects completely dominate transport and observables. Or the intermediate case can occur in which coherence effects are of some importance but not completely dominant. A system can also be in the incoherent limit in which coherence effects are completely absent. Well-defined phase relationships between lattice sites do not exist. Excitations are localized on individual molecules, and wave-like exciton transport does not occur. Transport can still occur by incoherent hopping of an excitation from one molecule to another in a manner completely analogous to excited state transport in dye solutions as mentioned above.

To understand excitation transport requires examination of those processes which affect the wave-like character of the exciton. The periodic potential and intermolecular interactions lead to delocalized eigenstates. But processes such as exciton–impurity scattering, exciton–defect scattering and exciton–phonon scattering can destroy exciton coherence, i.e. change the mode of transport from wave-like to incoherent hopping. In this chapter some aspects of the problem will be examined. An attempt will be made to illustrate basic concepts and to try to illuminate the gulf which still separates us from detailed understanding of the exciton transport problem.

2. Excitons in one dimension

Two of the most studied exciton systems are the one-dimensional first triplet, T^1, bands of 1,2,4,5-tetrachlorobenzene (TCB) (Francis and Harris 1971, Burland et al. 1977a, Botter et al. 1978, Dlott et al. 1978) and 1,4-dibromonaphthalene (DBN) (Burland et al. 1977b) molecular solids. These systems are considered one-dimensional because the intermolecular interactions responsible for exciton transport are overwhelmingly between molecules arrayed along one translational direction in the crystal. Molecules composing the exciton system form linear chains and exciton transport occurs along these one-dimensional chains. In what follows, virtually everything can be readily extended to multidimensions. However, certain experiments are peculiar to one-dimensional systems and therefore these systems make a convenient focus for this section.

Let the ground state of the mth molecule in a linear array of molecules be $|0_m\rangle$. An excited state of the mth molecule is $|1_m\rangle$. We are usually interested in the first excited state, either singlet, S^1, or triplet, T^1, since these states are metastable. Then the ground state of the system is

$$|0\rangle = |\ldots 0_{m-1}0_m0_{m+1}\ldots\rangle = \ldots |0_{m-1}\rangle|0_m\rangle|0_{m+1}\rangle\ldots, \tag{1}$$

and the state of the system in which the mth molecule is excited is

$$|m\rangle = |\ldots 0_{m-1}1_m0_{m+1}\ldots\rangle = \ldots |0_{m-1}\rangle|1_m\rangle|0_{m+1}\rangle\ldots. \tag{2}$$

If we consider the system to be effectively infinite in extent or to have a cyclic boundary condition, then the translational symmetry of the system results in a periodic potential. The state of the system $|m\rangle$ is degenerate with the state $|i\rangle$, i.e. the system looks the same regardless of which of the identical molecules is excited. The Bloch theorem of solid-state physics (Tinkham 1964) tells us that such a system containing N molecules will have N eigenstates of the form

$$|k\rangle = \frac{1}{\sqrt{N}} \sum_m e^{ikam} |m\rangle. \tag{3}$$

k is the wave vector or quantum number which labels the exciton band states, and a is the lattice spacing. In three dimensions kam would be replaced by $k \cdot r$, three-dimensional vectors. The above states have been written in terms of molecular eigenstates. Actually, they should be ortho-gonalized site functions, Wannier functions (Ziman 1972), in which the finite overlap of the molecular functions has been removed. In certain situations the distinction can be important, e.g. in the exciton spin–orbit coupling problem (Cooper and Fayer 1978). For our purposes, the molecu-lar functions are taken to be the orthogonal site functions.

The exciton Hamiltonian can be written in the site occupation number representation as (Davydov 1971)

$$\mathcal{H}_{ex} = \sum_j (\Delta E + D) B_j^+ B_j + \sum_{i,j}' \beta_{ij} B_i^+ B_j. \tag{4}$$

ΔE is the gas phase energy of the excitation, i.e. the energy of the excited state of an isolated molecule. D is the change in excitation energy of a molecule in the lattice due to the differences in Van der Waals interactions of a molecule with its surroundings in the ground and excited states. Since excited states tend to be more polarizable, Van der Waals interactions are stronger for excited states. The net effect is to lower the excitation energy in the crystal as compared to the gas phase. This reduction in energy is called the crystal shift. The second term in eq. (4) is the resonance interaction. It delocalizes the excitation and is reponsible for exciton transport. Combining ΔE and D as E_0, and considering nearest neighbor resonance interactions only (which is accurate for triplet states since the exchange interactions giving rise to β_{ij} for triplets are very short ranged) gives

$$\mathcal{H}_{ex} = \sum_j E_0 B_j^+ B_j + \sum_j \beta B_j^+ B_{j\pm1}, \tag{5}$$

where β is the nearest neighbor resonance interaction. The operators in eq. (5) operate on the site functions, $|j\rangle$, which are fermions since a double excitation of a single site is not a possible state of the system. However, the exciton eigenstates $|k\rangle$ obey commutator relationships which more closely resemble those of bosons. [See Davydov (1971) for a detailed discussion of this point.]

From the Bloch theorem we know that the $|k\rangle$'s are eigenstates of $\mathcal{H}_{ex}$. Therefore we can calculate the energy. The first term in eq. (5) is diagonal and gives E_0, since

$$B_j|m\rangle = |0\rangle\delta_{mj}, \qquad B_j^+|0\rangle = |j\rangle, \tag{6}$$

and the functions $|j\rangle$ and $|l\rangle$ are orthonormal. The second term in eq. (5) is

$$\langle k|\beta \sum_{j,j\pm1} B_j^+ B_{j\pm1}|k\rangle = \frac{\beta}{N}\left[\sum_l e^{-ikal}\langle l| \sum_{j,j\pm1} B_j^+ B_{j\pm1} \sum_m e^{ikam}|m\rangle\right]. \tag{7}$$

This gives N pairs of terms

$$= \frac{\beta N}{N}(e^{ika} + e^{-ika}) = 2\beta \cos ka. \tag{8}$$

Therefore the energy of the one-dimensional exciton state $|k\rangle$ with nearest neighbor interactions is

$$E(k) = E_0 + 2\beta \cos ka. \tag{9}$$

This is a band of states of width 4β resulting from the resonance interaction and centered around energy E_0. For any state $|k\rangle$ the probability of finding the excitation on any site j is equal, i.e.

$$\langle k|j\rangle\langle j|k\rangle = 1/N. \tag{10}$$

The eigenstates have the excitation delocalized over the entire system.

When a crystal is excited, excitation in general is not of a single eigenstate $|k\rangle$, but rather a superposition of k states is generated. Since there is a dense set of closely spaced k states, the superposition can be written as

$$|\omega\rangle = \int_{-\pi/a}^{\pi/a} f(k_0 - k)|k\rangle\, dk. \tag{11}$$

$f(k_0 - k)$ is a weighting function centered around the state k_0. The integral is over the wave vectors of the first Brillouin zone. The ket $|\omega\rangle$ is a wave packet. It is more or less localized in space, depending on the nature of the weighting function which is determined by the details of the excitation process.

Unlike the eigenstates $|k\rangle$ which are delocalized over the crystal, the wave packet $|\omega\rangle$ is localized in a particular region of the crystal and can move through the crystal. The velocity of the wave packet is the group velocity, V_g, which is given by

$$V_g(k) = \left(\frac{\partial \omega(k)}{\partial k}\right)_{k_0} = \frac{1}{\hbar}\left(\frac{\partial E(k)}{\partial k}\right)_{k_0} \tag{12a}$$

$$= \frac{2\beta a}{\hbar} \sin ka. \tag{12b}$$

The group velocity is maximum for wave packets comprised of states at the center of the band $k = \pm \pi/2a$, while it is zero for $k = 0$ or $k = \pm \pi/a$. For small k, the group velocity is approximately

$$V = \frac{2\beta a^2 k}{\hbar}. \tag{13}$$

The quantity $\hbar^2/2\beta a^2$ is a mass. The effective mass of the exciton can be defined as

$$M_{eff} = \hbar^2/2\beta a^2 \tag{14}$$

and

$$M_{eff} V = \hbar k. \tag{15}$$

If $M_{eff} V$ is taken to be the "momentum" of the exciton, then it is related to the wave vector in the same manner as for a free particle.

The introduction of the exciton wave packet brings forward the close analogy between the description of a photon in a box and an exciton in a crystal. The photon eigenstates are delocalized waves, but we generally consider a photon as a wave packet more or less localized in some region of space and moving with a group velocity given by eq. (12a). The eigenstates of an exciton are also delocalized waves, and an exciton wave packet is more or less localized in some region of the crystal and moves with a group velocity given by eq. (12). As in any wave packet, the exciton moves due to changing regions of constructive and destructive interference of the waves (k states) which form the packet. It is the time evolution of the well-defined phase relationships among the states which results in wave-like exciton transport. Thus the motion of the exciton wave packet as described above represents the coherent limit of exciton transport.

In principle coherent exciton transport can be very rapid. As examples, consider the band center ($k = \pi/2a$) group velocities associated with TCB triplet excitons which have a narrow band width of $1.4\,\text{cm}^{-1}$, DBN triplet excitons which have a wide band of width $29.6\,\text{cm}^{-1}$, and a singlet exciton band with a moderate width of $200\,\text{cm}^{-1}$. Using a value for the lattice spacing $a = 4\,\text{Å}$, these excitons have velocities at the band center of $5.4 \times 10^3\,\text{cm/s}$, $1.2 \times 10^5\,\text{cm/s}$ and $7.9 \times 10^5\,\text{cm/s}$, respectively.

Coherent transport is only expected to occur at low temperature. At high temperatures strong interactions with highly populated lattice phonons destroy the necessary phase relations of the exciton k states and therefore render exciton transport incoherent. However, even at low temperature, weak interactions with lattice phonons can bring the distribution of exciton k state populations into thermal equilibrium with the lattice temperature. To get an idea of the actual rates of coherent exciton transport it is necessary to examine the ensemble thermal average group velocity as a function of band width and temperature (Fayer and Harris 1974). For a thermal distribution of excitons in a band, both the probability, $P(k)$, of finding an exciton in a particular k state having group velocity $V_g(k)$, and the average group velocity $V_g(T)$, can be calculated:

$$P(k) = \frac{e^{-E(k)/KT}}{Z(T)} \tag{16a}$$

$$= e^{-y \cos ka}/\pi I_0(y) \tag{16b}$$

and

$$\langle V_g(T)\rangle = \frac{\Sigma_k V_g(k) e^{-E(k)/KT}}{Z(T)} \tag{17a}$$

$$= \left(\frac{2\beta a}{\hbar}\right)\left(\frac{2KT}{\pi\beta}\right)^{1/2} [I_{1/2}(y)/I_0(y)], \tag{17b}$$

where $Z(T)$ is the partition function, $I_0(y)$ and $I_{1/2}(y)$ are modified Bessel functions and $y = 2\beta/KT$. The summation in eq. (17a) is restricted to positive values of the wave vector k so that $\langle V_g(T)\rangle$ is the average velocity in one direction. $E(k)$ and $V_g(k)$ are given by eqs. (9) and (12), respectively. Table 1 gives the thermal average group velocity at three temperatures, 1.4 K, 4.2 K and 10 K, for the three band widths considered above with the lattice spacing $a = 4\text{ Å}$. Notice for the narrow band width, the thermal average group velocity is basically independent of temperature due to the almost uniform distribution of population among the k states even at 1.4 K. However for the wider bands, the thermal average group velocity increases considerably with temperature as states near the band center gain population.

Table 1
$\langle V_g(T)\rangle$ (cm/s).

T (K)	4β		
	$1.4\,\text{cm}^{-1}$	$29.6\,\text{cm}^{-1}$	$200\,\text{cm}^{-1}$
1.4	4.6×10^3	3.4×10^4	8.8×10^4
4.2	4.8×10^3	5.7×10^4	1.5×10^5
10	4.9×10^3	8.2×10^4	2.3×10^5

A thermal distribution of population among the k states is brought about by exciton–phonon scattering, i.e. interactions of the excitons with the lattice heat bath which produce changes of k state. These scattering processes can bring about a loss of exciton coherence. In the nearly coherent limit the k to k' scattering is fast enough to maintain thermal equilibrium, i.e. fast compared to the excited state lifetime but slow enough to permit significant wave-like transport. Kenkre and Knox have used a generalized master equation approach to describe the nature of exciton transport in the coherent, intermediate and incoherent regimes (Kenkre and Knox 1974a,b). Briefly, a memory function is used to describe how rapidly an exciton loses memory of its initial motion. For an exponentially decaying memory with decay constant α, the exciton is basically incoherent if $\alpha \gg \beta/h$, where β/h is on the order of the site-to-site coherent transfer rate. That is, if the system loses memory of its initial motion on a time scale rapid compared to the time required to move a single lattice site, then there will be no correlation between the site-to-site steps, and transport will occur as a random walk. In the other extreme, $\alpha \ll \beta/h$ and there is a good deal of correlated or wave-like transport before loss of memory occurs. Kenkre and Knox have obtained an approximate form for the "slow transfer" or incoherent stepping rate, ν_{inc}:

$$\nu_{inc} = 16\pi^2\beta^2/\alpha h^2. \tag{18}$$

Equation (18) and eq. (17), the thermal average group velocity, can be used to compare the rates of coherent and incoherent transfer. Taking TCB as an example, table 1 gives the thermal average group velocity as $\sim 5 \times 10^3$ cm/s, and with a lattice spacing of approximately 4 Å, this gives 8 ps/lattice site. If α is chosen to give a 1/e time of 1 ps, then $1/\nu_{inc}$ will be about 60 ps. Even for a value of α which gives $1/\nu_{inc}$ equal to 8 ps, long range transport is substantially slower via incoherent hopping, since the exciton executes a random walk instead of propagating with a well-defined group velocity. For example, a coherent exciton with 8 ps site-to-site travel time will require 0.8 ns to be displaced 100 lattice sites. Moving incoherently with the same step time, an rms displacement of 100 lattice sites requires 80 ns.

Strictly coherent or strictly incoherent transport are limiting cases which are useful conceptionally. Transport in real systems at low temperatures can have components of both modes of transport. An exciton may move with a well-defined group velocity for some distance before losing memory of its velocity. The distance traveled is referred to as the coherence length. Under these circumstances, at long times, macroscopic transport will be diffusive in nature. However, the rate of diffusion will be much greater than in the incoherent limit. Transport will occur as a random walk, but the step size in the walk will be on the order of the coherence length rather than the lattice spacing. This larger step size will result in a greatly increased macroscopic diffusion coefficient.

Grover and Silbey (1970) have presented a theoretical approach which gives a unified treatment of exciton transport in coherent and incoherent limits as well as the intermediate regime. They employ a model Hamiltonian which involves linear coupling of the excitation to optical phonons. A transformation is used and in the transformed basis set the excitation is described as a "clothed" (Mattuck 1976) exciton, i.e. an electronic excitation accompanied by a lattice distortion which changes the center of nuclear motion of the crystal from that of the ground state vibrational potential to that of the excited state potential. The transformed exciton states are stationary with respect to the major part of the exciton–phonon interaction. This permits the remaining exciton–phonon interaction terms to be treated perturbatively.

Using this approach Grover and Silbey obtained an expression for the exciton mean-square-displacement,

$$\langle R^2(t) \rangle = a^2 [(2\alpha + 4\tilde{J}^2/3\alpha)t + (4\tilde{J}^2/9\alpha^2)(e^{-3\alpha t} - 1)]. \tag{19}$$

In this expression a is the lattice spacing. $\tilde{J}$ is the intermolecular interaction in the clothed exciton basis set. It is the equivalent to β used in eq. (9) and accounts for both the magnitude of the pure electronic interactions of a bare excitation and the reduction of this interaction arising from the lattice distortion which accompanies excitation. This is effectively a Franck–Condon factor reduction of the excitation transfer matrix element which would occur in an undistorted lattice. Here the parameter α is given by

$$\alpha = 2\tilde{J}^2\gamma, \tag{20}$$

where γ is a parameter involving the strength of the exciton–phonon coupling which depends strongly on the dispersion of the phonon bands. At very low temperatures for which α is expected to be very small and for times such that αt is also small, Grover and Silbey find

$$\langle R^2(t) \rangle \approx 2\tilde{J}a^2 t^2. \tag{21}$$

The t^2 dependence of the mean-square-displacement is characteristic of wave-like or coherent transport. For high temperatures at which $\tilde{J}$ is small and α is large, they find

$$\langle R^2(t) \rangle = 2\alpha t a^2. \tag{22}$$

The linear t dependence is characteristic of diffusive or incoherent transport.

Like the Generalized Master Equation approach of Knox and Kenkre, the Green's Function approach of Grover and Silbey and other theoretical treatments (e.g. Haken and Reineker 1972, Reineker and Haken 1972) demonstrate that the mode of exciton transfer can be of mixed character with coherent and incoherent transport being the limiting cases. The problem then is to obtain well-defined experimental observables which permit examination of the microscopic details of exciton transport.

3. Trapping experiments in the quasi-one-dimensional exciton system 1,2,4,5-tetrachlorobenzene

In this section we will focus our attention on the time dependence of trap optical emission following impulse excitation of the host exciton band since this experimental observable can be related to the macroscopic dynamics of exciton transport and is strongly influenced by the processes mentioned above. In a recent theoretical study, a detailed model of the effect of impurity scattering on the time dependence of trapping of excitons undergoing basically one-dimensional transport was presented (Wieting et al. 1978). Both the coherent (wave-like) and incoherent (diffusive) microscopic modes of transport were considered and the effects of deviations from strictly one-dimensional transport topology (quasi-one-dimensional) were treated in detail. In one-dimensional systems, well-defined impurity scattering sites, i.e. impurities having excited states with higher energy than the corresponding host molecule excited state and which are not amalgamated into the host exciton band, can severely inhibit transport by "caging" a mobile exciton (Dlott et al. 1977, 1978). That is, the exciton is restricted to a chain of molecules bounded by scattering sites until it either tunnels past a scattering site at one of the ends or takes a non-one-dimensional step to an adjacent linear chain. Exciton trapping in this type of system is governed by a time-dependent trapping rate function which has a form dependent on the microscopic mode of transport, the topology of transport and various physical parameters of the system. Only in the case of nearly isotropic transport does this time-dependent trapping rate function reduce to a time-independent trapping rate constant.

Several recent experimental studies (Shelby et al. 1976, Guettler et al. 1977) of exciton trapping in one-dimensional systems have employed phenomenological trapping rate constants and rate equations which neglect the effects of scattering impurities and deviations from a strictly one-dimensional transport topology. This led to incomplete interpretations of observed results. Although not nearly as visible as their trap counterparts, scattering impurities can play the dominant role in exciton transport and in trapping experiments in one-dimensional systems by forcing the macroscopic mode of transport to be diffusive regardless of the microscopic transport mode. The influence of scattering impurities on observables associated with trapping in one-dimensional systems was suggested in a preliminary study on time-dependent trap emission in the 1,2,4,5-tetra-chlorobenzene triplet exciton system (Dlott et al. 1977). In that experiment, the time-dependent optical emission from the x-trap found in neat crystals of TCB was shown to be explicable in terms of the effects of impurity scattering on exciton transport. The scattering impurity was taken to be the naturally occurring isotopic impurity of monodeutero TCB. In that preliminary study the TCB system was assumed to be strictly one-dimensional and transport was assumed to be strictly coherent. Figure 1 demonstrates that using these assumptions, reasonable agreement between the preliminary theory and experiment was obtained without recourse to adjustable parameters. Thus, even in so called "pure crystals," intrinsic scattering species, which may be isotropic impurities, lattice defects, or difficult to remove chemical impurities, cannot be neglected.

In this section a series of experiments is described which demonstrate the applicability of the model of exciton migration, impurity scattering, and

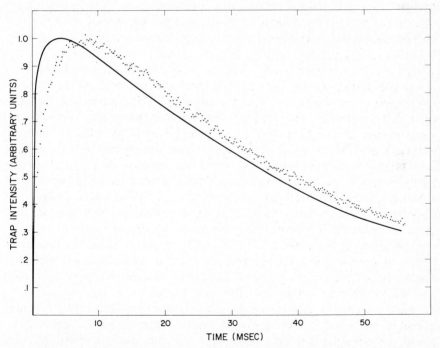

Fig. 1. The points are the experimentally determined time-dependent emission from the x-trap in h_2-1,2,4,5-tetrachlorobenzene (h_2-TCB) following impulse optical excitation. The solid line is the calculated trap intensity assuming the excitons are scattered only by the naturally occurring (0.03%) isotopic impurity hd-TCB. The calculated curve does not utilize adjustable parameters. [From Dlott et al. (1977).]

trapping in one-dimensional systems. These experiments measure the time dependence of optical emission from the triplet x-trap found in crystals of 1,2,4,5-tetrachlorobenzene (TCB) following impulse excitation of the TCB triplet exciton bands. A series of samples containing a range of known concentrations of the isotopic scattering impurity d_2-TCB is employed. Excellent overall agreement is shown between theory and experiment. The observed trapping rate was found to be proportional to the inverse square root of scattering impurity mole fraction. This concentration dependence is in agreement with theoretical prediction (Wieting et al. 1978), and demonstrates that impurity scattering determines the rate of exciton trapping in the TCB systems studied. From the observed concentration dependence and the physical parameters associated with the TCB system, it is possible to place an upper limit of $3 \times 10^3 \, s^{-1}$ on the frequency of cross-chain steps in this system.

The net result is that independent of the microscopic mode of transport, impurity scattering forces exciton transport in quasi-one-dimensional systems to be macroscopically diffusive. Macroscopic transport is characterized by the exciton cage-to-cage stepping frequencies which are obtained from the analysis of the trapping concentration dependence and the time-dependent trapping curves.

3.1. The trapping equations

Well-defined scattering sites have a transport blocking effect which causes an exciton to remain confined to a linear chain of molecules between two such sites, a cage, for a relatively long time before tunneling past an impurity site onto an adjacent chain of molecules. During this time the exciton probability will become uniformly distributed in the initial cage, ensuring equal probability of a step to either adjacent cage. Thus, the transport will describe a random walk between cages on the infinite linear chain when viewed on a time scale long relative to the time required to tunnel out of a particular cage. Such transport is strictly one-dimensional.

A more complete treatment of one-dimensional systems includes the effects of very small interactions leading to transport between adjacent linear chains (Wieting et al. 1978). This motion will be incoherent in nature since local potential fluctuations will certainly be much larger than cross-chain interactions in systems near the one-dimensional limit. The frequencies of interchain steps will be quite small compared to the frequency of site-to-site motion along a given linear chain. However, an interchain step is a step to a different cage, while on-chain a great number of site-to-site steps occur before enough encounters with the caging scattering impurities permit a single cage step. Thus, the frequency of steps between cages on adjacent chains may be comparable to or greater than the frequency of steps between cages on a single chain. This results in a change in the effective exciton transport topology. Transport will be a two- or three-dimensional random walk (not necessarily isotropic) between linear cages, supersites on the superlattice, i.e. the lattice of all linear cages. This will have a significant effect on trapping since multidimensional random walks greatly increase the total number of distinct lattice sites sampled (Montroll 1964) with a concomitant increase in the probability that an exciton will sample a low concentration trap site.

The caging effect of well-defined scattering impurities also has an important consequence for the trapping event. An exciton in such a system is confined to a small set of molecules in a cage for a relatively long time, so that if a trap site is present in that cage the exciton–trap interaction time is greatly extended and the exciton will trap on its first visit to the cage, even if the single encounter trapping probability is quite small.

The model presented below is the result of the above considerations applied to the time evolution of an exciton population ensemble interacting with dilute scattering and trapping impurities in a one-dimensional system. The time-dependent populations of the band states $E(t)$ and the trap states $T(t)$ are described by the rate equations

$$\dot{E}(t) = -[K_E + K_L(t)]E(t), \tag{23a}$$

$$\dot{T}(t) = -K_T T(t) + K_L(t)E(t). \tag{23b}$$

K_E is the decay rate constant (inverse lifetime) for band states and K_T is the decay rate constant for trap states. Thermally assisted promotion from a localized trap state is not included in this scheme, since it is negligible at sufficiently low temperatures (Brenner et al. 1975). $K_L(t)$ is the instantaneous rate of exciton localization per unit population, the time-dependent trapping rate function, the form of which depends on the effective transport topology. For impulse duration excitation of the system, the solutions to eq. (23) require only the form of $K_L(t)$. In what follows, N_T is the trap concentration, χ is the scattering impurity concentration, β is the intermolecular interaction matrix element responsible for on-chain one-dimensional transport, and S is the energy difference between the scattering impurity site excited state energy and the exciton band center.

3.1.1. Strictly one-dimensional transport
The time-dependent trapping rate function for strictly one-dimensional systems is given by

$$K_L(t) = At^{-1/2}, \tag{24}$$

and is independent of the microscopic mode of transport (Wieting et al. 1978). The value of the trapping rate coefficient, A, does depend on the

mode of transport and can be evaluated using the parameters of the
system, all of which are amenable to experimental determination.

For an ensemble of coherent excitons (exciton–phonon scattering is slow
relative to exciton–impurity scattering) at temperature T, the trapping
coefficient is

$$A_{coh} = \frac{2^{3/2}\Gamma(5/4)I_{3/4}(y)}{(y/2)^{3/4}I_0(y)} \frac{N_T\chi}{[\ln(1/(1-\chi))]^{3/2}} \frac{|\beta|^{3/2}}{h^{1/2}S}, \tag{25}$$

where $y = |2\beta/KT|$ and I_0 and $I_{3/4}$ are modified Bessel functions. For
incoherent excitons (fast exciton–phonon scattering) with an on-chain
site-to-site stepping frequency ν_{inc}, the trapping coefficient is given by

$$A_{inc} = \frac{N_T|\beta|}{S}\left[\frac{2(\chi^{-1}-1)\nu_{inc}}{\pi}\frac{1-P_e}{P_e}\ln\left(\frac{1+P_e}{1-P_e}\right)\right]^{1/2}, \tag{26a}$$

$$A_{inc} = \frac{4\pi\beta^2 N_T}{hS}\left[\frac{2(\chi^{-1}-1)}{\alpha\pi}\frac{1-P_e}{P_e}\ln\left(\frac{1+P_e}{1-P_e}\right)\right]^{1/2}, \tag{26b}$$

where eq. (26b) has used the explicit form of ν_{inc} given by Kenkre and
Knox (1974), eq. (18). The parameter α is the decay rate constant for the
loss of coherence of the exciton state, and in principle can be determined
from spectroscopic information (see section 2). The terms involving P_e in
eq. (26) correct for the probability of escape or "leakage" from the initial
cage before a uniform distribution is achieved in that cage. If P_e becomes
large, the results lose accuracy. In practice this somewhat limits the
diluteness of impurities for which a calculation can be made. P_e is given by

$$P_e = 1 - (1 - 2\beta^2/S^2)^{\langle n\rangle+1}, \tag{27a}$$

$$\langle n\rangle = (2/\pi)^{1/2}(\chi^{-1}-1)[1 + \tfrac{3}{8}(\chi^{-1}-1)^{-2} - \tfrac{7}{128}(\chi^{-1}-1)^{-4} + \cdots], \tag{27b}$$

where $\langle n\rangle$ is the mean number of exciton collisions with the impurity site it
last traversed before a uniform exciton probability distribution is
achieved in the cage (Zwemer and Harris 1978).

Equations (25) and (26) show that the dependence of the rate of exciton
trapping on scattering impurity concentration is basically identical for the
coherent and incoherent microscopic modes of exciton transport. For
coherent excitons, $A_{coh} \propto \chi/[\ln(1/(1-\chi))]^{3/2}$. However, for reasonably dilute
impurities $\ln(1/(1-\chi)) = \chi$, so $A_{coh} \propto \chi^{-1/2}$, which is also the dependence
manifested by A_{inc}, eq. (26), for incoherent excitons. Thus independent of
the microscopic mode of transport, the concentration dependence can be
used to access the transport topology and, as shown below, determine the
frequency of cross-chain steps. It is also worth noting that the rate of
trapping varies relatively slowly with scattering impurity concentration.

Using eqs. (23) to (27) the time-dependent exciton and trap populations
are obtained for an exciton system which is strictly one-dimensional in its

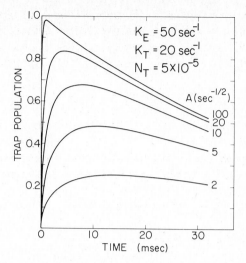

Fig. 2. Calculated trap populations as a function of time are used to illustrate the effect of increasing the coefficient A of the time-dependent trapping rate function for strictly one-dimensional transport on the time-dependent intensity of trap emission following impulse optical excitation of the exciton band. The rate constants for the exciton and trap decay, K_E and K_T, respectively, are 50 and 20 s^{-1}. These rates are typical of triplet systems. As A is increased, the population maximum is shifted to shorter times and the integrated population of the trap increases. [From Wieting et al. (1978).]

transport. A is obtained from eqs. (25) or (26).

$$E(t) = \exp(-K_E t - 2At^{1/2}), \tag{28a}$$

$$T(t) = \exp(-K_T t) \left(\frac{\pi A^2}{K_E - K_T}\right)^{1/2} \exp\left(\frac{A^2}{K_E - K_T}\right)$$
$$\times \left[\text{erf}\left([(K_E - K_T)t]^{1/2} + \frac{A}{(K_E - K_T)^{1/2}}\right) \right.$$
$$\left. - \text{erf}\left(\frac{A}{(K_E - K_T)^{1/2}}\right)\right], \tag{28b}$$

where erf(x) is the error function of argument x. Figure 2 illustrates the type of results obtained using eq. (28b).

3.1.2. Quasi-one-dimensional transport

For systems in which transport is close to but not strictly one-dimensional, an exciton undergoes a multidimensional random walk among the cages formed by the scattering impurities. Each cage, which is composed of many lattice sites, is a single supersite in the superlattice composed of all cages. An exciton performs a macroscopic random walk among the sites of the superlattice. This results in an identical time-dependent form of the trapping rate function for both microscopic modes of exciton migration (Wieting et al. 1978),

$$K_L(t) = K_L + B_L t^{-1/2}. \tag{29}$$

The values of the trapping parameters K_L and B_L depend on all of the physical parameters of the strictly one-dimensional problem and on the rate and relative anisotropy of the exciton walk on the three-dimensional superlattice.

The frequency of cage steps along the linear direction, denoted by ν_L, is given by the strictly one-dimensional cage stepping frequency. For an ensemble of coherent excitons at temperature T, the thermal average frequency of cage steps is given by

$$\nu_L(\text{coh}) = \frac{16\pi^{1/2}}{\chi^{-1}-1} \frac{|\beta|^3}{hS^2} \frac{I_{3/2}(y)}{(y/2)^{3/2} I_0(y)}, \tag{30}$$

where $y = |2\beta/KT|$ and I_0 and $I_{3/2}$ are modified Bessel functions. If the microscopic mode of transport is incoherent with a site-to-site stepping frequency ν_{inc}, the frequency of cage steps is given by

$$\nu_L(\text{inc}) = \frac{\nu_{\text{inc}}}{\chi^{-1}-1} \frac{\beta^2}{S^2} \frac{1-P_e}{P_e} \ln\left(\frac{1+P_e}{1-P_e}\right), \tag{31a}$$

or, using eq. (18),

$$\nu_L(\text{inc}) = \frac{16\pi^2\beta^4}{(\chi^{-1}-1)\alpha h^2 S^2} \frac{1-P_e}{P_e} \ln\left(\frac{1+P_e}{1-P_e}\right). \tag{31b}$$

Note that the dependence on χ, the scattering impurity mole fraction, is identical in eqs. (30) and (31) when P_e in eq. (31) is small, i.e. when P_e is in its regime of usefulness.

Generalized three-dimensional arrangements of one-dimensional molecular chains employed in this model included cross-chain motion through the use of two parameters, ν_c and $\nu_{c'}$, the frequencies of interchain steps in the two orthogonal off-chain directions. The relative probabilities for steps in different directions in the three-dimensional array of molecular chains are

on chain: $L = \nu_L/\nu_{\text{tot}}$,

interchain: $\begin{cases} C = \nu_c/\nu_{\text{tot}}, \\ C' = \nu_{c'}/\nu_{\text{tot}}, \end{cases}$

$$\nu_{\text{tot}} = \nu_L + \nu_c + \nu_{c'}. \tag{32}$$

From these relative probabilities the two leading terms of the random walk Green's function can be obtained (Montroll 1964). The remaining terms are negligible. The first term is the convergent integral

$$u_0 = \frac{1}{\pi^3} \int \int_0^\pi \int \frac{d\phi_1 \, d\phi_2 \, d\phi_3}{1 - (L \cos \phi_1 + C \cos \phi_2 + C' \cos \phi_3)}. \tag{33}$$

In all cases this may be evaluated by numerical integration. However, if $\nu_c = \nu_{c'}$, i.e. $C = C'$, a convenient closed form expression is obtained:

$$u_0 = \frac{1}{C} I([L/C]^{1/2}), \tag{34}$$

where $I(\alpha)$ is given by

$$I(\alpha) = 4[(\gamma+1)^{1/2} - (\gamma-1)^{1/2}]K(k_2)K(k_3)/\alpha\pi^2,$$
$$k_2 = \tfrac{1}{2}[(\gamma-1)^{1/2} - (\gamma-3)^{1/2}][(\gamma+1)^{1/2} - (\gamma-1)^{1/2}],$$
$$k_3 = \tfrac{1}{2}[(\gamma-1)^{1/2} + (\gamma-3)^{1/2}][(\gamma+1)^{1/2} - (\gamma-1)^{1/2}],$$
$$\gamma = (4 + 3\alpha^2)/\alpha^2, \tag{35}$$

and $K(k)$ is the complete elliptic integral of the first kind of modulus k. The second term is

$$u_1 = 1/(2\pi^2 LCC')^{1/2}. \tag{36}$$

The quasi-one-dimensional trapping rate parameters, K_L and B_L, are given by

$$K_L = N_T(\chi^{-1} - 1)\frac{\nu_{tot}}{u_0}, \tag{37}$$

$$B_L = \frac{N_T(\chi^{-1} - 1)\nu_{tot}^{1/2}u_1}{\pi^{1/2}u_0^2}. \tag{38}$$

The concentration dependence of K_L and B_L contains information pertaining to the relative anisotropy of the random walk and hence about the magnitude of the interchain interactions. If $\nu_L \gg \nu_c, \nu_{c'}$, then K_L and B_L are proportional to $(\chi^{-1} - 1)^{1/2}$. This is essentially the one-dimensional behavior and indicates that such a system is very close to the strictly one-dimensional limit. However, if $\nu_L \approx \nu_c$ a stronger dependence on scattering impurity concentration results, and in the limit that $\nu_L \ll \nu_c, \nu_{c'}$ the dependence goes as $(\chi^{-1} - 1)$. The net effect can be to produce trapping which behaves topologically multidimensional even though the intermolecular interactions and the total extent of excitation motion is virtually one-dimensional.

The trapping rate parameters are used with eq. (29) to solve the population rate equations, yielding the time-dependent populations

$$E(t) = \exp[-(K_E + K_L)t - 2B_L t^{1/2}],\qquad(39a)$$

$$T(t) = \exp(-K_T t)$$

$$\times \left\{ \frac{K_L}{K_L + K_E - K_T} (1 - e^{-(K_L + K_E - K_T)t - 2B_L t^{1/2}}) \right.$$

$$+ \left(\frac{B_L^2 \pi}{K_L + K_E - K_T} \right)^{1/2} e^{B_L^2/(K_L + K_E - K_T)} \left(1 - \frac{K_L}{K_L + K_E - K_T} \right)$$

$$\times \left[\mathrm{erf}\left([(K_L + K_E - K_T)t]^{1/2} + \frac{B_L}{(K_L + K_E - K_T)^{1/2}} \right) \right.$$

$$\left. \left. - \mathrm{erf}\left(\frac{B_L}{(K_L + K_E - K_T)^{1/2}} \right) \right] \right\},\qquad(39b)$$

where $\mathrm{erf}(x)$ is the error function of argument x. In the strictly one-dimensional limit, the expression of eq. (28) should be used. Figure 3 illustrates the effect that small multidimensional interactions can have on trapping in a basically one-dimensional system.

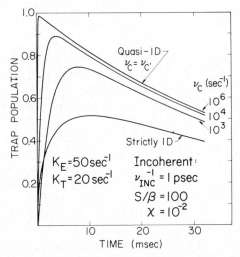

Fig. 3. The effect of very slow cross-chain transport on the time dependence of the trap population following impulse excitation of the exciton band is displayed. The bottom curve is for strictly one-dimensional transport with a site-to-site incoherent stepping time $\nu_{inc}^{-1} = 1$ ps. The exciton and trap decay rate constants (inverse life time) are 50 and 20 s^{-1}, respectively. The scattering impurity and trap concentrations are 10^{-2} and 5×10^{-5}, respectively. The other three curves are calculated using an identical set of parameters but including varying degrees of cross-chain stepping, i.e., 10^3, 10^4 and 10^6 s^{-1} moving from bottom to top. These cross-chain frequencies are 6 to 9 orders of magnitude smaller than the on-chain site-to-site stepping frequency, yet have a very large influence. [From Wieting et al. (1978).]

3.2. Experimental

The concentration of d_2-TCB in each TCB sample was determined by quantitative preparation. Single crystals were then grown from each mixture using the Bridgeman technique. All the time-dependent trapping experiments were performed at 1.35 K. A 3/4 meter monochromator was used to characterize the phosphorescent x-trap and exciton spectra. In the time-resolved measurements, the initial triplet exciton population was prepared either by direct triplet excitation with a 20 ns doubled ruby laser pulse or via the singlet manifold with a 3 μs xenon flash lamp filtered to pass light in the 2500 Å region of the spectrum. The results were the same with either excitation method. The time-decaying x-trap emission signals were detected at right angles with a photomultiplier tube and the monochromator set to the x-trap origin, digitally recorded and signal averaged with a transient recorder interfaced to a computer.

3.3. Results

The experiments employed TCB crystals doped with a range of concentrations of the isotopic scattering impurity, d_2-TCB. The time-dependent phosphorescent emission from the intrinsic TCB x-trap was monitored following impulse optical excitation. Typical data are displayed in Fig. 6 (see page 25). At short times the trap phosphorescence intensity increases as the exciton population flows into the trap. At longer times, all the population has trapped, and the trap emission decays exponentially. As the scattering impurity concentration increases, transport is hindered. The buildup of the trap phosphorescence becomes more gradual and the maximum is shifted to longer time. The physical constants which are necessary to use the models described above are known. The trap rate constant for decay to the ground state is $K_T = 25.5 \text{ s}^{-1}$ and the appropriate exciton decay rate constant in the absence of trapping is $K_E = 35.3 \text{ s}^{-1}$. The trap concentration is $N_T = 1/22\,000$. The TCB one-dimensional intermolecular interaction matrix element was previously determined to have a value of $\beta = 0.35 \text{ cm}^{-1}$. A reasonable value of the impurity–band center energy difference, S, can be obtained by using the difference between triplet state energies of h_2-TCB and d_2-TCB in a d_2-TCB host crystal. Spectroscopic measurements yield a value of $S = 20.9 \text{ cm}^{-1}$. The concentration of d_2-TCB scattering impurities is known from the sample preparation. All experiments were performed at 1.35 K.

The functional dependence of the observed time-resolved data on scattering impurity concentration is the fundamental test of the importance of exciton–impurity scattering and of the effective exciton trapping topology. As discussed in subsection 3.1, if the transport topology in TCB is strictly one-dimensional or deviates only slightly from the limit, the observed trapping rate function will vary with scattering impurity concentration in a manner which is directly proportional to $(\chi^{-1} - 1)^{1/2}$, or approximately the inverse square root of the scattering impurity concentration, χ. The concentration dependence of the trapping rate function is determined by the behavior of its time-independent coefficients. Using eq. (28b) and the known decay rate constants, K_E and K_T, an observed trapping rate function coefficient, A, is obtained from the optimal fit to data from each sample. [Equation (39b) could be employed to obtain identical results.] These experimentally determined trapping coefficients are plotted versus $(\chi^{-1} - 1)^{1/2}$ in fig. 4. The solid line through the data (crosses) shows that the theoretically predicted proportionality occurs over a thirty-fold range of

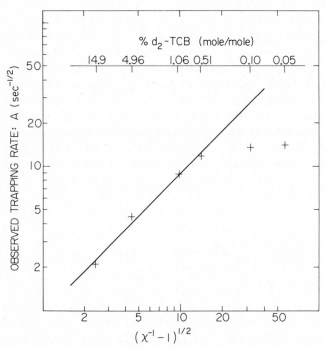

Fig. 4. The experimentally determined trapping rate coefficients, A, of the time-dependent trapping rate function $K_L(t) = At^{-1/2}$ (crosses) are plotted as a function of $(\chi^{-1} - 1)^{1/2}$, where χ is the mole fraction of the scattering impurity d_2-TCB doped in various concentrations in h_2-TCB host crystals. The predicted linear dependence, i.e. the solid line, is observed for all but the lowest concentrations. The crosses corresponding to concentrations less than 10^{-3} M/M fall below the line, demonstrating that in this case the d_2-TCB scattering impurities are dominated by residual impurities and intrinsic defects present in the h_2-TCB. [From Dlott et al. (1978).]

concentrations and at all but the lowest scattering impurity concentrations. (The lowest concentration regime is discussed below.) This observation confirms the importance of the role played by scattering impurities in the transport and trapping of triplet excitons in TCB crystals, and demonstrates the applicability of the model to real systems. The deviation from the predicted concentration dependence at low concentrations (doped-in scattering impurities ≤0.1%) suggests that additional impurities are also present in the samples and that these dominate at low concentrations. This is not an unexpected result since there are, at the very least, additional scattering impurities in the form of the naturally occurring isotopic impurity hd-TCB which is 0.03% abundant. Additionally, it is known that trichlorobenzene and tetrachlorobenzene isomers may remain in very low concentration even after extensive zone refining (Guettler et al. 1977). Other scattering sites, such as low concentration crystal lattice defects (Baughman and Turnbull 1971) (the high energy counterparts of x-traps), may also hinder transport when the doped-in impurity concentration is sufficiently low to unmask their effects.

It is clear from the agreement between the experimental and the theoretical scattering impurity concentration dependence that exciton–impurity scattering is the dominant factor in controlling the rate of triplet exciton trapping in TCB. Since all the required physical constants are known for TCB, a realistic upper bound can be placed on the deviation from strict one-dimensionality. First consider the strictly one-dimensional model of subsection 3.1.1. This model was employed to calculate the trapping rate coefficient, A, without the use of adjustable parameters for both the coherent [eq. (25)] and incoherent [eq. (26)] cases. The resulting

Table 2

The theoretical trapping rate coefficients, A, for strictly one-dimensional transport in TCB doped with d_2-TCB. A_{coh} is given by eq. (25). A_{inc} is given by eq. (26) using the theoretically predicted site-to-site hopping time of 1 ps.

% d_2-TCB	A_{coh} (s$^{-1/2}$)	A_{inc} (s$^{-1/2}$)	
14.9	0.486	2.10	(0.510)[a]
4.96	0.913	3.83	(0.930)
1.06	2.03	8.31	(2.02)
0.51	2.95	11.8	(2.86)

[a]The numbers in parentheses are the results of a calculation which treated the incoherent hopping time as an adjustable parameter, set to 17 ps in eq. (26).

values of A_{coh} and A_{inc} for the various concentrations of the d_2-TCB scattering impurity are given in table 2. These two models predict quite different rates of trapping. The incoherent calculation used eq. (26b), i.e., the Kenkre–Knox formalism was employed in calculating the microscopic hopping frequency, ν_{inc}. For TCB this approach yields a value of $\nu_{inc} = 1\,ps^{-1}$. However, this value was obtained using eq. (18) outside of its range of applicability for this system. (See discussions below and in sections 2 and 4.)

   The experimental data points and these calculations are plotted in fig. 5, where it can be seen that the values of A_{inc} (circles) agree exceedingly well with the observed data. The coherent transport calculations predict values of A_{coh} (triangles) which are rather too slow to account for the observed

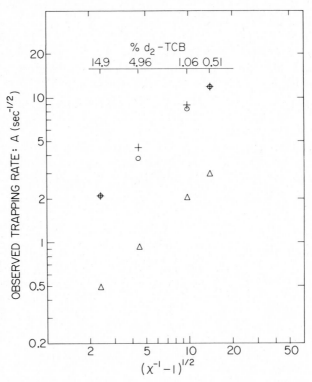

Fig. 5. Theoretically predicted trapping rate constants for TCB in the strictly one-dimensional limit are plotted along with the data (crosses) versus $(\chi^{-1} - 1)^{1/2}$, where χ is the mole fraction of scattering impurity. The open circles are calculated for a model of microscopically incoherent exciton migration using the Kenkre–Knox formalism for the site-to-site stepping time. The triangles are calculated for a model of microscopically coherent exciton migration. If small multidimensional interactions are considered, agreement between the data and the coherent model is greatly improved. [From Dlott et al. (1978).]

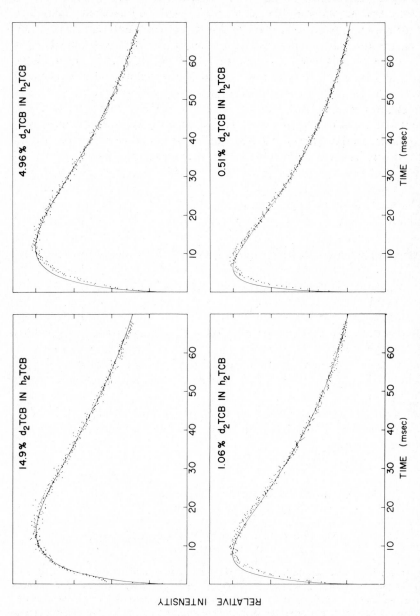

Fig. 6. Time-resolved trap emission from TCB crystals doped with various concentrations of scattering impurity. The solid lines are theoretical curves obtained without recourse to adjustable parameters using the model of microscopically incoherent one-dimensional transport and employing the Kenkre–Knox formalism for the incoherent site-to-site stepping frequency. [From Dlott et al. (1978).]

results. The actual trapping data and incoherent theoretical curves are
displayed in fig. 6 for each impurity concentration. It can be seen that the
overall agreement is good. However, the close agreement obtained from
the model of incoherent transport is due to a fortuitous calculation of ν_{inc}
using the Kenkre–Knox formula. This formula is not applicable for the
TCB physical situation. If the incoherent site-to-site hopping time is chosen
to be $\nu_{inc}^{-1} = 17$ ps, virtually identical results are obtained from the coherent
and incoherent models. The 1 ps site-to-site incoherent hopping time was
obtained using a memory function decay time which is much longer than
the coherent site-to-site transit time. In this situation the system is coherent
and eq. (18) does not apply. The 17 ps hopping time results from a memory
function decay which is a factor of ~ 2 faster than the coherent site-to-site
transit time. These results are given in parenthesis in table 2. Furthermore,
both calculations neglect the increase in trapping rate which results if
interchain interactions do not vanish.

In the TCB system, the $\chi^{-1/2}$ concentration dependence of the trapping
shows that any cross-chain interaction must be small indeed. If these
interactions are not negligible, TCB will behave as a quasi-one-dimensional
system and the model of subsection 3.1.2 can be used to interpret the
observed time-resolved x-trap phosphorescence. The necessary cage-to-
cage stepping frequency can be calculated from eqs. (30) and (31) for
coherent and incoherent transport, respectively. The results of these calculations for the various scattering impurity concentrations are given in
table 3. Cage stepping frequencies in the linear direction are orders of
magnitude slower than the inverse time for site-to-site motion along that
direction, typically $\sim 10^{11}$–10^{12} s^{-1}. Thus, even very low frequency cross-
chain steps can cause a significant deviation from strictly one-dimensional
transport. As discussed in subsection 3.1.2, the proportionality to $(\chi^{-1} - 1)^{1/2}$
will be observed only if a system is very near the strictly one-dimensional

Table 3

The intrachain cage stepping frequency, ν_L, of excitons in TCB doped with d_2-TCB. For coherent transport eq. (30) was used. For incoherent transport eq. (31) was employed with a site-to-site hopping time $\nu_{inc}^{-1} = 1$ ps.

% d_2-TCB	ν_L(coh) (s^{-1})	ν_L(inc) (s^{-1})
14.9	1.02×10^7	9.79×10^7
4.96	3.04×10^6	2.90×10^7
1.06	6.25×10^5	5.76×10^6
0.51	2.99×10^5	2.64×10^6

limit, i.e. $\nu_L \gg \nu_c, \nu_{c'}$. If interchain steps were not much slower than intrachain cage-to-cage steps, the transport topology will be much more isotropic and the concentration dependence will be stronger, i.e. it will approach χ^{-1}. Thus, for TCB the requirement that $\nu_L \gg \nu_c$ implies that interchain steps occur at a rate no greater than $\sim 10^5\,\mathrm{s}^{-1}$.

In the absence of an estimate of the anisotropy in the interchain interactions, it is assumed that all such nearest neighbor interchain interactions are equal. In this case the closed form expression describing exciton motion on the three-dimensional superlattice, eq. (34), may be employed to calculate the trapping rate function, eq. (29), for each scattering impurity concentration. The different on-chain cage stepping frequencies, ν_L, are given in table 3. Therefore the only unknown parameter is ν_c, and a single choice of ν_c should be able to reproduce the experimental exciton trapping curves for the entire range of scatter impurity concentrations.

Values of ν_c were combined with the values of $\nu_L(\mathrm{coh})$ in table 2 to evaluate the trapping parameters k_L and B_L. These parameters were then used in eq. (39b) to calculate the time-dependent exciton trapping curves. For coherent transport, the value $\nu_c = 2.5 \times 10^3\,\mathrm{s}^{-1}$ gave the best overall agreement between theory and the different trapping experiments. The values of K_L and B_L are listed in table 4 and the calculated curves along with the experimental data are displayed in fig. 7. Once again, agreement between observation and theory is generally quite good, although for this model significant disagreement is observed at the greatest impurity concentration. Pairs of scattering impurities which will be present at the

Table 4

The theoretical trapping rate coefficients, K_L and B_L, for quasi-one-dimensional transport in TCB doped with d_2-TCB. The intrachain one-dimensional transport is microscopically coherent with the cage stepping frequencies given in table 3. The interchain transport is assumed to be isotropic with a cross-chain step time $\nu_c = 2.5 \times 10^3\,\mathrm{s}^{-1}$.

% d_2-TCB	$K_L\,(\mathrm{s}^{-1})$	$B_L\,(\mathrm{s}^{-1/2})$
14.9	66.0	0.261
4.96	121	0.478
1.06	267	1.06
0.51	387	1.53

M.D. Fayer

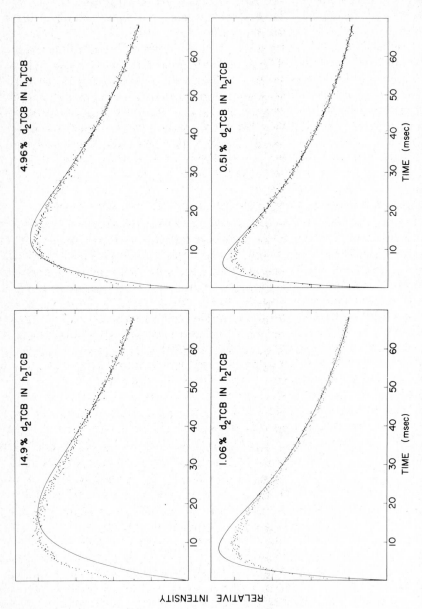

Fig. 7. Time-resolved trap emission from TCB crystals doped with various concentrations of scattering impurity and the best fit obtained from the microscopically coherent migration model. In this calculation the extent of multidimensional transport in this basically one-dimensional system was adjusted. The calculated curves were obtained using a cross-chain stepping frequency of $\nu_c = 2.5 \times 10^3 \ s^{-1}$. [From Dlott et al. (1978).]

highest concentration may be responsible for the deviations. However, the overall agreement between experiment and both models is sufficiently good that a clear choice between them is not possible. Thus, coherent transport is consistent with the observed trapping data.

Observing the effects of scattering impurity concentration on trapping in TCB demonstrates that the TCB triplet exciton system is virtually one-dimensional and that in the samples studied, exciton–impurity scattering is the dominant influence on the macroscopic rate of exciton transport. Furthermore, these results give additional strength to the argument that exciton–impurity scattering strongly influences energy migration in pure TCB crystals as well. From the functional form of the concentration dependence it is possible to put an upper bound on the cross-chain stepping frequency of $\sim 10^5\,\mathrm{s}^{-1}$. Detailed analysis of the time dependence of the trapping data gives a better estimate of $<3 \times 10^3\,\mathrm{s}^{-1}$. Comparing this to on-chain site-to-site tranfer frequencies of 10^{11}–$10^{12}\,\mathrm{s}^{-1}$ shows that the cross-chain frequency is down some eight orders of magnitude. Taking the cross-chain frequency to be proportional to the square of the cross-chain intermolecular interaction would put that interaction at $<10^{-4}$ of the on-chain interaction, corresponding to less than $10^{-4}\,\mathrm{cm}^{-1}$. Thus the impurity scattering concentration study provides a sensitive probe of the dimensionality of the exciton transport system.

Both the coherent and incoherent microscopic modes are consistent with the experimental trapping data. Impurity scattering forces long-range transport to be a diffusive process involving a random walk among cages, irrespective of the microscopic mode of transport. Thus a trapping experiment in this type of system examines the long-distance macroscopic rate of transport. Analysis of the trapping data provides the trapping rate parameters once the question of the transport topology has been sorted out. These parameters give directly the cage-to-cage stepping frequency which is the basic quantity characterizing long range energy transport in these quasi-one-dimensional systems.

4. Spectroscopic experiments in the 1,2,4,5-tetrachlorobenzene triplet exciton system

The previous section demonstrated that trapping experiments in quasi-one-dimensional systems can provide considerable information on the macroscopic nature of exciton transport provided essential features such as scattering impurities, transport topology, and the microscopic mode of exciton transport are properly included in the theoretical analysis. Oversimplified models can be quite misleading even when there is apparent agreement between experiment and calculation. Problems associated with trapping experiments in three-dimensional systems can be as severe as those encountered in the one-dimensional situation. For example, knowledge of the probability of exciton trapping on an encounter with a trap is avoided in the analysis of the experiments discussed in section 3, since an exciton remains in a cage for a considerable time. When the exciton enters a cage with a trap, it has a sufficient number of encounters with the trap to be trapped, regardless of the first-encounter trapping probability. However, in a multi-dimensional system this is not the case. A trapping experiment involves two processes, transport and trapping, and it is difficult to separate the effects of these two processes on the experimental observable.

Spectroscopic observables can in principle address the question of the extent of exciton coherence. Triplet excitons can be studied by both optical spectroscopy and electron spin resonance techniques. TCB has been particularly well studied by a variety of investigators and techniques (Francis and Harris 1971, Burland et al. 1977a, Botter et al. 1978).

Low temperature optical spectroscopy of an exciton band origin basically results in the examination of the $k = 0$ band state. This is because the wave vector, k, associated with an exciton band state defines the exciton's "pseudomomentum" (see section 2). Since the wave vector of the incoming photon has $k \approx 0$ when compared to the range of k values in the exciton band, i.e. $k = 0$ to $k = \pi/a$, conservation of momentum requires that the exciton produced must also have $k \approx 0$. A variety of effects can give width to the exciton $k = 0$ optical absorption line. T_2^* processes, i.e. pure dephasing processes, cause fluctuations in energy without inducing a change of state. T_1 processes result in a change in the state of the system. The most common of these is relaxation of the exciton to the ground state by either radiative or nonradiative pathways. In the absence of all other processes, the excited state life time will determine the absorption line width. In the presence of both types of processes, the line width is

$$1/\pi T_2 = 1/\pi T_2^* + 1/2\pi T_1. \tag{40}$$

As mentioned in previous sections, exciton–phonon interactions can cause scattering of an exciton from one wave vector state to another. This is a T_1 process since it involves a state change. For optical absorption experiments on the exciton origin, phonon scattering of the $k = 0$ state can contribute to the line width by shortening the $k = 0$ life time. Another way of describing this is to say that the exciton–phonon terms in the system's Hamiltonian admix the zeroth-order k states, eq. (3), obtained in the absence of the exciton–phonon terms. Thus $k = 0$ character is admixed into states other than the zeroth-order $k = 0$ state, giving some transition probability to a range of k states and consequently a broadened absorption line. In one-dimensional systems, $k = 0$ is either at the top or the bottom of the band. Therefore the final states in a scattering process (or the states into which

$k = 0$ character is admixed) are all displaced in energy to one side of the $k = 0$ state. This can result in asymmetric absorption lines at low temperature. The asymmetric line shape has been observed in both 1,4-dibromonaphthalene (DBN) (Burland et al. 1977b) and 1,2,4,5-tetra-chlorobenzene (TCB) (Burland et al. 1977a) one-dimensional triplet exciton systems. (In DBN, exciton–impurity scattering rather than exciton–phonon scattering has been proposed as the source of the low temperature asymmetric line broadening.) Then by examining the exciton optical absorption line it is possible to obtain information about exciton scattering and by inference information on exciton coherence.

The ESR experiments (Francis and Harris 1971, Botter et al. 1978) take advantage of the three triplet spin sublevels associated with the T_1 electronic state. These levels are split in zero magnetic field with typical splitting of 1–10 GHz. There are, then, three triplet exciton bands whose states are orthogonal due to the orthogonality of the spin functions. The energy difference between any two of the three bands is the spectrum of that band-to-band transition. To conserve momentum, $\Delta k = 0$. But since two bands are involved in an ESR band-to-band spectrum, it is possible to examine transitions among all the k states. This is in contrast to optical spectroscopy where only $k = 0$ is allowed. (Optical hot band experiments involving a vibrational exciton of the ground electronic state can examine all exciton k states.)

For one-dimensional systems like TCB, the energy difference between two of the states, x and y, as a function of the wave vector k is

$$\Delta E_{xy}(k) = (E_0^x - E_0^y) + 2(\beta_x - \beta_y) \cos ka. \tag{41}$$

If $\beta_x = \beta_y$, the intermolecular interactions are independent of spin, then ΔE_{xy} is independent of k and the spectrum would not display the k-dependent band structure. However, this is contrary to experiments performed on TCB discussed below. Spin–orbit coupling is responsible for the difference in β_x and β_y. For a molecule of high point group symmetry, like TCB, the three spin sublevels have different symmetries. They will only spin–orbit couple to states of the same symmetry. Therefore a triplet state of x symmetry on site m, $|T_{1m}^x\rangle$, will only couple to x symmetry singlet states, $|S_m^x\rangle$. If we assume that each triplet spin sublevel couples predominantly to a single singlet state, then in the derivation of the band dispersion eqs. (1)–(9), the ket $|1_m\rangle$ is replaced by

$$|1_m^x\rangle = |T_{1m}^x\rangle + \gamma_x |S_m^x\rangle \tag{42a}$$

for the x sublevel. γ_x is the probability amplitude for the singlet state spin–orbit admixed into the triplet state x spin sublevel. For the y spin sublevel,

$$|1_m^y\rangle = |T_{1m}^y\rangle + \gamma_y |S_m^y\rangle. \tag{42b}$$

Using eqs. (42) in eqs. (1)–(9) gives (Francis and Harris 1971)

$$\Delta E_{xy}(k) = (E_0^x - E_0^y) + 2(\gamma_x \beta_x^s - \gamma_y \beta_y^s) \cos ka$$
$$= \Delta E_0^{xy} + 2f_{xy} \cos ka. \tag{43}$$

β_x^s and β_y^s are the intermolecular interactions associated with the singlet bands involving $|S^x\rangle$ and $|S^y\rangle$, respectively. Singlet bands are generally much wider than triplet bands. However, γ_x, $\gamma_y < 10^{-3}$, so $4f_{xy}$ is very much smaller than 4β, the triplet band width. In TCB, $f_{xy} \approx 10\,\text{MHz}$, while $\beta \approx 10\,\text{GHz}$. The ESR spectrum, given by $\Delta E_{xy}(k)$, reflects the exciton band dispersion on a reduced scale. The full density-of-states function can be examined, and questions pertaining to exciton–phonon scattering, pure dephasing and exciton coherence can be addressed.

It is important to keep in mind that an optical absorption experiment and an ESR band-to-band experiment are inherently different. An optical experiment involves the ground electronic state and the exciton band. An ESR experiment involves transitions between two closely spaced triplet bands, differing only by a spin sublevel quantum number.

4.1. Temperature-dependent optical absorption experiments

In this subsection, temperature-dependent optical absorption experiments on the triplet exciton origin of 1,2,4,5-tetrachlorobenzene (TCB) are discussed (Burland et al. 1977a). As already mentioned, TCB is a particularly interesting compound to study for several reasons. First, it is an example of a basically one-dimensional exciton transport system. As such, investigations of many of its properties are simplified. Second, extensive ESR line shape experiments have been conducted on this system. They will be discussed in more detail in subsection 4.2. It is·of interest to compare the correlation times measured in ESR experiments with the corresponding times measured optically. Third, TCB has the interesting features that: (i) the optically accessible $k \approx 0$ level is at the top rather than at the bottom of the band (Dlott and Fayer 1976), and (ii) the TCB triplet exciton band width is only $1.4\,\text{cm}^{-1}$ (Francis and Harris 1971, Dlott and Fayer 1976) compared to band widths of $> 10\,\text{cm}^{-1}$ in other reported systems.

4.1.1. Experimental

TCB was purified by zone refining and crystals were grown using the Bridgeman method. During an experiment a crystal was immersed in He vapor and its temperature varied by varying the rate of flow of cooled He gas past the sample. The temperature was monitored near the sample with a Si diode, and the quoted temperatures are estimated to be accurate to within $\pm 0.3\,\text{K}$.

The spectra were recorded using a spectrometer (Jobin–Yvon THR 1500) with a measured resolution of better than 300 000. In no case did the instrumental line width contribute significantly to the observed line widths or line shapes. Care was taken to ensure that the absorption was small enough so that the measured line shape indeed reflected the true absorption line shape.

Crystals from several different boules were examined. In addition, the crystals were subjected to very different temperature cycling processes. In some cases the crystals were cooled to below 77 K in minutes; in others this process took over 24 hours. In only one case were we able to detect any effect of either the origin of the crystal or the manner of cooling on the line shape, position or width. It was observed that if a particular sample was taken to low temperatures, allowed to warm and then again cooled, significant distortion of both the optical and ground state Raman line shapes resulted (Schmidt and Macfarlane, 1977). Crystals subjected to this temperature cycling process were not used to obtain the results cited below.

4.1.2. Results

In fig. 8 the low temperature absorption line shape is shown. The first important point to note is that the line is an asymmetric lorentzian. Asymmetric exciton absorption line shapes have also been reported for DBN and were briefly discussed in the introduction to section 4. The asymmetry is attributed to the asymmetric position of the $k \approx 0$ level at one edge of the exciton band. Unlike DBN, the lorentzian broadening in TCB is on the low energy side of the line (fig. 8b), due to the fact that the $k \approx 0$ level is at the top of the band in TCB. The high energy side of the line falls off like a gaussian (fig. 8c).

As the temperature increases the line becomes more symmetric until, at around 14 K, it has become a symmetric lorentzian as indicated in fig. 9a. Upon further increasing the temperature, the line shape changes and at 22 K it has a gaussian shape as shown in fig. 9b.

Figure 10 shows the temperature dependence of the full width at half maximum for the TCB triplet exciton. At low temperatures (below 12 K) the width is constant within experimental uncertainty and equal to $1.3 \pm 0.2 \, \text{cm}^{-1}$. Note that while the overall width remains constant in this temperature region, the shape of the line as discussed in the previous subsection begins to change. Above 20 K the width begins to broaden rapidly.

The low temperature asymmetric lorentzian line shape arises from the asymmetric position of the optically accessible $k \approx 0$ level at one edge of the exciton band. In the DBN triplet exciton the $k \approx 0$ level is known to be at the bottom of the band (Hochstrasser and Whiteman 1972) and the asymmetric lorentzian broadening occurs on the high energy side of the absorption line. Dlott and Fayer (1976) have shown that for TCB the $k \approx 0$

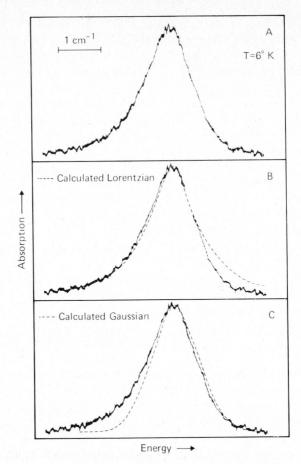

Fig. 8. TCB triplet exciton absorption line shape at 6 K. (a) Absorption line only. (b) Data and calculated lorentzian. (c) Data and calculated gaussian. The procedure in (b) and (c) was to measure the absorption peak. height and the full width at half height and to construct a lorentzian or gaussian from these two parameters only. The spectrum is lorentzian in shape on the red side of the line. This asymmetry has been predicted theoretically. [From Burland et al. (1977a).]

level occurs at the top of the band. We thus expect and in fact observe that the asymmetry in TCB occurs on the low energy side of the absorption line.

As shown in fig. 8c, the high energy side of the absorption line is gaussian in appearance. In the DBN system, the overall line is considerably narrower and the non-lorentzian side falls off extremely sharply. The line appears as if it is almost one half of a lorentzian. It has been suggested that when DBN molecules contain one or more ^{13}C atoms, this results in

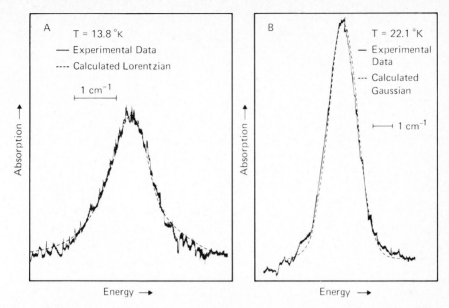

Fig. 9. TCB triplet exciton absorption line shape at (a) 13.8 K and at (b) 22.1 K. The procedure for the calculated curves is described in the caption of fig. 8. At high temperature the line shape (b) is gaussian while at lower temperatures the line shape (a) undergoes a transition to lorentzian. This change in shape has been predicted theoretically and indicates a transition from incoherent to coherent exciton transport as the temperature is reduced. [From Burland et al. (1977a).]

isotopic impurities that are amalgamated into the DBN band and are responsible for an exciton scattering process which makes a major contribution to the homogeneous line width (Burland et al. 1977b). However, similar calculations for the TCB triplet exciton system indicate that scattering by ^{13}C impurities will, in this case, make a negligible contribution when compared to the observed TCB line width. This fact, and the gaussian "high energy" shape, strongly suggest that the low temperature line is to some extent inhomogeneously broadened. If this is the case, it is reasonable to consider the possibility that the TCB absorption line is composed of a gaussian distribution of half-lorentzian lines. Computer simulation studies show that if the half-lorentzians have a width 40 to 50% of the gaussian width, the observed line shape, "low energy" lorentzian and "high energy" gaussian, could be reproduced. Thus, as a *rough estimate*, the optical homogeneous line width is between 0.4 and 0.5 of the observed 1.3 cm^{-1} width. This corresponds to an optical T_2 on the order of 15 to 20 ps.

As in the DBN case, the TCB triplet exciton absorption line shape becomes more symmetric as the temperature increases and at 14 K the line

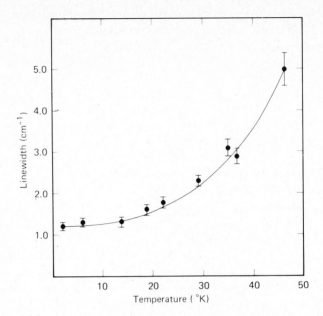

Fig. 10. Absorption line width, the full width at half maximum, as a function of temperature.
[From Burland et al. (1977a).]

is a lorentzian with full width at half maximum of 1.3 cm^{-1}, corresponding to a T_2 of 8 ps for the processes responsible for dephasing the $k \approx 0$ exciton.

It is important to compare the 15–20 ps low temperature optical T_2 we have observed in TCB with the 1 μs T_2 obtained for this exciton system from ESR line widths (Botter et al. 1978). Several points need to be made here. First, optical and magnetic resonance experiments measure different consequences of the exciton scattering process. The optical experiment is related to the electric dipole correlation function and the magnetic resonance experiment to the magnetic dipole correlation function (Gordon 1968). In principle there is no reason to believe that the two functions are affected in the same way by these scattering processes, or that they will yield identical correlation times. Second, even if this were the case, the relationship between the correlation time associated with a particular scattering process and line width or T_2 is not straightforward. One requires a model of the exciton scattering process to obtain such a relationship. Models relating optical line widths with microscopic exciton scattering processes have been proposed in the past (Toyozawa 1958, Grover and Silbey 1971, Morris and Sceats 1973, Brillante and Dissado 1976). These models all relate the optical line width to the scattering of the $k \approx 0$ exciton into other optically inaccessible k states, although the models do not yield values for the extent to which k itself is changed. The interpretation of the ESR experiments is more complicated since a direct relationship between the exciton scattering mechanism and the ESR line width has not been definitely established.

The most interesting aspect of the TCB absorption line shape is its change from lorentzian to gaussian in the region around 20 K. This behavior has been predicted theoretically (Toyozawa 1958), but the TCB system provides the first experimental observation of such a line shape change. The change corresponds to what Toyozawa has called the transition from weak to strong coupling. Its origin can best be seen by considering first the high temperature gaussian line shape. This line shape arises from the effect of a Boltzmann distribution of site energies on the energy of the electronic excited state. The line shape in this temperature region is given by the expression

$$I(E) = C \exp[-(E - E_a)^2/2D^2], \tag{44}$$

where E_a is the position of the absorption maximum and C is a constant. D is the thermally averaged value of the change in the exciton energy due to random lattice fluctuations. Note that for the gaussian lines the width cannot in any way be related to exciton scattering times.

Toyozawa (1958) has shown that eq. (44) will describe the exciton line shape provided $D \gg B$, where B is the exciton band half width (0.7 cm^{-1} for TCB). In the other extreme, where $B \gg D$, exciton transport is sufficiently rapid so that motional narrowing occurs, i.e., the exciton is not affected by the details of the local potential fluctuations. In this case the line shape should be lorentzian.

In fig. 11, we have extracted the values of D from the gaussian absorption lines. It is clear from this figure that in the region where $B \approx D$, the line changes shape from gaussian to lorentzian as predicted by theory. This change may be seen as a transition between the high temperature self-trapped exciton and the low temperature delocalized exciton, i.e. a transition from incoherent to coherent transport. Line shape measurements are always fraught with uncertainties. Small errors in baseline introduce uncertainties into widths. However, in this case there is a clean change in spectral shape with temperature and the change occurs in the temperature range predicted by theory.

Even at the lowest temperature, the dephasing time is fast, corresponding to the time required for an average exciton to travel only two or three lattice sites. This will be the case if T_2 is due to a process which results in complete loss of memory of the initial wave packet group velocity, e.g. exciton–phonon scattering in which the final state is a random exciton band state. However, this T_2 could arise from a more restricted form of scatter-

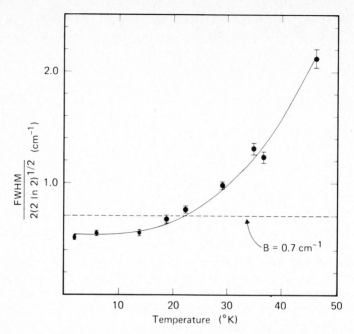

Fig. 11. The full width at half maximum (fwhm) divided by $2(2\ln 2)^{1/2}$ as a function of temperature. The ordinate is thus equal to the parameter D in eq. (44) when the line shape is gaussian, that is, above the dotted line in the figure. The dotted line locates the energy of the exciton band half width B. [From Burland et al. (1977a).]

ing, or from a pure dephasing process (T_2^*). These questions will be discussed in more detail in subsection 4.2 and alternate experimental methods will be suggested in section 5.

4.2. The relationship between ESR and optical experiments

In ESR investigations of triplet spin sublevel exciton band-to-band transitions in TCB, a dephasing time of approximately $1\,\mu s$ was obtained, independent of the band state $|k\rangle$ (Botter et al. 1978). In these studies the dephasing time was taken to imply that TCB is in the coherent limit at low temperatures. However, high resolution optical absorption measurements on the same system (Burland et al. 1977a), discussed in subsection 4.1, appeared to contradict these conclusions. In those experiments the optically accessible $k = 0$ triplet exciton band transition was found to dephase in $\sim 20\,ps$, implying a scattering rate four to five orders of magnitude faster than that determined by ESR techniques. Although some degree of coherence is indicated by the optical work, the question of the detailed nature of exciton transport in this system can only be answered by resolving the apparent conflict between the optical and ESR experiments. If scattering from an initial state $|k\rangle$ to a new state $|k'\rangle$ is fast and Markovian (Feller 1966), an exciton wave packet's group velocity fluctuates rapidly on the time scale of intermolecular motion and the transport occurs as a random walk between lattice sites. In the limit of slow scattering, the initial state $|k\rangle$ persists on a time scale long compared to intermolecular motion and transport has a wave-like component.

A possible consistent explanation for both the ESR and optical experiments can be found in a careful analysis of the relationship between exciton scattering and the spectroscopic dephasing times. (An alternative explanation involving pure dephasing will be discussed briefly at the end of this section.) Exchange of a species between two different states, or environments, causes the superposition associated with the spectral transition under observation to dephase, broadening the transition line width. This effect is well known from the classic example of the exchange of a spin between environments in which it has different resonance frequencies (Carrington and McLachlan 1967). The broadening of the observed line shape is a combined function of the exchange (scattering) rate, the rate of return to the initial state, and the difference in Larmor frequency of the spin transition in the two states.

An exciton system is a generalization of the two-state case to N states. In the former, each state forms an environment for the transition which is characterized by a unique resonance frequency. In an exciton band each state $|k\rangle$ is also characterized by a unique resonance frequency ω_0^k. If a low power applied field is used to probe the transition, then this type of system is equivalent to N independent two-level systems interacting only through exchange. In the absence of pure dephasing, dephasing of a transition in any state $|k\rangle$ is likewise a combined function of the rates of scattering to $|k'\rangle$, the time for return to $|k\rangle$, and the differences in Larmor frequency for the transition in the different band states. The form of the scattering, given by the probability distribution of final states $|k'\rangle$ following scattering from $|k\rangle$, is as important as the overall rate of leaving $|k\rangle$. Thus, the dephasing time may be quite different from the scattering time.

In the following, a two-part model of the dephasing of spectroscopic transitions by exciton scattering is presented and used to analyze the TCB triplet exciton system (Wieting and Fayer 1980). The first part employs an analytically tractable heuristic representation of an exciton band which permits direct insight into the problem. It is shown that the dephasing of optical transitions can yield the exciton scattering rate and that, in contrast, ESR dephasing times do not reflect the rate of scattering but rather are more closely related to the distribution of final states in the scattering event. In the second part the full exciton band is treated in the limit of a low power steady-state applied field and the complete exchange equations are formally solved for any distribution of final states in the scattering between band states. These results permit the quantitative calculation via numerical methods of the steady-state ESR line shape and the direct comparison of experiment with theory.

Applied to the TCB exciton system, this analysis identified the exciton scattering time with the optical dephasing time of 20 ps. The steady-state ESR experimental line shape is reproduced with excellent agreement for several different forms of the exciton scattering distribution. However, only in the case that exciton scattering occurs primarily to nearby $|k\rangle$ states is this theoretical agreement consistent with the 20 ps optical dephasing time and a reasonable number of states in the TCB exciton band. This result is in accord with independent stimulated spin echo measurements on this system which indicate that spectral diffusion occurs over a few percent of the total band width (Botter et al. 1978). However, very recent laser, ESR double resonance experiments appear to be in contradiction to both the theory and the previous stimulated spin echo experiments (Van Strien et al. 1980). These will be discussed below.

If the theoretical model presented here is correct, then a detailed description of exciton transport in TCB emerges. Exciton scattering occurs with high frequency, but each scattering event results in only a small change in the exciton wave vector. While not in the strict coherent limit, transport in this system is essentially wave-like. The only other possibility reasonably consistent with all experiments is that a pure dephasing mechanism is operative in the optical experiments only, and therefore the exciton system is even closer to the coherent limit. This scenario is discussed at the end of this section.

4.2.1. A heuristic model

In an exciton system with interactions along only one crystallographic axis (e.g. TCB), the wave vectors k lie on a one-dimensional periodic (inverse) lattice, i.e. a ring. Stochastic exciton scattering can be represented as a random walk on the k-space ring, where a change in the exciton state is represented by a step to a new (not necessarily adjacent) ring state. A fundamental theorem of random walks states that in a walk with a mean step rate k_S on a ring of N sites, the mean rate of return to the initial site is $k_B = k_S/N$, independent of the form of the walk (Lakatos-Lindenberg and Schuler 1971). Thus, an exciton initially in $|k\rangle$ which is scattered to other band states $|k'\rangle$ at rate k_S will return to $|k\rangle$ at rate k_B. This suggests a simple model for the T_1 dephasing of exciton band transitions (Wieting and Fayer 1980).

Consider the exchange of an excitation between a pair of two-level systems $|S\rangle$ and $|B\rangle$. The system $|S\rangle$ represents a single exciton band state

$|k\rangle$. The system $|B\rangle$ is a "bath" which represents the remaining $N-1$ states $|k'\rangle$ of the band which exchange population with $|k\rangle$. The transition frequency is ω_0 in $|S\rangle$ but is $\omega_0 + \Delta$ in $|B\rangle$, where Δ measures the average shift in Larmor frequency experienced during the time of residence in the bath. As discussed above, stochastic scattering from $|S\rangle$ with rate constant k_S implies that the rate constant for return from $|B\rangle$ will be $k_B = k_S/N$. Equilibrium of population between the state and the bath thus requires a bath population of N for unit population of $|S\rangle$.

We shall employ the Feynman, Vernon and Hellwarth (FVH) geometrical representation of the density matrix equation (Feynman et al. 1957). Viewed in a frame rotating (Slichter 1963) at the frequency ω of the applied field the components of a pseudospin vector $r = (r_1, r_2, r_3)$ contain linear combinations of the elements of the density matrix. The in-plane components r_1 and r_2 are equivalent to the in-plane magnetization of a spin system.

Consider the dephasing of $|S\rangle$ in the following gedanken experiment. The experiment begins with an applied pulse of sufficient power to rapidly establish an in-plane component and then the field remains off. With the field applied at $\omega = \omega_0$ and along r_1 in a frame rotating at ω, the initial conditions become $r_2^S(0) = 1$ and $r_2^B(0) = N$, with all other components equal to zero. Since the applied field is zero for $t > 0$, the in-plane components become decoupled from the out-of-plane (population) components which may then be neglected. The equations of motion are

$$\dot{r}_1^S = -k_S r_1^S + k_B r_1^B, \tag{45a}$$

$$\dot{r}_1^B = +k_S r_1^S - k_B r_1^B - \Delta r_2^B, \tag{45b}$$

$$\dot{r}_2^S = -k_S r_2^S + k_B r_2^B, \tag{45c}$$

$$\dot{r}_2^B = +\Delta r_1^B + k_S r_2^S - k_B r_2^B. \tag{45d}$$

For an exciton band $N \gg 1$, so $k_S \gg k_B$. With this condition, eqs. (45) may be solved subject to the initial conditions. We focus on the time dependence of the r_2 component of $|S\rangle$, $r_2^S(t)$. The decay of this initially prepared component yields the rate of phase loss in the state $|S\rangle$ *via* exchange. The general result is

$$r_2^S(t) = \frac{1}{D} \{ [\Delta^2(k_S^2 + \Delta^2) + k_S^3 k_B] e^{-k_S t}$$

$$+ k_S(k_S^2 + k_S k_B + \Delta^2)[(k_S + k_B)\cos\Delta t + \Delta \sin\Delta t] \}, \tag{46}$$

where $D = (k_S^2 + \Delta^2)^2 + k_S k_B(2k_S^2 + k_S k_B + 2\Delta^2)$ and $k_S \gg k_B$. Two special cases of the greatest interest are discussed below.

When the shift in Larmor frequencies Δ is much greater than the scattering rate, i.e. $\Delta \gg k_S$, eq. (46) reduces to an extremely simple form,

$$r_2^S(t) = e^{-k_S t}. \tag{47}$$

This result means that the superposition is completely dephased by the change of environment before it can return to the initial state, and the phase loss thus occurs simply with the scattering rate. This limit can be found in optical transitions where Δ is on the order of the exciton band width, typically 10^{10}–10^{12} Hz in triplet systems. Thus, optical spectra have a straightforward interpretation:

$$T_2^{opt} = 1/k_S. \tag{48}$$

If $\Delta \approx k_S$, dephasing occurs on a similar time scale.

When the scattering rate is much greater than the Larmor shift, i.e. $k_S \gg \Delta$, a more complicated expression is obtained,

$$r_2^S(t) = \cos \Delta t + \frac{\Delta}{k_S} \sin \Delta t + \frac{\Delta^2}{k_S^2} e^{-k_S t}. \tag{49}$$

The oscillatory behavior is an artifact of the oversimplified choice of a single resonance frequency $\omega_0 + \Delta$ for the bath. In order to more realistically model scattering in an exciton band, an average over a distribution of frequencies $\{\Delta\}$ should be taken, reflecting the average distribution of states $|k'\rangle$ occupied during exchange. A variety of distributions are plausible depending upon the form of scattering. A resonable choice is a gaussian function centered on the initial state, as might result from scattering only to nearby states. If this distribution ranges over only a small fraction of the band, an average over frequency yields

$$r_2^S(t) = e^{-(Et)^2/2}, \tag{50}$$

where E is the standard deviation of $\{\Delta\}$. The limit $k_S \gg \Delta$ is applicable in ESR band-to-band transitions, where the exciton band dispersion is mirrored but greatly reduced as discussed in the beginning of section 4. T_2^{ESR} cannot be obtained directly from this result. However, since the effective band width $B_E > E$, it is evident that dephasing of an ESR transition will occur on a time scale of B_E^{-1} or longer. These results suggest that T_2^{ESR} values measured in the $k_S \gg \Delta$ limit reflect the frequency spread of states involved in the exchange rather than the scattering rate between them.

4.2.2. The full exchange problem

The preceding heuristic model permits qualitative insights by reducing the problem to the elementary one of exchange between two different environments. In the following, quantitative results are derived from a treatment of the full exciton band exchange problem (Wieting and Fayer 1980). Exchange of an excitation between N independent two-level systems will be considered, where each system corresponds to a distinct exciton band state $|k\rangle$. A solution for the steady-state low power spectrum is obtained.

Again the FVH representation is employed. Since a low power coupling field is assumed, the population components r_3 are decoupled from the in-plane components and are constants proportional to the populations of the corresponding states. With an applied field at frequency ω along the r_1 axis in the rotating frame, the sum of the r_2 components of the N states yields the absorption at that frequency (Slichter 1963). The equations of motion for $|m\rangle$, with resonance frequency ω_0^m, are

$$\dot{r}_1^m = -\Delta\omega^m r_2^m - \left(\sum_{j=1}^{N}{}' k_{mj}\right) r_1^m + \sum_{j=1}^{N}{}' k_{jm} r_1^j, \tag{51a}$$

$$\dot{r}_2^m = -\omega_1 r_3^{0m} + \Delta\omega^m r_1^m - \left(\sum_{j=1}^{N}{}' k_{mj}\right) r_2^m + \sum_{j=1}^{N}{}' k_{jm} r_2^j, \tag{51b}$$

where the prime means that the $j = m$ term is excluded from the sums. The rate constant for scattering from $|m\rangle$ to $|j\rangle$ is k_{mj}, $\Delta\omega^m = \omega - \omega_0^m$, and r_3^{0m} is the steady-state value of r_3^m, proportional to the equilibrium population of $|m\rangle$. Equation (51) expresses just the mth pair of N pairs of equations. Defining the sum of rate constants in the "loss" terms as

$$k_{mm} = -\sum_{j=1}^{N}{}' k_{mj}, \tag{52}$$

the full set of $2N$ equations can be compactly written in matrix form,

$$\dot{r}_1 = -\Delta\omega\, r_2 + K^T r_1, \tag{53a}$$

$$\dot{r}_2 = -\omega_1 r_3^0 + \Delta\omega\, r_1 + K^T r_2, \tag{53b}$$

where K^T is the transpose of the scattering matrix K. The mth elements of the column vectors r_1, r_2 and r_3^0 are r_1^m, r_2^m and r_3^{0m}, respectively, and the matrix $\Delta\omega$ is diagonal with mth diagonal element $\Delta\omega_0^m$.

The central term in these equations is the scattering matrix K, whose elements k_{mj} embody a complete description of the exciton scattering process. For each state $|m\rangle$ the associated diagonal element k_{mm} measures the total rate of loss from $|m\rangle$ via all scattering paths. From the discussion surrounding eq. (48), we associate this total loss rate with the optical dephasing time of the state in question, i.e. $k_{mm} = -1/T_2^{\text{opt}}(m)$.

The specific form of any given scattering process determines the value of the off-diagonal elements. Exciton scattering may occur exclusively to states which are nearly isoenergetic with the initial state or to a broad distribution. At one extreme is the limit of scattering to nearest neighbor $|k\rangle$ states only. In this situation $k_{mj} = \frac{1}{2}k_S$ for $j = m \pm 1$ and zero for all other

$j \neq m$. At the opposite extreme scattering to all states is equally probable, i.e. $k_{mj} = k_S/(N - 1)$ for all $j \neq m$.

Scattering in real exciton systems probably lies intermediate to these two extremes. A useful model employs a distribution in which the scattering rate falls off exponentially as the difference in wave vector between final and initial state increases, i.e. $k_{mj} = \kappa \exp[-D(m, j)/R]$, where κ is a normalized rate constant chosen so that scattering into any $|j\rangle$ equals scattering out of $|j\rangle$. Since an exciton band has periodic boundary conditions, the wave vector difference $D(m, j) = |m - j|$ when $|m - j| \leq N/2$, but $D(m, j) = N - |m - j|$ when $|m - j| > N/2$. The parameter R is equal to the "average range" of a scattering event and characterizes the widths of the exponential distribution, which can be continuously varied between the two limits discussed above. Thus, while arbitrary, this form provides an instructive means of considering intermediate situations.

Equation (53) can be formally solved in steady state for r_2 to yield

$$r_2(\omega) = \omega_1(\Delta\omega + \boldsymbol{K}^{\mathrm{T}}\Delta\omega^{-1}\boldsymbol{K}^{\mathrm{T}})^{-1}(\boldsymbol{K}^{\mathrm{T}}\Delta\omega^{-1})r_3^0. \tag{54}$$

The absorption spectrum may then be calculated by evaluating the sum of the elements of eq. (54) at a range of frequencies.

4.2.3. Application to triplet excitons in 1,2,4,5-tetrachlorobenzene

In order to apply eq. (54) to a calculation of the TCB triplet exciton ESR spectrum, the elements of $\Delta\omega$, r_0^3 and $\boldsymbol{K}$ must be evaluated. It is first necessary to know the number of states which compose the exciton band, for the value of N sets the dimensionality of the matrix calculation. This number is equivalent to the mean number of molecules lying in a one-dimensional chain segment over which the excited states are not severely disrupted, suggesting that N will be of the order of the inverse number density of chemical impurities or major lattice defects. Since residual major impurities may remain in this system in concentrations $\leq 10^{-3}$–10^{-4} mole/mole, a rough estimate yields $N \sim 10^3$–10^4 (Dlott et al. 1978). N is also employed in the one-dimensional nearest neighbor exciton band dispersion along with the effective band width of the transition to evaluate the resonance frequencies ω_0^m of the states $|m\rangle$, which are used in $\Delta\omega$.

The population vector r_3^0 may be evaluated from the Boltzmann distribution in the band (Francis and Harris 1971, Dlott et al. 1978). Only this term contains the full band width 4β and the experimental temperature.

Finally, the scattering rate and its form must be chosen and incorporated into $\boldsymbol{K}$. As shown above in subsection 4.2.2, for this model the rate constant for leaving $|m\rangle$ is equal to the inverse of the dephasing time of the optical transition. Only $|k = 0\rangle$ is optically accessible from the ground state with $T_2^{\mathrm{opt}} = 20$ ps, but, since the ESR dephasing times are independent of the band state, it seems plausible to assume the same behavior in the optical case. Thus, it remains only to choose a specific form of the scattering distribution, and the steady-state ESR line shape may be calculated for TCB. However, to perform the full numerical solution, it would be necessary to solve a system of simultaneous equations of order $N \sim 10^3$–10^4. While possible in principle, it is more practical (and less costly) to perform the calculation for smaller values of N and to extrapolate the observed trends.

There are thus two variable parameters in the line shape calculation: (i) the number of band states, N, and (ii) the ratio of the scattering rate to the effective band width, k_S/B_E. Given a specific choice of (i) and (ii), the ESR spectrum is calculated on a frequency scale calibrated in units of B_E. Comparison of the calculated to the experimental spectrum provides in effect a measurement of this quantity and thereby yields the absolute scattering rate used in the calculation. The TCB spectrum is calculated over a range of N, optimizing the agreement to experiment for each choice of N by varying the value of k_S/B_E. The absolute scattering rates k_S versus N are plotted.

The optically detected ESR $|D|+|E|$ transition line shape in TCB was calculated for several different scattering models in the manner discussed above. Figure 12 displays the experimental spectrum, obtained at 3.2 K (Francis and Harris 1971), along with a pair of theoretically calculated curves which are optimized to bracket the experimental data. The bracketing procedure allows us to place an error bar on each calculation. These

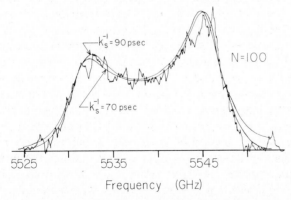

Fig. 12. The experimental ESR exciton band-to-band $|D|+|E|$ transition line shape for 1,2,4,5-tetrachlorobenzene at 3.2 K (Francis and Harris 1971). The theoretical curves were calculated from eq. (54) using the nearest neighbor scattering distribution, with $N = 100$. For this choice, best agreement was obtained with exciton scattering times $k_S^{-1} = 70$ to 90 ps (with the effective band width $B_E = 18$ MHz). This figure illustrates that the experimental data can be reproduced using exciton scattering times in the picosecond range. [From Wieting and Fayer (1980).]

curves were calculated using the nearest neighbor ($\Delta k = \pm 1$) scattering limit, with $N = 100$ and $k_S^{-1} = 70$–90 ps. The agreement is very good and typical of the fit obtained in all of the calculations. Figure 12 illustrates an important point. *It is possible to reproduce the experimental data using exciton scattering times in the picosecond range when exchange is properly included.*

A variety of scattering distributions was considered. In the order of increasing "average range" of scattering, the models employed were: (i) nearest neighbor, (ii) exponential, $R = 5\%$ of the exciton band; (iii) exponential, $R = 20\%$ of the exciton band; and (iv) Markovian, equal probability of scattering to any state in the band. For each of these models a set of calculations was performed in which k_S/B_E was optimized for $N = 20$, 50 and 100. The relationships between these optimized scattering rates and the number of band states is shown in fig. 13, a log–log plot of k_S^{-1} versus N. In each case the data fit a straight line with good precision, permitting extrapolation to large N. Consider the steepest line, labeled NN (for Nearest Neighbor). For a nearest neighbor scattering distribution, $N = 20$ band states requires a 2 ns scattering rate to fit the experimental ESR data, while $N = 50$ requires 320 ps. As discussed above the high resolution optical spectrum implies a 20 ps scattering rate for TCB at these temperatures if pure dephasing is absent. Extrapolating the nearest neighbor scattering requires $N \approx 200$, a number slightly below the possible range. Extrapolating the exponential $R = 5\%$ distribution to the 20 ps scattering line yields $N \approx 40\,000$. This value of N falls in the range consistent with experiment. In contrast, the longer ranged $R = 20\%$ exponential distribution requires an enormously greater number of band states, $N \approx 10^{10}$, a value which is orders of magnitude too large to be consistent with experiment. Finally, in the Markovian scattering limit it is impossible altogether to reconcile the ESR spectrum and a 20 ps scattering rate from the optical spectrum.

The results of these calculations suggest that TCB triplet exciton scattering at liquid helium temperatures occurs primarily to nearly isoenergetic states. Within the context of this model this conclusion is necessary to reconcile the optical and ESR experiments. Independent evidence also suggests this scattering behavior. Schmidt and co-workers (Botter et al. 1978), using stimulated spin echo measurements, observed that frequency instabilities of different sizes occur at quite different rates. They inferred that spectral diffusion due to scattering of the exciton $|k\rangle$ states is limited to a few hundred kHz in a total line width of about 10 MHz for the $|D| - |E|$ transition, or a few percent of the exciton band width. Thus, the optical, ESR, and stimulated echo experiments combined with the exchange calculations presented here provide a consistent picture of exciton scattering in the TCB system.

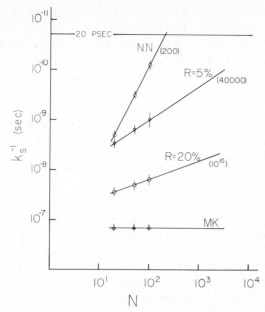

Fig. 13. The dependence of the exciton scattering rate k_S on the number of exciton band states N and on the distribution of final states in the scattering process, calculated via eq. (54) for triplet excitons in 1,2,4,5-tetrachlorobenzene at 3.2 K. The four scattering distributions employed, in order of increasing width of the scattering distribution, were: nearest neighbor scattering (NN); exponential ($R = 5\%$ of band); exponential ($R = 20\%$ of band); and Markovian (MK), equal probability of scattering to any final band state. For each model a linear log–log relationship is observed between k_S and N. The number in parentheses near each line is the value of N obtained from an extrapolation to $k_S^{-1} = 20$ ps, the exciton scattering time obtained from optical absorption experimental data. The error bars reflect the uncertainty in the optimization to the experimental ESR spectrum, e.g. see fig. 12. [From Wieting and Fayer (1980).]

In summary, this analysis may offer a resolution of the apparent conflict between the optical and ESR dephasing times which differ by orders of magnitude in TCB. Furthermore, it permits a semi-quantitative deter-mination of the exciton scattering distribution in this system assuming pure dephasing of the optical transition is absent (see below). These results show that exciton scattering is fast but that it occurs primarily to a small region of k space surrounding the initial state. In this situation the exciton group velocity undergoes small rapid fluctuations but is basically constant. Thus, exciton migration in TCB at low temperatures, while not strictly coherent, is nevertheless essentially wave-like.

The entire development above is based on the absence of pure dephasing in the optical absorption line. In terms of the possibility of coherent exciton transport in the TCB system, this is the worst case situation, i.e. the k-to-k scattering rate is as large as possible. However, some or all of the optical line width could be due to a pure dephasing process. It is possible that a mechanism exists which causes the optical transition energy to fluctuate without inducing k scattering. This would contribute to the width of the optical absorption line, and therefore the ~ 20 ps T_2 would not be the scattering rate. If such a mechanism exists, it is likely that the three triplet spin sublevel bands would undergo energy fluctuations in unison. Thus the optical pure dephasing would not manifest itself in the ESR band-to-band spectrum.

The net result could be that the scattering rate is considerably slower than 20 ps. It is clear from the detailed analysis presented above, that the ESR spectrum can be obtained regardless of the scattering rate. If the scattering rate is considerably slower than 20 ps, to fit the data, the scattering must occur to a broader range of final k states, i.e. a larger value of R in fig. 13.

The stimulated echo experiments mentioned above (Botter et al. 1978) would seem to indicate that k scattering is restricted to a very small region of k space and therefore support the association of the optical line width with a 20 ps scattering time. However, very recent laser ESR double resonance experiments, also by Schmidt and co-workers (Van Strien et al. 1980), are in contradiction to the earlier stimulated echo results. In these experiments a narrow band pulsed laser is tuned to the $k = 0$ optical absorption. ESR spectra are then recorded at various delay times after optical excitation. At short times, population is observed in the ESR spectrum only around $k = 0$. However, at later times, population increases in other parts of the band *uniformly across the band* on a microsecond time scale. Although there are still questions pertaining to the interpretation of these recent experiments, the Markovian form of scattering implied by the experiments is only consistent with the ESR data if the k-to-k scattering time is very long (see fig. 13), and if the optical line width is primarily due to a pure dephasing process.

The net result is that exciton transport in TCB is essentially wave-like, i.e. coherent. For the 20 ps scattering time, the analysis presented above demonstrates that scattering must be restricted to a very narrow region of k space surrounding the initial k state. This results in small, rapid fluctuations in the group velocity. In the other extreme, scattering is very slow, $\sim 1\ \mu$s, but occurs with equal probability to any final state. The change in optical absorption spectral shape (subsection 4.1) from gaussian at high temperature to lorentzian at low temperature adds additional strength to the argument that TCB triplet excitons are coherent, regardless of which scattering mechanism is correct. In terms of macroscopic transport, the end result is the same, long range wave-like transport. However, what is needed at this point, is a detailed and unified theoretical treatment of the optical and ESR experimental observables in terms of microscopic models of scattering and pure dephasing processes.

Since it is established that exciton transport in TCB is essentially coherent, the trapping models and experiments of section 3 can be reconsidered. Transport is basically one-dimensional with a small degree of cross-chain stepping which is eight orders of magnitude slower than on-chain site-to-site transfer. In "pure" crystals of TCB, intrinsic scattering impurities such as monodeutero-TCB determine the time dependence of exciton trapping. In quasi-one-dimensional systems, impurity scattering can turn microscopically coherent transport into macroscopically diffusive transport.

5. New experimental approaches

In this section, two new approaches to the study of exciton transport and exciton coherence, the picosecond transient grating method (Nelson and Fayer 1980, Nelson et al. 1979, Salcedo et al. 1978) and picosecond time scale photon echo experiments (Wiersma 1981, Cooper et al. 1980) will be discussed. Both techniques have been employed in the study of mixed molecular crystals but have not been applied to the study of exciton coherence in pure crystals.

5.1. The picosecond transient grating experiment in the study of exciton coherence

The transient grating method is a unique experimental technique for the investigation of the energy transport properties of solids. It will permit exciton migration to be investigated at the microscopic level by providing an accurate time scale and by introducing into the bulk crystal an accurate distance scale against which exciton migration can be measured. The method involves the diffraction of a probe pulse from a grating pattern induced in the sample. The grating pattern is generated by the interference between two coherently related optical beams in the following manner. A coherent pulse of light which has a duration in the picosecond time range is split in two. The resulting beams are arranged to have a known angle between them and to intersect in the sample (fig. 14). The path lengths traversed by the two pulses are adjusted so that the pulses arrive at the point of intersection in the sample simultaneously. Interference between the two coherently related exciting beams creates an optical interference pattern in the sample such that the intensity of light varies sinusoidally in

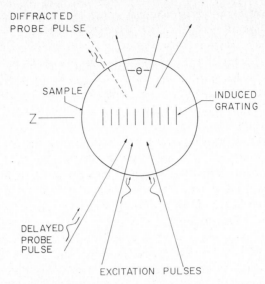

Fig. 14. Schematic illustration of the transient grating experiment. Two coherently related excitation beams cross in the sample and produce an optical interference pattern. Optical absorption by the sample results in a spatially sinusoidally varying concentration of excited states. This acts as a diffraction grating. A variably delayed probe pulse is diffracted from the excited state grating. Exciton transport which destroys the grating is monitored by the time-dependent intensity of the diffracted light. [From Nelson and Fayer (1980).]

the beam overlap region. The interference fringe spacing is determined by the angle between the beams and by the wave length of the light.

When the frequency of the exciting light coincides with an absorption band of the sample molecular crystal, excitations are produced. These excitations will have the same spatial distribution as the sinusoidal optical interference pattern, i.e. there will be a continuous oscillatory variation in the concentration of excited states. First, consider the case in which exciton migration does not occur, i.e. the excited states produced are fixed in their initial positions and only vanish via decay to the ground state. After a suitable time delay, a third "probe" pulse (which may differ in wave length from the exciting pulses) is directed into the sample along a third path. This probe pulse will experience an inhomogeneous optical medium in that the alternating regions of high and low concentration of excited states will have different complex indices of refraction. Thus the probe pulse encounters what amounts to a *diffraction grating* which causes it to diffract into one or more orders (see fig. 14). The diffracted beam leaves the sample along a unique direction determined by the relative impinging beam angles and has an intensity proportional to the square of the difference between the concentration of excitations at the grating peaks and the concentration at grating nulls. As the excitations which form the induced grating decay to the ground state, the difference in the concentrations and therefore in the indices of refraction of the grating peaks and nulls will be reduced. Since the intensity of the diffracted probe pulse depends on the square of the peak–null difference in excitation concentration, the diffracted beam will decrease in intensity as the time between excitation and probing becomes longer.

Now consider the effect on the diffracted signal beam produced by excitations which are mobile and free to migrate through the crystal. Exciton migration will carry excitations from areas of high exciton concentration, grating peaks, to areas of low concentration, grating nulls. Thus the exciton motion itself will fill in the grating nulls and reduce the peak heights (see fig. 15). Destruction of the grating pattern by redistribution of the excitations forming the grating at time $t = 0$, will lead to a decrease in the intensity of the diffracted probe pulse as the probe delay time is increased. Thus the time dependence of the loss of the grating pattern is determined both by the life time of the excited state and by the time dependence of exciton migration. If redistribution of the excitations in the pattern occurs on a time scale comparable to or faster than the excited state life time, then the time dependence of the intensity of the diffracted probe pulse will yield information on the time dependence of the exciton migration.

First consider the incoherent limit where exciton transport is diffusive. Taking the sinusoidal distribution of the grating pattern to occur along the x direction, at time $t = 0$ the spatial distribution of excitons, for delta function in duration excitation, is

$$N(x, 0) = \tfrac{1}{2}(1 + \cos \Delta x), \tag{55}$$

where

$$\Delta = 2\pi/d. \tag{56}$$

d is the grating fringe spacing given by

$$d = \frac{\lambda}{2 \sin \theta/2}, \tag{57}$$

where λ is the wavelength of the excitation beams and θ is the angle between the excitation beams. Only motion along the grating axis results in a change in signal, therefore the following one-dimensional diffusion equation with decay can be employed:

$$\frac{\partial N(x, t)}{\partial t} = \gamma \frac{\partial^2 N(x, t)}{\partial x^2} - \frac{N(x, t)}{\tau}. \tag{58}$$

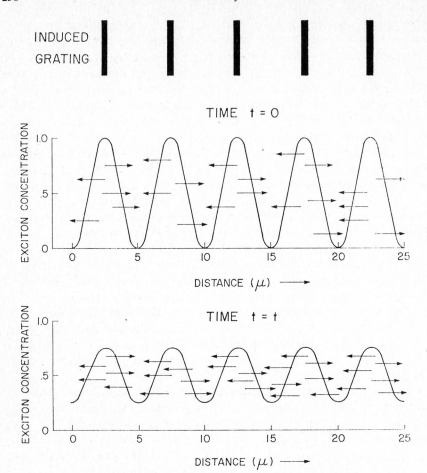

Fig. 15. A schematic illustration of the initial exciton distribution which forms the grating ($t = 0$) and the effect of transport on the grating pattern ($t > 0$). As the difference in peak–null exciton concentration decreases due to transport, the diffracted signal decreases.

$N(x, t)$ is the concentration of excitons, τ is the excited state life time and γ is the diffusion coefficient. The Green's function solution for impulse excitation of this equation at $t = t_0$, i.e. for

$$N_0(x, t_0) = \delta(x - x_0), \tag{59}$$

is

$$g(x, t; x_0, t_0) = \left(\frac{1}{4\pi\gamma(t - t_0)}\right)^{1/2} \exp\left(-\frac{(x - x_0)^2}{4\gamma(t - t_0)} - \frac{t - t_0}{\tau}\right). \tag{60}$$

For the initial exciton distribution given by eq. (55), the time-dependent concentration of excitations in the grating pattern is given by

$$N(x, t) = \int_{-\infty}^{\infty} \tfrac{1}{2}(1 + \cos \Delta x_0) \frac{1}{4\pi\gamma t} \exp\left(-\frac{(x - x_0)^2}{4\gamma t} - \frac{t}{\tau}\right) dx_0. \tag{61}$$

Evaluation of this integral yields

$$N(x, t) = \tfrac{1}{2} e^{-t/\tau}(1 + e^{-\Delta^2\gamma t} \cos \Delta x). \tag{62}$$

Note that at $t = 0$ this reduces to the initial condition, eq. (55), and the condition $\tau \gg t \gg \Delta^2\gamma$ gives $N(x, t) = 0.5 \exp(-t/\tau)$. The excitations are uniformly distributed and decaying with the excited state life time. Once the excitations are uniformly distributed the signal in the grating experiment is zero.

The time-dependent signal is proportional to the square of the difference, D, in the exciton concentrations of the grating peaks and nulls:

$$\begin{aligned} D(t) &= N(0, t) - N(d/2, t) \\ &= e^{-(\Delta^2\gamma + 1/\tau)t}. \end{aligned} \tag{63}$$

Therefore the signal is

$$S(t) \propto D(t)^2 = e^{-2(\Delta^2\gamma + 1/\tau)t}. \tag{64}$$

For diffusive processes, the signal decays exponentially. It will be shown below that this is in contrast to the coherent limit where the decay should be highly non-exponential.

Equation (64) demonstrates that the transient grating method can be used to obtain the diffusion constant γ directly given the excited state life time τ and the fringe spacing d. By plotting the grating signal decay constant $K = 2(\Delta^2\gamma + 1/\tau)$ versus Δ^2, the diffusion constant γ can be obtained from the slope and the intercept is $2/\tau$. This provides a rigorous test of the method in the diffusive limit.

Now consider the coherent limit in which exciton transport is wave-like and characterized by well-defined group velocities. The exciton population making up $N(x, 0)$, eq. (55), could in general be composed of any distribution of exciton k states. However, if the time for dressing of the bare exciton is short relative to the time required for the exciton to travel any significant distance, then thermal equilibrium with the lattice can be established without changing the initial spatial distribution of excitons. Theoretical calculations indicate that this is indeed the case (Munn 1974). In what follows, only the case in which there is a thermal equilibrium distribution of k states will be considered. Thus, the contribution of the kth exciton state to $N(x, 0)$ is

$$N^k(x, 0) = \tfrac{1}{2}(\cos \Delta x + 1)P(k), \tag{65}$$

where $P(k)$ is given by eq. (16). If vibrational relaxation and dressing do not result in a thermal distribution or if the band is selectively excited around $k = 0$, then the appropriate distribution function $P(k)$ is employed.

For $t \neq 0$, the microcanonical ensemble of excitons comprising $N^k(x, t)$ is moving with velocity $V_g(k)$,

$$N^k(x, t) = \tfrac{1}{2}\{\cos \Delta[x + V_g(k)t] + 1\}P(k), \tag{66}$$

where $V_g(k)$ is given by eq. (12). Then the time-dependent distribution of excitons comprising the grating pattern is obtained by summing the time-dependent distribution of each k ensemble, i.e.

$$N(x, t) = \sum_k N^k(x, t)$$

$$= \sum_k \frac{1}{2} \frac{\{\cos \Delta[x + t(2\beta a/\hbar) \sin ka] + 1\} e^{-y \cos ka}}{\pi I_0(y)}. \tag{67}$$

If there are a sufficient number of closely spaced k states the sum can be converted into an integral, and including decay due to radiative and nonradiative deactivation of the excited state,

$$N(x, t) = e^{-t/\tau} \left(\frac{1}{2} + \frac{1}{4\pi I_0(y)} \int_{-\pi}^{\pi} \cos \Delta(x + t\alpha \sin \theta) e^{-y \cos \theta} \, d\theta \right), \tag{68}$$

$$\alpha = 2\beta a/\hbar, \qquad \theta = ka, \qquad y = 2\beta/KT, \qquad \tau = \text{exciton life time.}$$

The time dependence of the signal, $S(t)$, observed in the transient grating experiment is proportional to the square of the difference, D, in the number of excitations at the peak and the null of the fringe pattern:

$$S(t) \propto D(t)^2 = [N(0, t) - N(d/2, t)]^2$$

$$= \left(e^{-t/\tau}[2\pi I_0(y)]^{-1} \int_{-\pi}^{\pi} \cos(\Delta t\alpha \sin \theta) e^{-y \cos \theta} \, d\theta \right)^2. \tag{69}$$

The integral can be evaluated by expanding $\cos(\Delta t\alpha \sin \theta)$ in powers of its argument. This results in a series in even powers of $\sin \theta$ multiplied by $\exp(-y \cos \theta)$. The integral of this series involves the sum of the integer order modified Bessel functions. Thus the difference, D, is given by

$$D(t) = e^{-t/\tau} \left(\frac{1}{2\pi I_0(y)} \sum_{n=0} (-1)^n \frac{(\Delta t\alpha)^{2n}}{(2n)!} \frac{2\pi^{1/2}\Gamma(n + \tfrac{1}{2})I_n(y)}{(\tfrac{1}{2}y)^n} \right). \tag{70}$$

The $I_n(y)$ are the modified Bessel functions and the other parameters are defined above. Although this is an infinite series, it converges quite rapidly. Figure 16 shows a plot of D^2, which is proportional to the observed signal, versus time for several sets of physical parameters.

Expression (70) for the time-dependent signal in the transient grating experiment can be directly related to the rate of exciton transport. The first

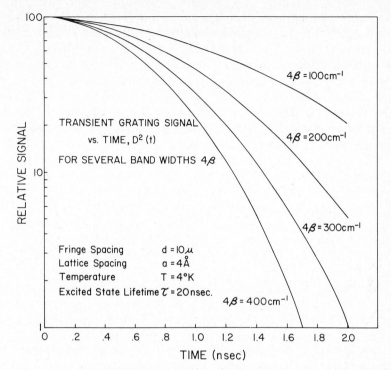

Fig. 16. Calculated transient grating decays for coherent exciton transport using several different band widths, 4β. The decays are highly non-exponential. Incoherent exciton transport should give rise to exponential decays.

term in the series is 1. The second term is

$$\frac{-(\Delta t)^2}{2} \frac{\alpha^2 I_1(y)}{y I_0(y)} = \frac{-(\Delta t)^2}{2} \frac{\alpha^2}{y} \frac{\int \sin^2 \theta\, e^{-y \cos \theta}\, d\theta}{Z(t)}. \tag{71}$$

$\alpha^2 \sin^2 \theta$ is the square of the group velocity $[V_g(k)]^2$, and $Z(t)$ is the partition function. So the second term in the series involves the thermal average value of the square of the group velocity, i.e.

$$\frac{-(\Delta t)^2}{2} \langle V_g^2(T) \rangle. \tag{72}$$

In an analogous manner the other terms can be identified with the average value of the even powers of the group velocity. Thus expression (70) for the transient grating signal can be written as

$$D(t) = \left(e^{-t/\tau} \sum_{n=0} \frac{(-1)^n (\Delta t)^{2n}}{(2n)!} \langle V_g^{2n}(T) \rangle \right). \tag{73}$$

From this expression, the direct connection between the observable in the grating experiment and the coherent nature of exciton transport can be seen.

The results of the above calculations demonstrate that the observable in a transient grating experiment can be directly related to the rate of exciton transport in the incoherent and coherent limits. In the incoherent limit, diffusive transport will produce an exponential decay of the grating signal. From the decay constant the exciton diffusion constant can be obtained. A test of the method involves varying the grating wave vector $\Delta = 2\pi/d$. The rate of transport is obtained along the direction of the grating wave vector, so spatial anisotropy in the rate of exciton transport can be examined. In the coherent limit, the signal decay in the grating experiment is predicted to be highly non-exponential and related to the thermal average group velocity. The difference in the functional form of the decay for coherent versus incoherent transport provides a useful test of the mode of transport.

Recently, Kenkre (1980) has applied the Generalized Master Equation approach in a detailed calculation of the effects of exciton migration on the transient grating observable for the incoherent and coherent limits as well as intermediate situations. This calculation involved the assumption that the initial density matrix was site diagonal. Kenkre also found that in the incoherent limit the signal decay is exponential and that in the coherent limit the decay is highly non-exponential. However, his results in the coherent limit have a different functional form than the results presented above. The differences may arise from the different choices of initial conditions.

5.2. Picosecond photon echo experiments on optical dimers

The optical homogeneous line shape can provide information on basic processes affecting an electronic excitation. However, inhomogeneous broadening can frequently mask the homogeneous line. In addition, it is not possible to distinguish T_2^* (pure dephasing) processes from T_1 (life time) processes. Application of optical coherence experiments such as photon echoes and stimulated photon echoes can be used to remove the effects of inhomogeneous broadening and distinguish T_1 effects, such as spectral diffusion, from T_2^* processes.

Until recently picosecond optical coherence experiments on large molecules have examined the states of isolated molecules in dilute mixed crystals. In recent experiments we have examined pentacene dimers in concentrated pentacene in p-terphenyl crystals. In this crystal, some of the wide variety of possible types of nearest neighbor dimers have large spectral shifts from the monomer $S_0 \to S_1$ origin due to differences in Van der Waals interactions and resonant dipole–dipole interactions which delocalize the excitation. The dipole interaction between the pentacene monomer components of the dimer splits the S_1 state into a doublet. Take the ground state of the system to be $|00\rangle$. $|10\rangle$ and $|01\rangle$ are the two states in which, respectively, the first molecule and the second molecule are excited. There is a fourth state of the system, the doubly excited state, $|11\rangle$. The pair eigenstates of the system including the dipolar interactions for monomers having identical energies are

$$|00\rangle, \tag{74a}$$

$$|+\rangle = (1/\sqrt{2})(|10\rangle + |01\rangle), \tag{74b}$$

$$|-\rangle = (1/\sqrt{2})(|01\rangle - |10\rangle), \tag{74c}$$

$$|11\rangle. \tag{74d}$$

The energies of these states are given by

$$E_{00} = 0, \tag{75a}$$

$$E_+ = E + \beta - D, \tag{75b}$$

$$E_- = E - \beta - D, \tag{75c}$$

$$E_{11} = 2E - D_{11}. \tag{75d}$$

E is the monomer excitation energy. In the pentacene in p-terphenyl host there are four well-defined site energies for pentacene, arising from the four symmetry inequivalent positions in the host unit cell. Thus some nearest neighbor dimers arising from translationally equivalent pairs are composed of monomers with degenerate electronic states. Equations (74) and (75) are correct for these dimer states. However, many of the monomer pairs have $E_{10} \neq E_{01}$. Even in this situation, for some molecular pairs, the dipolar interaction is so large that the excited state is nonetheless basically delocalized over the two molecules. The parameters D and D_{11} account for the differences in crystal shift of the delocalized dimer and the doubly excited pair as compared to the crystal shift of an isolated monomer which is absorbed into E, the monomer energy.

In fig. 17 the transmission spectrum of the spectral region to the red of the pentacene in p-terphenyl origin is shown. The sample is so highly concentrated that the monomer peaks are 100% absorbing over a broad spectral region. In a nonsaturated monomer spectrum, the monomer peaks are $\sim 1\,\text{cm}^{-1}$ wide (Olson and Fayer 1980). In this spectrum (fig. 17) five peaks are clearly visible to the red of the monomer origin. These are labeled R_1 through R_5. R_1 and R_2 appear as shoulders on the monomer spectrum while R_3, R_4 and R_5 are well resolved.

The dimer spectra can be calculated approximately using the p-terphenyl crystal structure and an approximate atom–atom potential calculation to

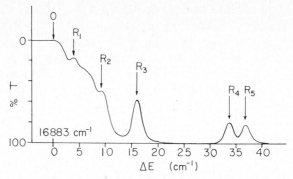

Fig. 17. Pentacene in *p*-terphenyl transmission spectrum showing the region to the red of the pentacene monomer line in a highly concentrated sample. The monomer absorption is so great that it appears as a very broad, 0% transmitting peak. The lines labeled R_1 through R_5 arise from pairs of pentacene molecules, i.e. they are delocalized dimer states. Photon echo experiments were performed on R_3, R_4 and R_5.

obtain the orientations of the pentacene molecules in the four crystallographic sites in the *p*-terphenyl lattice. These calculations demonstrate that most red shifted dimer lines have small splittings of less than 2 cm^{-1}. These will be obscured under the intense monomer peak. Attention must be focused on the red side of the monomer peak as intense absorption by the monomer phonon side band dominates the spectrum to the blue of the monomer peak. For a number of the different crystallographic dimer pairs, the dipolar splittings are large enough to produce the observed peaks R_1 through R_5. An exact calculation of the spectrum is not possible for two reasons. One, the exact orientations of the pentacenes in the lattice are unknown. Two, the differences between monomer and dimer Van der Waals interactions with the host lattice (*D* terms) are unknown. Although the spectrum cannot be calculated exactly, the observed peaks, R_1 through R_5, are entirely consistent with good estimates of the energies of the red shifted dimer states.

In addition to the energetics indicating that R_1 through R_5 are delocalized dimer states, the concentrations are correct. By examining the integrated absorption of the dimer lines their concentrations can be calculated. It is found that within experimental uncertainties the concentrations correspond to the square of the monomer concentration, as is to be expected for dimers, over a wide range of concentrations. Figure 18 is a simplified energy level diagram showing the energy of a monomer and the pair of states $|+\rangle$ and $|-\rangle$ arising from one crystallographic type of dimer. The dimer states are truly delocalized entities. They have been referred to as "mini-excitons" as they are the smallest multimolecule system which can exhibit delocalization.

There have been a number of experiments which examine spin

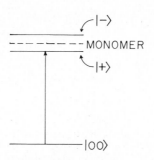

Fig. 18. A schematic dimer energy level diagram for one of the many crystallographically distinct pentacene dimers. The delocalized dimer states are split from the monomer state due to dipole–dipole interactions and changes in Van der Waals interactions. The photon echo experiments are performed on the red dimer state. The blue state is obscured by the highly absorbing monomer phonon side band.

coherence in triplet dimer systems. However, as discussed in subsection 4.2, it can be difficult to relate spin coherence measurements to the basic questions of optical coherence times. By performing photon echo experiments on the optical dimer states described above, we have been able to unambiguously measure the optical coherence time associated with a delocalized state.

5.2.1. Experimental

In working with molecular systems, the fast decay times necessitate the use of high energy, tunable picosecond light pulses. These pulses are provided by a synchronously pumped, mode locked and cavity dumped dye laser driven by the frequency doubled output of a cw pumped, acousto-optically mode locked and q-switched Nd:YAG laser. This dye laser system has been described elsewhere (Cooper et al. 1980) and provides a high repetition rate (0.4 kHz), stable source of 30 ps, 10 μJ pulses.

The dye laser output pulses pass through an attenuator and a beam splitter and are directed into a motorized optical delay line to form the desired pulse sequence (fig. 19). The beams are recombined in the appropriate intensity ratio, made collinear and sent to the sample in the liquid He Dewar. The echo signal which emerges from the sample is crossed with a single, 80 ps IR pulse from the YAG laser in an angle-tuned, type II ammonium dihydrogen phosphate (ADP) sum generation crystal. The resulting sum pulse is wave length and spatially filtered and detected by a

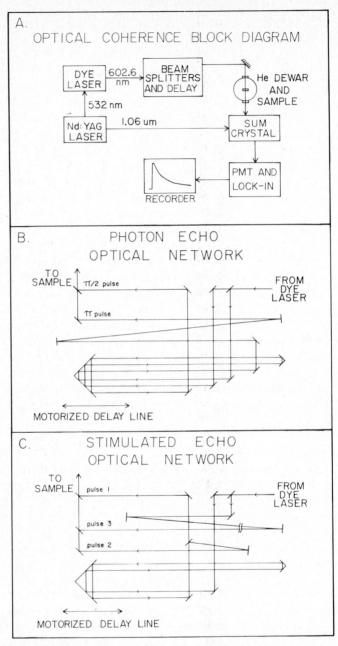

Fig. 19. (a) Block diagram of optical coherence experiments. (b) Arrangement of beam splitters and delay line to perform the photon echo decay experiment. (c) Arrangement of beam splitters and delay line to perform the stimulated echo decay experiment. [From Cooper et al. (1980).]

cooled EMI 6256B phototube and lock-in amplifier. The signal provides the
y-axis input to an *XY* recorder. The *x*-axis input is derived from a voltage
proportional to the optical delay line position. This method of detection
provides temporal, spatial and frequency separation of the signal from the
excitation pulses giving extremely good signal to noise ratio.

Echo decay data are generated by running the motorized optical delay
line. For the photon echo experiments, the π pulse traverses the round-trip
delay line path once, whereas the $\pi/2$ pulse makes two round trips, thus
ensuring that, for a given delay line travel, the $\pi/2$ pulse receives exactly
twice the temporal displacement of the π pulse (fig. 19b). For stimulated
echo experiments, one beam is double-passed down the delay line before
being separated into the first two pulses of the sequence. The second beam
is delayed a fixed amount external to the delay line to form the third pulse
of the sequence (fig. 19c). Echo decay curves are obtained several times for
each crystal to verify reproducibility. Often a set of echo decays is
repeated after $\sim$12 h to ensure long term reproducibility of the results.

The dye laser was tuned to the absorption maxima of the R_3, R_4 and R_5
lines and photon echo decay curves were recorded. The R_1 and R_2 lines are
not sufficiently well resolved from the monomer peak to obtain unam-
biguous results. The spectral spread of the dye laser and the Rabi
frequency, ω_1, are sufficiently small to assure that only the selected line
was excited in a given experiment. This permits the system to be analyzed
as a two-level system, since other transitions, including the transition from
the delocalized state to the doubly excited $|11\rangle$ state, are far off resonance.
The life times of the dimer lines R_3, R_4 and R_5 were obtained by selectively
exciting each line and measuring its fluorescence decay time using a fast
photomultiplier tube (rise time <2 ns) and a computer interfaced Tektronix
R7912 transient digitizer.

The pentacene in *p*-terphenyl crystals are immersed in liquid helium and
the temperature of the samples is controlled by the rate of pumping on the
liquid helium. The temperature was measured using a calibrated digital
manometer.

5.2.2. Results

Table 5 gives the results of the photon echo decay (T_2) and fluorescence
decay (T_1) measurements on the three dimer lines R_3, R_4 and R_5 in addition
to the pure dephasing times T_2^*. T_2^* is obtained from

$$\frac{1}{T_2} = \frac{1}{T_2^*} + \frac{1}{2T_1}. \tag{76}$$

The life time of the pentacene monomer is 24 ns. The life times of the three
crystallographically distinct delocalized dimer states are all somewhat
shorter than this and differ somewhat among themselves. This is not

Table 5
Dimer results.

Dimer lines	Wavelength (Å)	Life time T_1 (ns)	Echo decay time T_2 (ns)	Pure dephasing time T_2^* (ns)
R_3	5928	19.5	32	178
R_4	5934	16.5	22	67
R_5	5935	15.5	14	26

unreasonable. Both the radiative and nonradiative rates for decay to the ground state of a delocalized dimer state can be different from the monomer rates. The transition dipole of a dimer state is the appropriate vector sum of transition dipoles of the constituent monomers. The electronic excitation–phonon coupling for a delocalized dimer state also differs from that of a monomer state. This can affect the rate of vibrational relaxation to the ground state.

The values of T_2 obtained from the photon echo decay measurements in table 5 are all significantly shorter than $2T_1$, but all have $T_2 > 10$ ns. The corresponding homogeneous line widths, $1/\pi T_2$, are more than a thousand times narrower than the spectroscopically observed dimer widths, showing that the absorption line shapes arise from inhomogeneous broadening and do not reveal information on the dimer dynamics. The important fact uncovered by the photon echo experiments is that *the delocalized dimer state optical coherence time is very long.* These dimer "mini-excitons" are very close to the coherent limit.

The splittings (tens of cm^{-1}) between $|+\rangle$ and $|-\rangle$ states of each of the dimers that give rise to the red lines R_3, R_4 and R_5 are large relative to KT ($KT \approx 1\,cm^{-1}$). Thus phonon scattering between the $|+\rangle$ and $|-\rangle$ states of a dimer cannot occur since only the lower energy of the two states is excited and there is vastly insufficient thermal energy available for the scattering process. Since $T_2 \neq 2T_1$ the additional dephasing must arise from a pure dephasing process characterized by T_2^*. We are currently pursuing temperature-dependent photon echo studies of the dimer dephasing. Furthermore, recent theoretical work (Skinner et al. 1981) has demonstrated that for multilevel systems the relationship between T_2 and T_1 may not be straightforward. This arises because the traditional transition dipole line shape correlation function is not necessarily appropriate in describing photon echo results from multilevel systems.

In addition to showing that the coherence times are long, the echo measurements reveal a very interesting and possibly important feature. Geometrically different dimers can have substantially different T_2^*'s, i.e. pure dephasing rates. R_3 has a value of ~ 180 ns, R_4 has a value of ~ 70 ns while R_5 has a value of ~ 20 ns. Thus the coupling of the dimer delocalized state to the acoustic phonons must have a substantial dependence on the details of the dimer geometry. That is, the dimer electronic excitation–acoustic phonon coupling is spatially anisotropic. This could have implications for our understanding of coherence in exciton bands of pure crystals. Since the exciton band dispersion is based on pairwise interactions, the rate of loss of exciton coherence could also be spatially anisotropic. In a sufficiently anisotropic situation, coherent exciton propagation could effectively be limited in spatial direction, not because the energy dispersion is anisotropic but because the rate of exciton coherence loss is anisotropic.

Acknowledgments

Several of the sections in this article are based on extensive work performed in collaboration with a number of investigators. I would like to take this opportunity to thank them and say what a pleasure it has been working with them. Section 3 is based on work performed with R.D. Wieting and D.D. Dlott. Subsection 4.1 is based on work performed with D.M. Burland, D.E. Cooper and C.R. Gochanour. Subsection 4.2 is based on work performed with R.D. Wieting. Subsection 5.2 is based on work performed with R.W. Olson, H.W.H. Lee and F.G. Patterson. In addition I would like to thank the following agencies and organizations for providing support for much of the research described here: National Science Foundation, Division of Materials Research; Department of Energy, Division of Materials Research; the Dreyfus Foundation; The American Chemical Society, Petroleum Research Fund; The Sloan Foundation; and The Research Corporation.

References

Baughman, R.H. and D. Turnbull, 1971, J. Phys. Chem. Solids **32**, 1375.

Botter, B.J., A.I.M. Dicker and J.J. Schmidt, 1978, Mol. Phys. **36**, 129.

Brenner, H.C., J.C. Brock, M.D. Fayer and C.B. Harris, 1975, Chem. Phys. Lett. **33**, 471.

Brillante, A. and L. Dissado, 1976, Chem. Phys. **12**, 297.

Burland, D.M., D.E. Cooper, M.D. Fayer and C.R. Gochanour, 1977a, Chem. Phys. Lett. **52**, 279.

Burland, D.M., U. Kunzelmann and R.M. Macfarlane, 1977b, J. Chem. Phys. **67**, 1926.

Carrington, A. and A.D. McLachlan, 1967, Introduction to Magnetic Resonance (Harper and Row, New York).

Cooper, D.E. and M.D. Fayer, 1978, J. Chem. Phys. **68**, 229.

Cooper, D.E., R.W. Olson and M.D. Fayer, 1980, J. Chem. Phys. **72**, 2332.

Davydov, A.S., 1971, Theory of Molecular Excitons (Plenum Press, New York).

Dlott, D.D. and M.D. Fayer, 1976, Chem. Phys. Lett. **41**, 305.

Dlott, D.D., M.D. Fayer and R.D. Wieting, 1977, J. Chem. Phys. **67**, 2752.

Dlott, D.D., M.D. Fayer and R.D. Wieting, 1978, J. Chem. Phys. **69**, 2752.

Fayer, M.D. and C.B. Harris, 1974, Phys. Rev. **B9**, 748.

Feller, W., 1966, An Introduction in Probability Theory and Its Applications, vols. I–II (Wiley, New York).

Feynman, R.P., F.L. Vernon, Jr. and R.W. Hellwarth, 1957, J. Appl. Phys. **28**, 49.

Francis, A.H. and C.B. Harris, 1971, Chem. Phys. Lett. **9**, 188.

Gallus, G. and H.C. Wolf, 1966, Phys. Stat. Sol. **16**, 277.

Gochanour, C.R., H.C. Andersen and M.D. Fayer, 1979, J. Chem. Phys. **70**, 4254.

Godzik, K. and J. Jortner, 1980, J. Chem. Phys. **72**, 4471.

Gordon, R.G., 1968, in: Advances in Magnetic Resonance, vol. 3, ed. J.S. Waugh (Academic Press, New York) p. 1.

Grover, M. and R. Silbey, 1970, J. Chem. Phys. **52**, 2099.

Grover, M. and R. Silbey, 1971, J. Chem. Phys. **54**, 4843.

Guettler, W., J.O. von Schuetz and H.C. Wolf, 1977, Chem. Phys. **24**, 159.

Haken, H. and P. Reineker, 1972, Z. Physik **249**, 253.

Hochstrasser, R.M. and J.D. Whiteman, 1972, J. Chem. Phys. **56**, 5945.

Kenkre, V.M., 1980, Phys. Rev. **B22**, 3072.

Kenkre, V.M. and R.S. Knox, 1974a, Phys. Rev. **B9**, 5279.

Kenkre, V.M. and R.S. Knox, 1974b, Phys. Rev. Lett. **33**, 804.

Lakatos-Lindenberg, K. and K.E. Schuler, 1971, J. Math. Phys. **12**, 633.

Maradudin, A.A., 1966, Solid State Physics, eds. F. Seitz and D. Turnbull (Academic Press, New York).

Mattuck, R.D., 1976, A Guide to Feynman Diagrams in the Many-Body Problem (McGraw-Hill, New York).

Montroll, E.W., 1956, Proc. 3rd Berkeley Symp. on Mathematical Statistics and Probability, vol. III (Univ. of California, Berkeley, CA) p. 209.

Montroll, E.W., 1964, Proc. Symp. Appl. Math. Am. Math. Soc. **16**, 193.

Morris, G.C. and M.G. Sceats, 1973, Chem. Phys. **1**, 120.

Morris, G.C. and M.G. Sceats, 1974, Chem. Phys. **3**, 332, 342.

Munn, R.W., 1974, Chem. Phys. **6**, 469.

Nelson, K.A. and M.D. Fayer, 1980, J. Chem. Phys. **72**, 5202.

Nelson, K.A., D.D. Dlott and M.D. Fayer, 1979, Chem. Phys. Lett. **64**, 88.

Olson, R.W. and M.D. Fayer, 1980, J. Phys. Chem. **84**, 2001.

Powell, R.C., 1971, Phys. Rev. **B4**, 628.

Powell, R.C., 1973, J. Chem. Phys. **58**, 920.

Reineker, R. and H. Haken, 1972, Z. Physik **250**, 300.

Salcedo, J.R., A.E. Siegman, D.D. Dlott and M.D. Fayer, 1978, Phys. Rev. Lett. **41**, 131.

Schmidt, J. and R.M. Macfarlane, 1977, unpublished result.

Shelby, R.M., A.H. Zewail and C.B. Harris, 1976, J. Chem. Phys. **64**, 3192.

Simpson, O., 1956, Proc. R. Soc. **A238**, 402.

Skinner, J.L., H.C. Andersen and M.D. Fayer, 1981a, J. Chem. Phys. **75**, 3195.

Skinner, J.L., H.C. Andersen and M.D. Fayer, 1981b, Phys. Rev. **A24**, 1994.

Slichter, C.P., 1963, Principles of Magnetic Resonance (Harper and Row, New York).

Tinkham, M., 1964, Group Theory and Quantum Mechanics (McGraw-Hill, New York) p. 38.

Toyozawa, Y., 1958, Progr. Theoret. Phys. **20**, 53.

Van Strien, A.J., J.F.C. van Kooten and J. Schmidt, 1980, Chem. Phys. Lett. **76**, 7.

Wiersma, D.A., 1981, Adv. Chem. Phys., vol. XLVII (Wiley, New York) p. 421.

Wieting, R.D. and M.D. Fayer, 1980, J. Chem. Phys. **73**, 744.

Wieting, R.D., M.D. Fayer and D.D. Dlott, 1978, J. Chem. Phys. **69**, 1996.

Ziman, J.M., 1972, Principles of the Theory of Solids (Cambridge Univ. Press, Cambridge) p. 172.

Zwemer, D.A. and C.B. Harris, 1978, J. Chem. Phys. **68**, 2184.

CHAPTER 6

Theory and Experimental Aspects of Photon Echoes in Molecular Solids

WIM H. HESSELINK and DOUWE A. WIERSMA

Picosecond Laser and Spectroscopy Laboratory
State University of Groningen
Nijenborgh 16, 9747 AG Groningen
The Netherlands

Spectroscopy and Excitation Dynamics
of Condensed Molecular Systems
Edited by
V.M. Agranovich and R.M. Hochstrasser

Contents

1. Introduction

All echo phenomena, including the ones we generated as a youngster, have something fascinating. Maybe it is the *illusion* of time reversal in these effects which moves us. Whatever it may be, an increasing number of scientists make a living out of echoes.

The first echoes of scientific interest were *spin* echoes, observed and explained in the early fifties by Hahn (1950a,b). The theoretical foundation for the search for *photon* echoes was laid by Feynman et al. (1957), who showed that the Schrödinger equation for electric dipole transitions in a two-level system could be cast in a form identical to that of a spin $\frac{1}{2}$ particle interacting with a magnetic field. The first direct suggestion for the detection of a photon echo was made by Kopvillem and Nagibarov (1963), while Kurnit et al. (1964) made the first photon echo observation in 1964 in ruby.

With the development, in the early seventies, of tunable dye lasers (Hänsch 1968), interest in coherent optical effects as a tool to study relaxation processes has grown rapidly. Next to photon echoes, optical free induction decay (Brewer and Shoemaker 1972, Personov et al. 1974) and (photochemical) hole burning (Szabo 1975, de Vries and Wiersma 1976) have become useful techniques to study relaxation.

Recently a number of new nonlinear laser techniques, such as resonance Rayleigh scattering (Yajima and Souma 1978), polarization (Song et al. 1978) and four-wave mixing (Liao et al. 1977) have been used to study dynamical processes.

What specifically makes the photon echo a powerful technique is the fact that the Fourier transform of its decay gives the *homogeneous* optical line shape of the transition under study. With the other techniques this is true to a lesser extent. The main present interest in echo phenomena derives from the fact that, as stated (*vide retro*), the true line shape of a transition may be obtained, void from trivial inhomogeneity effects. This line shape, which contains all information on the dynamics of the system, is usually determined by a number of relaxation processes, which we generally divide in two types, population (T_1) and pure dephasing (T_2^*). Both these relaxation modes contribute to the homogeneous line width:

$$\Delta \nu_h = (\pi T_2)^{-1} \quad \text{(FWHM)},$$

where

$$T_2^{-1} = \tfrac{1}{2}T_1^{-1} + T_2^{*-1}.$$

It is known from Hahn's work (1950a,b) that two- and three-pulse photon echoes can be used to separate T_2^* from T_1. In this chapter we will show how this applies to optical transitions in molecular mixed crystals. Since the detection of the first photon echoes a number of reviews have appeared, e.g. Hartmann (1969), Abella (1969) and Brewer (1977). An excellent complete review of all the literature up to 1978 on coherent optical transient effects was recently given by Shoemaker (1979). A progress report on coherent optical studies in molecular mixed crystals was prepared by Wiersma (1981).

The intention of this chapter is therefore *not* to present a detailed survey or discussion of all photon echo experiments done in the past. In view of the recent work mentioned (Shoemaker 1979, Wiersma 1981) this seemed superfluous.

We have instead concentrated on some aspects of the photon echo technology which were less emphasized in Wiersma (1981). A general theoretical framework for the description of photon echoes in molecules is provided, as well as an overview of the most advanced experimental set-ups. In addition some new aspects of the (Redfield) relaxation theory, applicable to optical transitions in molecular solids, is presented. The discussion of experimental results has been restricted to what we consider to be the highlights, and in particular to the most recent ones not thoroughly discussed in Wiersma (1981). This (brief) review will show that the state of the art in generation and detection of photon echoes is now such that dependable results may be obtained, even for picosecond echo relaxation times.

The future of this branch of spectroscopy is hard to predict, be it that study of the optical homogeneous line shape of molecular excitons by coherence effects remains a great challenge.

2. Experimental

In this section we will describe some experimental set-ups which have been used in our and other laboratories to generate and detect photon echoes. Presently there is no set-up capable of measuring the whole range of echo relaxation times from picoseconds to minutes. The problem of interest therefore determines the set-up to be used. In the following we will describe three different set-ups centered around (1) a ns pulsed dye laser system, (2) a synchronously pumped ps dye laser system, and (3) a cw dye laser system.

2.1. Nitrogen- or YAG-pumped dye laser system

The basic set-up that we employ is shown in fig. 1. The excitation pulses are generated by two nitrogen- or YAG-pumped dye lasers in order to be able to electronically control the time separation between the pulses. To generate a stimulated photon echo, one dye laser pulse is split and one part delayed through an optical delay line [White cell (White 1942)]. These pulses then form the first and second excitation pulse, while the third excitation pulse is supplied by the other dye laser.

Crucial in these experiments is the fact that the timing jitter between the two dye laser pulses is negligible. This can be realized with commercially available pump lasers. In our set-up a computer-controlled scan–delay generator determines the speed and range over which the second laser pulse is delayed with respect to the first one. To generate a photon echo it is extremely important that the excitation pulses overlap in the crystal. To ensure this, both excitation pulses are focused through a 100 μm pin hole, mounted on the sample holder.

In this set-up the echo is most easily detected by choosing a collinear

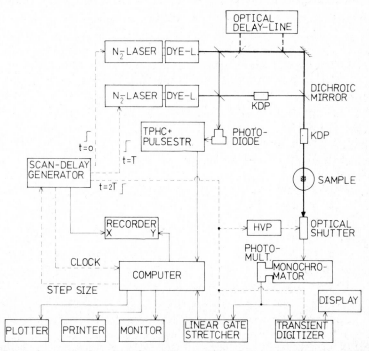

Fig. 1. Set-up for generating, detecting and processing of nanosecond photon echoes. The thick solid lines represent laser beams, the thin solid lines signal beams and the dashed lines trigger signals. HVP stands for high-voltage pulser.

excitation geometry: $k_e = k_1 = k_2$. This makes it relatively easy to align the optical shutter, which is synchronized via the scan–delay generator such that it only opens at the time of the echo passage. In our set-up the optical shutter consists of three Pockels cells in series. We note that if one of the pump lasers is a YAG laser it might be more convenient to detect the photon echoes by frequency mixing, a technique discussed in the next section. The photon echo is then detected via a monochromator, to suppress the incoherent fluorescence signal, on a photomultiplier. The signal is further fed into the computer via a linear gate stretcher. The information is stored permanently on magnetic disc, and can also be read out into a recorder.

As an example of the results obtainable with this system we show in fig. 2 a 3PSE decay curve for the system naphthalene in durene at 1.5 K. This curve was obtained with a repetition rate of 0.1 Hz and the total scan in this case was complete in 30 min. Note that every data point represents an average of over 20 echoes.

The strong point of this photon echo set-up is that it can measure echo decay times from minutes down to 2 ns. Also the great flexibility in the choice of exciting wavelength makes it attractive. We further note that it

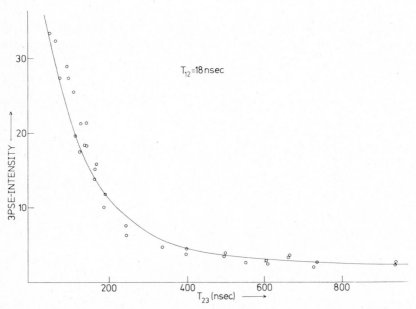

Fig. 2. Decay curve of the stimulated photon echo (3PSE) of naphthalene in durene. The open circles are the experimental data points, the solid line is a fit to eq. (10), with $k_{23}/(k_{21} + k_{23}) = 0.45$ and $k_{31} = 0$ ($k_{31} \ll k_{21}, k_{23}$). After Morsink (1982).

has been shown (Wallenstein and Hänsch 1975) that these dye laser pulses can be made close to transform limited (within a factor of 3) which makes a photon echo study as a function of position in the homogeneous line feasible.

2.2. *Synchronously pumped ps dye laser system*

In this set-up the ps excitation pulses are generated by synchronous pumping (Chan and Sari 1974) of a dye laser by a mode-locked Ar-ion laser. The shortest pulses that have been produced with this system are ≈ 0.5 ps.

If the accumulated grating echo (section 3) can be used, the set-up is very simple as shown in fig. 3. Important in this case is the reduction of amplitude noise in the dye laser output. This is achieved by inserting a Pockels cell in the beam, which compensates intensity fluctuations, monitored by a photodiode (PD), via rotation of the plane of polarization (P) of the incoming light. The rest of the set-up is identical to a normal pump–probe set-up. Note that a small part of the beam is used to monitor the cross-correlation of the exciting beams. For an example of a decay measured by this technique, see fig. 4, where the vibrational relaxation of a $747\,\text{cm}^{-1}$ mode in the upper B_{2u} state of pentacene is monitored via the accumulated grating echo.

If this method is not applicable, a more complex set-up is needed, which is shown in greater detail in figs. 5 and 6. To generate sufficient coherence

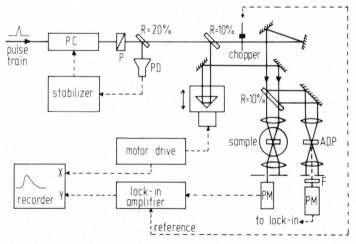

Fig. 3. Set-up for generating and detecting accumulated picosecond photon echoes. PC: Pockels cell, P: polarizer, PD: photodiode, F: filter, PM: photomultiplier.

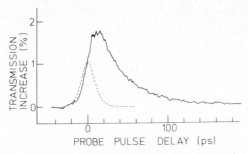

Fig. 4. Decay curve of the accumulated photon echo of the 747 cm^{-1} vibronic transition of pentacene in naphthalene. The dotted curve shows the cross-correlation of the excitation pulse. The pump–probe configuration of fig. 7c was used. After Hesselink and Wiersma (1979a).

in a transition in a single shot, the ps pulses from the dye laser need to be amplified. This can be achieved by transversely pumping two amplifier states by a Nd–YAG or nitrogen laser. Routinely we achieve ca 10 μJ in a single pulse with a duration as short as 2 ps. The second synchronously pumped dye laser in fig. 5 is used to generate a probe pulse that can be used to upconvert (Hesselink and Wiersma 1978, 1979a) the echo. The upconversion is achieved by phase-matched mixing of the probe pulse with the echo in an ADP crystal. The idea to use frequency upconversion for picosecond optical sampling was first suggested by Dugay and Hanson (1968) and recently used by Mahr and Hirsch (1975) to measure ultra-short fluorescence lifetimes. Note that the time resolution of this ps optical gate

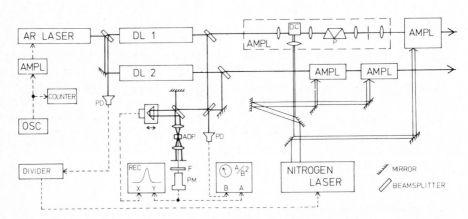

Fig. 5. Schematic diagram of the synchronously pumped ps dye laser set-up and inter-ferometer, used for measuring the auto- and cross-correlation. PD: photodiode, PM: pho-tomultiplier, F: filter, DC: dye cell, P: prism.

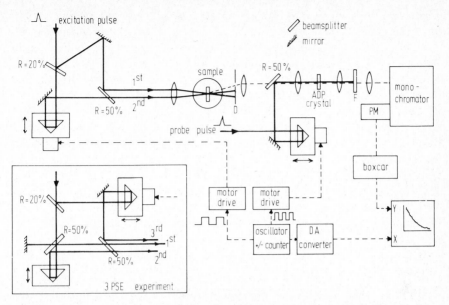

Fig. 6. Set-up for generating and detecting picosecond photon and stimulated photon echoes,
F: filter, PM: photomultiplier, D: diaphragm.

heavily relies on the synchronization (Jain and Heritage 1978) between the
two synchronously pumped dye lasers.

We are presently in the stage of also computer controlling this set-up.
We finally remark that this ps laser system can also be employed to study
other nonlinear coherence effects such as time-resolved CARS (Hesp and
Wiersma 1980).

For a study of photon echoes in molecular solids, Fayer and co-workers
(Cooper et al. 1979) have very successfully used a synchronously pumped
dye laser system based on a mode-locked and Q-switched Nd–YAG laser
(Kuizinga and Siegman 1970a,b). The high repetition rate ($\approx 1\,\mathrm{kHz}$) and
superb amplitude stability of this laser make it an ideal tool to study ps
relaxation effects. Unfortunately this Nd–YAG laser system is presently
not commercially available but, in our opinion, is highly competitive with a
system based on a mode-locked Ar-ion laser. The only limitation of this
system presently seems to be the dye laser pulse width of ca 30 ps, which
precludes study of very fast (≈ 5 ps) relaxation processes.

2.3. Gated cw dye laser system

For the measurement of photon echoes with a long lifetime ($\sim \mu$s) a cw dye
laser seems most suitable. Liao and Hartmann (1973a) were the first to use

a gated cw ruby laser to study photon echo relaxation in ruby. Macfarlane et al. (1979) recently showed that by a heterodyne detection technique this method can be made very sensitive. Basically their set-up is very simple. The excitation pulses are generated by acousto-optic switching of a frequency-stabilized dye laser. The acousto-optic switching introduces a frequency shift of the excitation pulses and the echo, equal to the rf frequency which drives the modulator. The important trick that Macfarlane et al. now apply is that just before the echo appearance they again shift the laser beam in frequency with an rf pulse slightly different from the modulator driving frequency in the excitation cycle. The echo then is heterodyne detected by interference between the echo polarization and the shifted laser beam. This detection technique increases the size of the echo signal by a factor 10^2–10^3. With the commercial availability of cw frequency-stabilized ring dye lasers this method will become important, especially for the study of long-lived states in ions and molecules.

Concluding remark. Next to a reliable set-up the optical quality of the material to be studied is of vital importance in a photon echo study. It is worthwhile therefore to devote much time to the growth of crystals (samples) with a high optical quality.

3. Photon echo formation

3.1. Introduction

The formation of a photon echo after two excitation pulses is presently well understood (Abella et al. 1966) and documented (Allen and Eberly 1975). Basically the first excitation pulse creates a macroscopic polarization in the medium, which quickly decays because of the inhomogeneous distribution in the transition excited. A second pulse a time τ later effectively induces time reversal in the system, which implies that at $t = 2\tau$ the macroscopic polarization in the medium is restored and a coherent light pulse, the *photon echo*, is emitted. The rephasing is complete only when during this time no phase-disturbing events (T_2^* effect) have occurred. This picture, while being completely satisfactory from a descriptive point of view, cannot account for some of the details of the photon echo formation, especially in multi-level systems.

A more quantitative description of photon echo formation relies on the density matrix. In this formulation the decay constants are introduced phenomenologically, which implies that a description in terms of relevant states (system) (states involved in the echo formation) and a heat bath is appropriate. Note that this implies that the decay of the coherence in the system is slow compared to the rate of fluctuations in the bath (Markov approximation).

In the following we will give a density matrix description for photon echo and stimulated photon echo formation in open and closed two-level systems and in a pseudo three-level system. At this point we remark that there are a variety of other new and exciting echo phenomena such as two-photon echoes (Flusberg et al. 1978), tri-level echoes (Mossberg et al. 1977), optical rotary echoes (Muramoto et al. 1980), notched and radiation-locked echoes (Liao and Hartmann 1973b), frequency grating (Morsink et al. 1979) and spatial grating echoes (Mossberg et al. 1979), which we will not discuss. The interested reader is referred to the references given.

3.2. Two-level systems

In a closed (a) two-level system the upper state decays radiatively and/or radiationless back into the ground state. This situation occurs, e.g., when a pure electronic $S_1 \leftarrow S_0$ transition of a guest molecule in a host lattice is excited, while the intersystem crossing plays no role on the timescale of the experiment. In an open (b) two-level system the upper state does not relax back into the ground state during the time of the experiment, which implies the presence of a bottleneck. This situation occurs, e.g., when a vibronic transition is excited where, on a ps time scale, the vibrationless excited state acts as a bottleneck.

The time development of the density matrix is given by

$$i\hbar\dot{\rho} = [H, \rho] + \text{relaxation terms}, \tag{1}$$

where

$$H = H_0 - \boldsymbol{\mu} \cdot \boldsymbol{E} \cos(\Omega t - \boldsymbol{k} \cdot \boldsymbol{r}).$$

In a frame rotating at the optical frequency Ω (indicated by a tilde) the individual elements of the density matrix evolve as follows:

$$\dot{\rho}_{22} = \tfrac{1}{2}i\chi(\tilde{\rho}_{12} \, e^{-i\boldsymbol{k} \cdot \boldsymbol{r}} - \tilde{\rho}_{21} \, e^{i\boldsymbol{k} \cdot \boldsymbol{r}}) - \rho_{22}/T_1,$$

$$\dot{\tilde{\rho}}_{12} = \tfrac{1}{2}i\chi(\rho_{22} - \rho_{11}) \, e^{-i\boldsymbol{k} \cdot \boldsymbol{r}} + (i\Delta - 1/T_2)\tilde{\rho}_{12}, \tag{2}$$

$$\tilde{\rho}_{21} = \tilde{\rho}_{12}^{*},$$

$$\rho_{11} = 1 - \rho_{22} \qquad \text{(case a),}$$

$$\dot{\rho}_{11} = \tfrac{1}{2}i\chi(\tilde{\rho}_{21} \, e^{i\boldsymbol{k} \cdot \boldsymbol{r}} - \tilde{\rho}_{12} \, e^{-i\boldsymbol{k} \cdot \boldsymbol{r}}) \qquad \text{(case b).}$$

The terms $e^{i\boldsymbol{k} \cdot \boldsymbol{r}}$ are purposely retained in the expressions to acquire the directional properties of the photon echo. In eq. (2) χ denotes the Rabi frequency ($\boldsymbol{\mu} \cdot \boldsymbol{E}/\hbar$), Δ the detuning of the absorbing centers due to inhomogeneous broadening, T_1 the population relaxation time of the *upper* state, and T_2 the phase relaxation time. Note that the so-called pure phase relaxation time T_2^{*} is related to T_2 by $T_2^{-1} = (T_2^{*})^{-1} + (2T_1)^{-1}$. The density matrix equations in (2) can be compactly written as

$$\dot{\rho} = -iL\rho, \tag{3}$$

where L is the Liouville operator. The formal solution to eq. (3) is simply

$$\rho(t) = e^{-iLt}\rho(0), \quad \rho(0) = \begin{pmatrix} \rho_{11}(0) \\ \rho_{12}(0) \\ \rho_{21}(0) \\ \rho_{22}(0) \end{pmatrix}.$$

With the usual assumptions of (i) resonant excitation and (ii) negligible relaxation during excitation, the exponential operator e^{-iLt} may be calculated using Putzers method (Putzer 1966).

We further distinguish between the situation where the exciting laser field is on (A) and off (B). After some algebraic manipulations we obtain (Hesselink 1980, Hesselink and Wiersma 1980) the following:

$$e^{-iL_A t} = \frac{1}{2} \begin{pmatrix} 1+\cos\chi t & -i\sin\chi t \, e^{ik\cdot r} & i\sin\chi t \, e^{-ik\cdot r} & 1-\cos\chi t \\ -i\sin\chi t \, e^{-ik\cdot r} & 1+\cos\chi t & (1-\cos\chi t)\,e^{-2ik\cdot r} & i\sin\chi t \, e^{-ik\cdot r} \\ i\sin\chi t \, e^{ik\cdot r} & (1-\cos\chi t)\,e^{2ik\cdot r} & 1+\cos\chi t & -i\sin\chi t \, e^{ik\cdot r} \\ 1-\cos\chi t & i\sin\chi t \, e^{ik\cdot r} & -i\sin\chi t \, e^{-ik\cdot r} & 1+\cos\chi t \end{pmatrix}, \tag{4a}$$

$$e^{-iL_B t} = \begin{pmatrix} 1 & 0 & 0 & 1-e^{-t/T_1} \\ 0 & e^{(i\Delta - 1/T_2)t} & 0 & 0 \\ 0 & 0 & e^{(-i\Delta - 1/T_2)t} & 0 \\ 0 & 0 & 0 & e^{-t/T_1} \end{pmatrix} \quad \text{(case a),} \tag{4b}$$

$$e^{-iL_B t} = \begin{pmatrix} 1 & 0 & 0 & 0 \\ 0 & e^{(i\Delta - 1/T_2)t} & 0 & 0 \\ 0 & 0 & e^{(-i\Delta - 1/T_2)t} & 0 \\ 0 & 0 & 0 & e^{-t/T_1} \end{pmatrix} \quad \text{(case b).} \tag{4c}$$

With these operators it is straightforward to calculate the echo intensities for the two-pulse and three-pulse (stimulated) photon echo (fig. 7) by simple matrix multiplication and collection of the appropriate terms. For optical thin media ($\alpha L \ll 1$, where α is the absorption cross section and L the crystal length) the photon echo amplitude can be shown (Hesselink 1980) to be proportional to:

$$\text{Tr}[\mu\rho(t)] = 2\mu_{12}\,\text{Re}(\tilde{\rho}_{12}\,e^{i\Omega t}), \tag{5}$$

where μ_{12} is the transition dipole. This expression explicitly shows that the

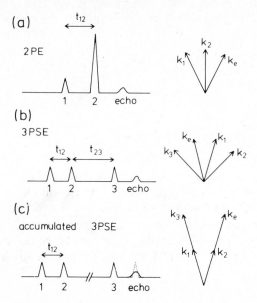

Fig. 7. Pulse sequences and phase matching geometries for various types of photon echoes.

photon echo formation is related to the off-diagonal element in the density matrix. Note that in a stimulated photon echo experiment the optical phase information temporarily is stored in the *diagonal* elements of the density matrix. We return to this point later. Using the Liouville operator techniques we recover the well-known results for the two-pulse echo (2PE) (Abella et al. 1966)

$$I_e(2t_{12}) = A_0 \sin^2(\chi\delta t_1) \sin^4(\tfrac{1}{2}\chi\delta t_2)\, e^{-4t_{12}/T_2},$$

$$k_e = 2k_2 - k_1,$$

(6)

where δt_1 and δt_2 are the duration of the first and second excitation pulse, respectively, and t_{12} is the exciting pulse separation. A_0 is a constant which depends on the system. The above results are valid for both open and closed systems.

At this point there are a number of interesting points to note. First, the well-known result that from the decay of the 2PE, the homogeneous line width ($\Delta\nu_h$) may be determined via the relation $\Delta\nu_h = (\pi T_2)^{-1}$. Secondly, when the excitation pulses are made shorter, then in order to generate the same degree of coherence in the system the *intensity* of the pulses should increase quadratically. Finally, from the intensity variation of the echo versus exciting field intensity, in principle, the transition dipole of the optical transition studied may be obtained. In fact the feasibility of this idea was recently demonstrated by Cooper et al. (1979).

We now proceed by presenting the results for the stimulated photon echo (3PSE):

$$I_e(2t_{12} + t_{23}) = A_e \sin^2(\chi \delta t_1) \sin^2(\chi \delta t_2) \sin^2(\chi \delta t_3)$$
$$\times e^{-4t_{12}/T_2} e^{-2t_{23}/T_1} \qquad \text{(case a)},$$
$$\times e^{-4t_{12}/T_2}(e^{-t_{23}/T_1} + 1)^2 \quad \text{(case b)},$$

$$k_e = k_3 + k_2 - k_1, \tag{7}$$

where t_{23} is the separation between the second and third pulse. Eqs. (7) show the important difference between the decay of the 3PSE in a closed and open system. In a closed two-level system (a) the 3PSE decays with a single relaxation time $\frac{1}{2}T_1$ when t_{23} is varied. In an open system (b) however, the 3PSE exhibits bi-exponential decay with relaxation times T_1 and $\frac{1}{2}T_1$, to a constant background. The background is due to the fact that the optical phase information, after two pulses, is stored independently in

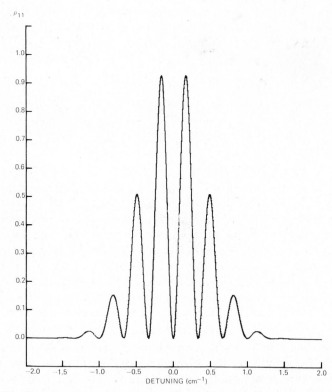

Fig. 8. Ground state population versus detuning from line center after two resonant $\pi/2$ pulses. Note that in the text this population distribution is referred to as "frequency grating".

the population distribution of the excited *and* ground state. After two excitation pulses of area θ_1 and θ_2 ($\chi\delta t_2$) separated in time by t_{12} we obtain:

$$\rho_{11}(t_{12}^+) = \tfrac{1}{2}\{1 + \cos\theta_1 \cos\theta_2$$
$$- \sin\theta_1 \sin\theta_2 \, e^{-t_{12}/T_2} \cos[\Delta t_{12} - (\boldsymbol{k}_1 - \boldsymbol{k}_2) \cdot \boldsymbol{r} + (\varphi_1 - \varphi_2)]\},$$
$$\rho_{22}(t_{12}^+) = 1 - \rho_{11}(t_{12}^+) \quad (t_{12} \ll T_1). \tag{8}$$

$\varphi_1 - \varphi_2$ is the phase difference between the first and second excitation pulse. The $\cos\Delta t_{12}$ modulation gives rise to a 3PSE when a third pulse is applied. Figure 8 shows the modulation in ρ_{11} for $\theta_1 = \theta_2 = \pi/2$. As long as this "grating in frequency space" persists, a 3PSE may be formed. This explains the "constant" background in case (b) when the excited state does not decay directly into the ground state.

3.3. Pseudo three-level systems

In many photon echo experiments on pure electronic transitions of molecular crystals, the two-level system approach fails to describe the observed stimulated photon echo decay. In this situation it is quite often necessary to take the lowest lying triplet state into account explicitly. We again distinguish between two different cases: (1) "single-shot"-type experiments, and (2) cw-type experiments, where coherent accumulation effects occur.

The distinction between these two cases is made on the basis of a comparison between the repetition rate of the experiment and the slowest relaxation rate in the optical pumping cycles of the system.

3.3.1. Echo and stimulated echo decay

We wish to describe the expected photon echo decay in the level scheme depicted in fig. 9. $|1\rangle$ is the ground state, $|2\rangle$ the electronically excited state and $|3\rangle$ the triplet state. The levels are connected by the indicated rate constants, where k_{21} is the fluorescence, k_{23} the intersystem crossing and k_{31} the phosphorescence rate constant. We note at this point that there is no basic difficulty in taking all triplet spin sublevels into account, but the important points to be made are clear by considering only three levels. As the exciting laser field is assumed to be (near) resonant with the optical transition $\langle 2| \leftarrow |1\rangle$, all off-diagonal elements of the density matrix involving level $|3\rangle$ may be neglected. We will further assume that during excitation no relaxation occurs, which implies that the field affects only the two-level system $\langle 2| \leftrightarrow |1\rangle$. In between the pulses the time evolution of the system is governed by the inhomogeneous width (Δ) and the optical decay parameters. The time evolution of the density matrix in this time span (field off) is, in the Liouville representation, governed by the following

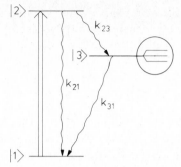

Fig. 9. Three-level structure representing the optical pumping cycle in a typical organic molecule. $|1\rangle$ is the ground state, $|2\rangle$ the excited singlet state and $|3\rangle$ the triplet state. Note that k_{21} and k_{31} include the effect of both radiative and radiationless decay. k_{23} only represents radiationless decay (intersystem crossing).

matrix equation (Hesselink and Wiersma 1981a):

$$\rho(t) = B(t)\rho(0) =$$

$$
\begin{pmatrix}
1 & 0 & 0 & \begin{matrix}1 - \beta\, e^{-k_{31}t} \\ + (\beta - 1)\, e^{-t/T_1}\end{matrix} & 1 - e^{-k_{31}t} \\
0 & e^{(i\Delta - 1/T_2)t} & 0 & 0 & 0 \\
0 & 0 & e^{(-i\Delta - 1/T_2)t} & 0 & 0 \\
0 & 0 & 0 & e^{-t/T_1} & 0 \\
0 & 0 & 0 & \beta\, e^{-k_{31}t} - \beta\, e^{-t/T_1} & e^{-k_{31}t}
\end{pmatrix}
\begin{pmatrix}
\rho_{11}(0) \\
\rho_{12}(0) \\
\rho_{21}(0) \\
\rho_{22}(0) \\
\rho_{33}(0)
\end{pmatrix}
\quad (9)
$$

where $\beta = k_{23}/(k_{21} + k_{23} - k_{31})$. T_1 is the population relaxation time of level $|2\rangle$, equal to $(k_{21} + k_{23})^{-1}$, and T_2 is the phase relaxation time of the $\langle 2| \leftarrow |1\rangle$ transition. Note that the expressions for $B(t)$ given in eqs. (4) may be recovered in the appropriate limits. Following now the procedure described earlier, the echo polarization for each pulse cycle may be obtained. For the decay of the 2PE the same result as given in eq. (6) is obtained.

The 3PSE intensity as a function of t_{23} is different and contains information on the intramolecular decay parameters:

$$I_{3PSE} = A(\chi_i, \delta t_i)\, e^{-4t_{12}/T_2} \left[\frac{k_{23}\, e^{-k_{23}t_{23}}}{k_{21} + k_{23} - k_{31}} + \left(1 + \frac{k_{21} - k_{31}}{k_{21} + k_{23} - k_{31}}\right) e^{-(k_{21} + k_{23})t_{23}} \right]^2$$

$$(10)$$

$A(\chi_i, \delta t_i)$ is identical to the one in eq. (7). Note again that the two-level results are recovered in the appropriate limits $k_{23} = 0$ (a) and $k_{21} = k_{31} = 0$ (b). In a system then where intersystem crossing is appreciable (as in most molecules) the decay of the stimulated photon echo is expected to be quite complex. However, quite surprisingly from the decay of the 3PSE quantitative information may be obtained on the intersystem crossing yield. We will discuss some results obtained by this method in subsection 5.4.

3.3.2. Coherent pumping effects

When the repetition rate in a photon echo experiment exceeds the lifetime of the bottleneck in the optical pumping cycle, coherent pumping effects may occur. Next to this effect one also should note that the system becomes "diluted" through population storage in the bottleneck. This implies that in photon echo experiments where the effect of the guest intermolecular interaction is studied one should adjust the repetition rate such that the bottleneck is not pumped. We will restrict ourselves here to a discussion of the accumulated grating echo phenomenon (Hesselink and Wiersma 1979b) which is induced by coherent pumping using a mode-locked cw dye laser. A simple physical explanation is obtained by considering eq. (8). This equation shows that two excitation pulses produce a grating $\cos[\Delta t_{12} - (k_1 - k_2) \cdot r + (\varphi_1 - \varphi_2)]$ in *both* the ground and excited states, from which a third pulse, independently, may stimulate an echo.

An important point to note here is that this grating, via the amplitude factor e^{-t_{12}/T_2}, contains information on the optical phase loss during the time between the first two pulses.

We now turn to a discussion of the accumulation effect. As time proceeds, a fraction $k_{23}/(k_{21} + k_{23})$ of the molecules decay into the triplet state which has a long lifetime, $k_{31} \ll k_{23} + k_{21}$. Therefore, at the time the upper state effectively is depleted, but the triplet state is not, $\rho_{11}(t) = 1 - \rho_{33}(t)$. By applying then a train of twin excitation pulses, with a separation between the twins short compared to the triplet state lifetime, one can easily imagine that the *modulated* part of $\rho_{11}(t)$ builds up in time. This is demonstrated in fig. 10, where results are shown from a computer simulation of the accumulation effect. The dashed line is the inhomogeneous line shape after one pair of excitation pulses. It is not distinguishable from the original distribution. After 300 pulse pairs, however, given by the solid line, the modulation in the ground state population becomes clearly visible. At this point it is important to realize that the grating can only grow provided that there is no phase jitter between consecutive twin excitation pulses in the preparation stage. This is clear from eq. (8), where a random phase difference in this period, $\langle\cos(\Delta t_{12} - k_{12} \cdot r + \varphi_{12})\rangle_a = 0$, destroys the grating. This phase stability requirement is similar to the one encountered in optical Ramsey fringe experiments (Salour 1978).

Hesselink and Wiersma (1981a) have recently described this echo phenomenon in greater detail and showed that under the excitation conditions shown in fig. 7c an interesting interference effect occurs between

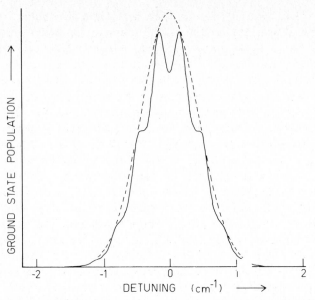

Fig. 10. Computer simulation of the accumulated grating, calculated by repeated application of eqs. (4) and (9). The dashed line represents the original Gaussian inhomogeneous distribution. The solid line shows the effect of 300 pulse pairs with a pulse area of 5 degrees. The parameters used are $t_{12} = 100$ ps, $T = 12.2$ ns, $k_{21} = k_{23} = 5 \times 10^7$ s^{-1}, $k_{31} = 10^6$ s^{-1}. After Hesselink (1980).

the echo polarization and probe beam. A quantitative description of this transparency effect (McCall and Hahn 1969, Slusher and Gibbs 1972) required consideration of both the optical Bloch and Maxwell equations. This is a unique situation, as normally in optically thin media coherence effects can be described in terms of the optical Bloch equations only.

Experimentally the stimulated photon echoes are detected as a change in intensity of the probe beam in a pump–probe set-up as discussed in section 2. Hesselink and Wiersma (1981a) calculated, under the condition k_{21}, k_{23}, $k_{31} \ll t_{12}^{-1}$, for the change in probe pulse intensity:

$$I_2 = I_2(0) \exp[-\alpha_{\text{eff}} \cos \theta_1 - (\cos \theta_1 + 2 \, e^{-2t_{12}/T_2})\theta_1^2 \gamma/2], \qquad (11)$$

where

$$\alpha_{\text{eff}} = \tau_p/2\pi g(0), \qquad (12)$$

and α is defined as

$$\alpha = (\pi^2 \omega N/\mu^2)/c\eta\hbar\epsilon_0, \qquad (13)$$

where $g(0)$ is the inhomogeneous distribution $g(\Delta)$ at $\Delta = 0$. γ is the enhancement factor produced by the accumulation effect. We will discuss

γ in more detail later. We note that in the above equation we assumed $\theta_2 \ll \theta_1 \ll 1$ and used the relation $\theta_2 = \chi_2 \tau_p$ for the area of the square wave.

In the absence of the first excitation pulse ($\theta_1 = 0$), eq. (11) simply gives the absorption law for the probe pulse. The reduction of the absorption coefficient [eq. (12)] is due to the fact that we assumed the pulse width to be much shorter than the inverse inhomogeneous width. Therefore the pulse has appreciable spectral components outside the absorption line. The second θ_1-dependent term in eq. (11) consists of two physically distinct components. These are an incoherent part (ordinary saturation) and a coherent part (the echo), which contains information about the phase loss during t_{12}. An interesting point is that both the saturation and coherent terms depend in the same way on the intensity of the first pulse and on the enhancement factor γ. From an experimental point of view this means that the echo is just as easily detectable as the saturation. The nice result of eq. (11) is that the observed change in probe pulse transmission is quantitatively correlated with the echo amplitude. The enhancement factor γ was also calculated by Hesselink and Wiersma (1981a) and in a realistic situation (applicable for the systems studied) turns out to be:

$$\gamma = \beta(1 - e^{-T/T_1})/2k_{31}T, \tag{14}$$

where $\beta = k_{23}/(k_{21} + k_{23})$ is the intersystem crossing yield.

Finally it is easy to show that the relative difference between the probe pulse intensity with the excitation beam on (I_{on}) and off (I_{off}) is given by:

$$S(t_{12}) = \frac{I_{on} - I_{off}}{I_{off}} \approx \alpha_{eff}[\tfrac{1}{2}\theta_1^2(\gamma + 1) + \theta_1^2\gamma \, e^{-2t_{12}/T_2}], \tag{15}$$

assuming θ_1, $S(t_{12}) \ll 1$. We note here that eq. (15) predicts the coherent signal to be twice as intense as the background signal. In practice we observe a ratio of approximately 1.3. This discrepancy may be due to several factors. First, our pulses are not short enough to entirely justify the assumption of negligible inhomogeneous dephasing during the excitation pulses. This means that the coherent excitation of the wings of the absorption line is not as efficient as that of the center of the lines, which may disturb the wings of the grating. Secondly, our picosecond pulses may be somewhat chirped, which would also affect the grating, but not the saturation.

4. Theories of optical line shape

4.1. Photon echo relaxation and optical line shape

Photon echo relaxation measurements are in fact line shape measurements, be it of the *homogeneous* line shape, which may be hidden under an inhomogeneous distribution. The formal relationship between the optical line shape and photon echo decay can be made as follows. The general expression for the absorption line shape is (Berne and Harp 1970):

$$I(\omega) \sim \int_{-\infty}^{\infty} dt\, e^{-i\omega t} \langle\langle \mu(0)\mu(t)\rangle\rangle, \tag{16}$$

where ω is the photon frequency, μ the transition dipole, and $\langle\langle\ \rangle\rangle$ denotes averaging over the initial ensemble states. We may further write (Mukamel 1979) for the unsaturated line shape $\langle\langle\mu(0)\mu(t)\rangle\rangle$:

$$\langle\mu_{if}(0)\mu_{if}(t)\rangle = \mu_{if}(0)\langle\mu_{fi}(t)\rangle = \rho_{fi}(t)|\mu_{if}|^2, \tag{17}$$

where we have used the fact that $\mu_{if}(0)$ is independent of the bath states (*vide infra*).

Combining eqs. (16) and (17) leads to the desired result, namely

$$I(\omega) \sim 2\,\mathrm{Re} \int_{0}^{\infty} dt\, e^{i\omega t} \rho_{fi}(t). \tag{18}$$

This equation shows that the line width is determined by $\mathrm{Re}[\rho_{fi}(t)]$. In the case of exponential decay of the photon echo amplitude with lifetime T_2, the homogeneous linewidth $\Delta\nu_h$ (FWHM) of the Lorentzian absorption line is given by the relation:

$$\Delta\nu_h = (\pi T_2)^{-1}. \tag{19}$$

Understanding of photon echo relaxation thus basically involves line shape theories of optical transitions.

Until recently it was assumed (Burke and Small 1974a,b) that the McCumber–Sturge theory (McCumber and Sturge 1963), which assumed quadratic electron–band phonon coupling to be important, adequately explained the line shape of optical transitions in organic impurity centers. Coherent optical experiments, however, showed (Aartsma and Wiersma 1976, Aartsma 1978) that the quasi-exponential activation of the line width observed at low temperature could not be understood on the basis of this theory. A number of attempts were then made either to refine the McCumber–Sturge theory (Small 1970, Jones and Zewail 1978) or to suggest other mechanisms for the observed line shape. Harris (1977), e.g., suggested the importance of exchange, an effect well known in magnetic resonance spectroscopy. De Bree and Wiersma (1979) later showed that for optical transitions this effect is expected to be less important than for Raman transitions. These authors, however, pointed out that inelastic phonon scattering in the ground and excited states via pseudolocal phonons (called local phonons hereafter) could explain the observed exponential activation of the optical line width. Sapozhnikov (1976) was the first to suggest that in organic crystals the anharmonic nature of the lattice phonons would be important. A decade earlier Lubchenko et al. (1964, 1973), and independently Krivoglaz (1964), had already worked out in detail the theory for this effect. Prior to this work it had been McCumber (1964) who noticed that the description of the optical line shape should extensively parallel the theory developed for magnetic resonance spectra. It is no surprise then that de Bree and Wiersma (1979) and de Bree (1981a) used Redfield theory (Redfield 1957, 1965) to describe optical relaxation.

We note that we have no intention of reviewing all the theoretical work that has been done on line shapes of optical transitions; rather we will focus on some recent results obtained using the Redfield theory of relaxation in molecular solids.

4.2. Redfield relaxation theory

Basic to the success of any theory in describing an experiment is a realistic choice of the levels to be considered. Based upon the outcome of the most recent photon echo experiments we claim that the level scheme shown in fig. 11 should be applicable to the description of the zero-phonon line shape in most molecular mixed crystals. In this figure, $|1\rangle$ is the ground state and $|2\rangle$ the electronically excited state of an isolated guest impurity in a host lattice. Levels $|3\rangle$ and $|4\rangle$ are associated with localized phonons (librations) of the guest molecule in the ground and excited state. The levels $|k\alpha\rangle$ are acoustic phonons of branch α with wave vector k. In connection with fig.

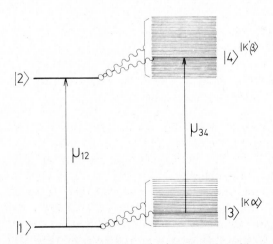

Fig. 11. Level structure of an optical transition $\langle 2| \leftrightarrow |1\rangle$ interacting with pseudolocalized levels $|3\rangle$ and $|4\rangle$, and delocalized crystal states $(k\alpha, k'\beta)$. Note that the levels $|3\rangle$ and $|4\rangle$ are projected out of the bath and included in H_A. Further note that this picture only holds at low temperature, where two-quantum excitations of the local phonon play no role.

11 a number of remarks are in order. First, coupling of the electronic transition to the vacuum electromagnetic field is omitted, which means that spontaneous photon emission from level $|2\rangle$ to level $|1\rangle$ is included as a decay constant. Secondly, other states of the guest molecule, e.g. the triplet state, are omitted, which implies that radiationless relaxation processes are also taken into account via relaxation constants. Finally, only one (local phonon) of the guest molecule is considered. This may not always be realistic; however, in that case it will be easy to generalize the relaxation expressions including the effect of the other levels.

Anticipating now the usage of Redfield relaxation theory (Redfield 1957, 1965) in a form worked out by de Bree and Wiersma (1979) we write the mixed crystal hamiltonian which describes the level structure of fig. 11 in the following form:

$$H = H_A + H_R + V_{AR},$$

$$H_A = \sum_f \left(\epsilon^f + \sum_\kappa V^f_{\kappa\kappa}(\bar{n}_\kappa + \tfrac{1}{2}) + \hbar(\Omega + \Delta\Omega^f)(B^+B + \tfrac{1}{2}) \right) a_f^+ a_f,$$

$$H_R = \sum_\kappa \hbar\omega_\kappa (b_\kappa^+ b_\kappa + \tfrac{1}{2}) + \frac{1}{3!} \sum_{\kappa,\kappa',\kappa''} (U_{\kappa\kappa'\kappa''} b_\kappa b_{\kappa'}^+ b_{\kappa''}^+ + cc) + \cdots,$$

$$V_{AR} = V_a + V_{e-p},$$

$$V_a = \frac{1}{3!} \sum_{\kappa,\kappa'} (U_{\kappa\kappa'\lambda} b_\kappa b_{\kappa'}^+ B^+ + cc) + \cdots,$$

$$V_{e-p} = \sum_{f,\kappa} V^f_\kappa (b_\kappa + b_\kappa^+) a_f^+ a_f + \sum_f V^f_B (B + B^+) a_f^+ a_f$$

$$+ \sum_{f,\kappa} V^f_{\kappa\kappa} (b_\kappa^+ b_\kappa - \bar{n}_\kappa) a_f^+ a_f + \sum_{\substack{f,\kappa,\kappa' \\ (\kappa \neq \kappa')}} V^f_{\kappa\kappa'} b_\kappa^+ b_{\kappa'} a_f^+ a_f$$

$$+ \sum_{f,\kappa} V^f_{\kappa\lambda} (b_\kappa^+ B + b_\kappa B^+) a_f^+ a_f, \tag{20}$$

with $\Delta\Omega^f = V^f_{\lambda\lambda}/\hbar$ and $\Delta\Omega^g = V^g_{\lambda\lambda}/\hbar = 0$. The index λ is used when the pseudolocal phonon is involved.

In eqs. (20) a_f^+ (a_f) denotes the creation (annihilation) operator for electronic state f with energy ϵ^f. b_κ^+ (b_κ) and B^+ (B) are the creation (annihilation) operators for the band phonons (acoustic) and local phonons, respectively. $\Delta\Omega^f$ represents the frequency change of the pseudolocal phonon in going to the excited state. V^f_κ, $V^f_{\kappa\kappa}$ and $V^f_{\kappa\kappa'}$, represent the linear, diagonal and nondiagonal electron–(acoustic) phonon interaction matrix elements. V^f_B and $V^f_{\kappa\lambda}$ are the corresponding quantities for the electron–(local) phonon coupling. $U_{\kappa\kappa'\kappa''}$ and $U_{\kappa\kappa'\lambda}$ represent cubic phonon anharmonic interactions and $\bar{n}_\kappa$ the average phonon occupation number: $\bar{n}_\kappa =$

$[\exp(\hbar\bar{\omega}_\kappa/kT) - 1]^{-1}$, where $\bar{\omega}_\kappa = \omega_\kappa + \delta\omega_\kappa$ and $\delta\omega_\kappa$ is a small renormalization factor of the phonon frequency due to anharmonicity, which is neglected.

We note here that the electronic energy has been redefined (Hubbard 1961) to include the thermal average of the electron–phonon coupling energy. As is obvious from this hamiltonian, the electronic states are treated on equal footing with the localized levels, as seems requested from a spectroscopic point of view (Aartsma et al. 1977). These localized levels (H_A) are interacting with the *anharmonic* phonon bath (H_R) through the electron–phonon coupling (V_{AR}) and anharmonicity effects. In their study de Bree and Wiersma (1979) focused on the role of the local and band phonons in the temperature dependence of the dephasing process, while de Bree (1981a) has also worked out the effect of band phonons on the low-temperature relaxation.

Let us consider in some more detail the perturbations in V_{AR}. The term V_a takes the local phonon–band phonon anharmonicity into account. The physical picture here is that a local phonon decays into two band phonons in either a frequency difference or sum generating process. The first two terms in V_{e-p} describe the linear electron–band phonon and electron–local phonon coupling. These terms are in fact responsible for most of the phonon side bands in absorption or emission. The third and fourth terms describe the quadratic diagonal and nondiagonal electron–band phonon coupling, while the last term in V_{e-p} is the mixed band–local quadratic electron–phonon coupling term. Note that the quadratic electron–local phonon coupling is included in H_A and redefines the excitation energies of the local phonon. We also wish to remark that in de Bree and Wiersma (1979) the linear electron–*local* phonon coupling was assumed to be zero.

We now continue our discussion of Redfield relaxation theory in connection with the hamiltonian of eq. (20). The basic assumption in the Redfield theory is that a partitioning of the total hamiltonian in terms of $H_A + H_R + V_{AR}$ is made such that the following equation of motion may be written:

$$\partial\rho_A/\partial t = -(i/\hbar)[H_A, \rho_A(t)] - \Lambda \cdot \rho_A(t), \tag{21}$$

where ρ_A is the reduced density matrix of the system, whose relaxation is described by

$$\rho_A = \mathrm{Tr}_R(\sigma(t)), \tag{22}$$

where $\sigma(t)$ is the density matrix for the complete system and the trace is taken over all degrees of freedom belonging to the subsystem described by H_R. Λ is a Liouville relaxation superoperator which works on $\rho_A(t)$. Redfield assumes now that the coupling between the states of H_A and H_R are so weak that Λ can be calculated using perturbation theory. Implicitly assumed is then that the fluctuations in the bath (H_R) are fast in comparison with the decay of the coherence in the system. This separation of time scales was discussed in detail by de Bree and Wiersma (1979).

This partitioning of the hamiltonian in $H_A + H_R + V_{AR}$ also implies a second important point, first discussed by Mukamel (1978), namely, that by a particular choice of H_A only scattering processes that are diagonal in the system states lead to pure T_2^*-type dephasing processes, while the other ones are found to be of T_1-type. This terminology, however, is intimately related to the *initial* choice of system (H_A) and bath states (H_R) and has only semantic value. At any rate, both T_1- and T_2^*-type processes lead to dephasing via the relation (valid for a two-level system interacting with a bath)

$$T_2^{-1} = (T_2^*)^{-1} + \tfrac{1}{2}T_1^{-1}. \tag{23}$$

We now apply Redfield relaxation theory in the form given by Cohen-Tannoudji (1975). We further assume no optical radiation field to be present and then find for the density matrix of the system (ρ_A) in the interaction representation

$$\dot{\tilde{\rho}}_A = -\frac{1}{\hbar^2} \int_0^t \mathrm{d}t' \, [\tilde{V}(t), [\tilde{V}(t'), \tilde{\rho}_A(t')\rho_R(0)]], \tag{24}$$

where the tilde ($\sim$) stands for interaction representation:

$$\tilde{V}(t) = \mathrm{e}^{\mathrm{i}(H_A+H_R)t/\hbar} V(t) \, \mathrm{e}^{-\mathrm{i}(H_A+H_R)t/\hbar}$$

and $\rho_R(0)$ is the density matrix of the bath states at time zero.

4.3. Linear electron–phonon coupling effects

The effect of *linear* coupling between the electronic transition and an anharmonic phonon bath on the impurity optical line shape can now easily be calculated by a choice of the relevant terms in eq. (20). The most simple hamiltonian that describes this effect consists of

$$H_A = \epsilon^f a_f^+ a_f,$$
$$H_R = \sum_\kappa \hbar\omega_\kappa (b_\kappa^+ b_\kappa + \tfrac{1}{2}) + \frac{1}{3!} \sum_{\kappa,\kappa',\kappa''} U_{\kappa\kappa'\kappa''} b_\kappa b_{\kappa'}^+ b_{\kappa''}^+ + \cdots, \tag{25}$$
$$V_{AR} = \sum_\kappa V_\kappa^f (b_\kappa + b_\kappa^+) a_f^+ a_f.$$

Skinner et al. (1981) used this hamiltonian to calculate the relaxation operator $\Lambda_{12,12}$ and found, using eq. (24), that

$$\Lambda_{12,12} = +\frac{1}{\hbar^2} \int_0^\infty \mathrm{d}\tau \, \langle \Delta(0)\Delta(\tau) \rangle, \tag{26}$$

where $\Delta = \Sigma_\kappa V_\kappa^f (b_\kappa + b_\kappa^+)$. In general, the integral in eq. (26) is a complex function. The imaginary part of $\Lambda_{12,12}$ leads to a renormalization of the bare electronic excitation and is of no interest. The real part of $\Lambda_{12,12}$, however, leads to loss of coherence and thus to a contribution to the homogeneous line width. It can be shown that (de Bree 1981a)

$$\frac{1}{\hbar^2} \operatorname{Re} \int_0^\infty \langle \Delta(\tau)\Delta(0) \rangle \, \mathrm{d}\tau = \frac{1}{T_2^*}$$

$$\approx \frac{1}{2\hbar^2} \lim_{\omega \to 0} \sum_\kappa |V_\kappa^f|^2 \int_{-\infty}^\infty \mathrm{d}\tau \, \mathrm{e}^{\mathrm{i}\omega\tau} [\langle b_\kappa^+(\tau)b_\kappa \rangle + \langle b_\kappa(\tau)b_\kappa^+ \rangle], \tag{27}$$

where the approximation stems from neglect of the terms $\langle b_\kappa^+(\tau)b_\kappa^+ \rangle$ and $\langle b_\kappa(\tau)b_\kappa \rangle$.

Equation (27) shows that pure dephasing effects are connected with the zero-frequency Fourier transform of the phonon correlation functions.

In the presence of anharmonicity, the phonon correlation function may be characterized by the following type of spectral functions:

$$\int_{-\infty}^\infty \mathrm{d}\tau \, \mathrm{e}^{-\mathrm{i}\omega\tau} \langle b_\kappa^+(\tau)b_\kappa \rangle = L_\kappa(\omega)n(\omega), \tag{28a}$$

with $n(\omega)$ the phonon occupation number and

$$L_\kappa(\omega) = \frac{2\gamma_\kappa(\omega)}{[\omega - \omega_\kappa - \pi_\kappa(\omega)]^2 + \gamma_\kappa^2(\omega)}. \tag{28b}$$

Here $\tilde{\omega}_\kappa = \omega_\kappa + \pi_\kappa(\omega)$ and $\gamma_\kappa(\omega)$ are the renormalized phonon frequency and (frequency-dependent) phonon lifetime, respectively. With the help of eqs. (28a) and (28b) the final result for $1/T_2^*$ is obtained:

$$\frac{1}{T_2^*} = \frac{1}{\hbar^2} \sum_\kappa \frac{|V_\kappa^f|^2}{\tilde{\omega}_\kappa^2 + \gamma_\kappa^2(0)} \lim_{\omega \to 0} \gamma_\kappa(\omega)[2n(\omega) + 1]. \tag{29}$$

Equation (29) shows that the zero-temperature limit for $1/T_2^*$ is determined by $\lim_{T \to 0} \gamma_\kappa(\omega = 0, T) = 0$.

The important conclusion then is that at zero temperature *pure dephasing* is absent, as was also recently pointed out, in another context, by Wertheimer and Silbey (1981). For *harmonic* phonons $\gamma_\kappa(\omega) = 0$ from the outset and therefore no contribution to optical dephasing is found at any temperature. According to Krivoglaz (1964), for cubic phonon anharmonicities the low temperature dephasing rate goes as T^9, and therefore vanishes, as it should, at $T = 0$ K. We end this subsection by noting that in the derivation of eq. (29) it is crucial to take into account the frequency dependence of the phonon damping parameter $\gamma_\kappa(\omega)$. If this is not done a finite pure dephasing constant is obtained at $T = 0$ K, which is incorrect! The conclusion of Sapozhnikov (1976) that anharmonic lattice phonons play an important role in the low-temperature optical dephasing, therefore most likely is incorrect.

4.4. Quadratic electron–phonon coupling effects

At higher temperature the effect of quadratic electron–phonon coupling becomes important. In principle, the effect of quadratic electron–phonon coupling on the Redfield relaxation operator could be dealt with in a similar fashion as described earlier. We will not do this but only quote the results obtained earlier by McCumber–Sturge (1963) and Krivoglaz (1964). In our terminology they find:

$$\Lambda^{b}_{12,12} = \pi \sum_{\kappa,\kappa'} |V^{f}_{\kappa\kappa'}|^2 \bar{n}_{\kappa}(\bar{n}_{\kappa}+1)\delta(\omega_{\kappa}-\omega_{\kappa'}), \qquad (30)$$

where $V^{f}_{\kappa\kappa'}$ is the nondiagonal quadratic electron–phonon coupling term and κ and κ' run over all energy conserving band phonon states.

We note here that essentially the same result is obtained when the phonon anharmonicity is taken into account (Krivoglaz 1964). This contribution to the line width from Raman scattering processes, which at low temperature ($T \ll \hbar\omega_{D}/k$) exhibits the well-known T^7 dependence (McCumber and Sturge 1963), however, vanishes at 0 K. De Bree and Wiersma (1979) calculated, using Redfield theory, the effect of quadratic electron–*local* phonon coupling on the line shape and arrived at the following simple expression (one local phonon in ground and excited state):

$$\Lambda^{\ell}_{12,12} = \tfrac{1}{2}[\Gamma_3(T)\,e^{-\omega_{31}/kT} + \Gamma_4(T)\,e^{-\omega_{42}/kT}], \qquad (31)$$

where $\Gamma^{-1}_{3,4}$ and ω_{31}, ω_{42} are the lifetimes and frequencies of the pseudolocal phonons in the ground and excited state.

Note that this contribution to the optical line shape also vanishes at 0 K. It is interesting to note that Krivoglaz (1964) also calculated the effect of localized phonons on the line shape, using a correlation function approach. His result, however, is not identical to the one we obtain using Redfield theory. The reason for the difference is that in the correlation function approach only adiabatic scattering processes are taken into account. For a more detailed discussion of this problem we refer to de Bree and Wiersma (1979).

In conclusion of this section we note that in the formalism presented the different contributions to the Redfield relaxation operator are additive

$$\Lambda_{12,12} = \Lambda^{b}_{12,12} + \Lambda^{\ell}_{12,12}. \qquad (32)$$

The photon echo thus probes these processes simultaneously. However, in a certain temperature range one of these processes may become dominant.

5. Results

5.1. Introduction

The first successful photon echo experiment on a molecular mixed crystal (Aartsma and Wiersma 1978a) was performed on the system pyrene in biphenyl, where it was shown that even at 2 K the optical dephasing time (T_2) could be shorter than $2T_1$ and varies from site to site. Since then a number of other preliminary photon echo studies have been reported (Wiersma 1981), which showed that coherent optical experiments could add much to our understanding of relaxation processes in molecular solids. As of today, however, the number of groups active in this field is very limited; in fact, so far only Fayer and coworkers and Wiersma and coworkers have reported results of photon echo experiments in molecular solids.

Coherent optical experiments other than echoes fortunately have been used to study relaxation in molecular solids. In particular we wish to mention here the FID work of Zewail and coworkers (Orlowski and Zewail 1979) and the photochemical hole burning studies of Völker et al. (1977), Haarer et al. (Friedrich et al. 1980, and references therein), and Small and coworkers (Hayes and Small 1978). This work is discussed in detail in this book by Small (Chapter 9) and Burns et al. (Chapter 7).

To date the molecular crystal most extensively studied by the photon echo technique is the *pentacene in naphthalene* mixed crystal. This system comes close to being the *"organic ruby"*, even qua colour.

In this chapter we will restrict ourselves to a discussion of the most recent photon echo studies of pentacene in several mixed crystals and of naphthalene in durene. We believe that these systems are typical examples in this class of solids and that the results obtained on other systems will only be qualitatively different.

As it becomes increasingly clear that a detailed understanding of coherence effects in molecular crystals necessitates a thorough understanding of the spectroscopy of these systems, we will first discuss the optical spectra of pentacene and naphthalene.

5.2. Pentacene absorption spectrum

5.2.1. Vibronic features

The optical absorption spectrum of the lowest $^1B_{2u} \leftarrow {}^1A_{1g}$ transition of pentacene in a supersonic jet (ter Horst and Kommandeur 1982) is displayed in fig. 12. The spectrum of pentacene in naphthalene at 1.5 K is shown in fig. 13. Comparison of these spectra shows that besides the large

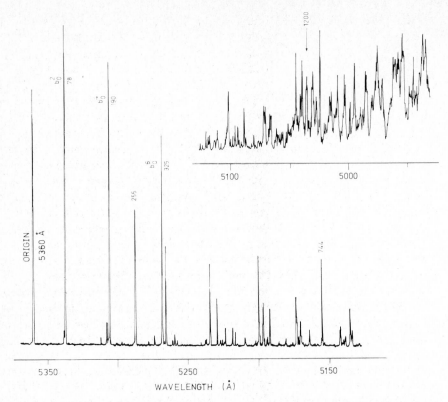

Fig. 12. Optical absorption spectrum of pentacene in a supersonic jet. Estimated rotational temperature is 2 K. After ter Horst and Kommandeur (1982).

gas-to-crystal shift of $\approx 2000 \, \text{cm}^{-1}$ there are large differences in the optical spectra. In particular, the butterfly (b_0^n) motion of pentacene, which dominates the gas phase spectrum (Amirav et al. 1980), is completely absent in the napthalene mixed crystal. Obviously the out-of-plane motion in pentacene is severely hindered by the naphthalene cage. A second interesting effect is observed in the vibrational congested region starting at $\approx 1200 \, \text{cm}^{-1}$ above the origin. While in the gas phase we observe in this region sharp structures in the spectrum, in the crystal the sharpness seems to be washed out, presumably by interaction with the lattice.

In other host lattices, such as p-terphenyl and benzoic acid (de Cola et al. 1980), the absorption spectrum, though shifted, looks similar to the spectrum in fig. 13. However, some differences may occur in the relative intensity of the vibronic absorptions versus the origin. As coherence effects basically probe the dynamics of the interaction of the excited species with its surroundings it is no surprise that these dynamics are host

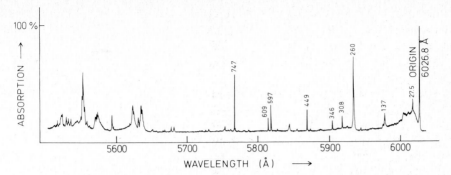

Fig. 13. Optical (*b*-polarized) absorption spectrum of pentacene in naphthalene at 1.5 K. Marked are the frequencies (in cm^{-1}) of the studied vibronic transitions, relative to the electronic origin. After Hesselink and Wiersma (1980).

specific. Of course the general picture of the interaction dynamics is expected to be transferable. The spectral data of the $^1B_{2u} \leftarrow {}^1A_{1g}$ transition of perproto- and perdeuteropentacene in naphthalene are collected in table 1. This table also contains information on the transition frequencies of the pure electronic transition of pentacene in *p*-terphenyl and benzoic acid.

5.2.2. Phonon side band

The optical spectrum of pentacene in naphthalene was previously studied by Prikhotko et al. (1969), Brillante and Craig (1975) and more recently by Lambert and Zewail (1980). As these studies were not accurate enough for our purposes and some of the mixed crystals were too concentrated (Lambert and Zewail 1980) we (Hesselink and Wiersma 1980) decided to re-investigate this spectrum under the highest possible optical resolution and most carefully around the region of the electronic origin. Figure 14 displays the origin region of the pentacene in naphthalene absorption and emission. In the best crystals the origin line shape was Gaussian with a line width of $0.6 \, cm^{-1}$. In none of our mixed crystals did we observe other sites near the origin as reported by Lambert and Zewail (1980).

We now focus attention on the phonon side band accompanying the pure electronic transition (fig. 14). Particularly noteworthy is the sharp line at $27.5 \, cm^{-1}$ in absorption ($26.1 \, cm^{-1}$ for pentacene-d_{14}) and a broader band at $\approx 36 \, cm^{-1}$ in emission ($35 \, cm^{-1}$ for deuteropentacene). Hesselink and Wiersma (1980) showed that these phonon features may be interpreted as optical transitions to overtones $(2 \leftarrow 0)$ of harmonic pentacene librations (localized phonons) in an undisplaced potential well. The librational frequency in the ground state is $17.8 \, cm^{-1}$ and in the excited state it is $13.8 \, cm^{-1}$. For deuteropentacene these numbers are 17.5 and $13.3 \, cm^{-1}$, respectively. Figure 15 shows that the hot $1 \leftarrow 1$ phonon transition may be

Table 1

Spectral data of the $^1B_{2u} \leftarrow {}^1A_{1g}$ transition of pentacene in naphthalene (NT), p-terphenyl (pTP) and benzoic acid (BA)

PTC-h₁₄/NT-h₈		PTC-d₁₄/NT-h₈		PTC-h₁₄/NT-d₈	
ν (vac. cm⁻¹)	intensity[a]	ν (vac. cm⁻¹)	intensity[a]	ν (vac. cm⁻¹)	intensity[a]
16 587.7 (0–0)	1.0	16 613.4 (0–0)	1.0	16 589.8 (0–0)	1.0
16 615.2 (27.5)	0.013	16 640.0 (26.6)	0.02	16 617.9 (28.1)	0.016
16 724.4 (136.7)	0.036	16 745.1 (131.7)	0.06	16 725.7 (135.9)	0.034
16 731.0 (143.3)					
16 819.6 (231.9)	0.002				
16 848.1 (260.4)	0.17	16 866.0 (252.6)	0.23	16 849.5 (259.7)	0.21
16 895.3 (307.6)	0.012	16 902.1 (288.7)	0.023	16 897.2 (307.4)	0.008
16 934.1 (346.4)	0.007	16 928.1 (314.7)	0.014	16 934.1 (344.3)	0.005
17 036.2 (448.5)	0.015	17 028.8 (415.4)	0.014	17 032.2 (447.4)	0.014
17 109.3 (521.6)	0.016	17 120.3 (506.9)	0.03	17 110.1 (520.3)	0.018
17 184.4 (596.7)	0.024	17 188.1 (574.7)	0.06	17 186.2 (596.4)	0.020
17 196.8 (609.1)	0.009	17 203.6 (590.2)	0.006	17 197.8 (608.0)	0.007
17 331.1 (743.4)					
17 334.9 (747.2)	0.044	17 328.5 (715.1)	0.09	17 337.3 (747.5)	0.046
17 350.6 (762.9)					
17 352.9 (765.2)					
17 376.5 (788.8)		17 447.1 (833.7)[b]	0.11		
17 594.9 (1007.2)		17 573.5 (960.1)[b]	0.06		
17 607.3 (1019.6)	0.008	17 810 (1197)[b]			
17 636.1 (1048.4)	0.004	17 956 (1343)[b]			
17 740 (1152)[c]	0.002	17 980 (1367)[b]			
17 790 (1202)[c]		18 023 (1410)[b]			
17 871.7 (1284.0)[c]	0.026	18 032 (1419)[b]			
17 940 (1352)[c]		18 043 (1430)[b]			

PTC-h₁₄/pTP
ν (vac. cm⁻¹)

16 882.7 (0–0, O₁)
16 886.5 (0–0, O₂)
17 005.2 (0–0, O₃)
17 064.7 (0–0, O₄)

PTC/BA
ν (vac. cm⁻¹)

16 998.9 (0–0)

[a]Relative to the 0–0. [b]Asymmetric bands, probably containing unresolved structure. [c]Very broad structures.

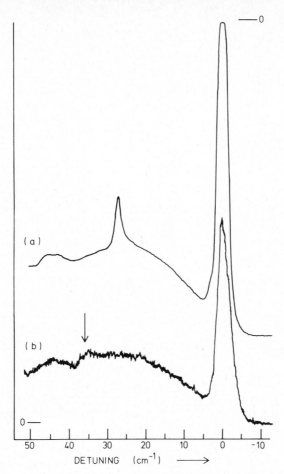

Fig. 14. Detailed absorption (a) and fluorescence (b) spectrum of the phonon side band of the origin of pentacene in naphthalene at 1.5 K. Note that in (a) the phonon side band is situated at higher energy and in (b) at lower energy relative to the zero-phonon line. Also note that the intensity of the zero-phonon line in emission is drastically reduced by the effect of re-absorption in this (concentrated) crystal. After Hesselink and Wiersma (1980).

observed at elevated temperature. We further note that the absence of the $1 \leftarrow 0$ phonon transition means, in terms of electron–phonon coupling, that the *linear* electron–phonon coupling for this (localized) phonon vanishes.

The broad underlying background on which the sharp line is superimposed is due to linear coupling of the electronic transition to the host acoustic phonon branch. The strength of this coupling is usually defined in terms of a Debye–Waller factor $\exp(-M)$, where $\exp(-M)$ is the ratio of the intensity of the zero-phonon line to the total intensity of the zero-

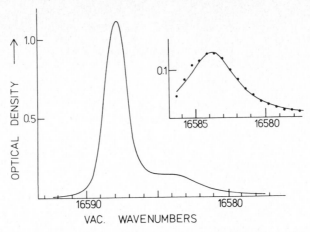

Fig. 15. Absorption spectrum of the zero-phonon line of pentacene in naphthalene at 16 K. The insert shows the spectrum of the side band obtained by subtracting the main (0–0) peak from the spectrum. It was assumed that the 0–0 band is symmetrical. The solid line is a fit to a Voigt line profile with $\Delta\nu_G = 2.0\,\mathrm{cm}^{-1}$ and $\Delta\nu_L = 2.0\,\mathrm{cm}^{-1}$. After Hesselink and Wiersma (1980).

phonon line and associated phonon side band:

$$M = \sum_s \xi_s^2 (\omega_s^i / \omega_s^f)^2 (2\bar{n}_s + 1), \tag{34}$$

where ξ_s is the shift in equilibrium position of phonon s upon excitation, ω_s^i and ω_s^f are the phonon frequency in the initial and final state of the guest, respectively, and $\bar{n}_s$ is the average phonon occupation number. In a following section we will show that both these host acoustic phonons and localized guest librations play an important role in the dephasing.

We conclude this subsection by noting that a similar situation occurs in the system pentacene in p-terphenyl, be it that the guest librational frequency is much higher ($\approx 30\,\mathrm{cm}^{-1}$, de Bree 1981b).

5.3. Naphthalene spectrum

In the early fifties McClure (1954, 1956) studied in great detail the absorption and emission spectrum of naphthalene in durene. As photon echo studies have only been performed on the pure electronic transition, we will limit ourselves to a discussion of the zero-phonon region. Figure 16 shows the absorption spectrum of a very concentrated crystal of naphthalene in durene. The arrow in the spectrum points to a phonon feature ($\approx 18.2\,\mathrm{cm}^{-1}$) whose frequency matches the observed activation energy of the photon echo relaxation (Aartsma and Wiersma 1978b). We believe that this phonon again is a guest libration, but in this case the local phonon oscillator is

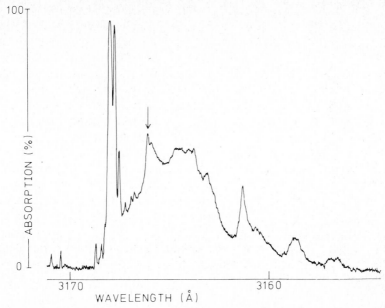

Fig. 16. Absorption spectrum of naphthalene in durene at 1.5 K. The sharp lines to the high energy side of the 0–0 line arise from ^{13}C isotopic impurities. After Aartsma (1978).

displaced upon excitation. This may be of importance for the interpretation of the low-temperature echo relaxation as remarked in subsection 4.3. Chereson et al. (1976) have reported observation of a $19\,cm^{-1}$ localized phonon in the ground state of naphthalene in durene.

Careful optical experiments are needed on both proto- and deuteronaphthalene to further characterize the nature of the phonon side band structure in this system before a more detailed interpretation of the photon echo results (than that given by Aartsma and Wiersma 1978b) becomes possible.

5.4. Photon echo relaxation

5.4.1. Low-temperature relaxation
The first detailed low-temperature photon echo relaxation study in molecular crystals was made by Morsink et al. (1977), of the system pentacene (PTC) and deuteropentacene in p-terphenyl. In very dilute mixed crystals (10^{-8} mole/mole) they measured at 1.5 K for PTC-h_{14} a fluorescence lifetime of 23.5 ± 1 ns (O_2) and a photon echo decay time of 45 ± 1 ns. For PTC-d_{14} the corresponding numbers were 27.5 ± 1 ns and 54 ± 1 ns. Morsink et al. conclude that for *isolated* pentacene molecules in p-terphenyl at low temperature $T_2 = 2T_1$. This implies that the low-temperature homogeneous

line width is only determined by radiative and radiationless decay. For PTC-h_{14} and PTC-d_{14} these homogeneous line widths are 7.1 and 5.9 MHz, respectively, which is almost a factor of 10^4 less than the inhomogeneous width.

Cooper et al. (1980) recently reported results of a photon echo relaxation study on pentacene in naphthalene. Their results are shown in fig. 17. The data presented clearly indicate that in this case at 1.4 K the photon echo decay is *faster* than expected from the fluorescence decay only: $T_2 < 2T_1$. A similar observation was recently made in our laboratory (Morsink et al. 1982) on the system pentacene in benzoic acid. In fig. 18 the low-temperature photon echo and fluorescence decay of this system are shown. Here the echo decay time is about half the fluorescence lifetime. Also in the system of naphthalene in durene the optical T_2 is much shorter than $2T_1$. Table 2 contains the relevant optical and relaxation data of these mixed crystals.

The original interpretation of this effect offered by Fayer and co-workers (Cooper et al. 1980), namely quadratic electron–phonon coupling to har-

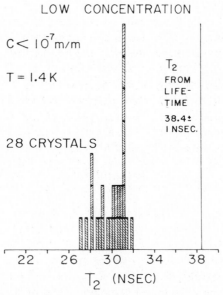

Fig. 17. Low concentration limit T_2 measurements on 28 pentacene in naphthalene mixed crystals with $<10^{-7}$ m/m concentration. The line at the right is the S_1 lifetime, i.e. the T_2 value which would be observed if the lifetime were the only broadening mechanism. The average T_2 value is 30 ns. After Cooper et al. (1980).

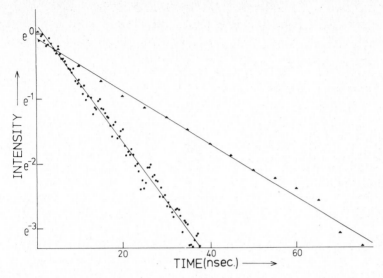

Fig. 18. Photon echo (●) and fluorescence (▲) decay of pentacene in benzoic acid at 1.5 K. After Morsink et al. (1982).

monic acoustic phonons, was not confirmed by their most recent experiments. Theoretically a T^7 temperature dependence is expected (subsection 4.4), while, in the range 1.4–2.2 K, none is observed. Skinner et al. (1981) recently speculated that the remaining low-temperature discrepancy between T_2 and $2T_1$ is caused by the fact that a two-level-coupled-to-a-phonon-bath description for these optical excitations is inadequate. They propose that a more realistic description is achieved by considering a

Table 2

Low-temperature (1.5 K) fluorescence (τ_{fl}) and photon echo ($\frac{1}{2}T_2$) decay times of pentacene and naphthalene in mixed crystals.

	Pentacene			Naphthalene durene[d]
	p-terphenyl[a]	naphthalene[b]	benzoic acid[c]	
transition ν (vac. cm^{-1})	16 886.5(O₂)	16 587.7	16 998.9	31 556.5
τ_{fl} (ns)	23.5 ± 1	19.2 ± 0.5	23 ± 2	180 ± 10
$\frac{1}{2}T_2$ (ns)	22.5 ± 1	15 ± 1	10 ± 1	80 ± 5

[a]Morsink et al. (1977). [b]Cooper et al. (1980). [c]Morsink et al. (1982). [d]Morsink et al. (1980).

quasi-two-level system, where the ground and excited states consist of multiplets. These multiplets arise from coupling of the electronic excitation to acoustic phonon levels. For this case they obtain the following correlation function for the photon echo amplitude:

$$\bar{p}(2t_1) \propto \langle \hat{\mu}(-t_1)\hat{\mu}(0)\hat{\mu}(t_1)\hat{\mu}(0)\rangle_0, \tag{35}$$

where $\hat{\mu}$ is the electronic transition dipole.

For times long compared to the characteristic phonon correlation times, this three-times correlation function can be decomposed into a product of two-times correlation functions with the result:

$$\bar{p}(2t_1) \propto \langle \hat{\mu}(-t_1)\hat{\mu}(0)\rangle_0 \langle \hat{\mu}(t_1)\hat{\mu}(0)\rangle_0, \tag{36}$$

which implies that

$$\bar{p}(2t_1) \propto \left| \int_{-\infty}^{\infty} d\omega\, e^{-i\omega t_1} I(\omega) \right|, \tag{37}$$

in agreement with the relation derived earlier (for a two-level system) in subsection 4.1. The important result obtained here is that only for very short excitation pulses ($\tau\omega_D \ll 1$, where ω_D is the Debye frequency) will the Fourier transform of the photon echo decay be identical to the homogeneous line shape.

The important point of the Skinner paper is that, if the optical excitation studied is not a real two-level system, there is no a priori reason why the photon echo decay at $T = 0\,K$ should be identical to the fluorescence decay. Another way of phrasing this is, that the inequality of the two marks the breakdown of the Markov approximation for the initial (two-level system) reduction of the degrees of freedom of the system (mixed crystal) into "molecule" and "bath" variables. What certainly remains to be done is to generate a "mixed crystal Hamiltonian" that relates the low-temperature discrepancy between T_2 and $2T_1$ to measurable (optical) properties of the mixed crystal.

In ending this section it is interesting to note that Skinner et al. (1981) for a quasi-two-level system also derive an expression for the amplitude of the stimulated photon echo:

$$\bar{p}(2t_1 + t_2) \propto [\langle \hat{\mu}(-t_1-t_2)\hat{\mu}(-t_2)\hat{\mu}(t_1)\hat{\mu}(0)\rangle_0$$

$$+ \langle \hat{\mu}(-t_1-t_2)\hat{\mu}(0)\hat{\mu}(t_1)\hat{\mu}(-t_2)\rangle_0], \tag{38}$$

where t_1 is the separation between the first and second and t_2 between the second and third excitation pulse. The decay of the stimulated echo is thus related to a four-times correlation function. Interestingly enough, in the mixed crystals, e.g. naphthalene in durene, where at low temperature $T_2 < 2T_1$, the decay of the stimulated echo is identical to the fluorescence lifetime. This, in our opinion, shows that the mystery of the low-temperature remnant pure dephasing has not been completely resolved.

5.4.2. *Stimulated photon echo and intersystem crossing*

Whereas the two-pulse photon echo (2PE) measures the loss of coherence, the stimulated photon echo (3PSE) probes the flow of population in the optical pumping cycle. In subsection 3.4 we showed that, in the case the optical pumping cycle involved three levels, the decay of the 3PSE could become quite complex. On the other hand, analysis of this decay yields new information on the relaxation parameters of the system.

In our laboratory the 3PSE decay of pentacene in *p*-terphenyl and naphthalene and of naphthalene in durene was measured. In the mixed crystal of pentacene in *p*-terphenyl (O_1, O_2) the 3PSE decay was exactly half the fluorescence lifetime as expected for a two-level system [eq. (7), case (a)]. De Vries et al. (1977), de Vries and Wiersma (1979) and Orlowski and Zewail (1979), using narrow band laser excitation, showed that in these sites the $T_1 \leftarrow S_1$ intersystem crossing yield is extremely low (<0.4%), which confirms the "two-level" nature of the optical transitions in these sites. This situation sharply contrasts with the case of pentacene in naphthalene. The decay of the 3PSE in this system (Morsink 1982) is shown in fig. 19. This experiment was done under single-shot conditions

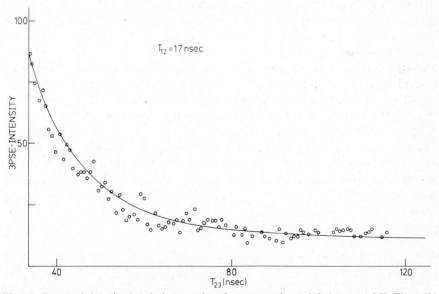

Fig. 19. Decay of the stimulated photon echo of pentacene in naphthalene at 1.5 K. The solid line is a fit to eq. (10) with $k_{23}/(k_{21} + k_{23}) = 0.15$ ($k_{31} \ll k_{21}, k_{23}$). After Morsink (1982).

(section 3), which establishes that the observed tail at long probe pulse delay times is not caused by an effect of accumulation.

We note that the 15 μs lifetime of the tail in the 3PSE decay, not shown in fig. 19, is in agreement with the EPR measurements of Van Strien and Schmidt (1980), who found dominant intersystem crossing to the Z spin substrate with 18 μs lifetime.

Using the known numbers of k_{21} and k_{31} we have attempted to fit the observed decay to the expression given in eq. (10). A reasonable fit is obtained by taking for $k_{23} = 8.7 \times 10^6\,\mathrm{s}^{-1}$. The total yield for intersystem crossing in this system is then found to be 15%.

We note that this number far exceeds our original estimate (de Vries and Wiersma 1979) and that of Lambert and Zewail (1980). Using the values obtained by van Strien and Schmidt (1980) for the relative intersystem crossing rates we can calculate the intersystem crossing yield for each spin substate.

Using the decay of the 3PSE as a probe, Morsink (1982) recently also measured the intersystem crossing yield in the system naphthalene in durene. As the triplet state lifetime is very long ($\approx$2.5 s, McClure 1949), in order to avoid accumulation effects, the repetition rate of the experiment needed to be taken $\leqslant$0.1 Hz. The decay of the 3PSE in this system was shown in fig. 2 together with a fit based on eq. (10). With $k_{21} = 5.3 \times 10^6\,\mathrm{s}^{-1}$ (Morsink 1982) we calculate $k_{23} = 2.4 \times 10^6\,\mathrm{s}^{-1}$ or that the intersystem crossing yield is 45%. We conclude that 3PSE decay measurements in combination with pulsed EPR measurements are a powerful tool to determine the intersystem crossing yield.

5.4.3. Electronic dephasing

While the effect of the anharmonic acoustic phonons on dephasing seems to be minor at low temperature, the local phonons seem to dominate the relaxation at elevated temperature. The importance of local phonons in dephasing was first recognized by Aartsma and Wiersma (1976) (see also Aartsma 1978) and then demonstrated in a number of other systems (Gorokhovski and Rebane 1977, Völker et al. 1978). With picosecond photon echoes (Hesselink and Wiersma 1978, 1979a) it has recently become possible to study dephasing in the system pentacene in naphthalene up to 20 K (Hesselink and Wiersma 1980), which, in combination with an optical study, revealed some new aspects of the local phonon induced relaxation process.

The most pertinent new result is shown in fig. 20, obtained by the technique of the accumulated 3PSE (Hesselink and Wiersma 1979b, 1981a). In a separate experiment it was established that the fluorescence lifetime of pentacene (19.5 ± 1 ns) was independent of temperature up to 20 K. This ascertains that the observed temperature effect on the echo relaxation is a

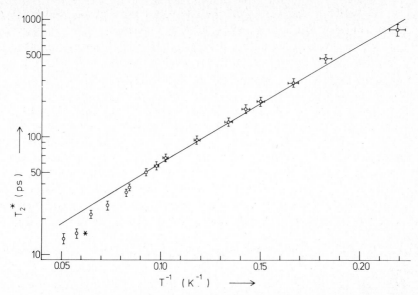

Fig. 20. Pure dephasing time T_2^* of the electronic origin of pentacene in naphthalene as a function of the inverse temperature. The results were obtained with the 3PSE experiments. The solid line is a fit to eq. (31) with numbers specified in the text. The asterisk at 16 K is calculated with $\Gamma_3^{-1} = 2.2$ ps and $\Gamma_4^{-1} = 6.8$ ps obtained from spectroscopic measurements. After Hesselink and Wiersma (1980).

genuine dephasing effect. In addition we have to note here that the echo relaxation constant $T_2^*(T)$ was not corrected for the anharmonic phonon coupling effect discussed in subsection 5.4. Inspection of eq. (29) shows that this effect is of minor importance in the temperature range studied.

We assume then that the observed dephasing basically may be under-stood on the basis of the four-level structure in fig. 11. In section 4.4 we pointed out that on the basis of the theory of de Bree and Wiersma (1979) (dBW) the echo relaxation constant for this four-level system should be of the form

$$\Gamma_{12}(T) = \tfrac{1}{2}[\Gamma_3(T) \exp(-\omega_{31}/kT) + \Gamma_4(T) \exp(-\omega_{42}/kT)], \qquad (31)$$

where Γ_3^{-1}, Γ_4^{-1} and ω_{31}, ω_{42} are the local phonon lifetimes and frequencies in the lower and upper state, respectively. The straight line in fig. 20 is a fit to eq. (31) for the data below 10 K, where Γ_3 and Γ_4 are assumed to be temperature independent. From the fit we calculate $\Gamma_3^{-1} = 3.5$ ps and $\Gamma_4^{-1} = 11$ ps. We note here that the calculated drastic difference in local phonon lifetimes in the upper and lower electronic state is in agreement with the spectral observation displayed in fig. 14, where the local phonon band in absorption $(2 \leftarrow 0)$ is much sharper than in emission. It may be useful to remark here that, in a limited temperature range, an excellent fit to the echo relaxation can also be obtained by a single exponential with an activation energy of $16 \, cm^{-1}$! This again points to the importance of combining echo relaxation data with optical data.

Obviously above 10 K this simple picture breaks down and the temperature dependence of the local phonon lifetime has to be considered. Information on the temperature dependence of the lifetime of the local phonon may be obtained from a study of the line width of the hot $1 \leftarrow 1$ phonon side band. Hesselink and Wiersma (1980) showed that by taking this temperature dependence into account, the dephasing could be described up to 20 K, consistently, by eq. (31).

The effect of (quadratic) coupling to the acoustic phonons seems to be unimportant in this temperature range. The only example we know of, where quadratic electron–acoustic phonon coupling seems to play a role, is the system of triphenylmethyl in triphenylamine (Hesselink and Wiersma 1977; see also Wiersma 1981). In this mixed crystal system the guest and host are so similar (in fact isoelectronic) that local guest librations are not formed and therefore the coupling to the acoustic phonons becomes important.

We end this section by commenting on the lifetime of the pentacene librational mode. In the dBW theory the local modes are expected to decay through local phonon–band phonon anharmonic interactions. In pure naphthalene, Hesp et al. (1981) showed that the lowest frequency librational mode of $58 \, cm^{-1}$ has a lifetime of 250 ps. For the corresponding pentacene librational mode of $18 \, cm^{-1}$ in the ground state we obtain a lifetime of 3.5 ps. As lower frequency modes are expected to decay slower (*ceteris paribus*) we conclude that the phonon-induced anharmonicity of the pentacene libration exceeds that of the naphthalene host libration by, at least, a factor of 10. This clearly shows that the insertion of an impurity in a host lattice introduces a local strain. The lifetimes of the localized librations thus, in a sense, probe the local lattice distortion.

5.4.4. Vibronic and vibrational dephasing

With the accumulated 3PSE technique, dephasing in a number of vibronic transitions was also studied by Hesselink and Wiersma (1980). Here we will only discuss the temperature dependence of vibronic dephasing. The interesting observation is that for all *vibronic* transitions studied the observed dephasing could be described with the same parameters as in the pure electronic transition.

This first of all implies that the vibrational relaxation process itself (T_1) is not dependent on temperature (or only weakly so) in the temperature region studied. The second implication is that the local phonon frequency is independent of vibrational excitation. An important consequence of this fact is that *exchange* will be important in Raman (vibrational) dephasing (Harris 1977, Marks et al. 1980) of ground and excited state vibrations of pentacene in naphthalene. We remind the reader here that optical exchange becomes important when the relation $\tau\Delta\omega \ll 1$ holds. Here $\Delta\omega$ is the frequency difference between the coupled transitions and τ the lifetime of the local phonon. When the diagonal quartic local phonon–vibron coupling, which determines $\Delta\omega$, is negligible, the effect of optical exchange on dephasing also become negligible. It seems certainly of interest to pursue a study of Raman dephasing in these systems.

5.5. *Intermolecular interaction dynamics*

So far we have, implicitly, been assuming that the impurity studied was isolated, in the sense that the intermolecular guest–guest interaction is negligible. Morsink et al. (1977) reported that in the system pentacene in *p*-terphenyl at higher guest concentrations the echo decay becomes faster. They interpreted this as evidence for intermolecular energy transfer. Cooper, Olson and Fayer (1980) (COF) made the first systematic 2PE and 3PSE study of pentacene in naphthalene as a function of the pentacene concentration. Figure 21 shows their result, which indicates a linear relationship between the concentration-dependent part of the homogeneous line width $(\pi T_2^c)^{-1}$ and the pentacene concentration. COF further reported that the decay of the stimulated echo was independent of concentration. In this experiment the time separation between the first and second excitation pulses was approximately 1 ns.

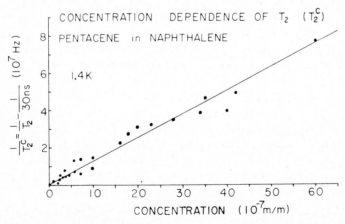

Fig. 21. Concentration-dependent photon echo decay rate $(T_2^c)^{-1}$ vs concentration. Decay rate (line width) is a linear function of concentration over two orders of magnitude. After Cooper et al. (1980).

A similar study was recently made on the system naphthalene in durene by Morsink, Kruizinga and Wiersma (1980) (MKW). The results of this experiment are given in fig. 22. The data, which show significant scatter, seem to indicate that at higher guest concentrations saturation of the dephasing effect occurs. MKW also studied the effect of higher guest concentrations on the decay of the 3PSE and, up to the onset of saturation in the 2PE decay, found none. At higher guest concentrations they were not able to measure the decay of the 3PSE.

Three different interpretations of the observed effects have been proposed. The first one, offered by Cooper et al. (1980) assumes that phonon-induced fluctuations of the transition dipole–dipole intermolecular interaction causes the observed effect. MKW criticized this model and in the meantime Fayer and co-workers, after a more detailed evaluation of its effect on the 2PE and 3PSE photon echo decay, have dropped it. We will therefore not go into the details of this mechanism.

MKW suggested that the *static* dipole–dipole coupling was responsible for the concentration-dependent dephasing. A necessary assumption was that the inhomogeneous broadening is mainly macroscopic, in the sense that the crystal consists of large *domains* of resonant guest molecules. In the absence of guest–guest dipole–dipole coupling the resonant molecules are isolated from each other. With dipole–dipole coupling the degeneracy among the states is removed and the optical transition is dipolar broadened.

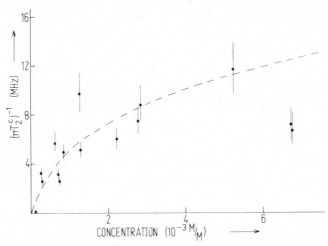

Fig. 22. Concentration-dependent contribution to the homogeneous line width $(\pi T_2^c)^{-1}$ of naphthalene in durene as a function of concentration at 1.5 K. The dashed curve is the *predicted* theoretical behaviour assuming a Gaussian band shape for the quasi-exciton band. After Morsink et al. (1980).

This dipolar-broadened line may also be looked at as a quasi-exciton band of the randomly distributed guest system.

This quasi-exciton band is not quite comparable to the usual exciton band, where the number of excitations is assumed to be much smaller than the number of lattice sites. On the contrary, in the quasi-exciton band we deal with here, the number of excitations may be comparable to the number of participating lattice guest sites. The dephasing then is due to destructive interference in the emission of all *coherently excited* clusters, containing random frequency spacings, in the mixed crystal. We note that a guest quasi-exciton band structure picture was earlier proposed by Port et al. (1975) in their study of energy transfer in isotopically mixed crystals of naphthalene.

A quantitative calculation of this dipolar broadening was attempted by MKW, which showed that, under certain conditions, agreement between theory and experiment was possible. The calculation made, however, was only correct for small rotations of the Bloch vector. Warren and Zewail (1981) recently improved the theoretical framework of the theory used by MKW to calculate the optical dipolar broadening effect. They arrive, for an initial pulse flip angle of $\pi/2$, at the following simple result for the concentration-dependent dephasing parameter T_2^c:

$$1/T_2^c = 1.51 f\mu^2 a^{-3}\hbar^{-1} \quad (f \ll 1),$$

where f is the impurity fraction, μ the transition dipole and a an average lattice parameter. For the case of pentacene in p-terphenyl the authors find good agreement between theory and experiment.

A word of caution, however, here is necessary. In a recent paper by the Fayer group (Olson et al. 1982) it was shown that their earlier obtained concentration-dependent dephasing had to be re-interpreted as an optical density effect. Whether this also applies to the results reported by Morsink et al. (1980) remains to be established. It seems that there is an urgent need for reliable photon echo measurements on concentrated mixed crystals of low enough optical density to probe into the mechanism of intermolecular interactions. This is of great importance, not the least because of the possibility to obtain new information on the magnitude of the local and macroscopic inhomogeneous broadening and its role in energy transport.

5.6. Vibrational relaxation

A process of fundamental importance in molecular crystals is vibrational relaxation. Rebane and co-workers (Rebane and Saari 1978) were the first to study vibrational relaxation in mixed crystals by measuring the un-relaxed (hot) fluorescence. De Vries and Wiersma (1976) showed that photochemical hole burning can also be used to study this process. Völker and Macfarlane (1979) made the first systematic study of vibrational relaxation in porphin in n-octane using this technique.

In this section we will present and discuss results obtained by Hesselink and Wiersma (1981b) on vibrational relaxation by measuring the photon echo decay of vibronic transitions in pentacene mixed crystals. The results were obtained by using the accumulated 3PSE method.

As in the accumulated 3PSE method *phase* relaxation times are measured, it first was necessary to determine the contribution of pure dephasing to the phase relaxation time. At very low temperature ($\leqslant 1.5$ K) we can assume this contribution to be negligible for the following reasons:

(1) In subsection 5.4.1 we have shown that at low temperature no pure dephasing processes occur on a picosecond time scale.

(2) For one particular vibration, the $747\,\mathrm{cm}^{-1}$ mode of pentacene in naphthalene, the population relaxation time (T_1) was measured directly, using the 3PSE, and the result is consistent with the situation where $T_2 = 2T_1$. The experimentally obtained decay curve is shown in fig. 23, and as the excited singlet state acts here as a bottleneck, eq. (7) case (b) was used to fit the decay curve. Note that the expected 4:1 ratio between the intensity at $t_{23} = 0$ and $t_{23} \to \infty$ is actually observed.

We thus conclude that at low temperature the phase relaxation times are completely determined by population relaxation.

In tables 3 and 4 the measured relaxation times of the transitions studied are summarized. Relaxation times longer than 2 ps could be measured

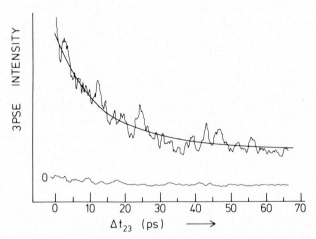

Fig. 23. Decay curve of the stimulated photon echo (single-shot type) of the $747\,\mathrm{cm}^{-1}$ vibronic transition of pentacene in naphthalene, at 1.5 K. The solid line is a fit to eq. (7) case (b) with $T_1 = 33$ ps.

Table 3

Phase relaxation times at 1.5 K of the vibronic transitions of pentacene in naphthalene.

PTC-h_{14}/NT-h_8		PTC-d_{14}/NT-h_8		PTC-h_{14}/NT-d_8	
ν^a (cm^{-1})	$\frac{1}{2}T_2$ (ps)	ν^a (cm^{-1})	$\frac{1}{2}T_1$ (ps)	ν^a (cm^{-1})	$\frac{1}{2}T_2$ (ps)
136.7	2.3 ± 0.4^b	131.7	1.5 ± 0.5^c	135.9	7.5 ± 2.5^c
260.4	2.4 ± 0.4^b	252.6	3.7 ± 0.7^b	259.7	3.5 ± 1^c
307.6	19 ± 2	288.7	14 ± 2	307.4	20 ± 2
346.4	8.5 ± 1	314.7	13 ± 1	344.3	14 ± 4
448.5	15 ± 1.5	415.4	30 ± 4	447.4	19 ± 2
521.6	1.5 ± 0.2^c	506.9	1.4 ± 0.2^c	520.3	1.7 ± 0.2^c
596.7	19.5 ± 1.3	574.7	13 ± 1.5	596.4	25 ± 2
609.1	59 ± 3	590.2	15 ± 1.5	608.0	47 ± 4
747.2	33 ± 1.5	715.1	26 ± 2	747.5	27 ± 2
		834	11 ± 2^d		
		960	10 ± 1.5		
		1197			
		1342			
		1367	<2		
		1409			
		1418			
		1430			

^{a}Relative to the 0–0.
^{b}From line width and photon echo experiments.
^{c}From line width measurements.
^{d}Decay showing beat pattern.

Table 4

Phase relaxation times at 1.5 K of the vibronic transitions of pentacene in *p*-terphenyl at the O_3 and O_4 sites.

O_3		O_4	
ν^a (cm^{-1})	$\frac{1}{2}T_2$ (ps)	ν^a (cm^{-1})	$\frac{1}{2}T_2$ (ps)
267.1	2.1 ± 0.5^b	267.8	2 ± 0.5^b
599.4	18 ± 3	599.4	10 ± 1
608.3	19 ± 3	606.9	32 ± 5
746.9	18 ± 3	744.9	31 ± 3

^{a}Relative to the 0–0.
^{b}From photon echo and line width measurements.

W.H. Hesselink and D.A. Wiersma

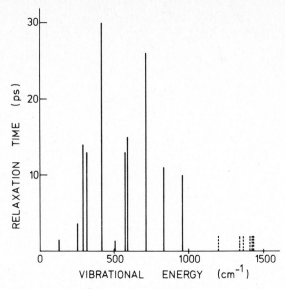

Fig. 24. Phase relaxation times $(\frac{1}{2}T_2)$ versus excess vibrational energy in the system pentacene-d_{14} in naphthalene at 1.5 K.

directly and were, within experimental error, exponential. Relaxation times shorter than 2 ps were calculated from the spectral line width, with the aid of the tabulated values for the Voigt profile (Posener 1959).

Figure 24 shows a plot of the measured vibrational relaxation times versus excess vibrational energy for the system pentacene-d_{14} in naphthalene. Notice the erratic behaviour of the relaxation times versus excess vibrational energy. We will only briefly comment on the results here and refer to Hesselink and Wiersma (1981b) for a more thorough discussion. The main conclusion is that there are basically three regions of excess vibrational energy (ΔE_v) which need to be considered.

In region I where $\Delta E_v \lesssim 2\omega_D$, with ω_D the Debye frequency, vibrational relaxation is fast and occurs by a direct one- or two-phonon relaxation process. In region II, where $2\omega_D < \Delta E_v < 1000 \text{ cm}^{-1}$, the vibrational relaxation time varies in an erratic way as a function of excess vibrational energy. Relaxation occurs here through mixed phonon–vibron anharmonic interactions in the sense that one vibration decays into another vibration plus lattice phonon. This process is very much dependent on the precise level structure and *local* anharmonicity. In region III $(\Delta E_v \gtrsim 1000 \text{ cm}^{-1})$ vibrational relaxation is a fast process again. The density of molecular "background" accepting states is so high that it approaches a continuum. In this case it is therefore the *intramolecular* anharmonicity that governs

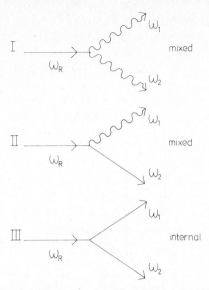

Fig. 25. Pictorial representation of the break-up (decay) of an internal vibrational mode (ω_R) into phonons ($\sim$) and/or other internal vibrations (—) through cubic anharmonic vibron–phonon (I, II) or vibron–vibron (III) interactions.

the decay. The decay mechanisms for the three regions of excess vibrational energy are shown in a pictorial way in fig. 25.

We note that this model also explains the observed relaxation pattern in porphin (Völker and Macfarlane 1979). Obviously, however, the range of region II is determined by the size of the molecule. The explanation of the precise variation of the vibrational relaxation time versus excess vibrational energy in region II, however, remains a challenge. It seems that study of a vibration-wise simpler molecule than pentacene offers better prospects for success.

6. Epilog

It seems a legitimate question at this point to ask: what did we *learn* from photon echoes and what are the future prospects of this technique for the understanding of relaxation phenomena in molecular solids.

The main accomplishment of photon echo studies in this area so far, in our opinion, has been the elucidation of the role that phonons play in the destruction of optical memory. In particular, the role of pseudolocal phonons (librons) in the dephasing process has been established. More work in this area, however, remains to be done, specifically on dephasing in mixed molecular crystals, where local phonons can *not* be formed.

Another area where important progress can be made using photon echoes is the field of energy transport in molecular solids. The photon echo is a much more sensitive probe here than fluorescence to reveal communication among different sites. The interpretation of the results here is hampered by a lack of knowledge on the nature of the inhomogeneous broadening in optical transitions of molecular mixed crystals.

Photon echoes may also play an important role in the study of vibrational relaxation. We emphasize here that stimulated photon echoes may be employed to identify the decay channels. A prerequisite is that the inhomogeneous broadening of different vibronic transitions is correlated. If so, a "grating" formed on a particular transition will relax downwards and a probe pulse at lower lying vibronic transitions, involved in the relaxation, will stimulate an echo. The rise and fall of such "delayed grating echoes" will inform us on the relaxation path. So far such studies have only been done using hot band emission spectroscopy, the results of which are not easy to interpret.

Study of exciton and polariton scattering by coherent optical techniques seems also within reach. There is a fundamental problem here, however. To generate a photon echo, the system must be excited in a macroscopic "coherent state". With a $\pi/2$ pulse as the first excitation pulse, 50% of *all* molecules will be excited, which in a pure solid implies that we are outside the usual "single-particle excitation" limit which is used to describe Frenkel excitons. The band structure of these highly excited solids will thus become excitation-degree dependent. At high excitation densities exciton–exciton annihilation/scattering may dominate the coherence decay. To really study exciton–phonon scattering of Frenkel-type excitons other techniques, such as optical free induction decay and CARS free induction decay, seem more suitable than photon echoes. The reason is that observation of free induction decay, in contrast to photon echoes, is possible for very small rotations of the Bloch vector (low excitation degree).

Finally we propose that photon echoes may be used to study certain aspects of solid state photochemistry, especially when metastable products are formed.

In conclusion of this review we confess that the artistry aspects of photon echoes remain a joyful element in this type of work.

Acknowledgments

We are greatly indebted to our co-workers for their invaluable support in preparing this chapter.

In particular we are very much indebted to Jos B.W. Morsink and Philip de

Bree for their help in formulating the theory of photon echo formation and Redfield relaxation in molecular crystals. The permission of professor Fayer to reproduce figs. 17 and 21 of this text is gratefully acknowledged. We are also thankful to professor Skinner for correspondence on the dephasing, induced by linear coupling to anharmonic acoustic phonons.

References

Aartsma, T.J., J.B.W. Morsink and D.A. Wiersma, 1977, Chem. Phys. Lett. **47**, 425.

Aartsma, T.J. and D.A. Wiersma, 1976, Chem. Phys. Lett. **42**, 520.

Aartsma, T.J., 1978, thesis (Univ. of Groningen, The Netherlands).

Aartsma, T.J. and D.A. Wiersma, 1978a, Phys. Rev. Lett. **36**, 1360.

Aartsma, T.J. and D.A. Wiersma, 1978b, Chem. Phys. Lett. **54**, 415.

Abella, I.D., N.A. Kurnit and S.R. Hartmann, 1966, Phys. Rev. **141**, 391.

Abella, I.D., 1969, in: Progress in Optics, ed. E. Wolf, vol. VII (North-Holland, Amsterdam) p. 158.

Allen, L. and J.H. Eberly, 1975, Optical Resonance and Two-Level Atoms (Wiley, New York).

Amirav, A., U. Even and J. Jortner, 1980, Chem. Phys. Lett. **72**, 21.

Berne, B.J. and G.D. Harp, 1970, Adv. Chem. Phys. **17**.

Brewer, R.G., 1977, Phys. Today **30**, 50.

Brewer, R.G. and R.L. Shoemaker, 1972, Phys. Rev. A **6**, 2001.

Brillante, A. and D.P. Craig, 1975, J. Chem. Soc. Faraday II, **71**, 1457.

Burke, F.P. and G.J. Small, 1974a, J. Chem. Phys. **61**, 4588.

Burke, F.P. and G.J. Small, 1974b, Chem. Phys. **5**, 198.

Chan, C.K. and S.O. Sari, 1974, Appl. Phys. Lett. **25**, 403.

Chereson, P.H., P.S. Friedman and R. Kopelman, 1976, J. Chem. Phys. **56**, 3716.

Cohen-Tannoudji, C., 1977, in: Frontiers in Laser Spectroscopy, vol. 1, Les Houches 1975, eds. R. Balian, S. Haroche and S. Liberman (North-Holland, Amsterdam).

Cooper, D.E., R.D. Wieting, R.W. Olson and M.D. Fayer, 1979, Chem. Phys. Lett. **67**, 41.

Cooper, D.E. R.W. Olson and M.D. Fayer, 1980, J. Chem. Phys. **72**, 2332.

de Bree, P., 1981a, Thesis (Univ. of Groningen, The Netherlands).

de Bree, P., 1981b, this laboratory, unpublished results.

de Bree, P. and D.A. Wiersma, 1979, J. Chem. Phys. **70**, 790.

de Cola, P., R.M. Hochstrasser, A.G. McGhie and H.P. Trommsdorff, 1980, J. Chem. Phys. **73**, 4695.

Dugay, M.A. and J.W. Hansen, 1968, Appl. Phys. Lett. **13**, 178.

de Vries, H. and D.A. Wiersma, 1976, Phys. Rev. Lett. **36**, 91.

de Vries, H., P. de Bree and D.A. Wiersma, 1977, Chem. Phys. Lett. **52**, 399; 1978, Erratum **53**, 418.

de Vries, H. and D.A. Wiersma, 1979, J. Chem. Phys. **70**, 5807.

Feynman, R.P., F.L. Vernon and R.W. Hellwarth, 1957, J. Appl. Phys. **28**, 49.

Flusberg, A., T. Mossberg, R. Kachru and S.R. Hartmann, 1978, Phys. Rev. Lett. **41**, 305.

Friedrich, J., J.D. Swalen and D. Haarer, 1980, J. Chem. Phys. **73**, 705.

Gorokhovski, A.A. and L.A. Rebane, 1977, Opt. Commun. **20**, 144.

Hahn, E.L., 1950a, Phys. Rev. **77**, 297.

Hahn, E.L., 1950b, Phys. Rev. **80**, 580.

Hänsch, T.W., 1968, Appl. Opt. **11**, 895.

Harris, C.B., 1977, J. Chem. Phys. **67**, 5607.
Hartmann, S.R., 1969, in: Proc. of the Intl. School of Physics Enrico Fermi, course XLII, ed. R.J. Glauber (Academic Press, New York) p. 532.
Hayes, J.M. and G.J. Small, 1978, Chem. Phys. **27**, 151.
Hesp, B.H. and D.A. Wiersma, 1980, Chem. Phys. Lett. **75**, 423.
Hesp, H.M.M., K. Duppen and D.A. Wiersma, 1981, Chem. Phys. Lett. **79**, 399.
Hesselink, W.H., 1980, Thesis (Univ. of Groningen, The Netherlands).
Hesselink, W.H. and D.A. Wiersma, 1977, Chem. Phys. Lett. **50**, 51.
Hesselink, W.H. and D.A. Wiersma, 1978, Chem. Phys. Lett. **56**, 227.
Hesselink, W.H. and D.A. Wiersma, 1979a, Chem. Phys. Lett. **65**, 300.
Hesselink, W.H. and D.A. Wiersma, 1979b, Phys. Rev. Lett. **43**, 1991.
Hesselink, W.H. and D.A. Wiersma, 1980, J. Chem. Phys. **73**, 648.
Hesselink, W.H. and D.A. Wiersma, 1981a, J. Chem. Phys., submitted for publication.
Hesselink, W.H. and D.A. Wiersma, 1981b, J. Chem. Phys., to be published.
Hubbard, P.S., 1961, Rev. Mod. Phys. **33**, 249.
Jain, R.K. and J.P. Heritage, 1978, Appl. Phys. Lett. **32**, 41.
Jones, K.E. and A.H. Zewail, 1978, Advances in Laser Chemistry, ed. A.H. Zewail (Springer, New York) p. 196.
Kopvillem, U.K. and V.R. Nagibarov, 1963, Fiz. Met. Metalloved. **15**, 136.
Krivoglaz, M.A., 1964, Sov. Phys. Solid State **6**, 1340.
Kuizinga, D.J. and A.E. Siegman, 1970, IEEE J. Quantum Electron. **QE-6**, 694, 709.
Kurnit, N.A., I.D. Abella and S.R. Hartmann, 1964, Phys. Rev. Lett. **13**, 567.
Lambert, W.R. and A.H. Zewail, 1980. Chem. Phys. Lett. **69**, 270.
Liao, P.F. and S.R. Hartmann, 1973a, Phys. Lett. **44A**, 361.
Liao, P.F. and S.R. Hartmann, 1973b, Opt. Commun. **8**, 310.
Liao, P.F., N.P. Economu and R.R. Freeman, 1977, Phys. Rev. Lett. **39**, 1473.
Lubchenko, A.F. and I.I. Fishchuk, 1973, Dokl. Akad. Nauk SSSR **211**, 319.
Lubchenko, A.F. and B.M. Pavlik, 1964, Phys. Stat. Sol. 7, 105, 433.
Macfarlane, R.M., R.M. Shelby and R.L. Shoemaker, 1979, Phys. Rev. Lett. **43**, 1726.
Mahr, H. and M.D. Hirsch, 1975, Opt. Commun. **13**, 96.
Marks, S., P.A. Cornelius and C.B. Harris, 1980, J. Chem. Phys. **73**, 3069.
McCall, S.L. and E.L. Hahn, 1969, Phys. Rev. **183**, 457.
McClure, D.S., 1949, J. Chem. Phys. **17**, 905.
McClure, D.S., 1954, J. Chem. Phys. **22**, 1668.
McClure, D.S., 1956, J. Chem. Phys. **24**, 1.
McCumber, D.E. and M.D. Sturge, 1963, J. Appl. Phys. **34**, 1682.
McCumber, D.E., 1964, Phys. Rev. **A133**, 163.
Morsink, J.B.W., 1982, this laboratory, unpublished result.
Morsink, J.B.W., T.J. Aartsma and D.A. Wiersma, 1977, Chem. Phys. Lett. **49**, 34.
Morsink, J.B.W., W.H. Hesselink and D.A. Wiersma, 1979, Chem. Phys. Lett. **64**, 1.
Morsink, J.B.W., B. Kruizinga and D.A. Wiersma, 1980, Chem. Phys. Lett. **76**, 218.
Morsink, J.B.W., H.P. Trommsdorff and D.A. Wiersma, 1982, unpublished result.
Mossberg, T., A. Flusberg, R. Kachru and S.R. Hartmann, 1977, Phys. Rev. Lett. **39**, 1523.
Mossberg, T.W., R. Kachru, E. Whittaker and S.R. Hartmann, 1979, Phys. Rev. Lett. **43**, 851.
Mukamel, S., 1978, Chem. Phys. **31**, 327.
Mukamel, S., 1979, Chem. Phys. **37**, 33.
Muramoto, T., S. Nakanishi, O. Tamura and T. Hashi, 1980, Jpn. J. Appl. Phys. **19**, L211.
Olson, R.W., H.W.H. Lee, F.G. Patterson and M.D. Fayer, 1982, J. Chem. Phys. **76**, 31.
Orlowski, T.E. and A.H. Zewail, 1979, J. Chem. Phys. **70**, 1390.
Personov, R.I., E.I. Al'shits, L.A. Bykovskaya and B.M. Kharlamov, 1974, Sov. Phys. JETP **38**, 912.

Port, H., D. Vogel and H.C. Wolf, 1975, Chem. Phys. Lett. **34**, 23.

Posener, D.W., 1959, Austral. J. Phys. **12**, 184.

Prikhotko, A.F., A.F. Skorobogatka and L.I. Tsikora, 1969, Opt. Spectrosc. **26**, 115.

Putzer, E.J., 1966, Am. Math. Monthly **73**, 2.

Rebane K. and P. Saari, 1978, J. Lumin. **16**, 223 and refs. therein.

Redfield, A.G., 1957, IBM J. Res. Dev. **1**, 19.

Redfield, A.G., 1965, Adv. Magn. Res. **1**, 1.

Salour, M.M., 1978, Appl. Phys. **15**, 119.

Sapozhnikov, M.N., 1976, Phys. Stat. Sol. **B75**, 11.

Shoemaker, R.L., 1979, Ann. Rev. Phys. Chem. **30**, 239.

Skinner, J.L., H.C. Andersen and M.D. Fayer, 1981, Phys. Rev. A **24**, 1994.

Slusher, R.E. and H.M. Gibbs, 1972, Phys. Rev. A **5**, 1634.

Small, G.J., 1978, Chem. Phys. Lett. **57**, 501.

Song, J.J., J.H. Lee and M.D. Levenson, 1978, Phys. Rev. A **17**, 1439.

Szabo, A., 1975, Phys. Rev. **B11**, 4512.

ter Horst G. and J. Kommandeur, 1982, this laboratory, unpublished result.

van Strien, A.J. and J. Schmidt, 1980, Chem. Phys. Lett. **70**, 513.

Völker, S., R.M. Macfarlane, A.Z. Genack, H.P. Trommsdorff and J.H. van der Waals, 1977, J. Chem. Phys. **67**, 1759.

Völker, S., R.M. Macfarlane and J.H. van der Waals, 1978, Chem. Phys. Lett. **53**, 8.

Völker, S. and R.M. Macfarlane, 1979, Chem. Phys. Lett. **61**, 421.

Wallenstein, R. and T.W. Hansch, 1975, Opt. Commun. **14**, 353.

Warren, W.S. and A.H. Zewail, 1981, J. Phys. Chem. **85**, 2309.

Wertheimer, R. and R. Silbey, 1981, J. Chem. Phys. **74**, 686.

White, J.U., 1942, J. Opt. Soc. Am. **32**, 285.

Wiersma, D.A., 1981, in: Photoselective Chemistry, part 2, Advances in Chemical Physics, vol. XLVII, eds. J. Jortner, R.D. Levine and S.A. Rice (Wiley, New York) p. 421.

Yajima, T. and H. Souma, 1978, Phys. Rev. A **17**, 309.

CHAPTER 7

Nonlinear Laser Spectroscopy and Dephasing of Molecules: An Experimental and Theoretical Overview

M.J. BURNS*, W.K. LIU** and A.H. ZEWAIL†

Arthur Amos Noyes Laboratory of Chemical Physics§
California Institute of Technology
Pasadena, CA 91125
U.S.A.

*Present address: Science and Technology Division, Institute for Defense Analyses, 1801 N. Beauregard St., Alexandra, VA 22311.

**Present address: Department of Physics, University of Waterloo, Waterloo, Ontario, N2L 3G1 Canada; NSERC (Canada) University Research Fellow.

†Alfred P. Sloan Fellow and Camille and Henry Dreyfus Foundation Teacher–Scholar; to whom correspondence should be addressed.

§Contribution No. 6394.

Spectroscopy and Excitation Dynamics
of Condensed Molecular Systems
Edited by
V.M. Agranovich and R.M. Hochstrasser

Contents

1. Introduction

1.1. Objectives

With the 1970's having ended, it is perhaps useful to give an overview of the tremendous accomplishments made over the last decade in the field of nonlinear laser spectroscopy or, more specifically, the field of optical molecular spectroscopy with coherent lasers. Using these techniques, a number of new experimental observations have been made on small or intermediate sized molecules in gases, beams, liquids, and solids. More recently the techniques have been expanded to the territories of large molecules.

Unlike conventional optical spectroscopic methods, coherent molecular spectroscopy can, in principle, give precise information on the nature of intramolecular and intermolecular interactions and dynamics. However, in obtaining such information one incorporates certain approximations to link theory to observables. Once these approximations and assumptions are tested and justified, numerous valuable dynamical parameters can be obtained. Examples of these parameters are vibrational and electronic energy relaxation rates, dephasing times, intramolecular coherence transfer rates, and homogeneous and inhomogeneous local structure parameters.

This review is written with several objectives in mind. First, we will present the reader with a simple but rigorous description of the time-dependent molecule–laser coherent interactions. We then present the formal theory of relaxation to explain the kind of molecular dynamics that can be obtained from coherent optical experiments. Finally, in section 4 we will give an overview of the applications of this type of spectroscopy to a variety of problems in liquids, solids, and gases, making the connection between theory and experiments as clear as possible and indicating the different approximations and problems encountered. Illustrative examples will be given, but no extensive tabulation of the many experiments in the field will be made.

1.2. Why coherent laser spectroscopy?

There are two basic observables that can be obtained from conventional steady-state spectroscopic measurements: the effective resonance line position and the lineshape. Traditionally, it is the line center position and the width of the resonance transition that are measured. Line positions tell us the energies (electronic, vibrational, and rotational) of the transitions, while the linewidths tell us a composite number of things. If the width of a single two-level transition is $\Delta\nu$, then by the Heisenberg Uncertainty Principle,

$$\Delta\nu \, \Delta t \sim 1; \tag{1}$$

the width, then, is related to a time Δt.

Several processes contribute to Δt. First, there is the known lifetime contribution. If a molecule has been excited into an upper level, then through radiative and/or nonradiative decay channels the molecule changes its energy and returns back to the ground state. This energy or level population relaxation (with a time constant T_1, say) leads to a resonance broadening.

The usual treatment for energy relaxation follows the Wigner–Weisskopf approximation, where it is assumed that the amplitude of the wavefunction decays according to a simple exponential law:

$$C_b(t) \propto e^{-t/2T_1}, \tag{2}$$

where b designates the excited state. Several approximations are involved here. First, we have assumed a weak coupling semiclassical treatment for the interaction of the radiation field with matter. The semiclassical nature of the approximation makes the treatment rather simple because only the molecule is treated quantum-mechanically. The radiation field is described classically. Second, we assume that all molecules decay with the same T_1. So, if we flash an ensemble of molecules with a laser pulse and monitor the fluorescence from the excited level, the decay will be a single exponential. Therefore, if all molecules are identical, the T_1 of the fluorescence and the absorption linewidth of the transition will be directly related. For the particular case at hand, a Lorentzian will be observed in the frequency domain and an exponential decay will be seen in the time domain, as dictated by the Fourier transform from one domain to the other: if the time-dependent emission field resulting from an optically prepared sample is given by

$$E_p(t) = \operatorname{Re} \tilde{E}_p(t) = \operatorname{Re} A e^{i\omega_0 t} e^{-t/2T_1}, \tag{3}$$

where ω_0 is the transition frequency and A is independent of t, then the spectral steady-state line shape will be given by

$$I(\omega) \propto \operatorname{Re} \int_{-\infty}^{\infty} E(t)\, e^{i\omega t}\, dt \tag{4}$$

$$\propto A \frac{\Delta\omega_{1/2}}{(\omega - \omega_0)^2 + (\Delta\omega_{1/2})^2}.$$

$\Delta\omega_{1/2}$ is directly related to $(T_1)^{-1}$ and is the half width at half maximum of the Lorentzian lineshape in frequency units, in the low power limit. But, besides population (energy) decay, other effects can contribute to a lineshape or to a transient relaxation signal.

The application of radiation to a two-level system mixes the wavefunctions of the two connected levels, creating a "mixed state". If the incident radiation is coherent, then the mixed states of the molecules in the ensemble will have a well-defined phase relationship. As will be shown in section 2, this ensemble-averaged, mixed state gives rise to a macroscopic dipole moment and, consequently, an observable electric field of well-defined phase; "coherence" is introduced into the sample. The decay time constant for the phase relationship between the wavefunctions of the various molecules is called T_2, the dephasing time. So, dephasing processes can cause relaxation.

Furthermore, except for isolated molecules at uniform speed, the environments of different groups of sample molecules with respect to the radiation field will be different. This is true for all phases of matter. This inhomogeneity in the sample will also contribute to line broadening. So, the

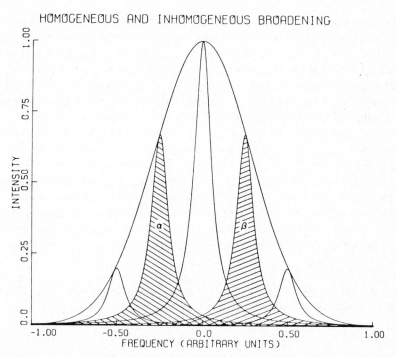

Fig. 1. Inhomogeneous and homogeneous broadening. The shaded areas, α and β, refer to distinct homogeneous packets of molecules.

observed resonance width might have nothing to do with the lifetime broadening and/or the actual energy dynamics of the system. The $\Delta \nu \, \Delta t$ relationship becomes ill-defined and the lineshape becomes a complicated function. To illustrate this point, let us consider the case of molecular gases.

Excited molecules in a gas can decay to the equilibrium state by T_1 or T_2 processes. If all the molecules have the same velocity, a Lorentzian will be obtained with a width determined by T_1 and T_2. Because there is a distribution of velocities in a gas, by virtue of the Doppler effect each velocity group or "packet" will have its own ω_0 and possibly different T_1 and T_2. As a result, the spectroscopic resonance will be a profile resonance made of many Lorentzians. Such a profile (a Doppler lineshape) is called an inhomogeneously broadened (IB) resonance. By contrast, the Lorentzians are the homogeneously broadened (HB) resonances (see fig. 1). The advantages of coherent optical spectroscopy are that HB can be disentangled from IB and that T_1 and T_2 can be measured for HB resonances. Hence the entire ensemble dynamics is measurable.

1.3. The approach

On the experimental side, there are two fundamentally different approaches. Basically, one can isolate the HB resonance by nonlinear frequency domain spectroscopy or by using a certain pulse sequence that allows one to resolve the energy and phase relaxation in the time domain. The former involves techniques (Demtroder 1980, Letokov 1976) such as Doppler-free spectroscopy, saturation spectroscopy, and many other derivative techniques. The latter makes use of NMR pulse methods, but instead of RF radiation, microwave or laser light is used. However, in the UV–visible spectral region molecules emit strongly and one can exploit this feature to monitor these processes (Zewail 1980b, for review). This is a feature that does not exist in NMR coherent spectroscopy (see section 4).

In parallel with NMR, optical free induction decay, optical nutation, and photon echoes are the transients that provide T_1 and T_2 relaxation times of optical transitions. In what follows, we define these and other terms that are commonly used in the field. They will be described in much greater detail in sections 2–4.

Optical nutation. The transient response of a molecular ensemble when resonant radiation is turned on is referred to as optical nutation. The molecules alternatively absorb and emit radiation in an oscillatory fashion until a steady state is achieved. The time necessary to achieve such a steady state depends upon the relaxation rates as well as the dipole moment of the transition and the incident power.

Optical free induction decay. After a pulse of radiation has been applied to a sample and the incident source light has been switched off (or off resonance), coherent emission will occur in approximately the same direction as the incident radiation. The decay of this emission, the free induction decay, is most sensitive to T_2 and inhomogeneous relaxation effects.

Incoherent resonance decay. If the emission during or after a pulse is monitored perpendicular to the laser propagation direction, instead of parallel to it, the spontaneous incoherent emission (e.g. fluorescence) is monitored. Such emission depends upon the energy, not the phase, relaxation of the system. Consequently, T_1 may be measured by this technique. By using multiple pulses, this emission can also give T_2 directly as we shall discuss later.

Photon echo. At a time after certain pulse sequences have been applied, a burst of coherent emission occurs in the forward direction. This optical analog of the spin echo is called the photon echo. As we shall see, the advantage of the echo is that the contribution to relaxation from the usual inhomogeneous broadening may be removed.

Spontaneous Raman scattering. A traditional Raman experiment is performed by monitoring the Stokes or anti-Stokes scattering perpendicular (usually) to the direction of the excitation beam. This scattering is called spontaneous Raman scattering.

Coherent Raman scattering. The Raman scattering in approximately the same direction as the incident coherent radiation is a stimulated radiation. This scattering radiation can be used to measure relaxation rates that destroy the coherence of Raman-excited levels. More details about these different coherent transients will be given later.

1.4. Techniques and historical developments of coherent transients

The optical analogue of magnetic resonance spectroscopy was demonstrated by the success of the Hartmann group (Kurnit et al. 1964) in observing the photon echo in ruby in 1964. The Columbia University group observed the echo as a spontaneous pulse of light from a ruby sample that had been irradiated by two (delayed in time) laser pulses. In 1969, Hocker and Tang (1969) observed an optical nutation in SF_6 gas when a pulse of a 10.6 μm CO_2 laser pumped the vibrational–rotational transition at this frequency. In 1971 Brewer and Shoemaker (1971) made an important observation; they used a continuous wave laser to observe coherent transients by Stark-switching the transition frequency of the sample into or out of resonance with the laser. In this clever way, the molecule in effect "sees" light only for the period that the laser is on resonance with the molecular transition (Stark modulation). Essentially all the optical analogues of spin resonance transients (nutation, free induction decay, echoes, etc.) have been observed by this method. The method, however, was limited to molecules with permanent dipole moments, because only for such materials can the transition frequencies be Stark-shifted.

In 1976, Brewer and Genack (1976) switched the single mode of a dye laser electro-optically and observed coherent transients for a iodine gas in the visible part of the spectrum. Hall (1973) observed the FID transient by frequency switching (electro-optically) a He–Ne laser operating at 3.39 μm earlier in 1973. In Brewer's experiment the sample was outside the laser cavity and the transients were observed in the forward direction of the laser beam by using the heterodyne detection method.

In 1976, Zewail et al. (Zewail 1980b, for review) exploited the *incoherent* spontaneous emission and used the electro-optic switching method to monitor the *coherent* transients at right angles to the exciting laser beam. The sample was also outside the laser cavity. Photon echoes, optical nutation and FID were observed in a variety of systems as discussed later. Also, instead of using the electro-optic modulator inside the dye laser cavity we have used an acousto-optic modulator outside the dye laser cavity and observed the photon echo in an iodine gas with nanosecond time resolution (Zewail 1980b, for review). Devoe and Brewer (1978) were successful in observing transients on the subnanosecond time scale (100 ps) for the sodium D_1 line using frequency switching.

The above methods of observing coherent transients have proven to be of considerable importance for studying dynamical processes in gases, solids and molecular beams. The group at IBM (since 1976) has focused on iodine and on Pr^{3+} in LaF_3 at 2 K. For the latter system, new and quite important effects have been found as we shall discuss later. Our efforts at Caltech (since 1976) have been focused on condensed phases (molecular crystals) and on molecular beams (see later sections). Molecular crystals have also been examined independently by Wiersma's group (since 1976) and by Fayer's group (since 1979). Wiersma's group has used pulsed lasers to observe echoes of impurity molecules in mixed crystals at low temperature, and hence to learn about the origin of T_2 and T_1. Fayer's group has used the transient grating technique and the photon echo method to study mainly energy transfer processes and coherence effects. The pioneering work of Hartmann's group (since 1964) and many other groups on echoes will not be surveyed here. Only those results pertinent to specific solids or molecules will be discussed in the text.

Finally, picosecond coherent transients in liquids have been observed via the coherent Raman effect by the Munich group (Von der Linde et al. 1971). These elegant experiments, which started in 1971, provide T_1 and T_2 directly in the liquid phase. More on this will be discussed later.

2. Elementary description of the time-dependent interaction of radiation with matter

2.1. Density matrix approach

Most time-resolved spectroscopic experiments involve the interaction of one or more radiation fields with an ensemble of atoms or molecules. Consequently, a density matrix approach is useful to describe such experiments. We will initially consider systems in which intermolecular interactions are ignored. Such effects will be included subsequently.

2.2.1. A two-level system

For a system of two-level atoms or molecules, the density matrix, ρ, has a particularly simple form. Fortunately, real molecular systems interacting with coherent radiation behave like two-level systems for many important cases when the radiation field is sufficiently weak so that multi-photon processes may be ignored. For example, this description is accurate when only one dipole-allowed molecular transition is within the effective bandwidth of the excitation source.

If the wavefunction of the τth two-level molecule is given by

$$|\psi_\tau\rangle = C_{i,\tau}|i\rangle + C_{j,\tau}|j\rangle, \tag{5}$$

where $|i\rangle$ and $|j\rangle$ are orthonormal time-independent eigenstates of the molecular Hamiltonian and $C_{i,\tau}$ and $C_{j,\tau}$ are arbitrary complex coefficients, then in this basis

$$\rho_{ij} = \langle C_{i,\tau}C_{j,\tau}^*\rangle_\tau. \tag{6}$$

The brackets indicate an ensemble average over the set of identical two-level molecules.

Noting that $|\langle i|\psi_\tau\rangle|^2$ is just the fractional population of state i of the τth molecule, we see that the diagonal elements of ρ are the fractional populations of the two levels in the ensemble. At equilibrium, when no radiation is present,

$$\rho_{ii}^0 = \langle i|e^{-H_0/kT}/Q|i\rangle = e^{-E_i/kT}/Q, \tag{7}$$

where H_0 is the molecular Hamiltonian, E_i is the energy of the ith state and Q is the canonical partition function. (The superscript 0 indicates equilibrium.) On the other hand, the off-diagonal elements are the "mixing" terms. At equilibrium, the phases of the states of the various molecules in the ensemble will be randomly distributed. Thus, for i not equal to j,

$$\rho_{ij}^0 = 0. \tag{8}$$

Equivalently, since $|i\rangle$ and $|j\rangle$ are eigenstates of the Hamiltonian, the

Hamiltonian must be diagonal in this representation. Consequently,

$$\rho^0_{ij} = \langle i| \, e^{-H_0/kT}/Q \, |j\rangle = 0, \tag{9}$$

since $H_{ij} = H_{ji} = 0$.

The normalization of $|\psi_\tau\rangle$ requires that

$$\mathrm{Tr} \, \rho = 1. \tag{10}$$

Furthermore, from eq. (6) it may be seen that the expectation value of any observable, A, is given by

$$\langle A \rangle = \mathrm{Tr}(\rho A). \tag{11}$$

Thus, ρ is a convenient tool for the calculation of macroscopic observables. The polarization, $P(t)$, for example, may be written as

$$P(t) = \langle \mu \rangle = N \, \mathrm{Tr}(\rho \mu), \tag{12}$$

where μ is the dipole moment operator and N is the number of absorbing (emitting) molecules.

The equation of motion for the density matrix is found by inserting eq. (5) into the Schrödinger equation and using eq. (6) to yield

$$i\hbar \frac{\partial \rho}{\partial t} = [H, \rho], \tag{13}$$

where H is the Hamiltonian for the system. The properties (10), (11), and (13) are general properties of density matrices for a system of molecules with an arbitrary number of states. Derivations may be found in many standard quantum-mechanics textbooks.

2.1.2. The Feynman–Vernon–Hellwarth representation

We now consider the form of the density matrix when a classical electromagnetic field of frequency ω, propagating in the z direction with wavevector k, is allowed to interact with a system of stationary, identical, two-level molecules. The field can be written as

$$E(t) = E_0 \cos(\omega t - kz), \tag{14}$$

where E_0 is the amplitude and may depend upon the x and y directions through the mode shape of the field. In matrix form, the Hamiltonian is given by (see, e.g., Orlowski and Zewail 1979)

$$H = \hbar\omega_{\mathrm{AV}} 1 + \tfrac{1}{2}\hbar\omega_0\sigma_z - \mu E_0\sigma_x \cos(\omega t - kz), \tag{15}$$

where $\hbar\omega_{\mathrm{AV}} = (E_a + E_b)/2$ and $\hbar\omega_0 = E_b - E_a$. E_a and E_b are the eigenvalues of the molecular Hamiltonian for the states a and b, and μ is the value of the dipole matrix element between states a and b. The phases of the wavefunctions are chosen such that $\mu_{ab} = \mu_{ba} = \mu$ (Liu and Marcus 1975).

In addition, we have assumed that the diagonal matrix elements of μ are zero. σ_x and σ_z are two of the well-known Pauli spin matrices:

$$\sigma_x = \begin{pmatrix} 0 & 1 \\ 1 & 0 \end{pmatrix}, \qquad \sigma_z = \begin{pmatrix} 1 & 0 \\ 0 & -1 \end{pmatrix}. \tag{16}$$

Immediately, we see that it is the term coupling the radiation field to the molecules through the dipole moment that contains off-diagonal elements. These elements will in turn create nonzero off-diagonal density matrix elements; thus, the states will become mixed as we expect from the usual qualitative picture of absorption.

Equation (13) is most easily solved if we move into a coordinate system rotating with the applied frequency ω (Orlowski and Zewail 1979), i.e.,

$$\rho' = \exp[\tfrac{1}{2}i(\omega t - kz)\sigma_z]\rho \exp[-\tfrac{1}{2}i(\omega t - kz)\sigma_z]. \tag{17}$$

Using the identity

$$e^{i\theta\sigma}z = 1 \cos\theta + i\sigma_z \sin\theta, \tag{18}$$

it may be shown that eq. (17) is equivalent to the transformation

$$\rho'_{aa} = \rho_{aa}, \qquad \rho'_{bb} = \rho_{bb},$$

$$\rho'_{ba} = \rho_{ba}\, e^{i(\omega t - kz)}, \qquad \rho'_{ab} = \rho_{ab}\, e^{-i(\omega t - kz)}. \tag{19}$$

In this representation the off-diagonal elements, which are given nonzero value by the applied field of frequency ω, rotate at frequency ω. The populations are the same as in the laboratory frame representation.

The Hamiltonian in this frame then becomes

$$H' = \hbar\omega_{\mathrm{AV}}\, 1 - \tfrac{1}{2}\hbar\, \Delta\omega\, \sigma_z - \tfrac{1}{2}\hbar\omega_1\sigma_x, \tag{20}$$

where $\Delta\omega$ is defined by $\omega - \omega_0$ and ω_1 is the circular Rabi frequency, $\mu E_0/\hbar$. In deriving eq. (20) we have also employed the rotating wave approximation by neglecting terms of the form $\exp(\pm 2i\omega t)$ with respect to unity.

Solving eq. (13) in the rotating frame yields the following equations:

$$\partial r_1/\partial t = \Delta\omega\, r_2, \tag{21}$$

$$\partial r_2/\partial t = \Delta\omega\, r_1 + \omega_1 r_3, \tag{22}$$

$$\partial r_3/\partial t = \omega_1 r_2, \tag{23}$$

where we have expanded in terms of the complete set of Pauli spin matrices (see, e.g., Steinfeld 1978),

$$\rho = r_1\sigma_x + r_2\sigma_y + r_3\sigma_z + r_4\, 1. \tag{24}$$

Furthermore, we define a vector r:

$$r \equiv \begin{pmatrix} r_1 \\ r_2 \\ r_3 \end{pmatrix} = \begin{pmatrix} \rho'_{ba} + \rho'_{ab} \\ i(\rho'_{ba} - \rho'_{ab}) \\ \rho_{bb} - \rho_{aa} \end{pmatrix}, \tag{25}$$

and

$$r_4 = \rho_{bb} + \rho_{aa} = \mathrm{Tr}(\rho) = 1,$$

from eq. (10).

This r vector was introduced by Dicke (1954) and by Feynman, Vernon, and Hellwarth (1957). Its use not only simplifies eqs. (21)–(23), but it also provides important connections between the well-known pulsed NMR experiments and time-dependent optical studies, as will be shown later. We note also that eqs. (21)–(23) may be written as

$$dr/dt = \Omega \times r, \tag{26}$$

where

$$\Omega = \begin{pmatrix} -\omega_1 \\ 0 \\ -\Delta\omega \end{pmatrix}. \tag{27}$$

Equation (26) is known as the Feynman–Vernon–Hellwarth (FVH) representation, and shows that, for two nondegenerate levels, one can write a torque equation that is of identical form to the Larmor expression for the precession of a spin 1/2 particle in a Zeeman field.

2.1.3. The relation of r to observables

The r vector elements may be related to the macroscopic polarization using eq. (12):

$$P(t) = N\mu(\rho_{ab} + \rho_{ba}). \tag{28}$$

Then, using eqs. (19) and (25)

$$P(z, t) = N\mu[r_1 \cos(\omega t - kz) - r_2 \sin(\omega t - kz)] \tag{29}$$

$$= \bar{P}_r(z, t) \cos(\omega t - kz) - \bar{P}_i(z, t) \sin(\omega t - kz), \tag{30}$$

where

$$\bar{P}_r(z, t) = N\mu r_1, \tag{31}$$

$$\bar{P}_i(z, t) = N\mu r_2. \tag{32}$$

We have written that the polarization is, in general, a function of z. The dependence of P on x, y has been temporarily suppressed. But the incident field is of the form

$$E(t) = E_0 \cos(\omega t - kz). \tag{33}$$

Therefore, this incident field has created both an in-phase component, $\bar{P}_r(z, t)$, and an out-of-phase component, $\bar{P}_i(z, t)$, of the polarization. The physical interpretation of these components will be discussed presently. [Note that in writing $E(t)$ in the form of eq. (33) we have tacitly assumed linearly polarized light. If circularly polarized light were used,

$$E(t) = E_0[\hat{x} \cos(\omega t - kz) - \hat{y} \sin(\omega t - kz)], \tag{34}$$

the resulting analysis would become only slightly more difficult.]

Equation (30) is often written in the equivalent complex form

$$
\begin{aligned}
P(t) &= \tfrac{1}{2}\bar{P}(z, t)\, e^{i(\omega t - kz)} + \tfrac{1}{2}\bar{P}(z, t)^*\, e^{-i(\omega t - kz)} \\
&= \mathrm{Re}(\bar{P}(z, t)\, e^{i(\omega t - kz)}),
\end{aligned}
\tag{35}
$$

where

$$\bar{P}(z, t) = \bar{P}_r(z, t) + i\bar{P}_i(z, t). \tag{36}$$

Thus, $\bar{P}_r(z, t)$ and $\bar{P}_i(z, t)$ are the real and imaginary parts, respectively, of the slowly varying part of the complex polarization. In terms of the complex susceptibility, χ,

$$\bar{P}(z, t) = \epsilon_0 E_0 \chi = \epsilon_0 E_0 (\chi' + i\chi''), \tag{37}$$

in MKS units. ϵ_0 is the electric permittivity of free space.

If the sample is optically thin so that only a small amount of the incident field is absorbed, we may neglect the effects of reabsorption or re-emission of the electric field created by $P(z, t)$. Then the polarization may be used as a source term in the usual Maxwell equations description of the electric field, E_p, created by that polarization. In MKS units,

$$\nabla^2 E_p - \frac{1}{c^2} \frac{\partial^2 E_p}{\partial t^2} = -\mu_0 \frac{\partial^2 P}{\partial t^2}, \tag{38}$$

for an isotropic medium. The constant c is the speed of light, and μ_0 is the permeability of free space. The dielectric constant ϵ for the medium (not counting the contribution from the resonant polarization P to ϵ) is taken to be that of vacuum for convenience. Equation (38) has been solved for the case at hand numerous times (McGurk et al. 1974, Shoemaker 1978). We will describe the method and results briefly.

An incident plane wave electric field of the form of eq. (33) is assumed. The electric field created by $P(z, t)$ [Eq. (30)] will be of the same form as that polarization. Therefore,

$$E_p = E_r(z, t) \cos(\omega t - kz) - E_i(z, t) \sin(\omega t - kz). \tag{39}$$

Equations (30) and (39) may be inserted into eq. (38), and the coefficients of the sines and cosines can be equated. Then the slowly varying envelope approximation is made. It consists of neglecting the time dependence of

$\bar{P}(z, t)$. This is reasonable since, from the form of eq. (35), $\bar{P}(z, t)$ will have a time dependence like $\exp(i \Delta \omega t)$. If ω is close to ω_0, this is certainly a small variation compared to $\exp(i\omega t)$. Similarly, we neglect rapidly varying parts of the created electric field amplitudes:

$$\frac{\partial^2 E_i(z, t)}{\partial z^2} \approx \frac{\partial^2 E_r(z, t)}{\partial z^2} \approx \frac{\partial^2 E_i(z, t)}{\partial t^2} \approx \frac{\partial^2 E_r(z, t)}{\partial t^2} = 0. \tag{40}$$

Consequently, eq. (38) reduces to

$$\frac{\partial E_r(z, t)}{\partial z} + \frac{1}{c} \frac{\partial E_r(z, t)}{\partial t} = \frac{\omega}{2\epsilon_0 c} \bar{P}_i(z, t) \tag{41}$$

$$\frac{\partial E_i(z, t)}{\partial z} + \frac{1}{c} \frac{\partial E_i(z, t)}{\partial t} = -\frac{\omega}{2\epsilon_0 c} \bar{P}_r(z, t). \tag{42}$$

Finally, the propagation delay of the field is taken into account by noting that, if $t_R = t - z/c$,

$$\frac{\partial}{\partial z} E_r(z, t_R) = \left(\frac{\partial}{\partial z} + \frac{1}{c} \frac{\partial}{\partial t} \right) E_r(z, t). \tag{43}$$

A corresponding equation holds for $E_i(z, t_R)$. Direct integration of eqs. (41) and (42) and transformation back into the t frame yields

$$E_r(z_D, t) = \frac{\omega z_D}{2\epsilon_0 c} \bar{P}_i(t) \tag{44a}$$

$$E_i(z_D, t) = -\frac{\omega z_D}{2\epsilon_0 c} \bar{P}_r(t), \tag{44b}$$

where z_D is the position of the detector at the end of the sample cell of length z_D. We note now that $\bar{P}(t)$ is, indeed, independent of z, although E_p depends on the length of the sample in the z direction.

From eqs. (14), (37), (39), and (44), the total field is given by

$$E_t(t) = [E_0 + E_r(t)] \cos(\omega t - kz) - E_i(t) \sin(\omega t - kz)$$
$$= E_0[1 - \tfrac{1}{2}\alpha(t)z_D] \cos(\omega t - kz) + \tfrac{1}{2}E_0\beta(t)z_D \sin(\omega t - kz), \tag{45}$$

where the inverse lengths α and β are introduced:

$$\alpha(t) = -\omega\chi''(t)/c, \qquad \beta(t) = -\omega\chi'(t)/c. \tag{46}$$

Inspection of eqs. (30), (32), (44), and (45) reveals that r_2 induces a polarization 90° out of phase to the incident field and thus r_2 is proportional to the absorption. On the other hand, r_1 creates an electric field component in phase with the incident field. Thus, r_1 is proportional to the dispersion. Since $\alpha(t)z_D$ is small by assumption,

$$E_t(t) \approx e^{-\alpha(t)z_D/2} E_0 \cos(\omega t - kz), \tag{47}$$

where we have assumed on-resonance absorption, which, as will be shown in section 2.4, makes β equal to zero. Thus, $\alpha(t)$ may be identified with the usual absorption coefficient, although in our case it is time dependent. Furthermore, since χ is the susceptibility to all orders in E_0 for a two-level system, $\alpha(t)$ will, in general, depend on E_0. To get the exact dependence of $E_t(t)$ on $\alpha(t)$, we would have had to solve the coupled Maxwell–Bloch equations. What we have done here is a very good approximation for optically thin samples.

Detectors are usually square-law, so they measure intensities, not electric fields. The total energy density at a point z_D is given in vacuo by

$$D = \epsilon_0 E_t^2. \tag{48}$$

An observation is usually on a time scale much slower than the optical cycle time. Hence we must consider the density averaged over many cycle times:

$$\bar{D} = \epsilon_0 \overline{E_t^2}$$
$$= \tfrac{1}{2}\epsilon_0\{[E_0 + E_r(t)]^2 + E_i(t)^2\}. \tag{49}$$

In MKS units,

$$\bar{D}(\text{J/m}^3) = 4.43 \times 10^{-12}\{[E_0 + E_r(t)]^2 + E_i(t)^2\}, \tag{50}$$

where the electric fields are in units of V/m. The irradiance or intensity (I) is simply the velocity of propagation times the total energy density at point z_D. Averaging over the optical cycle,

$$\bar{I} = \tfrac{1}{2}c\epsilon_0\{[E_0 + E_r(t)]^2 + E_i(t)^2\}. \tag{51}$$

Realizing that $E_r(t)$ and $E_i(t)$ are much less than E_0,

$$\bar{I} \approx \bar{I}_0 + c\epsilon_0 E_0 E_r(t), \tag{52}$$

where

$$\bar{I}_0 = \tfrac{1}{2}c\epsilon_0 E_0^2. \tag{53}$$

It is important to note from eq. (52) that, to first order, when the incident electric field is present, the change in intensity due to the small polarization is proportional to the imaginary (absorptive) part of the slowly varying part of the complex polarization. If one measured the emission with no incident field present, the intensity would be proportional to the square of the small, slowly varying parts of the polarization. It is possible to design experiments in which emission is detected in the presence of the incident field so that the first-order "heterodyne" between the incident field and $E_r(t)$ may be used to improve signal to noise. This idea will be discussed further in subsequent sections of this article.

If the intensity is in units of W/m^3 and the electric fields are in units of

V/m,

$$I = 1.327 \times 10^{-3}\{[E_0 + E_r(t)]^2 + E_i(t)^2\}. \tag{54}$$

For plane waves there is a magnetic field perpendicular to both the electric field and the direction of propagation of the wave. Therefore, these relations may be cast in terms of the magnetic field strength, H, using the fact that in vacuo

$$|E|/|H| = \mu_0 c = \sqrt{\mu_0/\epsilon_0} = 377 \ \Omega. \tag{55}$$

Thus, the intensity may also be found from the relation

$$\bar{I} = \overline{E \times H}, \tag{56}$$

which gives the result of eq. (51).

It is important to note that we have calculated the electric field propagating in the same direction (the z direction) as the incident field. If a detector is placed at a right angle to the z direction, the resulting intensity of radiation will be from spontaneous emission from excited states, not from the coherent field created by r_2. r_2 can be converted to r_3. This fact will be shown to be advantageous to the determination of T_1 and T_2 (Zewail 1980b, for review).

Equations (44) must be modified if the detector monitors only one mode of the E field propagating in the z direction, as would occur if the detector was attached to a microwave waveguide or an optical fiber. In this case it is useful to write

$$E_p = E_{mn}(x, y)[E_r(z, t) \cos(\omega t - kz) - E_i(z, t) \sin(\omega t - kz)], \tag{57}$$

where m and n signify the solution for the TE_{mn} or the TEM_{mn} mode that is detected. Then,

$$\left(\frac{\partial^2}{\partial x^2} + \frac{\partial^2}{\partial y^2}\right) E_{mn}(x, y) + \left(\frac{\omega^2}{c^2} - k^2\right) E_{mn}(x, y) = 0, \tag{58}$$

where the wavevector k is suitably defined for whichever mode is of interest. Then, eqs. (44) may be generalized to give

$$E_r(z_D, t) = \frac{\omega z_D}{2\epsilon_0 v_g B} \int dx \int dy \, E_{mn}(x, y)\bar{P}_i(x, y, t), \tag{59}$$

$$E_i(z_D, t) = \frac{-\omega z_D}{2\epsilon_0 v_g B} \int dx \int dy \, E_{mn}^*(x, y)\bar{P}_r(x, y, t), \tag{60}$$

where the radiation group velocity is given by

$$v_g = kc^2/\omega, \tag{61}$$

and

$$B \equiv \int dx \int dy \, |E_{mn}(x, y)|^2. \tag{62}$$

The x and y dependence of $\bar{P}_r$ and $\bar{P}_i$ have been written out explicitly.

Finally, in condensed phases, the polarization will be, in general, anisotropic. Consequently, the susceptibility becomes a tensor, not a scalar. The effects of this anisotropy include mode coupling and phase matching conditions. In addition, the fact that molecules contain more than two levels makes possible a number of multiphoton effects, including Raman scattering, that have not been discussed here. Detailed discussions of these anisotropy and multiphoton effects have been given by several authors (Laubereau and Kaiser 1978, Bloembergen 1965). A qualitative discussion of the effects of relaxation on Raman scattering will be given later.

2.2. Dynamical effects: phenomenological description of relaxation

It now remains that the effects of relaxation must be included in this treatment. Such relaxation may occur by inter- or intramolecular forces or by spontaneous emission. As will be shown in detail in the next section and as has been given by Liu and Marcus (1975), relaxation effects may be treated by adding a term to eq. (13) in the form of a generalized first-order rate equation:

$$(\partial \rho / \partial t)_{\text{relaxation}} = -\Lambda \rho. \tag{63}$$

Λ is called the relaxation or rate matrix. Physically, the diagonal elements, $\Lambda_{ii,ii}$, are proportional to the inelastic rate of population transfer out of level i. The elements of the form $\Lambda_{jj,ii}$ are proportional to the negative of the transfer rate from level i to level j. We assume that we have a many-level system, but that only two levels, a and b, are connected by the radiation field. Therefore, the only nonzero off-diagonal elements in the density matrix are ρ_{ba} and ρ_{ab}. Then, using eq. (63),

$$(\partial r_1 / \partial t)_{\text{relaxation}} = -\delta r_2 - r_1 / T_2, \tag{64a}$$

$$(\partial r_2 / \partial t)_{\text{relaxation}} = \delta r_1 - r_2 / T_2, \tag{64b}$$

where

$$T_2^{-1} + i\delta = \Lambda_{ba,ba}. \tag{65}$$

T_2^{-1} and δ are both real. We have also used the fact that elements of the form $\Lambda_{ij,ii}$ are zero by symmetry (Liu and Marcus 1975).

In order to derive a similar expression for the relaxation of the population difference, r_3, additional assumptions are necessary. First, we assume that the populations of the radiatively connected levels are the only populations that are significantly changed from their equilibrium values. This approximation clearly fails if there is some other level to which population is preferentially transferred during the time scale of the experiment. As will be shown in section 4, this assumption appears to be good for many systems. Then, using the fact that a collision must either result in the absorber being in the same state or be an inelastic collision, and using detailed balance,

$$\sum_i \Lambda_{jj,ii} \rho_{ii}^{(0)} = 0. \tag{66}$$

Thus, we have that

$$(dr_3/dt)_{\text{relaxation}} = \tfrac{1}{2}(\Lambda_{bb,aa} - \Lambda_{aa,aa} + \Lambda_{aa,bb} - \Lambda_{bb,bb})(r_3 - r_3^0)$$
$$-\tfrac{1}{2}(\Lambda_{bb,aa} - \Lambda_{aa,aa} - \Lambda_{aa,bb} + \Lambda_{bb,bb})(r_4 - r_4^0)$$
$$\equiv -k_{33}(r_3 - r_3^0) - k_{34}(r_4 - r_4^0), \tag{67a}$$

$$(dr_4/dt)_{\text{relaxation}} = \tfrac{1}{2}(\Lambda_{bb,aa} + \Lambda_{aa,aa} - \Lambda_{aa,bb} - \Lambda_{bb,bb})(r_3 - r_3^0)$$
$$-\tfrac{1}{2}(\Lambda_{bb,aa} + \Lambda_{aa,aa} + \Lambda_{aa,bb} + \Lambda_{bb,bb})(r_4 - r_4^0)$$
$$\equiv -k_{43}(r_3 - r_3^0) - k_{44}(r_4 - r_4^0). \tag{67b}$$

Thus, the inclusion of relaxation necessitates the coupling of r_4 to r, in general. We also note that since we are now considering a multi-level system, $\rho_{aa} + \rho_{bb} = r_4 \neq r_4^0$, as population is being transferred to other levels. In addition, the fundamental requirement of density matrices that $\text{Tr}\,\rho = 1$ is violated since $\rho_{ii} \equiv \rho_{ii}^0$ for i not equal to a or b. This should not be a serious problem if only levels a and b are isolated by the experiment. If emission from other levels is being monitored, a more elaborate treatment is necessary.

There are instances, though, in which eqs. (67) may be simplified. If the inelastic scattering dynamics out of each of the levels into the bath of levels are approximately equal, then k_{43} is about zero, and r_4 is decoupled from r_3. Thus, $r_4 = r_4^0$, and

$$dr_3/dt = -(r_3 + r_3^0)/T_1, \tag{68}$$

where

$$T_1^{-1} = \tfrac{1}{2}(\Lambda_{aa,aa} - \Lambda_{bb,aa} + \Lambda_{bb,bb} - \Lambda_{aa,bb}). \tag{69}$$

These assumptions are valid in the microwave and infrared regime for a gas at room temperature, for example. They are not generally valid for optical transitions between different electronic states, where the variation of r_4 should be included.

Combining eqs. (68), (64), (21), (22), and (23) yields the famous Bloch equations in the FVH picture:

$$\partial r_1/\partial t = \Delta\omega\, r_2 - r_1/T_2, \tag{70a}$$

$$\partial r_2/\partial t = -\Delta\omega\, r_1 + \omega_1 r_3 - r_2/T_2, \tag{70b}$$

$$\partial r_3/\partial t = -\omega_1 r_2 - (r_3 - r_3^0)/T_1, \tag{70c}$$

where

$$\Delta\omega \equiv \omega - \omega_0 - \delta. \tag{70d}$$

The rate T_1^{-1} is known as the population relaxation rate and is a measure of the rate of decay of the population difference between two levels to that difference's equilibrium value. From the form of eq. (69), we see that the population relaxation time is proportional to the average of the total inelastic rates out of the two levels plus the rate from state a to state b and the rate from state b to a, i.e., any process that knocks a molecule out of one of the radiatively coupled levels to some other level will cause a change in the population difference. But a process that deposits a molecule from the upper state to the lower state or vice versa will change the population difference by two "units", and therefore such processes are doubly counted. It was noted above that this model is generally inappropriate for high-frequency optical transitions. An exception occurs when relaxation within the upper and lower manifolds may be neglected (as might occur for "isolated" molecules in a beam) and relaxation is allowed back to the ground state only. Then,

$$T_1^{-1} = \Lambda_{aa,bb}, \tag{71}$$

and T_1^{-1} is the rate of spontaneous emission. It is important to emphasize that, although T_1 is defined only when very rigid restrictions are placed upon the allowed relaxation processes, all population relaxation events are commonly called "T_1 processes" in the literature.

The time constant T_2 is often called the "total dephasing time" or the "coherence relaxation time". Since this time is a measure of the decay of the off-diagonal density matrix elements, such a time will depend upon the relaxation of the optically induced macroscopic phase of the wavefunctions of the molecules in the ensemble. Consequently, both population relaxation and "pure phase changing" relaxation (in which there is not a change in the total energy of the system) will contribute to T_2.

As will be shown in section 3,

$$\frac{1}{T_2} = \tfrac{1}{2}(\Lambda_{aa,aa} + \Lambda_{bb,bb}) + \Gamma_{\text{dephasing}}$$

$$= \frac{1}{2}\left(\frac{1}{T_{1a}} + \frac{1}{T_{1b}}\right) + \frac{1}{T_2'}, \tag{72}$$

where $\Gamma_{\text{dephasing}}$ is the pure dephasing rate. We may consider some limits for the ratio of T_1 and T_2. When each of the probed levels is surrounded by many collisionally accessible, closely spaced levels with energy differences $\ll kT$, $\Gamma_{\text{dephasing}} \approx 0$. (This may be seen qualitatively by noting that any collision which has enough "strength" to significantly change the phase during one collision period will cause population transfer.) Also, $\Lambda_{aa,bb} \approx \Lambda_{bb,aa} \approx 0$ since population can be transferred to any of a large number of levels. Then, $T_1 = T_2$. On the other hand, if spontaneous emission is the predominant relaxation process, $\Lambda_{aa,aa}$, $\Lambda_{bb,aa}$, and $\Gamma_{\text{dephasing}}$ will all be nearly zero, and $\Lambda_{bb,bb} = -\Lambda_{aa,bb}$. Then, $T_2 = 2T_1$. Similarly, if there exists an isolated doublet with energy difference $\ll kT$, such as an inversion doublet of ammonia, the relaxation rates for the transition between the doublet levels will be dominated by population transfer between the two levels. Again, $T_2 = 2T_1$.

Finally, δ changes the apparent zero-field frequency of the molecules. In the gas phase, it is referred to as the pressure-induced frequency shift, and in the solid phase, the phonon-induced frequency shift. Physically, the fundamental difference between δ and $\Gamma_{\text{dephasing}}$ is that δ is due to a net static shift of the resonance frequency while the pure dephasing rate is caused by random and fast changes in the resonance frequency shifts.

2.3. Analogy to nuclear magnetic resonance

Equations (70) are the FVH analogs of the Bloch equations of NMR (Bloch 1946). The reason that these NMR equations can be used for optical transitions is that the electric dipole operator and the magnetic dipole operator for a spin 1/2 system both connect off-diagonal elements of the two-level wavefunctions in the appropriate basis sets. Consequently, the Hamiltonians for the two systems are the same, except for redefinition of constants. In NMR, ω_1 in eq. (20) must be defined as γH_1, where γ is the gyromagnetic ratio and H_1 is the amplitude of the applied oscillating magnetic field. Furthermore, ω_0 is γH_0, where H_0 is the static magnetic field applied in the z direction. Noting that in an NMR experiment it is the magnetization,

$$\langle \boldsymbol{\mu} \rangle = \text{Tr}(\boldsymbol{\mu}_{\text{m}}\rho), \tag{73}$$

where $\boldsymbol{\mu}_{\text{m}}$ is the magnetic dipole moment, that is measured, it is easily shown that eq. (26) for the optical case is equivalent to the Larmor expression for the precession of a classical magnetic moment about an external magnetic field in the rotating coordinate frame (Abragam 1961). One must remember, though, that for optical transitions eq. (26) does not correspond to any precession in real coordinate space as it does in the NMR case. But, as will be shown in the next section, the graphical representation of eq. (26) in the fictitious r space will be useful for considering optical pulsed experiments.

The use of denotations T_1 and T_2 for relaxation is taken directly from the language of NMR. Indeed, much of the work undertaken in coherent transient spectroscopy has been to take the well-known solutions of eqs. (70) for different NMR pulse sequences (Abragam 1961) and use them to find T_1 and T_2 for electric dipole systems. At present the reverse is also true; NMR spectroscopists are searching for the spin analogue of some optical effects.

2.4. Solutions to the Bloch equations

The general solutions to the Bloch equations (70) and the modified Bloch equations [in which eqs. (67) are used to describe population relaxation] are so complicated that they are generally not useful to describe experimental data. However, many simpler solutions relevant to experiments are possible when some terms may be neglected or treated as small.

2.4.1. Steady-state solutions

First, we consider the steady-state solutions to the Bloch equations, where the derivatives of the r vector elements are zero. A little algebra yields

$$r_1 = T_2 \Delta\omega \, r_2, \tag{74a}$$

$$\frac{r_2}{r_3^0} = \frac{\omega_1/T_2}{(\Delta\omega)^2 + (1/T_2)^2 + (T_1/T_2)\omega_1^2}, \tag{74b}$$

$$\frac{r_3}{r_3^0} = \frac{(1/T_2)^2 + (\Delta\omega)^2}{(\Delta\omega)^2 + (1/T_2)^2 + (T_1/T_2)\omega_1^2}. \tag{74c}$$

Equation (74b) is the usual expression for the Lorentzian saturated absorption lineshape, as we saw in the introduction. T_2^{-1} can be identified as the half width at half maximum of the low-power lineshape in circular frequency units.

If the modified Bloch equations are used, then T_1 is replaced by T_1^{eff}, where

$$(T_1^{\text{eff}})^{-1} = k_{33} - k_{34}k_{43}/k_{44}. \tag{75}$$

Also,

$$r_4 = r_4^0 - \frac{k_{43}}{k_{44}}(r_3 - r_3^0). \tag{76}$$

The same Lorentzian lineshape remains for r_2.

2.4.2. Transient nutation

In order to solve the Bloch equations for time-dependent phenomena, the method of Laplace transforms has been found to be useful. Solutions for several specific cases, some of which are reproduced here, have been given by Torrey (1949), McGurk et al. (1974), and Schenzle and Brewer (1976).

A time-dependent solution to the Bloch equations is easily found in the limit that ω_1 is sufficiently large that the relaxation terms may be neglected for times much less than the relaxation times. On resonance, then, the solutions for this transient absorption (or nutation) are straightforward:

$$r_1(t) = r_1(0), \tag{77a}$$

$$r_2(t) = r_2(0) \cos \omega_1 t + r_3(0) \sin \omega_1 t, \tag{77b}$$

$$r_3(t) = r_2(0) \sin \omega_1 t + r_3(0) \cos \omega_1 t. \tag{77c}$$

If the system is assumed to be at equilibrium at time zero, then, using eqs. (7) and (8),

$$r_1(t) = 0, \tag{78a}$$

$$r_2(t) = r_3^0 \sin \omega_1 t, \tag{78b}$$

$$r_3(t) = r_3^0 \cos \omega_1 t. \tag{78c}$$

If $\omega_1 t = \pi/2$, the population difference is zero, creating a maximally mixed state and thus, from eqs. (78b) and (32), a maximum polarization. In analogy to NMR experiments such a pulse is called a $\pi/2$ pulse. On the other hand, a π pulse will invert the populations of the two levels, but it will create no polarization. A graphical representation of the movement of the r vector when such pulses are applied is shown in fig. 2.

If we look at times comparable to T_1 and T_2, then the relaxation terms must be included in the solution of the Bloch equations. The solutions are much more complicated:

$$r_1(t) = r_1(0) \, e^{-t/T_2}, \tag{79a}$$

$$r_2(t) = r_3^0 A \left(1 - e^{-at} \cos bt - \frac{a}{b} e^{-at} \sin bt \right)$$
$$+ r_3(0)\omega_1 \, e^{-at} \frac{\sin bt}{b} + r_2(0) \, e^{-at} \left(\cos bt + C \frac{\sin bt}{b} \right), \tag{79b}$$

$$r_3(t) = r_3^0 A' \left[1 - e^{-at} \cos bt + \left(\frac{1}{T_1 A'} - a \right) e^{-at} \frac{\sin bt}{b} \right]$$
$$+ r_3(0) \, e^{-at} \left[\cos bt + \left(\frac{1}{T_2} - a \right) \frac{1}{b} \sin bt \right] - r_2(0)\omega_1 \, e^{-at} \frac{\sin bt}{b}, \tag{79c}$$

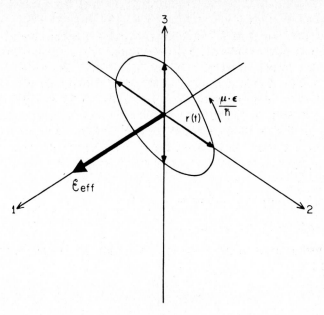

Fig. 2. A geometric representation of the precession of the vector $r(t)$ about an effective field in the rotating frame.

where

$$a = \frac{1}{2}\left(\frac{1}{T_1} + \frac{1}{T_2}\right), \tag{80a}$$

$$b^2 = \omega_1^2 - C^2, \tag{80b}$$

$$C = \frac{1}{2}\left(\frac{1}{T_1} - \frac{1}{T_2}\right), \tag{80c}$$

$$A = \frac{\omega_1}{T_1(a^2 + b^2)}, \tag{80d}$$

$$A' = \frac{A}{\omega_1 T_2}. \tag{80e}$$

These solutions are equivalent to those of McGurk et al. (1974).

Clearly, starting from equilibrium, the simple $\pi/2$ pulse will no longer maximize $r_2(t)$ and minimize $r_3(t)$ as before. Still, it is common to speak of a $\pi/2$ pulse as that time duration that maximizes $r_2(t)$ for fixed ω_1. Similarly, a π pulse minimizes $r_2(t)$.

Analytic solutions to the modified Bloch equations, even when $\Delta\omega = 0$, have not yet been derived, except in a formal sense (Schenzle and Brewer

1976). Consequently, these equations must be solved with further approximations or by numerical methods.

The solutions presented above, with $\Delta\omega = 0$, are appropriate for microwave transients in gases or for some optical experiments in isolated molecules. They are not appropriate for systems in which averaging over an inhomogeneous line is necessary, since in that case, as will be shown in the next section, the r vector elements must be averaged over $\Delta\omega$.

When $\Delta\omega \neq 0$, the Bloch equations for transient nutation have been solved analytically for the case where $T_1 = T_2$, which is appropriate for most microwave or infrared experiments, but not for optical work. A useful perturbative solution is possible, though, when

$$\left(\frac{1}{T_2} - \frac{1}{T_1}\right) \ll \omega_1. \tag{81}$$

In that case,

$$r_i = A\,e^{-at} + e^{-bt}(B\cos ct + C\cos ct) + D, \tag{82a}$$

where $i = 1, 2,$ or 3 and to second order in the difference in the relaxation times,

$$a = \frac{1}{T_2} + \left(\frac{1}{T_1} - \frac{1}{T_2}\right)\frac{(\Delta\omega)^2}{(\Delta\omega)^2 + \omega_1^2}, \tag{82b}$$

$$b = \frac{1}{T_2} + \left(\frac{1}{T_1} - \frac{1}{T_2}\right)\frac{\omega_1^2}{2[(\Delta\omega)^2 + \omega_1^2]}, \tag{82c}$$

$$c^2 = [(\Delta\omega)^2 + \omega_1^2] - \left(\frac{1}{T_1} - \frac{1}{T_2}\right)^2\frac{[\frac{1}{4}\omega_1^2 + (\Delta\omega)^2]\omega_1^2}{[(\Delta\omega)^2 + \omega_1^2]^2}. \tag{82d}$$

Torrey (1949) and Flygare and coworkers (1974) have given the general expressions for A, B, C, and D. A function $g(s)$ is defined for each of the r vector elements. Then, for each element r_i,

$$A = \frac{-g(-a)}{a[(b-a)^2 + c^2]}, \tag{83a}$$

$$B = -(A + D) + r_i(0), \tag{83b}$$

$$D = \frac{g(0)}{a(b^2 + c^2)}. \tag{83c}$$

For r_1,

$$C = aA + bB - \frac{r_1(0)}{T_2} + \Delta\omega\, r_2(0), \tag{84a}$$

and

$$g(s) = sr_1(0)\left[\left(s + \frac{1}{T_2}\right)\left(s + \frac{1}{T_1}\right) + \omega_1^2\right]$$
$$+ \Delta\omega\left[sr_2(0)\left(s + \frac{1}{T_1}\right) + \omega_1\left(sr_3(0) + \frac{r_3^0}{T_1}\right)\right]. \qquad (84b)$$

For r_2,

$$C = aA + aB - \Delta\omega\, r_1(0) - \frac{r_2(0)}{T_2} + \omega_1 r_3(0), \qquad (85a)$$

and

$$g(s) = s\left(s + \frac{1}{T_1}\right)\left[\left(s + \frac{1}{T_2}\right)r_2(0) - \Delta\omega\, r_1(0)\right] + \left(s + \frac{1}{T_2}\right)\omega_1\left(sr_3(0) + \frac{r_3^0}{T_1}\right). \qquad (85b)$$

For r_3,

$$C = aA + bB - \omega_1 r_2(0) - \frac{r_3(0)}{T_1} + \frac{r_3^0}{T_1}, \qquad (86a)$$

and

$$g(s) = \left(s + \frac{1}{T_2}\right)\left[\left(s + \frac{1}{T_2}\right)\left(sr_3(0) + \frac{r_3^0}{T_1}\right) - s\omega_1 r_2(0)\right]$$
$$+ \Delta\omega^2\left(sr_3(0) + \frac{r_3^0}{T_1} + sr_1(0)\right). \qquad (86b)$$

These perturbative solutions reduce to the exact results in the limit that $\Delta\omega = 0$ or that $T_1 = T_2$.

Although eqs. (82) are quite general in their description of transient nutation, they are so complicated that they are not generally useful for fitting experimental data. For many optical experiments, though, the laser source is of sufficiently high power that ω_1 is much greater than the relaxation times. Then the expression for r_2, which is related to the measured absorption, is particularly simple. When the initial conditions are the equilibrium conditions,

$$r_2(t) = \omega_1 r_3(0) \frac{\sin ct}{c} e^{-bt}. \qquad (87)$$

This is the commonly employed expression for laser-induced transient nutation.

2.4.3. Free induction decay and two-pulse delayed nutation
Experimentally, transient nutation has not been found to be the best way to extract relaxation data from the time-dependent signals. The power of the

source is not accurately known and may vary over distance. Also, as we see from eqs. (82), power effects may dominate relaxation effects. The usual case is to use transient absorption to create a macroscopic polarization and then watch the decay of the signal after the power has been switched off or switched far off resonance ($\Delta\omega \gg \omega_1$) from the excited molecular transition. In either case, we may take ω_1 to be zero in order to solve the Bloch equations or the modified Bloch equations. When ω_1 is zero, the coherence terms r_1 and r_2 are decoupled from the population relaxation terms. So, if the preparative pulse is said to end at time t',

$$r_1(t) = e^{-(t-t')/T_2}[r_1(t')\cos(\Delta\omega(t-t')) + r_2(t')\sin(\Delta\omega(t-t'))], \tag{88a}$$

$$r_2(t) = e^{-(t-t')/T_2}[r_2(t')\cos(\Delta\omega(t-t')) - r_1(t')\sin(\Delta\omega(t-t'))], \tag{88b}$$

where, if the power is turned completely off, we must put $\Delta\omega = 0$ (neglecting inhomogeneous broadening). If ω_1 cannot be completely neglected in comparison to $\Delta\omega$, perturbative solutions are available. [To first order in ω_1, $\Delta\omega$ is replaced by $[(\Delta\omega)^2 + \omega_1^2]^{1/2}$ as may be seen from eqs. (82).] Equations (88) describe what is known as a free induction decay (FID). The polarization relaxes with time constant T_2, modulated by the "heterodyne" of the emission at frequency ω_0 with the source at frequency ω. So, the FID has two main advantages. First, the functional dependence on T_2 is simple. Second, in the case of frequency switching, the resulting signal is heterodyned, making detection more efficient.

Similar expressions may be presented for the population relaxation. For the usual Bloch equations,

$$r_3(t) = r_3^0 + [r_3(t') - r_3^0]\, e^{-(t-t')/T_1}. \tag{89a}$$

Or, if the modified Bloch equations are employed,

$$\begin{aligned}
r_3(t) = r_3^0 + \frac{1}{k_2 - k_1}\{&r_3^0[(k_1 - k_{33})\, e^{-k_2 t} - (k_2 - k_{33})\, e^{-k_1 t}] \\
&+ r_3(0)[(k_2 - k_{44})\, e^{-k_2 t} - (k_1 - k_{44})\, e^{-k_1 t}] \\
&- [r_4^0 - r_4(0)]k_{34}(e^{-k_2 t} - e^{-k_1 t})\},
\end{aligned} \tag{89b}$$

where

$$k_1 = k_+ + k_t, \qquad k_2 = k_+ - k_t,$$

$$k_+ = \tfrac{1}{2}(k_{33} + k_{44}), \qquad k_t^2 = k_-^2 + k_{34}k_{43}, \qquad k_- = \tfrac{1}{2}(k_{33} - k_{44}).$$

The element $r_4(t)$ is of the same form as $r_3(t)$ and is found by making the following replacements:

$$r_3(0) \leftrightarrow r_4(0), \qquad r_3^0 \leftrightarrow r_4^0,$$

$$k_{44} \leftrightarrow k_{33}, \qquad k_{34} \leftrightarrow k_{43}.$$

If $r_4^0 = r_4(0)$ and the rate of relaxation between the two levels is negligible (as in many systems where the levels are closely spaced), then

$$r_3(t) = r_3^0 + \tfrac{1}{2}[r_3(0) - r_3^0](e^{-\gamma_a t} + e^{-\gamma_b t}), \tag{89c}$$

where $\gamma_a = \Lambda_{aa,aa}$ and $\gamma_b = \Lambda_b = \Lambda_{bb,bb}$ (Shoemaker 1978). This expression is much more tractable than eq. (89b).

The population decay may be measured by detecting the total emission from level b at right angles to the source $(\rho_{bb}(t) = \tfrac{1}{2}[r_3(t) + r_4(t)])$. Or, if $T_1 \gg T_2^*$ (where T_2^* is the effective dephasing time including inhomogeneous broadening, as discussed in the next section), a probe pulse can be used to measure T_1. Assuming that the normal Bloch equations are valid, if a second intense pulse is applied at time t'',

$$r_2(t) = \omega_1\{r_3^0 + [r_3(t') - r_3^0] e^{(t''-t')/T_1}\} \frac{\sin c(t - t'')}{c} e^{-b(t-t'')}, \tag{90}$$

where eqs. (79) and (82) have been used. By varying $(t'' - t')$, T_1 may be measured. Typically, an initial π pulse is applied to get a maximum population inversion. Then, at a variable time t'', a $\pi/2$ pulse is applied. The height of that pulse is a measure of T_1. The subsequent FID is used to measure T_2. A similar expression may be derived using the modified Bloch equations.

2.5. Inhomogeneous broadening and velocity effects

2.5.1. Inhomogeneous broadening

The treatment given so far has assumed that all molecules are "identical"; they are homogeneously broadened. Of course, real systems have molecules in different environments and are, therefore, inhomogeneously broadened in most instances. (Molecular beams can, in principle, eliminate inhomogeneous broadening.)

For simplicity, we consider inhomogeneous broadening in a dilute gas. Velocity changing collisions will be neglected. Molecules at position R at time t which still contribute to a signal at the detector were at position $R - v_1 t$ at time $t = 0$, where v_1 is the velocity of the absorber. Therefore (Coy 1975a, 1980),

$$P(R, t) = \int \int \int d^3 v_1 f(v_1) P(R - v_1 t, t), \tag{91}$$

where $f(v_1)$ is the Maxwell–Boltzmann velocity distribution. Motion in the x and y directions gives rise to effects such as moving outside the laser mode and wall broadening (Coy 1980). We will concern ourselves with movement in the z direction which causes Doppler broadening. From eq. (14) it may be seen that the effect of motion in the z direction is to replace

$\cos(\omega t - kz)$ in the electric field expression by $\cos(\omega t - kz + kv_1 t)$. But the traveling wave must be invariant to a transformation between inertial frames (Shoemaker 1978). Consequently, ω must be replaced by

$$\omega' = \omega - kv_{1z}. \tag{92}$$

Then, in the presence of Doppler broadening,

$$\Delta\omega = \omega - kv_{1z} - \omega_0, \tag{93}$$

where we have neglected the pressure-induced frequency shift. It is convenient to transform the v_{1z} integration in eq. (91) into an integral over $\Delta\omega$. For r_2, for example,

$$\langle r_2 \rangle = \frac{1}{\sqrt{\pi}\,\delta\omega_1} \int e^{-(\Delta\omega)^2/\delta\omega_1^2}\, r_2(\Delta\omega, t)\, d(\Delta\omega), \tag{94}$$

where $\delta\omega_1$ is the width of the Gaussian, Doppler-broadened line, and

$$\delta\omega_1 = kU, \tag{95}$$

where U is the most probable molecular speed.

Indeed, eq. (94) may be used to describe any inhomogeneous broadening mechanism which is Gaussian in nature. Strain field broadening in molecular crystals may be one such mechanism.

Equation (87) may be integrated over the inhomogeneous profile to yield the nutation signal in the limit of high power. In order to do the integration, certain additional approximations must be employed (Orlowski and Zewail 1979, Schenzle and Brewer 1976). First, terms of second order in the difference in the relaxation times are neglected. Second, we assume that the inhomogeneous width is much greater than ω_1. This is often the case experimentally. So, except for extremely short times, $(\sin ct)/c$ will be small except for a narrow range of frequencies around the center of the Doppler line. Consequently, $(\sin ct)/c$ may be averaged separately. Then,

$$\langle r_2 \rangle = \frac{\omega_1 r_3(0)}{\sqrt{\pi}\,\delta\omega_1}\, e^{-t/T_1} \left\langle \exp\left[-\frac{1}{2}\left(\frac{1}{T_1} - \frac{1}{T_2}\right)\frac{\omega_1^2}{(\Delta\omega)^2}\right]\right\rangle\!\left\langle\frac{\sin ct}{c}\right\rangle, \tag{96}$$

where ω_1^2 in the denominator of the averaged exponential has been neglected in comparison to $(\delta\omega_1)^2$. [This procedure has been checked by numerical integration (Orlowski and Zewail 1979).] Doing the integrals yields (see fig. 3):

$$\langle r_2(t) \rangle \approx \frac{\sqrt{\pi}\,\omega_1 r_3(0)}{\delta\omega_1}\, J_0(\omega_1 t)\, \exp\left(-\left\{\frac{t}{T_2} - \frac{\omega_1}{\delta\omega_1}\left[2t\left(\frac{1}{T_1} - \frac{1}{T_2}\right)\right]^{1/2}\right\}\right), \tag{97}$$

where J_0 is a Bessel function. There are several points to notice about eq. (97). First, the relaxation explicitly depends upon the power. Second, instead of a simple sine behavior as in eq. (78b), r_2 depends upon a Bessel

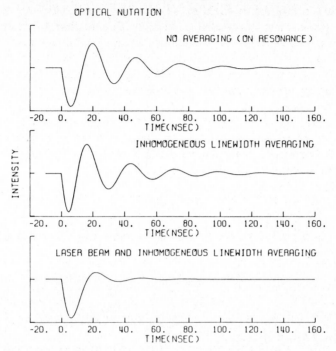

OPTICAL NUTATION

Fig. 3. A theoretical plot for the optical nutation of two-level systems (in this case using pentacene parameters, see text) with averaging and on resonance (top). The middle transient shows the results after averaging over the inhomogeneous linewidth. Here, the equation (see text) is multiplied by an error function to correct behavior near $t = 0$. Finally, at the bottom, is shown a plot which considers averaging over both the inhomogeneous lineshape and the laser beam spatial profile. In all three graphs, the Rabi frequency (36.6 MHz) is the same; only the amplitude and transient decay time are affected by the averaging process described in the paper by Orlowski and Zewail (1979). More recently, Orlowski has reconsidered the approximate averaging method, which uses approximation to integrals, and evaluated the exact expressions by numerical calculation on the computer. He found that (a) the error function correction for short times is not good as written in eq. (28) of the above reference, (b) the transient in the middle is in error but eq. (19) is valid; the figure should be a plot of eq. (19) by itself. So, the message is that all approximations we made earlier are very reasonable but one has to be careful in handling short times, 1 ns or so.

function of argument $\omega_1 t$. Consequently, the simple concept of a $\pi/2$ or a π pulse is only approximate when inhomogeneous broadening is included, even when T_1 and T_2 relaxation may be neglected.

If high-power pulses are employed, it is sometimes possible to achieve the condition that $\omega_1 \gg \delta\omega_1$. In that case, for $r_2(0) = r_1(0) = 0$,

$$\langle r_2 \rangle = r_3(0) \exp\left[-\frac{1}{2}\left(\frac{1}{T_1} + \frac{1}{T_2}\right)t\right] \sin \omega_1 t. \tag{98}$$

This is the same result as would be obtained by solving the Bloch equations for $\Delta\omega = 0$ and requiring that $\omega_1 \gg T_1^{-1}$, T_2^{-1}. In other words, all the inhomogeneous packets are rotated together towards the r_1-r_2 plane as power is applied. If we look at times short compared to the relaxation times, then eq. (98) reduces to eq. (78b). More generally, eqs. (77) can be used to describe this high-power, short-time situation, even in the presence of Doppler dephasing.

Another important example of an effect of inhomogeneous broadening that can be solved is that of a free induction decay following steady-state excitation. From eqs. (88b), (74), and (77b),

$$\langle r_2(t) \rangle = -\frac{\omega \sqrt{\pi}\, r_3^0}{\delta\omega_1} \left(1 - \frac{1/T_2}{(1/T_2)^2 + (T_1/T_2)\omega_1^2}\right)$$

$$\times \exp\left(-\left\{\frac{1}{T_2} + \left[\left(\frac{1}{T_2}\right)^2 + \frac{T_1}{T_2}\omega_1^2\right]^{1/2}\right\}t\right). \tag{99}$$

Note again that the inhomogeneous broadening has caused power-dependent relaxation. Qualitatively, the field causes many velocity groups to be excited. The various groups will emit at different frequencies causing interference and apparent relaxation. An example of an FID following a high-power, short pulse will be considered next in conjunction with the description of the photon echo.

2.5.2. Elimination of inhomogeneous relaxation—the photon echo

Often the experimentalist is not interested in exploring relaxation due to inhomogeneous broadening. This relaxation can be suppressed using the photon echo (Kurnit et al. 1964), the optical analog of the spin echo (Hahn 1950). We will present a simple description of this effect.

We consider the $\pi/2$-τ-π pulse sequence shown in fig. 4 in the limit of high power and short pulses so that eqs. (77) may be used to describe the r vector elements during the pulses. Equations (88) are used to describe the times between pulses. Then, at a time t after the last π pulse,

$$r_2(t) = -r_3^0\, e^{-[(t_2-t_1)+(t-t_3)]/T_2} \cos(\Delta\omega(t_2 - t_1 + t_3 - t)). \tag{100}$$

When $t - t_3 = t_2 - t_1$, r_2 is a maximum. On other words, a burst of emission will appear at time

$$t_4 = t_3 + t_2 - t_1. \tag{101}$$

This is the photon echo. An even more remarkable feature of this pulse train is apparent when r_2 is averaged over the inhomogeneous line:

$$\langle r_2 \rangle = -\frac{\sqrt{\pi}}{\delta\omega_1} r_3^0\, e^{-[(t_2-t_1)+(t-t_3)]/T_2} \exp\left[-\left(\frac{t_2 - t_1 + t_3 - t}{2\delta\omega_1}\right)^2\right]. \tag{102}$$

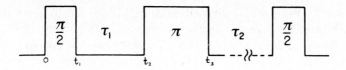

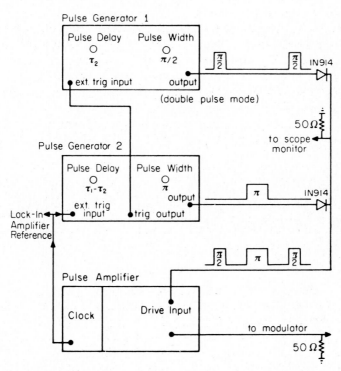

Fig. 4. Schematic of the pulse train required for the three-pulse photon echo (or two-pulse photon echo) and a pulse generator network devised to produce it.

At t_4,

$$\langle r_2 \rangle = -\frac{\sqrt{\pi}}{\delta\omega_1} r_3^0 e^{-2(t_2-t_1)/T_2}. \tag{103}$$

The exponential time dependence of r_2 on $\delta\omega_1$ has disappeared. So, if $(t_2 - t_1) = \tau$ is varied, and the amplitude of the signal is measured at time t_4 given by eq. (101), the homogeneous coherence relaxation time can be measured. We note that with the approximations employed here, the results are independent of the description of population transfer employed.

The reason that this pulse train works the way it does may be understood using the graphical representation of the FVH picture, as shown in fig. 5. At $t = 0$, the figure shows the equilibrium situation. Then, the $\pi/2$ pulse rotates the vector of all the molecules onto the r_2 axis, since $\omega_1 \gg \delta\omega_1$. Then, between the times t_1 and t_2, the molecules with different $\Delta\omega$'s will precess in the r_1–r_2 plane at different rates due to the inhomogeneous broadening. The disk in which they revolve will also shrink due to T_2 decay. This is shown in the figure. The π pulse flips the r vectors (in the r_1–r_2 plane) which inverts the order of the vectors. The molecules "farthest away" from the r_2 axis then have r vectors moving the fastest. So, at time t_4 they rephase at r_2, giving rise to a burst of coherent emission.

   The echo can also be detected using a *population difference* through monitoring fluorescence or some other means, rather than a polarization in the z direction. In fig. 5 we see that a $\pi/2$ pulse applied at t_4 will transform the r_2 polarization into a population difference, r_3 (Zewail et al. 1977). If the short pulse ends at time t_5,

$$r_3(t_5) = \frac{\sqrt{\pi}}{\delta\omega_1} r_3^0 \, e^{-2(t_2-t_1)/T_2}. \tag{104}$$

   In many experiments the conditions assumed for this derivation are not appropriate. Then the more complicated expressions derived in the last

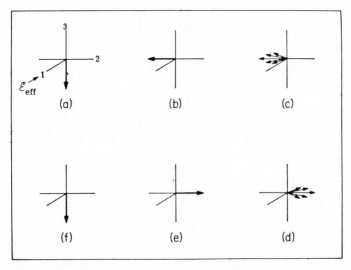

Fig. 5. A schematic for the geometric description of two-pulse and three-pulse (fluorescence detection) photon echoes: (a) all the molecules are in the ground state; (b) after a $\pi/2$ pulse (coherent superposition); (c) inhomogeneous dephasing; (d) after a π pulse (refocusing); (e) the echo, and (f) termination of the spontaneous emission from the upper level by the third $\pi/2$ pulse.

section must be used to describe the effects of the pulses (Schenzle and Brewer 1976). It should also be noted that a pure $\pi/2-\tau-\pi$ pulse sequence is not necessary to produce an echo. Any two pulses will, in general, cause an echo. Often it is not possible to achieve perfect $\pi/2$ or π pulses experimentally; yet, echoes may still be measured. The maximum intensity echo, though, is achieved through the use of $\pi/2$ and π pulses ignoring large IB effects.

Pulse trains containing more than two pulses have been employed to create different types of photon echoes (Shoemaker 1978). One of the most useful of these is the stimulated echo. The pulse train is shown at the top of fig. 6. Essentially this train is just the $\pi/2-\tau-\pi$ train with the last π pulse split up into two $\pi/2$ pulses. Again, an echo (the stimulated echo) will appear at $t_6 - t_5 = t_2 - t_1$. But other echoes due to the echoes created by sets of two pulses will also appear. The difference between this stimulated echo and a normal photon echo is that, for the stimulated echo, the r vector is pointing in the r_3 direction between the times t_3 and t_4; T_1 relaxation will occur during this time. By varying $t_2 - t_1$, T_2 may be measured; and by

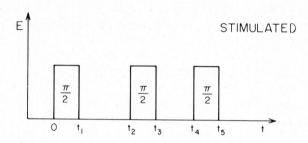

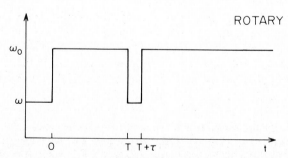

Fig. 6. Pulse sequences for observing the stimulated echo (electric field strength vs time) and the rotary echo (source frequency vs time; ω_0 is the resonance frequency).

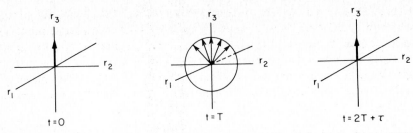

Fig. 7. The motion of the r vector during a rotary echo pulse train (at $t = 0$, we assume all molecules are up).

varying $t_4 - t_3$, T_1 is measured. These stimulated echo experiments suffer from the problem that some residual inhomogeneous broadening affects the echo amplitude unless $T_1 \gg T_2^*$.

Other, more complicated, echo sequences such as Carr–Purcell trains in which a $\pi/2 - \tau - \pi - \tau - \pi - \tau - \ldots$ pulse train is used have been discussed elsewhere (Shoemaker 1978).

A qualitatively different type of echo is the rotary echo (Wong et al. 1980, Rohart et al. 1977). We will discuss the elementary aspects of this echo; a more complete treatment has been given by Wong et al. (1980). Other echoes that we have discussed involved a rephasing of r along the r_2 coherence axis. During a rotary echo, on the other hand, r rephases along r_3, the population difference axis. Consequently, such an echo is most useful for eliminating inhomogeneities in ω_1, the Rabi frequency, rather than in ω_0. A pulse train as given by Wong et al. (1980) is shown at the bottom of fig. 6. In this application we consider changing the frequency of the source, not turning it on and off. We assume that we may switch completely out of resonance with the two-level transition ($\Delta E = \hbar \omega$) at frequency ω. When the source is brought into resonance at $t = 0$, the r vector rotates in the $r_3 - r_2$ plane with angle $\theta = \omega_1 t$ (see fig. 7). But since there is a distribution of ω_1's, there is a distribution of θ's. At time T, the frequency is switched such that $(\omega - \omega_0)\tau = \pi$; then $\omega_1 \to -\omega_1$. If $\tau \ll T_1, T_2$, T, at time $2T + \tau$ the r vectors will all rephase on the r_3 axis; and if the nutation frequency is greater than T^{-1}, a maximum in the nutation signal (r_2) will appear approximately centered about $2T + \tau$. Wong et al. (1980) have shown that, if Doppler broadening is taken into account, the echo heights decays exponentially with T with a rate given approximately by $\frac{1}{2}(T_2^{-1} + T_1^{-1})$ for many cases of interest.

2.5.3. Velocity-dependent effects

We conclude this section by mentioning some of the other phenomena that can contribute to relaxation in gaseous media.

First, we have tacitly assumed that T_1 and T_2 are independent of the absorber speed. In general, though, one would expect these relaxation rates to be quite speed dependent, depending upon the intermolecular potential operant in the collision. Calculations support this notion (Burns 1979). Consequently, the velocity integration in eq. (89) does not reduce to a simple integration over $\Delta\omega$ as in eq. (90). A model must be chosen to describe this speed dependence before the v_1 integration is performed. In the microwave region, the expression

$$T_2^{-1} = k_0 + k_1(v_1 - \bar{v}_1), \tag{105}$$

where $\bar{v}_1$ is the average absorber speed and where k_0 and k_1 are independent of v_1, has been used by Coy (1980) to describe the speed dependence of the total dephasing time. This choice has been supported by calculations (Burns 1979). For experiments in which $\omega_1 \ll \delta\omega_1$, essentially only a few closely spaced velocity groups are selected out by the experiment. In this case, the speed dependence of the relaxation times may be ignored.

Another collisional broadening mechanism is velocity changing collisions (VCC) (Berman and Lamb 1970, Berman et al. 1975). These collisions do not change the state of the molecule, but change the final velocity of the absorber relative to the initial velocity. Theoretically, these collisions are difficult to distinguish from pure phase changing collisions in the general case. The reason for this may be seen by considering that for the states a and b of a molecule there will be a classical trajectory corresponding to each of the states. But, if a coherent superposition of these states is formed by an electromagnetic source, no such classical trajectory exists. So the semiclassical approach that we have used in this section cannot be modified to include velocity changing collisions. An exception to this observation occurs when the form of the intermolecular interactions for states a and b are the same, such as that which occurs for those cases where the unmodified Bloch equation may be employed (e.g. rotational and vibrational transitions of molecules). Then the classical trajectories of the two states will be the same. Pure phase changing collisions will also be negligible since any collision that has the strength to change the phase appreciably has the strength to change the state of the molecule, since many closely spaced levels are near the excited levels. Then a velocity changing rate, Γ_{vc}, may be defined. Since this rate will couple different velocity groups, the Bloch equations for each of the velocity groups are all coupled. But this problem has been solved in the weak collision limit using Brownian motion theory (Berman et al. 1975).

How does one measure Γ_{vc}? The photon echo is one good method (Berman et al. 1975). This may be seen by looking back at fig. 5. After the $\pi/2$ pulse, a molecule with speed v_{1z} will have dephased by $kv_{1z}\tau$ at time t_3 (neglecting T_2 relaxation). Then at the time t_4, the phase will be $\pi - kv_{1z}t$

$+ kv_{1z}\tau = \pi$, and all molecules will have rephased to form the echo. But if, after the π pulse, there was a collision at time $t_4 - t'$ that changed the velocity by a small amount Δv_{1z}, then the phase at time t_4 would be $\pi - k\Delta v_{1z}(\tau - t')$. Thus, we see that a small velocity change can still create a large phase shift in the signal (since k is not small), causing relaxation. A detailed analysis (Berman et al. 1975) shows that the decrease in the echo due to VCC will go like $\exp(-A\tau^3)$ for small τ, where A depends linearly on Γ_{vc}. (This diffusion-type expression is expected since VCC are essentially diffusion within the inhomogeneous lineshape.) On the other hand, non-VCC relaxation decreases the echo like $\exp(-2\tau/T_2)$. So by fitting the decay of the echo amplitude with τ to a form

$$r_2 \quad \sim e^{-A\tau^3} \quad \text{short times,}$$

$$\sim e^{-\bar{A}\tau} \quad \text{long times,} \tag{106}$$

the VCC rate, Γ_{vc}, may be deduced, in principle, for small τ.

At long times, Berman et al. (1975) find that, for the photon echo,

$$r_2 \sim c' e^{-2c\tau}, \tag{107}$$

where $c = T_2^{-1} + \Gamma_{vc}$. Essentially, a new rate, c, is defined which includes the effects of population, phase, and velocity changes on the relaxation. Thus, if T_2^{-1} is known, Γ_{vc} may be deduced. How can T_2^{-1} be measured in the presence of VCC? Berman et al. (1975) have shown that the stimulated echoes resulting from a Carr–Purcell train are relatively insensitive to VCC if the pulses are closely spaced. This technique, as well as coherent Raman beats, may be used to measure T_2.

2.6. Multilevel and multiphoton processes

Coherent coupling of three or more levels exhibits new effects. The analogues to the Bloch equations for three levels have been given by several authors (Brewer and Hahn 1975, Vega and Pines 1977, Vega 1978, Feuillade et al. 1976). In general, eight coupled differential equations are necessary to describe such a system. Consequently, exact solutions are only possible in a few simple cases. There are numerous illuminating experiments, however, that may be done with multilevel or multiphoton coherence probing. We will discuss only a few of the basic sorts of experiments here:

Measurement of T_2 for dipole-forbidden levels. Consider the three-level system shown in fig. 8a, where the $a \to b$ and $b \to c$ transitions are electric dipole allowed, but the $a \to c$ transition is not. T_2 for the $a \to c$ transition may be measured by exciting the $a \to b$ transition and then transferring the coherence to the $a \to c$ levels by a pulse on the $b \to c$ levels. Applying

3-LEVEL SYSTEMS

(a)   (b)

Fig. 8. Three-level systems. The arrow indicates source radiation.

another pulse, after a delay, on the $b \to c$ transition transfers some or all of the $a \to c$ coherence back to the $a \to b$ transition, where it may be sampled (Hatanaka et al. 1975, Mossberg et al. 1977, Burns 1979, Burns and Coy 1983). Alternatively, the $a \to c$ transition may be excited by a two-photon process (Hatanaka and Hashi 1975, Loy 1977).

Quantum beats. If two transitions, $a \to b$ and $a \to c$, are so closely spaced that $\Delta\omega = |\omega_{ca} - \omega_{ba}|$ is less than or about equal to the inverse time of the pulse, then both the $a \to b$ and $a \to c$ transitions in fig. 8b may be excited by a single pulse. The subsequent FID (or, if multiple pulses are used, photon echo) will have a high frequency beat signal, $\sin(\Delta\omega\, t)$, superimposed on the usual FID (photon echo) (Brewer and Hahn 1975). The Fourier transform of the FID yields the frequency domain line positions and shapes. Such a procedure can be used to extract the frequency difference of closely spaced lines.

RAMAN BEAT LEVEL STRUCTURE

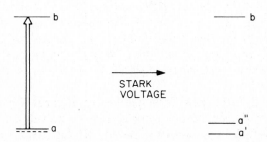

Fig. 9. The idea behind a Raman beat experiment. The arrow signifies the radiation source before the *a* level is split by the Stark voltage.

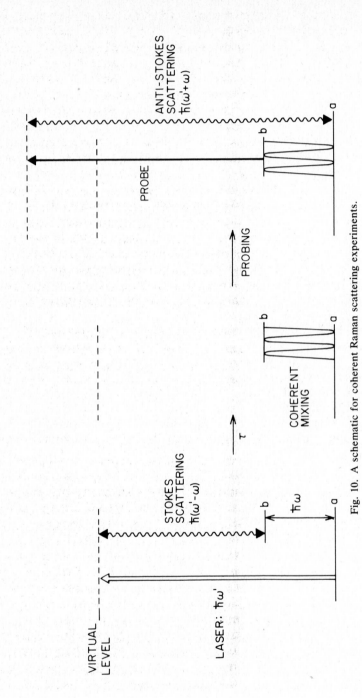

Fig. 10. A schematic for coherent Raman scattering experiments.

Coherent Raman beats. A useful variation of the coherent beats technique has been employed by Shoemaker and Brewer (1972) and Brewer and Hahn (1973). A degenerate two-level transition is excited by a laser as shown in fig. 9. The transition is then split to remove some degeneracy. (This may be done, for example, by resolving M levels using a Stark field.) Again a high frequency beat signal will appear on the FID. The decay of this signal has been shown to be quite insensitive to Doppler broadening and velocity changing collisions (Berman et al. 1975).

Stimulated Raman scattering. As shown in fig. 10, the application of a high-intensity pulse at frequency ω' will, in the presence of a Raman-active transition, create a coherent excitation of the Raman-active level as well as build up a Stokes field. After a time, τ, the application of a second (presumed weak) probe pulse of frequency ω' will "add" to the coherently excited vibration of frequency ω to yield a coherent anti-Stokes signal at frequency $\omega' + \omega$. The amplitude of this anti-Stokes signal as a function of time will depend upon the rate of loss of coherent excitation of the two-level system, i.e. on T_2 for the Raman-active transition. On the other hand, a measurement of the probe-induced incoherent (spontaneous) anti-Stokes scattering yields the population of the excited Raman level. Thus, this effect may be used as a measurement of T_1-like processes.

The excellent review article of Laubereau and Kaiser (1978) treats these stimulated Raman effects in detail. In particular, they treat the important point of how proper wavevector matching can yield T_2's for different groups inside an inhomogeneously broadened line.

3. Formal theory

3.1. Preliminaries

The macroscopic polarization $P(t)$ of a sample of optically active molecules infinitely diluted in a host of optically inactive molecules is given by [eq. (12), in slightly different notation]

$$P(t) = N_a \operatorname{Tr} \mu^{(s)} \rho(t), \tag{108}$$

where Tr denotes a trace taken over both the active and bath variables, and N_a and $\mu^{(s)}$ are the total number and dipole moment operator, respectively, of the optically active molecules. $\rho(t)$ is the density matrix of the entire system normalized to $\operatorname{Tr} \rho(t) = 1$ and satisfying the Liouville equation [eq. (13) with $\hbar \equiv 1$]

$$i \, \partial\rho/\partial t = [H, \rho] \equiv L\rho, \tag{109}$$

where

$$H = H^{(s)} + H^{(b)} + V. \tag{110}$$

$H^{(s)}$ and $H^{(b)}$ are the Hamiltonians for the active and bath molecules, respectively, and V is the interaction between them. It will be assumed throughout this paper that the eigenvectors of $H^{(s)}$, denoted below by $|i\rangle$, $|f\rangle, \ldots$, form a discrete set, while those of $H^{(b)}$, denoted by $|\alpha\rangle, |\beta\rangle, \ldots$, are dense and form a quasi-continuum. $L \equiv [H, \ ]$ is called a Liouville operator. Superscripts (s) and (b) will be used to denote quantities associated with the active molecules and the bath molecules, respectively. Density matrices associated with the active and bath molecules can be obtained from ρ by (Fano 1964)

$$\rho^{(s)}(t) = \mathrm{Tr}^{(b)}\rho(t), \qquad \rho^{(b)}(t) = \mathrm{Tr}^{(s)}\rho(t), \tag{111}$$

where $\mathrm{Tr}^{(s)}$ and $\mathrm{Tr}^{(b)}$ denote traces taken over the active and bath molecules, respectively. Since $\mu^{(s)}$ depends on the active molecule coordinates only, eq. (108) can be written as

$$P(t) = N_a \,\mathrm{Tr}\, \mu^{(s)}\rho_1(t) = N_a \,\mathrm{Tr}^{(s)}\mu^{(s)}\rho^{(s)}(t), \tag{112}$$

where

$$\rho_1(t) = \rho^{(b)} \,\mathrm{Tr}^{(b)}\rho(t) = \rho^{(b)}\rho^{(s)}, \tag{113}$$

and the normalization $\mathrm{Tr}^{(b)}\rho^{(b)} = 1$ has been used. The time development of $P(t)$ is hence determined by that of $\rho_1(t)$ or $\rho^{(s)}(t)$, a reduced density matrix, instead of $\rho(t)$.

Alternatively, according to the linear response theory, the response of the system to an optical excitation can be studied via the dipole autocorrelation function $C(t)$,

$$C(t) = \mathrm{Tr}\, \rho^0 \mu^{(s)}(0)\mu^{(s)}(t), \tag{114}$$

where [cf. eq. (7)]

$$\rho^0 = \mathrm{e}^{-\beta H}/\mathrm{Tr}\, \mathrm{e}^{-\beta H} \tag{115}$$

is the equilibrium density matrix for the entire system, $\beta = 1/k_B T$, and k_B, T are the Boltzmann constant and the temperature, respectively. The Heisenberg operator $\mu^{(s)}(t)$ is given by

$$\mu^{(s)}(t) = \mathrm{e}^{iHt}\mu^{(s)}\mathrm{e}^{-iHt}, \tag{116}$$

and $\mu^{(s)}(0) = \mu^{(s)}$. From the cyclic property of the trace of a product of operators, the autocorrelation function of eq. (114) can be written as

$$\begin{aligned} C(t) &= \mathrm{Tr}\, \mu^{(s)}\mathrm{e}^{-iHt}(\rho^0\mu^{(s)})\, \mathrm{e}^{iHt} \\ &= \mathrm{Tr}\, \mu^{(s)}D(t), \end{aligned} \tag{117}$$

where

$$D(t) \equiv e^{-iHt}(\rho^0\mu^{(s)})e^{iHt}$$
$$= e^{-iLt}(\rho^0\mu^{(s)}) \tag{118}$$

also satisfies the Liouville equation:

$$i\,\partial D(t)/\partial t = [H, D(t)] = LD(t), \tag{119}$$

and the initial condition

$$D(0) = \rho^0\mu^{(s)}. \tag{120}$$

Similar to eq. (112), eq. (117) can be written as

$$C(t) = \text{Tr}\,\mu^{(s)}D_1(t), \tag{121}$$

where

$$D_1(t) = \rho^{(b)}\,\text{Tr}^{(b)}D(t) \equiv \rho^{(b)}D^{(s)}(t). \tag{122}$$

Again, only an equation of motion for the reduced function $D_1(t)$ or $D^{(s)}(t)$ is required.

A powerful formalism, commonly known as the projection operator technique, was developed by Zwanzig (1960, 1961a,b, 1964, 1965) in his treatment of the theory of irreversibility. It subsequently found applications in many areas of nonequilibrium statistical mechanics (Mori 1965, Berne and Harp 1970, Boley 1975, Yip 1979). Zwanzig's theory will be discussed in detail in the next subsection; its relevance to the present problem can be recognized by introducing the Zwanzig–Fano projection operator (Fano 1963) defined by

$$PA = \rho^{(b)}\,\text{Tr}^{(b)}A \tag{123}$$

for an arbitrary operator A, where $\rho^{(b)}$ is the equilibrium density matrix for the bath molecules,

$$\rho^{(b)} = e^{-\beta H^{(b)}}/\text{Tr}\,e^{-\beta H^{(b)}}. \tag{124}$$

It can be seen from eq. (123) that P satisfies the requirement $P^2 = P$ of being a projection operator. From eqs. (112) and (122)

$$P\rho(t) = \rho_1(t) = \rho^{(b)}\rho^{(s)}(t), \tag{125}$$

$$PD(t) = D_1(t) = \rho^{(b)}D^{(s)}(t), \tag{126}$$

and an equation of motion for $\rho^{(s)}(t)$ ($D^{(s)}(t)$) can be obtained immediately from that of $P\rho(t)$ ($PD(t)$) by taking $\text{Tr}^{(b)}$.

3.2. The Zwanzig projection operator technique

The starting point is the Liouville equation for a density function $\bar{f}(t)$,

$$i\,\partial\bar{f}(t)/\partial t = L\bar{f}(t), \tag{127}$$

where $\tilde{f}(t)$ can be the density matrix $\rho(t)$ of eq. (109) or the function $D(t)$ defined by eq. (118). Note that $\tilde{f}(t)$ contains the dynamical information for all the particles in the entire system. In practice, only the dynamics of part of the entire system is of interest, which can be obtained from a "projected part" of $\tilde{f}(t)$. To be precise, a suitable projection operator P satisfying

$$P^2 = P \tag{128}$$

is introduced, and an equation of motion for $P\tilde{f}(t)$ is sought. Let

$$\tilde{f}_1(t) = P\tilde{f}(t), \tag{129}$$

$$\tilde{f}_2(t) = (1 - P)\tilde{f}(t) \equiv Q\tilde{f}(t). \tag{130}$$

Operating on eq. (127) by P and $Q \equiv (1 - P)$, respectively, generates the following coupled equations for $\tilde{f}_1(t)$ and $\tilde{f}_2(t)$:

$$i \, \partial\tilde{f}_1(t)/\partial t = PL\tilde{f}_1 + PL\tilde{f}_2, \tag{131}$$

$$i \, \partial\tilde{f}_2(t)/\partial t = QL\tilde{f}_1 + QL\tilde{f}_2. \tag{132}$$

These equations can readily be "solved" by the method of Laplace transform (Zwanzig 1960, 1961a,b, 1964). The complex Laplace transform of a function $\tilde{F}(t)$ can be defined by

$$F(z) = \int_0^\infty dt \, e^{izt} \, \tilde{F}(t), \quad z = \omega + i0^+, \tag{133}$$

where ω is real and the infinitesimal positive imaginary part of z ensures the convergence of the integral. The Laplace transforms of eqs. (131) and (132) become

$$zf_1(z) - i\tilde{f}_1(0) = PLf_1(z) + PLf_2(z), \tag{134}$$

$$zf_2(z) - i\tilde{f}_2(0) = QLf_1(z) + QLf_2(z). \tag{135}$$

$f_2(z)$ in eq. (135) can readily be solved in terms of $f_1(z)$ and substituted into eq. (134) to yield

$$zf_1(z) - i\tilde{f}_1(0) = PLf_1(z) + PL(z - QL)^{-1}i f_2(0) + PL(z - QL)^{-1}QLf_1(z), \tag{136}$$

which can be converted into an equation in the time domain by taking the inverse Laplace transform and applying the convolution theorem to the last term on its r.h.s., resulting in

$$i \frac{\partial\tilde{f}_1(t)}{\partial t} = PL\tilde{f}_1(t) + PL \, e^{-iQLt}\tilde{f}_2(0) - i \int_0^t dt' \, PL \, e^{-iQLt'} QL\tilde{f}_1(t - t'). \tag{137}$$

If the projection operator P is chosen such that initially at $t = 0$

$$P\tilde{f}(0) = \tilde{f}(0) \quad \text{and so} \quad \tilde{f}_2(0) = 0, \tag{138}$$

eq. (137) becomes a closed equation for $\tilde{f}_1(t)$. In most cases of interest, L can be separated into a unperturbed part L_0 and a perturbation L_1,

$$L = L_0 + L_1; \tag{139}$$

e.g., for H given by eq. (110), $L_0 = [H^{(s)} + H^{(b)}, \ ]$ and $L_1 = [V, \ ]$. If P has the further property that it commutes with L_0, i.e.,

$$[P, L_0] = 0, \qquad [Q, L_0] = 0, \tag{140}$$

then using the relation $QP = (1 - P)P = 0$ and observing that

$$QL\tilde{f}_1(t) = (1 - P)L_0 P\tilde{f}(t) + (1 - P)L_1\tilde{f}_1(t) = (1 - P)L_1\tilde{f}_1(t), \tag{141}$$

eq. (137) simplifies to (Zwanzig 1960, 1961a, 1964)

$$i\frac{\partial \tilde{f}_1(t)}{\partial t} = PL\tilde{f}_1(t) + \int_0^t dt' \ \check{K}(t')\tilde{f}_1(t - t'), \tag{142a}$$

where

$$\check{K}(t') = -iPL_1 e^{-iQLt'}QL_1. \tag{142b}$$

The corresponding complex Laplace transform of eq. (142a) is

$$zf_1(z) - i\tilde{f}_1(0) = PLf_1(z) + K(z)f_1(z), \tag{143a}$$

with

$$K(z) = \int_0^\infty dt \ e^{izt}\check{K}(t) = PL_1(z - QL)^{-1}QL_1, \tag{143b}$$

which can also be obtained directly from eq. (136) under the assumption of eqs. (138) and (140). Following Zwanzig, a memory function can be defined by (Zwanzig 1960)

$$zf_1(z) - i\tilde{f}_1(0) = P[L_0 + M_c(z)]f_1(z), \tag{144}$$

which, in the time domain, corresponds to

$$i\frac{\partial \tilde{f}_1(t)}{\partial t} = PL_0\tilde{f}_1(t) + P \int_0^t dt' \ \check{M}_c(t')\tilde{f}_1(t - t'). \tag{145}$$

$\check{M}_c(t)$ is related to $M_c(z)$ by an equation similar to eq. (133). [M_c is the same as T in eq. (13) of Zwanzig (1960).] Comparing eqs. (143) and (144) yields

$$M_c(z) = L_1 + L_1 \frac{1}{z - QL} QL_1. \tag{146}$$

From the useful operator identity

$$\frac{1}{A + B} = \frac{1}{A} - \frac{1}{A} B \frac{1}{A + B}, \tag{147}$$

and identifying A with $z - QL_0$ and B with $-QL_1$, the resolvent $(z - QL)^{-1}$ can be expressed as

$$\frac{1}{z - QL} = \frac{1}{z - QL_0 - QL_1} = \frac{1}{z - QL_0} + \frac{1}{z - QL_0} QL \frac{1}{z - QL}$$

$$= \frac{1}{z - L_0} Q + \frac{1}{z - L_0} QL_1 \frac{1}{z - QL}, \tag{148}$$

where the last equality follows from the assumption in eq. (140). Substituting eq. (148) into eq. (146) yields a Lippmann–Schwinger type of equation for the memory function (Zwanzig 1960):

$$M_c(z) = L_1 + L_1 \frac{1}{z - L_0} (1 - P) M_c(z). \tag{149}$$

It is worth noting that in arriving at eqs. (142), (143), and (149) the only assumptions made are: (i) eq. (138) concerning the initial value of $\bar{f}_1$, and (ii) the commutation relation eq. (140) between P and L_0. No approximation concerning the *dynamics* of the many-body system has been made. A perturbation scheme can readily be developed from eqs. (142) and (143) in powers of L_1 if the perturbation L_1 can be considered to be small (Zwanzig 1960, 1961a). If, on the other hand, the interaction which gives rise to L_1 is strong but short range, eqs. (144) and (149) serve as a convenient starting point for deriving a density expansion in which a binary collision expansion of $M_c(z)$ can be made (Fano 1963, Zwanzig 1961a).

3.3. The relaxation theory of Fano

One of the early and important applications of Zwanzig's projection operator approach was given by Fano in his treatment of pressure broadening in gases. Fano and later Ben-Reuven (1966a) took full advantage of the Liouville space formalism in which dynamical variables are considered to be vectors whose time development can be written in a very compact form. Following the notation of Ben-Reuven (1966a,b) an operator Φ in ordinary Hilbert space is considered to be a vector $|\Phi\rangle\rangle$ in the Liouville space in which an inner product is defined by

$$\langle\langle\Phi|\Psi\rangle\rangle = \mathrm{Tr}\, \Phi^+\Psi, \tag{150}$$

where Tr denotes a trace over the system variables. In this notation, the trace of Φ becomes

$$\mathrm{Tr}\, \Phi = \langle\langle 1|\Phi\rangle\rangle, \tag{151}$$

where 1 is the unit operator in ordinary Hilbert space. The statistical average of a dynamical quantity Φ can be written as

$$\langle\Phi\rangle = \mathrm{Tr}\, \rho\Phi = \langle\langle\rho|\Phi\rangle\rangle, \tag{152}$$

where ρ is the appropriate (Hermitian) density matrix of the system. In the following, the abbreviations L-space, L-vector, etc. will be used for Liouville space, Liouville vector, etc. If $(|i\rangle, |j\rangle, \ldots)$ form a basis in ordinary Hilbert space, a L-basis vector can readily be constructed from

$$|ij\rangle\rangle \equiv |i\rangle\langle j|, \qquad \langle\langle ij| = (|i\rangle\langle j|)^+ = |j\rangle\langle i|, \tag{153}$$

satisfying the orthonormal and completeness relations,

$$\langle\langle kl|ij\rangle\rangle = \delta_{ki}\delta_{lj}, \tag{154}$$

$$\sum_{i,j} |ij\rangle\rangle\langle\langle ij| = I, \tag{155}$$

where I is the identity L-operator. Using Eq. (155), any L-vector $|\Phi\rangle\rangle$ can be expanded in a convenient basis as

$$|\Phi\rangle\rangle = \sum_{i,j} \Phi_{ij}|ij\rangle\rangle, \tag{156}$$

where $\Phi_{ij} = \langle\langle ij|\Phi\rangle\rangle = \langle i|\Phi|j\rangle$. As mentioned earlier, the commutator relation

$$[H, \rho] = L\rho = L|\rho\rangle\rangle \tag{157}$$

defines a L-operator L, whose matrix element in this notation becomes

$$L_{ij,kl} = \langle\langle ij|L|kl\rangle\rangle. \tag{158}$$

If $|i\rangle$, $|f\rangle$ are eigenstates of the Hamiltonian H with eigenvalues E_i, E_f, respectively, it follows from eq. (157) that

$$L|if\rangle\rangle = (E_i - E_f)|if\rangle\rangle \equiv \hbar\omega_{if}|if\rangle\rangle. \tag{159}$$

$|if\rangle\rangle$ is thus an eigenvector of L with an eigenvalue $\hbar\omega_{if}$. The Hermitian conjugate of a L-operator $\mathscr{A}$, denoted by $\mathscr{A}^+$, is defined by

$$\langle\langle C|\mathscr{A}|D\rangle\rangle = \langle\langle \mathscr{A}^+C|D\rangle\rangle, \tag{160}$$

for any arbitrary L-vectors $|C\rangle\rangle$ and $|D\rangle\rangle$. It can be shown that if $\mathscr{A}$ is defined by a commutation operation

$$\mathscr{A} \equiv [A, \ ], \tag{161}$$

where A is a Hermitian operator in ordinary Hilbert space, then $\mathscr{A}$ is Hermitian, i.e., $\mathscr{A}^+ = \mathscr{A}$. Thus $L = [H, \ ]$, $L_0 = [H^{(s)} + H^{(b)}, \ ]$, and $L_1 = [V, \ ]$ are Hermitian L-operators.

The spectral lineshape function is given by the complex Laplace transform of the correlation function $C(t)$ defined by eq. (114) (Fano 1963):

$$F(z) = \frac{1}{\pi} \mathrm{Re} \int_0^\infty \mathrm{d}t \ e^{izt}C(t), \quad z = \omega + i0^+. \tag{162}$$

Using the L-space notation just introduced, it follows from eqs. (116)

and (117) that the lineshape function of eq. (162) can be written as

$$F(z) = \frac{1}{\pi} \text{Re}\langle\langle \mu^{(s)} | D(z)\rangle\rangle, \tag{163}$$

where

$$|D(z)\rangle\rangle = \int_0^\infty dt\ e^{izt} |D(t)\rangle\rangle = i\frac{1}{z-L} |\rho\mu^{(s)}\rangle\rangle. \tag{164}$$

If the Hamiltonian of the entire system is given by eq. (110) the L-operator L can be likewise split up as

$$L = L_0 + L_1, \qquad L_0 = L_0^{(s)} + L_0^{(b)}, \tag{165}$$

where $L_0^{(s)} = [H^{(s)},\ ]$, $L_0^{(b)} = [H^{(b)},\ ]$ and $L_1 = [V,\ ]$. The interaction L_1 can be disentangled from the resolvent $(z - L)^{-1} = (z - L_0 - L_1)^{-1}$ by using the operator identity eq. (147) and identifying A with $z - L_0$ and B with $-L_1$:

$$\frac{1}{z-L} = \frac{1}{z-L_0}\left(1 + L_1\frac{1}{z-L}\right). \tag{166}$$

An L-operator M, which formally contains all the effects due to L_1, can be defined by

$$\frac{1}{z-L} = \frac{1}{z-L_0}\left(1 + M(z)\frac{1}{z-L_0}\right). \tag{167}$$

An integral equation for $M(z)$ in terms of L_0 and L_1 can be obtained by the following formal manipulation. First, comparing eq. (167) with eq. (166),

$$M(z) = L_1\frac{1}{z-L}(z-L_0). \tag{168}$$

Again, from an operator identity similar to eq. (147):

$$\frac{1}{A+B} = \frac{1}{A} - \frac{1}{A+B}B\frac{1}{A}, \tag{169}$$

the resolvent in the r.h.s. of eq. (168) can be expressed as

$$\frac{1}{z-L} = \left(1 + \frac{1}{z-L}L_1\right)\frac{1}{z-L_0}, \tag{170}$$

and thus

$$M(z) = L_1 + L_1\frac{1}{z-L}L_1. \tag{171}$$

The desired integral equation for $M(z)$ can be obtained by substituting eq.

(166) for the resolvent on the r.h.s. of eq. (171):

$$M(z) = L_1 + L_1 \frac{1}{a - L_0} \left(1 + L_1 \frac{1}{z - L} \right) L_1$$

$$= L_1 + L_1 \frac{1}{z - L_0} M(z), \tag{172}$$

where eq. (171) is used in the last step to identify the $M(z)$ appearing on the r.h.s. Notice that no approximation has been made in deriving eqs. (171) and (172), and the results obtained are exact. Of course, the many-body problem has not been solved; all the interaction dynamics are contained in $M(z)$.

It will now be assumed that initial correlation between the system and the bath can be neglected. Thus the equilibrium density matrix ρ appearing on the r.h.s. of eq. (164) factors:

$$\rho \approx \rho^{(s)} \rho^{(b)}. \tag{173}$$

Consistent with the idea of a bath, it will further be assumed that $\rho^{(b)}$ is a canonical distribution among the bath variables and thus is diagonal in the eigenstates of $H^{(b)}$:

$$\langle \alpha | \rho^{(b)} | \beta \rangle = w_\alpha \delta_{\alpha\beta}, \tag{174}$$

where

$$H^{(b)} | \alpha \rangle = E_\alpha | \alpha \rangle \tag{175}$$

and

$$w_\alpha = e^{-E_\alpha / k_B T} / \sum_{\alpha'} e^{-E_{\alpha'} / k_B T}. \tag{176}$$

An immediate consequence of eq. (174) is that $\rho^{(b)}$ becomes an L-eigen-vector of $L_0^{(b)}$ with eigenvalue 0,

$$L_0^{(b)} | \rho^{(b)} \rangle\rangle = 0, \tag{177}$$

which can be verified by evaluating an explicit matrix element of the l.h.s. It must be emphasized that while eq. (173) holds in the theory of relaxation in the gas phase to lowest order in the density (Fano 1963), its validity in the general case remains to be ascertained (Fano 1964). However, eq. (173) will be treated as a working assumption in the following discussion.

Substituting eqs. (167) and (173) into eqs. (163) and (164), and observing that $\mu^{(s)}$ operates on the system variables only (i.e., $|\mu^{(s)}\rangle\rangle = |1^{(b)} \mu^{(s)}\rangle\rangle$, where $1^{(b)}$ is the unit operator in the space of the bath molecules), one obtains

$$F(\omega) = -\frac{1}{\pi} \text{Im} \langle\langle \mu^{(s)} | \frac{1}{z - L_0^{(s)}} \left(1 - \langle M(z) \rangle \frac{1}{\omega - L_0^{(s)}} \right) | \rho^{(s)} \mu^{(s)} \rangle\rangle, \tag{178}$$

where

$$\langle M(z)\rangle \equiv \langle\langle 1^{(b)}|M(z)|\rho^{(b)}\rangle\rangle = \mathrm{Tr}^{(b)}M(z)\rho^{(b)} \tag{179}$$

is still a L-operator on the active molecule variables. In deriving eq. (178), the property of eq. (177) has been used to eliminate $L_0^{(b)}$.

At this stage, the connection with Zwanzig's results of the last subsection will be made. A projection operator appropriate for the present discussions has been introduced in eq. (123), which can be written compactly in the L-space notation as (Jones and Zewail 1978, Diestler and Zewail 1979)

$$P = |\rho^{(b)}\rangle\rangle\langle\langle 1^{(b)}| = |\rho^{(b)}\rangle\rangle \sum_\alpha \langle\langle\alpha\alpha|. \tag{180}$$

As discussed in subsection 3.1 only the projected part of $D(t)$ is required to calculate the correlation function, and thus the lineshape function of eq. (162) can be written in terms of the complex Laplace transform of $D_1(t)$ by making use of eq. (121) as

$$F(\omega) = \frac{1}{\pi}\mathrm{Re}\langle\langle\mu^{(s)}1^{(b)}|D_1(z)\rangle\rangle, \quad z = \omega + \mathrm{i}0^+, \tag{181}$$

where

$$|D_1(z)\rangle\rangle = \int_0^\infty \mathrm{d}t\, \mathrm{e}^{\mathrm{i}zt}|D_1(t)\rangle\rangle. \tag{182}$$

To use the results of subsection 3.2 it must first be verified that P defined by eq. (180) satisfies eqs. (138) and (140). Because of the assumption of eq. (173), it follows from eq. (120) that

$$\begin{aligned} P|D(0)\rangle\rangle &= |\rho^{(b)}\rangle\rangle\langle\langle 1^{(b)}|\rho^{(b)}\rho^{(s)}\mu^{(s)}\rangle\rangle \\ &= |\rho^{(b)}\rho^{(s)}\mu^{(s)}\rangle\rangle \\ &= D(0). \end{aligned} \tag{183}$$

Furthermore, noting that $L_0^{(b)}|1^{(b)}\rangle\rangle = 0$ and that $L_0^{(s)}$ commutes with $\rho^{(b)}$ and $1^{(b)}$,

$$PL_0 = |\rho^{(b)}\rangle\rangle\langle\langle 1^{(b)}|(L_0^{(s)} + L_0^{(b)}) = L_0^{(s)}P. \tag{184}$$

From eq. (177),

$$L_0 P = (L_0^{(s)} + L_0^{(b)})|\rho^{(b)}\rangle\rangle\langle\langle 1^{(b)}| = L_0^{(s)}P, \tag{185}$$

and thus $[P, L_0] = 0$. Hence both conditions in eqs. (138) and (140) are satisfied, and identifying $f_1(z)$, $\tilde{f}_1(0)$ with $D_1(z)$ and $\rho^{(b)}\rho^{(s)}\mu^{(s)}$, respectively, it follows from eq. (144) that

$$zD_1(z) - \mathrm{i}\rho^{(b)}\rho^{(s)}\mu^{(s)} = [L_0^{(s)} + PM_c(z)P]D_1(z), \tag{186}$$

where eq. (184) and the fact $PD_1(z) = D_1(z)$ have been used. $M_c(z)$ is given explicitly by eqs. (146) or (149).

From eq. (186), $D_1(z)$ can be solved to yield

$$|D_1(z)\rangle\rangle = i \frac{1}{z - L_0^{(s)} - PM_c(z)P} |\rho^{(b)}\rho^{(s)}\mu\rangle\rangle. \tag{187}$$

From the definition in eq. (180),

$$PM_c(z)P = |\rho^{(b)}\rangle\rangle\langle\langle 1^{(b)}|M_c(z)|\rho)^{(b)}\rangle\rangle\langle\langle 1^{(b)}|$$
$$= |\rho^{(b)}\rangle\rangle\langle M_c(z)\rangle\langle\langle 1^{(b)}|. \tag{188}$$

Taking the trace over the bath variables on both sides of eq. (187), one obtains

$$|D^{(s)}(z)\rangle\rangle = \langle\langle 1^{(b)}|D_1(z)\rangle\rangle = i \frac{1}{z - L_0^{(s)} - \langle M_c(z)\rangle} |\rho^{(s)}\mu\rangle\rangle, \tag{189}$$

which, upon substitution into eq. (181), yields an alternative expression for the lineshape function (Fano 1963):

$$F(\omega) = -\frac{1}{\pi} \text{Im}\langle\langle\mu^{(s)}| \frac{1}{z - L_0^{(s)} - \langle M_c(z)\rangle} |\rho^{(s)}\mu^{(s)}\rangle\rangle, \quad z = \omega + i0^+. \tag{190}$$

Relations between $\langle M_c(z)\rangle$ and $\langle M(z)\rangle$ can readily be established by comparing eq. (190) with eq. (178) to give

$$\frac{1}{z - L_0^{(s)} - \langle M_c(z)\rangle} = \frac{1}{z - L_0^{(s)}}\left(1 + \langle M(z)\rangle \frac{1}{z - L_0^{(s)}}\right). \tag{191}$$

Applying the identity of eq. (147) with $A = z - L_0^{(s)}$, $B = \langle M_c(z)\rangle$, to the l.h.s. of eq. (191) one obtains, after cancelling common factors on both sides,

$$\langle M_c(z)\rangle \frac{1}{z - L_0^{(s)} - \langle M_c(z)\rangle} = \langle M(z)\rangle \frac{1}{z - L_0^{(s)}}. \tag{192}$$

Applying the identity of eq. (169) to the l.h.s. in a similar fashion, the following relations can be established:

$$\langle M(z)\rangle = \langle M_c(z)\rangle\left(1 + \frac{1}{z - L_0^{(s)} - \langle M_c(z)\rangle}\langle M_c(z)\rangle\right)$$
$$= \langle M_c(z)\rangle\left(1 + \frac{1}{z - L_0^{(s)}}\langle M(z)\rangle\right), \tag{193}$$

where the last equality is arrived at in the same manner as in the derivation of eq. (172). Finally, eq. (193) may be inverted to give (Fano 1963)

$$\langle M_c(z)\rangle = \langle M(z)\rangle\left(1 + \frac{1}{z - L_0^{(s)}}\langle M(z)\rangle\right)^{-1}$$
$$= \langle M(z)\rangle \sum_{n=0}^{\infty} \left(-\frac{1}{z - L_0^{(s)}}\langle M(z)\rangle\right)^n. \tag{194}$$

Since the zeros of the denominator of eq. (190) determine the resonance frequencies and widths of the spectrum, the quantity $\langle M_c(z) \rangle$ is of a more fundamental character than $\langle M(z) \rangle$. On the other hand, it appears from eqs. (149) and (172) that $M(z)$ has a simpler structure and is more manageable in actual calculation in some cases, as illustrated from the density expansion carried out by Fano (1963). Equations (193) and (194) then furnish the connection between $\langle M_c \rangle$ and $\langle M \rangle$.

The above considerations can also be applied directly to ρ_1 and $\rho^{(s)}$. In L-space notation, eq. (113) can be written as

$$|\rho^{(s)}(t)\rangle\rangle = \langle\langle 1^{(b)}|\rho_1(t)\rangle\rangle \quad \text{or} \quad |\rho^{(s)}(z)\rangle\rangle = \langle\langle 1^{(b)}|\rho_1(z)\rangle\rangle, \tag{195}$$

where $\rho^{(s)}(z)$, $\rho_1(z)$ are the complex Laplace transforms (cf. eq. (133)] of $\rho^{(s)}(t)$, $\rho_1(t)$, respectively.

Similar to eq. (173), initial correlation between the active molecules and the bath is assumed to be negligible. Thus

$$\rho(0) \approx \rho^{(s)}(0)\rho^{(b)}, \tag{196}$$

where $\rho^{(s)}(0)$ is the initial (nonequilibrium) density matrix for the active molecules and $\rho^{(b)}$ again obeys eq. (174). It then follows that for $f(z) = \rho(z)$, $f_1(z) = \rho_1(z)$, the condition eq. (138) is satisfied, and in exactly the same way eqs. (187) and (190) are derived one obtains

$$|\rho^{(s)}(z)\rangle\rangle = i\frac{1}{z - L_0^{(s)} - \langle M_c(z)\rangle}|\rho^{(s)}(0)\rangle\rangle, \quad z = \omega + i0^+. \tag{197}$$

Using the fact that the complex Laplace transform of $i\,\partial\rho^{(s)}(t)/\partial t$ is $z\rho^{(s)}(z) - i\rho^{(s)}(0)$ and the convolution theorem, an equation of motion for $\rho^{(s)}(t)$ emerges immediately from eq. (197) [cf. eq. (145)]:

$$i\frac{\partial\rho^{(s)}(t)}{\partial t} = L_0^{(s)}\rho(t) + \int_0^t dt' \langle \tilde{M}_c(t')\rangle\rho^{(s)}(t - t'), \tag{198}$$

where $\langle M_c(z)\rangle$ is the complex Laplace transform of $\langle \tilde{M}_c(t)\rangle$ [cf. eq. (133)].

Fano (1964, 1963) has observed that a well-resolved spectrum results if the condition

$$|\langle M_c\rangle_{f'i',fi}/(\omega_{f'i'} - \omega_{fi})| \ll 1 \quad \text{for } (f', i') \neq (f, i) \tag{199}$$

is satisfied. In this case, the off-diagonal elements of $\langle M_c\rangle$ can be ignored, and the matrix $[z - L_0^{(s)} - \langle M_c(z)\rangle]$ becomes diagonal. Recalling that $z = \omega + i0^+$, eq. (190) then simplifies to

$$F(\omega) = -\frac{1}{\pi} \text{Im} \sum_{f,i} |\mu_{fi}^{(s)}|^2 \rho_i^{(s)} \frac{1}{\omega - \omega_{fi} - \langle M_c(\omega + i0^+)\rangle_{fi,fi}}, \tag{200}$$

where f, i are the states of the active molecule coupled by the operator $\mu^{(s)}$, and $\hbar\omega_{fi} = E_f - E_i$ is the (unperturbed) energy difference between these states. For the simple case that only one line (f, i) is coupled by $\mu^{(s)}$, the lineshape becomes

$$F(\omega) = -\frac{1}{\pi} \operatorname{Im} |\mu_{fi}^{(s)}|^2 \left(\frac{1}{\omega - \omega_{fi} - \langle M_c(\omega + i0^+) \rangle_{fi, fi}} \rho_i^{(s)} \right.$$
$$\left. + \frac{1}{\omega - \omega_{if} - \langle M_c(\omega + i0^+) \rangle_{if, if}} \rho_f^{(s)} \right). \tag{201}$$

For $\omega_{fi} > 0$ and $\omega \sim \omega_{fi}$, the second term on the r.h.s. is usually much smaller than the first (since $\omega_{if} = -\omega_{fi}$) and can be neglected. This neglect of the nonresonant term is commonly known as "the rotating wave approximation" (Sargent et al. 1974). The resulting lineshape function is

$$F(\omega) = -\frac{1}{\pi} \operatorname{Im} |\mu_{fi}^{(s)}|^2 \rho_i \frac{1}{\omega - \omega_{fi} - \langle M_c(\omega + i0^+) \rangle_{fi, fi}}. \tag{202}$$

Applying the same assumptions to eq. (197), the off-diagonal density matrix element required in computing the macroscopic polarization $P(t)$ of eq. (112) obeys a similar equation:

$$\rho_{fi}^{(s)}(\omega) = i \frac{1}{\omega - \omega_{fi} - \langle M_c(\omega + i0^+) \rangle_{fi, fi}} \rho_{fi}^{(s)}(0), \tag{203}$$

whose equation of motion in the time domain is [cf. eq. (197)]

$$i \frac{\partial}{\partial t} \rho_{fi}^{(s)}(t) = \omega_{fi} \rho_{fi}^{(s)} + \int_0^t dt' \langle \tilde{M}_c(t') \rangle_{fi, fi} \rho_{fi}^{(s)}(t - t'), \tag{204}$$

or equivalently,

$$\rho_{fi}^{(s)}(t) = \int_{-\infty}^{\infty} \frac{d\omega}{2\pi} \frac{e^{-i\omega t}}{\omega - \omega_{fi} - \langle M_c(\omega + i0^+) \rangle_{fi, fi}} i\rho_{fi}^{(s)}(0). \tag{205}$$

3.4. Short-memory approximation

The frequency spectrum and the time dependence of the relaxation process will be very simple if $\langle M_c(z) \rangle$ is independent of z. Writing in this case

$$\langle M_c(z) \rangle = -i\Lambda, \tag{206}$$

where Λ can be identified with the relaxation matrix introduced earlier in eq. (63), the lineshape function of eq. (200) becomes a Lorentzian:

$$F(\omega) = \frac{|\mu_{fi}^{(s)}|^2 \rho_i}{\pi} \frac{\gamma}{(\omega - \omega_{fi} - \delta)^2 + \gamma^2}, \tag{207}$$

where

$$\gamma = \operatorname{Re} \Lambda_{fi, fi}, \qquad \delta = \operatorname{Im} \Lambda_{fi, fi}. \tag{208}$$

γ and δ give, respectively, the width and shift of the $i \to f$ spectral line. Furthermore, comparing eq. (208) with eq. (65), $\gamma = T_2^{-1}$.

Similarly, substituting eq. (206) into eq. (205) results in

$$\rho_{fi}^{(s)}(t) = \int_{-\infty}^{\infty} \frac{d\omega}{2\pi} \frac{e^{-i\omega t}}{\omega - \omega_{fi} - \delta + i\gamma} i\rho_{fi}^{(s)}(0)$$
$$= \rho_{fi}^{(s)}(0) e^{-i(\omega_{fi}+\delta)t} e^{-\gamma t}, \qquad (209)$$

where the integral is evaluated by closing the contour in the lower half ω-plane and applying the residue theorem. As mentioned earlier, γ is a positive quantity and eq. (209) represents an exponential decay of the initial coherent excitation. $\rho_{fi}^{(s)}(t)$ given by eq. (209) satisfies the equation of motion

$$i\frac{\partial}{\partial t} \rho_{fi}^{(s)}(t) = (\omega_{fi} + \delta)\rho_{fi}^{(s)}(t) - i\gamma\rho_{fi}^{(s)}(t). \qquad (210)$$

The second term on the r.h.s. of this equation, which describes the relaxation of $\rho_{fi}^{(s)}$, could be compared with the corresponding term in eq. (198). In the general case of eq. (198), the relaxation is described by a convolution integral and depends on the past history of the system. This indicates that the bath has a "memory", and such behavior is usually called "non-Markovian" (Fano 1964, Zwanzig 1960, 1961a). The relaxation term in eq. (210), on the other hand, depends only on the instantaneous behavior of the system. It thus has a short memory, and is called "Markovian". The validity of the short-memory (or Markovian) approximation for the case that the interaction between the active system and the bath is weak has been studied extensively by Van Hove (1955) (see also Zwanzig 1961a).

In order to investigate the conditions under which the short-memory approximation can be made, eq. (204) is first transformed into the rotating frame:

$$i\frac{\partial}{\partial t} \rho^R(t) = \int_0^t dt' \, M^R(t')\rho^R(t - t'), \qquad (211)$$

where

$$\rho^R(t) = e^{i\omega_{fi}t} \rho_{fi}^{(s)}(t) \qquad (212)$$

and

$$M^R(t) = e^{i\omega_{fi}t} \langle \tilde{M}_c(t) \rangle_{fi,fi}. \qquad (213)$$

If $M^R(t)$ is small, $\partial \rho^R/\partial t$ is small and $\rho^R(t - t')$ on the l.h.s. of eq. (211) can be replaced by $\rho^R(t)$. Furthermore, if we are only interested in the long-time behavior of $\rho^R(t)$ and the decay time or correlation time of $M^R(t)$ is short compared to the time scale of interest, the upper limit in the integral on the r.h.s. of eq. (211) can be extended to ∞. Then eq. (211) simplifies to

$$i\frac{\partial}{\partial t} \rho^R(t) = \langle M_c(\omega_{fi} + i0^+) \rangle_{fi,fi}\rho^R(t), \qquad (214)$$

which reduces to eq. (210) when transformed back to the laboratory frame and $\langle M_c(\omega_{fi} + i0^+) \rangle_{fi,fi}$ is identified with $-i\Lambda_{fi,fi} = \delta - i\gamma$, resulting in the simple behavior of eq. (209).

The same results can also be obtained by studying the problem in the frequency domain. From eq. (205), the time development of $\rho_{fi}^{(s)}(t)$ is given by the roots of

$$\omega - \omega_{fi} = \langle M_c(\omega + i0^+) \rangle_{fi,fi}. \tag{215}$$

If $\exp(i\omega_{fi}t)\langle \tilde{M}_c(t) \rangle_{fi,fi}$ has a short correlation time, $\langle M_c(z) \rangle_{fi,fi}$ is a slowly varying function about $z = \omega_{fi}$. Furthermore, if $\langle M_c(z) \rangle_{fi,fi}$ is small, eq. (215) can be solved iteratively (Wilson et al. 1968, Resibois 1969). The zeroth-order approximation is $z = \omega_{fi}$. The first-order result is obtained by substituting this zeroth-order result into the r.h.s. of eq. (215), resulting in

$$\omega = \omega_{fi} + \langle M_c(\omega_{fi} + i0^+) \rangle_{fi,fi}, \tag{216}$$

which can be written as $\omega = \omega_{fi} + \delta - i\gamma$ with

$$\delta = \text{Re}\langle M_c(\omega_{fi} + i0^+) \rangle_{fi,fi}, \tag{217}$$

$$\gamma = -\text{Im}\langle M_c(\omega_{fi} + i0^+) \rangle_{fi,fi}. \tag{218}$$

Putting this result into eq. (205) yields eq. (209) immediately.

Similar considerations can be given to the diagonal elements of $\rho^{(s)}(t)$. Under the condition of eq. (199), the following equation of motion for $\rho_{kk}^{(s)}(t)$ can be obtained from eq. (198):

$$i\frac{\partial \rho_{kk}^{(s)}(t)}{\partial t} = \sum_l \int_0^t dt' \, \langle \tilde{M}_c(t') \rangle_{kk,ll} \rho_{ll}^{(s)}(t - t'), \tag{219}$$

where the result $\langle\langle kk | L_0^{(s)} | ll \rangle\rangle = 0$ has been used. Employing the same arguments as in deriving eqs. (214) and (216), the short-memory approximation to eq. (219) is

$$i\frac{\partial \rho_{kk}^{(s)}(t)}{\partial t} = \sum_l \langle M_c(i0^+) \rangle_{kk,ll} \rho_{ll}^{(s)}(t), \tag{220}$$

where $M_c(i0^+)$ denotes $\lim_{\epsilon \to 0^+} M_c(i\epsilon)$. The matrix element $\Lambda_{kk,ll}$ introduced in section 2 is thus given by

$$\Lambda_{kk,ll} = i\langle M_c(i0^+) \rangle_{kk,ll} \tag{221}$$

and is related to T_1 by eq. (69), under the conditions given in subsection 2.2.

3.5. Perturbation expansion of $\langle M_c \rangle$

If the interaction between the active and bath molecules is weak, a perturbation expansion of $\langle M_c \rangle$ in powers of L_1 can be carried out. From eq. (149), $\langle M_c \rangle$ can be expanded to second order in L_1 as

$$\langle M_c(z) \rangle = \langle L_1 \rangle + \left\langle L_1 \frac{1}{z - L_0}(1 - P)L_1 \right\rangle. \tag{222}$$

As mentioned earlier, the linewidth of an isolated line and the T_1 and T_2 parameters measured in time-resolved experiments are related to the matrix elements of $\langle M_c \rangle$, expressions of which are derived in this subsection.

For many cases of interest, the interaction can be factored into a product of system and bath terms (Oxtoby and Rice 1976)

$$V = \sum_n V_n^{(s)} V_n^{(b)}. \tag{223a}$$

For simplicity we assume in this subsection that there is only one term in the expansion, i.e.,

$$V = V^{(s)} V^{(b)}. \tag{223b}$$

The results could readily be generalized for eq. (223a). The matrix element of L_1 can be written explicitly as

$$\langle\!\langle f\beta i\alpha | L_1 | k\mu l\lambda \rangle\!\rangle = \langle f\beta | V | k\mu \rangle \delta_{il}\delta_{\alpha\lambda} - \langle i\alpha | V | l\lambda \rangle^* \delta_{fk}\delta_{\beta\mu}. \tag{224}$$

The matrix elements of various terms of eq. (222) can readily be worked out (Chiu and Liu, unpublished). The first term in r.h.s. of eq. (222) gives

$$\langle L_1 \rangle_{fi,fi} = \sum_{\alpha\beta} \langle\!\langle f\alpha, i\alpha | L_1 | f\beta, i\beta \rangle\!\rangle w_\beta$$

$$= \sum_\beta w_\beta [\langle f\beta | V | f\beta \rangle - \langle i\beta | V | i\beta \rangle]$$

$$\equiv \Delta V. \tag{225}$$

The terms which are of second order in V are similarly given by

$$\left\langle L_1 \frac{1}{z - L_0} L_1 \right\rangle_{fi,fi} = \sum_{k,\beta,\alpha} w_\beta \left(\frac{|\langle f\beta | V | k\alpha \rangle|^2}{z - \omega_{ki} - \omega_{\alpha\beta}} + \frac{|\langle i\beta | V | k\alpha \rangle|^2}{z - \omega_{fk} + \omega_{\alpha\beta}} \right)$$

$$- \sum_{\alpha\beta} w_\beta \langle f\alpha | V | f\beta \rangle \langle i\alpha | V | i\beta \rangle^* \left(\frac{1}{z - \omega_{fi} - \omega_{\alpha\beta}} + \frac{1}{z - \omega_{fi} + \omega_{\alpha\beta}} \right), \tag{226}$$

$$\left\langle L_1 \frac{1}{z - L_0} P L_1 \right\rangle_{fi,fi} = \sum_k \left(\frac{|\langle V \rangle_{fk}|^2}{z - \omega_{ki}} + \frac{|\langle V \rangle_{ik}|^2}{z - \omega_{fk}} \right) - \frac{2\langle V \rangle_{ff} \langle V \rangle_{ii}}{z - \omega_{fi}}, \tag{227}$$

where

$$\langle V \rangle \equiv \sum_\beta w_\beta \langle \beta | V | \beta \rangle. \tag{228}$$

In eqs. (226) and (227), $\omega_{ki} = E_k - E_i$ and $\omega_{\alpha\beta} = E_\alpha - E_\beta$ are the energy differences of the active and bath molecules, respectively. For V given by eq. (223b), $\langle V \rangle$ of eq. (228) and ΔV of eq. (225) simplify to, respectively,

$$\langle V \rangle = V^{(s)} \langle V^{(b)} \rangle \tag{229a}$$

and

$$\Delta V = (V^{(s)}_{ff} - V^{(s)}_{ii}) \langle V^{(b)} \rangle, \tag{229b}$$

where

$$\langle V^{(b)} \rangle = \sum_{\beta} w_{\beta} \langle \beta | V^{(b)} | \beta \rangle, \qquad V^{(s)}_{kk} = \langle k | V^{(s)} | k \rangle.$$

Combining the results of eqs. (225) and (227), an explicit expression for $\langle M_c(z) \rangle_{fi,fi}$ can be obtained from eq. (222):

$$\langle M_c(z) \rangle_{fi,fi} = \Delta V + K(z), \tag{230}$$

where

$$K(z) = \sum_{\alpha,\beta} w_{\beta} (|\langle \alpha | V^{(b)} | \beta \rangle|^2 - \langle V^{(b)} \rangle^2 \delta_{\alpha\beta}) \Bigg[\sum_{k} \left(\frac{|V^{(s)}_{fk}|^2}{z - \omega_{ki} - \omega_{\alpha\beta}} + \frac{|V^{(s)}_{ik}|^2}{z - \omega_{fk} - \omega_{\beta\alpha}} \right)$$
$$- V^{(s)}_{ff} V^{(s)}_{ii} \left(\frac{1}{z - \omega_{fi} - \omega_{\beta\alpha}} + \frac{1}{z - \omega_{fi} + \omega_{\beta\alpha}} \right) \Bigg]. \tag{231}$$

In eq. (231), the form of eq. (223b) for V and eq. (229) have been used, and $V^{(s)}_{lm}$ stands for the matrix element $\langle l | V^{(s)} | m \rangle$. Since $\langle \Delta V \rangle$ is real, the imaginary part of $\langle M_c(z) \rangle_{fi,fi}$ is given by that of $K(z)$ (Mattuck 1976). From eq. (218), the linewidth, which is also T_2^{-1}, is given to second order in V by

$$\gamma = T_2^{-1} = -\text{Im } K(\omega_{fi} + i0^+), \tag{232}$$

while from eq. (217) the frequency shift is given to first order in V by

$$\delta = \Delta V. \tag{233}$$

Using the formula

$$\frac{1}{\omega + i0^+} = P \frac{1}{\omega} - i\pi\delta(\omega), \tag{234}$$

an explicit expression for T_2^{-1} can be obtained from eqs. (231) and (232):

$$\frac{1}{T_2} = \frac{1}{T_f^{in}} + \frac{1}{T_i^{in}} + \frac{1}{T_2^{ph}}, \tag{235}$$

where

$$\frac{1}{T_k^{in}} = \pi \sum_{\substack{\alpha,\beta \\ (\alpha \neq \beta)}} \sum_{\substack{l \\ (l \neq k)}} w_{\beta} |\langle \alpha | V^{(b)} | \beta \rangle|^2 |V^{(s)}_{lk}|^2 \delta(\omega_{lk} + \omega_{\beta\alpha}), \qquad k = i, f, \tag{236}$$

and

$$\frac{1}{T_2^{ph}} = \pi \sum_{\alpha,\beta} w_\beta (|\langle\alpha|V^{(b)}|\beta\rangle|^2 - \langle V^{(b)}\rangle^2 \delta_{\alpha\beta})|V_{ff}^{(s)} - V_{ii}^{(s)}|^2 \delta(\omega_{\alpha\beta}). \tag{237}$$

In eq. (236), the $\alpha = \beta$ term does not contribute because of the restriction of the δ-function. $(T_f^{in})^{-1}$ and $(T_i^{in})^{-1}$ represent the loss of coherence when the active molecule makes an inelastic transition out of the levels f and i, respectively, through interaction with the bath molecules. $(T_2^{ph})^{-1}$ given by eq. (237) arises from processes in which the active molecule returns to its original spectroscopic states i and f, with its phase being interrupted during interaction with the bath. This energetically elastic process (for the active molecule) is commonly known as the *pure dephasing process* (Jones and Zewail 1978; Diestler and Zewail 1979). Equations similar to eqs. (235)–(237) have been derived (Diestler 1976a,b) by applying the Zwanzig–Mori projection operator technique directly to the dynamical variables of a two-level system. The present derivation, however, does not require the assumption of a two level system.

The T_1-type matrix elements $\langle M_c(i0^+)\rangle_{kk,ll}$ of eq. (236) can similarly be worked out. One has

$$\Lambda_{kk,kk} = i\langle M_c(i0^+)\rangle_{kk,kk} = 2\pi \sum_{\substack{l \\ (l\neq k)}} \sum_{\substack{\alpha,\beta \\ (\alpha\neq\beta)}} w_\beta |\langle k\beta|V|l\alpha\rangle|^2 \delta(\omega_{lk} + \omega_{\beta\alpha}), \tag{238}$$

$$\Lambda_{kk,ll} = i\langle M_c(i0^+)\rangle_{kk,ll} = -2\pi \sum_{\alpha,\beta} w_\beta |\langle k\alpha|V|l\beta\rangle|^2 \delta(\omega_{kl} + \omega_{\beta\alpha}), \quad k \neq l. \tag{239}$$

Note that all these Λ matrix elements are real. Making use of the following property of the Boltzmann factor w_α:

$$w_\beta \delta(\omega_{kl} + \omega_{\beta\alpha}) = e^{\omega_{kl}/k_B T} w_\alpha \delta(\omega_{kl} + \omega_{\beta\alpha}), \tag{240}$$

the nondiagonal T_1-type relaxation matrix elements satisfy the detailed balance relation

$$\Lambda_{kk,ll} = e^{-\omega_{kl}/k_B T} \Lambda_{ll,kk}. \tag{241}$$

Comparing eq. (238) with eq. (236), we have

$$(T_k^{in})^{-1} = \tfrac{1}{2}\Lambda_{kk,kk}. \tag{242}$$

For the case of a *two-level system*, l and k in the above equation can take only the values i and f. It then follows from eqs. (238) and (239) that

$$\Lambda_{ff,ff} = -\Lambda_{ii,ff}, \qquad \Lambda_{ii,ii} = -\Lambda_{ff,ii} \quad \text{(two-level system)}, \tag{243}$$

where T_1^{-1} for a two-level system may be found from eq. (69):

$$T_1^{-1} = \tfrac{1}{2}(\Lambda_{ff,ff} + \Lambda_{ii,ii} - \Lambda_{ii,ff} - \Lambda_{ff,ii}). \tag{244}$$

From eqs. (238), (239), (241), (243), and (244), T_1^{-1} for a two-level system is explicitly given by

$$\frac{1}{T_1} = 2\pi(1 + e^{-\omega_{fi}/k_B T}) \sum_{\alpha,\beta} w_\beta |\langle f\beta|V|i\alpha\rangle|^2 \delta(\omega_{fi} + \omega_{\beta\alpha}). \tag{245}$$

Furthermore, from eqs. (238), (240), (242), and (245),

$$\frac{1}{T_f^{in}} + \frac{1}{T_i^{in}} = \frac{1}{2T_1} \quad \text{(two-level system)}, \tag{246}$$

and thus from eq. (235) the following well-known expression is obtained:

$$\frac{1}{T_2} = \frac{1}{2T_1} + \frac{1}{T_2^{ph}} \quad \text{(two-level system)}, \tag{247}$$

as was derived less generally in subsection 2.2 for an isolated two-level system such as the inversion doublets of ammonia.

3.6. Low-density expansion

When the interaction potential between the active and bath molecules is not weak, e.g. in a mixture of molecular gases, the interaction energy becomes very large when the molecules are close together, and the perturbation expansion of subsection 3.5 converges very slowly or not at all, and a new expansion procedure must be used. One of the major achievements of the relaxation theory of Fano is an expansion of $\langle M \rangle$ in powers of volume density n of the bath molecules interacting with the active molecule of interest. Fano showed that the lowest-order term

$$\langle M(z) \rangle = n\langle m(z) \rangle, \tag{248}$$

which is linear in the bath density n and is solely determined by the two-body interaction term, and corresponds to a binary collision expansion (Fano 1963, Zwanzig 1961a). Analogous to eq. (172), $m(z)$ obeys an integral equation,

$$m(z) = l_1 + l_1 \frac{1}{z - l_0} m(z), \tag{249}$$

where $l_1 = [v,\ ]$, $l_0 = [h_0,\ ]$ and h_0, v are the unperturbed Hamiltonian and the interaction for a two-body system consisting of one active and one bath molecule. By exploiting the similarity of eq. (249) with the corresponding Lippmann–Schwinger equation for the transition matrix t in ordinary Hilbert space,

$$t(E) = v + v \frac{1}{E - h_0} t(E), \tag{250}$$

Fano showed that in the "impact approximation", where only complete collisions are important, the matrix element of $\langle m \rangle$ can be expressed in terms of the on-the-energy-shell t-matrix elements as (Fano 1963)

$$
\begin{aligned}
n\langle m(\omega)\rangle_{f'i',fi} &\equiv -i\Lambda_{f'i',fi} \\
&= \sum_{\alpha,\alpha'} [t_{f'\alpha',f\alpha}\langle \alpha'i'|\alpha i\rangle - t^*_{i'\alpha',i\alpha}\langle \alpha'f'|\alpha f\rangle \\
&\quad + 2\pi i\delta(E_{i\alpha} - E_{i'\alpha'})t_{f'\alpha',f\alpha}t^*_{i'\alpha',i\alpha}]w_\alpha,
\end{aligned}
\tag{251}
$$

where $|i\rangle$, $|f\rangle$ denote the eigenstates of $h^{(s)}$ while $|\alpha\rangle$, $|\alpha'\rangle$ represent the eigenstates of $h^{(b)}$, and $E_{i\alpha}$ is the energy of the state $|i\alpha\rangle \equiv |i\rangle|\alpha\rangle$, and $h^{(s)}$, $h^{(b)}$ are the unperturbed Hamiltonian for a single active and a single bath molecule, respectively. Λ is the relaxation matrix introduced earlier in section 2. A succinct derivation of Fano's results and a discussion of the impact approximation has been given by Ben-Reuven (1975), to which the reader is referred for detail. The matrix elements of $\langle m(\omega)\rangle$ given by eq. (251) are now independent of ω. According to the discussion in subsection 3.4, the shift and width of an isolated line are given by the real part and the negative of the imaginary part, respectively, of $n\langle m\rangle_{fi,fi}$. In particular, the width, which is also the inverse of T_2, is given by (Jones and Zewail 1978, Diestler and Zewail 1979, Baranger 1958, Omont 1977)

$$
\gamma = \frac{1}{T_2} = -n \,\mathrm{Im}\left(\sum_\alpha w_\alpha(t_{f\alpha,f\alpha} - t^*_{i\alpha,i\alpha}) - 2\pi i \sum_{\alpha'} \delta(E_\alpha - E_{\alpha'})t_{f\alpha',f\alpha}t^*_{i\alpha',i\alpha}\right).
\tag{252}
$$

By applying the optical theorem to the first two terms on the r.h.s. of eq. (252), this expression can be written as

$$
\begin{aligned}
\gamma = \frac{1}{T_2} &= n\pi \sum_{\substack{k \\ (k\neq f)}} \sum_{\alpha,\alpha'} w_\alpha|\langle k\alpha'|t|f\alpha\rangle|^2\delta(\omega_{kf} + \omega_{\alpha'\alpha}) \\
&\quad + n\pi \sum_{\substack{k \\ (k\neq i)}} \sum_{\alpha,\alpha'} w_\alpha|\langle k\alpha'|t|i\alpha\rangle|^2\delta(\omega_{ki} + \omega_{\alpha'\alpha}) \\
&\quad + n\pi \sum_{\alpha,\alpha'} w_\alpha|\langle i\alpha'|t|i\alpha\rangle - \langle f\alpha'|t|f\alpha\rangle|^2\delta(\omega_{\alpha'\alpha}) \\
&\equiv \frac{1}{T_1} + \frac{1}{T_2'}\left(\frac{1}{T_2^{ph}}\right).
\end{aligned}
\tag{253}
$$

The three terms on the r.h.s. of eq. (253) can be identified with $(T_f^{in})^{-1}$, $(T_i^{in})^{-1}$ and $(T_2^{ph})^{-1}$ as in eqs. (236) and (237), and have the same physical interpretations. Equation (253) can also be expressed in terms of S-matrix elements which in some cases are more convenient for numerical computations. Using the relation between the on-the-energy-shell t-matrix and S-matrix elements,

$$
S_{i'\alpha',i\alpha} = \delta_{i'i}\delta_{\alpha'\alpha} - 2\pi \,it_{i'\alpha',i\alpha},
\tag{254}
$$

and transforming into the center-of-mass coordinates of the colliding pair, eq. (253) can be cast into the equivalent form (Liu and McCourt 1979)

$$\Lambda_{f'i',fi} = n\bar{v}\langle\sigma_{f'i',fi}\rangle = n\bar{v}\int_0^\infty dx\, x\, e^{-x}\sigma_{f'i',fi}(E_k), \quad x = E_k/k_B T, \tag{255}$$

and the complex cross section is

$$\sigma_{f'i',fi}(E_k) = \frac{\pi}{k^2}\sum_{\alpha,\alpha'} w_\alpha(\delta_{f'f}\delta_{i'i}\delta_{\alpha'\alpha} - S_{f'\alpha',f\alpha}S^*_{i'\alpha',i\alpha}), \tag{256}$$

where k and E_k are the wavevector and kinetic energy, respectively, of the relative motion of the colliding pair, $\bar{v} = (8k_B T/\pi\mu)^{1/2}$ is the average thermal velocity and i, α, etc. now denote the internal quantum numbers of the active and bath molecules. The separation of the contributions to the linewidth into inelastic and pure dephasing events is, however, not obvious in the S-matrix formulation.

The linewidth or the inverse of T_2 is given by the real part of $\Lambda_{fi,fi}$, while T_1 is related to the real relaxation matrix elements $\Lambda_{kk,ll}$, as discussed earlier. If the intermolecular forces between an active and a bath molecule are known, the two-body scattering problem can in principle be solved for values of various t- or S-matrix elements, allowing an ab initio calculation of T_1 and T_2 (Liu and Marcus 1975).

3.7. Examples

3.7.1. The width and shift of the zero-phonon line of impurity solids

As discussed before by Jones and Zewail (1978) and by Diestler and Zewail (1979), the perturbation theory of the relaxation matrix presented in subsection 3.5 can readily be applied to the study of the temperature dependence of the width and shift of the zero-phonon line (ZPL) in the optical spectra of impurities in inorganic solids and in molecular crystals. The "active system" is then the impurity center while the crystal vibration constitutes the "bath", and the electron–phonon interaction will be responsible for relaxation. The ZPL in these systems corresponds to a pure electronic transition in the impurity center without vibrational excitation of the system, thus leaving the number of phonons unchanged during the transition. At low temperature, very narrow lines corresponding to this transition can be observed in the absorption or emission spectra (Lax 1952, McCumber 1964, DiBartolo 1968, Rebane 1970, Sapozhnikov 1978). Because there exist interactions between the electrons of the impurity center and the phonons of the system, simultaneous multiphonon transitions can occur during the optical transition, giving rise to a "phonon sideband", and the intensity of the ZPL is reduced by a temperature-dependent factor similar to the "Debye–Waller factor" appearing in the neutron scattering of crystals. All these observations can be accounted for by assuming that the electron–phonon interaction is linear in the phonon normal coordinates, which corresponds physically to a shift of the equilibrium position of the potential energy function for the lattice vibration during an optical transition (Lax 1952, McCumber 1964, DiBartolo 1968, Rebane 1970, McCumber and Sturge 1963). Such a linear interaction alone, however, would predict an infinitely sharp ZPL, contrary to experimental observations. Broadening of ZPL results when quadratic coupling terms are included in the electron–phonon interaction. In this case a "Raman process", in which a simultaneous excitation and de-excitation of phonons of similar frequencies occurs, would limit the lifetime of the excited state and produce a width of the ZPL. The frequency of the ZPL is also shifted due to the average interaction of the electronic excitation with the phonons of the system. The width and shift are temperature dependent since they depend on the number of phonons present in the system. Hence, for spectra in which the ZPL is the dominant feature, a study of the temperature dependence of the width and shift of the ZPL would reveal information about the nature of the electron–phonon interaction.

The basic features of the temperature dependence of the width and shift of the ZPL can be explained by assuming the following model Hamiltonian for the system (in second quantized notation) (Orlowski and Zewail 1979, Krivoglaz 1964, de Bree and Wiersma 1979, Abram 1977):

$$H^{(s)} = \hbar\omega_{fi}a^+a, \qquad a^+|i\rangle = |f\rangle, \qquad a|f\rangle = |i\rangle, \tag{257}$$

$$H^{(b)} = \sum_\kappa \hbar\omega_\kappa b^+_\kappa b_\kappa, \tag{258}$$

$$V = V^{(s)}V^{(b)} = a^+a(V_1 + V_2), \tag{259}$$

$$V_1 = \sum_\kappa (V_\kappa b_\kappa + V^*_\kappa b^+_\kappa), \tag{260}$$

$$V_2 = \sum_{\kappa,\lambda'} (V'_{\kappa\lambda}b_\kappa b_\lambda + V'^*_{\kappa\lambda}b^+_\kappa b^+_\lambda + V_{\kappa\lambda}b^+_\kappa b_\lambda + V^*_{\kappa\lambda}b_\kappa b^+_\lambda), \tag{261}$$

where a^+, a are the creation and annihilation operators, respectively, for the impurity electronic excitation and obey Fermion anticommutation rules. The dipole moment operator for the electronic transition is thus given in terms of a^+ and a by

$$\mu^{(s)} = \mu_{fi}a^+ + \mu_{if}a. \tag{262}$$

The creation and annihilation operators for phonons of mode κ are denoted by b^+_κ and b_κ, respectively, and they obey Boson commutation rules. Equations (257)–(261) represent the Hamiltonian under the Born–Oppenheimer approximation, where nonadiabatic terms are neglected. The wavefunction of the system is approximated by the product of an electronic and a vibrational wavefunction; furthermore, the dependence of the electronic wavefunction on the vibrational coordinates is neglected, and thus a (or a^+) commutes with b_κ (or b_κ^+). The harmonic approximation is employed for the description of the phonon bath, yielding the expression of eq. (258) for $H^{(b)}$ and anharmonicity effects are neglected. V_1 and V_2 represent, respectively, the linear and quadratic coupling terms in the electron–phonon interaction. The phonon coordinate system is chosen such that V vanishes if the impurity is in the lower electronic state $|i\rangle$, a result expressed mathematically by the appearance of a^+a in eq. (259). b_κ^+ and b_κ are the phonon operators for the lower electronic state. Interactions linear in a and a^+, which mix the electronic states with the phonon coordinates, are neglected. Such terms would give rise to electronic "inelastic" transitions during interaction with the phonon bath, an unlikely event if the energy of the electronic transition greatly exceeds those of the phonons. Implicit in this model are the assumptions that: (1) the host crystal does not absorb near the optical absorption frequencies of the impurity center, and (2) the impurity centers are so dilute that interactions between themselves are negligible. These conditions are usually satisfied, and the case when they are not will be treated separately (subsection 4.4.2.1).

The spectral properties of the ZPL can be shown to be given by the long-time behavior of the dipole correlation function in the rotating frame with frequency ω_{fi} (Abram 1977). The results in subsection 3.4 are then applicable to the present situation. If V_1 and V_2 are small, expressions for the shift and width of the ZPL can be obtained from the perturbation treatment of subsection 3.5.

The state of the phonon bath is specified by the set of occupation numbers $\{n^\alpha\} = (n_1^\alpha, n_2^\alpha, \ldots, n_\kappa^\alpha)$:

$$H^{(b)}|\alpha\rangle = \sum_\kappa \hbar\omega_\kappa b_\kappa^+ b_\kappa |\{n^\alpha\}\rangle = \sum_\kappa n_\kappa^\alpha \hbar\omega_\kappa |\{n^\alpha\}\rangle. \tag{263}$$

Thus E_α of eq. (175) is given by $\sum_\kappa n_\kappa^\alpha \hbar\omega_\kappa$. Under the harmonic approximation of eq. (258) for $H^{(b)}$, the equilibrium density matrix for the phonon bath factors

$$\rho^{(b)} = \prod_\kappa \rho_\kappa, \tag{264}$$

where ρ_κ is the canonical distribution for phonons of mode κ,

$$\rho_\kappa = \exp(-\hbar\omega_\kappa b_\kappa^+ b_\kappa / k_B T)/z_\kappa, \tag{265}$$

with

$$z_\kappa = \mathrm{Tr}\,\rho_\kappa = (1 - e^{-\hbar\omega_\kappa/k_B T})^{-1}. \tag{266}$$

w_α of eq. (176) is thus given by

$$w_\alpha = \prod_\kappa w_\kappa^\alpha = \prod_\kappa \exp(-n_\kappa^\alpha \hbar \omega_\kappa / k_B T)/z_\kappa. \tag{267}$$

The average bath interaction $\langle V^{(b)} \rangle$ appearing in eq. (229) then becomes

$$
\begin{aligned}
\langle V^{(b)} \rangle &= \sum_\beta w_\beta \langle \beta | V_1 + V_2 | \beta \rangle \\
&= \sum_\kappa V_{\kappa\kappa} \sum_{n_1^\alpha, n_2^\alpha, \ldots}^\infty \left(\prod_{\kappa'} w_{\kappa'}^\alpha \right) \langle \{n^\alpha\} | b_\kappa^+ b_\kappa + b_\kappa b_\kappa^+ | \{n^\alpha\} \rangle \\
&= \sum_\kappa V_{\kappa\kappa} (2\bar{n}_\kappa + 1),
\end{aligned}
\tag{268}
$$

where

$$\bar{n}_\kappa = \sum_{n_\kappa^\alpha = 0}^\infty n_\kappa^\alpha w_\kappa^\alpha = (e^{\hbar\omega_\kappa / k_B T} - 1)^{-1} \tag{269}$$

is the average number of phonons in mode κ. In the second equality of eq. (268), we use the fact that the matrix element of V_1 and the nondiagonal terms $V_{\kappa\kappa'}$ ($\kappa \neq \kappa'$) in V_2 vanish, since they are nondiagonal in the phonon occupation number representation. The last step in eq. (268) follows from the commutation rule for b_κ and b_κ^+ that $b_\kappa b_\kappa^+ = b_\kappa^+ b_\kappa + 1$, and that $b_\kappa^+ b_\kappa$ is the number operator for phonons of mode κ, yielding $\langle \{n^\alpha\} | b_\kappa^+ b_\kappa | \{n^\alpha\} \rangle = n_\kappa^\alpha$. Note that $\langle V \rangle$ is still an operator for the impurity center. From eqs. (233), (229b) and (268), the frequency shift is then given by

$$\delta = \sum_\kappa V_{\kappa\kappa} (2\bar{n}_\kappa + 1). \tag{270}$$

An expression for the width (or T_2^{-1}) can be obtained from eqs. (235)–(237). Since there is no matrix element of the electron–phonon interaction $V_1 + V_2$ which connects the electronic states i and f, it follows from eq. (236) that $(T_i^{\text{in}})^{-1}$ and $(T_f^{\text{in}})^{-1}$ are zero for the model Hamiltonian of eqs. (257)–(261), and only the pure dephasing processes contribute to the linewidth. From eqs. (237), (263), and (267) we have

$$
\begin{aligned}
\frac{1}{T_2^{\text{ph}}} = \pi \sum_{\{n^\alpha\} \{n^\beta\}} \left(\prod_\nu w_\nu^\beta \right) &[|\langle \{n^\alpha\} | V^{(b)} | \{n^\beta\} \rangle |^2 \\
&- \langle V^{(b)} \rangle^2 \delta(\{n^\alpha\}, \{n^\beta\})] \delta \left(\sum_\kappa (n_\kappa^\alpha - n_\kappa^\beta)\omega_\kappa \right),
\end{aligned}
\tag{271}
$$

where

$$\delta(\{n^\alpha\}, \{n^\beta\}) = \prod_\kappa \delta_{n_\kappa^\alpha, n_\kappa^\beta}.$$

For $V^{(b)}$ given by eqs. (259)–(261), only V_2 contributes to $(T_2^{\text{ph}})^{-1}$ because of the δ-function. Using the commutator relations of b_κ^+ and b_λ, viz. $b_\kappa^+ b_\lambda = b_\lambda b_\kappa^+$ $(\lambda \neq \kappa)$ and $b_\kappa b_\kappa^+ = b_\kappa^+ b_\kappa + 1$, and eq. (268), eq. (271) can be cast into the form

$$\frac{1}{T_2^{\mathrm{ph}}} = \pi \sum_{\{n^\alpha\}} \sum_{\{n^\beta\}} \left(\prod_\nu w_\nu^\beta \right) \sum_{\kappa,\lambda} |\bar{V}_{\kappa\lambda}|^2 \tag{272}$$

$$\times [|\langle\{n^\alpha\}|b_\kappa^+ b_\lambda|\{n^\beta\}\rangle|^2 - \delta_{\kappa\lambda}\delta(\{n^\alpha\},\{n^\beta\})\bar{n}_\kappa^2]\delta\left(\sum_\mu (n_\mu^\alpha - n_\mu^\beta)\omega_\mu\right),$$

where $\bar{V}_{\kappa\lambda} = V_{\kappa\lambda} + V_{\lambda\kappa}^*$. The matrix element in eq. (272) can now be evaluated. For $\kappa \neq \lambda$, the only nonvanishing contribution comes from the phonon states

$$|\{n^\alpha\}\rangle = |n_1, \ldots, n_\kappa, \ldots, n_\lambda, \ldots\rangle,$$
$$|\{n^\beta\}\rangle = |n_1, \ldots, n_\kappa + 1, \ldots, n_\lambda - 1, \ldots\rangle,$$

yielding

$$|\langle\{n^\alpha\}|b_\kappa^+ b_\lambda|\{n^\beta\}\rangle|^2 = (n_\kappa + 1)n_\lambda,$$

and the argument of the δ-function

$$\sum_\mu (n_\mu^\alpha - n_\mu^\beta)\omega_\mu = \omega_\kappa - \omega_\lambda.$$

Averaging over the phonon bath then gives $|\bar{V}_{\kappa\lambda}|^2 \bar{n}_\lambda(\bar{n}_\kappa + 1)$. For $\kappa = \lambda$, only the phonon states

$$|\{n^\alpha\}\rangle = |\{n^\beta\}\rangle = |n_1, \ldots, n_\kappa, \ldots\rangle$$

contribute, yielding

$$|\langle\{n^\alpha\}|b_\kappa^+ b_\kappa|\{n^\beta\}\rangle|^2 = n_\kappa^2.$$

Taking the thermal average over the phonon states and using

$$\sum_{n_\kappa} w_\kappa n_\kappa^2 - \bar{n}_\kappa^2 = \bar{n}_\kappa(\bar{n}_\kappa + 1),$$

the diagonal term $\kappa = \lambda$ again gives $|\bar{V}_{\kappa\kappa}|^2 \bar{n}_\kappa(\bar{n}_\kappa + 1)$. Thus eq. (272) finally reduces to

$$\gamma = \frac{1}{T_2^{\mathrm{ph}}} = \pi \sum_{\kappa,\lambda} |\bar{V}_{\kappa\lambda}|^2 \bar{n}_\lambda(\bar{n}_\kappa + 1)\delta(\omega_\kappa - \omega_\lambda), \tag{273}$$

indicating that the process responsible for γ or $(T_2^{\mathrm{ph}})^{-1}$ $[(T_2')^{-1}]$ does not change the total number or energy of the phonon system, and its effect corresponds to a Raman scattering of the phonons. It is interesting to note that only the quadratic coupling term V_2 contributes to the width and shift of the ZPL *within the harmonic approximation*. Since $\bar{n}_\lambda$ vanishes as the temperature approaches zero, eq. (273) predicts that the linewidth vanishes at zero temperature. The final results on the temperature dependence have been derived previously using different methods (McCumber 1964, DiBartolo 1968, Rebane 1970, Sapozhnikov 1978, McCumber and Sturge 1963, Krivoglaz 1964). In Krivoglaz (1964), the anharmonicity of the lattice vibration has also been taken into account. For the impurity spectra of the ZPL (Sapozhnikov 1978, Krivoglaz 1964), it was found that while the contribution from Raman processes to the linewidth is virtually the same as in the harmonic case, an additional contribution to the linewidth of second order in the anharmonicity comes from the linear coupling terms. This latter contribution remains finite at zero temperature and has been used to account for the small residual broadening as $T \rightarrow 0$ K (Sapozhnikov 1978). In later work, however, Krivoglaz (1965) corrected an error and found that anharmonicities do not lead to pure dephasing at zero temperature, as we showed in this subsection. This point has also been confirmed more recently, at least for the weak coupling limit (Wertheimer and Silbey 1981).

The sums over the quasicontinuum of normal modes in eqs. (270) and (273) can be converted to an integral over the phonon frequency by introducing the following density functions:

$$p(\omega) = \sum_{\kappa} 2V_{\kappa\kappa}\delta(\omega - \omega_{\kappa}), \tag{274}$$

$$q(\omega, \omega') = \sum_{\kappa,\lambda} |\bar{V}_{\kappa\lambda}|^2 \delta(\omega - \omega_{\kappa})\delta(\omega' - \omega_{\lambda}). \tag{275}$$

From eq. (270) the temperature-dependent part of the shift can then be written as

$$\delta' = \int_0^{\infty} d\omega \, p(\omega)\bar{n}(\omega), \tag{276}$$

and the width, from eq. (273), can be expressed as

$$\gamma = \int_0^{\infty} d\omega \, q(\omega, \omega)\bar{n}(\omega)[\bar{n}(\omega) + 1], \tag{277}$$

where $\bar{n}(\omega)$ is given by

$$\bar{n}(\omega) = (e^{\omega/k_B T} - 1)^{-1}. \tag{278}$$

Equations (260) and (261) can be viewed as an expansion of the excited state vibrational potential in the lower state normal coordinates (Lax 1952) or the local strain (DiBartolo 1968). In either case, V_{κ} is proportional to $\omega_{\kappa}^{1/2}$ and $V_{\kappa\lambda}$ is proportional to $(\omega_{\kappa}\omega_{\lambda})^{1/2}$, and eqs. (274) and (275) simplify to

$$p(\omega) = A\omega g(\omega), \tag{279}$$

$$q(\omega, \omega') = \bar{A}\omega^2 [g(\omega)]^2, \tag{280}$$

where $g(\omega)$ is the phonon density of states and A, $\bar{A}$ are proportionality constants. If only acoustic phonons are coupled to the electronic transition, $g(\omega)$ may be approximated by the Debye density of states at low temperatures:

$$g(\omega) = 3N\omega^2/\omega_D^2, \quad \omega \le \omega_D,$$
$$= 0, \quad \omega > \omega_D, \quad (281)$$

where N is the total number of phonons and ω_D is the Debye frequency for the crystal. Substituting eqs. (279) to (281) into eqs. (276) and (277) then gives

$$\delta' \propto \int_0^{\omega_D} d\omega\, \omega^3/(e^{\omega/k_BT} - 1)$$
$$= (k_BT)^4 \int_0^{\omega_D/k_BT} dx\, x^3/(e^x - 1), \quad x = \omega/k_BT,$$
$$\simeq (k_BT)^4 \int_0^{\infty} dx\, x^3/(e^x - 1), \quad \text{for } T \to 0, \quad (282)$$

and

$$\gamma \propto \int_0^{\omega_D} d\omega\, \omega^6/(e^{\omega/k_BT} - 1) \simeq (k_BT)^7 \int_0^{\infty} dx\, x^7/(e^x - 1) \quad \text{for } T \to 0. \quad (283)$$

Thus, at low temperatures, the width is proportional to T^7 while the shift is proportional to T^4, a finding well supported by experiments on the impurity spectra of inorganic crystals (DiBartolo 1968, Rebane 1970).

On the other hand, if the electronic excitation is coupled to pseudolocalized phonons of frequency ω_0, $g(\omega)$ can be replaced by the Einstein density of states (the single-mode approximation (Abram 1977)).

$$g(\omega) = N\delta(\omega - \omega_0), \quad (284)$$

and eqs. (276) and (277) simplify immediately to

$$\delta' = \tilde{\alpha}\bar{n}(\omega_0), \quad (285)$$
$$\gamma = \tilde{\beta}\bar{n}(\omega_0)[\bar{n}(\omega_0) + 1], \quad (286)$$

where $\tilde{\alpha}$ and $\tilde{\beta}$ are temperature-independent constants. The temperature dependence of the width predicted by eq. (286) fits very well the results obtained from impurity spectra of mixed molecular crystals (Orlowski and Zewail 1979). The temperature dependence predicted by eq. (285) also fits the experiments, but only after a contribution to the total shift from thermal expansion has been subtracted out (Prasad and von Smith 1979, Prasad and Hess 1980).

3.7.2. T_1 and T_2 relaxation times of microwave transient experiments
The calculation of the spectral linewidth of gases from some known or assumed form of intermolecular potentials have been an active area of

research for many years (for a recent review, see Rabitz 1974). It has also been recognized that the relaxation theory of Fano (1963) and Ben-Reuven (1966a,b) is applicable not only to pressure broadening but also to other relaxation phenomena (Gordon 1968) such as NMR spin–lattice relaxation, sound absorption, Raman scattering, etc. With the development of efficient numerical methods to perform closed-coupling calculations of the S-matrix elements for small systems, a virtually exact ab initio computation of various relaxation parameters for the system H_2–He has been carried out by Shafer and Gordon (1973). Such a closed-coupling computation would not be feasible for larger molecular systems; however, advances in the coupled-state (Kouri 1979) and infinite-order sudden approximations (Kouri 1979) in collision dynamics renders the calculations of the relaxation matrix very promising.

Experiments on microwave transient phenomena (Schmalz and Flygare 1978, Schwendeman 1978) yield the relaxation times T_1 and T_2 directly. The relation of these relaxation times to various relaxation matrix elements is discussed in section 2 and subsection 3.6 of this chapter. Ab initio calculations of T_1 and T_2 had been performed by Liu and Marcus (1975) for the systems OCS–Ar and OCS–He. The semiclassical theory of Marcus (1973) for the S-matrix can be applied to eqs. (255) and (256) to give expressions for $\Lambda_{fi,fi}$ in terms of semiclassical probabilities, which in turn can be calculated from the classical trajectories of the colliding system. A semiclassical expression for the linewidth (and thus for T_2) was first derived by Fitz and Marcus (1973), and similar expressions for T_1 were later developed (Liu and Marcus 1975). Thus, T_2^{-1} is proportional to the real part of the complex cross section:

$$\langle \sigma_{fi,fi} \rangle = \mathrm{Av}\left(1 - \sum_{\mathrm{s.p.}} P_{j'j} D_{\delta\delta}^{k*}(\alpha, \beta, \gamma)\right), \tag{287}$$

$$T_2^{-1} = n\bar{v}\,\mathrm{Re}\langle \sigma_{fi,fi} \rangle \tag{288}$$

where Av( ) denotes an average over the relative kinetic energy of the colliding pair, the impact parameter and other relevant classical actions and angles. The sum in eq. (287) is over all "stationary phase points" or classical trajectories leading from the initial rotational angular momentum of the active molecule $j = \frac{1}{2}(j_f + j_i)$, to the final rotational angular momentum $j' = \frac{1}{2}(j_f + j_i)$, where j_f and j_i are the angular momenta of the active molecule coupled by the microwave field. $P_{j'j}$ is a classical probability for the $j \to j'$ collisional transition and $D_{\delta\delta}^{k*}(\alpha, \beta, \gamma)$ is the complex conjugate of the rotation matrix $D_{\delta\delta}^{k}(\alpha, \beta, \gamma)$, where β is the reorientation angle of the rotational angular momentum vector due to collision and α and γ are collision phase shifts. For microwave absorption experiments on rotating linear molecules, $k = \delta = 1$. The line shift is proportional to the imaginary part of eq. (287). An intermolecular potential of the form

$$V(R, \chi) = 4\epsilon\left[\left(\frac{\sigma}{R}\right)^{12} - \left(\frac{\sigma}{R}\right)^{6}[1 + a_p P_2(\cos \chi)]\right] \tag{289}$$

is assumed for both the systems OCS–Ar and OCS–He, where R and χ are the separation and relative orientation, respectively, of the colliding system, P_2 is a Legendre polynomial, and ϵ, σ and a_p characterize, respectively, the strength, range and anisotropy of the interaction. It was found that the calculated widths agree well with those obtained from experiments, and that the calculated shifts are very small, implying that the averaged reorientation and collisional phase shifts are small for the systems considered.

In the absence of an external field, T_1^{-1} is proportional to the following cross sections:

$$\langle\sigma_{kk, ll}\rangle = \mathrm{Av}\left(\delta_{kl} - \sum_{s.p.} P_{j_k j_l}\right). \tag{290}$$

The dipole radiation selection rule for linear molecules is $j_f - j_i = \pm 1$. For an intermolecular potential of eq. (289), there are *collisional* selection rules $j' - j = 0, \pm 2, \pm 4, \ldots$, and hence $\langle\sigma_{ff, ii}\rangle = 0$. For such systems T_1^{-1} is given by

$$T_1^{-1} = \tfrac{1}{2}n\bar{v}(\langle\sigma_{ff, ff}\rangle + \langle\sigma_{ii, ii}\rangle). \tag{291}$$

Again, the computed values for T_1-cross sections agree well with those from experiments. Recently, Green (1978) performed infinite-order sudden approximation calculations for these T_1- and T_2-cross sections for the same systems and obtained results in excellent agreement with the earlier semiclassical calculations of Liu and Marcus.

For both the OCS–Ar and OCS–He systems, it was found that T_1 and T_2 are about the same (Burns 1979). Furthermore, the frequency shift is found to be negligible (Coy 1980), indicating that reorientation and phase shift effects are small, and $D_{\delta\delta}^k$ in eq. (287) can be effectively replaced by unity. On comparing the subsequent expression with eqs. (290) and (291), and on account of the fact that the rotational levels of OCS are narrowly spaced so that the collisional dynamics of adjacent levels is expected to be similar, it is not surprising to find $T_1 \approx T_2$. For these systems the relaxation is dominated by inelastic collisions, and the effect of pure dephasing discussed in subsection 3.5, which in the present case is related to the reorientation and phase shift effects, is unimportant for the systems considered.

4. Experimental methods and results

4.1. Introduction

Many molecular systems have been studied using coherent transient techniques. This section encompasses a discussion of experimental methods as well as the results which have been derived from the experimental studies on *gases, beams, liquids*, and *solids*. Connections with theory will be made. The reader may also consult with several other reviews on microwave transients (Schmalz and Flygare 1978, Schwendeman 1978), infrared transients (Shoemaker 1978), electronic transients in the visible (Zewail et al. 1977, Zewail 1980), and coherent Raman transients (Laubereau and Kaiser 1978). Here, we will not discuss the coherent transients of atomic gases (Mossberg et al. 1979) and the large volume of NMR and ESR transients observed in ground (Jonas and Gutowsky 1980) and excited states (Harris and Breiland 1978).

4.2. Gases

In the gas phase, rotational, vibrational and electronic transitions have been studied by coherent transient techniques. Since the theory has been developed most completely for two and three levels, most of the experimental work has been performed on molecules where two (three) levels may be uniquely connected by the radiation field (fields). Consequently, small molecules, where the density of states is low, have been the samples for most reported gas phase transient experiments.

Demonstration of the effects discussed in section 2, such as FID, photon echo, etc., was the primary concern of the early gas phase work. In the course of this work, information was gathered on collision-induced dephasing rates (T_2^{-1}) and energy transfer rates (T_1^{-1}). Recently, more detailed studies have been undertaken in order to extract velocity dependences of rates and the effects of velocity changing collisions among other important results. These studies will be discussed presently.

The importance of measuring T_1 and T_2 for small molecules should be reiterated. Because the theory of relaxation for a simple two (or three-) level system is well developed, the measured relaxation rates may be compared to those calculated using different intermolecular potentials. Furthermore, the simplicity of the basic relation between T_1 and T_2, as discussed in subsection 2.2, allows qualitative trends to be understood. The technique of coherent transients appears to be the best way to measure these collisional relaxation rates. Steady-state lineshapes can, in principle, be used to measure T_1 and T_2, as seen from eqs. (74). Unfortunately, at high frequencies the lineshapes are dominated by uninteresting Doppler broadening. As we saw in section 2, the photon echo may be used to eliminate Doppler broadening in transient experiments. At lower frequencies, baseline and modulation problems make lineshape analysis difficult. In fact, only recently have high precision microwave lineshapes been recorded (Creswell et al. 1976, Amano and Schwendeman 1976). On the other hand, classical time-resolved fluorescence studies in the visible or IR region measure T_1-like rates, which may be convolutions of many rate processes, depending upon the frequency resolution of the detection apparatus employed. Saturated absorption measurements are much more sensitive to power effects than FID, for example.

Other advantages of coherent techniques in gases include the sensitive heterodyne nature of the detection, the availability of multipulse trains to measure relaxation rates inaccessible to classical experiments, and the sensitivity of some experiments to events such as velocity changing collisions.

4.2.1. Rotational transitions

Many dipole-allowed, low-J, rotational transitions fall in the microwave region (≈ 8–40 GHz) of the electromagnetic spectrum. This is a convenient region of the spectrum in which to study transients since the frequencies of transitions are accurately known and microwave technology is well developed compared to the newer laser technology. In particular, excellent, virtually noise-free, tunable sources are available. Using standard phase-lock techniques, these sources may be stabilized to better than 10 Hz FM. Furthermore, the theory of rotational relaxation in this region is quite mature (see section 3). The primary restraint on these microwave experiments is that, because there are many closely spaced levels with energy much less than kT, the value of the equilibrium population difference, r_3^0, is small. The amplitude of the transient absorption or emission depends linearly on r_3^0 [e.g. eqs. (78)], so the resulting microwave signal is small. Consequently, only those few small molecules that have relatively sparse level structures and/or large transition moments have been studied (e.g. linear molecules, symmetric tops, or near-symmetric tops). The low frequency of the radiation absorbed and emitted also makes for a rather low detection efficiency, removing the possibility of detecting the incoherent fluorescence easily. Finally, standard microwave sources have powers less than 40 W. Although this power is comparable to some CW laser sources, it is orders of magnitude less than the peak power achievable using pulsed laser sources. Picosecond switching of coherent transients in the microwave region has not yet been attained. Despite these drawbacks, probably the most accurate relaxation results on electric dipole allowed transitions have been achieved using microwave spectroscopy.

4.2.1.1. Experimental methods. A typical transient microwave spectrometer
is shown in fig. 11. The source is typically a klystron (≈ 1 W) or a backward
wave oscillator (≈ 100 mW) phase-locked to a harmonic of a stable local
oscillator. A counter is used to record the frequency. Solid state sources
have not yet been employed in any transient studies, possibly because of
the relatively low tunable power available and FM noise problems. The
source may be amplified to about 40 W using a traveling wave tube
amplifier (TWTA), at the expense of some increased noise. The
microwaves are propagated through a standard waveguide so that one
mode is present, usually the TE_{10} mode. The gas cell is also waveguide.
This has the advantages that the spatial mode shape is well known and that
there is little attenuation of the microwaves due to the cell.

Temporal modulation of the power may be achieved through frequency,
amplitude, or Stark modulation. The first transient experiments by Dicke
and Rohmer (1955) employed frequency modulation. They mixed a CW
microwave source with a modulated RF source. This technique has the
disadvantage that during the RF pulse two microwave frequencies are
propagated. More recently, frequency switching has been accomplished by
applying square wave voltages to the repeller of the klystron (Amano and
Shimizu 1973, Tanaka and Hirota 1976). Unfortunately, if the klystron is
phase locked, unwanted transients may arise.

Amplitude modulation of the signal can be effected through the use of
commercially available PIN diode switches (Ekkers and Flygare 1976,
Bestmann et al. 1980, Coy 1980). Less than 20 ns rise and fall times and
greater than 80 dB attenuation are possible with present technology. The
technique is simple and has become more popular in recent experiments
(Coy 1980, Bestmann et al. 1979, Schrepp et al. 1979). In these experiments,
a bridged cell is used (see fig. 11), in which the bridge radiation is used as a
local oscillator for heterodyne detection of the transient emission following
the PIN-modulated pulse into the cell. The radiation is typically a few MHz
off the molecular resonance in order to achieve efficient heterodyning of
the signal. Consequently, the power must be high enough so that the
inverse of the time of the pulse is $\geq$ the frequency offset. An additional
TWTA after the diode may be employed to boost the power. (The TWTA
must be placed after the switch because the diodes cannot, in general,
handle high powers.) The primary advantages of this scheme are that
efficient heterodyne detection is achieved and the problems of Stark field
inhomogeneities, to be discussed later, are avoided. A big disadvantage is
that, as of this writing, fast PIN switches are available only up to about
18 GHz. Also, for some applications, a TWTA, which is expensive and
introduces noise, is necessary.

For the last several years by far the most popular method of switching
has been the technique of Stark modulation (Harrington 1968, Schwen-

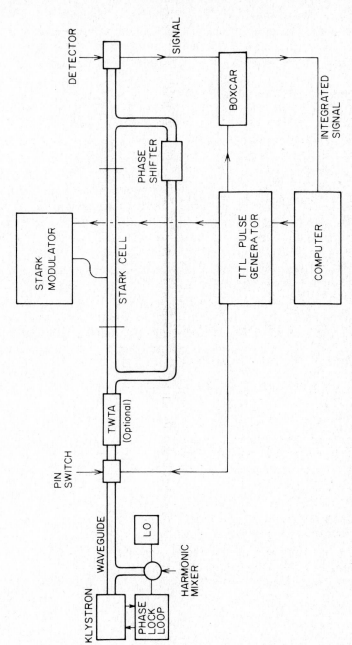

Fig. 11. A typical microwave transient spectrometer with options for amplitude or Stark modulation.

deman and Brittain 1970, McGurk et al. 1974, Coy 1975b). Much of this method's popularity results from the fact that Stark modulation is the most common detection scheme for conventional microwave spectroscopy. In transient experiments, the incident radiation is run in a CW mode and a large electric field is applied to the sample via a septum placed along the center of the cell. When the electric field is applied, the energy levels are shifted by the well-known Stark effect. Modulators have been built which can switch up to 1.5 kV with 200 ns rise and fall times. Two examples of Stark pulse trains are given in fig. 12. In the first example, the radiation field is set to be in resonance with the molecular transition when the voltage, V, is applied. When the voltage drops back to zero, the free induction decay will be a heterodyne beat signal between the molecular emission at frequency ω_0 and the source at frequency ω [see eq. (88)]. The Stark shift should be chosen such that the Rabi frequency, ω_1, is small compared to $\Delta\omega$ and such that heterodyning is efficient. [See the discussion following eq. (88).] For carbonyl sulfide at 10 mTorr pressure, a 10 MHz shift is typical. The molecules are allowed to relax at zero voltage because of the inhomogeneity of the Stark field which, when present, gives rise to inhomogeneous broadening of the transition. The second example of fig. 12 shows the two-pulse delayed nutation sequence used to derive a T_1 relaxation time (see subsection 2.4.3). Here the frequency of the radiation is equal to the zero-field molecular frequency. During the time τ the Stark field is nonzero, resulting in inhomogeneous phase destruction during this

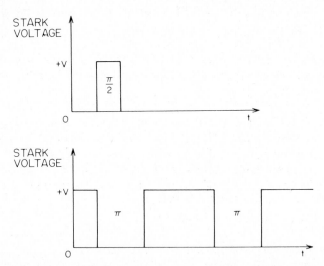

Fig. 12. Stark pulse trains; top: the Stark pulse for an FID T_2 experiment (the source is on resonance when the voltage is at V, and bottom: the Stark pulse train for a 2-PDN T_1 experiment (the source is on resonance when the voltage is zero).

time. But this is the effect that is desired, since for T_1 measurements it is easiest to interpret the results when $r_1(t_2) = r_2(t_2) = 0$. This fact is important because $r_1(t_2)$ and $r_2(t_2)$ are not in general zero, since the first π pulse which should make these elements zero is not a "pure" π pulse because of the microwave field strength distribution over the cell caused by the TE_{10} mode shape.

The primary advantage of the Stark technique is the automatic heterodyning of the FID signal. Also, since the Stark effect splits the rotational levels into their various M states, the effects of this projection quantum number on the relaxation times may be deduced (Hoke et al. 1975, 1976). For high-J states, though, this splitting is a disadvantage, since each M state will have only a small relative population. Despite its utility in T_1 measurements, the Stark field inhomogeneity presents problems in the interpretation of relaxation data because of the complexity of the distribution of excited molecules created during the pulse. Finally, only molecules with reasonably large Stark effects may be studied.

For Stark-modulated experiments, as well as amplitude modulation experiments, a phase-adjustable bridged cell is used so that maximum power in the cell may be achieved without burning out the detector. An isolated waveguide-mounted diode is typically used as the detector. The amplified signal is then averaged using a boxcar or a transient digitizer. The acquisition and storage of the data, as well as the triggering of a TTL level pulse generator which supplies the trigger pulses to the modulators, may be under computer control. Accurate knowledge of the timing of these pulses as well as elimination of TTL jitter are important for the success of these experiments.

Certainly all the schemes for temporal modulation have not been exhausted, and with the availability of new technology many more will arise. Even combinations of the afore mentioned techniques are useful for some experiments.

4.2.1.2. Results. The pioneer experiment in microwave transient spectroscopy was reported in 1955 by Dicke and Romer (1955). Yet, with only a couple of exceptions, work in this field did not begin in earnest until the early 1970's. Then several groups, most notably that of Flygare, began to experimentally verify the predictions of the Bloch equations. Because of its high line intensities and favorable line positions, the most commonly studied molecule was carbonyl sulfide (OCS).

OCS. This linear triatomic molecule possesses transitions at 12.162 GHz $(J = 0 \rightarrow 1)$ and at 24.324 GHz $(J = 1 \rightarrow 2)$, which are convenient for microwave spectroscopy. The coherence properties of these transitions, using the Stark-switching method, have been studied by transient nutation,

FID, two-pulse delayed nutation, adiabatic rapid passage and photon echoes (McGurk et al. 1974, Coy 1975b, Mäder et al. 1975, Tanaka and Hirota 1976, Hoke et al. 1976, Burns 1979, Macke and Glorieux 1972, Mäder 1979, Wang et al. 1973, Brittain et al. 1973, Rohart et al. 1977, Coy 1980). A typical FID is shown in fig. 13. Using amplitude modulation, free induction decays and two-pulse delayed nutations have been observed. The relaxation times at room temperature extracted from the various experiments are known for a variety of perturbers. For OCS self-broadened samples there is some scatter in the values of T_2. But, at least in terms of the reported accuracies, the FID data appear to give the best values. This is not surprising because for large heterodyne frequencies power effects do not affect FID as much as they do transient absorption or adiabatic rapid passage. Using eq. (91) the effects of Doppler and wall broadening may be handled exactly when the velocity dependence of T_2 is neglected. Photon echo results have the disadvantage that data cannot be collected until after two pulses have been applied, thereby losing the largest part of the T_2 emission. Also, the inhomogeneity that is operant is primarily the applied Stark inhomogeneity, not the small Doppler effect. The scatter in all the data may be due to the experiments having been done at different temperatures (Schwendeman 1978) (some reports only list the temperature as "room temperature") or to the velocity dependence of the relaxation rates

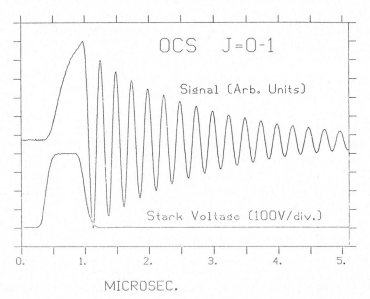

MICROSEC.

Fig. 13. A nutation and the resulting FID for the $J = 1 \leftarrow 0$ transition of OCS (self-broadened) at a pressure of 13.5 mTorr. Stark switching is employed (Coy 1980).

(Coy 1980). This latter effect will be discussed presently. Comparison of the T_2^{-1}'s with the best linewidth, $\Delta\omega_{1/2}$, shows reasonable agreement as expected from eq. (74). Calculations of T_2^{-1}, for the self-broadened OCS case, using Anderson theory (Anderson 1949, Frost 1976) or other perturbative approaches (Murphy and Boggs 1967a,b, 1968, Caltani 1971, Frost and MacGillivray 1977) yield results within about 10% of the experimental value when dipole-type and quadrupole-type intermolecular forces are included in the calculations. This is about the expected accuracy for perturbative theories. Similarly, the Infinite–Order Sudden Approximation (IOSA) (Green 1978) and semiclassical theories (Liu and Marcus 1977) have been used to calculate T_1 and T_2 for transitions of OCS colliding with He and Ar, when a simple potential [due to Gordon (1966)] was used. The results are given in table 1. The inaccuracy in the He case may be due to the simplicity of the potential.

Values of T_1 derived from different experiments are also given in table 1. Those results achieved by the two-pulse delayed nutation (2PDN) method are in good agreement with each other. The small M dependence of the population relaxation time for self-broadened OCS has been attributed to relaxation within the $\Delta J = 0$, M state manifold (Mäder 1979). For example, whereas the dipole-allowed $\Delta J = 0$, $\Delta M = \pm 1$ collisions contribute to T_1 for both the $|M| = 0$ and 1 transitions of $J = 1 \rightarrow 2$ (the radiation selection rule is $\Delta M = 0$), quadrupole-allowed $\Delta J = 0$, $\Delta M = \pm 2$ will not contribute to the $|M| = 1$ transition's T_1. This phenomenon has also been examined with the $J = 2 \rightarrow 3$ transition (Mäder 1979).

The important qualitative result of these studies is that $T_1 \approx T_2$ for all perturbers. As discussed in subsection 2.1, this behavior is expected for a

Table 1

The experimental and theoretical values[a),(b)] of T_1 and T_2 for the $J = 1 \leftarrow 0$ and $J = 2 \leftarrow 1$ lines of OCS.

Rate (NS^{-1} $Torr^{-1}$)	Transition ($J = $)	OCS–Ar			OCS–He		
		Expt.	IOSA	SC	Expt.	IOSA	SC
T_2^{-1}	$1 \leftarrow 0$	0.0233(2)	0.0247	0.0237	—	0.0151	0.0178
T_1^{-1}	$1 \leftarrow 0$	0.0223(8)	0.0242	0.0240	0.0188(5)	0.0138	0.0151
T_1^{-1}	$2 \leftarrow 1$, $|M| = 0$	0.0232(6)	0.0240	0.0235	0.0126	0.0133	0.0158
T_1^{-1}	$2 \leftarrow 1$, $|M| = 1$	0.0226(4)	0.0234	0.0231	0.0209	0.0121	0.0142

[a)]The experimental data are taken from Burns (1979) and Mäder (1979). Both the FID and the 2PDN were used. All values are at room temperature (adapted from Schwendeman 1978).
[b)]The calculated values using the infinite-order sudden approximations (IOSA) and the semiclassical (SC) methods are from Green (1978) and Liu and Marcus (1975b), respectively.

molecule like OCS that has numerous closely spaced levels with energies less than kT. In other words, T_2' (pure dephasing) processes do not appear to play a significant role in relaxation processes of the ground state of OCS.

NH_3. The rotational level structure of ammonia is completely different from that of OCS, consisting of "isolated" ·inversion doublets (see Oka 1973). Measurements of T_1 and T_2 for several transitions of this molecule (self-broadened) are now known (Hoke et al. 1976, Amano et al. 1973) for this prototype molecule. The transient emission results of Hoke et al. (1975, 1976, 1977) show that whereas T_2 is independent of M, the projection of the rotational angular momentum on the z axis, T_1 is highly M dependent. They have used a modified Anderson theory to show that, if only the predominant dipole–dipole forces are included in the calculation, the sum of the rates of transfer out of the two levels are M independent and that $(T_2')^{-1}$ is zero. From eq. (72), then, T_2 is M independent. But as argued in subsection 2.2, if the two radiatively connected levels are isolated, T_2 should approach $2T_1$. In this system, for some M states this is definitely the case. This modified Anderson theory, which takes into account effects such as resonance collisions (which were ignored in the discussion of subsection 2.2), can reasonably predict this M-dependent T_2/T_1 ratio.

Linewidth values give somewhat different results for T_1 and T_2 as a result of an averaging over M states that occurs. Schwendeman and Amano (1979), using an extension of the theory of Liu and Marcus (1975), considered this point in detail.

Asymmetric tops—formaldehyde and ethylene oxide. The level structures of ethylene oxide and formaldehyde are known. In ethylene oxide, there appears to be a small difference in T_1 and T_2 (Mäder et al. 1979b, Schrepp et al. 1979, Hom 1978, Coy 1980). The molecule CH_2O ($K = 1$) has an energy level structure for low J somewhat like that of NH_3 with "isolated" doublets (though less "isolated" than ammonia). Presumably, differences in T_1 and T_2 should be expected for this system but no T_1 measurements have been reported as of this writing. Formaldehyde is also of interest because of the postulated collisional mechanism for the anomalous absorption of CH_2O in interstellar space (Townes and Cheung 1969).

Other molecules. Several other molecules have been investigated by transient techniques: CH_3Cl (Hom 1978, McGurk et al. 1974, Frenkel et al. 1971), CH_3F (Jetter et al. 1973, McGurk et al. 1974, Glorieux et al. 1976), including rotary echoes (Rohart et al. 1977), SO_2 (Bestmann et al. 1979), HCN (Charron et al. 1980), CH_2CN (Mäder et al. 1979a), and CH_3CCH

Table 2
Microwave lineshifts of various self-broadened molecules at
room temperature[a].

Molecule	Transition	Lineshift (MHz/Torr)
OCS	$J = 0 \to 1$	$-0.015(12)$
	$J = 1 \to 2$	$0.00(10)$
H_2CO	$2_{12} \to 2_{11}$	$1.2(1)$
$CH_3^{35}Cl$	$3/2 \to 5/2$	$1.31(11)$
	$3/2 \to 3/2$	$1.17(16)$
	$3/2 \to 1/2$	$1.30(25)$
C_2H_4O	$2_{11} \to 2_{02}$	$-1.04(13)$
	$3_{30} \to 3_{21}$	$-0.37(14)$
	$4_{31} \to 4_{22}$	$0.37(10)$

[a]The $CH_3^{35}Cl$ transitions are all hyperfine transitions of the
$J = 1 \leftarrow 0$. From Coy (1980) and Hom (1978).

(Glorieux et al. 1975). Details on these specific molecules can be found in the original references.

Lineshifts. From eqs. (70d) and (88a), we see that the heterodyne beat frequency of an FID depends upon the pressure-induced lineshift. Lineshifts for several transitions of a few molecules have been measured in this way (Hom and Coy 1981, Coy 1980) and are given in table 2.

Velocity effects. Recently, nonexponential character (not due to Doppler, wall, or power broadening) in FID's have been seen in the mocrowave regime (Coy 1980). This nonexponentiality has been attributed to the absorber speed dependence of the coherence relaxation time. This effect has been discussed in the frequency regime (Luijendijk, 1977a,b) and has been seen in the infrared (Grossman et al. 1977), as will be discussed later in this section. In the microwave regime, this speed dependence has been measured in self-broadened OCS, NH_3, CH_2O, and ethylene oxide (Coy 1980), as well as OCS colliding with He and Ar (Burns 1979). The model of eq. (105) has been used. Anderson theory and IOSA calculations lend qualitative support to this choice of model (Burns 1979). Since this speed dependence increases the number of relaxation parameters that may be extracted, more information on the intermolecular potentials is derived. On the other hand, the results of previous exponential FID fits will depend upon the length and power of the preparation pulse as well as the decay sampling time because of this speed dependence. Some of the scatter in the FID T_2 results, especially for H_2CO, may be a result of this speed dependence. For example, the preparation time of the exciting pulse should affect the shape of the velocity distribution which is excited. Consequently, FID resulting from a long Stark excitation (1.1 μs) should have a different exponential relaxation time from that resulting from a 0.1 μs PIN diode pulse. This may be seen from the $J = 0 \to 1$ of OCS T_2 values from the work of Coy (1980). Similar velocity-dependent effects should be seen in T_1 results, which may mask the nonexponential character due to different energy relaxation rates to the various "bath" levels. Four-level pump and probe experiments (Oka 1973) predict such a difference in these rates.

4.2.2. Vibrational transitions

Many vibrational transitions in molecules are in the infrared portion of the electromagnetic spectrum. Since there are several high-power pulsed IR laser sources in this region (such as CO_2 lasers), this spectral regime would seem to be an ideal region in which to do transient experiments. Although pulsed lasers have been used successfully for several transient applications, there are problems with using them. It is difficult to control the length and power of the pulse to achieve $\pi/2$ or π pulses. Also, the heterodyne term of eq. (52) will be zero, meaning that the signal will be a small term proportional to the intensity of the molecular emission added to the huge source intensity term. This results in a great loss of sensitivity. Consequently, other forms of modulation, similar to those used in the microwave experiments, are frequently used.

Although these laser experiments give essentially the same information as microwave experiments (collisionally induced T_1 and T_2 relaxation rates), important differences are present. First, the effects of velocity changing collisions may, in some cases, be no longer negligible. Also, Doppler broadening is significant, so the photon echo pulse train takes on added importance.

4.2.2.1. Experimental methods.

Most of the transients measured in the IR have employed a CO_2 laser, working in either the pulsed or CW mode. Such a laser has the advantage of high power, but it has very narrow frequency tuning ranges. As described before, in many experiments the lasers are not run in the pulsed mode but in the CW mode. Thus, frequency-stable, single-mode CW CO_2 lasers (≈ 1–10 W) are necessary. The problems of short-term stability have been discussed by Freed (1968). For long-term stability, Grossman et al. (1977) have, on experiments with $^{13}CH_3F$, used a feedback loop with a second $^{13}CH_3F$ cell, and frequency locked the CO_2 laser to the Lamb dip of a $^{13}CH_3F$ transition using a piezoelectric tuner.

Modulation is generally provided by the Stark switching technique pioneered in the IR by Brewer and Shoemaker (1972). In the cell are two parallel Stark plates spaced from less than 1 to several cm. The gas handling cell encloses the plates. Furthermore, if a DC bias is placed on the Stark plates, the effect may be used to "tune" the molecular levels into resonance with the fixed laser frequency.

If the molecule is a symmetric top, which has a large Stark effect, it is necessary to switch only a hundred volts or so, which may be done with rise times of 40 ns or less. As in the microwave case, Stark switching yields an automatic heterodyne of the molecular signal, resulting in increased sensitivity.

A preliminary account of using interactivity frequency switching of a CO_2 laser as a modulation scheme has been given by Shoemaker et al. (1978). As of this writing, no detailed account has been given. The details of frequency switching will be discussed in the coming sections and in subsection 4.4.2.1 in conjunction with visible transients.

There is an apparent problem with frequency or Stark switching in the IR (or visible) regime. Many times one cannot switch completely out of a Doppler profile, so the switching is between one velocity group, v_{1z}, and another group, v'_{1z}. Then, for example after a $\pi/2$ pulse, one would have the FID from the v_{1z} group superimposed upon a nutation from the v'_{1z} group. This is often not a great problem because the effects can usually be separated during a fit of the data at different powers.

Recently, Genack et al. (1980) have reported the use of a CdTe crystal to phase shift the radiation of a CO_2 laser. They reported that they investigated SF_6 by this technique. Although no results were given, this method seems promising in that switching in less than 50 ps is possible.

Amplitude modulation has been achieved by using a Q-switched CO_2 laser (Hocker and Tang 1968) or, in the case of photon echoes (Patel and Slusher 1968), two such lasers. Some of the difficulties with using pulsed lasers have been stated previously. The advantages are the high power, the ability to look at molecules with no permanent dipole moment, and the availability of picosecond pulses.

A photodiode is usually the detector. The same type of signal processing and pulse generation as was described for the microwave experiments is commonly used.

4.2.2.2. Results. Several vibrational transitions of different gases have been studied by IR coherent transients. We will discuss here three of the most commonly studied molecules.

SF_6. Using a pulsed CO_2 laser or lasers, SF_6 was the first gaseous molecule for which optical nutation (Hocker and Tang 1968, 1969, Alimpiev and Karlov 1974), FID (Cheo and Wang 1970), and a photon echo (Patel and Slusher 1968, Gordon et al. 1969, Alimpiev and Karlov 1973, Meckley and Heer 1973, Heer and Nordstrom 1975, Gutman and Heer 1975, 1977) were

measured. Since it has no permanent dipole moment, Stark switching cannot be used. From eq. (78b), we see that the optical nutation signal can be used to measure the Rabi frequency ω_1. Consequently, from the discussion below eq. (20), the dipole matrix element for the transition may be measured if the incident electric field is known. This has been done by Hocker and Tang (1969) and by Alimpiev and Karlov (1974). Cheo and Wang (1970) have observed FID in SF_6, but the sample was optically thick so the interpretation of the results is difficult.

The photon echo has been the most common, and useful, technique for studying relaxation in SF_6. The values of T_2 derived from these studies are shown in table 3. By comparison with the microwave results on other molecules, we see that these SF_6 relaxation rates are of the correct size for rotational relaxation processes. No T_1 measurements have been reported. Two-photon transients by Doppler-free degenerate four-wave mixing have been recently observed by Steel and Lam (1979).

$^{13}CH_3F$. This molecule has been extensively studied by Brewer and colleagues (Schmidt et al. 1973, Berman et al. 1975, Grossman et al. 1977, Brewer and Shoemaker 1971) using a CW CO_2 laser and Stark modulation. The R(4) line of the fundamental ν_3 band of this symmetric top overlaps the P(32) line of the CO_2 laser.

T_1 for this self-broadened transition has been measured by the 2PDN method (discussed in subsection 2.4.3) by Schmidt et al. (1973). More recently, Berman et al. (1975) used a variation of this two-pulse method. Instead of a π pulse preparation, a steady-state initialization is used. When the Stark pulse is switched off resonance, the prepared molecules dephase quickly due to inhomogeneous broadening. Then, when the Stark voltage is

Table 3
The T_2 values of SF_6 with various perturbers, measured by the photon echo technique.

Perturber	T_2^{-1} (ns^{-1} Torr^{-1})	Reference
SF_6	0.045	(a), (b), (c)
	0.042 ± 0.009	(d)
He	0.030	(a)
Ne	0.021	(a)
H_2	0.061	(a)

[a] Patel and Slusher (1968).
[b] Alimpiev and Karlov (1974).
[c] Heer and Nordstrom (1975).
[d] Cheo and Wang (1970).

turned on resonance, the amplitude of the nutation depends only on T_1. The same equations as for the two-pulse delayed nutation describe this system, except that an $r_3(0)$ value for steady-state preparation should be used [see eq. (74c)]. The results of both these methods, as well as the rotational microwave transient results of Jetter et al. (1973) for the same transition, are presented in table 4. The numbers are quite close, indicating that rotational relaxation is probably the predominant mechanism of relaxation. Carr–Purcell trains and Raman echoes (see subsection 2.6) were used to measure T_2 (Berman et al. 1975), since these echoes are little affected by velocity changing collisions. From the echo studies, as expected for a molecule with many closely spaced rotational levels, Berman et al. (1975) found that $T_1 = T_2$ and no phase changing collisions are seen. Also, because $T_1 = T_2$, the semiclassical theory for VCC can be employed in the analysis of these data.

Methyl fluoride is the first molecule for which a velocity changing collision rate was measured. This was done by the two-pulse photon echo method described in subsection 2.5.2 (Berman et al. 1975). The results were somewhat difficult to interpret because of an intensity-dependent dephasing effect which has subsequently been said to be due to off-resonant power effects (Coy 1980). Still, this effect was accounted for empirically (Berman et al. 1975), and the value of Γ_{vc} from eqs. (106) and (107) was extracted (see table 4). Analysis of these data shows that Δv, the rms change in velocity per collision times the square root of two, is only 85 cm/s, a number very small compared to the most probable speed of the room temperature distribution. Much larger VCC parameters have been found by Hartmann et al. in atomic systems.

Grossman et al. (1977) have extended the two-pulse echo method in a very interesting experiment on $^{13}CH_3F$. Since $T_1 = T_2$ for the studied self-broadened transition of this molecule, the only contribution to relax-

Table 4
Relaxation rates for $^{13}CH_3F$.

Relaxation Rate	Value ($ns^{-1} Torr^{-1}$)	Spectral region	Reference
T_1^{-1}	0.095	MW	(a)
	0.076	IR	(b)
	0.089	IR	(c)
T_2^{-1}	0.089	IR	(c)
Γ_{vc}	0.074	IR	(c)

[a] Jetter et al. (1973).
[b] Schmidt et al. (1973).
[c] Berman et al. (1975).

ation from elastic collisions is through Γ_{vc}. In a two-pulse echo, the decay of the envelope of the FID after the second pulse as a function of pulse separation yields $T_1(=T_2)$. From eq. (107), the echo amplitude can be used to yield Γ_{vc}. If the sample is excited with the Stark voltage on, several M components will be excited. When the Stark voltage is turned off, they will decay with different zero-field frequencies. By Fourier transforming the FID and echo decays, T_1 and Γ_{vc} for each M state may be measured. Furthermore, by changing the Stark voltage, different v_{1z} groups are excited so these rates may also be measured as a function of absorber velocity, as was done for T_2 in the microwave region.

Grossman et al. (1977) found no M dependence for either of the rates for $^{13}CH_3F$. This corresponds well to the microwave data of Mäder (1979). He found only small M dependences in inelastic relaxation rates in OCS, a molecule with many closely spaced levels like $^{13}CH_3F$. The velocity dependence was fit to a Landau–Lifshitz elastic cross section. It was found that the inelastic data fit well to dipole–dipole forces, whereas Γ_{vc} fits well to an r^{-6} potential. It should be noted that, in the microwave region at least, this cross section has been found not to be adequate to even qualitatively describe the velocity dependence of rotational relaxation (Burns 1979). Better calculations than just a Landau–Lifshitz approach are probably necessary to interpret these IR data correctly.

NH_2D. The $(v_2, J, M) = (0, 4_{04}, |4|) \rightarrow (1, 5_{05}, |5|)$ and the $(0, 4_{04}, |4|) \rightarrow (1, 5_{14}, |5|)$ transitions of NH_2D can be Stark switched into resonance with the P(20) and P(14) lines, respectively, of a CO_2 laser. The molecule has been studied using optical nutation (Shoemaker and Van Stryland 1976, Shoemaker 1978), FID (Brewer and Shoemaker 1972) and photon echo (Shoemaker 1978, Brewer and Shoemaker 1971) techniques. In addition, quantum beats due to hyperfine splitting of the transitions have been seen in optical nutation and photon echo experiments (Shoemaker 1978, Shoemaker and Hopf 1974).

Self-broadened NH_2D is interesting in that 2PDN yields a sum of exponentials which from eq. (89c) implies that the inelastic rates (γ_a and γ_b) out of the two coupled levels are different (Shoemaker 1978). In such a system, the echo decays of a Carr–Purcell train need not yield T_2. Consequently, Van Stryland and Shoemaker (Shoemaker 1978, Van Stryland 1976) used a value of T_2 derived from pressure broadening (Plant and Abrams 1976). This T_2^{-1} result is within the experimental error of the value of $\frac{1}{2}(\gamma_a + \gamma_b)$ derived from the 2PDN measurements. From the long-time decay of the two-pulse echo, they were able to get an approximate value for Γ_{vc}.

Other molecules. Using a Q-switched laser, vibrational transients in BCl_3

(Alimpiev and Karlov 1973, 1974) and SiF_4 (Gutman and Heer 1977, Loy 1976) have been measured. Frequency switching has been used to measure FID in methane (Hall 1973). Stark switching has been used to study $^{15}NH_3$ (Shoemaker 1978).

4.2.3. Multiphonon and multilevel gas phase transients

Two-photon and three-level coherent transients have been seen with gaseous atomic systems in the visible (see e.g. Liao et al. 1977). Few gas phase molecular results have been given. Brown (1974) and Glorieux and Macke (1974) have used microwave double resonance techniques to look at T_2 for the $J = 1 \rightarrow 2$ transition of OCS. Burns and Coy (Burns 1979, Burns and Coy 1983) have used two-source microwave excitation to measure T_2 for the dipole-forbidden $J = 0 \rightarrow 2$ transition of OCS. Schwendeman (1978) discusses other MW–MW transient double resonance experiments. Loy has seen two-photon optical nutation (Loy 1976), FID (Loy 1976, 1977), and adiabaric inversion (Loy 1978) of a vibrational transition of NH_3 using CO_2 lasers. Shoemaker and Brewer (1973) have used coherent Raman beat signal decay to extract T_2's for self-broadened $^{13}CH_3F$. IR–microwave and other double resonance results have been discussed by Steinfeld and Houston (1978).

4.2.4. Electronic transitions

Coherent transients of transitions between two electronic states have been detailed for one gas phase molecule, I_2. This molecule has been chosen because it absorbs and fluoresces strongly in the visible part of the spectrum and many of its lines have already been assigned (Luc 1980). Various rovibronic transitions of the $X\,^1\Sigma_g \rightarrow B\,^3\Pi_{o+u}$ band have been studied.

4.2.4.1. Experimental methods. The radiation source for all the experiments has been a CW, argon-ion pumped, single-mode dye laser. As discussed later, both "linear" and ring dye laser configurations have been employed. These lasers deliver 100 mW (≈ 800 mW for the ring laser) in the 5800–6000 Å region using rhodamine 6G as the dye. The frequency jitter is less than 20 MHz and is probably 1 MHz or less for experiments of short time scale. In transient experiments, the quality of the single mode must be monitored by a high-resolution Fabry–Perot interferometer.

For experiments in the "*bulb*", a standard glass cell with a side arm is adequate, since I_2 gas is in equilibrium with its solid phase at most temperatures used in the experiments. The laser is along the longitudinal axis of the cell. For experiments in *beams*, an effusive I_2 beam crosses the laser beam at right angles (fig. 14) and the coherent and incoherent transients (Zewail et al. 1977) are detected either in the forward direction of the laser or at right angles to the laser and molecular beams.

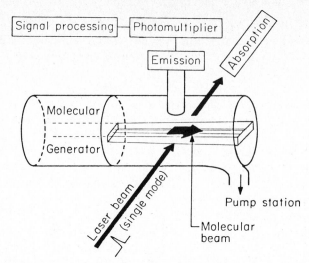

Fig. 14. A schematic of the apparatus used for observing coherent transients in beams.

There are numerous methods of modulating a CW laser to produce the desired pulses. One common method is the intracavity frequency switching of a dye laser (see subsections 1.4 and 4.4.2.1). The switching is done by placing an AD*P crystal in the laser cavity (fig. 15). Application of a voltage to the crystal changes the path length of the beam through the refractive crystal, thus yielding a different lasing frequency. Shifts of 0.6 MHz/V are typical for dye lasers used. Thus, only 50 to 100 V pulses, which are readily available with nanosecond rise times and kHz repetition rates from commercial instruments, are needed to shift out of a homogeneously broadened velocity group to another velocity group.

Orlowski et al. (Zewail et al. 1977, Orlowski et al. 1978) have used an extracavity dumper to *amplitude* modulate the laser to produce $\pi/2$ and π pulses. In this method, an RF pulse is applied to a transducer which will propagate sound waves through a fused quartz crystal. When the RF is on, a laser beam passing through the crystal (fig. 16) will be diffracted by the sound waves. Thus, pulses moving in a different direction from the incident laser beam are generated. Rise and fall times of 3 ns are achievable. This technique has the advantage that the effects of switching between two different velocity groups, as occur with the intracavity switching technique, and the necessity of deconvoluting these effects are avoided, especially in complicated molecular systems. Also, the voltage pulses are only 3–5 V. What is different from the electro-optic method is the lack of *intrinsic* heterodyne detection.

Instead of extracavity acousto-optic modulation, extracavity electro-

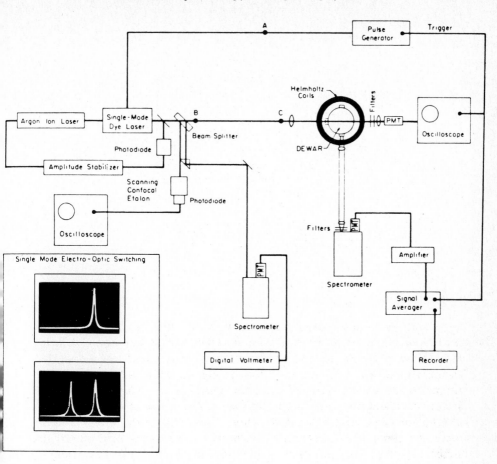

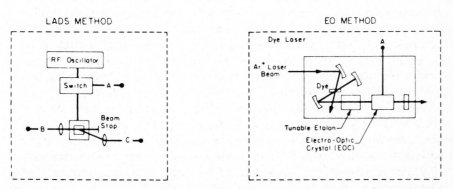

Fig. 15. The exprimental arrangements used for EO and LADS experiments (see Orlowski and Zewail 1979, and the text).

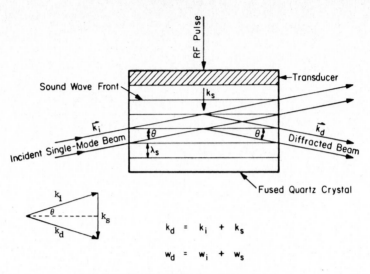

Fig. 16. The idea behind an AO modulator.

optic frequency modulation has also been reported on atomic systems
(DeVoe and Brewer 1978). A lithium tantalate crystal mounted in a strip-
line microwave transmission line was used. A voltage pulse was propagated
with the light down the length of the crystal, resulting in a change in the
refractive index and, thus, a frequency shift. Shifts in the range of 0 to
10 GHz (17 MHz/V) for 1.5 ns were achieved with a rise time of 100 ps. The
large frequency shifts attainable should make it possible to shift completely
out of the Doppler profile (similar to the RF switching method of AM) of
some molecules. For the large shifts, though, a detection system with a
very large bandwidth will be necessary in order to use heterodyne detection.

Genack et al. (1980) report the use of an AD*P crystal as a phase modulation device to study I_2 relaxation, although no detailed results for this
molecule were given. About a 300 V pulse was necessary to shift the phase
by π rad. This technique could also be capable of pico-second resolution.

Recently, Levenson (1979) has used polarization modulation of the laser
to observe FID. His apparatus is shown on fig. 17. Instead of the amplitude
or the frequency being switched, the polarization of the laser is modulated
using a Pockels cell, which produces left circularly polarized light when
voltage is applied and right circularly polarized light when the voltage is
zero. After the voltage is shut off, the FID from the sample will continue to
emit left circularly polarized light. The Pockels cell makes that light
linearly polarized. This light is then directed into the photomultiplier tube
(PMT) by a beam splitter. But the incident laser beam and the resulting

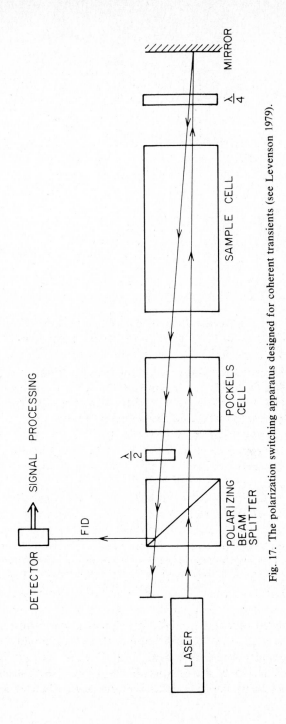

Fig. 17. The polarization switching apparatus designed for coherent transients (see Levenson 1979).

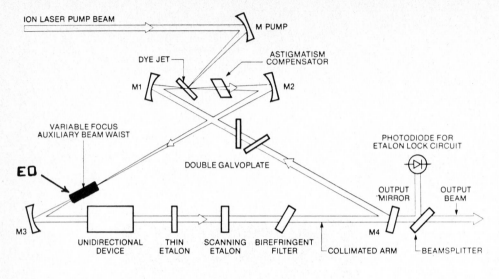

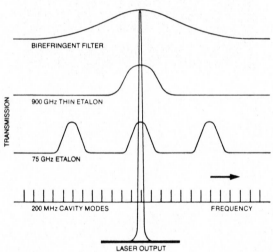

Fig. 18. Top: a schematic of a ring laser (Spectra Physics) *with* an EO crystal intracavity Lambert et al. (see Zewail 1980) used this configuration and observed the transients in fig. 19. The bottom figure shows the bandpass of the ring laser tuning elements (not to scale). (Courtesy of Spectra Physics.)

optical nutation after the pulse are directed into the beam stop. A half-wave plate is used to direct a small amount of the incident laser into the PMT, resulting in a heterodyned signal.

More recently Lambert and Burns (Zewail et al. 1980b for review) have obtained the coherent transients, FID and nutation, using a ring dye laser.

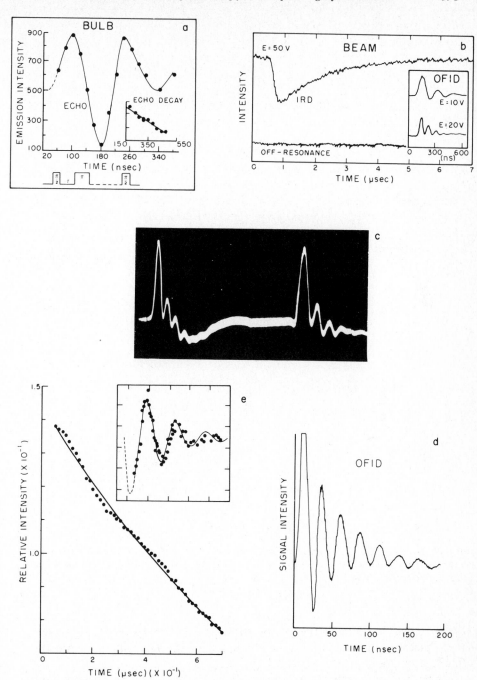

Fig. 19. The OFID and nutation of I_2 at 0°C, with 100 V of switching on the EO crystal shown in fig. 18 (see text).

The configuration of these experiments is shown in fig. 18, and the transients in fig. 19.

In general, detection of the transients is made with a fast photodiode and amplifier system. If the right-angle fluorescence is to be measured, a filtered, fast PMT is used. The pulse generation and data acquisition system is typically similar to that previously described for microwave and IR experiments.

4.2.4.2. Results—I_2 in a "bulb" and in a beam. I_2 was the first molecule for which coherent transients were observed in the forward direction of the laser (Brewer and Genack 1976) and at right-angles using the *incoherent* fluorescence (Zewail et al. 1976, 1980b for review). I_2 was also the molecule for which coherent transients in beams were observed for the first time (Zewail et al. 1977, 1980b for review). For the $(v, J) = (2, 59) \rightarrow (15, 60)$ transition T_1 and T_2 were measured (see figs. 20–24). Brewer et al. found that $T_1^{-1} = 0.71 + 0.029P \, \mu s^{-1}$, where the pressure, P, is in mTorr.

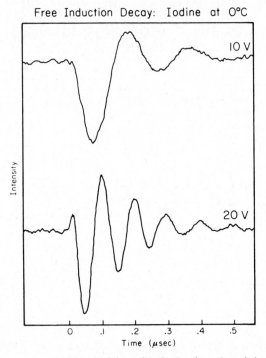

Fig. 20. The OFID of I_2 gas at 0°C coinciding with the leading edge of the EO crystal pulse. The switching frequency is 5 MHz for the top trace and 10 MHz for the bottom trace. Both transients were obtained at 500 μW laser power (Orlowski et al., see text).

SPONTANEOUSLY DETECTED PHOTON ECHO

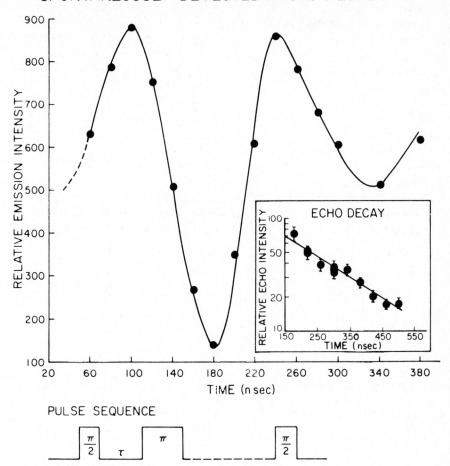

Fig. 21. The photon echo of I_2 at 10 mTorr, detected on the *fluorescence* using the three-pulse train (Zewail et al. 1977). Note the maximum amplitude is at $\tau_2 = 180$ ns, where the echo "burns" the spontaneous emission.

We measured T_1 by the IRD method and found $T_{1e}^{-1} = (0.775 \pm 0.029) + (0.0158 \pm 0.0004)P\,\mu s^{-1}$, where P is in mTorr. T_{1e}^{-1} is the decay rate of the excited level population (see figs. 25 and 26). Both experiments measure the relaxation due to inelastic collisions of the type $I_2 - I_2^*$. However, the difference in the pressure-dependent part of T_1 is due to the fact that the inelastic processes that can be sampled in transient experiments depend on the detection method and on the time scale of the experiments.

For I_2 in the excited electronic state, we may write the total population

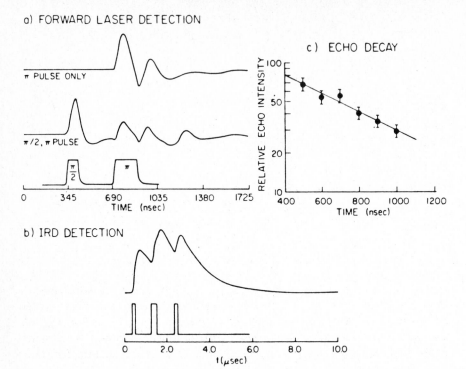

Fig. 22. The photon echo of I_2 monitored by using only two pulses (conditions are the same as those of fig. 21). Note the IRD at the bottom (Zewail et al. 1977).

decay rate as

$$T_{1e}^{-1} = T_{1PIe}^{-1} \text{ (pressure independent)} + T_{1PDe}^{-1} \text{ (pressure dependent).}$$

The term T_{1PI}^{-1} includes the radiative decay rate to all rotational–vibrational levels in the ground state, and spontaneous predissociation to nearby manifolds of levels. The term T_{1PD}^{-1} contains the pressure-dependent rates of relaxation to electronic (including pressure-induced predissociation), vibrational and rotational levels of the B-state. For the lower state ("ground" state)

$$T_{1g}^{-1} = T_{1PDg}^{-1} \equiv T_{1VRg}^{-1},$$

where only the pressure-dependent term (due to vibrational–rotational relaxation) is considered, since the radiative decay in the ground state manifold is relatively very weak.

In figs. 25 and 26, we show the IRD decay of I_2 as a function of pressure

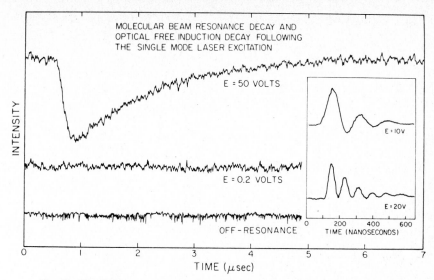

Fig. 23. The OFID and the IRD of I_2 in an effusive beam (Zewail et al. 1977).

and in an effusive beam. Firstly, we note that the zero-pressure value of the bulb experiment is in excellent agreement with the beam value. Secondly, the collisionless T_1 value gives directly T_{1PIe}, which translates to a homogeneous broadening of 128 ± 2 kHz. Thirdly, from the pressure-dependent measurement of T_1, we obtained a cross section (or more

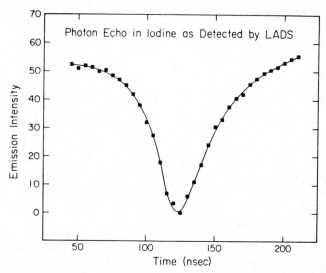

Fig. 24. The observed photon echo in I_2 using the LADS technique at 7 mTorr gas pressure (Orlowski et al. 1978).

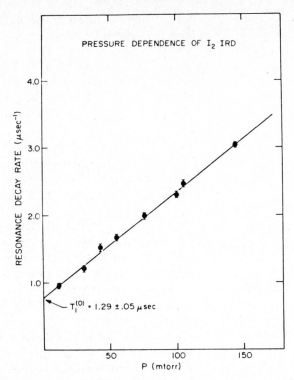

Fig. 25. Stern–Volmer plot of IRD in a iodine bulb. The error bars are standard deviations. The T_1 at zero pressure was obtained after least-squares fitting of all the data points (Zewail et al. 1977).

precisely a collision diameter squared) of 64 Å^2 for the inelastic electronic quenching. The IRD total decay gives the sum of T_{1PIe}^{-1} and T_{1PDe}^{-1} (rate of the process that takes the excited population to other electronic states including electronic predissociation). For excellent reviews on I$_2$ predissociation, see the papers by Lehmann (1977) and Broyer et al. (1975).

From the two-pulse nutation measurement of Brewer et al. on I$_2$ bulb, one obtains the total T_{1e}^{-1}. Thus by subtracting the T_{1e}^{-1} from the IRD rate, one obtains T_{1VRe}^{-1}, assuming that $T_{1VRe}^{-1} \approx T_{1VRg}^{-1}$. The collision diameter squared for the latter process is typically 10–40 Å^2 (cross section $\sigma = \pi D^2$, where D is the collision diameter). For the transition at 16 956.35 cm^{-1} (Wei and Tellinghuisen 1974), we already know the following (Chutjian et al. 1967, Kurzel et al. 1971):

D (electronic) = 70.0 Å^2,

D (vibration) = 10 Å^2,

D (rotation)   = 36 Å^2.

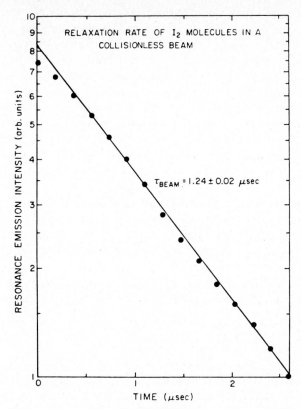

Fig. 26. The T_1 decay of I_2 in the beam. Note the agreement with the results of fig. 25 at zero pressure (Zewail et al. 1977).

For the same transition, Brewer's group obtained $D_e = 70.5 \text{ Å}^2$ and $D_{VR} = 40 \text{ Å}^2$.

The optical nutation gives the Rabi frequency ω_1. For I_2, $\mu = R_{el}R_{vib}R_{rot}$. By computer simulation of the I_2 nutation signal, Brewer's group obtained $R_{el}R_{vib} = 0.150 \pm 0.003$ Debye, after correcting for the degeneracy of the rotational magnetic quantum number M_J (the total wavefunction assumes the form $|\psi_{el}\psi_{vib}\psi_{JM}\rangle$) for the $(v, J) = (2, 59) \rightarrow (15, 60)$ transition. Using the calculated value of 0.2028 (Tellinghuisen 1978) for the vibrational overlap integral, $R_{el} = 0.740 \pm 0.015$ Debye. More on the use of nutation signals for obtaining μ will be discussed later.

The dephasing time of I_2 in a beam was obtained from the FID measurement of Zewail et al. (1977). In the bulb, T_2 was obtained directly from the two-pulse echo of Brewer et al. (1976) and the three-pulse echo of Zewail et al. (1977). As discussed before, T_2 provides the elastic cross section, which in the case of I_2 is a measure of the phase interrupting collisional cross section. The pressure-dependent rate is $T_2^{-1} =$ $0.79 + 0.071P\,\mu s^{-1}$, where P is in mTorr, for the same optical transition. After correcting for the T_1 contributions, T_2' can be obtained and related to the phase interrupting collisional processes ($D = 145\,\text{Å}^2$).

In conclusion, I_2 has been a prototype molecule for observing many transients using spontaneous emission detection methods, and forward heterodyne detection methods. OFID (in beams and bulbs), echoes (two-pulse and three-pulse), nutation (in the forward direction and at right angles) and more recently rotary echoes (Wong et al. 1980) have all been observed. These results are all consistent and separate T_1-, T_2- and T_2'-type processes. The results of the rotary echo experiments give $\frac{1}{2}(T_2^{-1} + T_1^{-1}) =$ $0.63 + 0.051P\,\mu s^{-1}$; the pressure-dependent part of this value is again in good agreement with the T_1 and T_2 previously measured. On the theoretical side there has not yet been any detailed calculation of T_2' for I_2, although our cross section data are in agreement with theoretical results of Jortner and his colleagues (Mukamel et al. 1976) on the effect of pressure broadening on scattering experiments. Finally it should be mentioned that the relaxation parameters above are good only for moderate pressures, since at very low pressures ground and excited state relaxation become very distinct.

4.3. Liquids

The measurement and theoretical interpretation of dephasing times in the liquid phase are much more difficult than in the gas phase. Experimentally, typical vibrational dephasing times are on the order of picoseconds. The impact approximation, which states that the length of time of inter-molecular interaction is short compared to the time between interactions, is not necessarily valid in the liquid phase, complicating the theoretical interpretation. In an exact sense, a many-body treatment is strictly needed to quantitatively describe these times. Several groups, however, have provided theoretical models of dephasing in liquids (see subsection 4.3.2).

Most of the experimental work on liquids has been done by the Munich group (see the excellent review by Laubereau and Kaiser 1978), primarily using coherent Raman spectroscopy. The lack (until recently) of a high-power, tunable, IR picosecond source dictated the use of Raman techniques for these experiments. Recently, the group in Amsterdam (Hesp et al. 1977) and Harris' group (1978) have used these techniques to examine some interesting dynamical problems in liquids.

4.3.1. Experimental methods

4.3.1.1. Stimulated Raman experiments. Kaiser and colleagues have employed a mode-locked Nd–glass laser operating at the $9455\,\text{cm}^{-1}$ line

(Laubereau and Kaiser 1978, for review). A single picosecond pulse was selected from the pulse train using a high-pressure spark gap and an optical Kerr cell. Since the decay times to be measured were on the order of the temporal pulse width, great care was taken to ensure that the pulses were well characterized. They achieved one gigawatt pulse, 6 ps long, in the TEM_{00} mode. The pulses cut from the train were then amplified with a gain of about 100, and the frequency was doubled to $18\,910\ cm^{-1}$ using a KDP crystal.

The rest of their apparatus is shown in figs. 27 and 28. The input intensity is monitored using one photodiode (P1). The probe pulse, about one hundredth the power of the incident laser pulse, is split from the main beam and then encounters a variable optical delay before transversing the sample cell. The gain or loss of the main beam, which excited the predominant Raman-allowed transition, is detected using a second photodiode (P2). The probe beam, which creates a coherent anti-Stokes signal from the excited vibrational state, is monitored at the correct phase-

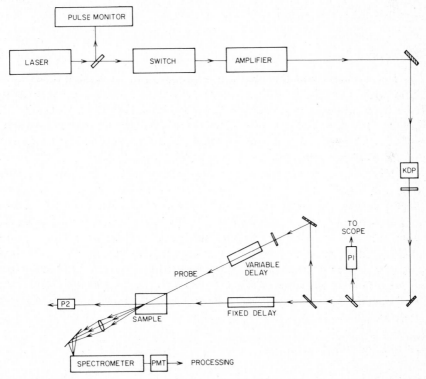

Fig. 27. Schematic representation of the noncollinear coherent Raman scattering experiments. (After Laubereau and Kaiser 1978).

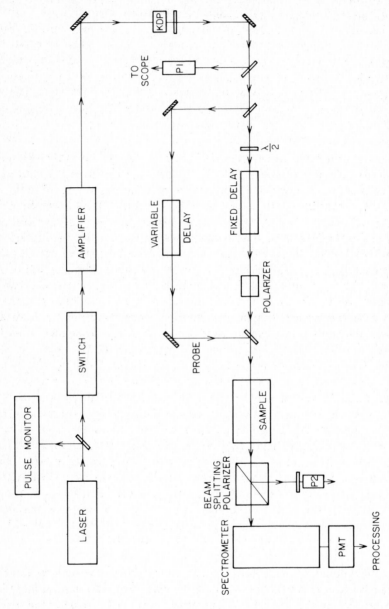

Fig. 28. The same as fig. 27, except the pump and probe pulses are spatially collinear, and polarizers are used to separate signals (see text). (After Laubereau and Kaiser 1978).

matched angle using a spectrometer (set to the correct anti-Stokes frequency) and a PMT. The signal is then normalized and displayed. By varying the probe pulse delay time while recording the probed anti-Stokes signal, T_2 can be deduced (see subsection 4.4.2.2).

The second set-up used collinear pump and probe beams, but in the pump beam a polarizer was placed. When the polarizer after the sample was crossed, only the coherent anti-Stokes Raman signal due to the probe was measured by the PMT after the spectrometer. Again, varying the delay time and measuring the anti-Stokes signal yield T_2. In addition, when an optical multichannel analyzer was used, frequency resolution was also possible.

4.3.1.2. CARS method. In the previous experiments, the stimulated Raman excitation is built from quantum noise. If, instead of using one excitation laser, two lasers with a frequency difference equal to that of the vibrational transition to be studied are used, then the build-up of selective Raman excitation is much more efficient. Consequently, relatively lower power lasers may be employed. The coherent vibrational excitation can be probed by splitting off a small amount of the excitation pulse and measuring the gain (or loss) of one of the weak probe pulses as a function of time delay of the probes. The advantages to this technique are that effects due to high power, such as self-focusing, may be avoided and that transitions other than the one with the largest Raman gain may be more conveniently studied. Also, the available laser pulse repetition rates for low-power lasers (> 1 MHz) are much larger than that of the high-power Nd–glass system ($\leqslant 10$ Hz). Thus, one can align the optics and the phase matching with ease.

Heritage (1979) used a single mode-locked CW argon-ion laser to synchronously pump two dye lasers operating in the rhodamine 6G region (≈ 5900 Å). Pulses of about 6 ps duration and about 100 W power were achieved. A variable optical delay and suitable optics were provided on one of the beam outputs to bring the lasers into temporal and spatial coincidence. As with the set-up described previously, a beam splitter and variable delays were used to generate the time-delayed probe pulses. After the sample, the probe pulses were collimated and then frequency resolved using prisms. The loss or gain in the desired probe beam was detected by a fast photodiode and a lock-in amplifier sensitive to the frequency at which the main beam was mechanically chopped (see fig. 29).

4.3.1.3. Vibrational T_1 measurements. The population relaxation time can be measured by Raman variations of the two-pulse delayed nutation method and the incoherent decay method, described previously. A powerful laser pulse at 9455 cm^{-1} generates vibrational excitation. A small part of the excitation beam is split off and doubled using a KDP crystal, creating a

M.J. Burns et al.

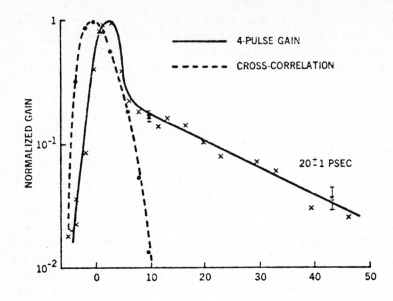

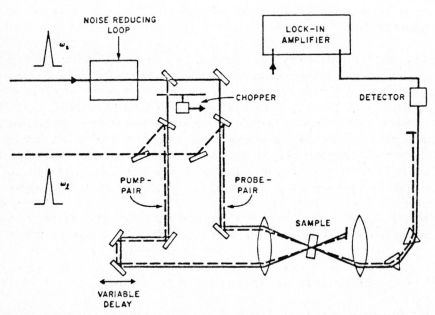

Fig. 29. Bottom: Experimental arrangements for dephasing measurements with CW mode-locked lasers (ω_ℓ and ω_s). The top figure shows the decay by dephasing in liquid CS_2. (After Heritage 1979).

probe pulse which can be variably delayed. The incoherent (spontaneous) anti-Stokes Raman signal perpendicular to the probe beam is measured using a monochromator and a photomultiplier tube. The decay of this signal as a function of time yields the population relaxation time of the vibrationally excited state.

Another technique for obtaining T_1 involves the use of direct IR pumping. Laubereau et al. (1974) have developed a method of generating high-power, tunable infrared pulses using parametric three-photon amplification in $LiNbO_3$ crystals. Exciting with the $9455 \, cm^{-1}$ line of the Nd–Glass laser, near bandwidth limited IR pulses from 6800 to $2700 \, cm^{-1}$ and of duration of 6 to 0.5 ps are achievable. The frequency may be monitored using a conventional infrared spectrometer.

Laubereau et al. (1975) have used these pulses to measure the population decay of excited vibrational levels of dye molecules. In their experiments, a vibrational transition was excited by the IR radiation. At a variable delay time, the doubled output of the glass laser was used to promote the molecules from the excited vibrational level into an excited singlet state, S_1. The energy of the singlet is such that molecules could not be promoted from the lowest IR connected vibrational state. Monitoring the probe-induced fluorescence as a function of time yielded T_1 for the excited vibrational state under certain reasonable approximations. Alternatively, incoherent anti-Stokes Raman scattering may be used as a probe, as discussed before.

4.3.1.4. Vibrational dephasing time measurements. Here, we will emphasize a few points pertinent to the measurements of T_2 that we described before. Firstly, since the interest is in monitoring the decay of vibrational coherence, the coherent anti-Stokes intensity (not the spontaneous) is measured as a function of the delay time between pump and probe pulses. Secondly, to produce coherent vibrational excitation, a well-defined phase matching relationship must exist between the Stokes, the anti-Stokes and the vibrational excitation of interest. This matching can be done if the wavelengths of the pump, probe, vibration, and the angles between the beams are known.

Finally, it must be remembered that for many of these experiments (especially the dephasing experiments) the desired decay signal must be deconvoluted from the temporal pulse shape (consequently, the pulse parameters must be accurately known and essentially constant throughout the experiment), and HB and IB effects must be separated.

4.3.2. Connections with theory
Information on the dynamics of molecular vibrations in liquids have been obtained from the analysis of lineshapes or from T_1- and T_2-type experi-

ments. In principle, these two kinds of experiments are related since the time evolution of molecule polarizability is related directly to the lineshape function $I(\omega)$:

$$I(\omega) \propto \sum_j \int_{-\infty}^{\infty} e^{i\omega t} \langle \alpha_i(t)\alpha_j(0) \rangle \, dt, \tag{292}$$

where α_i is the polarizability of molecule i. From the studies of $I(\omega)$, one can obtain details on (a) the nature of the lineshape [see the papers by Bratos and Marechal (1971) and by Rothschild (1976)], (b) the effect of *density* and *temperature* on the observed lineshapes (see Schroeder et al. 1977), and (c) the potential energy surfaces (Diestler 1978, Yarwood et al. 1977, 1979).

The picosecond experiments of the Munich group (Laubereau and Kaiser 1978) give direct information on the time evolution of the vibrational population and coherence through the measurements of T_1 and T_2. (Tables 5 and 6 show some of these T_1 and T_2 for different liquids.) The interpretation of these T_1 and T_2 in terms of microscopic parameters is, however, model dependent.

The theoretical models of Fischer and Laubereau (1975), Diestler (1976), Madden and Lynden-Bell (1976), Oxtoby and Rice (1976), Wang (1977), Oxtoby (1979), Harris et al. (1977, 1978), and the computer simulations by Oxtoby et al. (1978, 1980) have related T_1 and T_2 to parameters like

Table 5
The T_1 relaxation times in carbon tetrachloride[a].

Sample	Mole fraction	Vibration	T_1' (ps)	T_1 (ps)	Excitation method
CH_3CH_2OH	0.04	a C–H stretch	2 ± 1	40	IR
	1.0	a C–H stretch	0.5	22	Raman
$CHCl_3$	0.1	ν_1	—	2.5	IR
CH_2Cl_2	0.1	ν_1	4 ± 2	40 ± 10	IR
	1.0	ν_1	4	40	IR
CH_3Cl_3	0.2	ν_1	4	100 ± 30	IR
	0.4	ν_1	—	29	Raman
	0.6	ν_1	—	15	Raman
	0.8	ν_1	—	8	Raman
	1.0	ν_1	—	5.2	Raman
CH_3I	0.01	ν_1	—	1.5 ± 0.5	IR
	0.05	ν_1	—	1.0	IR
	0.05	ν_4	—	0.5	IR
coumarin 6	0.0004	—	1.3	8	IR

[a] All measurements were done at room temperature. T_1' is the fast relaxation time, and T_1 is the slow one. (Adapted from Laubereau et al. 1978a.)

Table 6
The T_2 relaxation times for various neat liquids[a].

Sample	$\bar{\nu}_0$ (cm^{-1})	T_2 (ps)	
		measured	spontaneous Raman linewidth
N_2(77 K)	2326	150 ± 16	158 ± 16
CCl_4	459	7.0 ± 0.8	7.6 ± 1.0
$SiCl_4$	425	6.0 ± 1.0	5.6 ± 1.0
$SnCl_4$	368	5.6 ± 0.6	5.0 ± 1.0
CH_3CCl_3	2939	2.4 ± 0.4	2.4 ± 0.2
$SnBr_4$	221	6.0 ± 0.6	>3.2
CH_3OH	2835	4.6 ± 1.0	>0.5
$(CH_2OH)_2$	2935	6.0 ± 1.0	>0.2
CS_2	657	20 ± 1	21

[a]Measurements were done at room temperature unless otherwise noted. $\bar{\nu}_0$ is the zero-field vibrational frequency. Adapted from Laubereau and Kaiser (1978). The values for CS_2 are from Heritage (1979).

the anharmonicity and exchange coupling between modes. There are, however, some problems in treating liquids. Firstly, the concept of homogeneous and inhomogeneous broadening is somewhat ill defined. Not every Lorentzian resonance is homogeneous in the sense of the two-level–bath problem discussed before. The weak coupling approximation, which yields Lorentzian resonances, and the binary collision approximation of gases is not straightforwardly transferable to liquid treatments. If the coupling in liquids is so strong as to produce Gaussian resonances, which will be confused with inhomogeneous broadenings, the interpretation of T_1 and T_2 must be done cautiously. These problems of correlation time changes are well known in other fields, and can be tested if careful consideration of the transient decay (exponential etc.) is given. In what follows, we shall consider a rather *simple* model given by Diestler et al. (1978) to describe vibrational dephasing in liquid N_2. We shall then comment on the changes of correlation times by discussing the dephasing of liquid mixtures of N_2 and Ar.

In the simplest version of the model Diestler (1978) considered only the homogeneous linewidth case, as described above, assuming that the diatomic has only two vibrational levels and that the coupling ΔV of eq. (237) does not depend on orientation of the diatomic. Furthermore, he also ignores the resonant ($V-V$) mechanism, which has been estimated (Kakimote and Fujiyama 1974) to be substantial in the case of $N_2(\ell)$ and $O_2(\ell)$. Hence, only the translational contribution to T_2' is considered. Clearly, this calculation of translational dephasing provides an upper bound on T_2 on account of neglecting the various other mechanisms of dephasing discussed before.

To determine the "translational" states of the liquid, they invoked the Lennard-Jones–Devonshire cell model in which the reference diatomic moves in an effective potential U obtained by averaging V, the interaction potential, over the orientation of the diatomic and smearing the nearest neighbors over the surface of a sphere. The effective potential, which is a function of both the intramolecular vibrational coordinate r and the displacement R of the center of mass of the diatomic from the center of the LJD cell, can be expanded as

$$U(R, r) = U(R, r_e) + \left(\frac{\partial U}{\partial r}\right)_{r=r_e} (r - r_e) + \frac{1}{2}\left(\frac{\partial^2 U}{\partial r^2}\right)_{r=r_e} (r - \acute{r}_e)^2 + \cdots, \quad (293)$$

where r_e is the equilibrium value of r. Using the dumbbell model, $U(R, r)$ can be evaluated. The unperturbed translation-like states of the LJD cell are determined in closed form by invoking the harmonic approximation (in R) for $U(R, r_e)$. Again, a harmonic fit of the R-dependent ΔV allows one to evaluate the expression for T_2' in closed form. The result is (Diestler 1978)

$$\frac{1}{\tau} = \frac{9\pi}{16} \frac{\Delta\omega_d^4}{\omega_d^3} [1 + \tfrac{4}{3} \sinh^{-2}(\hbar\omega_d/2k_B T)], \quad (294)$$

where ω_d is the frequency of oscillation of the diatomic in its cell and $\Delta\omega_d$ is the effective frequency obtained from the harmonic fit of ΔV. Note that this expression is expected to be applicable only at temperatures in the liquid phase range. In table 7 we compare the calculated values of τ for $N_2(\ell)$ and $O_2(\ell)$ with measured dephasing times and with other theoretical estimates. The results, which do not match the experimental τ, clearly indicate that one can obtain an order-of-magnitude estimate of τ by using certain potentials in the T_2' equation.

Mixtures of liquid N_2 and Ar exhibit a very interesting dephasing behavior. Hesp et al. (1977) have shown that the vibrational dephasing time of N_2 in liquid N_2/Ar mixtures decreases as the mole fraction of argon increases. They attributed this effect to changes in the correlation time (τ_c) of the molecular motion initiating the dephasing process. This time changes from 2.2 ps in pure N_2 to 4.6 ps in pure Ar. The correlation time and the dephasing times are related, in the fast modulation limit, by

$$\tau_{deph}^{-1} = 2\langle[\omega(0)]^2\rangle\tau_c, \quad (295)$$

where $\langle[\omega(0)]^2\rangle$ is the mean square frequency displacement of the frequency transitions due to intermolecular perturbations. These displacements also change in these liquid mixtures from $2.5 \times 10^{21}(\text{rad/s})^2$ in pure N_2 to $3.12 \times 10^{21}(\text{rad/s})^2$ in pure Ar. The conclusion of this work is that

Table 7

Comparison of experimental τ and calculated contribution to τ (in ps) for liquid N_2 and O_2 at their normal boiling points.

System	τ_{exp}[a]	τ[b]	τ[c]	τ[d]	τ[e]
N_2	79	1062	175	441	62
O_2	45	810	68	238	—

[a] Clements and Stoicheff (1968), Scotto (1968), Laubereau (1974). See also table 6, and note that τ and T_2 are related by a factor of 2.

[b] From eq. (294).

[c] Fischer and Laubereau (1975).

[d] Using a collinear potential and not the three-dimensional treatment of eq. (294).

[e] Using molecular dynamics simulation (Oxtoby 1979).

the dephasing results from the average force field determined by the number of nearest neighbor molecules, which is an interesting finding. We believe that more work on mixed liquids might prove fruitful in unravelling density and local inhomogeneity effects.

Finally, there are other mechanisms of dephasing (Oxtoby 1979, for review) that have been advanced recently. Among these is the modes exchange model of Harris et al. (1977, 1978), in which the vibrations are treated as anharmonic oscillators exchanging energy by the well-known Anderson (1954) mechanism.

4.4. Solids

In solids, essentially all the above-mentioned techniques of coherent transients have been successfully applied. In this section we shall focus our attention to the following:

(a) organic mixed crystals (guest electronic dephasing),

(b) organic "pure" crystals (electronic exciton dephasing),

(c) organic "pure" crystals (vibrational exciton relaxation),

(d) impurity ion in solids (magnetic dipolar interactions),

(e) other systems (color center dephasing and glasses).

The techniques used in investigating dephasing in the above-mentioned cases utilize either a low-power CW laser and a switch or high peak power pulsed lasers (see subsection 1.4). Photon echoes, OFID, nutation and IRD were all observed at relatively low temperatures.

4.4.1. Experimental techniques

4.4.1.1. The "CW laser and switch" methods. In 1976 we have reported on the first observation of optical nutation (fig. 30) in molecular crystals (Zewail and Orlowski 1976, 1977), using the intracavity frequency switching method (subsection 1.4). The system studied was pentacene in p-terphenyl mixed crystals at 1.7 K. Using the same technique, OFID was observed (fig. 31) in the same system (Zewail et al. 1976, 1977, 1980, de Vries et al. 1977, de Vries and Wiersma 1979). The extracavity acousto-optic modulation method (Orlowski et al. 1978) was used for measuring and studying the IRD in details for both pentacene/terphenyl and pentacene/naphthalene systems (Zewail et al. 1977, Orlowski and Zewail 1979, Lambert and Zewail 1980). Subsequently, the Stark switching method (subsection 1.4) of Brewer et al. was employed in organic crystals (Burland et al. 1979) using a CW laser. In a sense the switch in this case is the molecules in the solid, as described previously.

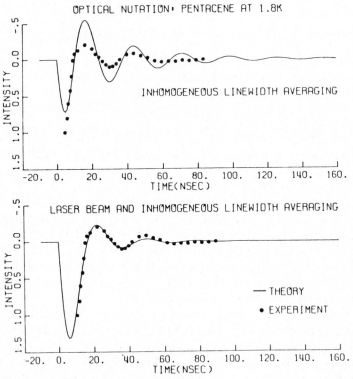

Fig. 30. Optical nutation of pentacene in p-terphenyl at 1.8 K (Orlowski and Zewail 1979); see also fig. 3.

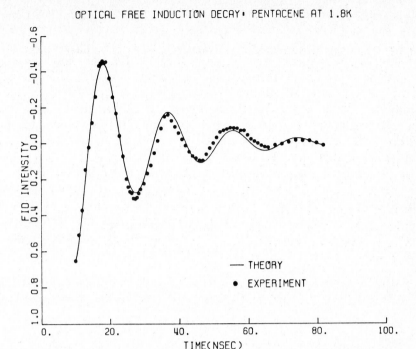

OPTICAL FREE INDUCTION DECAY: PENTACENE AT 1.8K

Fig. 31. OFID of pentacene in *p*-terphenyl at 1.8 K. The power of the laser in these experiments was 400 μW. T_2 is 45 ± 2 ns (Orlowski and Zewail 1979).

For the intracavity electro-optic (EO) switching method and the extracavity laser acoustic diffraction spectroscopic (LADS) method the following experimentation was used (see fig. 15). The single-mode dye laser amplitude was stabilized. A small portion of the main beam is sent to a scanning confocal etalon and spectrometer to ensure that the mode structure is single and to know λ of the radiation. For LADS experiments the modulator was placed in the optical path such that the diffracted beam continued on to the sample while the transmitted beam was blocked. Thus the sample was excited only when an RF pulse was supplied to the acousto-optic (AO) modulator. As discussed before, the AO modulator operates as follows: the RF pulses excite phonons in the transducer, which is bonded to a quartz crystal, creating travelling acoustic waves which diffract the incident beam (up to 50%) at the Bragg angle and shift its frequency by the acoustic phonon frequency (470 MHz). The characteristics of the laser pulse that this technique provides are determined by (1) the RF pulse width and its rise/fall times, and (2) the transit time of the acoustic wave across the focused laser beam in the crystal. We have measured laser pulse rise/fall times of 3 ns (1/e time) using a 75 mm focal length lens to focus the laser beam into the crystal.

There are advantages to both the EO and LADS techniques. The EO technique is most useful for observing OFID since extremely sensitive heterodyne detection techniques can be employed. The LADS method offers the following advantages. Firstly, it removes the influence of off-resonance effects, since the light is turned on and off at one frequency (fig. 32). Secondly, the optical pulses are generated by supplying low-voltage (TTL level) pulse trains to an RF oscillator instead of the high-voltage pulses that may be required by the EO technique.

For the detection, we use fast biased photodiodes (HP 5082-4203) for the coherent transient detection in the direction of the laser beam, and photomultipliers for coherent transient detection at right angles using spontaneous emission as a probe. The transients are displayed directly on a sampling scope, or in some cases, the signal will be averaged using a boxcar integrator. For the right-angle detection a lock-in amplifier is used for scanning the echo shape.

Finally, we would like to mention two additional developments related to the switching of CW dye lasers. These are the EO switching of a ring laser

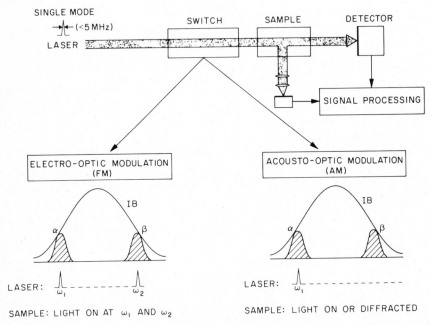

Fig. 32. A schematic showing differences between the frequency switching (EO) and the AO modulation method.

and the extension of the time resolution to the subnanosecond time regime. Lambert and Burns of this laboratory (see Zewail 1980) switched a Spectra Physics ring dye laser and observed the OFID and the nutation of a iodine gas. The set-up and the transients are shown in figs. 18 and 19. As for the extension of the switching to the subnanosecond time regime, this was accomplished (DeVoe and Brewer 1979) by using a travelling wave EO modulator outside the laser cavity, as discussed before.

4.4.1.2. The pulsed laser methods. Wiersma and co-workers (Aartsma et al. 1976, 1978) observed the echo of pentacene in p-terphenyl using the Hartmann group's pulse method (see subsection 1.4). In their configuration, shown in fig. 33, a nitrogen-pumped dye laser produces 5 ns duration pulses with a bandwidth of $0.4\,cm^{-1}$. This 10 Hz repetition rate laser produces 100–200 kW of visible power per pulse. The approximate $\pi/2$ and π pulses necessary for the photon echo experiments are generated and directed using beam splitters. An optical delay line allows the time between the pulses to be varied. After the sample, three Pockels cells are used as a fast optical shutter to attenuate the pump pulses and to pass only the photon echo. This shutter, as well as the detection apparatus, are triggered by a pulse from a photodiode which receives a small part of the input pulse. A monochromator is used to separate the photon echo from the nonresonant

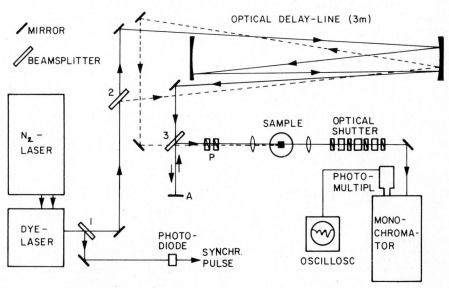

Fig. 33. A schematic drawing of the nanosecond time scale photon echo apparatus (Aartsma et al. 1976).

fluorescence. For some experiments (such as those on naphthalene) the laser pulses were doubled into the UV using KDP crystals.

Morsink et al. (1977) have used two nitrogen-pumped dye lasers for their photon echo experiments as shown in fig. 34. In this configuration, the two pulses, as well as the shutter and the boxcar, are triggered by TTL-generated pulses. Due to jitter in the lasers, though, the pulse separation is measured by allowing the two pulses to impinge upon a fast photodiode, which feeds into a time-to-peak-height converter. The sample holder, shutter, and detection apparatus are as just described except that a boxcar integrator is used to time average the signal.

Extension of the echo technique to the picosecond domain was made by using a mode-locked Ar^+ laser to synchronously pump two dye lasers operating at different frequencies (Hesselink et al. 1978). Transform-limited pulses of about 18 ps duration are produced. The average power is 25 mW, and the time jitter between the two pulses is typically 25 ps. These pulses are amplified using 400 kW nitrogen-pumped dye cells, synchronized by monitoring the Ar^+ pulses with a fast photodiode. A pulse from one of the amplified lasers is split into two (or more) pulses using (a) beam splitter(s). The pulses may be delayed using optical delay lines to create the necessary pulse timing for the photon echo (or stimulated echo) experiments. The pulse from the other laser is delayed and mixed with the photon echo signal

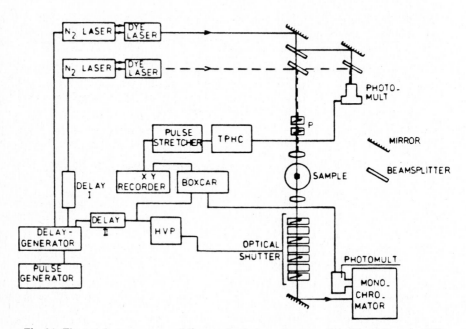

Fig. 34. The *two*-laser nanosecond time scale photon echo apparatus (Morsink et al. 1977).

from the sample in an ADP crystal, creating a sum frequency output whenever the probe and echo signal coincide. The pump laser is filtered out and the resulting sum echo signal transverses a monochromator and is detected by a PMT–boxcar system. This detection system (Dugay and Hansen 1968, Mahr and Hirsch 1975, Halliday and Topp 1977) allows photon echoes to be measured for excitation pulse separations down to 50 ps (Hesselink et al. 1978).

Cooper et al. (1979) have used a frequency-doubled, Q-switched, and mode-locked Nd:YAG laser to synchronously pump a cavity-dumped dye laser. Transform-limited, 6 μJ pulses of 40 ps duration were obtained. The pulses were split and delayed in the usual way to create photon echo pulse trains. The power of this laser was so high that "exact" $\pi/2$ and π pulses were achieved. The echoes were detected by the same above-mentioned nonlinear frequency mixing technique, though the echo was mixed with a delayed IR pulse (1.06 μm) from the Nd:YAG laser.

4.4.1.3. *The picosecond CARS method.*

In a series of experiments, Hochstrasser's group (Abram et al. 1979, 1980; Hochstrasser et al. 1980; De Cola et al. 1980) have used CARS to study relaxation of vibrational excitons in crystals at low temperatures. Also, Hesp and Wiersma (1980) measured the time-resolved CARS of naphthalene and found relaxation times ranging from <10 ps to 280 ps for different vibrational modes.

In these time-resolved experiments a synchronously pumped dual dye laser system is used. Basically, a mode-locked Ar^+ ion laser with pulses of ≈ 130 ps duration pumps two dye lasers (lasing at different frequencies ω_1 and ω_2) in parallel. The wavelengths are adjusted such that $2\omega_1 - \omega_2$ is matched to the frequency of the vibration. The dye laser pulses are typically 10 ps in duration. The CARS signal can then be detected after filtration from background incoherent light by using a monochromator with very high stray light rejection (ca. 10^{-18}). To measure the time constant for the decay, one plots the CARS signal as a function of the delay time between the pulses.

4.4.1.4. *Time-resolved laser line narrowing.*

The first report on the use of laser line narrowing (LLN) in molecular crystals was made a few years ago (Smith and Zewail 1979). In these time- and frequency-resolved experiments the narrow band laser is tuned to the allowed state of the exciton (from the ground state). After the pulse is turned off, the emission from the entire exciton band to another band in the ground state is then monitored (fig. 35). The delay between the time of observation and the time at which the laser pulse is turned off is determined using gating techniques.

In these LLN experiments we use a nitrogen laser pumped dye laser and

TRANSIENT BAND-TO-BAND SCATTERING

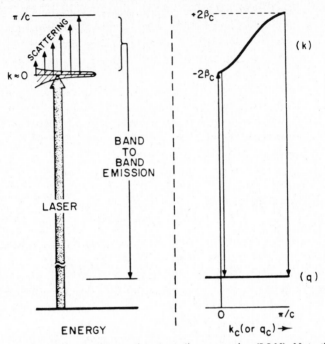

Fig. 35. The scheme of the transient exciton laser line narrowing (LLN). Note the selective preparation of $k \approx 0$ (Smith and Zewail 1979).

amplifier. The bandwidth of the laser is a few GHz and the pulse duration is ≈ 6 ns. The emission is gated by the use of an optical delay line and a boxcar integrator. Unwanted scattered light and "laser pulse tail" can be diminished by using a good spectrometer of high stray light rejection, and a gated circuit that turns off the photo-multiplier when the laser is on. Details can be found in the thesis of Smith (1981) from this laboratory. The set-up is shown in fig. 36.

4.4.2. Results and discussion

4.4.2.1. Pentacene in mixed molecular crystals. Two groups independently (Aartsma et al. 1976, 1980; Zewail et al. 1976, 1980) have examined the dephasing of pentacene in mixed crystals in detail. Subsequently, the energy transfer in these systems has been studied by the photon echo and the grating techniques (Cooper et al. 1979, 1980). Because the O_1 origin of the $^1A_g \rightarrow {}^1B_{2u}$ electronic transition of pentacene is conveniently near the

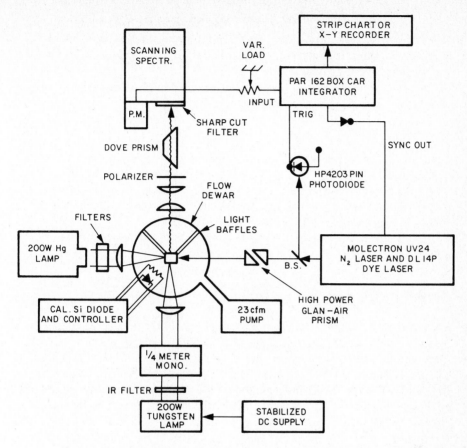

Fig. 36. A block diagram of LLN and absorption experiments (Smith 1981).

peak of the rhodamine 6G gain curve, many transients have been observed in this system. These are photon echoes, optical nutation, OFID, stimulated echoes and IRD.

From fluorescence and coherent transient studies in the nanosecond time regime, it is found that at very low temperatures (≤ 1.7 K), $T_2 \approx 2T_1 \approx 44$ ns in the p-terphenyl host. Thus, this very large and complicated molecule has an O_1 site transition that behaves just like a simple two-level system [see eq. (247)] when $(T_2^{ph})^{-1}$ or $(T_2')^{-1} \to 0$ on this time scale. At longer times (μs) intersystem crossing from the excited singlet to the triplet state takes place as evident from the IRD spectra (Zewail et al. 1977, Orlowski and Zewail 1979, de Vries and Wiersma 1979).

Nanosecond and picosecond transients have also been observed at different temperatures up to ≈ 20 K. While T_1 of the fluorescence is unchanged, the total T_2 changes with temperature. Thus both groups, Wiersma et al. and Zewail et al., explained the change in T_2 with temperature as due to temperature-dependent T_2' processes. *But* different mechanisms were provided.

Wiersma and co-workers used a four-level model to obtain an "activation-type" dephasing rate dependence on temperature:

$$(T_2')^{-1} = \beta \, e^{-\omega_0/kT}. \tag{296}$$

Basically, this is a kinetic argument where the population is distributed among the four levels, two of which constitute the transition of interest, while the other two are due to a low-frequency mode. Wiersma et al. found $\beta^{-1} \approx 3 \, \text{ps}$ (22 ps) and the phonon frequency $\approx 30 \, \text{cm}^{-1}$ ($21 \, \text{cm}^{-1}$) for the mixed crystals (Hesselink and Wiersma 1978, 1980).

Zewail and co-workers used the *two*-level–bath picture and obtained the following expression, which is an "activation-type" rate, in the relatively low-temperature regime (apart from a constant):

$$(T_2')^{-1} = |\langle \Delta V_2 \rangle|^2 \, e^{-\Delta/kT}, \quad \Delta > kT. \tag{297}$$

This equation, which was derived from eqs. (253) and (273) (Jones and Zewail 1978, Diestler and Zewail 1979), contains the physics of the pre-exponential: the difference in the coupling strength of the ground and excited states of the optically excited impurity to the bath. Furthermore, it only assumes that the phonon density of states peaks around a frequency Δ (quasi-local mode). The general expression for T_2 has been successfully used recently to describe dephasing of molecular glasses (Small 1981) and dephasing probed by light scattering experiments (Friedman 1979).

As recognized by Jones and Zewail (1978) four-level formal exchange does not give rise to population transfer but to a coherent exchange coupling by which the levels shift in energy. So, to invoke a T_1-type population transfer the four levels must couple to a bath, and our point is that two levels are sufficient to give the observed behavior. The point about the coherent exchange coupling was also recognized in a later paper by de Bree and Wiersma (1979). Our results (Orlowski and Zewail 1979) give a Δ that is close to $20 \, \text{cm}^{-1}$, a "pre-exponential time" close to 100 ps, and the transition dipole moment $\mu = 0.7 \pm 0.1$ Debye. Similar studies were done on the O_1 site of pentacene in naphthalene, and on the vibronic transitions of the same system (Hesselink and Wiersma 1979, 1980; Lambert and Zewail 1980). The long-time behavior in the IRD and its temperature dependence is detailed somewhere else (Orlowski and Zewail 1979). The important point to be mentioned here is that at relatively short times and at very low concentrations pentacene behaves as a two-level system. On a longer time scale, however, we must consider the nearby states, like the triplets. This was considered by us (Zewail et al. 1977; Orlowski and Zewail 1979) and by the Wiersma group (de Vries and Wiersma 1978) to account for the observed IRD and OFID. From these results, the rate of intersystem crossing and the lifetime of the nearby triplet state were obtained.

Cooper et al. (1979, 1980) have measured the concentration dependence of T_2 of pentacene in naphthalene using photon echoes and stimulated photon echoes. Pentacene concentrations ranged from 10^{-9} to 10^{-5} mole per mole of host. The temperature was 1.4 K for most of the studies. They found that the exponential photon echo decay rate was linear in concentration over the region studied. They also found that even at the lowest concentration of pentacene at 1.4 K, T_2 was not equal to $2T_1$ ($T_2 = 30$ ns and $T_1 = 19.2$ ns). Thus, pure dephasing (not due to apparent concentration effects) was seen even at this low temperature. Cooper et al. (1980) explained these results of pentacene in naphthalene in terms of phonon-induced fluctuations in distances between the transition dipoles of guest molecules. They replaced the lattice with a continuous distribution, and integrated the dipole–dipole interactions down to the nearest neighbor distance. Later, Morsink et al. (1980) pointed out that nearest neighbors should be excluded, because dimers have much different optical resonance frequencies than do monomers. Excluding dimers reduces the calculated effect by Cooper et al. (1980) substantially, thus inferring that this proposed mechanism is unimportant. Morsink et al. (1980) asserted that dipole–dipole energy transfer will cause line broadening which they estimate by *direct* analogy with magnetic resonance. However, because one is dealing with multilevel systems and *intense* pulses the treatment cannot satisfactorily explain dephasing of such optical transitions.

Warren and Zewail (1981) provided a description of bath-independent optical dephasing in multilevel systems. Specifically, we derived expressions for optical dephasing from transition multipole interactions, borrowing heavily from the theory of NMR dipolar line width in solids, but making crucial modifications to reflect differences between the *optical* and *NMR* Hamiltonians. The findings are consistent with all experimental data to date, and point out several important new features of optical experiments with multilevel systems. These include (a) the dependence of dephasing rates on pulse flip angles (θ), and (b) the existence of zero-temperature dephasing, which is absent in pure two-level systems.

Warren and Zewail (1981) calculated the second, M_2, and fourth M_4, moments for these optical systems. If M_2 and M_4 are known, the dephasing time (T_2^c) due to these multilevel interactions may be obtained by assuming a shape for the resonance curve. If the lineshape is Gaussian, $M_4 = 3M_2^2$, so this assumption might be reasonable for fractional occupation probability, $f \approx 1$. But for dilute mixed crystals M_4/M_2^2 is large. In this case a Lorentzian lineshape which is truncated far from resonance is a reasonable assumption. The full width at half maximum $\Delta\nu$ of such a curve is:

$$\Delta\nu = \frac{1}{\sqrt{12}} M_2^{1/2}\left(\frac{M_2^2}{M_4}\right)^{1/2} = \frac{1}{\pi T_2^c}, \tag{298}$$

and consequently,

$$(T_2^c)^{-1} = 1.51 f\mu^2 a^{-3}\hbar^{-1} \quad (f \ll 1). \tag{299}$$

The above equations indicate that the dephasing time can be related to the *transition dipole moment* and the *concentration* of such multilevel systems. (a^3 is the unit cell volume.)

For pentacene in naphthalene $\mu \approx 1$ Debye. The density at 78 K is 1.2433 g/cm³, and using this value at 0 K gives $a = 5.54$ Å. Then for $f = 10^{-6}$, $(T_2^c)^{-1} = 8 \times 10^6 \, \mathrm{s}^{-1}$, compared to the experimental value of $12 \times 10^6 \, \mathrm{s}^{-1}$. A more exact calculation requires an exact value for μ, and summations over the known lattice structure including the two interchange equivalent sites, but we can conclude that the agreement is good. Also, the dephasing rate is linearly proportional to the concentration and the decay is exponential, as observed experimentally.

For naphthalene in durene $\mu = 0.02$ Debye, so we predict $(T_2^c)^{-1} = 2.7 \times 10^6 \, \mathrm{s}^{-1}$ for $f = 10^{-3}$ (density of 1.03 g/cm³ is assumed). However, because μ is small, higher-order interactions (such as transition octopole–octopole) are two orders of magnitude larger than the dipole–dipole inter-action for near neighbors (Hanson 1970, Hong and Kopelman 1970), and are expected to be larger out to about 30 Å, assuming the interaction goes as r^{-7}. Their contribution can be roughly estimated. Similar to the above calculation, we obtain $(T_2^c)^{-1} = 8 \times 10^6 \, \mathrm{s}^{-1}$. The experimental value is $(T_2^c)^{-1} = 15 \times 10^6 \, \mathrm{s}^{-1}$ (Morsink et al. 1980).

The data of Morsink et al. (1980) show that $(T_2^c)^{-1}$ ceases to be a linear function of concentration for $f \approx 2 \times 10^{-3}$, and apparently decreases for $f \geqslant 5 \times 10^{-3}$. At these concentrations we still predict a linear dependence. *But* the reported maximum crystal optical density of 0.82 gives a pulse intensity which decreases by a factor of 7 across the sample. For this relatively optically thick sample the flip angle is not constant. The flip angle dependence of M_2 was derived to be (Warren and Zewail 1981):

$$M_2 = f \sum_j V_{ij}^2 \sin^2\theta, \tag{300}$$

and the dominant term in M_4 for $f \ll 1$ is

$$M_4 = f\left(\sum_j V_{ij}^4\right)(2\sin^2\theta + 3\sin^4\theta), \tag{301}$$

where V_{ij} is the coupling matrix element. These expressions assume that $f\sum V_{ij} = 0$. This means, for example that dimer peaks are ignored. It is clear that the lineshape is strongly dependent on the flip angle. The above equations would imply that for small flip angles $(T_2^c)^{-1}$ is proportional to the laser power. In a concentrated sample the observed signal will be a weighted sum of exponentials, so it is difficult to calculate an exact correction factor although from above we expect *the apparent dephasing rate to decrease as samples become optically thicker,* as indeed was observed experimentally. More details can be found in paper by Warren and Zewail (1981).

Several other mixed organic systems have been studied using basically the same techniques as those employed to study pentacene in various hosts. These systems included tetracene in a *p*-terphenyl host, naphthalene in durene, naphthalene in perdeuteronaphthalene, and pyrene in biphenyl (Aartsma et al. 1976, 1977, 1978). Coherent transient techniques have also been applied to radical species in organic hosts. Hesselink and Wiersma (1977) and Morsink et al. (1979) have investigated triphenylmethyl radical in a triphenylmethyl host. Burland et al. (1979) used duryl radical in durene as their impurity. In the latter work, the radicals were produced by X-ray irradiation of durene crystals at room temperature. The laser source was a standard argon-ion pumped single-mode dye laser operating around 4900 Å with mixed coumarin dyes. Optical nutation was observed using Stark switching, and OFID was measured using intracavity frequency switching, all at 2 K. From the OFID, T_2 was measured to be 212 ns, whereas T_1 was found to be 530 ns. The fact that $T_2 \neq 2T_1$ in this system was explained by the fact that the duryl radical has a complex hyperfine structure. Consequently, numerous lines were excited, giving rise to an apparent shortening of T_2.

4.4.2.2. Excitons

Electronic excitons. Dephasing of excitons in "pure" crystals has usually been inferred from lineshape analysis of their optical transitions. For example, for triplet excitons, a typical linewidth for absorption at low temperatures (2 K or so) is a few cm^{-1} or less, implying a dephasing time in the picosecond time range (Burland and Zewail 1979, for review). Exciton dephasing could be as complicated as the large molecule case (Jortner and Kommandeur 1978) and as the many-channels dephasing of condensed phases, treated formally by Mukamel (1978, 1979, 1980).

A system that has been studied in some detail is 1,4-dibromonaphthalene (DBN) solid, where the molecules in the crystal stack in a unique way, forming a quasi-one-dimensional exciton. In other words, the intrachain interaction matrix element ($6.2 \ cm^{-1}$) is orders of magnitude larger than the cross-chain interactions ($<0.01 \ cm^{-1}$), as was first established by Hochstrasser's group (Hochstrasser and Whiteman 1972, Hochstrasser and

Zewail 1974). The lineshape of the 0,0 transition at $20\,192\,\mathrm{cm}^{-1}$ was analysed by Burland et al. (1977). The shape is "half Lorentzian" at low temperatures and Lorentzian at high temperatures (see fig. 37). The Lorentzian fit was taken to mean a homogeneous broadening for the singlet–triplet exciton transition.

Smith and Zewail (1979) observed laser line narrowing (LLN), described in subsection 4.4.1.4, by tuning the laser to this same transition of DBN at $20\,192\,\mathrm{cm}^{-1}$, while observing the emission from the entire electronic exciton band to a vibrational exciton band in the ground state. It was found that even at 20 K the rate at which exciton wavevectors are altered by phonon processes is relatively slow. Similar rates have been found recently in other systems (Wolfrum et al. 1979, van Strien et al. 1981).

Smith and Zewail (1979) concluded that while the inelastic T_1-type dephasing is slow, elastic-type dephasing or impurity scattering must be dominating the width of the transition. Warren and Zewail (1981, 1983) have shown that for pure 1-D systems with exchange coupling (matrix element $= V$) and inhomogeneous broadening ($\Delta\omega_i$) the first six moments (see subsection 4.4.2.1) can be readily derived. The calculation shows that

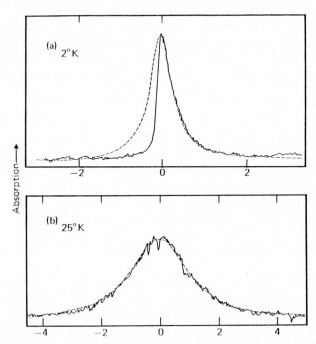

Fig. 37. The high-resolution 0,0 absorption of the DBN triplet exciton band at two temperatures. (After Burland et al. 1977)

the absorption lineshape should be asymmetrical at 0 K, reflecting *cross terms* between V and $\Delta\omega_i$, which would not be refocused in a conventional photon echo experiment. The usual $\Delta\omega_i$ effect can, of course, be eliminated by a two-pulse photon echo. In the absence of $\Delta\omega_i$ there will be no residual width in experiments like absorption where $\theta \to 0$. We have estimated the zero-temperature dephasing in such multilevel systems (see subsection 4.4.2.1) to be in the order of $\langle \Delta\omega_i^2 \rangle / \sqrt{6}\, V$ in the limit where $V > \Delta\omega_i$ and in the absence of resonance frequency correlations between adjacent sites. For DBN, $|V| = 6.2\ \mathrm{cm}^{-1}$, which means that the residual 0 K width is $\approx 0.6\ \mathrm{cm}^{-1}$ for inhomogeneous broadening of $3\ \mathrm{cm}^{-1}$ (close to the mixed crystal limit). This narrowing of IB due to the cross terms of V and $\Delta\omega_i$ (*motion* and *localization*) accounts for the observation and provides a new way of looking at these exciton dephasing dynamics. The origin of HB and IB (isotope effect, diagonal disorder, etc.) and their impact on pulsed laser spectra is currently under examination by us.

Vibrational excitons. In this section we present results on the relaxation of vibrational excitons obtained by Hochstrasser's group (1981) using CARS techniques (see subsection 4.4.1.3). Specifically, we shall consider the benzene crystal case which shows interesting new results.

Ho et al. (1981) have observed the coherent Raman signal of the 991 cm^{-1} vibrational mode of benzene crystal at low temperatures (see fig. 38). The

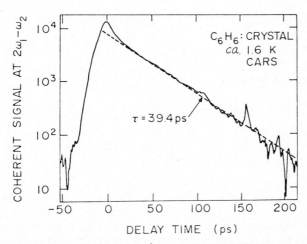

Fig. 38. CARS signal from benzene (991 cm^{-1} mode) crystal (1.6 K) as a function of the delay time (see text). The CARS signal from room temperature liquid benzene is symmetric relative to $t = 0$ and has a FWHM of 21.7 ps. (After Ho et al. 1981).

benzene crystal contains four molecules per unit cell, and the four factor group states are in the following order ($0 \, \mathrm{cm}^{-1}$ lowest in energy): $A_g(0)$, $B_{3g}(0.47)$, $B_{2g}(0.5)$ and $B_{1g}(1.05 \, \mathrm{cm}^{-1})$. In these CARS experiments the $A_g(k \approx 0)$ region is selectively populated, and the subsequent decay of the coherent state is probed. The data of Ho et al. are best fit to a decay constant of $39 \pm 2 \, \mathrm{ps}$ (see fig. 38).

Ho et al. (1981) have excluded IB as causing dephasing by estimating the effect of disorder on the overall width and by comparing the benzene results with other data on N_2 crystals (Abram et al. 1979). Comparing with the results of Smith and Zewail (1979) they also infer that intraband scattering is negligible. The conclusion is therefore that the observed decay corresponds to a spontaneous process. Since the A_g state is the lowest in energy, the $39 \pm 2 \, \mathrm{ps}$ decay time was assigned to a vibrational relaxation.

One interesting feature of these results is the relatively long (39 ps) relaxation time for an "extended" state when compared with the relaxation time of modes localized in nature. The lifetimes of the lattice $k = 0$ states are very long compared with those for a local lattice excitation such as occurs in mixed crystals. Ho et al. presented an interesting discussion of this point. Optical excitation of lattice modes in neat solids results in the preparation of a state that is more a "quasi"-eigenstate of the crystal Hamiltonian than would result from the excitation of lattice modes coupled to an impurity transition. In the latter case, the deformation potential may be directly involved whereas for the neat crystal nuclear kinetic or anharmonic interactions would be required. They, in fact, point out that specific anharmonic terms and modes may be involved in the relaxation of the $991 \, \mathrm{cm}^{-1}$ vibrational exciton. It will be extremely interesting to confront these experimental results with theoretical models invoking certain potentials, and rates for state-to-state energy relaxation from the $991 \, \mathrm{cm}^{-1}$ mode to all modes allowed by symmetry and energy considerations.

The dephasing of high-energy CH-stretch vibrational overtones in benzene-type molecules in solids at low temperatures has been studied by frequency domain spectroscopy (Perry and Zewail 1979). This intramolecular vibrational dephasing of molecules in solids will not be addressed here, and the reader is referred to the articles by Perry and Zewail (1979, 1981), Zewail and Diestler (1979), Smith and Zewail (1979) and Zewail (1980) for more details.

4.4.2.3. Impurity ion solids. The system that we shall discuss here is praseodymium (Pr^{3+}) impurity ions in lanthanum trifluoride (LaF_3) host crystals (the level structure is shown in fig. 39). Brewer and colleagues (De Voe et al. 1979, Rand et al. 1979, De Voe et al. 1981), and Shelby and Macfarlane (1978, 1979, 1980) have observed several transients in this system and obtained detailed and important information on dephasing by

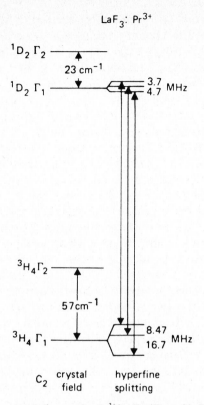

Fig. 39. Energy level diagram for Pr^{3+}/LaF_3. (From Shelby et al. 1980).

magnetic interactions. Hartmann and his group (Chen et al. 1979) have used their technique of photon echoes to measure relaxation in this system. Also, Szabo and Takeuchi (1975) observed self-induced transparency. The work of Erickson (1977) resulted in the discovery of an optical pumping cycle which leads to optical hole burning.

Brewer and co-workers reported on the OFID at 4 K using the frequency switching method. Because the experiments were done at low laser powers (10 mW in a 0.5 mm beam diameter), the OFID is not dominated by power broadening effects, and T_2 can be extracted. Optical free-induction dephasing times as long as $16 \mu s$, corresponding to an optical homogeneous linewidth of 10 kHz, have been reported for the $^3H_4 \leftrightarrow \,^1D_2$ transition of Pr^{3+} ions at 2 K. The measurements were facilitated by a frequency-locked CW dye laser. They also investigated the Zeeman effect and concluded that a Pr–F dipole–dipole dephasing is taking place.

Brewer et al.'s experiments on Pr^{3+}/LaF_3 yield important information on the magnetic interactions in this system. For example, they have shown that both homonuclear ^{19}F–^{19}F and heteronuclear ^{141}Pr–^{19}F local magnetic dipolar interactions are responsible for the broadening of the line. In a later study De Voe et al. (1981) carried out a Monte Carlo line broadening calculation which takes into account the role of the dipolar mechanism in the OFID problem. This allowed them to consider the LaF_3 as a discrete lattice structure and not as a continuum, and to include the preparation by the laser which burns a hole in the inhomogeneous distribution. The results of these investigations are summarized in table 8.

Chen et al. (1979) observed deeply modulated photo echoes from the Pr^{3+} ions which they ascribe to hyperfine interactions in both the ground $^3H_4$ and excited $^3P_0$ states. Splittings as small as $750\,kHz$ and $1.13\,MHz$ were inferred for the $^3P_0$ state. Also, the concentration dependence of the photon echo and a long-lived (3 min) stimulated photon echo were reported by the same authors. One important result emerges from the echo modulation experiments, namely the relative *orientation of the principal axes* associated with the ground and excited state Hamiltonians. Typical results from the work of Hartmann's group (Chen et al. 1979) is shown in fig. 40.

Macfarlane et al. (1979), using a delayed heterodyne detection of the photon echo, obtained a linewidth of $5\,kHz$ for the $^1D_2 \leftrightarrow {}^3H_4$ transition, which is narrower than the width obtained by OFID. They attribute this to the sensitivity of OFID to laser frequency jitter, power broadening and intensity-dependent dephasing arising from the presence of the heterodyne

Table 8

Linewidths of ^{141}Pr in LaF_3 due to magnetic inhomogeneous broadening[a],[b]

Transition	Method	Linewidth FWHM (kHz)
RF ($\Delta I_z'' = \pm 1$)	Monte Carlo theory[c]	82
	VanVleck second moment[c]	84.5
	CW RF–optical double resonance	
	$I_z'' = 1/2 \rightarrow 3/2$	$180 \pm 10^{[d]} (\approx 100)^{[e]}$
	$I_z'' = 3/2 \rightarrow 5/2$	$200 \pm 10^{[d]} (\approx 100)^{[e]}$
	optically detected RF transients	
	$I_z'' = 3/2 \rightarrow 5/2$	$230 \pm 25^{[d]}$
optical ($^3H_4 \leftrightarrow {}^1D_2$)	Monte Carlo theory[c]	
	$I_z'' \rightarrow I_z' = 1/2 \rightarrow 1/2$	42
	$I_z'' \rightarrow I_z' = 3/2 \rightarrow 3/2$	126
	$I_z'' \rightarrow I_z' = 5/2 \rightarrow 5/2$	210

[a]This table was taken from the paper by DeVoe et al. (1981).
[b]The smallest *homogeneous* broadening observed so far is $\approx 2\,kHz$ (from magic-angle line narrowing experiments; see Rand et al. (1979).
[c]The work of DeVoe et al. (1981).
[d]Earth's magnetic field.
[e]Static external field of $\geq 16\,G$.

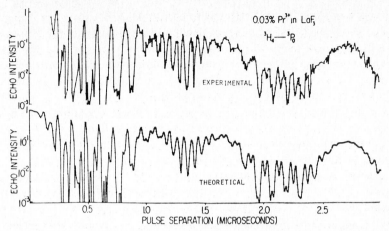

Fig. 40. Comparison of the experimental and theoretical photon echo modulating pattern. The data were taken for the $^3H_4$–$^3P_0$ transition of a 0.03 at% crystal of Pr^{3+}/LaF_3. (After Chen et al. 1979, 1980.)

field*. The same group have used the Pr^{3+}/LaF_3 system to make a number of interesting observations; optically detected coherent transients in the nuclear hyperfine levels of the 16.70 MHz transition (Shelby et al. 1978), coherent transients by optical phase switching (Genack et al. 1980), pseudo-Stark effect of the $^1D_2 \leftrightarrow {}^3H_4$ transition (Shelby et al. 1978), optical line narrowing by nuclear spin decoupling (Macfarlane et al. 1980), and optical measurements of spin–lattice relaxation (Shelby et al. 1980). In conclusion, it is clear that different magnetic interactions and dephasing mechanisms have been sorted out in the Pr^{3+}/LaF_3 systems by the use of the powerful coherent optical techniques.

4.4.2.4. Other systems. Other molecular systems have been investigated by hole burning techniques (Small 1981, for review) and by double resonance methods. Of particular interest to the authors is the development of a dephasing theory for experiments dealing with hole burning of molecules in *glasses* at low temperatures. This has been reported by Hayes and Small (1979), Hayes et al. (1980) and others. The pure dephasing part is still the same as that derived by Jones and Zewail (1978), while the population transfer part is modeled by using double-minimum potential energy surfaces in the ground and excited state of the molecules in the glass.

   The problem of *inhomogeneous broadening* for molecules (in different electronic states) in glasses has been addressed in a series of papers by Lemaistre and Zewail (1979). The effect of correlation between IB in

*Note added in proof. From a private communication by R. Brewer we learned that the 10 kHz value (and not 5 kHz) was obtained, when better laser frequency locking was used.

different electronic states and between homogeneous broadening, in-homogeneous broadening and laser line narrowing was discussed and related to measurements done on glassy systems.

Finally, we would like to conclude by mentioning briefly the results of two recent experiments on color centers which utilize optical–microwave double resonance and hole burning techniques. Glasbeek et al. (1980) studied the spin dephasing in an ensemble of dipolar coupled A and B spins in CaO crystals, where A-spin species are photoexcited ($S = 1$) F_2^{2+} centers and the group of B spins consists of ($S = 1/2$) F^+ centers. The interesting findings is that the A-spin dephasing as a function of temperature is not determined by the direct coupling of A spins with lattice vibrations, but instead is primarily governed by the temperature effect on the B-spin dephasing. This was established from the fact that the dephasing rate, obtained from the echo measurement, decreases as the temperature in-creases. For the first time, these results show that the A-spin dephasing is determined by exchange narrowing in the B-spin ensemble. The results are in quantitative agreement (Glasbeek et al. 1980) with the theoretical results of Jones and Zewail (1978).

In the color center F_3^+ in NaF, a reversible hole burning in the IB zero-phonon line was recently reported (Macfarlane and Shelby 1979). From these results, the homogeneous linewidth (17 MHz) was determined. The recovery of the hole shows two time components, 2.5 s and 70 min. The former time was assigned to triplet state relaxation, while the latter was assigned to electron tunneling from an adjacent center. These experi-ments, when completed, can provide valuable information on processes such as tunneling, which takes place on a very long time scale and usually is difficult to ascertain.

5. Concluding remarks

We have presented in this paper theoretical and experimental results on the techniques of nonlinear laser spectroscopy and on the dephasing of mole-cules in gases, beams, liquids and solids. We have provided an elementary description followed by a formal dynamical theory of optical dephasing. Two-level and multilevel systems were considered. Applications to a variety of systems were made and spanned isolated small molecules, gases (excited to rotational, vibrational and electronic states), impurity molecules or ions in solids, and vibrational and electronic excitons. The techniques used in each case were also discussed.

The field is still growing and more studies are needed before one knows the microscopic origin of dephasing in a general sense. Large molecules with complex level structure may exhibit different dephasing that simply probes different nonradiative and state-to-state processes. Exciton dephasing is also an important process that needs further examination as it might teach us more about disorder and transport in solids. Lasers with high resolution in the frequency and time domains will, hopefully, allow us to prove these new and exciting "avenues" that are amenable to coherent laser spectroscopy.

6. Postscript—some recent developments

While this chapter was in press several new experiments and techniques have been published. Here, we will highlight some, but not all, of the reported work.

The technique of acousto-optic modulation developed by us in 1977 (Zewail 1980b for review) for observing coherent transients has now been extended to vary independently the *phase* of optical pulses in a given laser pulse train; the phase of *each* pulse is independent of time delays between pulses or the duration of the pulse. With this technique, we (Warren and Zewail 1981b) observed the optical analogues of NMR multiple pulse spectroscopy. Specific multiple phase coherent pulse trains were developed to suppress unwanted background signals, enhance photon echoes and other transients by using composite pulses, and to overcome problems due to inhomogeneous broadening (Warren and Zewail 1981b for preliminary report, more detailed papers currently in press).

As for studies of dephasing processes in mixed crystals a number of new reports have appeared. These papers examine the effect of guest–guest interactions on dephasing near zero temperature (Warren and Zewail 1981, 1983), optical density effects on photon echo experiments (Olson et al. 1982), and the effect of 4-level coupling and strong phonon coupling (to guest molecules) on dephasing (Skinner et al. 1981, Skinner 1983).

In frequency domain experiments new focus has been on (i) the development of four-wave mixing techniques to provide ground *and* excited state vibrational frequencies (Decola et al. 1980, Hochstrasser et al. 1981); (ii) stabilization of narrow-band lasers to achieve very high resolution data (homogeneous widths) on solids (DeVoe et al. 1981); and (iii) studies of hole burning dynamics (homogeneous broadening, and hole burning mechanisms) in glasses and crystals (see chapter 9 in this book).

As mentioned in the text, a variety of photon echoes can be observed. The physics behind the observation of these echoes is somewhat different in that there are spatial and/or temporal requirements for each one of them. Hartmann and his group have recently developed an elegant model, called the billiard-ball model, to describe these photon echo phenomena (Beach et al. 1982).

Finally, we wish to conclude this chapter by mentioning a recent development concerning coherence effects in *large isolated molecules*. Coherence effects in anthracene were recently observed in this laboratory (Lambert et al. 1981) by exciting a molecular anthracene beam, under supersonic conditions (rotational temperature <1 K, vibrational temperature ~ 20 K), by picosecond pulses. Extension of the method to other large molecules, like stilbene and pyrazine, has also been successful (Syage et al. 1982, Felker et al. 1982). This development is interesting in that we are now dealing with collision-free molecules in the beam, and these molecules have their own heat bath, which induces the dephasing.

Acknowledgments

This material is based upon work supported by the National Science Foundation under Grants DMR8105034 and CHE8112833. One of us (AHZ) would like to thank Mrs. Tina Wood for typing this manuscript. Without her sincere and kind considerations, this chapter would never have been finished on time. Also, we like to thank colleagues who communicated their unpublished work and allowed us to use figures, tables, etc. Finally, we like to thank Dr. R. Brewer and Prof. J. Skinner for their useful comments and interest in this chapter.

References

Aartsma, T.J. and D.A. Wiersma, 1976a, Phys. Rev. Lett. **36**, 1360.
Aartsma, T.J. and D.A. Wiersma, 1976b, Chem. Phys. Lett. **42**, 520.
Aartsma, T.J. and D.A. Wiersma, 1978, Chem. Phys. Lett. **54**, 415.
Aartsma, T.J., J. Morsink and D. Wiersma, 1977, Chem. Phys. Lett. **47**, 425.
Abragam, A., 1961, The Principle of Nuclear Magnetism (Oxford, London), for example.
Abram, I.I., 1977, Chem. Phys. **25**, 87.
Abram, I.I. and R.M. Hochstrasser, 1980, J. Chem. Phys. **72**, 3617.
Abram, I.I., R. Hochstrasser, J.E. Kohl, M.C. Semack and D. White, 1979, J. Chem. Phys. **71**, 153.
Alimpiev, S.S. and N.V. Karlov, 1973, Sov. Phys. JETP **36**, 255.
Alimpiev, S.S. and N.V. Karlov, 1974, Sov. Phys. JETP **39**, 260.
Amano, T. and R.H. Schwendeman, 1976, J. Chem. Phys. **65**, 5133.
Amano, T. and T. Shimizu, 1973, J. Phys. Soc. Jpn **35**, 237.
Anderson, P.W., 1949, Phys. Rev. **76**, 647.
Anderson, P.W., 1954, J. Phys. Soc. Jpn **9**, 316.
Baranger, M., 1958, Phys. Rev. **112**, 855.
Beach, R., S. Hartmann and D. Friedberg, 1982, Phys. Rev. **A25**, 2658.
Ben-Reuven, A., 1975, Adv. Chem. Phys. **33**, 235.
Ben-Reuven, A., 1966a, Phys. Rev. **141**, 34.
Ben-Reuven, A., 1966b, Phys. Rev. **145**, 7.

Berman, P.R., 1975, Comm. Atom. Mol. Phys. **5**, 19.

Berman, P.R. and W.E. Lamb, 1970, Phys. Rev. **A2**, 2435.

Berman, P.R., J.M. Levy and R.G. Brewer, 1975, Phys. Rev. **A11**, 1668.

Berne, B.J. and G.D. Harp, 1970, Adv. Chem. Phys. **27**, 63.

Bestmann, G., H. Drietzler and H. Mäder, 1979, Z. Naturforsch. **34a**, 1330.

Bestmann, G., H. Drietzler, H. Mäder and U. Andresen, 1980, Z. Naturforsch. **35a**, 392.

Bloch, F., 1946, Phys. Rev. **70**, 460.

Bloembergen, N., 1965, Non-Linear Optics (Addison-Wesley, Reading, MA).

Boley, C.D., 1975, Phys. Rev. **A11**, 328.

Bratos, S. and E. Marechal, 1971, Phys. Rev. **A4**, 1078.

Brewer, R.G. 1977, Nonlinear Spectroscopy, in: Proc. Intl. School of Physics Enrico Fermi, Course LXIV, ed. N. Bloembergen (North-Holland, Amsterdam) p. 87.

Brewer, R.G. and A.Z. Genack, 1976, Phys. Rev. Lett. **36**, 959.

Brewer, R.G. and E.L. Hahn, 1973, Phys. Rev. **A8**, 464.

Brewer, R.G. and E.L. Hahn, 1975, Phys. Rev. **A11**, 1641.

Brewer R.G. and S.S. Kano, 1979, in: Nonlinear Behavior of Molecules, Atoms and Ions in Electric, Magnetic or Electromagnetic Fields (Elsevier, Amsterdam) p. 45.

Brewer, R.G. and R.L. Shoemaker, 1971, Phys. Rev. Lett. **27**, 631.

Brewer, R.G. and R.L. Shoemaker, 1972, Phys. Rev. **A6**, 2001.

Britt, C.O. and J.E. Boggs, 1966, J. Chem. Phys. **45**, 3877.

Brittain, A.H., P.J. Manor and R.H. Schwendeman, 1973, J. Chem. Phys. **58**, 5735.

Brown, S.R., 1974, J. Chem. Phys. **60**, 1722.

Brueck, S.R. and R.M. Osgood, Jr., 1976, Chem. Phys. Lett. **39**, 568.

Burland, D.M., U. Konzelmann and R.M. Macfarlane, 1977, J. Chem. Phys. **67**, 1926.

Burland, D.M. and A.H. Zewail, 1979, Adv. Chem. Phys. **40**, 369.

Burland, D.M., F. Carmona and E. Cuellar, 1979, Chem. Phys. Lett. **64**, 5.

Burns, M.J., 1979, Ph.D. Thesis (Harvard University).

Burns, M.J. and S.L. Coy, 1983, to be published.

Calaway, W.F. and G.E. Ewing, 1975, Chem. Phys. Lett. **30**, 485.

Caltani, M., 1971, J. Chem. Phys. **54**, 2291.

Charron, M., T.G. Anderson and J.I. Steinfeld, 1980, J. Chem. Phys. **73**, 1494.

Chen, Y.C., K. Chiang and S.R. Hartmann, 1979, Opt. Commun. **29**, 181.

Chen, Y.C., K. Chiang and S.R. Hartmann, 1980, Phys. Rev. **B21**, 40.

Cheo, P.K. and C.H. Wang, 1970, Phys. Rev. **A1**, 225.

Chiu, L.C. and W.K. Liu, unpublished.

Chutjian, A., J.K. Link and L. Brewer, 1967, J. Chem. Phys. **46**, 2666.

Clements, W.R.L. and B.P. Stoicheff, 1968, Appl. Phys. Lett. **12**, 246.

Cooper, D.E., R.W. Olson, R. Wieting and M. Fayer, 1979, Chem. Phys. Lett. **67**, 41.

Cooper, D.E., R.W. Olson and M.D. Fayer, 1980, J. Chem. Phys. **72**, 2332.

Coy, S.L., 1975a, Ph.D. Thesis (Harvard University).

Coy, S.L., 1975b, J. Chem. Phys. **63**, 5145.

Coy, S.L., 1980, J. Chem. Phys. **73**, 5531.

Creswell, R.A., S.R. Brown and R.H. Schwendeman, 1976, J. Chem. Phys. **64**, 1820.

DeBree, P. and D.A. Wiersma, 1978, Opt. Commun. **26**, 248.

DeBree, P. and D.A. Wiersma, 1979, J. Chem. Phys. **70**, 790.

DeCola, P.L., R.M. Hochstrasser and H.P. Trommsdorff, 1980, Chem. Phys. Lett. **72**, 1.

DeCola, P.L., J. Andrews, R.M. Hochstrasser and H.P. Trommsdorff, 1980, J. Chem. Phys. **73**, 4695.

Demtroder, W., 1980, Laser Spectroscopy: Basic Concepts and Instrumentation (Springer, Berlin), for example.

DeVoe, R.G. and R.G. Brewer, 1978, Phys. Rev. Lett. **40**, 862.

DeVoe, R.G., A. Szabo, S.C. Rand and R.G. Brewer, 1979, Phys. Rev. Lett. **42**, 1560.
DeVoe, R.G., A. Wokaun, S.C. Rand and R.G. Brewer, 1981, Phys. Rev. **B23**, 3125.
de Vries, H. and D.A. Wiersma, 1978, J. Chem. Phys. **69**, 897.
de Vries, H. and D.A. Wiersma, 1979, J. Chem. Phys. **70**, 5807.
de Vries, H., P. DeBree and D.A. Wiersma, 1977, Chem. Phys. Lett. **52**, 399.
Di Bartolo, B., 1968, Optical Interactions in Solids (Wiley, New York) Ch. 15, p. 341.
Dicke, R.H., 1954, Phys. Rev. **93**, 99.
Dicke, R.H. and R.H. Romer, 1955, Rev. Sci. Instrum. **26**, 915.
Diestler, D.J., 1976a, Chem. Phys. Lett. **39**, 39.
Diestler, D.J., 1976b, Mol. Phys. **32**, 1091.
Diestler, D.J. in: Advances in Laser Chemistry, 1978, ed. A.H. Zewail (Springer, Berlin) p. 258.
Diestler, D.J. and A.H. Zewail, 1979a, J. Chem. Phys. **71**, 3103.
Diestler, D.J. and A.H. Zewail, 1979b, J. Chem. Phys. **71**, 3113.
Dugay, M.A. and J.W. Hansen, 1968, Appl. Phys. Lett. **13**, 178.
Ekkers, J. and W.H. Flygare, 1976, Rev. Sci. Instrum. **47**, 448.
Erickson, L.E., 1977a, Phys. Rev. **B16**, 4731.
Erickson, L.E., 1977b, Opt. Commun. **21**, 147.
Fano, U., 1963, Phys. Rev. **131**, 259.
Fano, U., 1964, in: Lectures on the Many-Body Problem, vol. 2, ed. E.R. Cianello (Academic Press, New York) p. 217.
Felker, P.M., Wm.R. Lambert and A.H. Zewail, 1982, Chem. Phys. Lett. **89**, 309.
Feuillade, C., J.G. Baker and C. Bottcher, 1976, Chem. Phys. Lett. **40**, 121.
Feynman, R.P., F.L. Vernon, Jr. and R.W. Hellwarth, 1957, Appl. Phys. **28**, 49.
Fischer, S.F. and A. Laubereau, 1975, Chem. Phys. Lett. **35**, 6.
Fischer, S.F. and A. Laubereau, 1978, Chem. Phys. Lett. **55**, 189.
Fitz, D.E. and R.A. Marcus, 1973, J. Chem. Phys. **59**, 4380.
Forster, D., 1976, Hydrodynamic Fluctuations, Broken Symmetry, and Correlation Functions (Benjamin, New York).
Freed, C., 1968, IEEE J. Quantum Electron QE-**4**, 404.
Frenkel, L., H. Marantz and T. Sullivan, 1971, Phys. Rev. **A3**, 1640.
Friedman, J.M., 1979, J. Chem. Phys. **71**, 3147.
Frost, B.S., 1976, J. Phys. **B9**, 1001.
Frost, B.S. and W.R. MacGillivray, 1977, J. Phys. **B10**, 3649.
Genack, A.Z. and R.G. Brewer, 1978, Phys. Rev. **A17**, 1463.
Genack, A.Z., D.A. Weitz, R.M. Macfarlane, R.M. Shelby and A. Schenzle, 1980, Phys. Rev. Lett. **45**, 438.
Gilbert, J., 1970, Rev. Sci. Instrum. **44**, 1050.
Glasbeek, M., R. Hond and A.H. Zewail, 1980, Phys. Rev. Lett. **45**, 744.
Glorieux, P. and B. Macke, 1974, Chem. Phys. **4**, 120.
Glorieux, P., J. Legrand and B. Macke, 1975, J. Physique **36**, 643.
Glorieux, P., J. Legrand and B. Macke, 1976, Chem. Phys. Lett. **40**, 287.
Gordon, R.G., 1966, J. Chem. Phys. **44**, 3083.
Gordon, R.G., 1968, Adv. Magn. Reson. **3**, 7.
Gordon, J.P., C.W. Wang, C.K.N. Patel, R.E. Slusher and W.J. Tomlinson, 1969, Phys. Rev. **179**, 294.
Green, S., 1978, J. Chem. Phys. **69**, 4076.
Grossman, T.S.B., A. Schenzle and R.G. Brewer, 1977, Phys. Rev. Lett. **38**, 275.
Gutman, W.M. and C.V. Heer, 1975, Phys. Lett. **A51**, 437.
Gutman, W.M. and C.V. Heer, 1977, Phys. Rev. **A16**, 659.
Hahn, E.L., 1950, Phys. Rev. **80**, 580.

Hall, J., 1973, Atomic Physics, vol. 3, eds. S.J. Smith, G.K. Walters and L. Volsky (Plenum Press, New York).

Hallidy, L.A. and M.R. Topp, 1977, Chem. Phys. Lett. **46**, 8.

Harrington, H.W., 1968, in: Symp. Mol. Structure, Spectrosc., Ohio State Univ, Columbus, OH, paper S4.

Harris, C.B. and W. Breiland, 1978, in: Laser and Coherence Spectroscopy, ed. J. Steinfeld (Plenum Press, New York) p. 373.

Harris, C.B., R. Shelby and P. Cornelius, 1977, Phys. Rev. Lett. **38**, 1415.

Harris, C.B., P. Cornelius and R. Shelby, 1978, in: Advances in Laser Chemistry, ed. A.H. Zewail (Springer, Berlin), p. 223.

Hatanaka, H. and T. Hashi, 1975, J. Phys. Soc. Jpn **39**, 1139.

Hatanaka, H., T. Lerao and T. Hashi, 1975, J. Phys. Soc. Jpn **39**, 835.

Hayes, J.M. and G.J. Small, 1979, J. Lumin. **18/19**, 219.

Hayes, J.M., R.P. Stout and G. Small, 1980, J. Chem. Phys. **73**, 4129.

Heer, C.V. and R.J. Nordstrom, 1975, Phys. Rev. **A11**, 536.

Heritage, J.P., 1979, Appl. Phys. Lett. **34**, 470.

Hesp, H. and D.A. Wiersma, 1980, Chem. Phys. Lett. **75**, 423.

Hesp, H.M.M., J. Langelaar, D. Bebelaar and J.D.W. van Voorst, 1977, Phys. Rev. Lett. **39**, 1376.

Hesselink, W.H. and D.A. Wiersma, 1977, Chem. Phys. Lett. **50**, 51.

Hesselink, W.H. and D.A. Wiersma, 1978, Chem. Phys. Lett. **56**, 227.

Hesselink, W.H. and D.A. Wiersma, 1979, Chem. Phys. Lett. **65**, 300.

Hesselink, W.H. and D.A. Wiersma, 1980, J. Chem. Phys. **73**, 648.

Hill, R.M., D.E. Kaplan, G.F. Herrmann and S.K. Ichiki, 1967, Phys. Rev. Lett. **18**, 105.

Ho, F., W.-S. Tsay, J. Trout and R.M. Hochstrasser, 1981, Chem. Phys. Lett., to be published.

Hochstrasser, R.M. and C. Nyi, 1979, J. Chem. Phys. **70**, 1112.

Hochstrasser, R.M. and J. Whiteman, 1972, J. Chem. Phys. **56**, 5945.

Hochstrasser, R.M. and A.H. Zewail, 1974, Chem. Phys. **4**, 142.

Hochstrasser, R.M., G.R. Meredith and H.P. Trommsdorff, 1980, J. Chem. Phys. **73**, 1009.

Hocker, G.B. and C.L. Tang, 1968, Phys. Rev. Lett. **21**, 591.

Hocker, G. and C. Tang, 1969, Phys. Rev. **184**, 356.

Hoke, W.E., J. Ekkers and W.H. Flygare, 1975, J. Chem. Phys. **63**, 4075.

Hoke, W.E., D.R. Bauer, J. Ekkers and W.H. Flygare, 1976, J. Chem. Phys. **64**, 5276.

Hoke, W.E., D.R. Bauer and W.H. Flygare, 1977, J. Chem. Phys. **67**, 3454.

Hom, R.H., 1978, Ph.D. Thesis (Harvard University).

Hom, R.H. and S. Coy, 1981, J. Chem. Phys. **74**, 5453.

Hong, H.K. and R. Kopelman, 1970, J. Chem. Phys. **55**, 724.

Jetter, H., E.F. Pearson, C.L. Norris, J.C. McGurk and W.H. Flygare, 1973, J. Chem. Phys. **59**, 1796.

Jonas, J. and H.S. Gutowsky, 1980, Ann. Rev. Phys. Chem. **31**, 1.

Jones, K.E. and A.H. Zewail, 1978, in: Advances in Laser Chemistry vol. 3, ed. A. Zewail (Springer, Berlin) p. 196.

Jones, K.E., A. Nichols and A.H. Zewail, 1978, J. Chem. Phys. **69**, 3350.

Jortner, J. and J. Kommandeur, 1978, Chem. Phys. **28**, 273.

Kakimoto, M. and T. Fujiyama, 1974, Bull. Chem. Soc. Japan **47**, 1883.

Kasuga, T., T. Amano and T. Shimizu, 1976, Chem. Phys. Lett. **42**, 278.

Kouri, D.J., 1979, in: Atom–Molecule Collisions Theory: A Guide for the Experimentalist, ed. R.B. Bernstein (Plenum Press, New York) p. 301.

Krishnaji, S.L. Srivastava and P.D. Pandey, 1972, Chem. Phys. Lett. **13**, 372.

Krivoglaz, M.A., 1964, Sov. Phys. Solid State **6**, 340.

Krivoglaz, M.A., 1965, Sov. Phys. JETP **21**, 204.

Kurnit, N., I. Abella and S. Hartmann, 1964, Phys. Rev. Lett. **13**, 567.

Kurzel, R.B., J.I. Steinfeld, D.A. Hatzenbuhler and G.E. Leroi, 1971, J. Chem. Phys. **55**, 4822.

Lambert, Wm. and A.H. Zewail, 1980, Chem. Phys. Lett. **69**, 270.

Lambert, Wm.R., P.M. Felker and A.H. Zewail, 1981, J. Chem. Phys. **75**, 5958.

Laubereau, A., 1974, Chem. Phys. Lett. **27**, 600.

Laubereau, A. and W. Kaiser, 1975, Ann. Rev. Phys. Chem. **26**, 83.

Laubereau, A. and W. Kaiser, 1977, in: Chemical and Biochemical Applications of Lasers, vol. II, ed. C.M. Moore (Academic Press, New York) p. 87.

Laubereau, A. and W. Kaiser, 1978, Rev. Mod. Phys. **50**, 607.

Laubereau, A., D. Von der Linde and W. Kaiser, 1972, Phys. Rev. Lett. **28**, 1162.

Laubereau, A., L. Greiter and W. Kaiser, 1974, Appl. Phys. Lett. **25**, 87.

Laubereau, A., A. Seilmeier and W. Kaiser, 1975a, Chem. Phys. Lett. **36**, 232.

Laubereau, A., G. Wochner and W. Kaiser, 1975b, Opt. Commun. **14**, 75.

Laubereau, A., S.F. Fischer, K. Spanner and W. Kaiser, 1978a, Chem. Phys. **31**, 335.

Laubereau, A., G. Wochner and W. Kaiser, 1978b, Chem. Phys. **28**, 363.

Lax, J., 1952, J. Chem. Phys. **20**, 1752.

Lee, C.H. and D. Ricard, 1978, Appl. Phys. Lett. **32**, 168.

Legan, R.L., J.A. Roberts, E.A. Rinehart and C.C. Lin, 1965, J. Chem. Phys. **43**, 4337.

Lehman, J.C., 1977, Atomic Physics **5**, 167.

Lemaistre, J.P. and A.H. Zewail, 1979a, Chem. Phys. Lett. **68**, 296.

Lemaistre, J.P. and A.H. Zewail, 1979b, Chem. Phys. Lett. **68**, 302.

Letakov, V.S., 1976, in: Topics in Applied Physics, vol. 13, ed. K. Shimoda (Springer, Berlin).

Levenson, M.D., 1979, Chem. Phys. Lett. **64**, 495.

Liao, P.F., J. Bjorkholm and J.P. Gordon, 1977, Phys. Rev. Lett. **39**, 15.

Liu, W.-K. and R.A. Marcus, 1975a, J. Chem. Phys. **63**, 272.

Liu, W.-K. and R.A. Marcus, 1975b, J. Chem. Phys. **63**, 290.

Liu, W.-K. and F.R. McCourt, 1979, J. Chem. Phys. **71**, 3750.

Loy, M.M.T., 1976, Phys. Rev. Lett. **36**, 1454.

Loy, M.M.T., 1977, Phys. Rev. Lett. **39**, 187.

Loy, M.M.T., 1978, Phys. Rev. Lett. **41**, 473.

Luc, P., 1980, J. Mol. Spectrosc. **80**, 41.

Luijendijk, S.C.M., 1977a, J. Phys. **B10**, 1735.

Luijendijk, S.C.M., 1977b, J. Phys. **B10**, 1741.

Lynden-Bell, R., 1977, Mol. Phys. **33**, 907.

Macfarlane, R.,R. Shelby and R. Shoemaker, 1979, Phys. Rev. Lett. **43**, 1726.

Macfarlane, R., C.S. Yannoni and R. Shelby, 1980, Opt. Commun. **32**, 101.

Macke, B. and P. Glorieux, 1972, Chem. Phys. Lett. **14**, 85.

Madden, P.A. and R. Lynden-Bell, 1976, Chem. Phys. Lett. **38**, 163.

Mäder, H., 1979, Z. Naturforsch. **34a**, 1170.

Mäder, H., J. Ekkers, W. Hoke and W.H. Flygare, 1975, J. Chem. Phys. **62**, 4380.

Mäder, H., H. Bomsdorf and U. Andresen, 1979a, Z. Naturforsch. **34a**, 850.

Mäder, H., W. Lalowski and R. Schwarz, 1979b, Z. Naturforsch. **34a**, 1181.

Mahr, A. and M.D. Hirsch, 1975, Opt. Commun. **13**, 96.

Marcus, R.A., 1973, J. Chem. Phys. **59**, 5135, and references therein.

Mattuck, R.D., 1976, A Guide to Feynman Diagrams in the Many-Body Problem, 2nd Ed. (McGraw-Hill, New York).

McCumber, D.E., 1964, J. Math. Phys. **5**, 21.

McCumber, D.E. and M.B. Sturge, 1963, J. Appl. Phys. **34**, 1682.

McGurk, J.C., T.G. Schmalz and W.H. Flygare, 1974a, Adv. Chem. Phys. **25**, 1.

McGurk, J.C., R.T. Hofmann and W.H. Flygare, 1974b, J. Chem. Phys. **60**, 2922.

McGurk, J.C., C.L. Norris, T.G. Schmalz and W.H. Flygare, 1974c, in: Laser Spectroscopy, eds. R.G. Brewer and A. Mooradian (Plenum Press, New York).

McGurk, J.C., H. Mäder, R.T. Hofmann, T.G. Schmalz and W.H. Flygare, 1974d, J. Chem. Phys. **61**, 3759.

Meckley, J.R. and C.V. Heer, 1973, Phys. Lett. **46A**, 41.

Mori, H., 1965, Prog. Theor. Phys. **33**, 423.

Morsink, J.B.W., T.J. Aartsma and D.A. Wiersma, 1977, Chem. Phys. Lett. **49**, 34.

Morsink, J.B.W., W.H. Hesselink and D.A. Wiersma, 1979, Chem. Phys. Lett. **64**, 1.

Mossberg, T., A. Flusberg, R. Kachru and S.R. Hartmann, 1977, Phys. Rev. Lett. **39**, 1523.

Mossberg, T., A. Flusberg, R. Kachru and S.R. Hartmann, 1979a, Phys. Rev. Lett. **42**, 1665.

Mossberg, T., R. Kachru, E. Whittaker and S.R. Hartmann, 1979b, Phys. Rev. Lett. **43**, 851.

Mukamel, S., 1978, Chem. Phys. **31**, 327.

Mukamel, S., 1979a, Chem. Phys. **37**, 33.

Mukamel, S. 1979b, J. Chem. Phys. **71**, 2884.

Mukamel, S., 1980, J. Chem. Phys. **73**, 5322.

Mukamel, S., A. Ben-Reuven and J. Jortner, 1976, J. Chem. Phys. **64**, 3971.

Murphy, J.S. and J.E. Boggs, 1967a, J. Chem. Phys. **47**, 691.

Murphy, J.S. and J.E. Boggs, 1967b, J. Chem. Phys. **47**, 4152.

Murphy, J.S. and J.E. Boggs, 1968, J. Chem. Phys. **49**, 3333.

Nelson, K., D. Dloff and M. Fayer, 1979, Chem. Phys. Lett. **64**, 88.

Nitzan, A., M. Shugard and J. Tully, 1978, J. Chem. Phys. **69**, 2525.

Nordstrom, R.J., W.M. Gutman and C.V. Heer, 1974, Phys. Lett. **50A**, 25.

Oka, T., 1973, Adv. At. Mol. Phys. **9**, 127.

Olson, D.S., C.O. Britt, V. Prakash and J.E. Boggs, 1973, J. Phys. **B6**, 206.

Olson, R., H.W.H. Lee, F. Patterson and M. Fayer, 1982, J. Chem. Phys. **76**, 31.

Omont, A., 1977, Prog. Quantum Electron. **5**, 69.

Orlowski, T.E. and A.H. Zewail, J. Chem. Phys. **70**, 1390.

Orlowski, T.E., K.E. Jones and A.H. Zewail, 1978, Chem. Phys. Lett. **54**, 197.

Oxtoby, D.W., 1979a, J. Chem. Phys. **70**, 2605.

Oxtoby, D.W., 1979b, Adv. Chem. Phys. **40**, 1.

Oxtoby, D.W., 1981a, J. Chem. Phys. **74**, 1503.

Oxtoby, D.W. 1981b, J. Chem. Phys. **74**, 5371.

Oxtoby, D.W. and S.A. Rice, 1976, Chem. Phys. Lett. **43**, 1.

Oxtoby, D.W., D. Levesque and J.J. Weis, 1978, J. Chem. Phys. **68**, 5528.

Oxtoby, D.W., D. Levesque and J.J. Weis, 1980, J. Chem. Phys. **72**, 2744.

Patel, C.K.N. and R.E. Slusher, 1968, Phys. Rev. Lett. **20**, 1087.

Perry, J.W. and A.H. Zewail, 1979a, J. Chem. Phys. **70**, 582.

Perry, J.W. and A.H. Zewail, 1979b, Chem. Phys. Lett. **65**, 31.

Perry, J.W. and A.H. Zewail, 1981, J. Phys. Chem. **85**, 933.

Plant, T.K. and R.L. Abrams, 1976, J. Appl. Phys. **47**, 4006.

Prasad, P.N. and L. Hess, 1980, J. Chem. Phys. **72**, 573.

Prasad, P.N. and R. Von Smith, 1979, J. Chem. Phys. **71**, 4646.

Rabitz, H., 1974, Ann. Rev. Phys. Chem. **25**, 155.

Rand, S.C., A. Wokaun, R. DeVoe and R.G. Brewer, 1979, Phys. Rev. Lett. **43**, 1868.

Rebane, K.K., 1970, Impurity Spectra of Solids (Plenum Press, New York).

Resibois, P., 1969, in: Elementary Excitations in Solids, eds. A.A. Maradudin and G.F. Nardelli (Plenum Press, New York) p. 340.

Rogers, D.V. and J.A. Roberts, 1973, J. Mol. Spectrosc. **46**, 200.

Rohart, F., P. Glorieux and B. Macke, 1977, J. Phys. **B10**, 3835.

Rothschild, W.G., 1976, J. Chem. Phys. **65**, 455.

Sapozhnikov, M.N., 1978, J. Chem. Phys. **68**, 2352.

Sargent, M., III, M.O. Scully and W.E. Lamb, Jr., 1974, Laser Physics (Addison-Wesley, Reading, MA).

Schenzle, A. and R.G. Brewer, 1976, Phys. Rev. **A14**, 1756.

Schmalz, T.G. and W.H. Flygare, 1978, in: Laser and Coherence Spectroscopy, ed. J.I. Steinfeld (Plenum Press, New York) p. 125.

Schmidt, J., P.R. Berman and R.G. Brewer, 1973, Phys. Rev. Lett. **31**, 1103.

Schrepp, W., G. Bestmann and H. Drietzler, 1979, Z. Naturforsch. **34a**, 1467.

Schroeder, J., V. Schiemann and J. Jonas, 1977, Mol. Phys. **34**, 1501.

Schwendeman, R.H., 1978, Ann. Rev. Phys. Chem. **29**, 537.

Schwendeman, R.H. and T. Amano, 1979, J. Chem. Phys. **70**, 962.

Schwendeman, R.H. and A.H. Brittain, 1970, Symp. Mol. Structure Spectrosc., Ohio State Univ., Columbus, OH, paper Q4.

Scotto, M., 1968, J. Chem. Phys. **49**, 5362.

Shafer, R. and R.G. Gordon, 1973, J. Chem. Phys. **58**, 5422.

Shelby, R.M. and R. Macfarlane, 1978, Opt. Commun. **27**, 399.

Shelby, R.M., C.S. Yannoni and R. Macfarlane, 1978, Phys. Rev. Lett. **41**, 1739.

Shelby, R.M., R. MacFarlane and C.S. Yannoni, 1980, Phys. Rev. **B21**, 5004.

Shoemaker, R.L., 1978, in: Laser and Coherence Spectroscopy, ed. J.I. Steinfeld (Plenum Press, New York) p. 197.

Shoemaker, R.L., 1979, Ann. Rev. Phys. Chem. **30**, 239.

Shoemaker, R.L. and R.G. Brewer, 1972, Phys. Rev. Lett. **28**, 1430.

Shoemaker, R.L. and F.A. Hopf, 1974, Phys. Rev. Lett. **33**, 1527.

Shoemaker, R.L. and E.W. Van Stryland, 1976, J. Chem. Phys. **64**, 1733.

Shoemaker, R.L., R.E. Scotti and B. Comaskey, 1978, J. Opt. Soc. Am. **68**, 1388.

Skinner, J., H.C. Andersen and M.D. Fayer, 1981, J. Chem. Phys. **75**, 3195.

Skinner, J., 1983, to be published.

Slichter, C.P., 1978, Principles of Magnetic Resonance, 2nd Ed. (Springer, Berlin).

Small, G.J., 1983, chapter 9 in this volume.

Smith, D., 1981, Ph.D. Thesis, Calif. Inst. Tech.

Smith, D. and A.H. Zewail, 1979, J. Chem. Phys. **71**, 3533.

Smith, D. and A.H. Zewail, 1979, J. Chem. Phys. **71**, 540.

Srivastava, G.P. and A. Kumar, 1974, J. Phys. **B7**, 2578.

Steel, D.G. and J.F. Lam, 1979, Phys. Rev. Lett. **43**, 1588.

Steinfeld, J.I., 1978, Molecules and Radiation: An Introduction to Modern Molecular Spectroscopy (The MIT Press, Cambridge, MA).

Steinfeld, J.I. and P. Houston, 1978, in: Laser and Coherence Spectroscopy, ed. J.I. Steinfeld (Plenum Press, New York) p. 1.

Syage, J.A., Wm.R. Lambert, P.M. Felker, A.H. Zewail and R.M. Hochstrasser, 1982, Chem. Phys. Lett. **88**, 266.

Szabo, A. and N. Takeuchi, 1975, Opt. Commun. **15**, 250.

Tanaka, K. and E. Hirota, 1976, J. Mol. Spectrosc. **59**, 286.

Tellinghuisen, J., 1978, J. Quant. Spectrosc. Radiat. Transfer **19**, 149.

Torrey, H.C., 1949, Phys. Rev. **76**, 1059.

Townes, C.H. and A.C. Cheung, 1969, Astron. Astrophys. J. (Lett.) **157**, L103.

Van Hove, L., 1955, Physica **21**, 517.

Van Strien, A.J., J. van Kooten and J. Schmidt, 1980, Chem. Phys. Lett. **76**, 7.

Van Stryland, E.W., 1976, Ph.D. Thesis (University of Arizona).

Vega, S., 1978, J. Chem. Phys. **68**, 5518.

Vega, S. and A. Pines, 1977, J. Chem. Phys. **66**, 5624.

Von der Linde, D., O. Bernecke and W. Kaiser, 1970, Opt. Commun. **2**, 149.

Von der Linde, D., A. Laubereau and W. Kaiser, 1971, Phys. Rev. Lett. **26**, 954.

Wang, C.H., 1977, Mol. Phys. **33**, 207.

Wang, J.H.-S., J.M. Levy, S.G. Kuholich and J.I. Steinfeld, 1973, Chem. Phys. **1**, 141.

Warren, W. and A.H. Zewail, 1981a, J. Phys. Chem. **85**, 2309.

Warren, W. and A.H. Zewail, 1981b, J. Chem. Phys. **75**, 5956.

Warren, W. and A.H. Zewail, 1983, J. Chem. Phys., to be published.

Wei, J. and J. Tellinghuisen, 1974, J. Mol. Spectrosc. **50**, 317.

Wertheimer, R. and R. Silbey, 1981, J. Chem. Phys. **74**, 686.

Wilson, R.S., W.T. King and K.S. Kim, 1968, Phys. Rev. **175**, 1164.

Wolfrum, H., K. Renk and H. Sixl, 1979, Chem. Phys. Lett. **68**, 90.

Wong, N.C., S.S. Kano and R.G. Brewer, 1980, Phys. Rev. **A21**, 260.

Yarwood, J., R. Andt and G. Döge, 1977, Chem. Phys. **25**, 387.

Yarwood, J., R. Andt and G. Döge, 1979, Chem. Phys. **42**, 331.

Yip, S., 1979, Ann. Rev. Phys. Chem. **30**, 547.

Zewail, A.H., 1977, Opt. Eng. **16**, 206.

Zewail, A.H., 1978, J. Opt. Soc. Am. **68**, 696.

Zewail, A.H., 1979, J. Chem. Phys. **70**, 5759.

Zewail, A.H., 1980a, Physics Today **33**, 27.

Zewail, A.H., 1980b, Acc. Chem. Res. **13**, 360 for a review of work started in 1976.

Zewail, A.H. and D.J. Diestler, 1979, Chem. Phys. Lett. **65**, 37.

Zewail, A.H. and Wm. Lambert, 1979, J. Lumin. **18/19**, 205.

Zewail, A.H. and T.E. Orlowski, 1977, Chem. Phys. Lett. **45**, 399.

Zewail, A.H., T.E. Orlowski and D.R. Dawson, 1976, Chem. Phys. Lett. **44**, 379.

Zewail, A.H., D.E. Godar, K.E. Jones, T.E. Orlowski, R.R. Shah and A. Nichols, 1977a, in: Advances in Laser Spectroscopy I, vol. 113, Proc. SPIE Conf. San Diego, CA, 1977, ed. A.H. Zewail, (SPIE Publ. Co., Bellingham, WA) p. 42.

Zewail, A.H., T.E. Orlowski, K.E. Jones and D.E. Godar, 1977b, Chem. Phys. Lett. **48**, 256.

Zewail, A.H., T.E. Orlowski, R.R. Shah and K.E. Jones, 1977c, Chem. Phys. Lett. **49**, 520.

Zwanzig, R., 1960, J. Chem. Phys. **33**, 1338.

Zwanzig, R., 1961a, in: Lectures in Theoretical Physics, vol. 4, ed. W.E. Brittain (Wiley, New York) p. 106.

Zwanzig, R., 1961b, Phys. Rev. **124**, 983.

Zwanzig, R., 1964, Physica **30**, 1109.

Zwanzig, R., 1965, Ann. Rev. Phys. Chem. **16**, 67.

CHAPTER 8

Theory of Light Absorption and Emission by Organic Impurity Centers

I.S. OSAD'KO

V.I. Lenin State Pedagogical Institute
119435 Moscow
U.S.S.R.

Translated from the Russian by Nicholas Weinstein.

Spectroscopy and Excitation Dynamics
of Condensed Molecular Systems
Edited by
V.M. Agranovich and R.M. Hochstrasser

Contents

1. Introduction

Spectroscopic research on polyatomic molecules in gaseous, liquid and solid media has developed rapidly in the last decade. An organic molecule is an impurity center in a solid solvent and, consequently, the theory of impurity centers is applicable to such molecules. The development of the modern theory of impurity centers is associated, to a considerable extent, with attempts to explain the optical properties of color centers in inorganic alkali-halide crystals (Huang and Rhys 1950, Pekar 1950, 1953). For many years the theory of impurity centers was applied mainly to inorganic crystals. This inevitably affected the theory. Hence, the application of many results of the theory to organic impurity centers leads to appreciable difficulties. Certain aspects of the theory, the Jahn–Teller effect, for instance, were found to be inessential because, as a rule, organic impurity centers have a very low symmetry. It is impossible, as yet, to apply certain other aspects of the theory, such as the calculation of the adiabatic potential, to organic matter because of the present lack of knowledge of electron wave functions for polyatomic organic molecules. The infeasibility of solving the direct problem of spectroscopy as applied to organic impurity centers has stimulated the development of methods to solve the inverse problem, namely, the reconstruction of electron–vibrational interaction on the basis of spectroscopic data. Solving the inverse problem can prove to be effective only after clarification of the quantitative relation between optical bands on the one hand, and the various parameters of the total electron–vibrational interaction on the other. This, exactly, is the problem in whose solution appreciable advances have been made in recent years.

The Pekar–Huang theory satisfactorily explained the basic features of wide structureless bands of color centers by drastic changes in the polarizing effect of the color center on neighboring ions when the electron in the center is excited. This change in the polarizing effect is proportional to the displacement a of the equilibrium positions of the ions surrounding the center. It was quite clear even at that time that the total electron–vibrational interaction is not restricted to the displacement a of the equilibrium positions (Kubo and Toyozawa 1955). Nevertheless, the role of the remaining part of the electron–vibrational interaction was subject to much less intensive research because it was less essential

439

for wide bands and because of the formidable mathematical difficulties encountered.

Intensive research conducted in the last decade on high-resolution optical spectra of various mixed crystals and, especially, solid solutions of organic matter revealed, however, that the part of the electron–vibrational interaction not accounted for by the Pekar–Huang theory plays a cardinal role in certain effects, for example, the temperature broadening and shift of optical zero-phonon lines (ZPL) (Silsbee 1962, McCumber and Sturge 1963, McCumber 1964, Krivoglaz 1964, 1965). This gave impetus to further development of impurity center theory. The results of these investigations are described in the books by Maradudin (1966), Rebane (1968), Kristofel (1974), Perlin and Tsukerblat (1974) and in a review by the present author (1979).

The quadratic Franck–Condon interaction, the Herzberg–Teller interaction and the existence of several minima in the adiabatic potential all combine to cause the violation of mirror symmetry of conjugate absorption and fluorescence bands. Substantial advances have been made in the last 5 or 7 years in devising a theory that expresses conjugate bands in terms of the same parameters of the electron–vibrational interaction. This made it possible to convert the lack of mirror symmetry into a highly important source of information on electron–vibrational interaction. An account of these advances is the primary aim of the present survey.

2. Adiabatic and diabatic approaches

The electron–vibrational properties of a system of interacting electrons and nuclei can be calculated by either the adiabatic method (Born and Oppenheimer 1927, Born and Huang 1954) or the so-called diabatic method (Longuet-Higgins 1961, O'Malley 1971, Gregory et al. 1976). For the same of convenience we shall consider them concomitantly.

The Hamiltonian operator $H(r, R)$ of a system of interacting electrons and nuclei is written in the following forms for the adiabatic and diabatic treatments, respectively:

$$H(r, R) = H_0(r, R) + \hat{T}(R), \tag{1a}$$

$$H(r, R) = H_0(r, 0) + V(r, R) + \hat{T}(R). \tag{1b}$$

Here $\hat{T}(R) = -\frac{1}{2}\Sigma_n \partial^2/\partial R_n^2$ is the kinetic energy operator of the nuclei, and the equilibrium position of the nuclei of an electronically nonexcited system is taken as 0. It follows from eqs. (1a) and (1b) that $V(r, R) = H_0(r, R) - H_0(r, 0)$. The perturbation in the adiabatic approach is $\hat{T}(R)$; in diabatic treatment it is $V(r, R)$. Accordingly, the electron wave functions (zeroth-order approximation) are determined by the equations

$$[H_0(r, R) - \epsilon^f(R)]\varphi^f(r, R) = 0,$$ (2a)

$$[H_0(r, 0) - \epsilon^f]\varphi^f(r) = 0.$$ (2b)

The total function $\Psi(r, R)$ of the system is written in the form of an expansion in the zeroth-order wave functions:

$$\Psi(r, R) = \sum_f \Phi_a^f(R)\varphi^f(r, R),$$ (3a)

$$\Psi(r, R) = \sum_f \Phi_d^f(R)\varphi^f(r).$$ (3b)

Substituting eq. (3a) into the Schrödinger equation with the total Hamiltonian $H(r, R)$, we obtain a system of equations for the adiabatic vibrational wave functions $\Phi_a^f(R)$:

$$[\hat{T}(R) + U^f(R) - E]\Phi_a^f(R) + \sum_{f'(\neq f)} \hat{\Lambda}^{ff'}(R)\Phi_a^{f'}(R) = 0,$$ (4a)

where

$$U^f(R) = \epsilon^f(R) + \hat{\Lambda}^{ff}(R)$$ (5a)

and

$$\hat{\Lambda}^{ff'}(R) = -\sum_n \int \varphi^f(r, R) \left(\frac{\partial}{\partial R_n} \varphi^{f'}(r, R) \frac{\partial}{\partial R_n} + \frac{1}{2} \frac{\partial^2}{\partial R_n^2} \varphi^{f'}(r, R) \right) dr;$$ (6a)

likewise, upon substitution of eq. (3b), we obtain for the diabatic vibrational wave functions $\Phi_d^f(R)$:

$$[\hat{T}(R) + V^{ff}(R) - E]\Phi_d^f(R) + \sum_{f'(\neq f)} V^{ff'}(R)\Phi_d^{f'}(R) = 0,$$ (4b)

where

$$V^{ff'}(R) = \int \varphi^f(r)V(r, R)\varphi^{f'}(r)\, dr + \epsilon^f\delta_{ff'}.$$ (5b)

The systems of equations (4a) and (4b) are the ones we proceed from in solving any specific electron–vibrational problem.

There is a simple relationship between the adiabatic and diabatic approaches. This is established by means of the matrix $\hat{A}(R)$, which diagonalizes the matrix $\hat{V}(R)$, i.e. it obeys the matrix equation

$$\hat{A}^{-1}(R)\hat{V}(R)\hat{A}(R) = \hat{\epsilon}(R),$$ (7)

where $\hat{\epsilon}(R)$ is the diagonal matrix with the elements $\epsilon^f(R)$. The matrix $\hat{A}(R)$, determined by eq. (7), relates the diabatic and adiabatic vibrational functions to each other:

$$\Phi_d^f(R) = \sum_{f'} A^{ff'}(R)\Phi_a^{f'}(R). \tag{8}$$

This can be checked by substituting eq. (8) into the system of equations (4b). Taking eq. (7) into account we obtain

$$[\hat{A}^{-1}(R)\hat{T}(R)\hat{A}(R) + \hat{\epsilon}(R) - E]\Phi_a(R)$$
$$= [\hat{T}(R) + \hat{\epsilon}(R) + \hat{\Lambda}(R) - E]\Phi_a(R) = 0, \tag{9}$$

where

$$\hat{\Lambda}(R) = -\sum_n \left(\hat{A}^{-1}(R)\frac{\partial}{\partial R_n}\hat{A}(R)\frac{\partial}{\partial R_n} + \tfrac{1}{2}\hat{A}^{-1}(R)\frac{\partial^2}{\partial R_n^2}\hat{A}(R)\right). \tag{10}$$

It is readily evident that $\hat{\Lambda}(R)$ is the nonadiabaticity operator with the matrix elements (6a). As a matter of fact, after substituting eq. (8) into eq. (3b) we obtain

$$\Psi(r, R) = \sum_{f,f'} \varphi^f(r)A^{ff'}(R)\Phi_a^{f'}(R) = \sum_{f'} \varphi^{f'}(r, R)\Phi_a^{f'}(R), \tag{11}$$

i.e., the adiabatic and diabatic electronic wave functions are related to each other by means of the same matrix $A(R)$:

$$\varphi^{f'}(r, R) = \sum_f \varphi^f(r)A^{ff'}(R). \tag{12}$$

Next, substituting eq. (12) into eq. (6a) we obtain

$$\hat{\Lambda}^{ff'}(R) = -\sum_n \sum_{f''} (A^{-1})^{ff''}(R) \left(\frac{\partial}{\partial R_n} A^{f''f'}(R)\frac{\partial}{\partial R_n} + \frac{1}{2}\frac{\partial^2}{\partial R_n^2} A^{f''f'}(R)\right). \tag{13}$$

This expression is eq. (10) in terms of the elements of the matrix $A(R)$. Hence, the matrix $A(R)$, determined by eq. (7), establishes relations (8) and (12) between the vibrational and electronic wave functions in the diabatic and adiabatic approximations, and determines the form of the nonadiabaticity operator.

The adiabatic and diabatic approaches both have their merits and shortcomings. An advantage of the diabatic approach is the relative simplicity of eq. (2b) as compared to (2a). Therefore, the diabatic approach is resorted to in all investigations in which the electronic wave function φ^f is to be calculated explicitly. It is also widely employed in computing the electron–vibrational properties of impurity centers in alkali-halide and other inorganic crystals. If, however, the impurity center is a polyatomic organic molecule in an organic matrix, the calculation of the function φ^f is practically impossible at present. Under these conditions the adiabatic approach has an advantage over the diabatic one because in the framework of the adiabatic approximation the mutual influence of the electronic levels splits, in a natural way, into two parts, called the Herzberg–Teller (HT) interaction and the nonadiabatic interaction. This can be readily shown by solving the system of equations (7) according to perturbation theory. In the first-order approximation with respect to $\hat{V}(r, R)$ we obtain

$$\epsilon^f(R) = V^{ff}(R),\tag{14}$$

$$A^{ff'}(R) = \delta_{ff'} + (1 - \delta_{ff'})\frac{V^{ff'}(R)}{\epsilon^{f'} - \epsilon^f}.\tag{15}$$

Substituting eq. (15) into eqs. (8) and (12) we obtain

$$\Phi_d^f(R) \approx \Phi_a^f(R) + \sum_{f'(\neq f)} \frac{V^{ff'}(R)}{\epsilon^{f'} - \epsilon^f} \Phi_a^{f'}(R),\tag{16}$$

$$\varphi^f(r, R) \approx \varphi^f(r) + \sum_{f'(\neq f)} \frac{V^{f'f}(R)}{\epsilon^f - \epsilon^{f'}} \varphi^{f'}(r).\tag{17}$$

The electronic wave function (17) can also be obtained by solving the system of equations (2a) according to perturbation theory (Herzberg and Teller 1933). Consequently, the dependence of the electronic function $\varphi^f(r, R)$ on the vibrational coordinates R accounts for the effect of the electronic levels on one another, which can be taken into consideration in calculating the matrix $\hat{A}(R)$ according to perturbation theory (HT interaction). Neglecting off-diagonal matrix elements in the systems of equations (4a) and (4b) we obtain

$$\Psi_a^f(r, R) = \varphi^f(r, R)\Phi_a^f(R),\tag{18}$$

$$\Psi_d^f(r, R) = \varphi^f(r)\Phi_d^f(R).\tag{19}$$

The adiabatic function $\Psi_a^f(r, R)$ partly accounts for the influence of the electronic levels on one another, whereas the diabatic function $\Psi_d^f(r, R)$ does not. Therefore, all the effects that can be obtained by taking into consideration the influence of the electronic levels on one another according to perturbation theory can be attributed to HT interaction and they can be dealt with by resorting to the functions (18). All other effects of the mixing of electronic states should be attributed to nonadiabatic interaction, due to the off-diagonal elements of the system of equations (4a). This natural partitioning of the total interaction is the main advantage of the adiabatic approach over the diabatic one. Hence, in the present paper we shall mainly employ the adiabatic approach.

The part of the total electron–vibrational interaction taken into account in the wave functions (18) is said to be adiabatic. Adiabatic interaction manifests itself in two ways. Firstly, in the modulation of the electronic motion by nuclear vibrations, i.e. in the dependence of the electronic

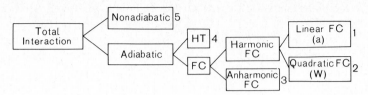

Fig. 1. Partitioning of the total electron–vibrational interaction into its component parts within the scope of the adiabatic approach.

function $\varphi^f(r, R)$ and, consequently, of the electronic matrix element

$$M_{ff'}(R) = \int \varphi^f(r, R)(dE)\varphi^{f'}(r, R)\, dr \tag{20}$$

on the vibrational coordinates. We shall conditionally call the $M(R)$ dependence the HT interaction. Secondly, the adiabatic interaction manifests itself in the dependence of the adiabatic potential $U^f(R)$ and, consequently, of the vibrational functions $\Phi_a^f(R)$, determined from the equation

$$[\hat{T}(R) + U^f(R) - E]\Phi_a^f(R) = 0, \tag{21}$$

on the subscript or superscript f of the electronic state. The change of the adiabatic potential upon electronic excitation $U^e(R) - U^0(R)$, is called the Franck–Condon (FC) interaction. In the harmonic approximation the FC interaction is a quadratic function of the coordinates R:

$$
\begin{aligned}
U^e(R) - U^0(R) &= (R - a)\frac{U^e}{2}(R - a) - R\frac{U^0}{2}R \\
&= a\frac{U^e}{2}a - (aU^e)R + R\frac{W}{2}R,
\end{aligned} \tag{22}
$$

i.e., the FC interaction is the sum of the linear and quadratic interactions, characterized by the parameters a and $W = U^e - U^0$. The diagram in fig. 1 shows the partitioning of the total electron–vibrational interaction into its component parts in the framework of the adiabatic approach (approximation). The present paper is based primarily on this approach.

3. Optical spectra in the adiabatic and harmonic approximations

The theory presented below takes into account interactions 1, 2 and 4 of fig. 1; symbolically they can be written as

$$a \neq 0; \qquad W \neq 0; \qquad M(R) \neq \text{const.} \tag{23}$$

The theory is applicable to impurity centers in which the second excited electronic state is separated from the first by a gap several times larger than the vibrational frequencies that are clearly displayed in the optical spectrum; in addition, the first excited state should not be degenerate. Optical spectra described by the given theory should have the following features: the conjugate absorption and fluorescence spectra should either be mirror symmetric or have small violations of mirror symmetry, whereas the 0–0 transitions of the conjugate spectra should be resonance ones.

Substantial advances have been made in the last 5 or 7 years in investigating the roles played by the HT interaction and the quadratic FC interaction. Present-day theory is capable of taking into account their effect on optical spectra without restrictions imposed by the magnitude of the matrix W and the function $M(R)$. Since it is precisely these two interactions that are responsible for the violation of mirror symmetry in the adiabatic approximation, the investigation of the lack of mirror symmetry is the most important source of information on the matrix W and the function $M(R)$. The method of deriving information on W and $M(R)$ from the lack of mirror symmetry was devised by the author (1973, 1977a, 1979). It proved quite efficient because the quadratic FC interaction produces a "lack of mirror symmetry with respect to shape", whereas the HT interaction produces a "lack of mirror symmetry with respect to intensity", i.e. departures from mirror symmetry of different kinds (Osad'ko 1979). These two kinds are discussed in detail in subsections 3.1 and 3.2.

3.1. Quadratic Franck–Condon interaction

Since the vector R in eq. (22) may have components corresponding to both intra- and intermolecular vibrational degrees of freedom, the quadratic FC interaction may manifest itself both in vibronic and in electron–phonon spectra. Let us begin by considering the simpler of these: vibronic spectra.

3.1.1. Vibronic spectra
When only the linear and quadratic FC interactions are taken into account the shape of the light absorption band is described by the equation*

$$I^a(\omega) = \sum_v |\langle v|0\rangle|^2 \delta(\omega - v\Omega_e), \tag{24}$$

where

$$\langle v|0\rangle = \int_{-\infty}^{\infty} dR\, \Phi_v^e(R)\Phi_0^0(R) \tag{25}$$

is the multidimensional Franck–Condon integral. The energy of the 0–0

---

*We take $\hbar = 1$ everywhere.

transition is taken as zero, and the frequencies $\Omega_e(j)$ of the normal modes of the excited molecule are the components of the vector $\mathbf{\Omega_e}$. If the shape of the molecule remains unchanged upon electronic excitation, a unified system of normal coordinates exists for an excited and a nonexcited molecule. But if the shape changes (the Dushinsky effect), the normal coordinates q^e and q^0 are related by the equation

$$q^e = \hat{S}q^0 + a, \tag{26}$$

where $\hat{S}$ is a matrix, which becomes the unit matrix for $W = 0$. The quadratic FC interaction has one specific property that fundamentally distinguishes it from the linear FC interaction. If, at a large displacement a of the equilibrium positions, the 0–0 transition can be less than the 0–1 transition, and the latter can be less than the 0–2 transition, then such a situation cannot, in principle, be produced by the quadratic interaction because, even for $W = \infty$, the most intense peak is the 0–0 peak and each subsequent peak is lower than the preceding one. For this reason the quadratic interaction has a weak effect on the vibronic spectra of molecules that have no degenerate frequencies. This can be readily shown, using the one-mode model. It has been investigated in detail in many papers. The following equations were obtained by the author (1973):

$$I^a(\omega) = I(0) \sum_{v=0}^{\infty} \frac{1}{2^v v!} \left(\frac{\Omega_0 - \Omega_e}{\Omega_0 + \Omega_e}\right)^v H_v^2\left(-a \frac{\sqrt{\mu \Omega_e}\,\Omega_0}{(\Omega_0^2 - \Omega_e^2)^{1/2}}\right) \delta(\omega - v\Omega_e), \tag{27}$$

where

$$I(0) = \frac{2\sqrt{\Omega_0 \Omega_e}}{\Omega_0 + \Omega_e} \exp\left(-a^2 \mu \frac{\Omega_0 \Omega_e}{\Omega_0 + \Omega_e}\right) \tag{28}$$

is the intensity of the 0–0 transition, $H_v(x)$ is the complete Hermitian, μ is the mass of the normal oscillator, Ω_e and Ω_0 are its frequencies, and a is the displacement from the equilibrium position. The equation for the emission band is obtained from eq. (27) by the substitutions: $\Omega_e \rightleftarrows \Omega_0$, $a \rightarrow -a$ and $\omega \rightarrow -\omega$. The coefficient of the δ-function describes the intensity of the vth vibronic peak. When $a \neq 0$, the vibronic progression contains peaks both with even and with odd v values. The peaks with the same v value in conjugate spectra are not only located at different distances from the 0–0 peak, but have different intensities as well. For instance, it follows from eq. (27) that for one- and two-quantum peaks

$$\frac{I_1^a}{I_1^e} = \frac{\Omega_0}{\Omega_e}, \qquad \frac{I_2^a}{I_2^e} = \left(\frac{\Omega_e^2 - \Omega_0^2 + 2a^2 \mu \Omega_e \Omega_0^2}{\Omega_e^2 - \Omega_0^2 - 2a^2 \mu \Omega_e^2 \Omega_0}\right)^2. \tag{29}$$

Consequently, the quadratic interaction leads to different intensity distributions in the progressions of the absorption and fluorescence spectra. We have called this lack of mirror symmetry "lack of mirror symmetry in shape" because the integrated intensity of the electron–vibrational part of the spectrum, referred to the intensity of the 0–0 transition, is the same for conjugate spectra. This follows from the general equation (24) and the equation

$$I^e(\omega) = \sum_v |\langle 0|v \rangle|^2 \delta(\omega + v\Omega_0) \tag{30}$$

for the emission spectrum. Integrating these equations with respect to ω, we obtain $I^e = I^a = 1$, and this leads to the statement made above.

If $a = 0$ the odd vibronic peaks vanish because $H_{2m+1}(0) = 0$. Taking into consideration the fact that $H_{2m}(0) = 2^m(2m-1)!!(-1)^m$, we obtain from eq. (27) the following expression for the intensity of the even peaks:

$$I^a_{2m} = I^e_{2m} = \frac{2\sqrt{\Omega_e \Omega_0}}{\Omega_e + \Omega_0} \frac{(2m-1)!!}{(2m)!!} \left(\frac{\Omega_e - \Omega_0}{\Omega_e + \Omega_0} \right)^{2m}. \tag{31}$$

It follows from this equation that even for $\Delta\Omega = \Omega_e - \Omega_0 \to \infty$, the intensity of the peaks decreases with increasing m. Normally, $\Delta\Omega/\Omega < 0.1$ in vibronic spectra. In this case, according to eq. (31), the contribution of the quadratic FC interaction to the intensity of the vibronic peaks can be neglected. Exceptions are vibronic peaks corresponding to degenerate or quasi-degenerate vibrations.

Next we shall consider the multimode case. When there are many modes, degeneracy of vibrations (in high-symmetry molecules) or accidental proximity of two or more vibrational frequencies (quasi-degeneracy) is possible. With degeneracy or quasi-degeneracy the Dushinsky effect may lead to an appreciable redistribution of the intensities of the corresponding peaks. It is necessary in this case to take into account the intermixing of the normal coordinates, i.e. the off-diagonality of the matrix $\hat{S}$. As a result, the multidimensional Franck–Condon integral (25) does not factorize into the product of one-dimensional integrals. Attempts have been made to circumvent this obstacle by means of approximate factorization of the integral (25) (Coon et al. 1962, Sharp and Rosenstock 1964). Using the coherent states of the harmonic oscillator, Doktorov et al. (1975, 1976, 1977) worked out a method for calculating the nonfactorizable integral (25). This method is based on the recurrence formulas that relate various Franck–Condon integrals, for example (Doktorov et al. 1977):

$$\langle v_k|0 \rangle = \langle 0|0 \rangle (M^{v_k}_{kk}/2v_k!)^{1/2} H_{v_k}(\langle 1_k|0 \rangle /(2M_{kk})^{1/2}), \tag{32}$$

where $H_{v_k}(x_k)$ is the Hermite polynomial in one variable, and M_{kk} is the matrix element of matrix M, which is related to the matrix $\hat{S}$ by the equations

$$M = 1 - 2I(1 + \tilde{I}I)^{-1}\tilde{I}, \quad I = \lambda_{\Omega_e}\hat{S}\lambda_{\Omega_0}^{-1},$$

$$\lambda_\Omega = \mathrm{diag}\{\Omega^{1/2}(1), \Omega^{1/2}(2), \ldots, \Omega^{1/2}(N)\}. \tag{33}$$

Not only can the elements of the matrix M be calculated by eq. (33) for a given matrix $\hat{S}$, but they can be determined experimentally by means of the equations

$$M_{kk} = \langle 1_k | 0 \rangle^2 - 2\langle 2_k | 0 \rangle, \qquad M_{kl} = \langle 1_k | 0 \rangle \langle 1_l | 0 \rangle - \langle 1_k 1_l | 0 \rangle. \tag{34}$$

Eqs. (32) and (34) enable one to calculate the intensity distribution in the kth progression on the basis of only the intensities of the first and second peaks. The following equation has been derived for phototransitions due to the excitation of two modes, k and l:

$$\langle v_k v_l | 0 \rangle = \langle 0 | 0 \rangle (v_k! v_l!)^{-1/2} H_{v_k v_l}(\tau_k, \tau_l), \tag{35}$$

where

$$\begin{pmatrix} \tau_k \\ \tau_l \end{pmatrix} = \begin{pmatrix} M_{kk} & M_{kl} \\ M_{kl} & M_{ll} \end{pmatrix}^{-1} \begin{pmatrix} \langle 1_k | 0 \rangle \\ \langle 1_l | 0 \rangle \end{pmatrix}. \tag{36}$$

Here $H_{vv'}(x, y)$ is the Hermite polynomial in two variables. The properties of multidimensional Hermite polynomials are expounded in Appell et al. (1926). The theory outlined above was applied in the calculation of the intensity distribution in the vibronic progression of the SO_2 molecule (Doktorov et al. 1975). Eqs. (34) and (33) make it feasible to consider the problem of determining the matrix $\hat{S}$ from spectroscopic data. But the effectiveness of this procedure has not yet been demonstrated in practice.

3.1.2. Electron–phonon bands

The quadratic interaction plays a more important role in electron–phonon spectra than it does in vibronic spectra. This is so, primarily, because the phonon frequencies form a continuous frequency range and are therefore always quasi-degenerate. Taking into account the effect of the quadratic interaction on the shape of the electron–phonon band is a more complex problem, because the phonon frequencies form a continuum. Hence, attempts were made to obtain an approximate solution of the problem by neglecting the intermixing of the normal coordinates (Ratner and Zilberman 1959; Kelley 1972, 1973; Mostoller et al. 1971), by applying the moment method (Kristofel et al. 1963) or by taking only part of the quadratic interaction into account (Small 1972, Osad'ko 1973). A new method was devised recently (Osad'ko 1977a) with which the effect of the total quadratic interaction on the shape of the electron–phonon spectra can be taken into account.

Since quadratic interaction affects both the frequencies of vibrations and the intensities of phototransitions, it does not prove expedient, in calculating the shape of the electron–phonon band, to deal separately with the Franck–Condon integrals. It proves more advantageous to consider the complete expression (24) for the band shape; this expression can be presented in the form (Lax 1952)

$$I^a(\omega) = \int_{-\infty}^{\infty} dt \ e^{i\omega t} \langle e^{iH^0 t} e^{-iH^e t} \rangle, \tag{37}$$

where

$$H^0 = \hat{T}(R) + R \frac{U^0}{2} R, \tag{38}$$

$$H^e = \hat{T}(R) + (R - a) \frac{U^e}{2} (R - a)$$

are the vibrational Hamiltonians of the nonexcited and the excited impurity center, respectively. Equation (37) can be readily transformed to

$$I^a(\omega) = \int_{-\infty}^{\infty} dt \ e^{i(\omega - Va/2)t} \langle \hat{S}(t) \rangle, \tag{39}$$

where

$$\hat{S}(t) = \hat{T} \exp \left[-i \int_0^t d\tau \left(-VR(\tau) + R(\tau) \frac{W}{2} R(\tau) \right) \right], \tag{40}$$

and the notation $V = aU^e$ has been used. The time dependence of the operators is determined by the Hamiltonian H^0, whereas the averaging $\langle \ \rangle$ is carried out with a density matrix which also depends on H^0. After cumulant expansion of the average of the operator $\hat{S}$, we obtain (Kubo 1962)

$$\langle \hat{S}(t) \rangle = \exp(\langle \hat{S}(t) - 1 \rangle_c) = \exp g(t), \tag{41}$$

where $\langle \ \rangle_c$ denotes averaging with only connected pairs taken into account. The procedure for calculating the cumulant function $g(t)$ is extremely simple if we neglect quadratic interaction.

Assume that $W = 0$. Then, under the integral in eq. (40) we have only linear interaction and there is only a single connected pair:

$$g(t) = - \int_0^t d\tau \int_0^\tau d\tau' \langle VR(\tau)R(\tau')V \rangle$$

$$= i \frac{Va}{2} t + \sum_q (\tfrac{1}{2}a_q)^2 [(n_q + 1) e^{-i\nu_q t} + n_q e^{i\nu_q t} - (2n_q + 1)]. \tag{42}$$

Here $a_q/\sqrt{2}$ is the shift of the equilibrium position of mode q with respect to the amplitude of the zero-point vibrations, ν_q is the phonon frequency, and $n_q = [\exp(\nu_q/kT) - 1]^{-1}$. Since the phonon frequencies form a continuous frequency range, it proves expedient in eq. (42) to go over to a continuous function of the frequency,

$$f(\nu) = \sum_q (\tfrac{1}{2}a_q)^2 \delta(\nu - \nu_q), \tag{43}$$

which is nonzero in the one-phonon frequency range and can therefore be called the one-phonon function. In terms of this function the cumulant function $g(t)$ is expressed as

$$g(t) = i\frac{Va}{2}t - f(0, T) + f(t, T),\qquad(44)$$

where

$$f(t, T) = \int_{-\infty}^{\infty} d\nu\, e^{-i\nu t}f(\nu, T),\qquad(45)$$

$$f(\nu, T) = [n(\nu) + 1]f(\nu) + n(-\nu)f(-\nu).\qquad(46)$$

According to eq. (46), the temperature-dependent one-phonon function $f(\nu, T)$ is the sum of a Stokes and an anti-Stokes part. Substituting eq. (44) into eq. (41), and the latter into eq. (39), we obtain, after expansion of $\exp[f(t, T)]$ into a series in terms of $f(t, T)$ and term-by-term calculation of the integrals,

$$I^a(\omega) = 2\pi\, e^{-f(0,T)}[\delta(\omega) + \Phi(\omega)],\qquad(47)$$

in which the first term describes the zero-phonon line (ZPL) and the second the phonon wing (PW):

$$\Phi(\omega) = \sum_{k=1}^{\infty}\frac{1}{k!}\int_{-\infty}^{\infty} d\nu_1 \ldots \int_{-\infty}^{\infty} d\nu_k\, f(\nu_1, T)\ldots f(\nu_k, T)\delta(\omega - \nu_1 - \cdots - \nu_k).$$
$$(48)$$

The expression for the light emission band is obtained from eqs. (47) and (48) by the substitution $\omega \rightarrow -\omega$, i.e., the conjugate bands are mirror symmetric for $\hat{W} = 0$. According to eqs. (47) and (48), the temperature dependence of the integrated intensity of the ZPL and the shape of the PW are determined by a single function, the one-phonon function $f(\nu)$. Consequently, there must be a relationship between them that can be observed experimentally. For this purpose it is only necessary to "extract" the function $f(\nu)$ from the measured PW. The method of "extraction" proposed by Kukushkin (1963, 1965) has been applied many times in practice (Loorits and Rebane 1967, Personov et al. 1971, Osad'ko et al. 1974, Ranson et al. 1976).

Assume that $W \neq 0$. The difficulty of calculating the cumulant function $g(t)$ by eq. (41) grows drastically in this case because there exist an infinite number of connected averages with different powers of W. It can be shown that this infinite sum is a resolvent of a Fredholm integral equation of the second kind with a difference kernel (Levenson 1971, Osad'ko 1972). Recently an efficient method was devised (Osad'ko 1977a) to solve this equation in analytic form. The results of this investigation are given below.

For $W \neq 0$ the shape of the conjugate absorption and emission bands is described, at zero temperature, by the expression

$$I^s(t) = \int_{-\infty}^{\infty} dt \, \exp[\pm i(\omega - \Delta)t + f^s(t) - f(0)], \quad s = a, e \tag{49}$$

where the superscript s is to be taken as a or e according to whether light is absorbed or emitted. The upper sign applies for $s = a$ and the lower one for $s = e$. The quantity

$$f(0) = \int_{-\infty}^{\infty} dt \, f^s(t) \tag{50}$$

is independent of the superscript s. It determines the integrated intensity of the ZPL. The quantity Δ determines the shift of the 0–0 transition due to the quadratic interaction. It is evident that eq. (49) can also be converted to the form of eq. (47), in which case the PW will also be described by eq. (48), when $f^s(\nu)$, the Fourier transform of the function $f^s(t)$, is substituted for $f(\nu, T)$. Since $f^a(\nu) \neq f^e(\nu)$, the form of the PW of the conjugate bands is different.

The functions $f^s(\nu)$ are not one-phonon functions as in the case of $W = 0$. They can be represented as an infinite series:

$$f^s(\nu) = f_1^s(\nu) + f_2^s(\nu) + f_3^s(\nu) + \cdots, \tag{51}$$

where the function $f_n^s(\nu)$ is different from zero in the n-phonon frequency range of the finite electronic state. The terms of the series (51) decrease in magnitude with increasing n for all W values (even infinitely large ones). At actually attainable values of W the series converges rapidly and therefore we only need to take the one-phonon functions $f_1^s(\nu)$ into account. These functions, multiplied by $\exp[-f(0)]$, represent the probability of a one-phonon transition and are given by

$$f_1^a(\nu) = \frac{a^2}{\pi} (\nu + 2\Delta)^2 \Gamma^a(\nu) \exp[2\psi(\nu)], \tag{52}$$

$$f_1^e(\nu) = \frac{a^2}{\pi} (\nu - 2\Delta)^2 \Gamma^e(\nu) \exp[-2\psi(\nu)], \tag{53}$$

where

$$\Delta = \int_0^{\infty} -\frac{d\nu}{2\pi} \Delta(\nu), \qquad \psi(\nu) = \int_0^{\infty} \frac{d\omega}{\pi} \frac{\Delta(\omega)}{\nu + \omega},$$

$$\Delta(\nu) = \arctan \frac{W\Gamma^e(\nu)}{1 - W\Omega^e(\nu)}, \quad \Omega^e(\nu) = \int_0^{\infty} \frac{d\omega}{\pi} \frac{2\omega\Gamma^e(\omega)}{\nu^2 - \omega^2}. \tag{54}$$

Equations (52), (53) and (54) were obtained under the assumption that upon electronic excitation of the impurity center, a change occurs in the force constant, which binds the impurity only to the closest molecule, thereby changing their relative distance by an amount a. Such a simplified model does not imply, of course, that we are considering the one-mode case.

The functions $\Gamma^a(\nu)$ and $\Gamma^e(\nu)$ are not independent; they are related by the equation

$$\Gamma^a(\nu) = \frac{\Gamma^e(\nu)}{[1 - W\Omega^e(\nu)]^2 + [W\Gamma^e(\nu)]^2}. \tag{55}$$

Consequently, both one-phonon functions f_1^a and f_1^e are expressible in terms of the single function $\Gamma^e(\nu)$, and the parameters a and W of the linear and quadratic FC interaction. Assuming that the interaction is with acoustic phonons, the following expression can be obtained for the function $\Gamma^e(\nu)$:

$$\Gamma^e(\nu) = \frac{\pi}{2\nu} \sum_q [u(1, q) - u(0, q)]^2 \delta(\nu - \nu_q). \tag{56}$$

Here ν_q are the phonon frequencies of the nonexcited impurity crystal, and $u(n, q)$ are the coefficients of the resolution of the displacement of the nth molecule in normal coordinates,

$$R_n = \sum_q \frac{u(n, q)}{\sqrt{2\nu_q}} R_q. \tag{57}$$

A numerical computation of the one-phonon functions was carried out by means of eqs. (52) and (53) for various values of the dimensionless parameter $b = W/\nu_D^2$, using the following model function for $\Gamma^e(\nu)$:

$$\Gamma^e(x) = 13\pi x^3 (1 - x)^{1/2}, \quad 0 \leq x = \nu/\nu_D \leq 1. \tag{58}$$

The result of this calculation is presented in fig. 2. The departure from mirror symmetry, progressing with increasing quadratic interaction, is evident in the graphs. Weakening of the binding of the impurity molecule with its surroundings when the impurity is excited corresponds to negative values of the parameter b, strengthening to positive values. Accordingly, a low-frequency peak due to a quasi-localized vibration appears in the one-phonon absorption function at $b = -0.2$. There is no such peak in the one-phonon fluorescence function $f_1^e(\nu)$, because such a quasi-localized vibration can only exist together with an impurity molecule in an electronically excited state. In a similar way a localized vibration, which is absent in the fluorescence function $f_1^e(\nu)$, appears in the one-phonon absorption function at $b = 0.2$. The conjugate pairs of functions shown in fig. 2 practically coincide with the PW for small linear FC interaction $(a^2\nu_D < 1)$. But for $a^2\nu_D > 1$, the PW can be calculated from eq. (48), into which the one-phonon functions shown in fig. 2 should be substituted. It is obvious that the mirror symmetry of the PW of the conjugate bands will be violated.

Also calculated was the contribution of the quadratic interaction to the

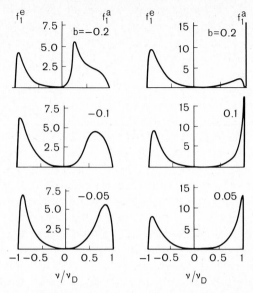

Fig. 2. Violation of mirror symmetry of one-phonon functions due to quadratic FC interaction ("lack of mirror symmetry with respect to shape"); $b = W/\nu_D^2$ and $a^2\nu_D = 1$.

intensity of the ZPL. The correction to the ZPL intensity reaches values of only 15 to 20% even for values of $|b| = 0.2$. Such values of b are to be regarded as large because at these values the phonon spectrum is changed. The shift of the ZPL is described by the approximate formula

$$\Delta \approx b\nu_D/2. \tag{59}$$

3.1.3. Temperature broadening of the zero-phonon lines
The quadratic FC interaction is also responsible for the temperature broadening and shift of the ZPL. At nonzero temperature T the cumulant function contains, besides the terms presented in eq. (44), another term: $-|t|\gamma(T)/2$ (Krivoglaz 1964). Therefore, adding this term to the exponent of eq. (49) and putting $f^s = 0$, we obtain the following expression for the ZPL:

$$I_{ZPL}^s(t) = e^{-f(0,T)} \frac{\gamma(T)}{[\omega - \Delta(T)]^2 + [\gamma(T)/2]^2}. \tag{60}$$

It follows from this equation that the ZPL has a Lorentz profile with the position of its maximum $\Delta(T)$ and its half width $\gamma(T)$ depending upon the temperature. Some papers contend that quadratic interaction can lead to the violation of resonance between the fluorescence and absorption ZPL and also to their different broadening. This problem was investigated by the author. The following was established: (1) both ZPL are resonant and broaden equally [i.e. the superscript s can be discarded in eq. (60)]; (2) the linear FC interaction and HT interaction have absolutely no effect on the ZPL broadening and shift, i.e., the quadratic FC interaction shows itself in the shape of the ZPL in its pure form.

The half width $\gamma(T)$ cannot be explained by decay processes of the electronic excitation because there are no radiationless transitions in the adiabatic approximation. Consequently, the adiabatic half width $\gamma(T)$ is not related to the lifetime τ of the electronic excitation by the ordinary function $\gamma = h/\tau$. As a matter of fact, experiments indicate that the homogeneous half width of the ZPL at 4.2 K exceeds h/τ by more than one order of magnitude (Völker et al. 1977). It is entirely natural to associate this fact with the existence of an adiabatic half width of the ZPL.

Silsbee (1962) was evidently the first to draw attention to the vital role played by the quadratic FC interaction in the temperature broadening of the ZPL. Subsequently, equations were obtained for $\gamma(T)$ and $\Delta(T)$ in the framework of perturbation theory in terms of W (McCumber and Sturge 1963, McCumber 1964, Krivoglaz 1964) and without applying perturbation theory (Osad'ko 1972). The ZPL broadening and shift were investigated in many experiments [see the review by the author (Osad'ko 1979)]. It was established that the displacement of the electron term due to thermal expansion of the crystal (Fitchen 1968) makes a substantial contribution to the observed ZPL shift. Without taking this fact into consideration there is no point in comparing the adiabatic shift with the observed value. Hence in what follows we shall discuss in detail only ZPL broadening.

The equation describing the ZPL half width for an arbitrary strength of the quadratic interaction was derived by Osad'ko (1972) and can be presented in the form

$$\gamma(T) = \int_0^\infty \frac{d\nu}{2\pi} \ln\{1 + 4W^2 n(\nu)[n(\nu) + 1]\Gamma^e(\nu)\Gamma^a(\nu)\}. \tag{61}$$

The functions Γ^e and Γ^a are determined by eqs. (55) and (56), i.e., the ZPL half width and the probabilities $f_1^e(\nu)$ and $f_1^a(\nu)$ of the one-phonon transitions depend upon one and the same function $\Gamma^e(\nu)$. It follows from eq. (61) that $\gamma(0) = 0$ and that the half widths coincide for resonance ZPL. We point out that when $W \neq 0$, the equations for the emission band are obtained from the corresponding equations for the absorption band by the substitutions ω, a, W, $\Gamma^a(\nu) \rightarrow -\omega$, $-a$, $-W$, $\Gamma^e(\nu)$, where ω is the light frequency. This rule can be checked on the one-phonon functions, determined by eqs. (52), (53) and (54)*. The half width $\gamma(T)$ is invariant with respect to such a substitution.

---

*In this case it is necessary to take into account the identity $1 + W\Omega^e(\nu) - iW\Gamma^e(\nu) \equiv [1 - W\Omega^0(\nu) + iW\Gamma^0(\nu)]^{-1}$.

The great majority of experimental data on ZPL broadening were compared, not with the results obtained by means of eq. (61), but with those of the simpler equation

$$\gamma(T) = 2W^2 \int_0^\infty \frac{d\nu}{\pi} [\Gamma^e(\nu)]^2 n(\nu)[n(\nu) + 1],$$ (62)

which chronologically precedes eq. (61). It is the zeroth order term of the expansion of eq. (61) in terms of $W\Gamma^e(\nu)$. Consequently, the criterion of applicability of this equation for weak broadening is

$$W\Gamma_m^e \ll 1,$$ (63)

where Γ_m^e is the maximum of the function $\Gamma^e(\nu)$. Most frequently used are two versions of eq. (62). The first is obtained with $\Gamma^e(\nu) = 5\pi\nu^3/\nu_D^5$, where ν_D is the Debye frequency. In this case it is assumed that ZPL broadening is due to interaction with acoustic phonons and that eq. (62) takes the form

$$\gamma(T) = 50\pi \frac{W^2}{\nu_D^3} \left(\frac{T}{\theta}\right)^7 \int_0^{\theta/T} \frac{x^6 e^x \, dx}{(e^x - 1)^2},$$ (64)

where $\theta = \nu_D/k$ is the Debye temperature. The second version is obtained with

$$\Gamma^e(\nu) = \frac{1}{\nu_0} \frac{\gamma_0}{(\nu - \nu_0)^2 + \gamma_0^2}.$$ (65)

In this case it is assumed that the ZPL broadening is due to interaction with the localized vibration. Then

$$\gamma(T) \approx \nu_0 \frac{W}{\nu_0^2} \frac{W}{\nu_0 \gamma_0} n(\nu_0)[n(\nu_0) + 1].$$ (66)

The narrower the peak of the localized vibration (i.e., the lower the value of γ_0), the less the temperature broadening of the ZPL. This follows from eq. (61) (Osad'ko and Zhdanov 1977). When $\gamma_0 \to 0$, the ZPL half width, determined by eq. (66), tends to infinity. But if we take into account the fact that the criterion of applicability of eq. (66) is the condition

$$W/\nu_0\gamma_0 \ll 1,$$ (67)

which follows from eq. (63), it becomes obvious that the result obtained by Osad'ko and Zhdanov (1977) also follows from eq. (66).

The exact eq. (61) possesses the important property of "saturation", i.e. when $W \to \infty$, the function $\gamma(T)$ does not continue to grow infinitely, but tends to a limiting curve. Numerical calculation by means of eq. (61), employing the function $\Gamma^e(\nu)$ of eq. (58), yields the following condition (Osad'ko and Zhdanov 1977):

$$\Delta\gamma/\Delta T \leq 0.3\text{--}0.4 \text{ cm}^{-1}/\text{K}.$$ (68)

The approximate eqs. (64) and (66) do not possess the property of "saturation". According to the data obtained by Osad'ko and Zhdanov (1977) the approximate and exact equations for $\gamma(T)$ yield close results provided that the temperature broadening of the ZPL is not large, i.e.

$$\Delta\gamma/\Delta T \leqslant 0.05 \text{ cm}^{-1}/\text{K}. \tag{69}$$

This inequality may serve as a practical criterion for the applicability of eqs. (64) and (66). For the latter equation, the inequality should even be increased.

3.1.4. Relation of the temperature broadening of the ZPL to the lack of mirror symmetry of conjugate phonon wings

Although many papers have been published in which experimental data on ZPL broadening were compared with the results obtained from eqs. (64) and (66), no proof has yet been obtained that the observed ZPL broadening is really due to a change of W in the force matrix. This uncertainty is due to the fact that eqs. (64) and (66) contain two free parameters: the frequency ν_D or ν_0, characterizing the phonons, and the coupling constant W. Either one or both parameters were selected in all papers. But these two parameters give three equations, eqs. (61), (64) and (66), with extensive possibilities for approximation. Korotaev and Kaliteevskii (1980), for instance, have shown that their experimental results on ZPL broadening are quite satisfactorily described by eq. (66) for several selected sets of values for the pair of parameters ν_0 and W. But a further analysis by these authors indicated that eq. (66) was not applicable to their case (see below, fig. 5). The chance nature of the coincidence between the calculated and experimental results obtained in many papers is indisputable. A more detailed comparison of experimental data with theory can be found in a review by the author (Osad'ko 1979).

It is clear from the afore said that the parameters in the equations for $\gamma(T)$ should not be selected. They should be derived from experimental investigations. Such a possibility is opened by a theory relating the violation of mirror symmetry of conjugate PW to the temperature broadening of the ZPL (Osad'ko 1977a). Since both effects have the same origin, they should be correlated. And, as a matter of fact, correlation of a qualitative nature has recently been observed in experimental investigations (Sheka and Meletov 1977). From the quantitative point of view, the relation between these two effects is manifest from the fact that the one-phonon functions f_1^a and f_1^e, as well as $\gamma(T)$, the ZPL half width, are expressible, according to eqs. (52), (53), (54) and (61), in terms of the general function $\Gamma^e(\nu)$ and the parameter W. Consequently, by analyzing the PW of conjugate bands we can find $\Gamma^e(\nu)$ and W, after which no free parameters and functions remain in eq. (61). The only obstacle to the application of this

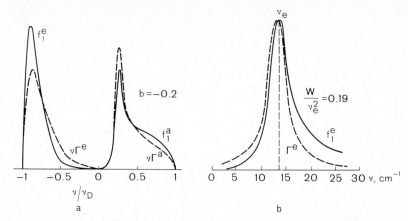

Fig. 3. Comparison of the shapes of exact and approximate one-phonon functions, corresponding to interaction with acoustic phonons (a) and with a quasi-localized phonon (b). The exact functions f_1^e and f_1^a (full lines) were calculated by means of eqs. (52) and (53). Their approximations, the functions $\nu\Gamma^{a,e}$ and Γ^e (dashed lines) have been normalized with respect to area to the exact functions $f_1^{a,e}$.

procedure in practice is that the relation between the functions f_1^a, f_1^e and $\gamma(T)$ is not an explicit one. It is therefore necessary to simplify this relation, and this can be done. Shown in fig. 3a are the exact functions f_1^a and f_1^e, calculated by eqs. (52) and (53), and the approximate functions, calculated by the simplified equation

$$f_1^s(\nu) \approx \frac{a^2}{\pi} \nu_D \nu \Gamma^s(\nu). \tag{70}$$

The main difference between the approximate and exact functions is in the normalization. This difference can easily be taken into account. This leads to the following procedure for comparing the experimental data with theoretical results: (1) the one-phonon functions $f_1^a(\nu)$ and $f_1^e(\nu)$ are derived from the PW of conjugate bands by a tested method (Personov et al. 1971); (2) applying eq. (70) we find the functions $\Gamma^s(\nu)$, which must be normalized by the condition

$$\int_{-\infty}^{\infty} \frac{d\nu}{\pi} \nu \Gamma^s(\nu) = 1, \quad s = a, e, \tag{71}$$

which follows from eq. (56); (3) the obtained and normalized functions $\Gamma^e(\nu)$ and $\Gamma^a(\nu)$ are substituted into eq. (55), from which the parameter W can be determined; (4) using the normalized function $\Gamma^e(\nu)$ and the determined parameter W, a numerical calculation of the function $\gamma(T)$ is carried out by means of eq. (61); this function is then compared to the measured temperature dependence of the ZPL half width.

This procedure is valid under the condition that the HT interaction makes no contribution to the observed lack of mirror symmetry. This can be checked by ensuring that the ratio of the integrated intensities. $\alpha^s = I_{ZPL}/(I_{ZPL} + I_{PW}^s)$, is the same in the conjugate bands being considered (see subsection 3.2). The procedure described above has recently been applied in practice by Korotaev and Kaliteevskii (1980). Figure 4 shows the conjugate spectra they obtained of the 3,4,6,7-dibenzpyrene molecule in *n*-octane at 4.2 K. In the given spectra $\alpha^a \approx \alpha^e = 0.71$ to 0.73. The spectrum in fig. 4 is inhomogeneously broadened; this can be readily seen from the ZPL. The half width of the ZPL is $3.8 \, \mathrm{cm}^{-1}$, whereas its homogeneous width at 4.2 K, according to theory and in agreement with experimental data (Kharlamov et al. 1974, Gorokhovski et al. 1976), should be less by at least one order of magnitude. It can therefore by presumed that the shape of the ZPL corresponds to an inhomogeneous broadening function. The peak with half width $7 \, \mathrm{cm}^{-1}$ being investigated is the convolution of a real peak with an inhomogeneous broadening function. Using standard methods, the homogeneous half width γ_0 of a one-quantum peak can be separated out. Korotaev and Kaliteevskii (1980) found that the half width γ_0 of the one-quantum peak equals $4.6 \, \mathrm{cm}^{-1}$. Assuming, in accordance with eq. (70), that the one-quantum peaks found with half width $\gamma_0 = 4.6 \, \mathrm{cm}^{-1}$

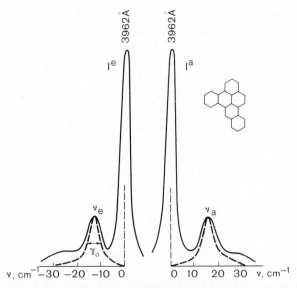

Fig. 4. Fluorescence (left) and absorption (right) bands for the 0–0 transition of 3,4,6,7-dibenzpyrene molecules in *n*-octane at 4.2 K. The full lines are the observed spectrum, whereas the dashed lines are the spectrum after exclusion of inhomogeneous broadening; $\nu_e = 13.1 \, \mathrm{cm}^{-1}$, $\nu_a = 15.6 \, \mathrm{cm}^{-1}$ and $\gamma_0 = 4.6 \, \mathrm{cm}^{-1}$.

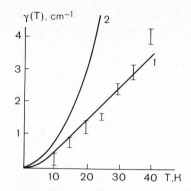

Fig. 5. Variation of the homogeneous half width of the ZPL at 3962 Å (vertical bars) and the temperature broadening calculated from eq. (61) (curve 1) and from eq. (66) (curve 2), with the parameters $\gamma_0 = 4.6\,\text{cm}^{-1}$ and $W = 36\,\text{cm}^{-2}$ that were determined experimentally (see text, subsection 3.1.4).

are proportional to the functions $\Gamma^s(\nu)$, according to eq. (55) and the more approximate equation

$$W \approx (\nu_a^2 - \nu_e^2)/2, \tag{72}$$

which follows from eq. (55) with $\gamma_0 = 0$, these investigators found that the dimensionless parameter $b = W/\nu_e^2$ is practically the same and equal to 0.19 for both methods of calculation. Substituting the functions found experimentally and the parameter $b = 0.19$ into eq. (61), Korotaev and Kaliteevskii obtained the theoretical curve 1 shown in fig. 5, which satisfactorily fits the experimentally determined points. Curve 2 in this figure was calculated with the same b and γ_0 values, but from eq. (66) for weak broadening. Of interest is the fact that by selecting the parameter b they could satisfactorily describe their experimental data by eq. (66) as well.

3.2. Herzberg–Teller interaction

If the dependence on the vibrational coordinates R is not neglected in eq. (20) for the electron matrix element $M(R)$, then the amplitude $\langle v|0\rangle$ of the electron–vibrational phototransition is of the form

$$\langle v|0\rangle = \int_{-\infty}^{\infty} dR\ \Phi_v^e(R)M(R)\Phi_0^0(R). \tag{73}$$

It is obvious that even in the absence of FC interaction, i.e. for $\Phi_v^e = \Phi_v^0$, this amplitude is nonzero and is determined by the derivatives of M with respect to R. These derivatives should be regarded as parameters of the HT interaction. This interaction is not large and therefore it is important to take it into account only to elucidate the nature of electron–vibrational transitions, especially in electronically forbidden spectra, i.e., when there is no 0–0 transition.

3.2.1. Lack of mirror symmetry of conjugate spectra

The HT interaction is manifest, however, in electronically allowed spectra as well, because it leads to the violation of mirror symmetry of conjugate spectra. This lack of mirror symmetry differs in principle from that due to the quadratic FC interaction. In contrast to the "lack of mirror symmetry with respect to shape", discussed in subsections 3.1.1 and 3.1.2, this can be called "lack of mirror symmetry with respect to intensity". To clarify the nature of this lack of mirror symmetry, we shall consider one-quantum conjugate phototransitions in the absence of quadratic FC interaction (fig. 6). Their amplitudes are determined by such expressions as

$$\langle 1|0 \rangle = \int_{-\infty}^{\infty} dR \, \Phi_1(R-a)M(R)\Phi_0(R)$$

$$= \sqrt{2} \int_{-\infty}^{\infty} d\rho \, (\rho - \tfrac{1}{2}a)\Phi_0(\rho - \tfrac{1}{2}a)M(\rho + \tfrac{1}{2}a)\Phi_0(\rho + \tfrac{1}{2}a),$$

$$\langle 0|1 \rangle = \int_{-\infty}^{\infty} dR \, \Phi_1(R)M(R)\Phi_0(R-a)$$ (74)

$$= \sqrt{2} \int_{-\infty}^{\infty} d\rho \, (\rho + \tfrac{1}{2}a)\Phi_0(\rho + \tfrac{1}{2}a)M(\rho + \tfrac{1}{2}a)\Phi_0(\rho - \tfrac{1}{2}a),$$

in which use in made of the relation $\Phi_1(R) = \sqrt{2}\, R\Phi_0(R)$ between the normalized functions of the harmonic oscillator, and the variable is changed according to $R = \rho + \tfrac{1}{2}a$. Since the product $\Phi_0(\rho + \tfrac{1}{2}a)\Phi_0(\rho - \tfrac{1}{2}a)$ is an even function of ρ, integrals containing odd powers of ρ vanish. Hence, if M is approximated, for example by a linear function, i.e. $M(\rho + \tfrac{1}{2}a) =$

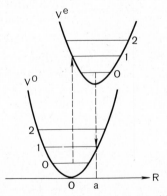

Fig. 6. One-quantum conjugate phototransitions.

$M_0 + M_1\rho$, then it follows from eq. (74) that

$$\langle 1|0\rangle^2 = 2M_0^2(-\tfrac{1}{2}a\langle\Phi^e\Phi^0\rangle + \alpha\langle\Phi^e\rho^2\Phi^0\rangle)^2,$$

$$\langle 0|1\rangle^2 = 2M_0^2(\tfrac{1}{2}a\langle\Phi^e\Phi^0\rangle + \alpha\langle\Phi^e\rho^2\Phi^0\rangle)^2, \tag{75}$$

where $\Phi^e = \Phi_0(\rho - \tfrac{1}{2}a)$, $\Phi^0 = \Phi_0(\rho + \tfrac{1}{2}a)$ and $\alpha = M_1/M_0$. It is quite clear that the amplitudes of the conjugate one-quantum transitions are the sum and the difference of the FC and HT amplitudes, because the parameters a and α, determining their intensity, are the parameters of the FC and HT interactions. If one of the interactions is absent, the intensities of the conjugate phototransitions are equal. Consequently, "lack of mirror symmetry with respect to intensity" is the result of the interference of the FC and HT amplitudes. On a qualitative level this effect was discussed many years ago by Sponer and Wollman (1941) in analyzing the vibronic spectra of benzene. Though the joint effect of the FC and HT interactions on optical spectra was dealt with in subsequent years in many papers (Markham 1959, Albrecht 1960, Rebane et al. 1964, Rebane 1968), the effect of the lack of mirror symmetry had been neglected. More recently, attention was again drawn to this effect in the investigation of vibronic spectra (Craig and Small 1969) and electron–phonon spectra (Osad'ko 1970). Since this effect is quite frequently employed today in analyzing vibronic spectra [see the review by the author (Osad'ko 1979)], the corresponding theory should be considered in more detail.

3.2.2. *Probability of phototransitions with an arbitrary dependence of the electronic matrix element $M(R)$ on the nuclear coordinates R*

Usually, in taking into account the HT interaction, i.e. the deviation from the Condon approximation, the electronic matrix element $M(R)$ is approximated by a first- or second-degree polynomial over the complete range of variation of the variable R. The components of this multidimensional vector correspond to nuclear displacements (intramolecular coordinates) as well as the motion of the molecule as a whole (intermolecular coordinates). Obviously, if in a phototransition the displacement a of the equilibrium positions is comparable to the interatomic (intermolecular) distances, the above-mentioned approximation is unsatisfactory. In this subsection we shall derive equations for the band shape that are valid for an arbitrary dependence of $M(R)$ on R.

The probabilities of absorption $I^a(\omega)$ and emission $I^e(\omega)$ of a photon are described by the equations

$$I^a(\omega) = \int_{-\infty}^{\infty} dt\, e^{i\omega t}\langle 0|\rho(H^0)\, e^{iH^0 t} M(R)\, e^{-iH^e t} M(R)|0\rangle,$$

$$I^e(\omega) = \int_{-\infty}^{\infty} dt\, e^{-i\omega t}\langle e|\rho(H^e)\, e^{iH^e t} M(R)\, e^{-iH^0 t} M(R)|e\rangle, \tag{76}$$

where ρ is the density matrix and the functions $|0\rangle$ and $|e\rangle$ are the eigenfunctions of the vibrational Hamiltonians H^0 and H^e of the nonexcited and the excited impurity crystal, respectively. Ignoring the change in the force matrix upon excitation of the impurity center, we can write

$$H^0 = H(R) = \hat{T}(R) + R\,\frac{U}{2}\,R,$$

$$H^e = H(R-a) = \hat{T}(R) + (R-a)\frac{U}{2}(R-a). \tag{77}$$

The eigenfunctions of these Hamiltonians and the Hamiltonians themselves are related by a unitary transformation:

$$|e\rangle = e^{-\hat{Q}}|0\rangle, \qquad e^{-\hat{Q}}H(R)\,e^{\hat{Q}} = H(R-a), \qquad e^{\hat{Q}}H(R)\,e^{-\hat{Q}} = H(R+a), \tag{78}$$

where

$$\hat{Q} = \sum_q a_q\,\partial/\partial R_q. \tag{79}$$

Making use of eqs. (78) we can transform the initial eqs. (76) to the form

$$I^a(\omega) = \int_{-\infty}^{\infty} \mathrm{d}t\, e^{i\omega t}\langle 0|\rho(H^0)\, e^{iH(R)t} M(R)\, e^{-iH(R-a)t} M(R)|0\rangle,$$

$$I^e(\omega) = \int_{-\infty}^{\infty} \mathrm{d}t\, e^{-i\omega t}\langle 0|\rho(H^0)\, e^{iH(R)t} M(R+a)\, e^{-iH(R+a)t} M(R+a)|0\rangle. \tag{80}$$

To calculate the averages in these equations it proves convenient to represent $M(R)$ in terms of a Fourier integral. Then eqs. (80) can be transformed to

$$I^s(\omega) = \int_{-\infty}^{\infty} \mathrm{d}k\, M(k) \int_{-\infty}^{\infty} \mathrm{d}k'\, M(k')\, \exp[-\tfrac{1}{2}i(1 \mp 1)(k + k')a]$$

$$\times \int_{-\infty}^{\infty} \mathrm{d}t\, e^{\pm i\omega t} I_{kk'}(\mp a; t), \tag{81}$$

where

$$I_{kk'}(\mp a, t) = \langle e^{-ikR(t)}\, e^{\mp \hat{Q}(t)}\, e^{\pm \hat{Q}}\, e^{-ik'R}\rangle. \tag{82}$$

In these and the following equations the upper sign is for an absorption and the lower for an emission spectrum. The averaging $\langle\ \rangle$ and the time dependence of the operators are determined by the Hamiltonian $H(R)$. The calculation of the average in eq. (82) is based on the two equations:

$$e^{\hat{A}}\,e^{\hat{B}} = e^{\hat{A}+\hat{B}}\,e^{[\hat{A},\hat{B}]/2}, \tag{83}$$

$$\langle e^{\hat{A}+\hat{B}}\rangle = \exp\langle\tfrac{1}{2}(\hat{A}+\hat{B})^2\rangle, \tag{84}$$

of which eq. (84) is only valid for operators $\hat{A}$ and $\hat{B}$ that depend linearly on the coordinates and momenta. Calculating eq. (82) we obtain

$$I_{kk'}(\mp a, t) = \exp\left(-f^{\mathrm{FC}}(T) \mp \tfrac{1}{2}\mathrm{i}(k + k')a + f_{kk'}(\mp a, t)\right.$$
$$\left. - \sum_q \tfrac{1}{2}(k_q^2 + k_q^2)(2n_q + 1)\right), \tag{85}$$

where

$$f_{kk'}(\mp a, t) = \sum_q [(\mp \tfrac{1}{2}a_q - \mathrm{i}k_q)\, \mathrm{e}^{-\mathrm{i}\nu_q t}(n_q + 1)(\mp \tfrac{1}{2}a_q - \mathrm{i}k_q')$$
$$+ (\mp \tfrac{1}{2}a_q + \mathrm{i}k_q)\, \mathrm{e}^{\mathrm{i}\nu_q t} n_q (\mp \tfrac{1}{2}a_q + \mathrm{i}k_q')] \tag{86}$$

is a one-phonon function that takes into account the joint effect of the HT interaction and the linear FC interaction. In the Condon approximation ($k = k' = 0$), the exponent in eq. (85) goes over into eq. (42) without the first term because

$$f^{\mathrm{FC}}(T) = \sum_q (\tfrac{1}{2}a_q)^2(2n_q + 1). \tag{87}$$

Substituting expression (85) into eq. (81) we obtain

$$I^s(\omega) = \mathrm{e}^{-f^{\mathrm{FC}}(T)} \int_{-\infty}^{\infty} \mathrm{d}k\, L(k) \int_{-\infty}^{\infty} \mathrm{d}k'\, L(k') \int_{-\infty}^{\infty} \mathrm{d}t\, \mathrm{e}^{\pm\mathrm{i}\omega t + f_{kk'}(\mp a, t)}, \tag{88}$$

where

$$L(k) = M(k) \exp\left(-\tfrac{1}{2}\mathrm{i}ka - \sum_q \tfrac{1}{2}k_q^2(2n_q + 1)\right). \tag{89}$$

The time integral in eq. (88) is calculated by expansion of $\exp f_{kk'}$ into a series in terms of $f_{kk'}$ and term-by-term integration. In such an expansion of $\exp f_{kk'}$, integrals of the following type appear in eq. (88):

$$\int_{-\infty}^{\infty} \mathrm{d}k\, L(k)k_1, \ldots, k_p = \left[\int_{-\infty}^{\infty} \mathrm{d}k\, L(k)\, \mathrm{e}^{-\mathrm{i}kx} k_1, \ldots, k_p\right]_{x=0}$$
$$= (+\mathrm{i})^p \left[\frac{\partial}{\partial x_1}, \ldots, \frac{\partial}{\partial x_p} L(x)\right]_{x=0}. \tag{90}$$

Proceeding from eq. (90) we can transform eq. (88) to its final form:

$$I^s(\omega) = 2\pi\, \mathrm{e}^{-f^{\mathrm{FC}}(T)} \left[\left(\delta(\omega) + \int_{-\infty}^{\infty} \mathrm{d}\nu\, \hat{f}^s(\nu, T)\delta(\pm\omega - \nu)\right.\right.$$
$$+ \frac{1}{2!} \int_{-\infty}^{\infty} \mathrm{d}\nu_1 \int_{-\infty}^{\infty} \mathrm{d}\nu_2\, \hat{f}^s(\nu_1, T)\hat{f}^s(\nu_2, T)\delta(\pm\omega - \nu_1 - \nu_2) + \cdots\right)$$
$$\left. \times L(x)L(x')\right]_{x=x'=0}, \tag{91}$$

where

$$\hat{f}^s(v, T) = [n(v) + 1] \sum_q (\tfrac{1}{2}a_q \mp \partial/\partial x_q)\delta(v - v_q)(\tfrac{1}{2}a_q \mp \partial/\partial x_q')$$

$$+ n(v) \sum_q (\tfrac{1}{2}a_q \pm \partial/\partial x_q)\delta(v + v_q)(\tfrac{1}{2}a_q \pm \partial/\partial x_q'). \tag{92}$$

According to eq. (91), the parameters of the HT interaction that appear in optical bands are derivatives of the function $L(x)$ at zero. The function $L(x)$ can be expressed in terms of $M(R)$ if relation (89) is used, as well as the equation

$$e^{-k^2\xi/2} = \int_{-\infty}^{\infty} \frac{\mathrm{d}R}{\sqrt{2\pi\xi}} \, e^{\mathrm{i}kR - R^2/2\xi}. \tag{93}$$

Calculating the Fourier component from both parts of eq. (89) and taking eq. (93) into account, we obtain

$$L(x) = \int_{-\infty}^{\infty} \frac{\mathrm{d}R(T)}{\pi^{N/2}} \, M(x + \tfrac{1}{2}a - R) \, e^{-R^2(T)}, \tag{94}$$

where

$$R_q(T) = \frac{R_q}{(4n_q + 2)^{1/2}} = R_q[\tfrac{1}{2} \tanh(v_q/2kT)]^{1/2}. \tag{95}$$

Here R_q is the dimensionless normal coordinate, which is the ratio of the normal coordinate to the corresponding amplitude $(\mu v_q)^{-1/2}$ of the zero-point vibrations. The decrease of $R_q(T)$ with increasing T is due to the increasing amplitude of the vibrations of oscillator q.

Due to the exponent, the main contribution to the integral (94) is made by the interval $|R(T)| < 1$. Consequently, according to eq. (94) the relation $M(R)$ is of interest over a range of the order of the amplitude of the vibrations and including the point $\tfrac{1}{2}a$. At low temperature this range is much less than the interatomic (intermolecular) distances. Hence, the function $M(\tfrac{1}{2}a - R)$ can be approximated over such a range by means of a polynomial. Using a quadratic polynomial and zero temperature, eq. (91) can be transformed to

$$I^s(\omega) = 2\pi M^2(\tfrac{1}{2}a) \, e^{-f^{FC}(0)} \left(\delta(\omega) + \sum_q (\tfrac{1}{2}a_q \mp \alpha_q)^2 \, \delta(\omega \mp v_q) \right.$$

$$+ \frac{1}{2!} \sum_{q,q'} (\tfrac{1}{2}a_q \cdot \tfrac{1}{2}a_{q'} \mp \alpha_q \cdot \tfrac{1}{2}a_{q'} \mp \alpha_{q'} \cdot \tfrac{1}{2}d_q + \alpha_{qq'})^2$$

$$\left. \times \delta(\omega \mp v_q \mp v_{q'}) + \cdots \right), \tag{96}$$

where

$$\alpha_q = \left[\frac{\partial M(R)}{\partial R_q} \bigg/ M(R) \right]_{a/2} \quad \text{and} \quad \alpha_{qq'} = \left[\frac{\partial^2 M(R)}{\partial R_q \, \partial R_{q'}} \bigg/ M(R) \right]_{a/2} \tag{97}$$

are the parameters of the HT interaction. They are determined by the derivatives of M with respect to R at the point located in the middle between the equilibrium positions of the ground and excited states (see fig. 6). It is especially important to take this into account for large displacements a.

Due to the interference of the FC and HT amplitudes, the electron–vibrational transitions in conjugate spectra have different intensities in accordance with eq. (96). The "lack of mirror symmetry with respect to intensity" appears, as shown in fig. 7. The bands shown in this figure were obtained from eq. (96) for $\alpha_{qq'} = 0$ and $\alpha_q = \eta a_q/2$. The main indication of the interference of the FC and HT amplitudes is the different PW intensity in the conjugate spectra with respect to the ZPL. In accordance with fig. 7, the degree to which mirror symmetry is violated depends only on the relative magnitudes of the FC and HT interactions, characterized by the parameter η. Since the HT interaction is not strong as a rule, it cannot cause appreciable lack of mirror symmetry in multiphonon bands, which are characterized by strong FC interaction.

Equation (96) is also applicable, of course, to the analysis of vibronic spectra. It is especially effective for analyzing quasi-line spectra of polyatomic molecules, because it makes it possible to determine the parameters a_q, α_q and $\alpha_{qq'}$ of the vibronic coupling from spectroscopic data [see the review by the author (Osad'ko 1979)].

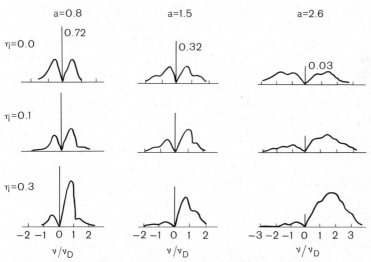

Fig. 7. "Lack of mirror symmetry with respect to intensity" in electron–phonon bands. The linear FC interaction increases from left to right, whereas the relative contribution of the HT interaction increases from top to bottom. The numbers alongside the ZPL are the Debye–Waller factors $\exp(-a^2/2)$.

3.2.3. Temperature effects

Like the linear FC interaction, the HT interaction does not affect the temperature broadening and shift of the ZPL. But it does influence the integrated intensity of the ZPL and the PW and the dependence of their intensity on the temperature. We obtain the temperature dependence of the integrated intensity $I_{ZPL}(T)$ of the zero-phonon line by means of eq. (91) (first term):

$$I_{ZPL}(T) = e^{-f^{FC}(T)} L^2(0, T), \tag{98}$$

where $L(0, T)$ is determined by eq. (94) for $x = 0$, and $f^{FC}(T)$ by eq. (87). It proves most convenient to obtain the temperature dependence $I^s(T)$ of the integrated intensity of the whole band by means of eq. (88). Integrating this equation over the frequency and making use of eq. (93), we obtain

$$I^a(T) = \int_{-\infty}^{\infty} \frac{d\boldsymbol{R}(T)}{\pi^{N/2}} M^2(\boldsymbol{R}) e^{-R^2(T)}, \tag{99}$$

$$I^e(T) = \int_{-\infty}^{\infty} \frac{d\boldsymbol{R}(T)}{\pi^{N/2}} M^2(\boldsymbol{R} + \boldsymbol{a}) e^{-R^2(T)}. \tag{100}$$

This leads to the well-known result that the integrated intensity of the bands depends upon the temperature, provided that the HT interaction is nonzero. Moreover, the integrated intensity of conjugate absorption and emission bands differs if the bands are plotted to a scale at which the intensities of the zero-phonon transitions are equal to each other. The most convenient quantity for comparison with the experimental data is the Debye–Waller factor, i.e. $\alpha^s(T) = I_{ZPL}(T)/I^s(T)$. Substituting eqs. (98), (99) and (100) into this expression we obtain the Debye–Waller factor with linear FC interaction and arbitrary HT interaction taken into account.

The effect of the HT interaction on the Debye–Waller factor was recently investigated by Korotaev and Kaliteevskii (1980). They measured the temperature dependence $\alpha^e(T)$ of the band of 3,4,6,7-dibenzpyrene shown in fig. 4. This temperature dependence is shown by the points in fig. 8. Since the spectra shown in fig. 4 have practically the same values of α^a and α^e, we have in the case being investigated either pure FC or pure HT interaction. In the former case eqs. (98), (99) and (100) yield

$$\alpha^{FC}(T) = \exp\left(-\int_{-\infty}^{\infty} d\nu\, f(\nu)[2n(\nu) + 1]\right), \tag{101}$$

in which the one-phonon function $f(\nu)$ is described by eq. (43). In the latter case

$$\alpha^{HT}(T) = L^2(0, T)/I(T), \tag{102}$$

in which L and I are determined by eqs. (94), (99) and (100) for $\boldsymbol{a} = 0$. A

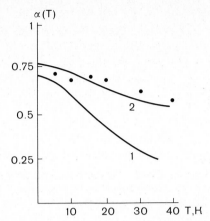

Fig. 8. Temperature dependence of the Debye–Waller factor for a band with a ZPL at 3962 Å for 3,4,6,7-dibenzpyrene molecules. Solid circles are experimental points. Curve 1 is for α^{FC}, curve 2 for α^{HT}.

calculation by means of eq. (101), in which the one-phonon peak shown by the dashed line of fig. 4 was taken instead of $f(\nu)$, yielded curve 1 of fig. 8, i.e., it led to a discrepancy between the theoretical and experimental data. Then a calculation using eq. (102) was carried out with the electronic matrix element taken in the form $M(R) = M_0 + M_1 R + M_2 R^2$. According to eq. (96), the intensities of the one- and two-quantum peaks for $a = 0$ are equal to $(M_1/M_0)^2$ and $(M_2/M_0)^2$, respectively. Using the curves of fig. 4, the following values were found for these parameters: $M_1/M_0 = 0.64$ and $M_2/M_0 = 0.17$. Then all the quantities in eq. (102) are known and it was employed to plot curve 2 of fig. 8. The coincidence of curve 2 with the experimental data indicates that the one- and two-phonon peaks in fig. 4 are produced by the HT interaction.

4. Adiabatic potentials with several minima

The adiabatic theory of phototransitions, discussed in the preceding section, includes the additional assumption that the adiabatic potentials have only one minimum each and that the vibrations occurring in the vicinity of these minima are harmonic. No fundamentally new effects appear when the small anharmonicity of the vibrations is taken into account. This situation changes dramatically, however, when the adiabatic potentials have more than one minimum. In this case, many basically new effects appear.

The existence of additional minima of the adiabatic potentials is usually associated with the degeneracy of the excited electronic levels, i.e. with the

Jahn–Teller effect. A review of the papers devoted to the Jahn–Teller effect is beyond the scope of the present paper, the more so because there are many reviews and books specially devoted to this problem (Longuet-Higgins 1961, Toyozawa 1967, Sturge 1967, Englman 1972, Bersuker 1976).

In the present review we shall discuss the possible existence of a multiple-minimum adiabatic potential in low-symmetry impurity centers, where there is no Jahn–Teller effect. We shall also consider the spectral effects of such a potential. To indicate more clearly the difference between this new approach and the traditional "Jahn–Teller approach", we shall briefly consider the traditional mechanism of the occurrence of a multiple-minimum potential in a system with a pair of closely spaced electronic levels. Denoting the corresponding two excited electronic states by the superscripts 1 and 2, and neglecting the influence of the other electronic states, we reduce the system of equations (4b) to a system of the second order:

$$[\hat{T}(R) + V^{11}(R) - E]\Phi_d^1(R) + V^{12}(R)\Phi_d^2(R) = 0,$$
$$V^{21}(R)\Phi_d^1(R) + [\hat{T}(R) + V^{22}(R) - E]\Phi_d^2(R) = 0. \tag{103}$$

According to eq. (8) the diabatic vibrational functions $\Phi_d^{1,2}$ are related to the adiabatic functions $\Phi_a^{1,2}$ by a simple orthogonal transformation which in this case is of the form

$$\begin{pmatrix} \Phi_d^1(R) \\ \Phi_d^2(R) \end{pmatrix} = \begin{pmatrix} \cos\theta & -\sin\theta \\ \sin\theta & \cos\theta \end{pmatrix} \begin{pmatrix} \Phi_a^1(R) \\ \Phi_a^2(R) \end{pmatrix}, \tag{104}$$

where

$$\tan 2\theta = 2V^{12}(R)/|V^{11}(R) - V^{22}(R)|. \tag{105}$$

This becomes readily evident when we substitute eq. (104) into (103). After this substitution we obtain instead of the system of equations (103):

$$[T(R) + U^1(R) - E]\Phi_a^1(R) + \Lambda^{12}(R)\Phi_a^2(R) = 0,$$
$$\Lambda^{21}(R)\Phi_a^1(R) + [T(R) + U^2(R) - E]\Phi_a^2(R) = 0, \tag{106}$$

where

$$U^{1,2}(R) = \frac{V^{11}(R) + V^{22}(R)}{2}$$
$$\pm \left\{ \left(\frac{V^{11}(R) - V^{22}(R)}{2} \right)^2 + V^{12}(R)V^{21}(R) \right\}^{1/2} + \Lambda^{11,22}(R) \tag{107}$$

is the adiabatic potential. The "diabatic" potentials V^{11} and V^{22} are usually approximated by polynomials of the second degree, whereas V^{12} and V^{21} are approximated by polynomials of the first degree. But even in this simplest case the adiabatic potential (107) can have more than one minimum. The occurrence of a two-well adiabatic potential is often shown

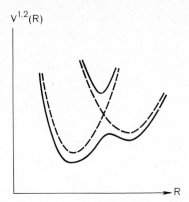

Fig. 9. Formation of a two-well adiabatic potential (full lines) from two one-well diabatic potentials (dashed lines).

schematically as in fig. 9, though a specific adiabatic potential may be substantially more complicated. For instance, a doubly degenerate electronic level of symmetry E, interacting with an e-mode, corresponds to the adiabatic potential shown in fig. 10.

Though the "Jahn–Teller approach" allows an explanation of the origin of the additional minima for the adiabatic potential of an excited center, it does not explain the fact that degeneracy or quasi-degeneracy is not a necessary condition for the appearance of such potentials. The existence of a two-well potential is possible, in principle, in a system without electronic quasi-degeneracy. There are various aspects in this case that are not explained in the "Jahn–Teller approach". We shall consider several of them. In the first place, a two-well potential can exist only with respect to a

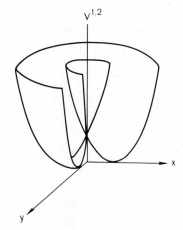

Fig. 10. Adiabatic potential of the E–e system.

single coordinate. Such a coordinate may, for example, be the angle determining the orientation of an impurity molecule in the crystal lattice. Secondly, there can be a two-well potential in the ground state as well as in the electronically excited state. Thirdly, with this approach, an electron–vibrational problem can be solved in the adiabatic approximation. This allows one to deal with the effect of the additional minima of the adiabatic potential on the optical spectra in its pure form, i.e. without the influence of the nonadiabatic interaction.

Without the concept of the two-well potential it proves impossible to explain a number of physicochemical phenomena. These include, for instance, the intra- and intermolecular proton transport in organic impurity centers of low symmetry, and the large conformation changes that occur in polyatomic molecules. Anomalies were discovered about ten years ago in the low-temperature properties of glasses. For example, the specific heat was found to be proportional to the first power of the temperature and the thermal conductivity to the second power for $T < 1$ K (Zeller and Pohl 1971). Phillips (1972) and Anderson et al. (1972) explained these anomalies on the basis of the two-well potential. In recent years other anomalies were discovered that confirmed the existence of two-well potentials in low-symmetry systems. Winterling (1975), for example, observed low-frequency Raman scattering in glass, and Hegarty and Yen (1979) detected anomalous low-temperature behavior of the half width of optical ZPL [see also the review by Jackle et al. (1976)]. The above-mentioned phenomena drew considerable attention to two-well systems. The present paper discusses theoretical investigations of impurity centers with two-well adiabatic potentials, carried out by the author together with S.A. Kulagin.

4.1. The vibrational system

The vibration of an impurity center can be found by solving the vibrational Schrödinger equation

$$[\hat{T}(x) + V(x) + \hat{T}(\varphi) + U(x, \varphi) - E]F(x, \varphi) = 0, \tag{108}$$

where $\hat{T}$ are the kinetic energy operators, and $V + U$ is the potential energy operator. The potential $V(x)$ is of the two-well type. Hence, in the following we shall call x the anharmonic coordinate. The set of coordinates φ describes the remaining vibrational modes.

Assume that we know the solution of the Schrödinger equation with the two-well potential

$$[\hat{T}(x) + V(x) - \epsilon_l]\Psi_l(x) = 0. \tag{109}$$

Then the complete vibrational function can be represented in the form of the expansion

$$F(x, \varphi) = \sum_l \Phi^l(\varphi)\Psi_l(x). \tag{110}$$

Substituting this expansion into eq. (108), we obtain a system of equations for the vibrational function $\Phi^l(\varphi)$:

$$[H_l(\varphi) + \epsilon_l - E]\Phi^l(\varphi) + \sum_{l'(\neq l)} U_{ll'}(\varphi)\Phi^{l'}(\varphi) = 0, \tag{111}$$

where

$$H_l(\varphi) = T(\varphi) + U_{ll}(\varphi), \tag{112}$$

and

$$U_{ll'}(\varphi) = \int dx\ \Psi_l(x)U(x, \varphi)\Psi_{l'}(x). \tag{113}$$

The function $U_{ll}(\varphi)$ is the adiabatic potential with respect to the coordinates φ. We shall assume that the corresponding vibrations are harmonic, i.e., that

$$H_l(\varphi) = T(\varphi) + (\varphi - a_l)\frac{\hat{U}}{2}(\varphi - a_l). \tag{114}$$

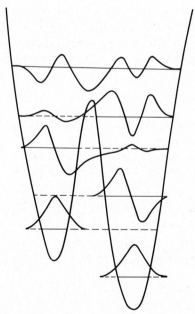

Fig. 11. Eigenfunctions and eigenvalues of eq. (109) with the potential (117) for $V = 2V_0\mu/(\hbar k)^2 = 1200$, $\xi = 0.37$ and $\kappa = 0.02$.

If the function $U_{ll'}$, off-diagonal with respect to the anharmonic mode subscripts, equals zero, then the vibrational wave functions $\Phi^l(\varphi) \equiv \Phi(\varphi - a_l)$ satisfy the harmonic oscillator equation

$$[H_l(\varphi) + \epsilon_l - E_{lv}]\Phi_v(\varphi - a_l) = 0, \qquad (115)$$

whereas the complete vibrational function and the vibrational energy are described by the equations

$$F_{lv}(x, \varphi) = \Phi_v(\varphi - a_l)\Psi_l(x), \qquad E_{lv} = \epsilon_l + v\nu. \qquad (116)$$

Shown in fig. 11 are the eigenfunctions and eigenvalues of eq. (109); they have been calculated with the two-well potential

$$V(x) = -V_0(\cos kx - \xi \cos 2kx + \kappa \sin kx), \qquad (117)$$

by a numerical method described by Osad'ko and Kulagin (1980). It is quite evident that the wave functions of the lower steady states are localized, even for a small value of the asymmetry parameter κ, either in a shallow or in a deep well. We shall therefore denote levels whose functions are localized in a shallow well by the subscript α, and use the subscript β for a deep well. We shall employ the sub- or superscript l to denote any level of a two-well system. If $U_{ll'} = 0$, the system cannot go over from one level to another because the function (116) describes a steady state.

4.2. *Probabilities of intrawell and interwell transitions*

Intrawell and interwell transitions are caused by the interaction between modes that is described by the function $U_{ll'}(\varphi)$. Hence, in the first nonzero order with respect to $U_{ll'}$, the expression for the transition probability per unit time is of the form:

$$\gamma_{l'l} = \int_{-\infty}^{\infty} dt \, e^{i\epsilon_{ll'}t} \, \text{Tr}\{\rho(H_l) \, e^{iH_lt} U_{ll'} \, e^{-iH_{l'}t} U_{l'l}\}, \qquad (118)$$

where $\epsilon_{ll'} = \epsilon_l - \epsilon_{l'}$, the Hamiltonians are determined by eq. (114), and $\rho(H_l)$ is the density matrix. Expression (118) describes both intrawell and interwell transitions. It looks very much like the first of eqs. (76) and is therefore calculated in a similar manner. Assuming $a_l = 0$ and $a_{l'} = a$, we obtain an equation analogous to eq. (91):

$$\gamma_{l'l} = 2\pi \, e^{-f^{FC}(T)} \left[\left(\hat{f}(\epsilon_{ll'}) + \frac{1}{2!} \int_{-\infty}^{\infty} d\nu \, \hat{f}(\nu)\hat{f}(\epsilon_{ll'} - \nu) \right. \right.$$

$$\left. + \frac{1}{3!} \int_{-\infty}^{\infty} d\nu_1 \int_{-\infty}^{\infty} d\nu_2 \, \hat{f}(\nu_1)\hat{f}(\nu_2)\hat{f}(\epsilon_{ll'} - \nu_1 - \nu_2) + \cdots \right)$$

$$\times \Lambda_{ll'}(\varphi)\Lambda_{l'l}(\varphi') \Bigg]_{\varphi = \varphi' = 0}, \qquad (119)$$

where

$$\hat{f}(\nu) = [n(\nu) + 1] \sum_q (\tfrac{1}{2}a_q - \partial/\partial\varphi_q)\delta(\nu - \nu_q)(\tfrac{1}{2}a_q - \partial/\partial\varphi'_q)$$

$$+ n(\nu) \sum_q (\tfrac{1}{2}a_q + \partial/\partial\varphi_q)\delta(\nu + \nu_q)(\tfrac{1}{2}a_q + \partial/\partial\varphi'_q), \qquad (120)$$

and the function $\Lambda_{l'l}(\varphi)$ is expressed in terms of the function $U_{l'l}(\varphi)$, which describes the interaction between the anharmonic mode x and the harmonic modes φ, by the equation:

$$\Lambda_{l'l}(\varphi) = \int_{-\infty}^{\infty} \frac{\mathrm{d}\mathbf{R}(T)}{\pi^{N/2}} U_{l'l}(\varphi + \tfrac{1}{2}\mathbf{a} - \mathbf{R}) \, e^{-R^2(T)}, \qquad (121)$$

where

$$R_q(T) = [\tfrac{1}{2}\tanh(\nu_q/2kT)]^{1/2} R_q. \qquad (122)$$

The temperature function $f^{FC}(T)$ is determined by eq. (87), but in this case a_q is the displacement of the equilibrium position of the harmonic mode φ_q in the transition $l' \leftarrow l$. The first term in eq. (119) describes one-phonon transitions between the levels ϵ_l and $\epsilon_{l'}$ for an arbitrary dependence of the interaction $U_{l'l}(\varphi)$ on the harmonic coordinates φ and arbitrary displacements a_q of the equilibrium positions of the harmonic oscillators. The second term describes the two-phonon transitions, the third term the three-phonon transitions, etc. There is a simple relation between the probabilities of the direct and reverse transitions:

$$\gamma_{l'l} = e^{(\epsilon_l - \epsilon_{l'})/kT}\gamma_{ll'}. \qquad (123)$$

This relation can be obtained, for instance, from eq. (119) considering that $-a_q$ corresponds to the reverse transition.

4.3. Interwell relaxation

The barrier separating the wells is the reason why the probabilities $\gamma_{\alpha\alpha'}$ and $\gamma_{\beta\beta'}$ of intrawell relaxation are much larger than the probability $\gamma_{\alpha\beta}$ of an interwell relaxation (see subsection 4.4). To stress this we introduce the new notation $\gamma_{\alpha\alpha'} = \Gamma_{\alpha\alpha'}$ and $\gamma_{\beta\beta'} = \Gamma_{\beta\beta'}$ for the probabilities of intrawell relaxation, retaining the old notation $\gamma_{\alpha\beta}$ only for the probability of interwell relaxation. Then the system of kinetic equations for the populations N_l of the levels ϵ_l of the anharmonic mode can be written in the new notation as

$$\dot{N}_l = \sum_{l'} (-\gamma\delta_{ll'} + \Gamma_{ll'} + \gamma_{ll'})N_{l'} + I(t)\delta_{ll_0}. \qquad (124)$$

Here $I(t)$ is the rate at which the level ϵ_{l_0} is populated by an external source, and γ is the probability of departure from the two-well system.

Since the subscript l numbers the levels of both wells, then, in accordance with what was stated at the beginning of this subsection, we should take $\Gamma_{\alpha\beta} = 0$ and $\gamma_{\alpha\alpha'} = \gamma_{\beta\beta'} = 0^*$. The following relation exists for the matrix elements of $\hat{\Gamma}$ and $\hat{\gamma}$:

$$\Gamma_{ll} + \sum_{l'(\neq l)} \Gamma_{l'l} = 0, \qquad \gamma_{ll} + \sum_{l'(\neq l)} \gamma_{l'l} = 0. \tag{125}$$

Making use of the inequality $\hat{\Gamma} \gg \hat{\gamma}$ we shall solve the system of equations (124), regarding $\hat{\gamma}$ as a small perturbation.

In the zeroth-order approximation, only the matrix $\hat{\Gamma}$ remains in eq. (124). This matrix can be diagonalized, going over from the l-basis to the j-basis by means of the matrix $\hat{n}$:

$$N_l = \sum_j n_{lj} N_j, \qquad N_j = \sum_l (n^{-1})_{jl} N_l. \tag{126}$$

The matrix elements of $\hat{n}$ and $\hat{n}^{-1}$ are determined from the systems of equations

$$\Gamma_j n_{lj} + \sum_{l'} \Gamma_{ll'} n_{l'j} = 0, \qquad \Gamma_j (n^{-1})_{jl} + \sum_{l'} (n^{-1})_{jl'} \Gamma_{l'l} = 0. \tag{127}$$

From the solvability condition for these systems of equations we find the roots Γ_j, which are the same for both systems. Using eqs. (125) and relation (123), we can ascertain by a direct check that the system of equations (127) has the root $\Gamma_0 = 0$, associated with the following linearly independent solutions:

$$n_{ls} = z_s \sum_\alpha e^{-\epsilon_\alpha/kT} \delta_{l\alpha}, \qquad (n^{-1})_{sl} = \sum_\alpha \delta_{l\alpha} \qquad \text{(s-solution)},$$

$$ \tag{128}$$

$$n_{ld} = z_d \sum_\beta e^{-\epsilon_\beta/kT} \delta_{l\beta}, \qquad (n^{-1})_{dl} = \sum_\beta \delta_{l\beta} \qquad \text{(d-solution)},$$

where

$$z_s^{-1} = \sum_\alpha e^{-\epsilon_\alpha/kT}, \qquad z_d^{-1} = \sum_\beta e^{-\epsilon_\beta/kT}. \tag{128a}$$

Recall that the subscript α only numbers the levels of the shallow well, and β only those of the deep well.

In the j-basis, in which the matrix $\hat{\Gamma}$ is diagonal, the system of equations

---

*We restrict ourselves to the consideration of sufficiently low temperatures at which above-barrier transitions play no appreciable role. If above-barrier levels are neglected, then $l = \alpha$ or β.

(124) is of the following form:

$$\dot{N}_s = -\gamma N_s + \sum_j \gamma_{sj} N_j + I_s(t),$$

$$\dot{N}_d = -\gamma N_d + \sum_j \gamma_{dj} N_j + I_d(t), \tag{129}$$

$$\dot{N}_j = (-\gamma + \Gamma_j) N_j + \sum_{j'} \gamma_{jj'} N_{j'} + I_j(t) \quad (j \neq s, d),$$

where

$$\gamma_{jj'} = \sum_{l,l'} (n^{-1})_{jl} \gamma_{ll'} n_{l'j'}, \tag{130}$$

$$I_j(t) = (n^{-1})_{jl_0} I(t) \quad \text{(including } j = s, d), \tag{131}$$

and $\Gamma_s = \Gamma_d = \Gamma_0 = 0$ was taken into account. Applying eqs. (128), as well as the second of eqs. (126), we find that

$$N_s = \sum_\alpha N_\alpha \quad \text{and} \quad N_d = \sum_\beta N_\beta \tag{132}$$

are the total populations of the shallow and the deep well, respectively. Instead of N_s and N_d we introduce two new variables $N = N_s + N_d$ and $D = N_s - N_d$ that describe the total and the difference populations of the two wells. By addition and subtraction of the first two of eqs. (129) we obtain a new system of equations in the new variables:

$$\dot{N} = -\gamma N + I_+(t),$$

$$\dot{D} = (-2\gamma + \gamma_{ss} - \gamma_{sd} - \gamma_{ds} + \gamma_{dd}) D/2 + (\gamma_{ss} + \gamma_{sd} - \gamma_{ds} - \gamma_{dd}) N/2$$

$$+ \sum_j{}'' (\gamma_{sj} - \gamma_{dj}) N_j + I_-(t), \tag{133}$$

$$\dot{N}_j = (-\gamma + \Gamma_j) N_j + \sum_{j'} \gamma_{jj'} N_{j'} + I_j(t),$$

where

$$I_\pm(t) = [(n^{-1})_{sl_0} \pm (n^{-1})_{dl_0}] I(t). \tag{134}$$

The double prime on the summation sign in the second equation indicates that terms with $j = s$ and $j = d$ are excluded from the summation. The first equation is separated from the other two. It can be used to find the time dependence of the total population of both wells. The second and subsequent equations of the system (133) describe the intrawell and interwell kinetics. In the first-order approximation with respect to the matrix $\hat{\gamma}$ we should discard all off-diagonal elements γ_{sj} and γ_{dj}. In this approximation, the second equation is also separated from the rest and we obtain an independent system of equations for the total and the difference populations N and D:

$$\dot{N} = -\gamma N + I_+(t),$$
$$\dot{D} = -(\gamma + p)D + rN + I_-(t),$$ (135)

where

$$p = p_{sd} + p_{ds}, \qquad r = p_{sd} - p_{ds},$$ (135a)

$$p_{sd} = z_d \sum_{\beta,\alpha} e^{-\epsilon_\beta/kT} \gamma_{\alpha\beta}, \qquad p_{ds} = z_s \sum_{\alpha,\beta} e^{-\epsilon_\alpha/kT} \gamma_{\beta\alpha}.$$ (135b)

Thus, we have found that in the first-order approximation with respect to $\gamma_{\alpha\beta}$ the interwell kinetics can be investigated on the basis of a two-level arrangement with the effective probabilities p_{sd} and p_{ds} described by eqs. (135b).

4.4. One-phonon transitions

When the distance between adjacent levels of one well is less than ν_D, the limiting frequency of acoustic phonons, and the displacements a_q of the equilibrium positions are small in the transition $l' \leftarrow l$, we can confine ourselves to taking only one-phonon transitions into account in the general eq. (119). Then

$$\gamma_{l'l} = e^{-f^{FC}(T)} \sum_q \left(\frac{\partial \Lambda_{l'l}}{\partial \varphi_q}\right)^2 [(n_q + 1)\delta(\epsilon_{ll'} - \nu_q) + n_q\delta(\epsilon_{ll'} + \nu_q)].$$ (136)

After taking into account relation (121) between $\Lambda_{l'l}$ and the interaction $U_{l'l}$, as well as the small amplitude of the intermolecular vibrations, we obtain the following expression for the coefficient in eq. (136):

$$\left(\frac{\partial \Lambda_{l'l}}{\partial \varphi_q}\right)_0 \approx \left(\frac{\partial U_{l'l}}{\partial \varphi_q}\right)_{a/2} = \int dx\ \Psi_{l'}(x) \frac{\partial U(x, \varphi)}{\partial \varphi_q} \Psi_{l'}(x)$$
$$= \frac{u_q}{\sqrt{N}} \int dx\ \Psi_{l'}(x)S(x)\Psi_l(x) = \frac{u_q}{\sqrt{N}} S_{l'l},$$ (137)

where $u_q/\sqrt{N}$ is the component of a unit vector. Substituting eq. (137) into (136) we obtain

$$\gamma_{l'l} = S_{l'l}^2 e^{-f^{FC}(T)} g_{l'l}(T),$$ (138)

where

$$g_{l'l}(T) = [n(\epsilon_{ll'}) + 1]\rho(\epsilon_{ll'}) + n(\epsilon_{l'l})\rho(\epsilon_{l'l}),$$ (139)

and

$$\rho(\nu) = \frac{1}{N} \sum_q u_q^2 \delta(\nu - \nu_q)$$ (140)

is the weighted density of phonon states. The temperature dependence of the $l' \leftarrow l$ transition is determined by the exponential factor, which decreases as the temperature is raised, and the increasing quantity $g_{l'l}(T)$. Consequently, $\gamma_{l'l}(T)$ as a function of temperature may pass through a maximum. The magnitude of the probability is mainly determined by the factor $S_{l'l}$ and strongly depends on which well the functions of the levels l' and l are localized in. When they are localized in a single well (intrawell transition), the factor $S_{l'l}$ is substantially larger than when they are localized in different wells (interwell transition). Numerical calculations were carried out for the factor $S_{l'l}$ with the function $S(x) = \sin kx$, $\cos kx$ or $\cos 2kx$. Results obtained for the case $S(x) = S_0 \sin kx$ are given in fig. 12 (the other cases show analogous behavior). It is evident that the time for interwell relaxation $1' \leftarrow 1$ and $1 \leftarrow 2'$ differs from the time for intrawell relaxation $1 \leftarrow 2$ by several orders of magnitude, even for a comparatively low barrier between the wells. Resonance between the levels of different wells (potential b) accelerates interwell and slows down intrawell relaxation, i.e. it facilitates, to a considerable extent, the thermalization of the excitations, not in separate wells, but over the whole two-well system. In

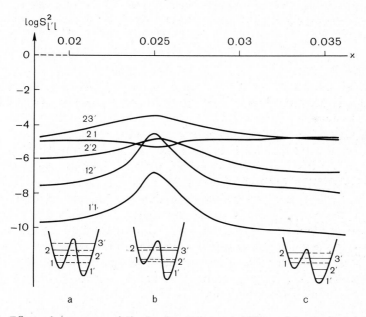

Fig. 12. Effect of resonance of the levels on the probabilities of interwell and intrawell transitions. Here $V = 1200$ and $\xi = 0.37$. The potentials are arranged under the corresponding κ values. The calculation of the factor $S_{l'l}$ in this case and in all numerical calculations in subsection 4 was carried out with $S(x) = S_0 \sin kx$. The constant S_0 was chosen in such a way that the probability $\gamma_{l'l}$ of relaxation between above-barrier levels equalled unity.

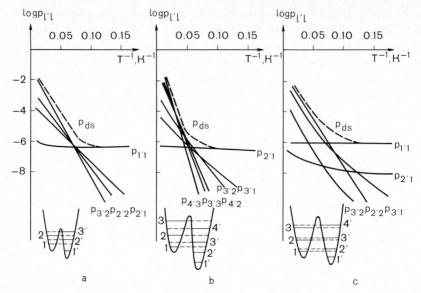

Fig. 13. Temperature dependence of the partial probabilities of transitions from levels of the shallow well to levels of the deep well (full lines) and the effective probability p_{ds} (dashed lines). The probability of transitions between high above-barrier levels has been taken as unity and $V = 1200$. (a) $\xi = 0.37$, $\kappa = 0.01$ (barrier height $U = 130\,\mathrm{cm}^{-1}$); (b) $\xi = 0.4$, $\kappa = 0.035$ ($U = 160\,\mathrm{cm}^{-1}$); (c) $\xi = 0.4$, $\kappa = 0.025$ ($U = 180\,\mathrm{cm}^{-1}$).

considering resonance situations it is necessary to take into account the two-phonon processes in eq. (119), because $\rho(\epsilon_{l'l})$ tends to zero when $\epsilon_{l'l}$ tends to zero.

Making use of the one-phonon approximation, we calculated the effective constants that determine the interwell kinetics from eq. (135b). The weighted density $\rho(\nu)$ in eq. (139) and in all calculations of section 4 was taken in the form $\rho(\nu) = \nu^3 \theta(\nu)\theta(\nu_D - \nu)$, where $\theta(\nu)$ is the Heaviside step function. Shown in fig. 13 is the temperature dependence of the partial probabilities $p_{\beta\alpha} = \gamma_{\beta\alpha} z_s \exp(-\epsilon_\alpha/kT)$ and the total probability p_{ds} (dashed line) for three potentials. Evidently there is no monotonic relation between the barrier height and p_{ds} at a fixed temperature. Moreover, the dependence of $\log p_{ds}$ on the inverse temperature is not linear. This indicates that the simple activation equation $p_{ds} \sim \exp(-U/kT)$, where U is the barrier height, is inapplicable to multi-level systems.

4.5. Phototransitions in systems with two-well potentials

Relaxation processes do not only determine the population of the levels ϵ_l of the anharmonic mode. They are also responsible for the broadening of optical and infrared lines that correspond to the transitions between these levels. But in real systems inhomogeneous broadening occurs, which often appreciably exceeds the homogeneous linewidth. Under these conditions the main factors that affect the shape of the luminescence band are the populations N_l of the levels of the anharmonic mode. Therefore, ignoring the processes of relaxation broadening, we shall take the wave function of an impurity center in the form:

$$F_{lv}^q(r, x, \varphi) = \chi^g(r, x, \varphi)\Psi_l^g(x)\Phi_v(\varphi - a_l^g), \tag{141}$$

where r, x and φ are the coordinates of the optical electron, the anharmonic mode and the harmonic modes, respectively. The superscript $g = 0$ refers to the nonexcited electron state, and $g = e$ to the excited electron state. The vibrational functions Φ and Ψ are found from eqs. (109) and (115), respectively.

In the first-order approximation with respect to the interaction with light, the shape of the optical absorption band is described by the expression:

$$I(\omega) = 2\pi \sum_{l,v} N_l(T)z\, e^{-vv/kT} \sum_{l',v'} \left| \int_{-\infty}^{\infty} d\varphi\, \Phi_v(\varphi - a_l^e)\Phi_v(\varphi - a_l^0) \right|^2$$

$$\times \left| \int_{-\infty}^{\infty} dx\, \Psi_{l'}^e(x)M^{e0}(x)\Psi_l^0(x) \right|^2 \delta(\omega + E_{lv}^0 - E_{l'v'}^e), \tag{142}$$

where $E_{lv}^g = E_g + \epsilon_l^g + vv$ is the electron–vibrational energy, which is the sum of the energy E_g of the electron, the energy ϵ_l^g of the anharmonic mode and the energy vv of the harmonic modes; $N_l(T)$ is the population of level l; $z^{-1} = \Sigma_v \exp(-vv/kT)$; and $M^{e0}(x)$ is the electronic matrix element of the transition $e \leftarrow 0$:

$$M^{e0}(x) = \int dr\, \chi^e(r, x, \varphi)(dE)\chi^0(r, x, \varphi), \tag{143}$$

in which we neglected the dependence on harmonic coordinates φ. Using the integral representation of the δ function, we can transform eq. (142) to the form

$$I(\omega) = \sum_l N_l(T) \sum_{l'} M_{l'l}^2 \int_{-\infty}^{\infty} dt\, e^{i(\omega - \epsilon_{l'l})t} \langle e^{iH_l^0 t} e^{-iH_{l'}^e t} \rangle$$

$$= \sum_l N_l(T) \sum_{l'} M_{l'l}^2 \int_{-\infty}^{\infty} dt\, \exp[i(\omega - \epsilon_{l'l})t + f_{l'l}(t) - f_{l'l}(0)], \tag{144}$$

where

$$M_{l'l} = \int dx\, \Psi_{l'}^e(x)M^{e0}(x)\Psi_l^0(x), \tag{145}$$

$$f_{l'l}(t) = \sum_q [\tfrac{1}{2}(a_{l'}^e - a_l^0)]_q^2[(n_q + 1)\, e^{-i\nu_q t} + n_q\, e^{i\nu_q t}], \tag{146}$$

and $\epsilon_{l'l} = \epsilon_{l'}^e - \epsilon_l^0$. The electron excitation energy is taken as zero. The equations for luminescence are obtained by changing ν_q to $-\nu_q$ in eq. (146). Applying term-by-term integration over time to the expansion of $\exp[f_{l'l}(t)]$ into a series in terms of $f_{l'l}(t)$, we obtain an equation analogous to eq. (47), i.e., the optical band consists of lines, corresponding to the electron–vibrational transitions $el' \leftarrow 0l$, and of the phonon wings (PW) accompanying these lines. If the displacements $(a_{l'}^e - a_l^0)_q$ of the equilibrium positions of the oscillators are sufficiently large, then each phototransition $el' \leftarrow 0l$ had a corresponding PW of Gaussian shape. It is obvious that the intensity distribution in an optical band depends essentially on the population N_l of the anharmonic mode level. And the populations N_l depend upon the probability of the relaxation processes and upon the optical excitation conditions.

4.6. Dependence of the population on the temperature and on the wavelength of the exciting light

In this subsection we consider the practically important special case of steady-state excitation $[I(t) = I]$. Under steady-state conditions we obtain from eq. (135):

$$N = I_+/\gamma, \qquad D = (I_- + rN)/(\gamma + p). \tag{147}$$

These equations establish the relationship between the populations of the two wells under steady-state conditions. Let us calculate the population distribution over the levels under steady-state conditions. In the zeroth-order approximation with respect to the probabilities $\gamma_{l'l}$ of interwell relaxation and under steady-state conditions the system of equations (124) is transformed into

$$-\gamma N_l + \sum_{l'} \Gamma_{ll'} N_{l'} + I\delta_{ll_0} = 0. \tag{148}$$

In this case the subscript l numbers the levels of the well whose level is excited optically with intensity I. Summing all the equations of the system of equations (148) and taking eq. (125) into account, we obtain $I = \gamma \sum_l N_l$. Substituting this relation into eq. (148) we find that

$$-\gamma N_l + \sum_{l'} (\Gamma_{ll'} + \gamma) N_{l'} = 0. \tag{149}$$

Since the probability γ of radiative decay of the excitation is much lower than the probability $\Gamma_{ll'}$ of intrawell relaxation, we obtain in the zeroth-order approximation with respect to γ:

$$\sum_{l'} \Gamma_{ll'} N_{l'} = 0. \tag{150}$$

Recalling eq. (123) and relation (125) we can ascertain by a direct check that the solution of the system of equations (150) is of the form

$$N_\alpha = A\, e^{-\epsilon_\alpha/kT} \quad \text{or} \quad N_\beta = B\, e^{-\epsilon_\beta/kT}, \tag{151}$$

depending upon the well in which the optically excited level is located. The constants A and B are determined by the condition that $\Sigma_l N_l$ is equal to the total population $N = I/\gamma$. The probabilities p and r in eqs. (147) depend upon the temperature and the arrangement of the energy levels in the shallow and deep wells. The intensities $I_\pm$ depend, according to eq. (134), on the number l_0 of the level being excited. Hence eqs. (147) permit a calculation of the dependence of population of the two wells on these factors. We shall consider three cases.

(a) Assume that the light induces a transition to the level $l_0 = \alpha$ of the shallow well. Then, according to eqs. (128) we have $(n^{-1})_{sl_0} = 1$ and $(n^{-1})_{dl_0} = 0$. Hence

$$N = I/\gamma, \qquad D = N(\gamma + r)/(\gamma + p). \tag{152}$$

(b) Assume that the light induces a transition to the level $l_0 = \beta$ of the deep well. Then $(n^{-1})_{sl_0} = 0$ and $(n^{-1})_{dl_0} = 1$. Therefore

$$N = I/\gamma, \qquad D = N(-\gamma + r)/(\gamma + p). \tag{153}$$

(c) Assume that the level l_0 is located above the barrier that separates the wells. Since the probability of a transition from this level to the first levels under the barrier is of the same order of magnitude as the probability of an intrawell relaxation, this case obviously reduces to the first two. It is equivalent to the excitation by light of the first subbarrier levels with intensities $I_- = I_s$, I_d, with $I_s + I_d = I$, where I is the intensity of excitation of level l_0.

According to eqs. (152) and (153), the population of a well does not depend on which level of this well is optically excited. These equations neglect the fact that for levels close to the top of the barrier, the probabilities of interwell and intrawell transitions are almost equal in magnitude. When this is taken into account, we obtain the dependence of the population of the well on the number of the level that is optically excited in the given well. This is demonstrated by Table 1. Listed in this table is the equilibrium population N_s of a shallow well at various temperatures and with various levels excited by light. The calculations have been carried out by means of the exact system of equations (124) with $I(t) = I = \gamma = 1$. The probabilities $\gamma_{l'l}$ and $\Gamma_{l'l}$ were calculated for a potential with the parameters $V = 1200$, $\xi = 0.37$ and $\kappa = 0.01$ (fig. 13a). The distance between the adjacent levels in the same well was taken equal to $80\ \text{cm}^{-1}$. Table 1 demonstrates the strong dependence of the population of the well at low temperature on the excitation wavelength, and the levelling out of this

Table 1
Dependence of the equilibrium population of a shallow
well on the number of the level being excited and on the
temperature.

Level N being excited	Temperature (K)				
	4.2	15	30	45	60
0 (1')*	0	0	0.08	0.22	0.29
1 (1)	1	0.99	0.62	0.34	0.34
2 (2')	0	0	0.08	0.23	0.29
3 (2)	1	0.99	0.60	0.34	0.33
4 (3')	0.08	0.09	0.15	0.25	0.30
5	0.69	0.68	0.42	0.30	0.32
6	0.57	0.57	0.37	0.29	0.32
7	0.48	0.48	0.34	0.28	0.31

*Given in parentheses are the numbers of the same levels as denoted in fig. 13a.

dependence as the temperature is raised. Next we analyze two specific experimental effects on the basis of the theory discussed above.

4.7. "Multiplets"

The optical spectra of polyatomic organic molecules, dissolved in normal paraffins, often have a "multiplet", i.e. a group of resonance lines. In discussing the nature of the multiplet, Bowen and Brocklehurst (1955) supposed that the lines of a multiplet correspond to spatially separated impurity centers. This assumption was soon confirmed experimentally by Svishchev (1963), using the coronene doublet as an example. Somewhat later, Korotaev and Personov (1972) discovered the existence of reversible changes in the lines of a multiplet due to the effect of laser irradiation. These changes contradicted the hypothesis of Bowen and Brocklehurst. It was therefore proposed (Osad'ko 1977b, 1979) to resort to two-well adiabatic potentials to explain the reversible changes occurring in impurity centers*. According to this new hypothesis, a number of lines of the multiplet can belong to a single impurity center, which, under certain conditions, can go over from one metastable state to another. Let us consider this question in more detail.

---

*The specific case of intramolecular proton transport in H_2-porphin by means of light (Solov'ev et al. 1973) has also previously been dealt with in terms of two-well potentials (Völker and van der Waals 1976).

In accordance with eq. (144) we carried out numerical calculations of the optical absorption and fluorescence spectrum for an impurity center having a two-well potential in the ground and excited electronic states. These potentials and levels of the anharmonic mode are shown in fig. 14 at the upper left. Also shown in fig. 14 are the calculated absorption spectrum a and three fluorescence spectra b, c and d, which correspond to various wavelengths of the exciting light. The distance between the adjacent levels 0 and 1 of the shallow well was taken equal to $80 \, \mathrm{cm}^{-1}$. Lines corresponding to electron–vibrational transitions have been artificially broadened; this was done to show the inhomogeneous broadening that always exists in real spectra. We ignored the effect of the phonon wings on the spectrum because phonon wings are actually small in real quasi-line spectra (Osad'ko et al. 1974). The populations N_l in eq. (144) were calculated by means of eqs. (151), (152) and (153), in which the probabilities p and r were calculated from eqs. (135a, b) and (138) with $f^{\mathrm{FC}}(T) = 0$.

At $T = 15 \, \mathrm{K}$, the interwell relaxation is slower than the optical transitions. Therefore, the two states of the impurity center in the shallow and the deep well are displayed spectroscopically as two spatially separated

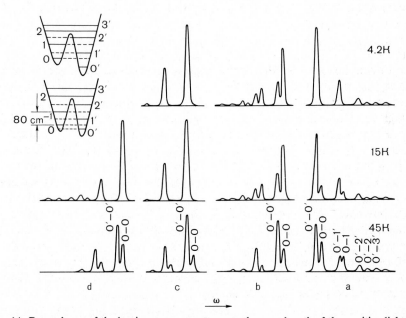

Fig. 14. Dependence of the luminescence spectrum on the wavelength of the exciting light and on the temperature. Spectra: (a) absorption; (b) luminescence upon excitation in the 0′–2 line; (c) in the 0′–0′ line; and (d) in the 0–0 line. Parameters of the upper and lower potentials: $V = 1200$, $\xi = 0.37$ and $\kappa = 0.01$ and 0.05, respectively.

impurity centers. Here, upon excitation with white light or upon laser excitation of high levels, a resonance doublet, fig. 14b, is observed in the fluorescence spectrum. When a component of the doublet is excited, fig. 14c and d, the fluorescence spectrum simplifies, as in the experiments of Svishchev that served as the main proof of the validity of the Bowen and Brocklehurst hypothesis. But when the temperature is raised to 45 K, the rate of interwell relaxation becomes higher than the rate γ of the radiative transition. Consequently, the fluorescence spectrum practically ceases to depend upon the wavelength of the exciting light. Hence, the vanishing dependence of the fluorescence spectrum on the wavelength of the exciting light, as the temperature is increased (see fig. 14 from the top downward), can serve as a qualitative proof of the two-well nature of the doublet.

4.8. "Hole burning" in absorption spectra

In investigating inhomogeneously broadened bands of impurity centers, formed by polyatomic organic molecules in a polycrystalline or vitrifiable solvent, the phenomenon of "hole burning" was discovered. It was found that holes could be burned in an absorption spectrum by laser irradiation with a power rating of several mW/cm^2 (see the review by Personov, chapter 10 of this volume). The hole was maintained for 24 h at 4.2 K, but quickly disappeared when the temperature was raised to 20 K or the specimen was irradiated with white light (Kharlamov et al. 1975). The hypothesis was advanced that "hole burning" in the spectra of aromatic organic substances was due to the transfer of the impurity centers from one potential well to the other (Hayes and Small 1978, Osad'ko 1979). It is, however, most difficult within the framework of this hypothesis to reconcile the very long lifetime of the hole at 4.2 K with its rapid disappearance when the temperature is raised to only 20 K. It is impossible to answer the question posed here without resorting to quantitative theory. The theory set forth above provides an answer to this "delicate" question as well.

   Assume that the impurity center has adiabatic potentials, in the ground and excited states, that have two minima with respect to either their orientational or translational coordinate (fig. 15). Contained in a real solid solution are centers with various energy gaps between the pair of potential curves shown in fig. 15. Hence, the optical spectrum of such a solution is inhomogeneously broadened. Assume further that the laser frequency coincides with the frequency of the transition $0'_d \leftarrow 0_d$. Then, for comparable probabilities of a radiative transition, γ, and tunneling, p'_{sd}, part of the excited impurity centers get into the other well and from there, by means of a phototransition, into the shallow well of the ground electronic state. They remain there for a long time due to the low probability p_{ds} of tunneling through a high barrier. In this manner, the population N_d of the deep well is

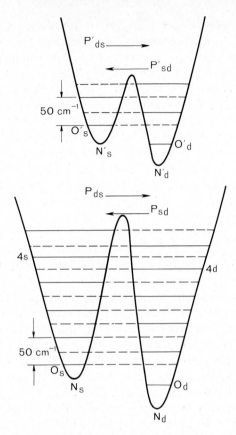

Fig. 15. Two-well potentials for bringing the "hole-burning" theory proposed in the text into agreement with the quantitative data obtained by Kharlamov et al. (1975).

reduced in the process of irradiation and the population N_s of the shallow well is increased. When N_d has decreased appreciably, a narrow hole of a lifetime determined by the probability p_{ds} is formed in the in-homogeneously broadened absorption band. Kharlamov et al. (1975) established the following quantitative facts: (1) the lifetime of the hole at 4.2 K is of the order of 24 h, i.e., $p_{ds} \approx 10^{-5}\,\text{s}^{-1}$; (2) raising the temperature to 20 K increases p_{ds} by four or five orders of magnitude; and (3) the "hole-burning" time for a laser power rating of 10^{16} photons/cm^2 s was 10^2 s. Let us consider these facts on the basis of fig. 15.

At 4.2 K the probability of a transition from the shallow to the deep well is determined by the magnitude of the overlap integral $S(0_d \leftarrow 0_s)$ of the anharmonic functions, eq. (137). A numerical calculation of this magnitude, making use of the potential (117), yielded a linear dependence of log $S(0_d \leftarrow 0_s)$ on the parameter ξ, which, other parameters being constant, determines the height of the barrier, i.e.

$$S(0_d \leftarrow 0_s) = 10^{-45(\xi-0.32)}.\tag{154}$$

The validity of this equation has been proved by direct calculation of the change in the parameter ξ within the range 0.34 to 0.42. Extrapolating to $\xi = 0.5$ we find that this value of the parameter corresponds to the integral $S(0_d \leftarrow 0_s)$ which yields $p_{ds} \approx 10^{-5}\,\mathrm{s}^{-1}$. This is precisely the value of the parameter ξ that the lower adiabatic potential in fig. 15 corresponds to.

According to the second of eqs. (135b), tunneling from the higher levels of the shallow well begins to make an increasing contribution to the probability p_{ds} when the temperature is raised. The contribution, for instance, of the transition $4d \leftarrow 4s$ at 20 K increases p_{ds} by five orders of magnitude, if the distance between adjacent levels in the same well is approximately $50\,\mathrm{cm}^{-1}$. Thus, the lower adiabatic potential in fig. 15 can explain facts (1) and (2) above with a barrier height of approximately $250\,\mathrm{cm}^{-1}$.

Let us consider the kinetics of "hole burning". The "hole-burning" time is determined by the probability of tunneling in the excited state and the intensity of the laser radiation. As a matter of fact, if we neglect tunneling in the ground electronic state of the center, we can write the following system of kinetic equations for the populations of the wells:

$$\dot{N}_d = -kN_d + \gamma N_d',$$

$$\dot{N}_s = \gamma N_s',$$

$$\dot{N}_d' = -(\gamma + p_{sd}')N_d' + p_{ds}'N_s' + kN_d,\tag{155}$$

$$\dot{N}_s' = p_{sd}'N_d' - (\gamma + p_{ds}')N_s'.$$

The prime denotes quantities referring to the excited state, N_d and N_s are the populations of the deep and shallow wells, respectively, γ is the probability of radiative decay of the electronic excitation, $k = \sigma I$, where σ is the effective cross section of the interaction between the impurity center and light, and I is the number of photons of the laser radiation incident on $1\,\mathrm{cm}^2$ per s. The system of equations (155) is essentially of the third order because the variable N_s appears only in the second equation. The attenuation constants Γ of the populations are determined from the condition of solvability of a system of three algebraic equations, i.e. from the equation

$$\Gamma = k\left(1 + \gamma \frac{\Gamma - \gamma - p_{ds}'}{(\Gamma - \gamma)(\Gamma - \gamma - p_{ds}' - p_{sd}')}\right).\tag{156}$$

In the zeroth-order approximation with respect to k, this equation has only the single root $\Gamma^{(0)} = 0$. Corresponding to this root is the solution $N_d =$

const. and $N'_s = N'_d = 0$. In the first-order approximation with respect to k we obtain

$$\Gamma^{(1)} = k \frac{p'_{sd}}{\gamma + p'_{sd} + p'_{ds}}. \tag{157}$$

This root describes the reduction of the population N_d under the effect of the laser radiation, i.e., it determines the kinetics of "hole burning".

Summarizing we can state the following: (1) A numerical calculation indicates that the probability of interwell transitions can be several orders of magnitude lower than that of intrawell transitions even at a barrier height of the order of the phonon frequency. (2) Resonance between levels belonging to different wells greatly increases the probability of interwell transitions and reduces that of intrawell transitions (see fig. 12). (3) Multi-level two-well systems can be dealt with on the basis of a two-level arrangement with the effective transition probability $p = p_{ds} + p_{sd}$ [see eqs. (135a,b)]. (4) The logarithm of the effective transition probability has no linear dependence on the inverse temperature, indicating, thereby, that the simple activation equation $\exp(-U/kT)$, where U is the height of the barrier separating the wells, is inapplicable for multilevel systems. (5) Two-well potentials provide a new approach to the problem of multiplets in quasi-line spectra. (6) On the basis of two-well potentials, phenomena accompanying the "hole-burning" effect, with respect to metastable holes in inhomogeneously broadened impurity center bands, can be explained in a quantitative manner.

5. Nonadiabatic interaction

Within the scope of the adiabatic approximation, the various electronic states interact with each other by means of the HT interaction, as previously mentioned in section 2. But certain effects due to the interaction of the electronic states with each other, for instance, radiationless transitions, cannot be explained by the HT interaction. This interaction is also found to be insufficient in calculating the absorption band of an impurity center having close electronic levels. In such cases also the nonadiabatic interaction, i.e. the solution of the system of equations (4a), must be taken into account. Here the wave function of the impurity center,

$$\Psi(r, R) = \sum_{f,v} C_v^f \varphi^f(r, R) \Phi_v^f(R), \tag{158}$$

is a superposition of the adiabatic functions, and the coefficients C_v^f are determined by a system of algebraic equations following from eq. (4a):

$$(E_{fv} - E)C_v^f + \sum_{f',v'} \Lambda_{vv'}^{ff'} C_{v'}^{f'} = 0, \tag{159}$$

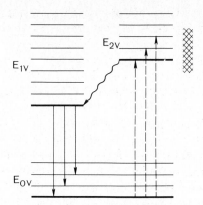

Fig. 16. Electron–vibrational levels of three electronic states. The energy region in which nonadiabatic interaction is essential is indicated by hatching. Luminescence (full arrows) is not affected by nonadiabatic interaction, whereas 0–2 absorption (dashed arrows) is. The wavy arrow indicates radiationless transitions.

where

$$\Lambda_{vv'}^{ff'} = \int_{-\infty}^{\infty} dR\ \Phi_v^f(R)\Lambda^{ff'}(R)\Phi_{v'}^{f'}(R) \tag{160}$$

are the matrix elements of the nonadiabaticity operator. Here E_{fv} and Φ_v^f are the eigenvalues and eigenfunctions of the adiabatic eq. (21).

As is well known, the coefficients C_v^f are determined mainly by quasi-resonant states, which satisfy the criterion

$$|E_{fv} - E_{f'v'}| < \Lambda_{vv'}^{ff'}. \tag{161}$$

If the gap between electronic levels is much larger than the vibrational frequency, as shown in fig. 16, the vibrational quantum numbers v and v' of the quasi-resonant states differ by many units. Since the magnitude of the matrix element $\Lambda_{vv'}^{ff'}$ decreases rapidly with increasing difference $v - v'$, the greater the distance between the electronic levels, the less condition (161) will be satisfied. This is why the ground electronic state of polyatomic organic molecules can practically always be dealt with in the adiabatic approximation. Very frequently the first excited state can also be considered in the adiabatic approximation, because the second electronic state is separated from the first by a gap of five to ten thousand cm^{-1}.

5.1. Nonadiabaticity operator

In practice it is most frequently necessary to take the nonadiabatic interaction between the first and second excited electronic states into account.

Let us discuss the feasibility of solving the system of equations (159) for this special case. The first question that arises in this connection is: can we express the nonadiabaticity operator in terms of quantities observed in experiment? This turns out to be possible, but only under certain conditions. The line of reasoning for this case is as follows. Assume that the levels E_{1v} and E_{2v} in fig. 16 represent the electron–vibrational levels of the first two excited states $f = 1$ and $f' = 2$. Then the nonadiabatic interaction intermixes the electron–vibrational states, whose energies are located in the hatched region. It is obvious that these states do not participate in the low-temperature luminescence. Hence, the shape of the luminescence spectrum can be dealt with in the adiabatic approximation. According to the results of subsection 3.2, the parameters a_q of the FC interaction and α_q and $\alpha_{qq'}$ of the HT interaction can be derived from the luminescence spectrum.

This derivation procedure is based on the application of eq. (96) to the experimental spectrum. It has been carried out repeatedly (Korotaeva and Naumova 1977, Nersesova and Shtrokirkh 1978). Since the HT interaction manifests itself in the dependence of the electronic function $\varphi(r, R)$ on the vibrational coordinates R, it becomes possible to express the nonadiabaticity operator in terms of the parameters of the HT interaction found from the fluorescence spectrum. Let us consider this procedure in more detail.

Within the framework of the assumption that the HT interaction takes the mutual influence of the electronic states 1 and 2 into account, the electronic functions of these two states are, according to eq. (12), of the following form:

$$\varphi^1(r, R) = \varphi^1(r) \cos \theta + \varphi^2(r) \sin \theta,$$
$$\varphi^2(r, R) = -\varphi^1(r) \sin \theta + \varphi^2(r) \cos \theta. \tag{162}$$

The explicit form of the matrix $\hat{A}$, relating the diabatic functions $\varphi(r)$ to the adiabatic functions $\varphi(r, R)$, follows from the unitarity condition

$$(\hat{A}^*)^{-1}\hat{A} = 1. \tag{163}$$

Substituting the functions (162) into eq. (6a) we obtain the following expression for the matrix elements of the nonadiabaticity operator:

$$\Lambda^{11}(R) = \Lambda^{22}(R) = \sum_q \left(\frac{\partial \theta}{\partial R_q}\right)^2, \tag{164}$$

$$\Lambda^{12}(R) = -\Lambda^{21}(R) = \Lambda(R) = \sum_q \left(\frac{\partial \theta}{\partial R_q}\frac{\partial}{\partial R_q} + \frac{1}{2}\frac{\partial^2 \theta}{\partial R_q^2}\right), \tag{165}$$

where R_q are the normal vibrational coordinates. The diagonal matrix elements $\Lambda^{11} = \Lambda^{22}$ can be included, according to eq. (5a), in the adiabatic potential. But we shall consider the off-diagonal matrix elements in more detail. To find the parameter θ we shall resort to the electronic matrix element $M(R)$, which appears in the fluorescence spectrum. Substituting eq. (162) into eq. (20), we obtain for the matrix element of the transition $0 \leftarrow 1$:

$$M(R) = M_{01} \cos \theta + M_{02} \sin \theta, \tag{166}$$

where M_{01} and M_{02} are the matrix elements of the transitions $0 \rightarrow 1$ and $0 \rightarrow 2$ in the Condon approximation (in the absence of HT interaction). The HT interaction is small, i.e. $\theta \ll 1$. Hence, for practical purposes it is sufficient to put

$$M(R) \approx M_{01} + M_{02}\theta. \tag{167}$$

It follows that

$$\frac{\partial M}{\partial R_q} = M_{02} \frac{\partial \theta}{\partial R_q}, \qquad \frac{\partial^2 M}{\partial R_q^2} = M_{02} \frac{\partial^2 \theta}{\partial R_q^2}, \tag{168}$$

i.e. the desired derivatives of θ are proportional to the derivatives of the electronic matrix element. These derivatives are related to the parameters α_q and $\alpha_{qq'}$ by eqs. (97). Expanding $M(R)$ into a Taylor series at the point $\frac{1}{2}a$ we obtain

$$M(R) = M(\tfrac{1}{2}a) \left(1 + \sum_q \alpha_q (R_q - \tfrac{1}{2}a_q) \right.$$
$$\left. + \frac{1}{2!} \sum_{qq'} \alpha_{qq'}(R_q - \tfrac{1}{2}a_q)(R_{q'} - \tfrac{1}{2}a_{q'}) + \cdots \right). \tag{169}$$

By this equation the derivatives $\partial M/\partial R_q$ and $\partial^2 M/\partial R_q \partial R_{q'}$ can be readily expressed in terms of the parameters α_q and $\alpha_{qq'}$. Considering that $M(0) = M_{01}$, we obtain

$$M(\tfrac{1}{2}a) = M_{01}(1 - \boldsymbol{\alpha} \tfrac{1}{2}a + \tfrac{1}{2}a \tfrac{1}{2}\hat{\alpha} \tfrac{1}{2}a - \cdots)^{-1}. \tag{170}$$

We now have all that is needed to express the nonadiabaticity operator $\Lambda(R)$ in terms of the parameters of the FC and HT interactions. Restricting ourselves to the quadratic approximation for $M(R)$, we obtain the following expression for the nonadiabaticity operator Λ:

$$\Lambda(R) = \frac{M(\tfrac{1}{2}a)}{M_{02}} \sum_q \left\{ \left(\alpha_q + \sum_{q'} \alpha_{qq'}(R_{q'} - \tfrac{1}{2}a_{q'}) \right) \frac{\partial}{\partial R_q} + \tfrac{1}{2}\alpha_{qq} \right\}, \tag{171}$$

where $M(\tfrac{1}{2}a)$ is determined by eq. (170), in which only the terms written out are to be taken into account.

Let us analyze eq. (171). Assume that $q = (n, s)$, where n denotes the not completely symmetric vibrations and s the completely symmetric ones. Then, in a linear approximation, the nonadiabaticity operator decomposes into two terms:

$$\Lambda(R) = \Lambda_N(R) + \Lambda_S(R), \tag{172}$$

where

$$\Lambda_N(R) = \frac{M_{01}}{M_{02}} \frac{\Sigma_n \, \alpha_n \, \partial/\partial R_n}{1 - \boldsymbol{\alpha} \frac{1}{2} a}, \qquad \Lambda_S(R) = \frac{M_{01}}{M_{02}} \frac{\Sigma_s \, \alpha_s \, \partial/\partial R_s}{1 - \boldsymbol{\alpha} \frac{1}{2} a}. \tag{173}$$

Account should be taken in eqs. (173) of the fact that the equilibrium position of not completely symmetric oscillators does not usually change in impurity centers with nondegenerate electronic levels ($a_n = 0$). If the symmetries Γ_1 and Γ_2 of the electronic levels 1 and 2 differ, $\alpha_s = 0$ and, consequently, $\Lambda_S = 0$. If $\Gamma_1 = \Gamma_2$, then $\alpha_n = \Lambda_N = 0$. Therefore, in a linear approximation, two electronic states are intermixed due to interaction either only with not completely symmetric vibrations or only with completely symmetric ones. Modes of different symmetries can participate in the intermixing of two electronic states only if $\alpha_{ns} \neq 0$ in the nonadiabaticity operator.

Equation (171) was obtained under the assumption that only the pair of electronic levels, 1 and 2, whose nonadiabatic interaction we intend to investigate contribute to the HT interaction. If other states also contribute to the HT interaction (Bolotnikova and El'nikova 1974), we obtain, instead of eq. (167),

$$M(R) \approx M_{01} + \sum_{j(\neq 1)} M_{0j} \theta_j, \tag{174}$$

and now additional parameters appear in the nonadiabaticity operator. At this stage we are obliged to consider these parameters as unknown. This is why only the operator (171) proves convenient for practical calculations.

5.2. Radiationless transitions

Nonadiabatic interaction is the cause of the radiationless transition from the electronic level f' to the lower lying electronic level f (see fig. 16). The problem of radiationless transitions deserves a special review. Lacking the opportunity to deal here with this problem in detail, we shall discuss only the main obstacle with which the theory of radiationless transitions is confronted, using the transitions between a pair of electronic states, 2 and 1, as an example.

The probability γ_{12} of a radiationless transition from electronic level 2 to electronic level 1 is determined, in the first-order approximation with respect to the nonadiabaticity operator $\hat{\Lambda}$, by the expression

$$\gamma_{12} = 2\pi \sum_v e^{-E_{2v}/kT} \sum_{v'} |\langle 2v|\hat{\Lambda}|1v'\rangle|^2 \delta(E_{2v} - E_{2v'}). \tag{175}$$

Here $|fv\rangle$ and E_{fv} are adiabatic electron–vibrational functions and energies. Using the linear approximation for the nonadiabaticity operator (171) we obtain

$$\Lambda = i \sum_q \lambda_q \, \partial/\partial R_q = \sum_q \lambda_q P_q, \tag{176}$$

where

$$\lambda_q = \frac{M(\tfrac{1}{2}a)}{M_{02}} \alpha_q. \tag{177}$$

Substituting eq. (176) into (175) and making use of the integral representation of the δ-function, we transform eq. (175) into the following form:

$$\gamma_{12} = \int_{-\infty}^{\infty} dt \, e^{i\epsilon_{21}t} \langle 2|\rho[H(R)] \, e^{iH(R)t} \boldsymbol{\lambda P} \, e^{-iH(R-b)t} \boldsymbol{\lambda P}|2\rangle, \tag{178}$$

where $\rho(H)$ is the density matrix, $\epsilon_{21} = E_{20} - E_{10}$, $H(R)$ is the adiabatic vibrational Hamiltonian of electronic state $|2\rangle$, and $b = a^{(1)} - a^{(2)}$ is the displacement of the equilibrium position of the first electronic state with respect to the second. Using the second of eqs. (78), we replace $H(R - b)$ by $H(R)$ and transform eq. (178) to the form

$$\gamma_{12} = \int_{-\infty}^{\infty} dt \, e^{i\epsilon_{21}t} \langle \boldsymbol{\lambda P}(t) \, e^{ibP(t)} \, e^{-ibP} \boldsymbol{\lambda P} \rangle$$

$$= \sum_{q,q'} \lambda_q \lambda_{q'} \left[\frac{\partial}{\partial b_q} \frac{\partial}{\partial b_{q'}} I(b, b'; \epsilon_{21}) \right]_{b=b'}, \tag{179}$$

where

$$I(b, b'; \epsilon_{21}) = \int_{-\infty}^{\infty} dt \, \langle e^{ibP(t)} \, e^{-ib'P} \rangle \, e^{i\xi_{21}t}. \tag{180}$$

Calculating the average in the integrand with the aid of eqs. (83) and (84) we obtain

$$\langle e^{ibP(t)} \, e^{-ib'P} \rangle = \exp[-\varphi(b, b') + f(b, b'; t)], \tag{181}$$

where

$$f(b, b'; t) = \sum_q \tfrac{1}{4} b_q b'_q [(n_q + 1) \, e^{-i\nu_q t} + n_q \, e^{i\nu_q t}], \tag{182}$$

$$\varphi(b, b') = \sum_q \tfrac{1}{8}(b_q^2 + b_q'^2)(2n_q + 1). \tag{183}$$

Substituting eq. (181) into (180) and calculating the value of the time integral, we have

$$I(b, b'; \epsilon_{21}) = 2\pi \, e^{-\varphi(b,b')} \left(\delta(\epsilon_{21}) + \int_{-\infty}^{\infty} d\nu \, f(b, b'; \nu)\delta(\epsilon_{21} - \nu) \right.$$

$$\left. + \frac{1}{2!} \int_{-\infty}^{\infty} d\nu_1 \int_{-\infty}^{\infty} d\nu_2 \, f(b, b'; \nu_1)f(b, b'; \nu_2)\delta(\epsilon_{21} - \nu_1 - \nu_2) + \cdots \right), \tag{184}$$

where $f(b, b'; \nu)$ is the Fourier component of the function (182). Substituting eq. (182) into (179) we obtain

$$\gamma_{12} = \int_{-\infty}^{\infty} d\nu \, \Psi(\nu) I(\epsilon_{21} - \nu)$$
$$+ \int_{-\infty}^{\infty} d\nu \, \chi(\nu) \int_{-\infty}^{\infty} d\nu' \, \chi(\nu')[I(\epsilon_{21}) - 2I(\epsilon_{21} - \nu) + I(\epsilon_{21} - \nu - \nu')],$$

$$(185)$$

where $I(\epsilon)$ is determined by eq. (184), with $b = b'$; this function describes the shape of the absorption band $2 \leftarrow 1$ when only the FC interaction is taken into consideration,

$$\Psi(\nu) = \sum_q (\tfrac{1}{2}\lambda_q)^2 f_q(\nu), \qquad \chi(\nu) = \sum_q \tfrac{1}{4}\lambda_q b_q f_q(\nu), \tag{186}$$

and

$$f_q(\nu) = (n_q + 1)\delta(\nu - \nu_q) + n_q\delta(\nu + \nu_q). \tag{187}$$

Equation (185) expresses the probability of the radiationless transition $1 \leftarrow 2$ in terms of the displacement $b_q = a_q^{(1)} - a_q^{(2)}$ of the equilibrium positions and of the parameters λ_q, characterizing the nonadiabatic interaction. According to eq. (177), the parameters λ_q depend on the quantities α_q, which can be found from the fluorescence spectrum and on the quantities $M(\tfrac{1}{2}a)/M_{02}$. The latter can be found from the ratio of the integrated intensities of the $1 \leftarrow 0$ and $2 \leftarrow 0$ transitions provided that the intensity of the second transition is substantially higher than that of the first. If the parameters λ_q can be determined approximately from spectroscopic data, the displacements b_q of the equilibrium positions can be found only if the absorption spectrum $2 \leftarrow 1$ is available. Considerable difficulties are encountered in obtaining such a spectrum. The unknown displacements b_q are the main obstacle in calculating γ_{12} from eq. (185).

Next we consider the probability γ_{01} of a radiationless transition from level 1 to the ground electronic state. If we neglect the effect of the excited electronic states on the electronic function of the ground state, then $\varphi^0(r, R) = \varphi^0(r)$. In this approximation the radiationless transitions $0 \leftarrow 1$ are absent. If the mutual influence of the ground and first excited states on each other is taken into account,

$$\varphi^0(r, R) = \varphi^0(r) \cos \theta + \varphi^1(r) \sin \theta,$$
$$\varphi^1(r, R) = -\varphi^0(r) \sin \theta + \varphi^1(r) \cos \theta. \tag{188}$$

In this approximation, the nonadiabaticity operator is still described by eq. (165), but the parameter θ is now related to the electronic matrix element $M(R)$ of the $0 \leftarrow 1$ transition by an equation of the form

$$M(R) = M_{01} \cos 2\theta(R). \tag{189}$$

Since $\theta(0) = 0$, eq. (189) does not contain a term linear in R. In this case we do not succeed in expressing the derivatives in eq. (165) in terms of the spectroscopic parameters, and therefore the nonadiabaticity operator contains unknown constants λ_q. This leads to the main difficulty in calculating the probability γ_{01}. The two cases discussed above well illustrate the main problem posed in calculating the probabilities of radiationless transitions: the probability equation always contains unknown quantities that cannot be extracted from experimental data.

5.3. Vibronic spectra

If the distance between the excited electronic states 1 and 2 is comparable to the frequencies Ω_q of intramolecular vibrations, it is necessary to take the nonadiabatic interaction into account in calculating the absorption spectrum in the region of the second electronic transition. To determine the vibronic spectrum it is necessary to calculate the electron–vibrational function (158), i.e., to find the coefficients C_v^f from the infinite algebraic system of equations (159). The solution of this system of equations is equivalent to the solution of the adiabatic system of two differential equations (106) or to the solution of the diabatic system of equations (103). The diabatic system of equations (103) is more frequently used in practical calculations; the matrix elements $V^{11}(R)$, $V^{22}(R)$ and $V^{12}(R)$ are then approximated by quadratic and linear polynomials, and only one vibrational mode is taken into account. The one-mode model allows one to solve the diabatic system of equations (103) numerically without resorting to such assumptions as the smallness of the perturbation V. Such calculations were carried out for Jahn–Teller systems by Longuet-Higgins et al. (1958) and Longuet-Higgins (1961), these systems being doubly and triply degenerate. The vibronic systems were calculated for dimers and molecules having two closely spaced electronic levels (pseudo-Jahn–Teller systems) by Fulton and Gouterman (1964), Gregory et al. (1976) and Henneker et al. (1978). In some papers (Hochstrasser and Marzzacco 1969, Orlandi and Siebrand 1972, Orr and Small 1973, Small 1975) electron–vibrational spectra were dealt with under the assumption of the smallness of the interaction $V(R)$. A vital shortcoming of these numerical calculations was the use of the one-mode model. Such a model greatly restricts the application of the theory to experiment.

As has been noted in section 2, the diabatic approach has advantages over the adiabatic approach only if the electronic functions $\varphi(r)$ can be calculated. These functions are unknown, as a rule, in organic impurity systems. This is why off-diagonal matrix elements $V^{12}(R)$ of a diabatic system introduce unknown parameters into the problem when they are approximated by polynomials. This is an important drawback of the diabatic approach. It makes generalization to a multimode model practically impossible.

Along these lines the adiabatic approach has a doubtless advantage over the diabatic one because, as has been shown in subsection 5.1, it makes it possible to express the nonadiabaticity operator, in certain cases, in terms of parameters extracted from the fluorescence spectrum. In this case, the off-diagonal matrix elements in the system of equations (159) are known. Hence, the coefficients C_v^f can be calculated by numerical methods on an electronic computer, using the multimode model. In this case, in order to calculate the vibronic absorption spectrum it is necessary to specify the relative intensity M_2/M_1 of the second and first electronic transitions, the adiabatic parameters a_q and $a_q^{(2)}$ (displacements of the equilibrium positions of the oscillators) in the $1 \leftarrow 0$ and $2 \leftarrow 0$ transitions, respectively, as well as the vibrational frequencies Ω_q and the parameters α_q of non-Condon behavior.

The parameters a_q, Ω_q and α_q can be determined from the fluorescence spectrum. The rest of the parameters have to be selected. Hence, in the adiabatic approach it is also impossible to eliminate all unknown parameters, but their number is substantially less than in the diabatic approach. The unknown parameters M_2/M_1 and $a_q^{(2)}$ are determined by a fit of the calculated spectrum to the measured one. A numerical calculation of a multimode model in accordance with the above-described procedure has been carried out recently by Chigirov (1979).

5.4. Electron–phonon spectra

Since the phonon frequencies ν_q form a continuum, a direct solution of the system of equations (159) proves inconvenient. Therefore calculations of electron–phonon optical bands were performed either by the semiclassical method (Toyozawa and Inoue 1966) or by the method of ordered operators, but with the neglect of retardation in the correlation functions (Vekhter et al. 1972, Khizhnyakov 1970). These calculations dealt with the case of large heat release, associated with multiphonon bands and strong FC interaction. Neither method is applicable to the description of high-resolution optical spectra at low temperatures. But these are precisely the type of spectra obtained for many organic mixed crystals. Therefore, in the following we shall consider electron–phonon bands under the conditions of the pseudo-Jahn–Teller effect and weak electron–phonon interaction, employing the quantum Green's function method for the calculations. This was the method used by Kogan and Suris (1966) [see also the review by Levinson and Rashba (1973)].

Let us introduce the operators B_f^+ and B_f, the creation and annihilation operators of the fth electron excitation in an impurity center, as well as the creation and annihilation operators b_q^+ and b_q of the phonon q. The operators B_f^+ and B_f satisfy Fermi commutation relations, and b_q^+ and b_q Bose commutation relations. Taking only into account the electron–phonon interaction that is linear in the phonon operators, we can write the impurity crystal Hamiltonian in the form

$$H = H_0 + V, \tag{190}$$

$$H_0 = \sum_f \epsilon_f B_f^+ B_f + \sum_q \nu_q b_q^+ b_q, \tag{191}$$

$$V = \sum_{f,f',q} [V(f,f';q)B_f^+ B_{f'} b_q + V^*(f,f';q)B_{f'}^+ B_f b_q^+]. \tag{192}$$

Here ϵ_f and ν_q are the energy of electron excitation and the phonon frequency. Terms with $f = f'$ in the electron–phonon interaction operator V determine the linear FC interaction, and terms with $f \neq f'$ determine the nonadiabatic and HT interactions. In the diabatic approach these two kinds of interaction are not separated.

In the dipole approximation the operator of the interaction between the impurity center and light can be written in the form

$$\hat{M} = i \sum_f M_f (B_f^+ - B_f), \tag{193}$$

where M_f is the electronic matrix element for the $0 \rightarrow f$ transition. In the first-order approximation with respect to operator $\hat{M}$, the probability $I^a(\omega)$ of the absorption of a photon with frequency ω is described by the expression

$$I^a(\omega) = \pi \sum_v e^{-E_{0v}/kT} \sum_j |\langle 0v|M|j\rangle|^2 \delta(\omega - E_j + E_{0v}), \tag{194}$$

where $|0v\rangle$ and E_{0v} are the eigenfunctions and eigenvalues of the electronically nonexcited crystal, and $|j\rangle$ and E_j are the eigenfunctions and energies of the excited crystal. Making use of the integral representation of the δ-function and taking into account that $B_f|0v\rangle = 0$, we can transform eq. (194) to

$$I^a(\omega) = -\mathrm{Im}\left(\sum_{f,f'} M_f G_{ff'}(\omega) M_{f'} \right), \tag{195}$$

where $G_{ff'}(\omega)$ is the Fourier component of the retarded Green's function,

$$\begin{aligned} G_{ff'}(t) &= -i\theta(t)\langle 0|e^{iHt} B_f\, e^{-iHt} B_{f'}^+|0\rangle \\ &= -i\theta(t)\langle 0|B_f(t)S(t)B_{f'}^+|0\rangle, \end{aligned} \tag{196}$$

where

$$S(t) = e^{iH_0 t}\, e^{-iHt} = \hat{T}\exp\left(-i\int_0^t d\tau\, V(\tau)\right), \tag{197}$$

and the time dependence of the operators $B_f(t)$ and $V(\tau)$ is determined by the zero-order Hamiltonian H_0.

With the aim of comparing the shapes of the electron–phonon absorption and fluorescence bands, occupying a small spectral range, we can write the following expression for the probability $I^e(\omega)$ of emission per unit time of a photon with frequency ω:

$$I^e(\omega) = \text{const.} \sum_j e^{(E_0 - E_j)/kT} \sum_v |\langle 0v|\hat{M}|j\rangle|^2 \delta(\omega - E_j + E_{0v})$$

$$= \text{const.} \sum_v e^{(E_0 - E_{0v} - \omega)/kT} \sum_j |\langle 0v|\hat{M}|j\rangle|^2 \delta(\omega - E_j + E_{0v})$$

$$= \text{const.} \, e^{(E_0 - \omega)/kT} I^a(\omega), \tag{198}$$

where the constant depends very weakly on the frequency. Equation (198), where E_0 is the lowest of the levels E_j, enables one to find the fluorescence band provided that the absorption band has been calculated. Let us now calculate the absorption band.

5.4.1. *Approximate calculation of the absorption band*

We shall calculate the Green's function (196) under the assumption that the interaction V is sufficiently weak that all effects due to this interaction, i.e. electron–phonon phototransitions and radiationless transitions, may be dealt with in the first nonvanishing order with respect to V. Accordingly, in calculating the Green's function we shall consecutively discard all terms of higher order than the approximation chosen. Then in the expansion (196) in terms of V we can certainly discard all terms that are described by Feynman diagrams with intersecting phonon lines. The approximate Green's function $G_{ff'}(\omega)$ obtained in this manner will satisfy the following system of equations [see, e.g., Agranovich (1968)]:

$$G_{ff'}(\omega) = G_f^0(\omega)\delta_{ff'} + G_f^0(\omega) \sum_{f''} \mu(f, f''; \omega)G_{f''f'}(\omega), \tag{199}$$

where

$$G_f^0(\omega) = (\omega - \epsilon_f + i0^+)^{-1}, \tag{200}$$

$$\mu(f, f''; \omega) = \sum_{f_1, f_2, q} V(f, f_1; q)[(n_q + 1)G_{f_1 f_2}(\omega - \nu_q)$$

$$+ n_q G_{f_1 f_2}(\omega + \nu_q)]V^*(f'', f_2; q), \tag{201}$$

and $n_q = [\exp(\nu_q/kT) - 1]^{-1}$. Equation (199) still contains terms of higher order than the accepted approximation. If, in the mass operator μ and in the right-hand side of eq. (199), we replaced function $G_{ff'}(\omega)$ by $G_f^0(\omega)\delta_{ff'}$, we obtain the Green's function in the first nonvanishing order with respect to V. This approximation is quite crude. In this approximation we can calculate the probability of optical electron–phonon transitions, but the probability of radiationless transitions will be equal to zero. Consequently, the HT interaction is taken into account in this approximation, but not the nonadiabatic interaction. In order to take nonadiabatic interaction into consideration, the system of equations (199) must be solved more accurately. Let us transfer the term with $f'' = f$ in eq. (199) from the right- to the left-hand side and solve the obtained equation by an iteration procedure. Restricting ourselves to the first iteration we obtain

$$G_{ff'}(\omega) = \frac{1}{\omega - \epsilon_f - \mu(f, f; \omega)} \left(\delta_{ff'} + \frac{(1 - \delta_{ff'})\mu(f, f'; \omega)}{\omega - \epsilon_{f'} - \mu(f', f'; \omega)} \right). \tag{202}$$

Since the mass operator μ is proportional to V^2, it is sufficient to substitute only the first term of eq. (202) into the right-hand side of eq. (201). Then we obtain for the mass operator

$$\mu(f, f''; \omega) = \sum_{f_1, q} V(f, f_1; q) \left(\frac{n_q + 1}{\omega - \epsilon_{f_1} - \nu_q - \mu(f_1, f_1; \omega - \nu_q)} \right.$$

$$+ \left. \frac{n_q}{\omega - \epsilon_{f_1} + \nu_q - \mu(f_1, f_1; \omega + \nu_q)} \right) V^*(f'', f_1; q). \tag{203}$$

We resort to the iteration method again to solve eq. (203). The first iteration yields

$$\mu^{(1)}(f, f'; \omega) = \Delta_{ff'}(\omega) - i\gamma_{ff'}(\omega), \tag{204}$$

in which the real part of the mass operator is

$$\Delta_{ff'}(\omega) = \sum_{f_1, q} V(f, f_1; q) \left(\frac{n_q + 1}{\omega - \epsilon_{f_1} - \nu_q} + \frac{n_q}{\omega - \epsilon_{f_1} + \nu_q} \right) V^*(f', f_1; q). \tag{205}$$

The function $\Delta_{ff}(\omega)$ determines the shift of the fth electronic level, due to the electron–phonon interaction. The new electronic level E_f is a root of the equation

$$\omega - \epsilon_f - \Delta_{ff}(\omega) = 0. \tag{206}$$

To first order the imaginary part of the mass operator is determined by the expression

$$\gamma_{ff'}(\omega) = \pi \sum_{f_1, q} V(f, f_1; q) f_q(\omega - E_{f_1}) V^*(f', f_1; q), \tag{207}$$

where

$$f_q(\nu) = (n_q + 1)\delta(\nu - \nu_q) + n_q \delta(\nu + \nu_q). \tag{208}$$

If we now substitute the mass operator $\mu^{(1)}$ into eq. (202), the latter will still contain terms of higher order than the accepted approximation. To get rid of them we should carry out the following simplification:

$$\frac{1}{\omega - \epsilon_f - \mu^{(1)}(f, f; \omega)} \approx \frac{1}{\omega - E_f + i\gamma_{ff}(\omega)} \approx \frac{1}{\omega - E_f + i\gamma_f} - i\frac{\gamma_{ff}(\omega) - \gamma_f}{(\omega - E_f + i\gamma_f)^2},$$

(209)

where

$$\gamma_f = \gamma_{ff}(E_f).$$

(210)

In the first simplification above we neglected corrections to the zero-phonon lines that are proportional to V^2, and in the second simplification, corrections to the intensity of the electron–photon phototransitions that are proportional to V^4. If we substitute eq. (209) into eq. (202), and the latter into eq. (195) for the shape of the absorption band, we obtain the expression that is the final result of the present theory:

$$
\begin{aligned}
I^a(\omega) &\approx \sum_f M_f^2 \frac{\gamma_f}{(\omega - E_f)^2 + \gamma_f^2} \\
&\quad + \sum_{f,f'} M_f \frac{\omega - E_f}{(\omega - E_f)^2 + \gamma_f^2} \gamma_{ff'}(\omega) \frac{\omega - E_{f'}}{(\omega - E_{f'})^2 + \gamma_{f'}^2} M_{f'} \\
&= \sum_f \left(M_f^2 \frac{\gamma_f}{(\omega - E_f)^2 + \gamma_f^2} + \pi \sum_q |A(f, q; \omega)|^2 f_q(\omega - E_f) \right),
\end{aligned}
$$

(211)

where

$$A(f, q; \omega) = \sum_{f'} M_{f'} \frac{\omega - E_{f'}}{(\omega - E_{f'})^2 + \gamma_{f'}^2} V(f, f'; q).$$

(212)

According to eq. (211), the absorption band is the sum of the partial bands that correspond to the $f \leftarrow 0$ transitions. The fth band consists of a Lorentzian ZPL of the half width $2\gamma_f$, which determines the probability of radiationless transitions, and a one-phonon PW, described by the summation over q. The amplitude of the one-phonon transitions is described by eq. (212). This amplitude, as well as the probability $2\gamma_f$ of radiationless transitions, have been calculated to the first nonzero order with respect to V.

5.4.2. Two close electronic levels

Let us apply eq. (211) to the case when two excited electronic levels, 1 and 2, are separated by the gap $\epsilon = E_2 - E_1$, which is less than the phonon frequency ν_D (fig. 17b). Considering that the terms diagonal with respect to f in the operator V described by eq. (192), determine the FC interaction, and that the off-diagonal terms determine the nonadiabatic and HT interactions, we introduce the following notation:

$$V(11; q) = F_1(q), \qquad V(22; q) = F_2(q),$$

$$V(12; q) = V^*(21; q) = H(q).$$

(213)

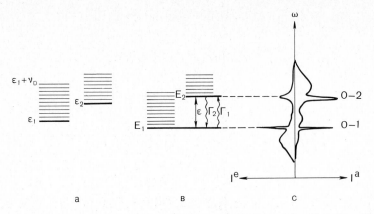

Fig. 17. Electron-phonon levels with (a) the electron–phonon interaction V neglected and (b) with the interaction V "switched on". (c) Schematic representation of the electron–phonon bands at $kT \sim \epsilon$.

Making use of this notation in eq. (212), which describes the amplitudes of one-phonon transitions, and taking the frequency E_1 of the $0 \rightarrow 1$ transition are zero, we have

$$A(1, q; \omega) = M_1 F_1(q) \frac{\omega}{\omega^2 + \gamma_1^2} + M_2 H(q) \frac{\omega - \epsilon}{(\omega - \epsilon)^2 + \gamma_2^2},$$

$$A(2, q; \omega) = M_2 F_2(q) \frac{\omega - \epsilon}{(\omega - \epsilon)^2 + \gamma_2^2} + M_1 H^*(q) \frac{\omega}{\omega^2 + \gamma_1^2}, \tag{214}$$

i.e., each one-phonon amplitude is the sum of two terms, the first of which is due to the FC interaction, and the second to the HT interaction. Phenomena due to the interference of the FC and HT amplitudes have been discussed previously in subsection 3.2.2. They lead to a violation of mirror symmetry (see fig. 7) even when level 2 is at an appreciable distance from level 1. Here we obtain a generalization of this phenomenon to the case of close electronic levels. To ascertain this, we write down the expressions for the probabilities $f_1^a(\omega)$ and $f_1^e(\omega)$ of conjugate one-phonon transitions at zero temperature. The function $f_1^a(\omega)$ is described by the summation over q in eq. (211):

$$f_1^a(\omega) = M_1^2 \pi \sum_q \left| \frac{F_1(q)}{\nu_q} + \frac{M_2}{M_1} H(q) \frac{\nu_q - \epsilon}{(\nu_q - \epsilon)^2 + \gamma_2^2} \right|^2 \delta(\omega - \nu_q). \tag{215}$$

We obtain the function $f_1^e(\omega)$, describing the probability of one-phonon transitions with light emission, from $f_1^a(\omega)$ by making use of relation (198). Since we have taken E_0 as zero, we have $f_1^e(\omega) = \exp[(-\omega/kT)f_1^a(\omega)]$. Since $\exp(\nu_q/kT)n_q = n_q + 1$, we find that $f_1^e(\omega)$ is obtained from $f_1^a(\omega)$ where n_q

and $n_q + 1$ are interchanged. Hence, for $f_1^e(\omega)$ at zero temperature we have the following expression:

$$f_1^e(\omega) = M_1^2 \pi \sum_q \left| \frac{F_1(q)}{\nu_q} + \frac{M_2}{M_1} H^*(q) \frac{\nu_q + \epsilon}{(\nu_q + \epsilon)^2 + \gamma_2^2} \right|^2 \delta(\omega + \nu_q). \tag{216}$$

Let us compare eqs. (215) and (216). The first term in the amplitude is the displacement of the equilibrium position of the qth oscillator, i.e. $F_1(q)/\nu_q = a_q/2$. The second term is the Herzberg–Teller parameter α_q. For $\epsilon \gg \nu_q$ and real $H(q)$ we obtain the following relations for conjugate amplitudes:

$$A^a \approx \tfrac{1}{2} a_q - \alpha_q, \qquad A^e \approx \tfrac{1}{2} a_q + \alpha_q, \tag{217}$$

which are characteristic of conjugate transitions to an isolated electronic level 1 [see the second term in eq. (96)]. In the case of close electronic levels, the interference of the FC and HT amplitudes still leads to a violation of the mirror symmetry of conjugate transitions, but the quantitative relation between the amplitudes of conjugate transitions is more complicated than eq. (217).

Let us next consider the half widths $\Gamma_1 = 2\gamma_1$ and $\Gamma_2 = 2\gamma_2$ of two ZPL. Applying eqs. (210) and (207), we have

$$\Gamma_1 = 2\pi n(\epsilon) \sum_q |H(q)|^2 \delta(\epsilon - \nu_q),$$

$$\Gamma_2 = 2\pi [n(\epsilon) + 1] \sum_q |H(q)|^2 \delta(\epsilon - \nu_q). \tag{218}$$

Both half widths are due to the off-diagonal terms in the interaction V, i.e. to the nonadiabatic interaction. The half width Γ_2 describes the transition from level 2 to level 1 with the creation of a phonon, whereas the half widths Γ_1 describes the probability of the reverse process, i.e. the transition from level 1 to level 2 with the absorption of a phonon from the crystal. The "decay" half width Γ_2 does not vanish as $T \to 0$, whereas the "activation" half width

$$\Gamma_1 = e^{-\epsilon/kT} \Gamma_2 \tag{219}$$

tends to zero as $T \to 0$. ZPL broadening of an activation-like nature was investigated by Krivoglaz (1965).

That the ZPL half width is determined by the nonadiabatic interaction only is a consequence of the fact that in the electron–phonon interaction V only the part linear in the phonon operators is taken into consideration. If quadratic interaction is added to eq. (192) it is still possible, with such a more complicated interaction, to carry out the calculations leading to eq. (211). In this case, the ZPL half width will contain, as a term, the adiabatic half width (Krivoglaz 1965), which has been discussed in subsections 3.1.3

and 3.1.4. The adiabatic half width makes the main contribution to the ZPL half width of $1 \leftarrow 0$ transitions to the isolated level 1. The results of the analysis given here are represented schematically in the absorption band I^a and fluorescence band I^e shown in fig. 17c.

5.5. Luminescence of shallow traps

Impurity levels located near the bottom of the exciton band (shallow traps) are of prime importance in the low-temperature luminescence of organic crystals. A free exciton, captured by a shallow trap, is localized close to an impurity molecule, thereby becoming a localized exciton. Whereas the bands of light absorption by free excitons have been the subject of many papers (Knox 1963, Agranovich 1968, Davydov 1968), the optical bands of localized excitons have been investigated to a much lesser extent. Localized exciton bands should have an intermediate form between the bands of deep traps, i.e. the impurity centers whose spectra we have been discussing up to now, and free exciton bands. This was the aspect of one-phonon luminescence of a localized exciton, captured by an isotopic impurity, that was discussed by Ochs et al. (1974) and by Meletov et al. (1979). We shall consider this problem on the basis of the paper by Osad'ko and Chigirov (1981).

The following Hamiltonian can be related to an impurity molecular crystal:

$$H = H_0 + V, \tag{220}$$

where

$$H_0 = \sum_\kappa \epsilon_\kappa B_\kappa^+ B_\kappa + \sum_q \nu_q b_q^+ b_q, \tag{221}$$

$$V = \sum_{\kappa,\kappa',q} [V(\kappa, \kappa'; q) B_\kappa^+ B_{\kappa'} b_q + \text{h.c.}]. \tag{222}$$

Here ν_q are the phonon frequencies, and ϵ_κ are the energies of free excitons, which are scattered by the impurity molecules ($\kappa = s$) or are localized ($\kappa = \ell$) (see fig. 18). The creation and annihilation operators B_κ^+ and B_κ of the elementary excitations of an impurity crystal are related to the creation and annihilation operators $B_{k\mu}^+$ and $B_{k\mu}$ of excitons of an ideal crystal by the unitary transformation

$$B_{k\mu}^+ = \sum_\kappa \chi_{k\mu}^*(\kappa) B_\kappa^+, \qquad B_{k\mu} = \sum_\kappa \chi_{k\mu}(\kappa) B_\kappa. \tag{223}$$

Formally, the Hamiltonian (220) resembles the Hamiltonian (190). Hence, without repeating the discussion in subsection 5.4, we can at once write

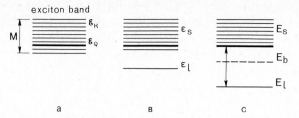

Fig. 18. Energy spectrum of (a) an ideal crystal with fixed molecules, (b) after doping with the impurity and (c) subsequent "switching on" of the electron–phonon interaction. E_b is the energy of the exciton bound to a phonon.

down the expression for the absorption band:

$$I^a(\omega) = -\mathrm{Im}\left(\sum_{\kappa,\kappa'} \lambda_Q(\kappa) G_{\kappa\kappa'}(\omega) \lambda_Q(\kappa')\right). \qquad (224)$$

Here $\lambda_Q^2(\kappa)$ is the integrated intensity of absorption of a photon with wave vector Q by the creation of an elementary excitation κ, and $G_{\kappa\kappa'}(\omega)$ is determined by eqs. (196) and (197), in which the subscript f has to be changed to κ. The approximate calculation of the Green's function is carried out similar to that in subsection 5.4.1 and the result is analogous to eqs. (211) and (212):

$$I^a(\omega) = \sum_{\kappa}\left(\lambda_Q^2(\kappa)\frac{\gamma_\kappa}{(\omega - E_\kappa)^2 + \gamma_\kappa^2} + \pi \sum_q |\lambda_Q(\kappa, q; \omega)|^2 f_q(\omega - E_\kappa)\right), \qquad (225)$$

where

$$\lambda_Q(\kappa, q; \omega) = \sum_{\kappa'} \lambda_Q(\kappa')\frac{\omega - E_{\kappa'}}{(\omega - E_{\kappa'})^2 + \gamma_{\kappa'}^2} V(\kappa, \kappa'; q), \qquad (226)$$

$$\gamma_\kappa = \pi \sum_{\kappa',q} |V(\kappa, \kappa'; q)|^2 f_q(E_\kappa - E_{\kappa'}), \qquad (227)$$

and the function $f_q(\nu)$ is determined by eq. (208).

The luminescence equation can be obtained from eq. (225) by making use of the relation

$$I^e(\omega) = \text{const.}\ e^{(E_\ell - \omega)/kT} I^a(\omega), \qquad (228)$$

where E_ℓ is the energy of a localized exciton. Here the following remark is in order. It has been established (Brodin et al. 1976, Ferguson 1976, Galanin et al. 1980) that the low-temperature exciton luminescence of an anthracene crystal can only be explained on the basis of polariton effects (Agranovich 1968, 1979, Agranovich et al. 1980) that are not taken into account by eq. (228). But polariton effects are appreciable only in crystals whose molecules have a high oscillator strength (for instance, anthracene).

For weakly absorbent crystals (such as naphthalene) polariton effects can be neglected and the luminescence can be described by eq. (228).

In deriving eq. (225) we used no specific concepts concerning the nature of the traps and their concentration. With a numerical calculation in mind, we shall extend eq. (225) by assuming: (a) that an isotopic impurity serves as the traps, (b) that the trap concentration is so low that the existence of an impurity band can be ignored, and (c) that the bottom of the lower exciton band corresponds to optically active excitons with wave vector Q (Q excitons), whereas the luminescence of the higher bands can be neglected. Such a crystal can be described by a model, according to which a single impurity molecule is located in the zero lattice point of the crystal. Then, taking the integrated absorption of light by the band of free excitons as unity, we have

$$\lambda_Q(\kappa) = \chi_Q(\kappa), \tag{229}$$

$$\chi_k(\ell) = \frac{1}{\sqrt{N}} \frac{1}{E_\ell - \mathscr{E}_k} \left[\frac{1}{N} \sum_k \left(\frac{1}{E_\ell - \mathscr{E}_k} \right)^2 \right]^{-1/2}, \qquad \chi_k(s) \approx \delta_{ks}, \tag{230}$$

where $\mathscr{E}_k$ are the energies of the excitons. The concentration in this model is $C = N^{-1}$. Taking into account eq. (228) and the fact that

$$V(\kappa, \kappa'; q) = \sum_k V(k, q)\chi_k(\kappa)\chi_{k-q}(\kappa'), \tag{231}$$

where $V(k, q)$ is the amplitude of the scattering of an exciton of an ideal crystal by the phonon $q = qi$, i being the number of the phonon branch, we can transform the amplitude $\lambda_Q(\kappa, q; \omega)$ of one-phonon luminescence to the form

$$\lambda_Q(\kappa, q; \omega) = \sum_k V(k, q)\chi_k(\kappa) \sum_{\kappa'} \chi_Q(\kappa') \frac{\omega - E_{\kappa'}}{(\omega - E_{\kappa'})^2 + \gamma_{\kappa'}^2} \chi_{k-q}(\kappa'). \tag{232}$$

By making use of the smallness of the concentration, the equation describing the shape of the luminescence band can be simplified. Substituting $\chi_k(s) = \delta_{ks}$ into eq. (232), we obtain the following expression, in the first nonzero order with respect to the concentration, for the shape of the luminescence band ($E_\ell = 0$, and E is the depth of the trap):

$$I^e(\omega) = \chi_Q^2(\ell) \left(\frac{\gamma_\ell}{\omega^2 + \gamma_\ell^2} + f_\ell^+(\omega, T) + f_\ell^-(\omega, T) \right)$$

$$+ e^{-E/kT} \left(\frac{\gamma_Q}{(\omega - E)^2 + \gamma_Q^2} + f_Q^+(\omega, T) + f_Q^-(\omega, T) \right). \tag{233}$$

The expression in parentheses in the first term describes the luminescence of the trap, that in the second term the free exciton luminescence. For deep traps $\chi_Q^2(\ell) \approx C$, i.e., the intensity of trap luminescence is proportional to the trap concentration. The intensity $\chi_Q^2(\ell)$ increases as the trap depth is reduced. The coefficient $\chi_Q^2(\ell)$ is responsible for the increase of light absorption by the localized level as it approaches the exciton band (Rashba 1957, 1962). This effect has been observed repeatedly in experimental investigations (Rashba and Gurgenishvili 1962, Sheka 1972). Band luminescence is small by the activation factor. The first term in each pair of parentheses describes the ZPL of Lorentz shape, whereas the function f^+ and f^- describe the Stokes and anti-Stokes one-phonon luminescence that accompanies each of the two ZPL. The one-phonon luminescence is described by the expressions

$$f_\ell^\pm(\omega, T) = \pi \sum_q (n_q + \tfrac{1}{2} \pm \tfrac{1}{2}) \left(\pm \frac{\nu_q}{\nu_q^2 + \gamma_\ell^2} \sum_k V(k, q) \chi_k(\ell) \chi_{k-q}(\ell) \right.$$
$$\left. + \frac{E \pm \nu_q}{(E \pm \nu_q)^2 + \gamma_Q^2} \frac{E}{\mathscr{E}_{Q+q} + E - \mathscr{E}_Q} V(Q + q, q) \right)^2 \delta(\omega \pm \nu_q),$$

(234)

$$f_Q^\pm(\omega, T) = \pi \sum_q (n_q + \tfrac{1}{2} \pm \tfrac{1}{2}) \sum_k V^2(k, q) \left(\frac{\mathscr{E}_{Q+q} - \mathscr{E}_Q \mp \nu_q}{(\mathscr{E}_{Q+q} - \mathscr{E}_Q \mp \nu_q)^2 + \gamma_Q^2} \delta_{k,Q+q} \right.$$
$$\left. + \frac{\mathscr{E}_k \mp \nu_q}{(\mathscr{E}_k \mp \nu_q)^2 + \gamma_\ell^2} \chi_Q(\ell) \chi_{k-q}(\ell) \right)^2 \delta(\omega - \mathscr{E}_k \pm \nu_q).$$

(235)

Here the upper sign refers to the function f^+ and the lower sign to f^-. The coefficient of the δ-function is the square of the amplitude of a one-phonon transition.

One-phonon transitions. Let us first deal with exciton–phonon transitions that are described by eq. (235). The first term in parentheses is the amplitude of the exciton–phonon transitions of a pure crystal. The second term is the correction to this amplitude due to the existence of a trap. This correction is proportional to $C = N^{-1}$, but increases with a reduction in depth E of the trap. Exciton–phonon transitions play a secondary role in the luminescence spectrum, because they are doubly small: due to the activation factor $\exp(-E/kT)$ and due to the weak electron–phonon coupling.

A more important role is played by the electron–phonon luminescence of the trap [see eq. (234) and fig. 19]. The first term with the double sign in eq. (234) is the FC amplitude and the second term in the parentheses is the HT amplitude. The latter is due to the HT interaction of a localized exciton with the free excitons. The HT amplitude tends to zero, not only as $E \to \infty$, but as $E \to 0$ as well. This last is an indication of the relative reduction of the probability of one-phonon phototransitions, due to the HT interaction, as compared to the zero-phonon transitions, which increases with a reduction of the depth of the trap. The FC amplitude has a different dependence on the depth of the trap (fig. 20). This amplitude decreases monotonically,

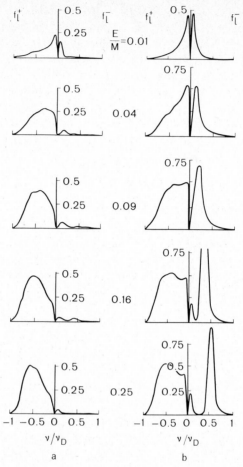

Fig. 19. Dependence of one-phonon luminescence of a trap on its depth E. (a) $kT/\nu_D = 0.05$, (b) $kT/\nu_D = 0.2$, and $\nu_D/M = 0.5$, where M is the width of the exciton band.

compared to the case when $E = \infty$, as the depth of the trap is reduced. As a matter of fact, we can present the FC amplitude in the form

$$\frac{\nu_q}{\nu_q^2 + \gamma_\ell^2} \sum_k V(k, q)\chi_k(\ell)\chi_{k-q}(\ell) \approx \frac{V(q)}{\nu_q} \sum_k \chi_k(\ell)\chi_{k-q}(\ell)$$

$$= \frac{V(q)}{\nu_q} \sum_n C_n^2(\ell)\, e^{-iqn}, \qquad (236)$$

in which the coefficient

$$C_n(\ell) = \frac{1}{\sqrt{N}} \sum_k \chi_k(\ell)\, e^{-ikn} \qquad (237)$$

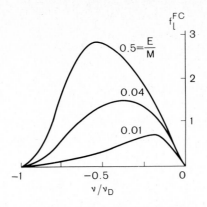

Fig. 20. Dependence of the one-phonon FC amplitude (in arbitrary units) on the trap depth.

describes the distribution of the electronic excitation in a localized exciton over the points of the crystal lattice. For a deep trap, $C_n \sim \delta_{n0}$ and the FC amplitude equals $V(q)/\nu_q = \frac{1}{2}a_q$, where a_q is the displacement of the equilibrium position of the qth mode. The shallower the trap, the larger the region of the crystal covered by the function $C_n^2(\ell)$ and the narrower its Fourier transform in q-space. Consequently, a shallow trap corresponds to less total displacement of the equilibrium position than does a deep trap. This reduction in FC interaction is a consequence of the delocalization of the electronic excitation. Figures 19 and 20 indicate that when the depth of the trap is reduced, the maximum of one-phonon luminescence is displaced to the low-frequency region, approaching the ZPL. In conclusion we point out that the FC and HT amplitudes have the same sign for Stokes luminescence and opposite sign for anti-Stokes luminescence. The resulting interference effects for deep traps have been discussed in subsection 5.4.2.

Zero-phonon transitions. The two ZPL in eq. (233) correspond to these transitions. Their half widths are due to radiationless transitions and can be calculated by eq. (227). Let us first consider the ZPL half width $\Gamma_\ell = 2\gamma_\ell$ of a localized exciton:

$$\Gamma_\ell = 2\pi \sum_{k,q} V^2(k, q)\chi_{k+q}^2(\ell)n_q\delta(E_\ell + \nu_q - \mathcal{E}_k). \tag{238}$$

It is evident from the argument of the δ-function that the half width Γ_ℓ is due to the upward transition of a localized exciton into the band with simultaneous annihilation of a phonon, i.e., the half width is of an activational nature. The ZPL half width $\Gamma_Q = 2\gamma_Q$ of a Q exciton is the sum of two terms:

$$\Gamma_Q = \Gamma_t + \Gamma_s, \tag{239}$$

where

$$\Gamma_t = 2\pi \sum_q V^2(\boldsymbol{Q}, q)\chi^2_{\boldsymbol{Q}-q}(\ell)(n_q + 1)\delta(\mathscr{E}_\boldsymbol{Q} - E_\ell - \nu_q), \tag{240}$$

$$\Gamma_s = 2\pi \sum_q V^2(\boldsymbol{Q}, q)n_q\delta(\mathscr{E}_\boldsymbol{Q} + \nu_q - \mathscr{E}_{\boldsymbol{Q}-q}), \tag{241}$$

where Γ_t is the probability of Q-exciton capture by a trap. The other part of the ZPL half width of Q excitons, i.e. Γ_s, determines the probability of scattering of a Q exciton by a phonon.

Employing the equations of this subsection, Osad'ko and Chigirov (1981) calculated the one-phonon luminescence of a localized exciton (fig. 19), the probability of exciton capture by a trap (fig. 21), and the probability of upward transition of a localized exciton into the band and the scattering of Q excitons (fig. 22). The quantities required for numerical calculations were taken in the form

$$\mathscr{E}_k = \mathscr{E}_0 + \frac{M}{2}\sum_{j=1,2} \sin^2 k_j a_j/2, \qquad \nu_q = \nu_D \left(\frac{1}{3}\sum_{j=1,2,3} \sin^2 q_j a_j/2\right)^{1/2}, \tag{242}$$

$$V(k, qi) = \frac{1}{\sqrt{\nu_q}}\sum_{j=1,2} e^i_j(q)[\Delta \sin q_j a_j + \mu \sin (k_j - q_j)a_j].$$

These calculations were based on anisotropic molecular crystals of the naphthalene type. Let us discuss the results of the calculations.

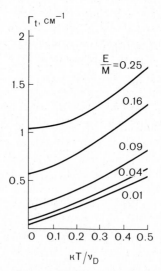

Fig. 21. Probability of free exciton capture by a trap for various trap depths E/M. Trap concentration $C = 0.01$, $\nu_D/M = 0.5$.

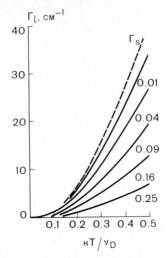

Fig. 22. Probability of thermal upward transition of a localized exciton into the band for various trap depths E/M. The dashed line indicates the probability of Q exciton scattering by a phonon; $\nu_D/M = 0.5$.

According to fig. 19, the intensity of one-phonon transitions decreases with a reduction in the trap depth in comparison to that of zero-phonon transitions, whose probability is taken as unity. The maximum of the Stokes function is displaced toward lower frequencies. The high-frequency peak in the anti-Stokes one-phonon luminescence is due to strong HT transfer of intensity from the exciton band. The dip is due to the subtraction of the FC amplitude from the HT amplitude. Shown in fig. 20 are calculations by means of eq. (234), but with only the FC amplitude taken into consideration. The reduction in intensity and the displacement of the maximum to low frequencies can be explained by the delocalization of the electronic excitation with a decrease in trap depth. Even when the constants Δ and μ have the same magnitude in eq. (242), the contribution of the μ term to the FC amplitude is an order of magnitude less than the contribution of the Δ term. This is due to the nonlocalized nature of the resonance interaction to which the μ term corresponds, and signifies that the contribution of resonance interaction to the FC amplitude can be neglected.

The probability Γ_t of an exciton capture by a trap decreases with the depth of the trap (fig. 21), whereas, on the contrary, the probability Γ_ℓ of the thermal upward transition of a localized exciton into the band increases (fig. 22). The numerical values along the ordinates in figs. 19, 21 and 22 depend upon the magnitude of the electron–phonon interaction. In the present case this magnitude was taken such that the Debye–Waller factor of the luminescence band for a deep trap at zero temperature would be equal to 0.7. This magnitude of interaction approximately corresponds to mixed crystals of naphthalene.

The complete luminescence band of an impurity crystal is shown in fig. 23. It was calculated by means of eq. (233) with the same magnitude of electron–phonon interaction and for a concentration $C = 0.01$. The bands in (a) are normalized to an area under the curves of one. The ratio of the ZPL intensities of the localized and the Q excitons depends to a high degree on the temperature and concentration of the traps.

The principal conclusions are as follows. When the trap depth is reduced: (a) a decrease is observed in the integrated intensity of the trap's electron–phonon luminescence relative to the trap's ZPL (fig. 19): (b) the relative intensity of the low-frequency one-phonon transitions increases (figs. 19 and 20); (c) there is a drastic increase in the rate of temperature broadening of the ZPL of the localized exciton, i.e. in the probability Γ_ℓ of the thermal upward transition of the localized exciton into the band (fig. 22); and (d) there is a decrease in the probability Γ_t of capture of a free exciton. Only at very low temperatures and relatively high concentrations can Γ_t compare with the probability Γ_s of scattering, i.e. be manifest in the ZPL half width of optically active Q excitons being observed (figs. 21 and 22).

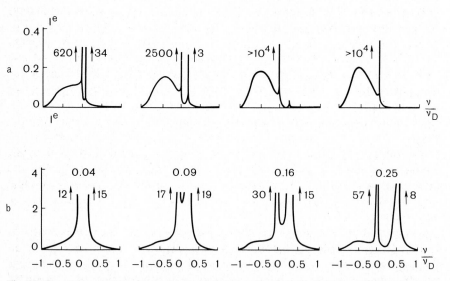

Fig. 23. Dependence of luminescence band shape on the trap depth and on the temperature. (a) $kT/\nu_D = 0.05$, (b) $kT/\nu_D = 0.2$; trap concentration $C = 0.01$ and $\nu_D/M = 0.5$. The numbers alongside the arrows indicate the peak intensity of the ZPL.

6. Conclusion

In the present paper we have dealt with the following effects displayed in optical spectra of an organic impurity center: adiabatic electron–vibrational interaction (section 3), adiabatic potentials with two minima (section 4), as well as nonadiabatic interaction (section 5). As is evident from this review, various parts of the theory have been developed to different degrees of completion.

The greatest possibilities for application to experimental investigations are undoubtedly offered by the theory in the adiabatic approximation. Combining conjugate absorption and luminescence spectra by quantitative relationships, this approach has converted the violation of mirror symmetry of spectra into a most important source of information on the quadratic FC interaction (subsection 3.1) and the HT interaction (subsection 3.2).

The achievements of other parts of the theory dealt with in this review are much less dramatic. So far only the first steps have been taken in elucidating specific features in the spectral manifestations of adiabatic potentials having two minima. For example, only recently have calculations been carried out for the optical spectra of impurity centers having one-dimensional adiabatic two-well potentials. There is still a large number of unsolved problems in this field. They include, for instance, taking the multidimensionality of potentials into account, calculation of the probabilities of above-barrier transitions and tunneling, taking the effects of the temperature into consideration, etc. The difficulties are aggravated by the fact that calculations employing two-well potentials require, as a rule, the application of an electronic computer and numerical techniques. It is quite clear today that a large number of phenomena accompanied by substantial changes in the shape of molecules or in their chemical composition should be described in terms of potentials having several minima. This should serve as a strong impetus for the development of the theory along these lines.

The effect of nonadiabatic interaction in the optical spectra of impurity centers has been investigated more comprehensively than the spectral effects of adiabatic potentials having several minima. But the theory in this field has not reached a stage at which it can be successfully applied to interpret the results of experiment. For example, the calculation of the vibronic spectra of dimers and impurity centers with close electronic levels is carried out using numerical techniques and strongly idealized (one-mode) models. A generalization of available calculation procedures to cover models with many vibrational modes encounters substantial difficulties.

Electron–phonon spectra are calculated at present either under the assumption of the smallness of the electron–phonon interaction or by making use, for a large electron–phonon interaction, of semiclassical calculation methods. Using weak electron–phonon coupling, the theory, as has been shown in the present review, can quantitatively describe the optical properties of shallow traps. An investigation of the luminescence of shallow traps can yield much new information on the electron–phonon interaction, especially in isotopic impurity crystals. This is so because in such crystals we can obtain traps of various depths, practically without changing the electronic properties of the impurity center.

References

Agranovich, V.M., 1968, Theory of Excitons (Nauka Publishers, Moscow) (in Russian).

Agranovich, V.M., 1979, Izv. Akad. Nauk SSSR Fiz. Ser. **43**, 1298.

Agranovich, V.M., S.A. Darmanyan and V.I. Rupasov, 1980, Zh. Eksp. Teor. Fiz. **78**, 656.

Albrecht, A.C., 1960, J. Chem. Phys. **33**, 156.

Anderson, P.W., B.I. Halperin and C. Varma, 1972, Phil. Mag. **25**, 1.

Appell, P. and J. Kampé de Feriet, 1926, Fonctions hypergeometriques et hyperspheriques Polynomes d'Hermite (Paris).

Bersuker, I.B., 1976, Electronic Structure and Properties of Coordination Compounds (Khimiya Publishers, Leningrad) (in Russian).

Bolotnikova, T.N. and O.F. El'nikova, 1974, Opt. Spektrosk. **36**, 292 [1974, Opt. Spectrosc. (USSR) **36**, 168].

Born, M. and K. Huang, 1954, Dynamical Theory of Crystal Lattices (Clarendon Press, Oxford).

Born, M. and R. Oppenheimer, 1927, Ann. der Phys. **84**, 457.

Bowen, E.J. and B. Brocklehurst, 1955, J. Chem. Soc. **99**, 4322.

Brodin, M.S., M.A. Dudinski, S.V. Marisova and E.N. Myasnikov, 1976, Phys. Stat. Sol. (b) **74**, 453.

Chigirov, A.R., 1979, Opt. Spektrosk. **46**, 683.

Coon, J.B., R.E. De Wames and C.M. Loyd, 1962, J. Mol. Spectrosc. **8**, 285.

Craig, D.P. and G.J. Small, 1969, J. Chem. Phys. **50**, 3827.

Davydov, A.S., 1968, Theory of Molecular Excitons (Nauka Publishers, Moscow) (in Russian).

Doktorov, E.V., I.A. Malkin and V.I. Man'ko, 1975, J. Mol. Spectrosc. **56**, 1.

Doktorov, E.V., I.A. Malkin and V.I. Man'ko, 1976, J. Phys. **B9**, 507.

Doktorov, E.V., I.A. Malkin and V.I. Man'ko, 1977, J. Mol. Spectrosc. **64**, 302.

Englman, R., 1972, The Jahn–Teller Effect in Molecules and Crystals (Wiley–Interscience, New York).

Ferguson, J., 1976, Z. Chem. Phys. **101B**, 46.

Fitchen, D.B., 1968, in: Physics of Color Centers, ed. W. Fowler (Academic Press, New York).

Fulton, R.L. and M. Gouterman, 1964, J. Chem. Phys. **41**, 2280.

Galanin, M.D., E.N. Myasnikov and Sh.D. Khan-Magometova, 1980, Izv. Akad. Nauk SSSR Fiz. Ser. **44**, 730.

Gorokhovski, A., R. Kaarli and L. Rebane, 1976, Opt. Commun. **16**, 282.

Gregory, A.R., W.H. Henneker, W. Siebrand and M.Z. Zgierski, 1976, J. Chem. Phys. **65**, 2071.

Hayes, J.M. and G.J. Small, 1978, Chem. Phys. **27**, 151.

Hegarty, J. and W.M. Yen, 1979, Phys. Rev. Lett. **43**, 1127.

Henneker, W.H., A.P. Penner, W. Siebrand and M.Z. Zgierski, 1978, J. Chem. Phys. **69**, 1884.
Herzberg, G. and E. Teller, 1933, Z. Phys. Chem. (Leipzig) **21**, 410.
Hochstrasser, R.M. and C.A. Marzzacco, 1969, Molecular Luminescence, p. 631.
Huang, K. and A. Rhys, 1950, Proc. R. Soc. **204**, 406.
Jäckle, J., L. Piché, W. Arnold and S. Hunklinger, 1976, J. Non-Cryst. Solids **20**, 365.
Kelley, G.J., 1972, Phys. Rev. **136**, 4112.
Kelley, G.J., 1973, Phys. Rev. **138**, 1806.
Kharlamov, B.M., R.I. Personov and L.A. Bykovskaya, 1974, Opt. Commun. **12**, 191.
Kharlamov, B.M., R.I. Personov and L.A. Bykovskaya, 1975, Opt. Spektrosk. **39**, 240.
Khizhnyakov, V.V., 1970, Izv. Akad. Nauk ESSR Fiz.-Mat. Ser. **19**, 144.
Knox, R.S., 1963, Theory of Excitons (Academic Press, New York).
Kogan, Sh.M. and R.A. Suris, 1966, Zh. Eksp. Teor. Fiz. **50**, 1279.
Korotaev, O.N. and M.Yu. Kaliteevskii, 1980, Zh. Eksp. Teor. Fiz. **79**, 439 [1980, Sov. Phys. JETP **52**, 220].
Korotaev, O.N. and R.I. Personov, 1972, Opt. Spektrosk. **32**, 900.
Korotaeva, E.A. and T.M. Naumova, 1977, Opt. Spektrosk. **42**, 912 [1977, Opt. Spectrosc. (USSR) **42**, 524].
Kristofel, N.N., 1974, Theory of Impurity Centers of Small Radius in Ionic Crystals (Nauka Publishers, Moscow) (in Russian).
Kristofel, N.N., K.K. Rebane, O.I. Sil'd and V.V. Khizhnyakov, 1963, Opt. Spektrosk. **15**, 569 [1963, Opt. Spectrosc. (USSR) **15**, 306].
Krivoglaz, M.A., 1964, Fiz. Tverd. Tela **6**, 1707 [1964, Sov. Phys. Solid State **6**, 1340].
Krivoglaz, M.A., 1965, Zh. Eksp. Teor. Fiz. **48**, 310.
Kubo, R., 1962, J. Phys. Soc. Jpn. **17**, 1100.
Kubo, R. and Y. Toyozawa, 1955, Prog. Theor. Phys. **13**, 160.
Kukushkin, L.S., 1963, Fiz. Tverd. Tela **5**, 2170 [1964, Sov. Phys. Solid State **5**, 1581].
Kukushkin, L.S., 1965, Fiz. Tverd. Tela **7**, 54 [1965, Sov. Phys. Solid State **7**, 38].
Lax, M., 1952, J. Chem. Phys. **20**, 1752.
Levenson, G.F., 1971, Phys. Stat. Sol. **43B**, 739.
Levinson, I.B. and E.I. Rashba, 1973, Usp. Fiz. Nauk **111**, 683.
Longuet-Higgins, H.C., 1961, Adv. Spectrosc. **2**, 429.
Longuet-Higgins, H.C., O. Öpik, M.H.L. Pryce and R.A. Sack, 1958, Proc. R. Soc. **244**, A1.
Loorits, L.A. and K.K. Rebane, 1967, Trudy IFA Akad. Nauk ESSR **32**, 3.
Maradudin, A.A., 1966, Solid State Phys. **18**, 273; **19**, 1.
Markham, J.J., 1959, Rev. Mod. Phys. **31**, 956.
McCumber, D.E., 1964, J. Math. Phys. **5**, 221, 508.
McCumber, D.E. and M.D. Sturge, 1963, J. Appl. Phys. **34**, 1682.
Meletov, K.P., E.I. Rashba and E.F. Sheka, 1979, Zh. Eksp. Teor. Fiz. Pis'ma **29**, 184.
Mostoller, M., B.N. Ganguly and R.F. Wood, 1971, Phys. Rev. **B4**, 2015.
Nersesova, G.N. and O.F. Shtrokirkh, 1978, Opt. Spektrosk. **44**, 102 [1978, Opt. Spectrosc. (USSR) **44**, 58].
Ochs, F.W., P.N. Prasad and R. Kopelman, 1974, Chem. Phys. Lett. **29**, 290.
O'Malley, T.F., 1971, Adv. At. Mol. Phys. **7**, 223.
Orlandi, G. and W. Siebrand, 1972, Chem. Phys. Lett. **15**, 465.
Orr, G. and G.J. Small, 1973, Chem. Phys. **2**, 60.
Osad'ko, I.S., 1970, Fiz. Tverd. Tela **12**, 2123 [1971, Sov. Phys. Solid State **12**, 1686].
Osad'ko, I.S., 1972, Fiz. Tverd. Tela **14**, 2927 [1973, Sov. Phys. Solid State **14**, 2252].
Osad'ko, I.S., 1973, Fiz. Tverd. Tela **15**, 2429 [1974, Sov. Phys. Solid State **15**, 1614].
Osad'ko, I.S., 1977a, Zh. Eksp. Teor. Fiz. **72**, 1575 [1977, Sov. Phys. JETP **45**, 827].
Osad'ko, I.S., 1977b, Phys. Stat. Sol. (b) **82**, K107.
Osad'ko, I.S., 1979, Usp. Fiz. Nauk **128**, 31 [1979, Sov. Phys. Usp. **22**, 311].
Osad'ko, I.S. and A.R. Chigirov, 1981, Fiz. Tverd. Tela **23**, 538.

Osad'ko, I.S. and S.A. Kulagin, 1980, Opt. Spektrosk. **49**, 290.

Osad'ko, I.S. and S.A. Zhdanov, 1977, Fiz. Tverd. Tela **19**, 1683 [1977, Sov. Phys. Solid State **19**, 982].

Osad'ko, I.S., E.I. Al'shits and R.I. Personov, 1974, Fiz. Tverd. Tela **16**, 1974 [1975, Sov. Phys. Solid State **16**, 1286].

Pekar, S.I., 1950, Zh. Eksp. Teor. Fiz. **20**, 510.

Pekar, S.I., 1953, Usp. Fiz. Nauk **50**, 197.

Perlin, Yu.E. and B.S. Tsukerblat, 1974, Effects of Electron–Vibrational Interaction in Optical Spectra of Impurity Paramagnetic Ions (Shtiintsa Publishers, Kishinev) (in Russian).

Personov, R.I., I.S. Osad'ko, E.D. Godyaev and E.I. Al'shits, 1971, Fiz. Tverd. Tela **13**, 2653 [1971, Sov. Phys. Solid State **13**, 2224].

Phillips, W.A., 1972, J. Low Temp. Phys. **7**, 351.

Ranson, P., P. Peretti, J. Laport and Y. Rousset, 1976, J. Chim. Phys. **73**, 545.

Rashba, E.I., 1957, Opt. Spektrosk. **2**, 568.

Rashba, E.I., 1962, Fiz. Tverd. Tela **4**, 3301.

Rashba, E.I. and G.E. Gurgenishvili, 1962, Fiz. Tverd. Tela **4**, 1029.

Ratner, A.M. and G.E. Zilberman, 1959, Fiz. Tverd. Tela **1**, 1697.

Rebane, K.K., 1968, Elementary Theory of the Vibrational Structure of Impurity Center Spectra in Crystals (Nauka Publishers, Moscow) [1970, English translation, Plenum Press, New York].

Rebane, K.K., O.I. Sil'd and I.Yu. Tekhver, 1964, Trudy IFA Akad. Nauk ESSR **27**, 23.

Sharp, T.E. and H.H. Rosenstock, 1964, J. Chem. Phys. **41**, 3453.

Sheka, E.F., 1972, in: Physics of Impurity Centers in Crystals, ed. G. Zavt (Tallinn) p. 431 (in Russian).

Sheka, E.F. and K.P. Meletov, 1977, Mol. Cryst. Liq. Cryst. **43**, 203.

Silsbee, R.H., 1962, Phys. Rev. **128**, 1726.

Small, G.J., 1972, Chem. Phys. Lett. **15**, 147.

Small, G.J., 1975, J. Chem. Phys. **62**, 4661.

Solov'ev, K.N., I.E. Zalesskii, V.N. Kotlo and S.F. Shkirman, 1973, Zh. Eksp. Teor. Fiz. Pis'ma **17**, 463.

Sponer, H. and S. Wollman, 1941, J. Chem. Phys. **9**, 816.

Sturge, M.D., 1967, Solid State Phys. **20**, 91.

Svishchev, G.M., 1963, Izv. Akad. Nauk SSSR Fiz. Ser. **27**, 696.

Toyozawa, Y., 1967, in: Tokyo Summer Lectures on Theoretical Physics, eds. R. Kubo and H. Kamimura, p. 461.

Toyozawa, Y. and M. Inoue, 1966, J. Phys. Soc. Jpn. **36**, 1663.

Vekhter, B.G., Yu.E. Perlin, V.Z. Polinger, Yu.B. Rozenfeld and B.S. Tsukerblat, 1972, in: Physics of Impurity Centers in Crystals, ed. G. Zavt (Tallinn) p. 461 (in Russian).

Völker, S. and J.H. van der Waals, 1976, Mol. Phys. **32**, 1703.

Völker, S., R.M. Macfarlane, A.Z. Genack, H.P. Tromsdorff and J.H. van der Waals, 1977, J. Chem. Phys. **67**, 1759.

Winterling, G., 1975, Phys. Rev. **B12**, 2432.

Zeller, R.C. and R.O. Pohl, 1971, Phys. Rev. **B4**, 2029.

CHAPTER 9

Persistent Nonphotochemical Hole Burning and the Dephasing of Impurity Electronic Transitions in Organic Glasses

GERALD J. SMALL

Ames Laboratory* and Department of Chemistry
Iowa State University
Ames, IA 50011
U.S.A.

*Operated for the U.S. Department of Energy by Iowa State University under contract No. W-7405-Eng-82. This research was supported by the Director for Energy Research, Office of Basic Energy Science, WPAS-KC-01-03-01-1.

Spectroscopy and Excitation Dynamics
of Condensed Molecular Systems
Edited by
V.M. Agranovich and R.M. Hochstrasser

Contents

1. Introduction

The burning of *persistent* holes (dips) in the site inhomogeneously broadened absorption bands of organic molecules imbedded in low temperature matrices was first observed by Kharlamov et al. (1974, 1975) and Gorokhovskii et al. (1974). The hole production resulted from irradiation with a narrow line laser in an impurity vibronic absorption band associated with the lowest singlet (S_1) absorption system. By the term persistent, it is generally meant that the holes last *indefinitely* with no profile change provided the sample is maintained at or below the temperature at which the holes are burned (the "burn temperature" T_B) and in the dark. Obviously, this type of solid state hole burning spectroscopy has its origin in mechanisms very different from that operative in transient saturation hole burning spectroscopy (Lee and Skolnick 1967, Szabo 1975). In this chapter, we will not be concerned with the latter and, unless otherwise mentioned, hole burning spectroscopy will be understood to be of the persistent type.

Generally in a hole burned spectrum, a (zero-phonon) hole is observed coincident with the laser burn frequency ω_B. Additionally, holes in vibronic bands other than the one coinciding with ω_B can be produced. The latter will be referred to as intramolecular satellite holes. In fig. 1 the arrows locate the positions of the most prominent satellite holes in the S_1 absorption system of tetracene in an EtOH:MeOH glass. The deepest hole occurs in the origin band at 476.5 nm, coincident with ω_B. Another common feature is the "phonon side band" hole which appears at low frequency and most prominently to the low energy side of the zero-phonon hole. An example is apparent in fig. 1 just to the right of the zero-phonon hole at 476.5 nm. Much more dramatic examples of the phonon side band hole are evident in the hole burned spectra of cresyl-violet in the EtOH:MeOH glass, fig. 2, where the zero-phonon holes coincident with the laser burn frequency are very weak in comparison.

The initial step of the hole burning process is the preparation of the isochromat (subset of impurity sites whose absorption profiles overlap the laser profile). This is also the initial step in fluorescence line narrowing which is discussed in chapter 9 of this volume by Personov. The relationship between hole burning and fluorescence line narrowing will be considered later in this chapter. It should be clear that hole production is due

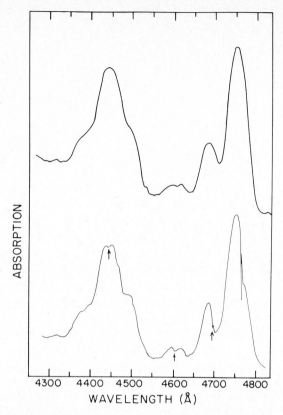

Fig. 1. Nonphotochemical hole burning of tetracene's S_1 absorption system, $T_B = 2$ K. The glass is 4:1 ethanol–methanol. The upper trace is the absorption spectrum before burning and the lower the spectrum after burning at 476.5 nm for 20 minutes with a laser flux of ≈ 50 mW/mm^2. The deepest hole is in the origin band at 476.5 nm. (From Hayes and Small 1979.)

to site selective photobleaching and that the elucidation of the mechanisms of such represents an interesting problem.

Before delineating which of the many aspects of hole burning spectroscopy will be emphasized in this chapter, it would be useful to consider several of its possible applications so that the current excitement in the field can be better appreciated. They are:

(i) vibronic and electronic state assignments in biologically important molecules (Kharlamov et al. 1977, Friedrich et al. 1980a, 1981);

(ii) kinetics of ultra-fast intramolecular photochemical reactions;

(iii) intramolecular vibrational relaxation (de Vries and Wiersma 1976, Voelker and Macfarlane 1979);

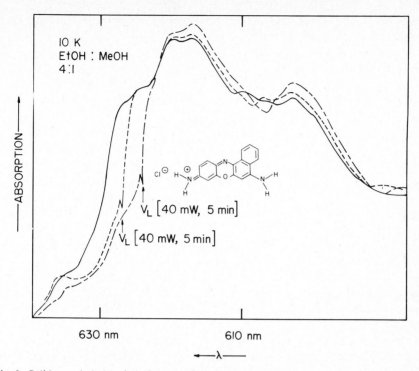

Fig. 2. Solid state hole burning of cresyl-violet in the ethanol–methanol glass, $T_B = 10$ K. The solid curve is the absorption spectrum (S_1) prior to burning. The dashed curve is the spectrum following the first burn at the frequency V_L furthest to the left. Note the shallow zero-phonon hole coincident with V_L and the very intense and broad phonon side band hole to lower energy of V_L. The interrupted dashed spectrum was recorded following a second burn at V_L (furthest to the right) with a laser flux of 40 mW/mm^2 and an irradiation time of 5 min. Note that this burn eliminates (fills) the first zero-phonon hole. (Unpublished data from our laboratory.)

(iv) dependence of impurity site energy distribution on electronic state;

(v) dephasing of excited electronic states due to the electron–phonon interaction (Voelker et al. 1977, 1978);

(vi) structure of amorphous solids and dephasing of impurity electronic transitions in glasses (Hayes and Small 1978a,b, Hayes and Small 1979, Hayes et al. 1980, 1981).

Hole burning has also been utilized for ultra-high resolution Stark and Zeeman studies (Macfarlane and Shelby 1979). With reference to point (i) above, we note that it is possible to perform two-photon excitation spectroscopy on an absorption system following one-photon hole burning (Edelson et al. 1979). Additionally, solid state hole burning may find important technological applications, e.g. fabrication of high density computer memory storage devices and ultra-narrow high transmission optical filters.

In the literature one finds reference to photochemical hole burning (PHB) spectroscopy and nonphotochemical hole burning (NPHB) spectroscopy, the implication being that in certain systems the hole burning mechanism is photochemical in nature while in others it is nonphotochemical. To date, the term NPHB has been restricted to *photostable* molecules, e.g. perylene, tetracene, anthracene, imbedded in glasses since hole burning for such molecules in crystalline matrices has not been observed. It will become apparent that the distinction between PHB arising from intermolecular photochemistry and NPHB in amorphous solids can be subtle. Originally, the term NPHB was introduced to make a distinction between it and PHB due to the intrinsic photoreactivity of the impurity (Hayes and Small 1978a).

Since the focus of this chapter is on NPHB spectroscopy and because PHB is discussed only when interesting parallels or contrasts with the former can be made, it is appropriate at this point to mention that PHB has been observed in several systems including: H_2-phthalocyanine in n-octane (Gorokhovskii et al. 1974, 1976, Gorokhovskii 1976); H_2-tetra-4-tert-butyl phthalocyanine in tetradecane (Gorokhovskii and Kikas 1977) and in n-nonane (Gorokhovskii and Rebane, 1977); dimethyl-s-tetrazine in durene (de Vries and Wiersma 1976) and in polyvinylcarbazole (Cuellar and Castro 1980); resorufin in polymethylmethacrylate (Marchetti et al. 1977); free-base porphyrin in n-octane (Voelker et al. 1977); the F_3^+ color center in NaF (Macfarlane and Shelby 1979); a variety of dihydroxyquinones in polymers and glasses (Graf et al. 1978, Drissler et al. 1980, Friedrich and Haarer 1980); and phycoerythrin and c-phycocyanin in glycerol/buffer glasses (Friedrich et al. 1980, 1981).

Our decision to emphasize NPHB spectroscopy was based on a number of considerations. The first, of course, is that it has been extensively studied in our laboratory. Another is that because NPHB is a manifestation of the disorder which characterizes the glassy state, the disorder is an underlying theme which ties together the various facets of NPHB. One has hope, therefore, that a well-knit story can be told. Yet another is that while PHB followed logically from an understanding of fluorescence line narrowing, NPHB was quite unexpected. As might be imagined, NPHB can shed light on the disorder of glasses and, furthermore, the framework for understanding hole formation and profiles (dephasing) of impurities in glasses is novel.

The organization of this chapter is as follows: In section 2 the early work on NPHB is reviewed and a mechanism for hole formation is discussed. Section 3 deals primarily with a theory for NPH profiles and, more generally, the dephasing of impurity electronic transitions in glasses. It will be emphasized that the theory may prove useful for understanding the dephasing of optical transitions of rare earth ions imbedded in "hard" inorganic glasses. In section 4, we consider NPHB data which, as of yet, have no clear interpretation, e.g. data on hole filling by thermal annealing. The primary purpose of this section is to underscore the fact that much work remains to be done before it can be said that our understanding of the glassy state or NPHB is satisfactory. Finally, in section 5, we consider a few of the many promising directions for NPHB spectroscopy in the years ahead.

2. The properties of nonphotochemical holes, and a mechanism of their production

The observation by Kharlamov et al. (1974, 1975) that following laser irradiation (15 mW/cm^2) persistent holes are burned into the origin band of the lowest singlet absorption systems of perylene and 9-aminoacridine in an ethanol glass and that the mechanism is of the one-photon type was very interesting since these two molecules are thought to be photostable in their lowest singlet state. The observed hole widths of ≈ 0.5 cm^{-1} at 4.2 K were interpreted as being homogeneous. However, no broadening mechanism was proposed. Although their results suggested that the mechanism for hole production might be nonphotochemical, the possibility that the hole production was photochemical (due to unknown photochemistry in ethanol) was not eliminated. Moreover, no mechanism for the non-photochemical production of holes was offered. Since our initial interest was in the study of the hole burning of molecules, known to undergo intramolecular photochemistry, imbedded in glasses, it was important to ascertain whether NPHB of organic molecules in disordered media can occur and, if so, what the mechanism for hole production is and how one can distinguish between a photochemical and a nonphotochemical hole.

In a series of early survey experiments performed at 2 K, hole burning was observed in the S_1 absorption systems of naphthalene, anthracene and pyrene (Hayes and Small 1977). The glasses employed were 4:1 ethanol–methanol and 5:2 ethyl ether–isopropanol. Hole burning for tetracene in these glasses was also observed (Hayes and Small 1978a)—see for example fig. 1—and for phenanthrene in the ethanol–methanol glass (Edelson et al. 1979). All five molecules are *not* known to be photoreactive in their S_1 states. The laser flux employed was typically in the range 10–30 mW/mm^2.

2.1. Some important properties of nonphotochemical holes

Prior to the formulation of a mechanism for hole production (Hayes and Small 1978a), the hole burning studies on the naphthalene, anthracene and tetracene impurity–glass systems at $T_B \approx 2\,K$ reveals that the holes have the following properties:

(i) Their production rate depends linearly on the light flux, that is, is consistent with a one-photon process. For tetracene in the ethanol–methanol glass, both a cw and pulsed dye laser of approximately the same average power were employed for hole burning. No significant variations in hole burning quantum yield were observed (Hayes and Small 1978a,b, 1980).

(ii) They exhibit saturation. That is, after a certain laser fluence at ω_B hole burning ceases. A recent example of the time evolution of a zero-phonon hole (and its phonon side band hole) is shown in fig. 3 for the tetracene in glycerol–dimethylsulphoxide (DMSO)–dimethylformadide (DMF) glass (1:1:1). The laser flux employed was $\approx 10\,mW/mm^2$ and the saturation time of $\approx 130\,min$ is fairly typical for many of the systems

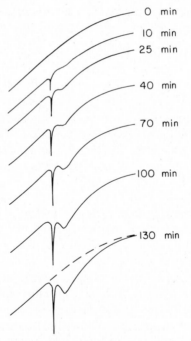

Fig. 3. Irradiation time dependence of the zero-phonon and associated low energy phonon side band hole for tetracene in the glycerol–DMSO–DMF glass (see text). The times labeling the curves give the burn duration with a laser flux of $10\,mW/mm^2$, $T_B = 1.9\,K$. The zero-phonon hole is saturated by 130 min. The laser burn was into the origin band of tetracene's S_1 absorption system at 476.5 nm. Note the increase in the intensity of the side band hole relative to the sharp zero-phonon hole with increasing irradiation time. (From Hayes et al. 1981.)

studied. Usually, the saturation corresponds to an optical density (OD) change in the range 10–20% for the zero-phonon hole. The hole burning quantum yields are very low, for example, about 10^{-4} for the tetracene in glycerol–DMSO–DMF glass (Hayes et al. 1981).

(iii) The holes persist indefinitely after their production, provided the sample is maintained at or below the burn temperature T_B and in the dark [irradiation with an intense white light source can irreversibly eliminate them (Kharlamov et al. 1975)]. For the tetracene in ethanol–methanol and glycerol–DMSO–DMF systems, the zero-phonon holes coincident with the laser burn frequency have been monitored for several hours following their production with no detectable change in the hole profiles.

(iv) The holes can be irreversibly removed by sample warming above T_B.

(v) The $T_B \approx 2\,\mathrm{K}$ width (FWHM) of the zero-phonon hole in the in-homogeneously broadened origin band of the S_1 absorption system and coincident with ω_B is quite broad, $\approx 1\,\mathrm{cm}^{-1}$ (not laser or instrument broadened). A recent example is shown in fig. 4 for the tetracene in glycerol–DMSO–DMF glass. Zero-phonon *photochemical* holes far nar-rower than this had been observed. Examples are shown in figs. 5 and 6.

(vi) The hole burning efficiency is not correlated to the linear electron-

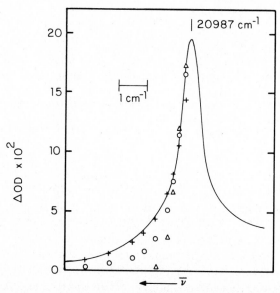

Fig. 4. Zero-phonon hole profile for tetracene in the glycerol–DMSO–DMF glass, $T_B = 1.8\,\mathrm{K}$. The hole is coincident with the laser burn frequency at $20\,987\,\mathrm{cm}^{-1}$ (476.5 nm), see caption to fig. 3. ΔOD denotes the change in optical density produced by the burn. The solid curve is experimental. The circles and triangles are Lorentzian and Gaussian fits while the crosses were calculated using eq. (44). (From Hayes et al. 1981.)

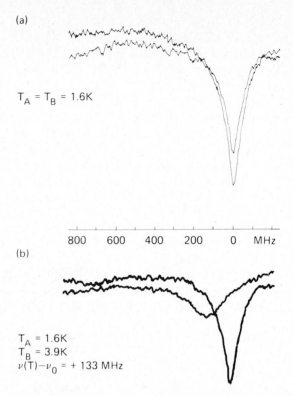

Fig. 5. Photochemical hole burning of free-base porphin in an n-octane crystal. The upper traces (a) show two zero-phonon holes corresponding to samples in two different cryostats, A and B. The holes are coincident with the laser burn frequency for a burn temperature of 1.6 K. The lower set of tracings (b) show the broadening and shift of the zero-phonon hole of sample B when its temperature is raised (following the burn at 1.6 K) to 3.9 K. The vibronic band being burned is the origin band of the S_1 absorption system. (From Voelker et al. 1978.)

phonon interaction as measured by fluorescence line narrowing studies (Hayes and Small 1978a). That is, the zero-phonon hole production mechanism does not depend on the excitation of the optically allowed phonon modes. This observation is consistent with the fact that a zero-phonon hole at ω_B is produced.

(vii) The $T_B \approx 2$ K saturation depth and width can show a dependence on cool-down procedure, see subsection 3.1.2 for further discussion.

Although the alleged photostability of the molecules discussed above, observation (ii), and the fact that hole burning for tetracene in p-terphenyl and Shpolskii crystals was not observed (Hayes and Small 1978a) pointed to the existence of a nonphotochemical mechanism for hole production in glasses, it was left to the hole filling experiment of Hayes and Small (1978a)

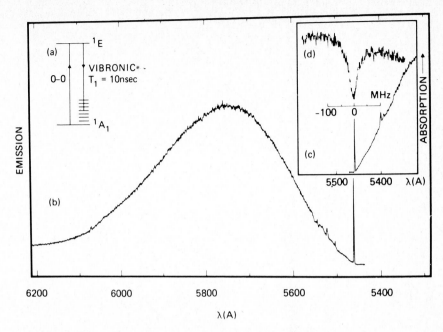

Fig. 6. Photochemical hole burning in the zero-phonon absorption band of the F_3^+ color center in NaF, $T_B = 1.5$ K. The transition being burned is indicated by (a) and a blow up of the zero-phonon hole, shown in (b), is given in (d). All traces are excitation spectra. There is also a population hole burning mechanism operative in this system. (From Macfarlane and Shelby 1979.)

to provide the most convincing case. The experiment involves burning a primary hole at ω_B, followed by the burning of additional holes at other burn frequencies ω'_B. The results for tetracene in the ethanol–methanol glass are shown in fig. 7 with $\omega_B = 21\,041$ cm^{-1} (laser linewidth = 0.6 cm^{-1}). Burning a secondary hole at $21\,038$ cm^{-1} was observed to have no effect on the primary hole. However, production of a third hole midway between the two resulted in their being partially filled. The analysis showed that $\approx 45\%$ of the tetracene sites burnt are shifted to sites differing in their excitation energies by <2 cm^{-1}. Given that any conceivable photochemistry of tetracene would most likely produce a photoproduct whose absorption is substantially removed from that of tetracene and that it would be irreversible, Hayes and Small (1978a) concluded that the hole production mechanism was nonphotochemical. It may well be, however, that the hole filling experiment is not a fool-proof .method for distinguishing between photochemical and nonphotochemical holes; see the discussion on annealing in section 4.

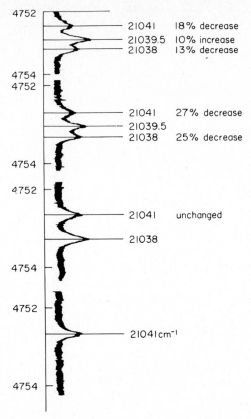

Fig. 7. Hole filling in the origin band of tetracene's S_1 absorption system. $21\,041\ cm^{-1}$ is the primary burn frequency, see text for discussion. (From Hayes and Small 1978a.)

2.2. The two-level systems structural model for nonphotochemical hole burning in glasses

It seems clear that the NPHB must arise from some type of impurity–glass "photoisomerization" process which permanently depletes the population of impurity sites whose excitation energies overlap the laser profile. The only mechanism which has been proposed (Hayes and Small 1978a,b, Hayes et al. 1980, 1981) is based on the so-called two-level systems (TLSs) structural model for the glassy state. In this model, the glass is considered to contain a distribution of asymmetric intermolecular double-well potentials (TLSs), see fig. 8, with varying barrier height V, zero-point energy difference ϵ and well separation d. There may also be a variety of distinctly different TLS coordinates q. In fig. 8, TLS_G and TLS_F are the potential

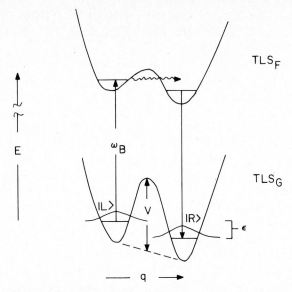

Fig. 8. Two-level system potential energy diagrams for a two-level system coupling to an impurity in its ground (TLS$_G$) and excited electronic state (TLS$_F$), see text. (From Hayes et al. 1981.)

energy curves for a particular TLS interacting with an impurity molecule in its ground (G) and excited (F) electronic states. Before discussing the hole production mechanism, we must emphasize that the TLS structural model was first proposed by Anderson et al. (1972) and Philips (1972) in order to explain the anomalous T-dependences of the specific heat and thermal conductivity of inorganic glasses observed at low temperatures, $T \leqslant 10$ K. Such measurements and others will be discussed in section 5. Suffice it to say now that TLSs appear to be a universal accoutrement of glasses (Hunklinger and Arnold 1976, Hunklinger 1978).

To explain NPHB at a given burn temperature, T_B, Hayes and Small (1978a) proposed that hole formation is due to a *subset* of the TLSs interacting with impurity molecules which, at the temperature T_B, have the following two properties:

(i) Relaxation (either by phonon assisted tunneling or thermal activation) between the two minima of the TLS$_G$ is sufficiently long (i.e. many hours) to account for the hole persistence times, see observation (iii) of subsection 2.1.

(ii) The impurity electron–TLS interaction is sufficiently strong to modify the double-well potential to an extent which makes relaxation between the TLS$_F$ minima competitive with the normal, e.g. radiative, decay processes of the excited electronic state of the impurity. Again, relaxation can occur by PAT (phonon assisted tunneling) or by a thermally activated process. Typical fluorescence lifetimes for the molecules studied thus far are in the range ≈ 100–1 ns.

With these two properties, one can readily see from fig. 8 that if the laser burn frequency ω_B is on resonance with the "left" transition but not with the "right" transition, hole burning of the persistent type can ensue. The details of hole formation are discussed in section 3. A key feature of the above mechanism is the assertion that at a given T_B only a subset of the TLSs interacting with the impurity is involved in hole formation. This was necessary in order to explain hole saturation or, in other words, why 100% OD changes are not generally observed. That this assertion is valid was first established by Hayes and Small (1978b) for the tetracene in ethanol–methanol glass system. They observed that for $T_B = 2$ K the zero-phonon hole coincident with ω_B (laser linewidth = 0.25 cm^{-1}) had a full width at half maximum (FWHM) of 0.9 cm^{-1} and that this hole was irreversibly eliminated by warming by 10 K. Under identical conditions, but with $T_B = 13$ K, a much broader hole (FWHM = 5.3 cm^{-1}) was burned. The profile of this high temperature hole suffered no change upon reduction of the temperature to 2 K. Subsequently, hole burning was repeated at 2 K and the end result was a hole clearly interpretable as the superposition of the 2 and 13 K holes. This was an interesting result because it proved that *by burning holes at different burn temperatures it is possible to probe different TLS subsets.* Since this early work, the temperature dependent hole burning of tetracene in ethanol–methanol has been extensively studied (Hayes et al. 1981) and the data are reproduced in fig. 9. The open data points correspond to the integrated OD of the zero-phonon hole as a function of T_B. The solid data points show the decrease in hole intensity with thermal annealing following production of the saturated hole. One sees, for example, that while a $T_B \approx 2$ K hole is annealed out by 5 K, a $T_B \approx 8$ K hole persists to a much higher temperature, roughly 30 K. Although experimental details are discussed elsewhere (Hayes et al. 1981), a few remarks on the method of studying the hole intensity as a function of T_B are in order. The open data points were obtained sequentially starting from the left (lowest T_B) and proceeding to the right. Following the production and measurement of a hole at T_B, the laser was shuttered and the hole completely eliminated by warming. The sample temperature was then decreased to the next $T'_B > T_B$ and the process repeated. This procedure is time consuming and it was subsequently determined that for this system the same data can be obtained without having to anneal prior to the next burn.

The data in fig. 9 and similar data for another system are more thoroughly discussed in section 4. A discussion of the hole profiles associated with these data is postponed until section 3.

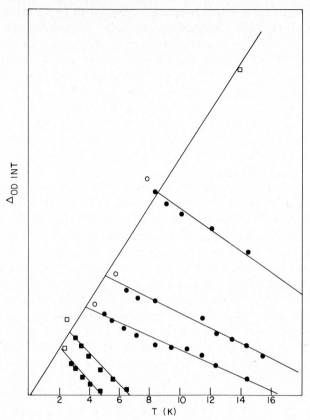

Fig. 9. Temperature dependence of the integrated optical density (ΔOD_{int}) for a zero-phonon hole of tetracene in the ethanol–methanol glass. Open data points show the dependence of the hole intensity (saturated) on burn temperature, T_B. Solid data points show the disappearance of the hole with warming above T_B. (From Hayes et al. 1981.)

We conclude this subsection by noting that the TLS mechanism for NPHB was conceived in ignorance of the early theoretical work of Anderson et al. (1972) and Philips (1972). Thus, the fact that very different data led to the same TLS structural model for soft and hard glasses adds to its credibility; nevertheless, it is just that—a (simple) model which has met with a great deal of success. As new types of data on glasses are obtained, it will very likely have to be improved upon, see section 4.

2.3. *The origin of the phonon side band holes*

The occurrence of phonon side bands building on the high energy side of zero-phonon lines in the optical absorption spectra of pure and mixed crystals is very common and need not be discussed here. In most cases, it is the result of the linear electron–phonon interaction and is interpretable in terms of the Condon approximation. The observation that the dominant phonon side band hole lies on the *low energy side* of the associated zero-phonon hole (see figs. 1–3) now appears to be very general and is most obvious following saturation of the zero-phonon hole. The first observation of this was for the tetracene in ethanol–methanol glass system (Hayes and Small 1978a). These authors interpreted the hole as being due to impurity sites with zero-phonon absorption lines to lower energy of ω_B which undergo hole burning by virtue of absorption by their high energy phonon wings which overlap the laser profile. No theoretical discussion of the total hole shape including zero-phonon and high and low energy phonon side band holes was given. An approximate, but quite satisfactory, theoretical discussion based on the above model was subsequently reported by Friedrich et al. (1980b). In what follows, we consider the essential features of their treatment and some of their results.

To a first approximation at least, the phonon sideband hole problem does not depend on whether hole burning proceeds by a photochemical or nonphotochemical mechanism. Friedrich et al. (1980b) were interested in the time evolution of the photochemical hole burning for 1,4-dihydroxy-anthraquinone (quinizarin) in the ethanol–methanol glass shown in fig. 10. Earlier work (Graf et al. 1978, Drissler et al. 1980) had shown that the photochemistry proceeds via intermolecular hydrogen bonding with the alcoholic glass molecules. The behavior shown in fig. 10, where the low energy phonon side band hole is more intense than its high energy counterpart, and steadily increases in intensity relative to the zero-phonon hole with increasing burn time, is qualitatively similar to that shown in fig. 3. The fact that the zero-phonon photochemical hole in fig. 10 saturates is interesting. It may be that an equilibrium between the intra- and intermolecular hydrogen bonded forms of quinizarin is being established and/or that a subset of the quinizarin sites does not have a configuration appropriate for creation of the intermolecular hydrogen bond.

Friedrich et al. (1980b) adopt a model in which there is a Gaussian distribution of site excitation energies with an inhomogeneous half width $\Gamma_{inh} \gg \Gamma_z$. Γ_z is the homogeneous half width of an individual site zero-phonon line. For glasses at low temperatures, this is certainly a valid assumption. Γ_z and its associated Debye–Waller factor

$$\alpha = e^{-S} \tag{1}$$

are *assumed* to be independent of site. In eq. (1),

$$S = \frac{1}{2} \sum_j \frac{\mu_j \omega_j}{h} \Delta q_j^2, \tag{2}$$

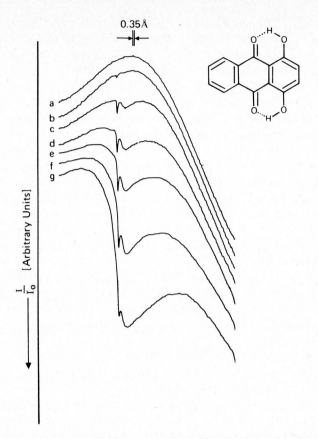

Fig. 10. Irradiation time dependence of the zero-phonon and associated phonon side band holes for quinizarin in the ethanol–methanol glass, $T_B = 2\,K$. Curve a is the origin absorption band of the S_1 state prior to burning. Curves b–g are spectra following burning at 514.5 nm for 1 s, 6 s, 16 s, 1 min, 5 min and 32 min, respectively. The laser power was 6 mW. (From Friedrich et al. 1980b.)

with μ_j and ω_j the appropriate mass and frequency for the jth phonon mode. The term Δq_j is the normal coordinate shift which results from impurity excitation. We assume that $S < 1$ so that we are in the weak linear electron–phonon coupling regime, i.e., one-phonon structure is dominant but weak relative to the zero-phonon line. The weak light source limit where ground state depletion effects can be neglected is also assumed.

For the case of a normal Gaussian distribution for site excitation energies, and the absorption line function

$$g(\omega - \omega') = \alpha z(\omega - \omega') + (1 - \alpha)p(\omega - (\omega' + \Delta)) \qquad (3)$$

for a single site, it follows that

$$\frac{N_\tau(\omega')}{N} = \left(\frac{\ln 2}{\pi}\right)^{1/2} \Gamma_{\text{inh}}^{-1} \exp\left(-\frac{\omega'^2 \ln 2}{\Gamma_{\text{inh}}^2}\right) \exp[-\sigma I \phi \tau g(\omega_B - \omega')] \tag{4}$$

is the fraction of absorbers left at frequency ω' after an irradiation time τ for an absorption cross section σ, a laser intensity I (photons $\text{cm}^{-2}\text{s}^{-1}$) and the hole burning quantum yield ϕ. Γ_{inh} is the half width of the Gaussian distribution, ω_B the burn frequency and in eq. (3) z and p are the zero-phonon and phonon side band line shape functions, respectively. Δ corresponds to the frequency of the single-phonon side band measured relative to the zero-phonon line. The absorption line shape function, $A_\tau(\omega)$, is the convolution of eq. (4) with the normalized line shape function $g(\omega - \omega')$, i.e.,

$$A_\tau(\omega) = \int_{-\infty}^{+\infty} \frac{N_\tau(\omega')}{N} g(\omega - \omega')\, d\omega'. \tag{5}$$

In the *short burn time limit*, where the last exponential term in eq. (4) is close to unity, it follows easily that

$$\begin{aligned}
A_{\tau=0}(\omega) - A_\tau(\omega) = \sigma\phi I \tau \alpha \bigg(&\int_{-\infty}^{+\infty} \alpha z(\omega_B - \omega') z(\omega - \omega')\, d\omega' \\
&+ (1-\alpha)[p(\omega - (\omega_B + \Delta)) \\
&\qquad + p(\omega_B - (\omega + \Delta))] \bigg)
\end{aligned} \tag{6}$$

is the hole line shape function after a burn time τ. In the derivation of eq. (6), it is assumed that the single-site zero-phonon line shape function is a deltoid relative to the phonon side band. The first term in eq. (6) is the

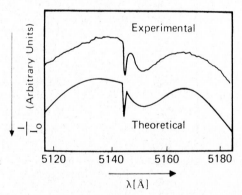

Fig. 11. Experimental and theoretical hole profiles for quinizarin in ethanol–methanol, $T_B = 2\ \text{K}$. See text for discussion. (From Friedrich et al. 1980b.)

zero-phonon hole line shape function, which is a Lorentzian with half width equal to $2\Gamma_z$ for the case z is a Lorentzian with half width Γ_z. The second and third terms describe the phonon side band holes building to the high and low energy side of ω_B, respectively. The high energy side band should be viewed as building on the zero-phonon hole. From eq. (6) one observes that in the short burn time limit, the intensities of the three types of holes increase linearly with τ and that the side band holes are of equal intensity.

Such is not the case for longer burn times and when saturation of the

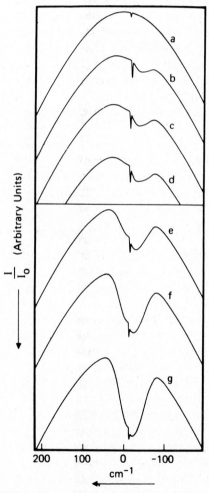

Fig. 12. Calculated irradiation time dependence of the hole profile for quinizarin in the ethanol–methanol glass, see text. (From Friedrich et al. 1980b.)

zero-phonon hole occurs prior to that of the low energy phonon sideband hole. In this case the hole structure must be simulated using eq. (5) appropriately modified to take into account saturation. Using this equation and experimentally determined values for Γ_z, Δ and Γ_{inh}, Friedrich et al. (1980b) have performed computer calculations for the quinizarin in ethanol–methanol glass system. Figure 11 shows their theoretical fit to an experimental profile (fig. 10d) for a Debye–Waller factor $\alpha = 0.58$. The calculation mirrors the main features of the observed spectrum. The upper half of fig. 12 shows the simulated time evolution of the hole profile corresponding to curves a–d in fig. 10. Again, the agreement is reasonable although the calculated zero-phonon hole saturates too early (by curve b in fig. 12).

In our opinion, this early saturation is probably the result of the assumption that even for a very narrow laser source a single quantum yield can be used for all (burnable) sites whose excitation energies overlap the laser profiles. The point is that in a glass, sites having the same excitation energy can have significantly different ground and excited state "lattice" configurations. Thus, it would be more appropriate to use a distribution of quantum yields chosen so as to match the observed saturation time. Additional improvements would follow by relaxing the requirement that the Debye–Waller factor is independent of site excitation energy. There seems little doubt, however, that the simple model of Friedrich et al. (1980b) correctly accounts for the main features of the hole growth.

2.4. Population bottle neck and power broadening effects on hole profiles

In the preceding subsection, we were only concerned with understanding the general features of the zero-phonon plus phonon side band hole profiles. As will be discussed in section 3 for NPHB in glasses, the zero-phonon hole profile can provide us with important information about the dynamical (dephasing) processes the coupled impurity–TLS "clusters" undergo during the hole formation process. It is important, therefore, that appropriate measures be taken to ensure that contributions to the hole profile, other than that from dephasing, be taken into account.

The importance of using a burn laser and a probe laser (or continuum source spectrometer) whose linewidths are narrow relative to the homogeneous width of the hole is rather obvious and need not be considered here. Another cautionary remark is that one must ensure that the flux of the burn laser is sufficiently low to eliminate sample warming and the resulting thermal broadening of the hole. In other words, the measured sample temperature should accurately reflect the actual sample temperature.

De Vries and Wiersma (1980) have recently given a detailed density matrix treatment on the effects of a population bottle neck and power broadening on (persistent) hole profiles. The reader is referred to this paper for details and, in addition, a very thorough listing of earlier related work. Because their treatment is not sufficiently complete for NPHB, in the sense that hole saturation, the site distribution of NPHB quantum yields and the electron–phonon interaction are not taken into account, we will limit our discussion of their work to a few pertinent remarks.

De Vries and Wiersma show that the contribution to the homogeneous width of the Lorentzian hole profile from power broadening will be negligible when

$$\chi^2 T_1 T_2 \ll 1, \tag{7}$$

in agreement with the earlier work of Sargent and Toschek (1976). The parameter χ is the on-resonance Rabi frequency equal to $eDE_0/\hbar$, where D is the transition dipole for the *vibronic* transition being burned and E_0 the electric field amplitude. T_2 and T_1 are the total dephasing and longitudinal relaxation times. From published NPHB experiments, we may take the following values as typical for the laser (cw) burn flux, D, T_1 and T_2: 10 mW/mm^2, 0.1 Å, 10^{-8} s and 10^{-12} s. One calculates that $\chi^2 \approx 10^{13} \text{ s}^{-2}$ so that the above inequality is easily satisfied. In general, one can assert that with cw laser fluxes sufficiently low to avoid sample heating, the power broadening is negligible. With pulsed lasers, this may not be the case so that a study of the hole width dependence or burn power is advisable.

For organic molecules pumped into a vibronic band of their S_1 absorption spectrum, relaxation by intersystem crossing to the T_1 state may occur prior to hole burning in S_1. Of course, hole burning may occur in the T_1 state. Nonetheless, one must consider the contribution to the persistent hole width from the T_1 population bottle neck which develops due to the long T_1 lifetime (T_1 here should not be confused with the longitudinal relaxation time of the state directly pumped by the burn laser). It is instructive first to consider the case where persistent hole burning is not operative, i.e. population bottleneck hole burning by itself. The steady state solution of the equations of motion for the appropriate density matrix elements yields (de Vries and Wiersma 1980)

$$\Gamma^h = \frac{2}{T_2}[1 + (1 + K^2 T_2^2)^{1/2}] \tag{8}$$

for the homogeneous hole width (circular frequency). In this equation

$$K^2 = \chi^2(2 + k'_{IX}k_{IX}^{-1})/2T_2\kappa, \tag{9}$$

with k'_{IX} and k_{IX} the rate constants corresponding to the $T_1 \leftrightsquigarrow S_1$ and $S_0 \leftrightsquigarrow T_1$ relaxation processes. The term κ is the total relaxation constant for the S_1 state. The hole depth relative to the unburned absorption coefficient is given by

$$\mathscr{D} = \frac{K^2 T_2^2}{(1 + K^2 T_2^2)^{1/2} + 1 + K^2 T_2^2}.$$ (10)

Equations (8) and (10) provide a useful guide for estimating when the population bottle neck may complicate a persistent hole burning experiment. Referring to the model calculation in our discussion of power broadening and applying the same parameter values to the problem at hand, one can readily see, from eq. (9), that the ratio k'_{IX}/k_{IX} would have to equal ca. 10^6 for the bottle neck to affect the homogeneous hole width.

Noting again that eq. (8) is only valid in the steady state condition, it is clear that Γ^h for burn times shorter than the time required to establish a steady state, will be narrower than the steady state value. It turns out, however, that the most reliable experimental method for extracting T_2 from a hole burning experiment is to determine the dependence of Γ^h on χ^2 for a fixed burn time and extrapolate to zero laser burn flux (de Vries and Wiersma 1980). This is also the case for persistent hole burning.

To end this subsection, we comment on an interesting possible complication to the problem of determining T_2 from PHB or NPHB experiments. It is this: the process of persistent hole formation must, by definition, change the structure of the solid. Thus, when there is communication between different impurity sites, those sites which are initially on resonance and undergo hole burning may "drive" other sites, initially off-resonance, into resonance. One expects that this long range communication would be most effective in the easily deformable organic glasses and polymers.

3. Nonphotochemical hole profiles and ultra-fast dephasing of impurity transitions in glasses

Within the last year, it has become apparent that the homogeneous hole widths of impurity molecules imbedded in organic glasses are anomalously large in comparison with the photochemical hole widths observed in crystalline hosts (Hayes et al. 1980). A comparison of fig. 4 and fig. 5 effectively illustrates the magnitude of the difference. Figures 13 and 14 show two other examples of a large hole width at low T_B (burn temperature) for quinizarin in the ethanol–methanol glass and dimethyl-s-tetrazine in the polyvinylcarbazole polymer. Recall also that in the early work of Kharlamov et al. (1974, 1975) on perylene and 9-aminoacridine in an ethanol glass, the homogeneous hole widths were reported to be $\approx 1\,\mathrm{cm}^{-1}$ for $T_B = 4\,\mathrm{K}$. To our knowledge, the dependence of the zero-phonon hole width on T_B has only been reported for the tetracene in

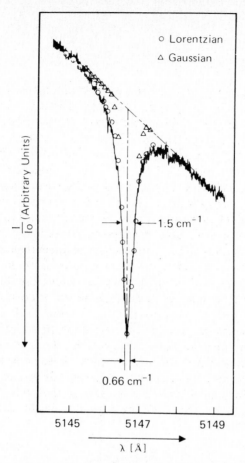

Fig. 13. The zero-phonon hole profile for quinizarin in the ethanol–methanol glass, $T_B = 2$ K and the laser power = 3 mW. 0.66 cm^{-1} is the spectrometer band pass used to measure the spectrum. The triangles and circles are Gaussian and Lorentzian fits. (From Friedrich and Haarer 1980.)

ethanol–methanol and glycerol–DMSO–DMF glasses (Hayes et al. 1981). These data are reproduced later in this section. Suffice it to say now that this dependence is linear with T_B for $T_B \gtrsim 4$ K and that the homogeneous hole width at $T_B = 4$ K is about 2 cm^{-1}. Such widths correspond to dephasing times 2–3 orders of magnitude shorter than those observed in photon echo experiments for polycyclic aromatic hydrocarbons imbedded in host crystals (Wiersma 1980).

It was pointed out by Hayes et al. (1980) that existing theories (Aartsma and Wiersma 1976, Wiersma 1980, Jones and Zewail 1978, Harris 1977, Small 1978, de Bree and Wiersma 1979, Burke and Small 1974a,b) of

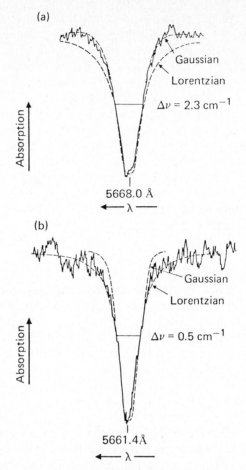

Fig. 14. Zero-phonon hole profiles (excitation) for dimethyl-s-tetrazine in a polyvinyl carbazole polymer. The lower hole (b) is nonphotochemical in origin and was burned with a cw
laser power of 10 mW, $T_B = 1.8$ K. The laser linewidth was 0.01 cm^{-1}. The upper hole (a) is
photochemical and biphotonic in nature and was produced using a pulsed dye laser, $T_B = 8$ K.
Notice that the lower hole profile is Lorentzian. (From Cuellar and Castro 1980.)

dephasing triggered by the electron–phonon interaction in mixed crystals
cannot account for the aforementioned T-dependence. Nor can they
readily account for the anomalously fast dephasing of impurity electron
transitions in glasses.

This is not surprising, perhaps, since these theories are not designed for
amorphous solids where disorder and the TLSs (two-level systems, see
subsection 2.2) come into play. Hayes et al. (1980, 1981) have developed a
theory based on a model which has the impurity interacting with several

TLSs. This "multiple" interaction was necessary because the hole formation process, by virtue of its low quantum yield (see observation (ii) of subsection 2.1), is too slow to account for the observed hole widths. Thus, it was proposed that hole formation was due to interaction with a TLS possessing the two properties discussed in subsection 2.2 and the dephasing due to another (or other) TLS(s) capable of ultra-fast phonon assisted tunneling (PAT) but with its PAT not leading to hole formation. It was argued that the lack of hole formation was due to the fact that the TLS_G and TLS_F barrier heights, fig. 8, are correlated so that if PAT for TLS_F occurs on the picosecond timescale, PAT for TLS_G would be rapid on the timescale of the experiment and, consequently, hole formation precluded.

Because the theory proposed by Hayes et al. (1981) is new and reveals how NPHB can be used to probe the structure of glasses, we proceed now to a detailed description of it.

3.1. Theory

In what follows, it will be shown that PAT between the tunnel states of TLSs can account for the linear dependence of the NPH width on T_B. Specifically, this dependence arises in the high T limit of the theory, where kT is greater than the width of the distribution function $f(\epsilon)$ for the TLSs, see fig. 8. It will also be shown that the low T limit of the theory yields a dependence which is close to quadratic.

3.1.1. Energetics and tunneling of a two-level system

Before considering the impurity–glass system and hole burning, it is instructive to consider the energetics of the glass TLSs themselves. Figure 8 depicts the intermolecular double-well potential for a TLS comprised of some "cluster" of glass molecules. The TLS potential energy is obtained with the Hamiltonian H_0 in

$$H = H_0 + \sum_i \left(\frac{\partial V_{int}}{\partial \xi_j}\right)_{\xi=0} \xi_i. \tag{11}$$

H_0 is the electronic Hamiltonian of the TLS with the medium molecules surrounding the TLS frozen in a suitable configuration. Thus, H_0 contains $V_{int}(x, q, \xi = 0)$, where V_{int} is the intermolecular potential energy operator which governs interactions between TLS molecules and others. The ξ_i are suitable displacement coordinates of the latter, q is the appropriate displacement coordinate for the TLS and x denotes electron coordinates. It is the last term in eq. (11) which is considered responsible for PAT. Higher order terms not included in eq. (11) can be taken into account as the need arises. The Hamiltonian H is viewed as the total electronic Hamiltonian H for the TLS.

Ignoring, for the moment, PAT, one proceeds to form the tunnel states

$$|A\rangle = a|L\rangle + b|R\rangle \tag{12a}$$

and

$$|B\rangle = -b|L\rangle + a|R\rangle \tag{12b}$$

from the localized oscillator states $|L\rangle$ and $|R\rangle$, cf. fig. 8. These states are eigenfunctions of the energy matrix (Jäckle 1972)

$$\frac{1}{2}\begin{pmatrix} \epsilon & \Delta \\ \Delta & -\epsilon \end{pmatrix}.$$

$\Delta/2$ is the tunneling frequency given, approximately, by

$$\Delta \approx \hbar\omega_0\, e^{-\lambda}, \tag{13}$$

with ω_0 on the order of the zero-point frequency of the localized oscillators and

$$\lambda \approx d(2mV)^{1/2}/\hbar. \tag{14}$$

In eq. (14), d is the distance between minima and m is the appropriate mass (moment of inertia). The exponential in eq. (13) is the overlap factor between oscillators. The tunnel states $|A\rangle$ and $|B\rangle$ are, in the absence of coupling of the TLS to the phonon bath via the second term in eq. (11), stationary states with energies

$$E_{A,B} = \pm E/2, \tag{15}$$

where

$$E = (\epsilon^2 + \Delta^2)^{1/2}. \tag{16}$$

For what follows, we note that $ab = \Delta/2E$. With regard to the potential barrier in fig. 8, it is reasonable to assume that it varies from close to zero to kT (glass transition temperature) $\approx 100\ \mathrm{cm}^{-1}$ in our glasses. From eqs. (13) and (14) one expects, therefore, that the tunneling frequencies vary over several orders of magnitude. Of course, the mass m and displacement coordinate q may also be different for different TLSs.

We are now in a position to derive an expression for the phonon assisted tunneling frequencies between tunnel states using the second term in eq. (11) as the transition operator, which is expressed in terms of the system phonon coordinates (Q_s) as

$$V_1 \equiv \sum \left(\frac{\partial V_{int}}{\partial Q_s}\right)_{Q=0} Q_s. \tag{17}$$

The derivatives in this equation are functions of the electron coordinates and q. Utilizing the Fermi–Golden Rule, we have, for example, that

$$\langle \Gamma_{em} \rangle_T = \frac{2\pi}{\hbar} \sum_n W_n \sum_m |\langle GBm|V_1|GAn \rangle|^2 \rho(E_{A,n} - E_{B,m}) \tag{18}$$

is the thermally averaged PAT frequency for one-phonon emission type transitions originating from the tunnel phonon states $|A(q)\rangle|n(Q)\rangle$ and terminating at $|B(q)\rangle|m(Q)\rangle$. $|G\rangle = |G(x, q, Q = 0)\rangle$ is the TLS electronic eigenfunctions of H_0. In the argument of the density of states function ρ, we have, for example, $E_{A,n} = E/2 + E_n$, where E is defined by eq. (16) and E_n is the energy of the phonon state $|n(Q)\rangle$. Finally, W_n is the probability that the initial phonon state $|n\rangle$ is thermally occupied. Integrating only over electron coordinates leads to

$$\langle \Gamma_{em} \rangle_T = \frac{2\pi}{\hbar} \sum_n W_n \sum_m \sum_s |\langle B|v_s^G(q)|A \rangle|^2 |\langle m|Q_s|n \rangle|^2 \rho(E_{A,n} - E_{B,m}), \tag{19}$$

where

$$v_s^G(q) = \langle G|(\partial V_{int}/\partial Q_s)_0|G \rangle. \tag{20}$$

As will be shown, the phonons involved in PAT are of the low frequency acoustical type and for this reason we employ the long wave approximation (Silsbee 1979) to write $v_s^G(q) = \omega_s c^{-1} v^G(q)$, where c is an average sound velocity for the acoustical branches. This approximation with the tunnel state wavefunctions, eqs. (12), allow Γ_{em} to be written as

$$\langle \Gamma_{em} \rangle_T = \frac{\pi c^{-2}}{2\hbar} \left(\frac{\Delta \mathcal{B}_G}{E} \right)^2 \sum_n W_n \sum_m \sum_s \omega_s^2 |\langle m|Q_s|n \rangle|^2 \rho(E - E_m + E_n), \tag{21}$$

with $E_m - E_n > 0$. $\mathcal{B}_G$ is the deformation potential defined as

$$\mathcal{B}_G = \langle R|v^G|R \rangle - \langle L|v^G|L \rangle. \tag{22}$$

The phonon matrix elements in eq. (21) are readily evaluated in the harmonic approximation where $|m\rangle$ and $|n\rangle$ are simple products of oscillator wavefunctions and the summations over m and n are over all possible oscillator quantum numbers. We have, for example, $\langle m|Q_k|n \rangle = (\hbar/2M\omega_k)^{1/2}(n_k + 1)^{1/2}$ for $|m\rangle$ and $|n\rangle$ differing only by one quantum oscillator k and zero otherwise. M is the mass of the sample. Equation (21) becomes

$$\langle \Gamma_{em} \rangle_T = \frac{\pi}{2\hbar^4} \left(\frac{\Delta \mathcal{B}_G}{\omega} \right)^2 \sum_s \frac{\hbar \omega_s}{2Mc^2} (\langle n_s \rangle_T + 1) \rho(\omega - \omega_s), \tag{23}$$

where $\hbar \omega = E$ and $\langle n_s \rangle_T = [\exp(\omega_s/kT) - 1]^{-1}$ is the thermal occupation number. Expressing the density of states function as $g(\omega_s)\delta(\omega - \omega_s)$, where

g is the phonon density of states, and integrating eq. (23) leads to

$$\langle \Gamma_{em} \rangle_T = \frac{3c^{-5}}{8\pi\rho\hbar^3} \Delta^2 \mathcal{B}_G^2 \omega (\langle n_\omega \rangle_T + 1), \tag{24}$$

where a Debye density of states $3V\omega^2/2\pi^2c^3$ has been employed. V is the sample volume and ρ in eq. (24) is the sample mass density. Again, $\hbar\omega$ is the tunnel state splitting.

For the situation where the initial tunnel state in the PAT process lies lower in energy, one-phonon annihilation (absorption) must be considered and one finds that

$$\langle \Gamma_{abs} \rangle_T = \langle n_\omega \rangle_T (\langle n_\omega \rangle_T + 1)^{-1} \langle \Gamma_{em} \rangle_T. \tag{25}$$

With the above background material, one can consider the dephasing problem for the impurity electronic transition.

3.1.2. Hole widths arising from phonon assisted tunneling of TLSs

As discussed at the beginning of this subsection, the anomalously fast dephasing, which gives rise to the large homogeneous hole widths at low T, is considered to arise from PAT of the TLSs surrounding and interacting with the impurity site. One can expect that the surrounding TLS structure will vary with the impurity site. We begin by considering the dephasing of an impurity (I) electronic transition due to PAT of a single TLS, TLS^α.

The impurity molecule I is considered to be part of the medium interacting with TLS^α and this interaction is now assumed to be included in V_{int}, vide supra. Thus, $|G\rangle = |G(x, q, Q = 0)\rangle$ represents the crude (with respect to phonon coordinates) electronic wavefunction of TLS^α for the ground state of the impurity (TLS_G^α) and yields the type of potential energy surface shown in fig. 8. The tunnel state wavefunctions, eqs. (12), utilized below are derived from this potential. We define $|F\rangle = |F(x, q, Q = 0)\rangle$ as the TLS^α electronic wavefunction for the case when the impurity is electronically excited to state F. Thus, $\langle F|H_0|F\rangle$ yields the "excited" state potential energy curve for TLS^α as a function of q. This curve need not be the same as that for TLS_G^α. We refer to the excited state TLS as TLS_F^α. The following expression for the dephasing frequency of the impurity electronic transition is based on the recent work of Jones and Zewail (1978):

$$\langle \Gamma_I^\alpha \rangle_T = \frac{2\pi}{\hbar} \sum_n W_n \sum_m |\langle FBm|\hat{T}|FAn \rangle - \langle GBm|\hat{T}|GAn \rangle|^2 \rho(E_{A,n} - E_{B,m}). \tag{26}$$

For notational simplicity, the dependence of the operator and wavefunctions on I and α is not explicitly indicated in eq. (26). The inverse of $\langle \Gamma_I^\alpha \rangle_T$ is the pure dephasing time T_2'. $\hat{T}$ is the transition operator and satisfies Dyson's equation, so that eq. (26) is accurate to all orders in the perturbation giving rise to the dephasing. The bra and ket wavefunctions are Born–Oppenheimer products of previously defined wavefunctions. From the energetics of the tunnel states, one sees that the PAT described in eq. (26) involves one-phonon emission. The case of one-phonon absorption need also be considered, *vide infra*. For reasons to be discussed, one should consider the two lowest order contributions to the matrix elements. For the problem at hand we can recast eq. (26) into the more convenient form

$$\langle \Gamma_I^\alpha \rangle_T = \frac{2\pi}{\hbar} \sum_n W_n \sum_m |\langle Bm|\hat{V}(q, Q)|An\rangle|^2 \rho(E_{A,n} - E_{B,m}), \tag{27}$$

where the perturbation associated with the Dyson transition operator $\hat{V}$ is

$$W(q, Q) = \delta U(q) + \delta v(q, Q). \tag{28}$$

Here

$$\delta U(q) = \langle F|H_0|F\rangle - \langle G|H_0|G\rangle \tag{29}$$

and

$$\delta v(q, Q) = \sum_s Q_s[v_s^F(q) - v_s^G(q)], \tag{30}$$

with, for example,

$$v_s^F(q) = \langle F|(\partial V_{int}/\partial Q_s)_0|F\rangle. \tag{31}$$

In eqs. (29)–(31), integration is over electron coordinates only. Returning to eq. (27), we have that for the problem at hand

$$\langle Bm|\hat{V}|An\rangle = \langle Bm|W|An\rangle$$
$$+ \left\{ \frac{\langle Bm|W|Am\rangle\langle Am|W|An\rangle}{\Delta E_{nm}} + \frac{\langle Bm|W|Bn\rangle\langle Bn|W|An\rangle}{\Delta E_{AB}} \right\} \tag{32}$$

is the appropriate expression for the scattering matrix element correct to second order. The leading term, with W defined by eq. (28), reduces to

$$\langle Bm|W|An\rangle = \langle B|\delta U|A\rangle\langle m|n\rangle + \sum_s \langle B|v_s^F(q) - v_s^G(q)|A\rangle\langle m|Q_s|n\rangle. \tag{33}$$

Since we are interested in inelastic scattering, the first term vanishes. With the long wave approximation and eqs. (12), the last equation becomes

$$\langle Bm|W|An\rangle = \frac{\Delta_{\alpha,I}}{2c}(\delta\mathcal{B})_{\alpha,I}E_{\alpha,I}^{-1}\sum_s \omega_s\langle m|Q_s|n\rangle, \tag{34}$$

with

$$(\delta\mathcal{B})_{\alpha,I} = \mathcal{B}_{\alpha,I}^F - \mathcal{B}_{\alpha,I}^G \tag{35}$$

the change in the deformation potential which accompanies electronic excitation of the impurity. In eq. (34), the α and I labels are included as a reminder that the tunnel frequency Δ, $\delta\mathcal{B}$ and tunnel state splitting E do depend on the impurity site I and the TLS it is coupled to. Using the same procedure employed to derive eq. (24), one finds that the first approximation to $\langle\Gamma_I^\alpha\rangle_T$ is

$$\langle\Gamma_I^\alpha\rangle_T^{(1)} = \frac{3c^{-5}}{8\pi\rho\hbar^3}\Delta_{\alpha,I}^2(\delta\mathcal{B})_{\alpha,I}^2\omega_\alpha(\langle n_{\omega_\alpha}\rangle_T + 1). \tag{36}$$

Since the model has the dephasing due to TLS_GS, which do not lead to hole formation, one may assume that their two tunnel states are in thermal equilibrium. The appropriate weighting (population) factor, $\exp(-\omega_\alpha/kT) \times[1+\exp(-\omega_\alpha/kT)]^{-1}$, must, therefore, be applied to eq. (36). For the case of one-phonon absorption, where in eq. (36) the last bracketed term is replaced by $\langle n_{\omega_\alpha}\rangle_T$, the analogous weighting is $[1+\exp(-\omega_\alpha/kT)]^{-1}$. For both cases,

$$\langle\Gamma_I^\alpha\rangle_T^{(1)} = \frac{3c^{-5}}{16\pi\rho\hbar^3}\Delta_{\alpha,I}^2(\delta\mathcal{B})_{\alpha,I}^2\omega_\alpha\,\mathrm{cosech}(\omega_\alpha/kT) \tag{37}$$

is the dephasing frequency.

Because the tunnel frequency in this equation is that of TLS_G^α, one sees that this first approximation does not take into account, for example, the change in tunnel frequency (potential barrier) which accompanies electronic excitation. This change is a consequence of the potential energy difference $\delta U(q)$, eq. (29). This difference appears in the second order contribution to the transition matrix element, i.e. the curly bracketed term in eq. (32). Denoting this term by { }, one can show that in the long wave approximation

$$\{\,\} = \frac{(\delta\mathcal{B})_{\alpha,I}}{2cE_{\alpha,I}}\Delta_{\alpha,I}'\sum_s\omega_s\langle m|Q_s|n\rangle, \tag{38}$$

where

$$\Delta_{\alpha,I}' = 2\langle B|\delta U|A\rangle_{\alpha,I}. \tag{39}$$

Finally, one has that

$$\langle\Gamma_I^\alpha\rangle_T^{(2)} = \frac{3c^{-5}}{16\pi\rho\hbar^3}(\delta\mathcal{B})_{\alpha,I}^2(\Delta_{\alpha,I}+\Delta_{\alpha,I}')^2\omega_\alpha\,\mathrm{cosech}(\omega_\alpha/kT) \tag{40}$$

is the dephasing frequency correct to second order. Our data indicate that Δ' cannot be assumed small relative to Δ.

We move now to a consideration of the averaging of $\langle\Gamma_I^\alpha\rangle_T^{(2)}$ over α and I. Exact averaging over α for a given I is clearly not possible. The data indicate that the impurity site is coupled to several TLSs. The assumption is made that the dephasing of site I is dominated on average by one TLS which has its PAT dephasing frequency lying in some "maximum" range or interval. With the above assumption, the α-averaging is circumvented and one may write

$$\langle \Gamma_I \rangle_T = \frac{3c^{-5}}{16\pi\rho\hbar^4} [(\delta\mathcal{B})_I(\Delta_I + \Delta_I')]^2_{max} \int_{\Delta_{min}}^{\infty} f(E)E \cosech(E/kT)\,dE \tag{41}$$

for the dephasing of impurity site I undergoing hole burning. $f(E)$ is the probability distribution for the TLSs. The averaging in eq. (41) takes into account the fact that there is a distribution of tunnel splittings that the TLS (dominating the dephasing) can have. The simplifying but, in our opinion, reasonable assumption is made that ϵ and $\Delta(\delta\mathcal{B})$ are not correlated. The lower limit on the integral is discussed later.

It is convenient to cast eq. (41) into a different form by utilizing the relationship $E^2 = \epsilon^2 + \Delta^2_{min}$, eq. (16), and a Gaussian distribution function $f(\epsilon)$:

$$\langle \Gamma_I \rangle_T = \frac{3c^{-5}}{16\pi\rho\hbar^4} [(\delta\mathcal{B})_I(\Delta_I + \Delta_I')]^2_{max} \left(\frac{\pi}{2\sigma}\right)^{1/2}$$
$$\times \int_0^{\infty} e^{-\sigma\epsilon^2}\epsilon \cosech((\epsilon^2 + \Delta^2_{min})^{1/2}/kT)\,d\epsilon. \tag{42}$$

For what follows, we note that when the distribution is narrow on the scale of kT, the high T limit of eq. (42), where $\langle \Gamma_I \rangle_T \propto T$, is attained. Importantly, the onset of this linear behavior can be related to the width of $f(\epsilon)$.

It should be noted that in using eq. (42) the upper limit in the integral is set by the maximum phonon frequency when it is smaller than ϵ corresponding to a nonnegligible $f(\epsilon)$ value.

The observed zero-phonon hole is a superposition of the Lorentzian profiles corresponding to different impurity sites. The single-impurity profile has a linewidth equal to $\langle \Gamma_I \rangle_T$, eq. (42). Defining Γ_{max} and Γ_{min} as the upper and lower limits for the "maximum" dephasing frequency interval introduced above, one has

$$g(\omega) \propto \int_{\Gamma_{min}}^{\Gamma_{max}} \frac{\langle \Gamma_I \rangle_T\, d(\langle \Gamma_I \rangle_T)}{(\omega - \omega_B)^2 + (\langle \Gamma_I \rangle_T/2)^2} \tag{43}$$

$$= \ln\left(\frac{(\omega - \omega_B)^2 + (\Gamma_{max}/2)^2}{(\omega - \omega_B)^2 + (\Gamma_{min}/2)^2}\right). \tag{44}$$

The linewidth of this function is $(\Gamma_{max}\Gamma_{min})^{1/2}$. Calculations show that for $\Gamma_{max}/\Gamma_{min} \lesssim 7.5$, $g(\omega)$ would be difficult to distinguish from a single Lorentzian with width $(\Gamma_{max}\Gamma_{min})^{1/2}$. For large values of this ratio $g(\omega)$ is more sharply peaked than a Lorentzian, falls off more slowly than a Lorentzian at intermediate $|\omega - \omega_B|$ values and converges to a Lorentzian at large $|\omega - \omega_B|$.

We consider now the application of the above theory to the data in figs. 15 and 16. They show the dependence of the NPH width on T_B for the tetracene in glycerol–DMSO–DMF and ethanol–methanol glasses (Hayes et al. 1980, 1981). The glass transition temperatures are ≈ 90 and ≈ 150 K, respectively. For the former glass, the T_B-dependence is linear down to the lowest burn temperature, $T_B \approx 1.8$ K, and it is observed that the hole profiles are close to Lorentzian particularly for $T_B \gtrsim 4$ K. For the latter glass, fig. 16 shows that the broadening curves do depend on the particular sample used (even for identical chemical composition). Evidently, variations in the cool-down procedure lead to variations in the microscopic structure of the glass. Nonetheless, the onset of the linear dependence near 4 K is sample independent. For the tetracene in ethanol–methanol glass system, the hole profiles were observed to possess a significant Gaussian component at $T_B \approx 2$ K and by $T_B \approx 3$–4 K to be close to Lorentzian. Thus, for both systems, the dephasing time $T_2 = 2(\pi \Gamma c)^{-1} \approx 10$ ps at $T \approx 4$ K, but see below.

For a consideration of the linear behavior, the pertinent equation is eq. (42). Noting that, when $(\epsilon^2 + \Delta_{min}^2)^{1/2} \ll kT$, the cosech function carries a *linear* dependence on T, one can see that when the TLS distribution function $f(\epsilon)$ is narrow relative to kT, $\langle \Gamma_l \rangle_T$ will also carry a linear dependence on T_B. The calculations of Hayes et al. (1981) show that the distribution function

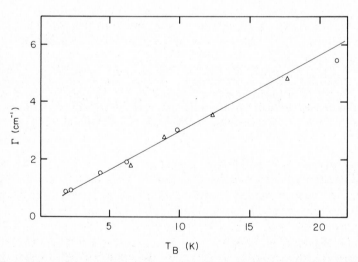

Fig. 15. Dependence of the zero-phonon hole width (Γ) on the burn temperature for tetracene in the glycerol–DMSO–DMF glass. The zero-phonon hole is coincident with the laser burn wavelength at 476.5 nm and in the origin band of the S_1 absorption system. The circles and triangles are data points from two different samples of the same chemical composition. (From Hayes et al. 1981.)

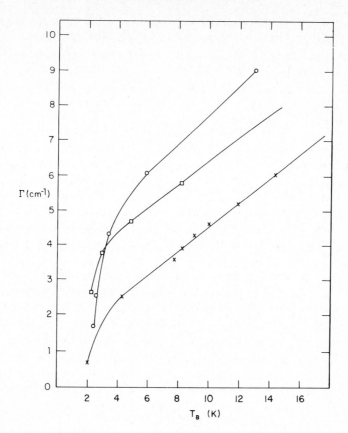

Fig. 16. Dependence of the zero-phonon hole width (Γ) on the burn temperature for tetracene in the ethanol–methanol glass. The zero-phonon hole is coincident with the laser burn wavelength at 476.5 nm and in the origin band of the S_1 absorption system. The squares, circles and crosses are data points from three different samples of the same chemical composition. (From Hayes et al. 1981.)

must be narrow ($<4\,\text{cm}^{-1}$) for both glasses and note that, for tetracene in ethanol–methanol, this is consistent with the hole filling data, see subsection 2.2. Such a narrow distribution width has been postulated more recently for ethanol and boric acid glasses on the basis of the thermal cycling properties of photochemical holes for quinizarin (Friedrich et al. 1982).

Finally, Hayes et al. (1981) point out that the above theoretical model is also supported by the data in fig. 4 for the tetracene in glycerol–DMSO–DMF glass, since eq. (44) allows for the possibility which is observed, namely, a deviation from Lorentzian behavior. This equation reveals that the homogeneous linewidth of a nonphotochemical zero-phonon hole is representative of a distribution of dephasing frequencies arising from the disorder inherent to the glass. A ratio of $\Gamma_{max}/\Gamma_{min} = 10$ was used to obtain the fit (crosses) in fig. 4. The fit was restricted to the high energy side of the hole since the low energy side suffers interference from the phonon side band hole. With reference to fig. 13, which shows an example of photochemical hole burning in a glass, it is apparent that the homogeneous widths of photochemical holes in glasses may frequently be dominated by TLS phonon assisted tunneling.

3.2. The low temperature limit for dephasing due to phonon assisted tunneling

In addition to hole burning, coherent transient spectroscopies and time resolved resonant fluorescence line narrowing spectroscopy can be used to measure the dephasing of impurity electronic transitions in glasses. Such measurements for organic glasses have not yet been reported, but FLN data for the $^3P_0-^3H_4$ transition of Pr^{3+} in a BeF_2 glass (Selzer et al. 1976, Hegarty et al. 1979) and the $^5D_0-^7F_0$ resonant transition of Eu^{3+} in silicate (Avouris et al. 1977) exist. The thermal broadening of the zero-phonon line follows quite closely a T^2-dependence from liquid helium to room temperature and the low T homogeneous linewidths are *anomalously large* in comparison to those measured in crystalline hosts. It may be fair to say at this point that the T^2-dependence over such an extended temperature range is due to a special set of circumstances since there are a variety of possible broadening mechanisms and the glass undergoes significant structural changes from liquid helium to room temperature.

Still it is instructive to consider whether the theory of Hayes et al. (1981) can explain the T^2-dependence. The low T limit of the theory corresponds to kT less than the width of the TLS distribution function $f(\epsilon)$. Because the maximum TLS barrier height V_{max}, fig. 8, is roughly equal to the glass transition temperature, it follows that V_{max} for the hard inorganic glasses should be a factor of about ten larger than for the organic glasses considered above. It is reasonable to apply the same scaling procedure to the width of $f(\epsilon)$. Thus, an estimate of the width for a "hard" glass is $\approx 100\ cm^{-1}$. Our numerical evaluation of the integral in eq. (42) for a number of wide distribution widths and a range of Δ_{min} values show that for a given width the onset of the linear behavior occurs near $kT \approx$ distribution width (insensitive to the choice of Δ_{min}). In the low temperature limit, the limiting power dependence on T varies between ≈ 2 and ≈ 3. For example, a distribution width of $400\ cm^{-1}$ yields a power dependence of 2 and 3 for Δ_{min} equal to 0.5 and $20\ cm^{-1}$, respectively. It appears, therefore, that our theory is capable of explaining the dephasing data for the aforementioned RE ion transitions.

We note that Orbach and Lyo (1980) have also invoked PAT of two-level systems to account for the above T^2-dependence. They find, for kT less than the maximum phonon frequency or the maximum ϵ (whichever is smaller), a power law of 2. In the high temperature limit, the homogeneous line broadening follows a linear dependence on T. Thus, although there are significant differences between the two theoretical approaches, the principal predictions are the same. One important difference is in the method used for averaging with the TLS distribution functions. Orbach and Lyo adopt the procedure of Anderson et al. (1972), which assumes that the density of states of the TLSs as a function of ϵ is constant on the temperature scale of the experiment. It is difficult to understand how this assumption can be valid when kT becomes comparable to the distribution width or greater than it in the high T limit. Originally, Anderson et al. (1972) utilized this approximation for $T \leqslant 10$ K.

In summary, there seems little doubt that PAT of two-level systems plays a very important role in the dephasing of impurity electronic transitions in both soft and hard glasses and, in particular, at low temperatures.

4. The thermal annealing of nonphotochemical holes

In addition to the dependence of the zero-phonon hole profile on the burn temperature, the temperature dependence of the irreversible hole disappearance as the sample is warmed above T_B is also determined by the structural and dynamical properties of the glass. Figure 9 shows, in part, annealing data for tetracene in the ethanol–methanol glass. Similar data for tetracene in the glycerol–DMSO–DMF glass are given in fig. 17. The dependence of the zero-phonon hole integrated optical density on T_B for the same system is shown in fig. 18. Considering this dependence first, we note that it is linear up to ≈ 10 K. Above this temperature, the curve bends over. Although the bending over is not observed in the temperature range employed for the ethanol–methanol glass, fig. 9, one expects that at a sufficiently high T_B it would. In the linear regime, the number of impurity–TLS sites undergoing hole burning in an interval dT_B about T_B must be constant. Furthermore, given the importance of PAT for dephasing, it is reasonable to assert that PAT is the initial step of hole formation. Thus, with Δ_F and $\mathscr{B}_F$ the TLS$_F$ tunneling and deformation potential frequencies, it would appear that the number of impurity–TLS sites with "infinitely slow" ground state relaxation and with $\Delta_F\mathscr{B}_F$ in the appropriate range for hole formation is also constant in the interval dT_B about T_B. It is clear, however, that the dependence of hole intensity on T_B cannot be further understood until the annealing data are.

The most obvious interpretation of the data in Figs. 9 and 17, which

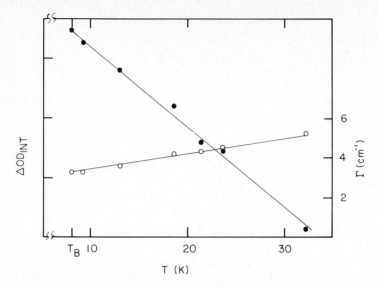

Fig. 17. Integrated zero-phonon hole intensity as a function of annealing temperature (solid circles) for tetracene in the glycerol–DMSO–DMF glass. Note the linear dependence. The open circles show the dependence of the zero-phonon hole width on the annealing temperature. The zero-phonon hole is as defined in the caption to fig. 18. (From Hayes et al. 1981.)

show that the loss of hole intensity with warming above T_B is roughly linear with T, is based on PAT of the TLS_Gs, fig. 8. That is, the annealing is due to PAT which returns the impurity–TLS_G sites, burned (permanently) at T_B, back to their original configurations with a site excitation energy coincident with ω_B. However, this interpretation cannot be correct in light of the following experimental facts; as noted earlier, subsection 2.1, the holes produced at T_B last indefinitely provided the sample temperature is maintained at or below T_B; warming of the sample above T_B and the first measurement of the hole at $T' > T_B$ takes about 20 minutes; subsequent measurements of the hole at $T = T'$ reveal no further change in the hole profile, i.e., the hole at $T = T'$ also lasts indefinitely! Given the weak T-dependence of PAT, or even the stronger T-dependence of a thermally activated (exponential) process, the intriguing persistence properties are not consistent with the above interpretation. Thus, the physical model of Hayes et al. (1980, 1981), which is consistent with the available data on hole formation and the dephasing of the impurity electronic transitions, cannot account for the annealing data.

This deficiency is most likely the result of the model not being sufficiently complete. One speculative but interesting possible explanation of the annealing data is that the hole burning produces a macroscopic state

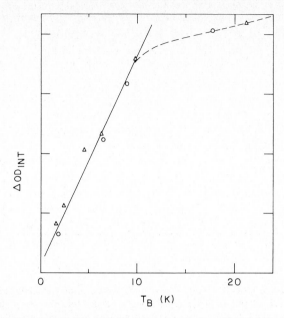

Fig. 18. The dependence of the integrated zero-phonon hole optical density for the tetracene in glycerol–DMSO–DMF glass. The zero-phonon hole is coincident with the laser burn wavelength at 476.5 nm. The circles and triangles are saturated hole intensities from two different samples of the same chemical composition. (From Hayes et al. 1981.)

of the glass more metastable than the original state and that the annealing involves cooperative (network) order–disorder transitions which are continuous.

5. Concluding remarks

We have attempted to bring the reader up-to-date on the status of nonphotochemical hole burning spectroscopy of organic molecules imbedded in host organic glasses. This spectroscopy is very new, but already it is apparent that the glass disorder, specifically the TLS, plays an important role in the hole formation process and the dephasing of the electronically excited impurity molecule. As a result, NPHB promises to become a valuable tool for probing the nature of disorder in glasses. There is little doubt, in our opinion, that the TLS structural model proposed by Hayes and Small (1978a,b; 1980, 1981) to explain hole formation, persistence and dephasing is basically correct. There are, however, indications (see preceding section) that it may need to be improved upon. Although much work remains to be done before the microscopic structures or even spatial extents of TLSs will be understood, it is fair to say that TLSs are a universal accoutrement of all glasses. For this reason, their study should lead to a very close interaction between workers in several different fields. Recalling that the NPHB data indicate that PAT of two-level systems occurs on a very short timescale even at temperatures close to 0 K, and noting that one can view hole formation and impurity dephasing as a manifestation of impurity–glass photoisomerization, it would seem that NPHB should be of interest to chemists interested in structural trans-formations of easily deformable molecular clusters.

We consider now a few of the more important experiments on or related to NPHB which, hopefully, will be pursued in the near future. First, hole burning has been studied primarily in alcoholic glasses and, to a lesser extent, in polymers, for example, polyvinylcarbazole and polymethyl-methacrylate. It is important, therefore, to determine the extent to which NPHB occurs in a wider variety of glasses (inorganic, hydrocarbon, rare earth ion–organic chelate, other polymers). If systems with high NPHB quantum yields can be found, time resolved measurements of hole for-mation would be possible and very interesting. In order that the theory for dephasing of impurity electronic transitions in glasses, section 3, be adequately tested, additional studies on the dependence of zero-phonon hole width on burn temperature are required. So too are coherent transient spectroscopic and resonant fluorescence line narrowing measurements. However, the relationship of the former experiment, for example photon echo, to hole burning is not simple, because a substantial component of the absorption at any frequency within the severely inhomogeneously broadened profile will often be due to one- or multi-phonon transitions. With regard to the relationship of the homogeneous linewidths from FLN measurements to those from hole burning, one must consider the pos-sibility that impurity–TLS sites which undergo hole burning, undergo more rapid dephasing than those which do not. Thus, it would be important to perform FLN studies prior to and after production of the saturated hole. The observation that for tetracene in ethanol–methanol the $T_B = 2$ K hole profile contains a significant Gaussian component, which is not laser limited, is interesting. The same observation was made for the photo-chemical hole of dimethyl-s-tetrazine in polyvinyl carbazole, upper trace in fig. 14. One possible explanation for this is discussed in the last paragraph of subsection 2.4. Concentration (impurity) dependent studies of the resi-dual Gaussian contribution to the hole width would establish whether or not this explanation is correct.

It was noted earlier that the anomalous low temperature behavior of the specific heat, thermal conductivity and ultrasonic attenuation in inorganic glasses is well known (Hunklinger and Arnold 1976, Hunklinger 1978). For example, the former two follow, respectively, a linear and a quadratic dependence on T for $T \leq 10$ K. The theories of Anderson et al. (1972) and Philips (1972) based on the TLS model explain these anomalies *but* with the assumption that the TLS density of states as a function of ϵ, fig. 8, is constant on the temperature scale of the experiment. According to Hayes et al. (1981), this assumption for the same temperature range would not be valid. Thus, specific heat, thermal conductivity and ultrasonic attenuation measurements on organic glasses and polymers would provide a critical test of their assertion.

References

Aartsma, T.J. and D.A. Wiersma, 1976, Chem. Phys. Lett. **42**, 520.

Anderson, P.W., G.I. Halperin and C.M. Varma, 1972, Philos. Mag. **25**, 1.

Avouris, P., A. Campion and M.A. El-Sayed, 1977, J. Chem. Phys. **67**, 3397.

Burke, F.P. and G.J. Small, 1974a, Chem. Phys. **5**, 198.

Burke, F.P. and G.J. Small, 1974b, J. Chem. Phys. **61**, 4588.

Cuellar, E. and G. Castro, 1980, Chem. Phys. **54**, 217.

De Bree, P. and D.A. Wiersma, 1979, J. Chem. Phys. **70**, 790.

De Vries, H. and D.A. Wiersma, 1976, Phys. Rev. Lett. **36**, 91.

De Vries, H. and D.A. Wiersma, 1980, J. Chem. Phys. **72**, 1851.

Drissler, F., F. Graf and D. Haarer, 1980, J. Chem. Phys. **72**, 4996.

Edelson, M.C., J.M. Hayes and G.J. Small, 1979, Chem. Phys. Lett. **60**, 307.

Friedrich, J., H. Scheer, B. Zickendraht-Wendelstadt and D. Haarer, 1980a, J. Chem. Phys. **72**, 705.

Friedrich, J., J.D. Swalen and D. Haarer, 1980b, J. Chem. Phys. **73**, 705.

Friedrich, J., H. Scheer, B. Zickendraht-Wendelstadt and D. Haarer, 1981, J. Am. Chem. Soc. **103**, 1030.

Friedrich, J. and D. Haarer, 1980, Chem. Phys. Lett. **74**, 503.

Friedrich, J., H. Wolfrum and D. Haarer, 1982, J. Chem. Phys., to appear.

Gorokhovskii, A.A., 1976, Opt. Spectrosc. **40**, 272.

Gorokhovskii, A.A. and J. Kikas, 1977, Opt. Commun. **21**, 272.

Gorokhovskii, A.A. and L.A. Rebane, 1977, Opt. Commun. **20**, 144.

Gorokhovskii, A.A., R.K. Kaarli and L.A. Rebane, 1974, JETP Lett. **20**, 216.

Gorokhovskii, A.A., R.K. Kaarli and L.A. Rebane, 1976, Opt. Commun. **16**, 282.

Graf, F., H.-K. Hong, A. Nazzal and D. Haarer, 1978, Chem. Phys. Lett. **59**, 217.

Harris, C.B., 1978, J. Chem. Phys. **57**, 501.

Hayes, J.M. and G.J. Small, 1978a, Chem. Phys. **27**, 151.

Hayes, J.M. and G.J. Small, 1978b, Chem. Phys. Lett. **54**, 435.

Hayes, J.M. and G.J. Small, 1979, J. Luminesc. **18/19**, 219.

Hayes, J.M. and G.J. Small, 1980, unpublished results.

Hayes, J.M., R.P. Stout and G.J. Small, 1980, J. Chem. Phys. **73**, 4129.

Hayes, J.M., R.P. Stout and G.J. Small, 1981, J. Chem. Phys. **74**, 4266.

Hegarty, J. and W.M. Yen, 1979, Phys. Rev. Lett. **43**, 1126.

Hunklinger, S., 1978, J. Physique C6-1444.

Hunklinger, S. and W. Arnold, 1976, in: Physical Acoustics, vol. 12, eds. R.N. Thurston and W.P. Mason (Academic Press, New York) p. 155.

Jäckle, J., 1972, Z. Phys. **257**, 212.

Jones, K.E. and A.H. Zewail, 1978, Theory of Optical Dephasing in Condensed Phases, in: Advances in Laser Chemistry, vol. 3, ed. A.H. Zewail (Springer, Berlin).

Kharlamov, B.M., R.I. Personov and L.A. Bykovskaya, 1974, Opt. Commun. **12**, 191.

Kharlamov, B.M., R.I. Personov and L.A. Bykovskaya, 1975, Opt. Spectrosc. **39**, 137.

Kharlamov, B.M., L.A. Bykovskaya and R.I. Personov, 1977, Chem. Phys. Lett. **50**, 407.

Lee, P.H. and L.S. Skolnick, 1967, Appl. Phys. Lett. **10**, 303.

Lyo, S.K. and R. Orbach, 1980, Phys. Rev. **B22**, 4223.

Macfarlane, R.M. and R.M. Shelby, 1979, Phys. Rev. Lett. **42**, 788.

Marchetti, A.P., M. Scozzafava and R.H. Young, 1977, Chem. Phys. Lett. **51**, 424.

Philips, W.A., 1972, J. Low Temp. Phys. **7**, 351.

Sargent III, M. and P.E. Toschek, 1976, Appl. Phys. **11**, 107.

Selzer, P.M., D.L. Huber, D.S. Hamilton, W.M. Yen and M.J. Weber, 1976, Phys. Rev. Lett. **36**, 813.

Silsbee, R.H., 1969, in: Optical Properties of Solids, eds. S. Nudelman and S.S. Mitra (Plenum Press, New York) p. 607.

Small, G.J., 1978, Chem. Phys. Lett. **57**, 501.

Szabo, A., 1975, Phys. Rev. **B11**, 4512.

Voelker, S., R.M. Macfarlane, A. Genack, H.P. Tromsdorff and J.H. van der Waals, 1977, J. Chem. Phys. **67**, 1759.

Voelker, S., R.M. Macfarlane and J.H. van der Waals, 1978, Chem. Phys. Lett. **53**, 8.

Voelker, S. and R.M. Macfarlane, 1979, Chem. Phys. Lett. **61**, 421.

Wiersma, D.A., 1981, Coherent Optical Transient Studies of Dephasing and Relaxation in Electronic Transitions of Large Molecules in the Condensed Phase, in: Advances in Chemical Physics, vol. 17, eds. J. Jortner, R.D. Levine and S.A. Rice (Wiley, New York).

CHAPTER 10

Site Selection Spectroscopy of Complex Molecules in Solutions and Its Applications

R.I. PERSONOV

Institute of Spectroscopy
USSR Academy of Sciences
142092 Troitsk, Moskovskaya Oblast
U.S.S.R.

Translated from the Russian by Nicholas Weinstein.

Spectroscopy and Excitation Dynamics
of Condensed Molecular Systems
Edited by
V.M. Agranovich and R.M. Hochstrasser

Contents

1. Introduction

The optical spectra of molecules are the most vital source of information on their structure and properties. Of particular importance among all kinds of molecular spectra are the electronic spectra, in which all types of motion in the molecule manifest themselves: electronic, vibrational and rotational. The relations between the optical properties of molecules and their physical, chemical and photochemical properties are most distinctly revealed in electronic spectra. Such spectra are the most sensitive indications of various kinds of infra- and intermolecular interactions and serve as a valuable means of investigating the interaction of a molecule with its environment.

It is a well-known fact, however, that in contrast to the spectra of atoms and simple molecules, electronic absorption and emission spectra of polyatomic organic molecules (most frequently investigated in solutions) are diffuse as a rule and consist of one or several broad bands (of a width from about 10^2 to $10^3 \, \mathrm{cm}^{-1}$). Such broad-band spectra cannot yield much information and this drastically reduces their usefulness for scientific and practical purposes. Hence the problem of the origin of broad bands in the spectra of complex organic molecules and the search for conditions that would enable one to observe finer details in spectra have always been a matter of attention for spectroscopists.

A breakthrough along these lines was due to Shpol'skii and co-workers (Shpol'skii et al. 1952, Shpol'skii 1960), who showed in 1952 (employing certain aromatic hydrocarbons as examples) that at low temperatures and with a definite type of solvent (crystallized short-chain n-paraffins) the fluorescence and absorption spectra consisted of dozens of comparatively narrow bands or quasi-lines, instead of broad bands. At the present time several hundreds of compounds have been found that produce such spectra, called "quasi-line" spectra, in n-paraffin matrices. In investigating the nature of quasi-line spectra it was established that the lines observed in them correspond to optical zero-phonon transitions in the molecules being studied and possessed all the typical features of these transitions (see e.g. Rebane and Khizhnyakov 1963, Persohov et al. 1971a, Richards and Rice 1971, Al'shits et al. 1972, Osad'ko et al. 1973).

But the fact that many molecules have quasi-line spectra in specially selected n-paraffin matrices does not eliminate the general problem of the origin of broad-band spectra of complex molecules. As a matter of fact, many compounds have low or practically no solubility in *n*-paraffins. Many compounds that can be introduced into frozen *n*-paraffin matrices continue, even under such conditions, to produce broad-band spectra. Taking into consideration the huge variety of organic compounds and the wide choice of available solvents, it can be stated that in the great majority of cases, the spectra of the solutions consist of broad bands even at liquid-helium temperatures. Consequently, the problem of band broadening in spectra of complex molecules retains its high theoretical and practical significance.

Substantial advances have been made in recent years in elucidating the nature of broad-band spectra of organic compounds. A promising development along these lines was the establishment in our laboratory of the fact that in many cases at sufficiently low temperatures the spectra of solutions are mainly broadened inhomogeneously, and possess a concealed, but useful, line structure (Personov et al. 1972a,b, 1973). This circumstance served as the starting point for initiating a new trend in the electronic spectroscopy of molecules: fine-structure selection spectroscopy (site selection spectroscopy) of complex molecules in frozen solutions. In the following years procedures have been worked out for eliminating in-homogeneous broadening and for revealing line structure in the fluores-cence, phosphorescence and absorption spectra of molecules in arbitrary solvents by means of selective laser excitation. All the principal features and characteristics have been studied in line spectra obtained by this procedure. Extensive opportunities were thus created available for fine-structure spectroscopic investigation of complex molecules and their in-teraction in a great variety of media.

The present review deals with the basic principles of site selection spectroscopy of complex molecules in solutions, and with certain ap-plications of the corresponding methods.

Before going over to a consideration and discussion of specific experi-mental results, we shall recall certain general principles of the theory that will be required further on.

2. Certain propositions of the theory of impurity center spectra of crystals

2.1. Introduction

Molecules in solutions (matrices) are similar in many aspects to impurity centers in crystals. For this reason we shall resort to the main concepts of impurity center theory to analyze the features of the electronic spectra of molecules in solutions. Here we shall assume (in accordance with the conditions commonly existing in experimental investigations) that the electronic excitations of the solvent require much higher energy than those of the molecules (impurities) being investigated. This means that in the spectral region that interests us only the electrons of the impurity molecules interact with light. Moreover, we assume that the concentration of impurity molecules is low enough to neglect all interaction between them.

The nature of the spectrum produced by an impurity molecule is determined by electron–vibrational interaction of two types: interaction of the molecule's electrons with its intramolecular vibrations (vibronic coupling) and interaction with the intermolecular vibrations of the solution (electron–phonon coupling). Vibronic coupling leads to the presence in the molecular spectrum of a series of vibronic bands, along with the band in the region of the purely electronic transition. The nature and shape of each vibronic band is determined by electron–phonon coupling. Each vibronic band may consist, in principle, of two parts: (a) a narrow zero-phonon line (ZPL), corresponding to transitions in the impurity without a change in the number of phonons in the matrix (an optical analogy of the resonance γ line in the Mössbauer effect) and (b) a relatively broad phonon wing (PW), due to phototransitions in the impurity molecule with the creation or annihilation of matrix phonons. What in particular is actually observed in the spectrum (ZPL, PW or ZPL + PW) depends upon the strength of the electron–phonon coupling and upon temperature. To realize how these statements follow from the theory and to render a further analysis of specific experimental data more convenient, we recall here certain of the basic propositions of the theory (see, e.g., Maradudin 1966, Rebane 1968). A more comprehensive and general treatment of the state-of-the-art in this theory can be found in a special review published as chapter 8 in the present volume.

2.2. General nature of a spectral band

Let us begin by analyzing a portion of the spectrum in the region of a purely electronic transition, ignoring vibronic coupling. We consider the transition of an impurity crystal from the fixed state $|i\rangle$ with energy E_i to the state $|f\rangle$ with energy E_f, accompanied by the absorption (or emission) of a light quantum of frequency $\omega = (E_f - E_i)/\hbar$. As is known from quantum mechanics, the probability of such a transition is proportional to the square of the matrix element: $|\langle f|\hat{V}|i\rangle|^2$, where $\hat{V}$ is the operator of the interaction between the electrons of the impurity and light (in the dipole approximation, this matrix element is proportional to the dipole moment of the transition). In order to write the expression for the intensity of the spectral band at frequency ω we should take into account the fact that at a nonzero temperature a contribution to the intensity at this frequency is made by the transitions between various states that have the same energy difference. Hence, we must take into consideration the contributions of all these transitions, bearing in mind the probability of each initial state being realized and the Bohr frequency condition. Then, for the intensity of the spectral band we can write the expression

$$I(\omega) = C \sum_i v_i \sum_f |\langle f|\hat{V}|i\rangle|^2 \delta(E_f - E_i - \hbar\omega), \tag{1}$$

where v_i is the probability that the initial phonon state i is populated (Boltzmann factor); the δ-function under the summation sign over the final phonon states ensures the selection of transitions with frequency $\omega = (E_f - E_i)/\hbar$ only; and C is a constant*.

We now make use of the adiabatic approximation; this is based on the substantial difference between the masses of the electrons and nuclei. In this approximation, the wave functions of the initial and final states are represented in the form of a product of the electronic, $|\alpha\rangle$, and vibrational, $|n\rangle$, wave functions:

$$|i\rangle = |\alpha_i\rangle|n_i\rangle, \qquad |f\rangle = |\alpha_f\rangle|n_f\rangle, \tag{2}$$

whereas the energy is the sum of the electronic, E_α, and phonon, E_n, parts:

$$E_i = E_{\alpha_i} + E_{n_i}, \qquad E_f = E_{\alpha_f} + E_{n_f}. \tag{3}$$

Substituting eqs. (2) and (3) into eq. (1), and introducing the additional assumption that the matrix element of the purely electronic transition $\langle \alpha_f|\hat{V}|\alpha_i\rangle$ is independent of the nuclear coordinates (the Condon approximation), we can obtain [see, e.g., the paper by Lax (1952)]:

$$I(\omega) = \frac{1}{2\pi} \int_{-\infty}^{\infty} dt\, e^{-i\omega t} S(t), \tag{4}$$

where

$$S(t) = \sum_{n_i} v_{n_i} \langle n_i | e^{i\hat{H}_f t}\, e^{-i\hat{H}_i t} | n_i \rangle. \tag{5}$$

Here $\hat{H}_f$ and $\hat{H}_i$ are the phonon Hamiltonians of the impurity crystal with and without an excitation of the impurity molecule. In deriving eqs. (4) and (5) from eq. (1), we put $\hbar = 1$; as the reference point for the frequency ω we selected the frequency of the purely electronic transition $E_f - E_i$; the integrated intensity of the band was normalized to unity: $\int_{-\infty}^{+\infty} I(\omega)\, d\omega = 1$.

It is evident from eqs. (4) and (5) that when there is no electron–phonon coupling (i.e., the phonon subsystem "does not sense" the change in the

*Strictly speaking, C depends upon ω, but for our purpose this weak dependence is inessential.

electronic state of the impurity and $\hat{H}_f = \hat{H}_i$), then $S(t) = 1$, and $I(\omega)$ is a ZPL having the form of a δ-function. The presence of such a coupling, however, leads to a "smearing" of the spectral band and to its specific shape, as well as to the dependence of the band shape on temperature.

It proves convenient for further discussion to introduce an auxiliary time and temperature function $g(t, T)$, defined by the relation

$$S(t) = \exp[g(t, T) - g(0, T)].\tag{6}$$

Then, making use of eq. (6), expression (4) can be rewritten in the form:

$$I(\omega) = \frac{1}{2\pi} \int_{-\infty}^{\infty} dt \, \exp[-i\omega t + g(t, T) - g(0, T)].\tag{7}$$

In order to determine the spectral band shape, it is obviously necessary at this point to find the form of the function $g(t, T)$, which is dictated by the nature of the electron–phonon coupling operator $\hat{\Lambda} = \hat{H}_f - \hat{H}_i$. The operator $\hat{\Lambda}$ depends upon the nuclear coordinates and can be represented in the form of an expansion in a series in terms of these coordinates. In the harmonic approximation the general form of this expansion is

$$\hat{\Lambda} = \hat{H}_f - \hat{H}_i = \sum_j V_j R_j + \sum_{j,j'} W_{jj'} R_j R_{j'}.\tag{8}$$

The term in expression (8) that is linear with respect to the coordinates is related to the change in the equilibrium position of the molecules in the crystal following the phototransition in the impurity center. The quadratic terms are due to the change in the normal coordinates and frequencies of the impurity crystal resulting from this transition.

The function $g(t, T)$ has been calculated, taking both linear and quadratic terms in eq. (8) into account, based on perturbation theory by Krivoglaz (1964) and, with a more general approach, by Osad'ko (1972, 1977). This function can be presented in the form

$$g(t, T) = i\Omega(T)t - \Gamma(T)|t| + f(t, T).\tag{9}$$

Here $\Omega(T)$ and $\Gamma(T)$ are real-valued functions of the temperature [whose explicit form is given, for instance, in Maradudin (1966)] associated with the presence of the quadratic terms in eq. (8). As will be shown below, these functions determine the ZPL temperature shift and broadening. The function $f(t, T)$, which tends to zero as $t \to \infty$, determines the shape of the PW.

Next, let us consider the shape of the spectral band in more detail, on the basis of relations (7) and (9). Substituting eq. (9) into (7), expanding $\exp[f(t, T)]$ in a series in terms of $f(t, T)$, and representing $f(t, T)$ in the form of a Fourier integral, we readily obtain

$$I(\omega) = \Phi_0(\omega, T) + \Phi(\omega, T),\tag{10}$$

$$\Phi_0(\omega, T) = \frac{1}{\pi} \frac{e^{-f(T)}\Gamma(T)}{[\omega - \Omega(T)]^2 + \Gamma^2(T)}, \tag{10a}$$

$$\Phi(\omega, T) = \sum_{m=1}^{\infty} \Phi_m(\omega, T), \tag{10b}$$

$$\Phi_m(\omega, T) = \frac{1}{m!} \int_{-\infty}^{\infty} d\nu_1 \int_{-\infty}^{\infty} d\nu_2 \ldots \int_{-\infty}^{\infty} d\nu_m f(\nu_1, T) \ldots f(\nu_m, T)$$
$$\times \Phi_0(\omega - \Omega + \nu_1 + \cdots + \nu_m, T), \tag{10c}$$

$$f(T) = f(t, T)_{t=0} = \int_{-\infty}^{\infty} f(\nu, T)\, d\nu, \tag{10d}$$

where the function $f(\nu, T)$ is the spectral density of the function $f(t, T)$.

Relations (10) and (10a–d) provide a means to analyze the general nature of the spectral band (fig. 1). It is evident from the above expressions that the band consists of two portions: ZPL, eq. (10a), and PW, eq. (10b). According to eq. (10a), the ZPL is of Lorentzian shape. Its width is determined by the function $\Gamma(T)$ and its temperature shift by the function $\Omega(T)$. [If $W_{jj'} = 0$ in eq. (8), then $\Gamma(T) = \Omega(T) = 0$ and expression (10a) is transformed into a δ-function.] From relations (10b) and (10c), which describe the PW, it is evident that the PW can occupy a sufficiently wide spectral range and is completely determined by the function $f(\nu, T)$. This function includes information on the phonon spectrum of an impurity crystal and on the nature of the electron–phonon coupling. We shall call it the phonon function in accordance with Osad'ko (1979). (In the case of $W_{jj'} = 0$, it is often called the Stokes loss function as well.)

Up to this point we have neglected the natural intramolecular vibrations of the impurity, as well as the vibronic coupling. When these factors are taken into account, the spectrum also displays a series of vibronic bands,

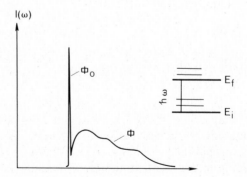

Fig. 1. The general nature of the spectral band of an impurity center (one vibronic transition).

consisting of ZPL and PW and similar to a purely electronic band. Their intensities may vary and are determined by the strength of the vibronic coupling.

2.3. Phonon wing structure

If the phonon function $f(\nu, T)$ is known, the shape of the PW can be determined, and vice versa. The corresponding relationship was investigated theoretically by Kukushkin (1963, 1965) and by Personov et al. (1971b), who established an integral equation by which the phonon function $f(\nu, T)$ can be obtained from the spectral curve for the PW. Without going into the details of this problem, we shall analyze here only a single question: what is the role played by the one-phonon and multi-phonon transitions in the formation of the PW and what spectral range can it occupy? Here, in our qualitative treatment, we may restrict ourselves to the linear approximation for electron–phonon coupling. [In this case, in relation (8) all $W_{jj'} = 0$, $\Gamma(T)$ and $\Omega(T) \to 0$, and $\Phi_0(\omega, T)$ is the δ-function.] In this approximation, the phonon function has a simple physical meaning and can be represented [see, e.g., Personov et al. (1971b), Osad'ko (1979)] in the form:

$$f(\nu, T) = [n(\nu) + 1]f_0(\nu) + n(-\nu)f_0(-\nu), \tag{11}$$

where

$$f_0(\nu) = \sum_q \xi_q^2 \delta(\nu - \nu_q). \tag{11a}$$

Here ν_q is the phonon frequency of the impurity crystal, ξ_q is the displacement of the equilibrium position of the corresponding oscillator upon a phototransition in the impurity, and $n(\nu) = [\exp(\nu/kT) - 1]^{-1}$.

The function $f_0(\nu)$, which is nonzero only for the ν values within the phonon band, can be represented in the form:

$$f_0(\nu) = 6N\xi^2(\nu)\rho(\nu), \tag{11b}$$

where $\rho(\nu)$ is the density of phonon states in the impurity crystal, $\xi^2(\nu)$ is the mean square displacement of the equilibrium positions of oscillators of frequency $\nu_q = \nu$, and N is the number of molecules in the crystal. On the basis of eq. (11b), the function $f_0(\nu)$ is frequently called the weighted density of phonon states, and $\xi^2(\nu)$ the electron–phonon coupling function.

In the approximation being discussed the phonon function $f(\nu, T)$ is proportional to the probability of one-phonon transitions [the first term in eq. (11) corresponding to the creation and the second to the annihilation of one phonon upon a phototransition in the impurity]. As a matter of fact, if we substitute eqs. (11) and (11a) into (10c), we can readily see that

$$\Phi_1(\omega, T) = e^{-f(T)} f(-\omega, T). \tag{12}$$

The function $f_0(\nu)$ is nonzero in the region from 0 to ν_q^{max} (where ν_q^{max} is the maximum frequency of the phonon spectrum, with which the electronic transition in the impurity interacts). Hence, the function $\Phi_1(\omega, T)$ is non-zero in the one-phonon range $(-\nu_q^{max}, +\nu_q^{max})$ [see eqs. (11) and (12)]. In the same manner, it is also readily evident that $\Phi_2(\omega, T)$ is nonzero in the region $(-2\nu_q^{max}, +2\nu_q^{max})$, and $\Phi_m(\omega, T)$ in the region $(-m\nu_q^{max}, +m\nu_q^{max})$. Thus, each $\Phi_m(\omega, T)$ is the probability of an m-phonon transition (expressed in terms of the probability of a one-phonon transition).

In order to assess the contribution to the PW from phototransitions with the creation of various numbers of phonons, we resort to the integrated intensities of the ZPL and PW:

$$I_{ZPL} = \int_{-\infty}^{\infty} \Phi_0(\omega, T) \, d\omega, \qquad I_{PW} = \int_{-\infty}^{\infty} \Phi(\omega, T) \, d\omega, \tag{13}$$

and introduce the parameter α, determined by the ratio

$$\alpha = I_{ZPL}/(I_{ZPL} + I_{PW}). \tag{14}$$

The quantity α indicates the share of the ZPL intensity in the integrated intensity of the spectral band, and may serve as a characteristic of the strength of the electron–phonon coupling. The parameter α (by analogy with the theory of neutron and X-ray scattering) is usually called the Debye–Waller factor.

Now, making use of eqs. (10) and (10a–d), we calculate the contribution to the PW intensity from the m-phonon transitions. Integrating eq. (10c) with respect to ω we obtain

$$\Phi_m(T) = e^{-f(T)} f^m(T)/m!, \tag{15}$$

where $f(T)$ is determined by eq. (10d). Further, on the basis of eqs. (10a), (10b) and (13), and with the normalization $\int_{-\infty}^{\infty} I(\omega) \, d\omega = 1$ taken into account, we readily obtain the following expression for the Debye–Waller factor:

$$\alpha(T) = e^{-f(T)}. \tag{16}$$

It is evident from eqs. (15) and (16) that a single-valued relation exists between $\Phi_m(T)$ and α. Knowing α we can easily find the contribution to the PW of the m-phonon transitions (and vice versa). The required calculations were carried out by Osad'ko et al. (1974) and their results are presented in fig. 2. From the curves given there it is evident, for example, that, while for $\alpha = 0.9$ the PW is practically determined by one-phonon transitions only, for $\alpha = 0.1$ a substantial contribution to the PW comes

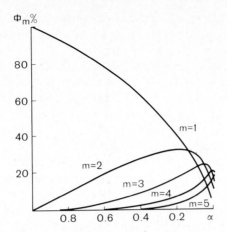

Fig. 2. Relative contribution to PW of transitions with the creation of various numbers of phonons as a function of α (from Osad'ko et al. 1974).

from two-, three- and even four-phonon transitions. Since the Debye frequency of molecular crystals lies in the range of 40 to 100 cm^{-1} [see, e.g., the Personov et al. (1971b), Osad'ko et al. (1974)], the PW can occupy a spectral range from several tens to several hundreds of reciprocal centimeters.

2.4. Temperature dependence of spectra

The temperature has a pronounced effect on the type of spectral band produced, on the ratio of the ZPL and PW intensities, on their widths and shapes, and also leads to temperature shifts of the ZPL.

In the case of linear electron–phonon coupling, when the ZPL are of zero (natural) width at all temperatures, the principal temperature effect is the redistribution of intensities between the ZPL and PW. In this approximation, with eqs. (10d) and (11) taken into account, we readily obtain the following expression for the Debye–Waller factor (16):

$$\alpha(T) = \frac{I_{ZPL}}{I_{ZPL} + I_{PW}} = \exp\left[-\int_0^\infty f_0(\nu)\left(\frac{2}{e^{\nu/kT}-1}+1\right)d\nu\right]. \qquad (17)$$

It is evident from eq. (17) that the ZPL intensity is determined .by the weighted density $f_0(\nu)$ of the phonon states and the temperature. The stronger the electron–phonon coupling, the lower the ZPL intensity. At a fixed coupling force an increase in the temperature leads to a drop in ZPL intensity and to the "transfer" of this intensity to .the PW (the integrated intensity $I_{ZPL} + I_{PW}$ is independent of the temperature).

When the quadratic terms in eq. (8) are taken into consideration, temperature broadening and shift occur along with a temperature drop in intensity. The corresponding theory has been developed in sufficient detail in Silsbee (1962), McCumber and Sturge (1963), Krivoglaz (1964) and Osad'ko (1975, 1979). But this problem is beyond the scope of the present paper.

3. Principles of site selection spectroscopy of complex molecules in solutions

3.1. Introduction

From the viewpoint of the concepts discussed in the preceding section, there are two possible extreme cases that can elucidate the origin of the broad bands in the spectra of organic molecules in solutions.

1. Strong electron–phonon coupling occurs between the molecules and the solvent. Here the ZPL intensity can be very low even at low temperatures [see eq. (17)], and the wide bands that are observed are well developed PW. These PW are due to both one- and multi-phonon transitions.

2. The electron–phonon coupling is weak and the spectra of the separate molecules consist of narrow ZPL. But, in this case, the investigated molecules are in somewhat differing local conditions in the field of the solvent. This leads to statistical straggling in the positions of their electronic levels and to relative displacements of their spectra along the frequency scale. Here each broad band observed in the spectrum is the envelope of a family of ZPL, and band broadening is inhomogeneous. (In the first case, the broadening is homogeneous.)

In order to ascertain which of the above-mentioned cases actually occurs, experiments were conducted with the excitation of fluorescence by means of monochromatic laser radiation. It could be expected that in the case of homogeneous broadening the change from ordinary to laser excitation would not lead to any essential alterations in the nature of the fluorescence spectrum, whereas in the case of inhomogeneous broadening the situation should change appreciably. In the latter case, monochromatic excitation (in the region of the 0–0 transition or the lowest vibronic transitions) affects (excites) mainly the molecules that have an absorption ZPL at the frequency of the laser. Then the emission spectrum displays narrow lines that belong only to these molecules.

As a result of these experiments, conducted in our laboratory on a number of different specimens, it was established that at sufficiently low temperatures, the decisive role, in many cases, is played by inhomogeneous broadening (Personov et al. 1972a,b, 1973, Personov and Kharlamov 1973, Bykovskaya et al. 1974). This was subsequently confirmed in many other papers (Avarmaa 1974a, Eberly et al. 1974, Cunningham et al. 1975, Abram et al. 1975). Later papers were devoted to the development of selective methods for eliminating inhomogeneous broadening and for revealing line structure in fluorescence, phosphorescence and absorption spectra, as well as to the investigation of this structure [see Personov (1978) and references therein]. Let us consider each of the three above-mentioned types of spectra separately.

3.2. Fluorescence spectra

3.2.1. Emergence of fine structure upon selective excitation
Experiments conducted on a number of aromatic hydrocarbons, porphyrins and dyes (at 4.2 K) in various solvents have indicated that changing over from conventional methods of excitation to monochromatic techniques leads to a drastic transformation of the fluorescence spectrum and to the emergence of a rich line structure. Some examples illustrating the afore said are given in fig. 3. It is obvious that when selective excitation is resorted to, narrow ZPL are displayed instead of a few broad bands. These ZPL are accompanied on the long-wave side by relatively wide wings. (This is especially clear in figs. 3a, b and c.) Such line spectra can be obtained for various compounds, both in the neutral and in the ionic forms, and in a great variety of solvents (glassy and crystalline, polar and nonpolar, etc.). These data unambiguously point to the essential role played by inhomogeneous broadening of the spectra and to the feasibility of its elimination by selective excitation.

Next we shall consider the principal characteristics of the line spectra obtained in this manner.

3.2.2. General nature of spectral bands
First of all, we note that the long-wave wings, observed close to the ZPL in monochromatically excited fluorescence spectra, are not simply phonon wings, but have a more complicated origin. They include the true PW of the resonantly excited centers, but also fluorescence bands (ZPL + PW) of the nonresonantly excited centers (i.e. excited not through the ZPL, but through the PW in absorption). To elucidate this point, let us consider fig. 4, showing the pattern of the spectrum in the 0–0 transition region in absorption and in fluorescence. The absorption bands of the centers, laser excited through the PW, are shown by dashed lines in the upper part of the figure. Since the intensity at a maximum PW is usually substantially less than that at a maximum ZPL, the fluorescence of each of these

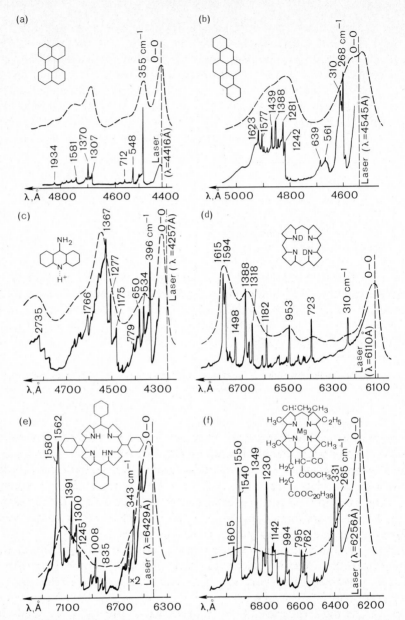

Fig. 3. Emergence of fine structure in fluorescence spectra of solutions of organic compounds upon selective laser excitation (T = 4.2 K): (a) perylene in ethanol, (b) 3,4-8,9-dibenzpyrene in polymethylmethacrylate, (c) protonated form of 9-aminoacridine in ethanol, (d) D$_2$-porphin in deuterated ethanol, (e) tetraphenylporphin in polystyrene, (f) protochlorophyll in ether (ordinary excitation is indicated by dashed lines, selective laser excitation by full lines). (From Bykovskaya et al. 1981.)

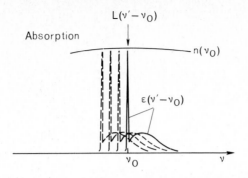

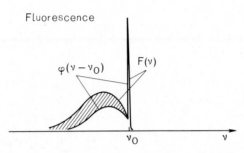

Fig. 4. Schematic diagram of the formation of a wing near a ZPL in a selectively excited fluorescence spectrum.

nonresonantly excited centers is relatively weak. But, since the width of the PW is much greater than that of the ZPL, the number of nonresonantly excited centers is large. All together they make an appreciable contribution to the integrated intensity of the long-wave wing close to the fluorescence line. (This contribution is represented by the hatched area in the lower part of fig. 4.) This leads to an apparent increase in the PW and reduction in the Debye–Waller factor as compared to the true values.

Let us make a simple quantitative assessment and establish a relationship between the true and "observed" Debye–Waller factors. Let $\epsilon(\nu' - \nu_0)$ and $\varphi(\nu - \nu_0)$ be the true spectral distribution curves of the 0–0 bands in the absorption and fluorescence spectra of one type of centers, having maximum ZPL at the frequency ν_0. We assume that the shape of the spectra of the various types of centers is the same, and that they differ only in the value of ν_0. Let $L(\nu' - \nu_\ell)$ be the laser line profile (where ν_ℓ is the frequency of the maximum), and $n(\nu_0)$ the distribution function of the number of centers with respect to their 0–0 transition frequency. Then the spectral curve for the 0–0 band in the fluorescence spectrum is expressed (up to a constant factor) in the form of the convolution

$$F(\nu) = \int_0^\infty \int_0^\infty d\nu' \, d\nu_0 \, \varphi(\nu - \nu_0)\epsilon(\nu' - \nu_0)n(\nu_0)L(\nu' - \nu_\ell). \tag{18}$$

We also assume that the half width of the distribution function of the centers $n(\nu_0)$ is considerably larger than the PW half width in absorption, and put $n = \text{const.}$ (This assumption is valid only for infinitely large inhomogeneous broadening and is very crude. It is, however, admissible for elucidation of the qualitative picture.) Moreover, we shall assume that the laser line $L(\nu' - \nu_\ell)$ is much narrower than the ZPL and shall take it as a δ-function. Then, instead of eq. (18), we can write for the fluorescence band:

$$F(\nu) \approx \int_0^\infty d\nu_0 \, \varphi(\nu - \nu_0)\epsilon(\nu_\ell - \nu_0). \tag{19}$$

Next we present the curves φ and ϵ in the form of two terms:

$$\varphi(\nu - \nu_0) = \varphi_1(\nu - \nu_0) + \varphi_2(\nu - \nu_0), \tag{20}$$

$$\epsilon(\nu_\ell - \nu_0) = \epsilon_1(\nu_\ell - \nu_0) + \epsilon_2(\nu_\ell - \nu_0), \tag{21}$$

where φ_1 and ϵ_1 describe the ZPL, and φ_2 and ϵ_2 the PW.

Substituting eqs. (20) and (21) into (19) we have

$$F(\nu) \approx F_1(\nu) + F_2(\nu), \tag{22}$$

where

$$F_1(\nu) = \int_0^\infty d\nu_0 \, \varphi_1(\nu - \nu_0)\epsilon_1(\nu_\ell - \nu_0), \tag{23}$$

$$F_2(\nu) = \int_0^\infty d\nu_0 \, \{\varphi_2(\nu - \nu_0)\epsilon_1(\nu_\ell - \nu_0) \\ + \epsilon_2(\nu_\ell - \nu_0)[\varphi_1(\nu - \nu_0) + \varphi_2(\nu - \nu_0)]\}. \tag{24}$$

Here $F_1(\nu)$ describes the observed ZPL and $F_2(\nu)$ the observed wing. Recalling the definition of the Debye–Waller factor, eq. (14), we introduce, by analogy, the "observed Debye–Waller factor" α'. Thus

$$\alpha' = \frac{I_{\text{ZPL}}^{(\text{obs})}}{I_{\text{ZPL}}^{(\text{obs})} + I_{\text{W}}^{(\text{obs})}} \approx \frac{\int_0^\infty F_1(\nu)\,d\nu}{\int_0^\infty F(\nu)\,d\nu}. \tag{25}$$

Substituting eqs. (19) and (23) into (25), we obtain

$$\alpha' \approx \frac{\int_0^\infty d\nu \, \varphi_1(\nu - \nu_0) \int_0^\infty d\nu_0 \, \epsilon_1(\nu_\ell - \nu_0)}{\int_0^\infty d\nu \, \varphi(\nu - \nu_0) \int_0^\infty d\nu_0 \, \epsilon(\nu_\ell - \nu_0)} = \alpha_{\text{fluor}}\alpha_{\text{absorp}}. \tag{26}$$

Assuming that the Debye–Waller factors are the same for the fluorescence and absorption bands, i.e. putting

$$\alpha_{\text{fluor}} \approx \alpha_{\text{absorp}} = \alpha, \tag{27}$$

we finally have

$$\alpha' \approx \alpha^2. \tag{28}$$

In addition we shall determine the share of the true PW in the integrated intensity of the observed wing. The integrated intensity of the observed wing is

$$I_W^{(\text{obs})} = \int_0^\infty F_2(\nu) \, d\nu. \tag{29}$$

The true PW is formed by the centers excited through the ZPL and is determined by the expression

$$I_{\text{PW}} = \int_0^\infty d\nu \int_0^\infty d\nu_0 \, \varphi_2(\nu - \nu_0) \epsilon_1(\nu_\ell - \nu_0). \tag{30}$$

On the basis of eqs. (29) and (30), and taking eq. (27) into account, we readily obtain

$$I_{\text{PW}}/I_W^{(\text{obs})} \approx \alpha/(\alpha + 1). \tag{31}$$

It is evident from eq. (31) that the intensity of the true PW constitutes less than one half of that of the observed wing (because $\alpha \leq 1$).

The approximate relations (28) and (31) indicate that the general appearance of the fluorescence spectra upon selective excitation (and, in particular, the possibility of revealing narrow ZPL in them) drastically depends on the quantity α, i.e. on the strength of the electron–phonon coupling. The fluorescence of nonresonantly excited centers, in this case, greatly "spoils" the line pattern of the spectrum. Thus, for instance, with a true Debye–Waller factor $\alpha = 0.05$ (with the multi-phonon transitions, in this case, making an essential contribution to the PW, see fig. 2) the observed factor is $\alpha' \approx 0.0025$, and it may prove difficult to detect ZPL in the spectrum.

The following is a specific example of how a variation in α affects the appearance of a spectrum. Figure 5 presents fluorescence spectra of the same dye (thioindigo) in three different solvents upon laser excitation in the 0–0 transition region (in the case of *n*-paraffin as solvent the excitation wavelength was taken somewhat shorter than in the other two cases, because of the general displacement of the spectrum in this solvent). Approximate estimates, based on measurements in these spectra [making use of eq. (28)], yield the following three values for α: 0.3, 0.15 and 0.07. It is evident from fig. 5c that any further small reduction of α would lead to the complete disap-

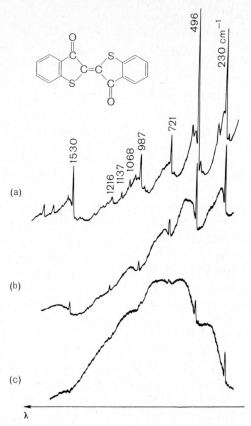

Fig. 5. Portion of the fluorescence spectrum of thioindigo in three different solvents upon monochromatic excitation by a tunable laser in the 0–0 transition region (T = 4.2 K): (a) in paraffin oil (λ_{exc} = 5550.6 Å), (b) in ethanol (λ_{exc} = 5621.5 Å), (c) in chloroform (λ_{exc} = 5621.5 Å). (From Al'shits et al. 1980.)

pearance of the ZPL in the spectrum and to the appearance of a structure-less band.

From the afore said it follows that when line structure in the spectrum is absent even at low temperatures and upon monochromatic excitation, the origin should primarily be looked for in the high value of the electron–phonon coupling. This can be weakened, in principle, by a proper selection of the medium.

3.2.3. Zero-phonon line width
A measurement of the ZPL widths in fluorescence spectra of organic molecules (at 4.2 K) upon excitation in the 0–0 transition region by narrow laser lines ($\Delta\nu \leqslant 0.1$ cm^{-1}) usually yields values within the range of 0.5 to

$5 \, \text{cm}^{-1}$. All these ZPL correspond to vibronic transitions, and their width is determined to a considerable degree by the relaxation times of the vibrational excitations of the molecules under investigation. This width cannot be appreciably reduced. The corresponding times, estimated from the width of these vibronic ZPL, are found to be of the order of 10^{-11} s. This agrees with data obtained by other methods (Personov and Solodunov 1968, Tamm and Saari 1974).

It should be pointed out, however, that there are data which indicate that upon selective excitation (in the 0–0 transition region) the width of the vibronic lines in a fluorescence spectrum may, in many cases, also include an appreciable share of inhomogeneous width. This width is of the same order of magnitude as the homogeneous width (Gorokhovski and Kikas 1977). This means that in an inhomogeneous system, molecules with a fixed frequency of the purely electronic transition may differ slightly in the frequencies of their intramolecular vibrations.

At low temperature, when the temperature broadening of the ZPL due to electron–phonon coupling is negligibly small, the homogeneous width of a purely electronic line should be substantially less than the homogeneous width of vibronic lines (for electronic states with a lifetime of 10^{-8} to 10^{-9} s). But this width cannot be determined from the fluorescence line spectra. As a matter of fact, observation is hindered by scattered laser radiation upon excitation in the 0–0 band. Upon excitation in the vibronic absorption band, the 0–0 fluorescence line will be inhomogeneously broadened. In the latter case, the profile of the 0–0 line in the fluorescence spectrum is a convolution of the comparatively wide vibronic absorption line profile with the profile of the 0–0 fluorescence line of one type of centers. In this case, therefore, the width of the 0–0 line in the fluorescence spectrum cannot be less than the homogeneous width of the vibronic absorption line.

The homogeneous width of the purely electronic line can be determined from experiments on "burning" of narrow holes in an inhomogeneously broadened absorption band (see subsections 3.5,6). The relevant measurements indicate that the homogeneous width of purely electronic lines is in fact very small. It is especially small in the case of crystalline matrices, in which case, at 2 to 4 K, it ranges from about 10^{-2} and $10^{-3} \, \text{cm}^{-1}$ (de Vries and Wiersma 1976, Gorokhovski et al. 1976, Voelker et al. 1977).

3.2.4. Temperature dependence of spectra

As has been mentioned previously (subsection 2.4), a typical feature of ZPL is a drop in intensity as the temperature is raised. This drop is distinctly observed in spectra obtained upon selective excitation. Shown in fig. 6 as a typical example is a portion of the fluorescence spectrum (one vibronic band) of perylene in two solvents with different properties and at different temperatures. The drastic temperature drop in ZPL intensity

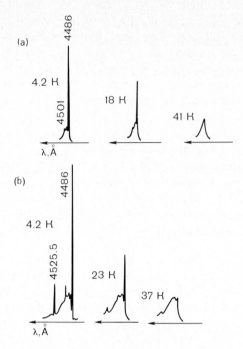

Fig. 6. Vibronic band $(0-355\,\text{cm}^{-1})$ of the fluorescence spectrum of perylene upon laser excitation $(\lambda_{\text{exc}} = 4415.6\,\text{Å})$ at various temperatures: (a) in ethanol, (b) in n-undecane. (From Personov et al. 1972a.)

is readily evident. In the case being considered, the ZPL practically disappear from the spectrum at a temperature between 40 and 50 K. [Note that the observed temperature weakening of ZPL in selective excitation spectra is larger than the true weakening, due to relation (28).]

It follows from temperature measurements that in order to reveal the fine structure of a spectrum, sufficiently low cooling of the solution is required along with selective excitation.

3.2.5. Dependence of spectra on the excitation wavelength

In the examples discussed above, fine-structure (high-resolution) fluorescence spectra were obtained with monochromatic excitation in the 0–0 transition region. The question which naturally arises concerns the dependence of the spectra on λ_{laser}. This was investigated experimentally by many researchers: Personov and Al'shits (1975), Al'shits et al. (1975), Rebane et al. (1975), Cunningham et al. (1975), Abram et al. (1975), etc. [A theoretical investigation of this dependence, using simple models, was carried out by Avarmaa (1974b).] The results of these investigations reduces to the following:

(a) Variation of the laser frequency within the limits of the in-homogeneous broadening (100 to $300\,\mathrm{cm}^{-1}$) in the 0–0 transition region causes a corresponding displacement of the spectrum along the frequency scale without changing its nature. This is quite obvious under conditions of inhomogeneous broadening.

(b) Upon a considerable increase in the laser frequency, when the excitation is performed in the region of high vibronic transitions of the excited electronic state (exceeding the 0–0 transition frequency by 3000 to $5000\,\mathrm{cm}^{-1}$ and more), the fluorescence spectrum is broad-bended. It does not, in this case, differ from spectra obtained with excitation by broad-band sources. Shown in fig. 7 as one such example are the fluorescence spectra of Zn-phthalocyanine in *n*-butyl alcohol obtained by various excitation methods. The transformation of the spectrum with increasing laser frequency can be seen clearly. The 6764 Å laser line falls in the long-wave

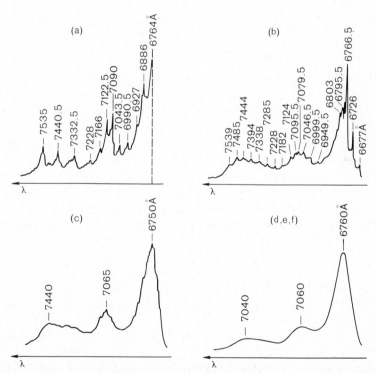

Fig. 7. Fluorescence spectra of Zn-phthalocyanine in *n*-butyl alcohol obtained by various methods of excitation ($T = 4.2\,\mathrm{K}$): (a) by a Kr laser with $\lambda_{\mathrm{Kr}} = 6764$ Å, (b) with $\lambda_{\mathrm{Kr}} = 6471$ Å, (c) with $\lambda_{\mathrm{Kr}} = 5682$ Å, (d) with $\lambda_{\mathrm{Kr}} = 5208$ Å, (e) by a cadmium laser with $\lambda_{\mathrm{Cd}} = 4415.6$ Å, (f) by a mercury–quartz lamp with $\lambda_{\mathrm{Hg}} = 4046$ and 4358 Å. (From Al'shits et al. 1975.)

part of the absorption band. Here the 0–0 transition line is not visible in the fluorescence spectrum (fig. 7a) because its frequency coincides with that of the laser line. Its vibrational recurrences, however, are indications of the elimination of inhomogeneous broadening. In passing over to excitation by the 6471 Å laser line (fig. 7b) there is still a fine structure in the spectrum. Here (in contrast to the preceding case) the region of the 0–0 transition is also visible. But a further increase in the frequency of the laser leads to substantial broadening of the spectrum. At $\lambda_{laser} = 5682$ Å, for instance, the spectrum has broad bands with hardly observable inflections (fig. 7c). When laser lines of higher frequencies (5208 and 4415.6 Å) are used for excitation, the fluorescence spectrum of Zn-phthalocyanine is the same as with excitation by ordinary sources (mercury discharge or xenon lamps) (figs. 7d,e,f).

Qualitatively, there is no difficulty in understanding the above-mentioned broadening of the spectrum upon a large increase in the laser frequency. As a matter of fact, high vibronic levels (in the region of overtones and combination tones) and the ZPL that correspond to them are considerably broader than the purely electronic ones. Moreover, there is a significant increase in the density of states in this region. Hence, there is appreciable

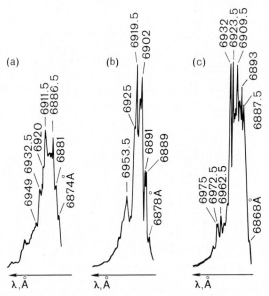

Fig. 8. "0–0 multiplets" in the fluorescence spectrum of H_2-phthalocyanine in paraffin oil upon excitation by various laser lines ($T = 4.2$ K): (a) $\lambda_{exc} = 6764$ Å (vibrational region from 200 to 400 cm^{-1} in the excited electronic state), (b) $\lambda_{exc} = 6471$ Å (vibrational region from 900 to 1100 cm^{-1}), (c) $\lambda_{exc} = 6328$ Å (vibrational region from 1200 to 1400 cm^{-1}). (From Personov and Al'shits 1975.)

overlap of the spectral bands of various centers in the region of high vibronic states in the absorption spectrum. In this region the laser excites practically all types of centers and inhomogeneous broadening is not eliminated.

(c) More important and interesting is the dependence of the fluorescence spectrum on λ_{laser} upon excitation to not particularly high vibronic levels (up to 2000 or 3000 cm^{-1}), when the spectrum is still of the line type. Under these conditions, the detailed structure of the spectrum essentially depends on the wavelength of the laser excitation. When the excitation is performed in the 0–0 transition region, the fluorescence spectrum consists of single vibronic lines. On the contrary, with excitation to the vibronic levels, the fluorescence spectrum becomes more complicated and consists of groups of lines: "multiplets". The number and arrangement of the lines in a multiplet depend to a large extent on λ_{laser}. This is illustrated in fig. 8, which shows the 0–0 transition region in the fluorescence spectrum of phthalocyanine, excited by three different laser lines. In each case, as is evident, the "0–0 multiplet" contains from 7 to 10 lines, and the multiplets differ

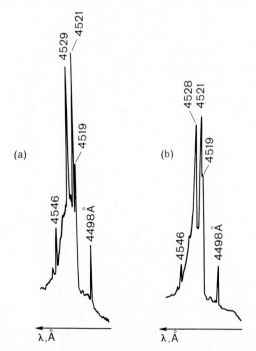

Fig. 9. "0–0 multiplets" in the fluorescence spectrum of 3,4-8,9-dibenzpyrene ($T = 4.2$ K) upon excitation by a laser with $\lambda = 4415.6$ Å: (a) in ethanol, (b) in paraffin oil. (From Personov and Al'shits 1975.)

considerably. (The most intense components of these multiplets are subsequently repeated at vibronic transition frequencies throughout the whole
fluorescence spectrum.)

An essential property of these multiplets is that they are independent of
the solvent. This is quite evident from the example in fig. 9, which shows
the 0–0 multiplet in the fluorescence spectrum of 3,4-8,9-dibenzpyrene in
two solvents: in the polar glassy solvent ethanol, and in the neutral
crystalline solvent paraffin oil. It is obvious that the number of lines in the
multiplet and their arrangement are the same in both cases.

The occurrence of these multiplets can be readily explained by referring
to the simple diagram in fig. 10. (First we ignore the presence of PW in the
spectra.) Assume that the molecules under investigation have in the excited
electronic state S_1 two close vibronic levels corresponding to the excitation
of vibrations with frequencies $\Delta \nu_1$ and $\Delta \nu_2$. Assume further that for part of
the dissolved molecules the laser frequency corresponds to excitation to
the vibronic level with frequency $\Delta \nu_1$. After vibrational relaxation to the
state S_1, these molecules will emit a quantum of fluorescence. The 0–0 line
in the fluorescence spectrum of these molecules is separated from the laser
line by a distance $\Delta \nu_1$ (fig. 10a). If the amount of inhomogeneous broaden-

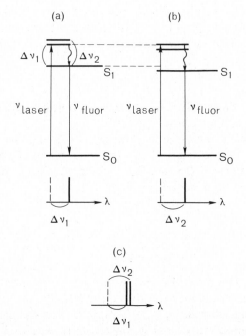

Fig. 10. Schematic diagram of the formation of "multiplets" in a fluoresence spectrum.

ing is larger than the difference $\Delta\nu_2 - \Delta\nu_1$, there will be molecules in the solution whose energy of the purely electronic S_1–S_0 transition lies exactly an amount $\Delta\nu_2 - \Delta\nu_1$ below that of the former molecules. (However, the vibrational quanta of all the molecules in an inhomogeneous system can, in this case, be assumed the same.) Molecules of the latter type are excited by the same laser radiation to the vibronic level $\Delta\nu_2$. In their fluorescence spectrum the 0–0 line will be separated from the laser line by a distance $\Delta\nu_2$ (fig. 10b). Since molecules of both types are present in the solution simultaneously, a doublet appears in the fluorescence spectrum (fig. 10c). If, in the excited state, the dissolved molecules have not two, but several closely spaced vibronic levels close to the laser frequency, the corresponding, more complex, "multiplet" appears in the 0–0 transition region of the fluorescence spectrum. The distances from the laser line to the various components of this multiplet are equal to the vibrational frequencies of the molecules in the excited state S_1. By varying the frequency of the exciting line we can "probe" various regions of the vibronic states and determine all the vibration frequencies in the excited state (that are active in the given electronic transition).

The mechanism presented above explains the dependence of the "multiplet structure" on the excitation frequency of the laser and its independence of the solvent. The validity of the above interpretation of "multiplets" was experimentally confirmed on a number of molecules, for which the vibration frequencies in the excited electronic state were determined by independent methods (Personov and Al'shits 1975, Al'shits et al. 1975).

It should be pointed out that, if an absorption spectrum contains sufficiently intense PW, with relatively narrow maxima, a supplementary structure, due to these PW, may appear in the multiplets. But, as a rule, PW do not contain such narrow peaks.

In this manner, by investigating the fine-structure fluorescence spectra upon excitation in the regions of various vibronic transitions, we can establish the vibrational frequencies of molecules, not only in the ground state, but in the excited electronic states as well.

3.2.6. Line polarization

A vital source of information on the properties and structure of molecules is the polarization of their radiation. The line structure of fluorescence spectra upon selective excitation enables one to measure the degree of polarization for separate vibronic lines (and not the total radiation, as is frequently done). If a molecule has a definite symmetry, the degree of polarization of the vibronic lines is associated with the types of symmetry of the corresponding vibrations. In some recent papers, Bykovskaya et al. (1978, 1979, 1980a) developed a method to determine the symmetry of molecular vibrations. It is based on obtaining fluorescence line spectra in isotropic media upon selective excitation and measuring the degree of polarization of separate lines.

A homogeneous nonscattering medium is required for making quantitative measurements of the degree of fluorescence polarization. But many organic solvents (and their mixtures) that form transparent glasses at 77 K become fissured at lower temperatures and scatter light appreciably. This leads to substantial depolarization of the radiation. Investigations have established that exceptionally convenient matrices for performing low-temperature polarization measurements are transparent polymer films. These films retain their good optical quality at the liquid-helium temperature and upon repeated cooling. In conducting polarization measurements by the method being discussed, fluorescence is excited by linearly polarized light (along the vertical Oz axis) of a tunable dye laser. The intensity of each vibronic line is measured in two polarizations: along the Oz and Ox axes (for which purpose a polarizing filter is set in front of the spectrometer slit). The degree of polarization P is calculated by the

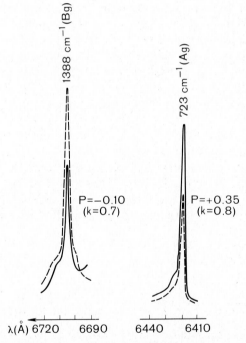

Fig. 11. Polarization of two vibronic lines in the fluorescence spectrum of porphin in polystyrene ($T = 4.2$ K and $\lambda_{exc} = 6134$ Å). Full lines indicate z-polarization, dashed lines x-polarization. (From Bykovskaya et al. 1980a.)

relation

$$P = (I_z - kI_x)/(I_z + kI_x), \tag{32}$$

where the factor k takes the polarizing effect of the elements of the optical installation into account and is determined for all wavelengths by preliminary measurements.

As an example, we cite certain data on the spectra of porphin in polystyrene. The porphin molecule belongs to the symmetry point group D_{2h}. Allowed in its fluorescence spectrum with respect to symmetry are the completely symmetrical vibrations A_{1g} and the not completely symmetrical ones B_g. Upon excitation of fluorescence in the 0–0 transition region by linearly polarized light, the limiting degrees of polarization of the cor-

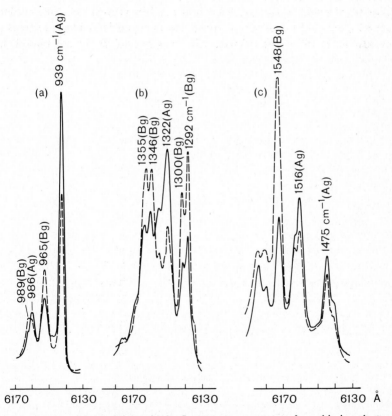

Fig. 12. Polarization of 0–0 multiplets in the fluorescence spectrum of porphin in polystyrene ($T = 4.2$ K) upon monochromatic excitation to the vibrational sublevels of the excited electronic state: (a) $\lambda_{exc} = 5807$ Å, (b) $\lambda_{exc} = 5688$ Å, (c) $\lambda_{exc} = 5634$ Å. Full lines indicate z-polarization, dashed lines x-polarization. (From Bykovskaya et al. 1979.)

responding vibronic lines are $+0.5$ and -0.33. Photoelectric recordings of two vibronic lines in the porphin spectrum are shown in fig. 11 for two polarizations. It can be established from these recordings that the $723\,\mathrm{cm}^{-1}$ vibration is completely symmetrical (A_{1g}), whereas the $1388\,\mathrm{cm}^{-1}$ vibration is not completely symmetrical (B_g)*.

In subsection 3.2.5 we demonstrated that multiplets appear in the fluorescence spectrum when we excite fluorescence by monochromatic radiation in the region of vibronic states. Making use of these multiplets we can determine the frequency of the molecular vibrations in the excited electronic state. Here we contend that we can also determine the symmetry of the vibrations of the excited state by measuring the degree of polarization of the separate components of the multiplets. The feasibility of the afore said is illustrated by fig. 12, which shows a recording of 0–0 multiplets in the fluorescence spectrum of porphin in polystyrene upon excitation by three different laser lines. The substantial differences in the intensity distribution in the multiplets at various polarizations is quite evident. These differences are indicative of the different symmetry of the vibrations in the spectrum. These types of symmetry can be established from the degree of polarization of the lines (see fig. 12).

3.3. Phosphorescence spectra

Along with fluorescence, many organic compounds also show phosphorescence. There is every reason to believe that the broad bands of phosphorescence (like those of fluorescence) at low temperature are mostly broadened inhomogeneously. It was certainly of interest, therefore, to attempt to eliminate this broadening and to reveal the fine structure of the spectrum in this case as well by means of selective excitation. But in the very first experiments, using phenanthrene in ethanol as an example, we established that the application of laser excitation in the region of the S_1–S_0 transition (at $4.2\,\mathrm{K}$), leading to the appearance of fine structure in the fluorescence spectrum, does not change the nature of the phosphorescence spectrum. The latter remains just as broad-banded as when the excitation is done by ordinary sources. A similar result was obtained by Cunningham et al. (1975) in their investigation of the spectra of 1-chloronaphthalene in

---

*Attention is drawn to the following fact. Although the differences in the degree of polarization of the lines enables one to distinguish reliably between completely symmetrical and not completely symmetrical vibrations, the degree of polarization does not reach its limiting values. The reason for this is that tautomerism, associated with the displacement of two central protons, readily occurs in the porphin molecule due to the light. During the laser irradiation anisotropy appears in the initially isotropic solution, and the degree of polarization is reduced. This phenomenon and methods for partial elimination of its effects are discussed in more detail in the original papers mentioned above.

ethanol. The lack of narrowing of the phosphorescence bands upon reduction of the spectral width of the exciting light (in excitation through a monochromator in the region of the S_1–S_0 transition) was noted by Tamm and Saari (1975a,b) for the case of 1,12-benzperylene and 1,2-benzpyrene. It has thus been established by employing a number of specimens that monochromatic S_1–S_0 excitation does not reveal fine structure in phosphorescence spectra. In the above-mentioned papers various hypotheses were proposed and various causes analyzed to explain this specific feature of phosphorescence spectra [see also Avarmaa and Suisalu (1975), Saari and Tamm (1976) and Al'shits et al. (1976a)]. Here we shall only indicate the cause which actually turned out to be the prevailing one.

The scatter in the energies of the S_1–S_0 and T_1–S_0 transitions of molecules in solution depends, in general, on various parameters characterizing the arrangement of the molecules of the dissolved substances and of the solvent relative to one another. Hence, monochromatic radiation in the region of the S_1–S_0 transition excites molecules for which the energies of this transition are identical, but the local conditions of these molecules are not completely identical. Although their S_1–S_0 transitions have the same energy, the T_1–S_0 transitions of these molecules may have different energies (and vice versa). To eliminate inhomogeneous broadening in this case it is necessary to provide selection with respect to the T_1–S_0 energy, i.e. to ensure direct monochromatic T_1–S_0 excitation of phosphorescence.

The main difficulty in such experiments is associated with the very low absorption coefficient in the region of the T_1–S_0 transition, which is forbidden with respect to spin. The absorption coefficient in this region is usually less by a factor of 10^6 to 10^7 than in the case of S_1–S_0 transitions. Moreover, in investigating radiation excited in the region of negligibly low absorption coefficients, there is the problem of separating this radiation from the interfering luminescence of impurities that are always present in negligible concentrations, and from Raman scattering of the solvent. In the investigations by Al'shits et al. (1976a,b), these difficulties were overcome by use of sufficiently intensive laser lines for excitation, the effect of an external heavy atom that increases the probability of T_1–S_0 transitions, special phosphoroscopes, etc. Phosphorescence spectra were obtained with laser T_1–S_0 excitation for a number of molecules and it was shown that under these conditions broad-band phosphorescence spectra actually were converted into line-type spectra. Two examples are given in fig. 13. As could be expected, a vibrational analysis of these line spectra yields the frequencies of the normal vibrations of the investigated molecules in the ground electronic state. It should be pointed out that the symmetry of vibrations can be determined, in principle, by measuring the polarization degree of the phosphorescence lines [Suter and Wild (1981)].

The cited data directly indicate that broad bands in phosphorescence

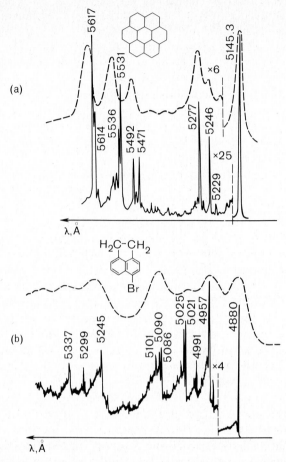

Fig. 13. Phosphorescence spectra of (a) coronene and (b) 5-bromacenaphthene in butyl bromide excited (at 4.2 K) by two methods: dashed lines indicate the spectrum upon ordinary S_1–S_0 excitation (λ_{Hg} = 313 and 365 nm), full lines the spectrum upon laser T_1–S_0 excitation (λ = 5145 and 4880 Å). (Spectra obtained by Al'shits and Kharlamov.)

spectra are inhomogeneously broadened and that this broadening can be eliminated by selective T_1–S_0 excitation.

In the examples illustrated in fig. 13, laser excitation was performed in the region of purely electronic transitions. Upon laser excitation to vibronic levels of the T_1 state (and not in the 0–0 transition region), "multiplets" depending on λ_{laser} are found in the phosphorescence spectrum. The mechanism for the occurrence of these multiplets is similar to that discussed above for fluorescence spectra (subsection 3.2.5). The only difference is that here the triplet state T_1 is involved instead of the singlet state S_1. An analysis of these multiplets at various λ_{laser} values enables one to determine the frequencies of molecular vibrations in the excited elec-

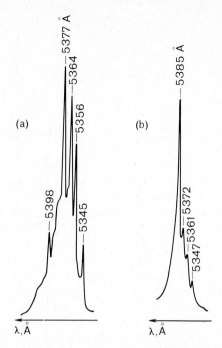

Fig. 14. "0–0 multiplets" in the phosphorescence spectrum of 1,2-benzpyrene in butyl bromide ($T = 4.2$ K) upon excitation by two different laser lines: (a) $\lambda_{exc} = 5145$ Å, (b) $\lambda_{exc} = 4965$ Å. (From Al'shits et al. 1976b.)

tronic state, but now in the triplet state. As an example, fig. 14 shows the region of the 0–0 transition in the phosphorescence spectrum of 1,2-benzpyrene upon excitation with two laser lines: 5145 and 4965 Å. The first line corresponds to excitation in the triplet state of vibrations in the frequency range of 700 to 900 cm^{-1}; the second line corresponds to the frequency range of 1400 to 1600 cm^{-1}. In the first case the multiplet consists of five lines (fig. 14a), in the second of four lines (fig. 14b) (the 5385 Å line being considerably more intense than the others). In accordance with the afore said, an analysis of the first multiplet, for instance, yields the following frequencies of vibrations for 1,2-benzpyrene in the T_1 state: 725, 765, 790, 835 and 910 cm^{-1}.

The feasibility of determining the frequencies of molecular vibrations in the triplet state by this method deserves special attention because very few data of this kind are available to date.

3.4. Excitation spectra

The methods of selective excitation discussed above enable one to obtain line spectra in emission. It is of no less interest to reveal an analogous fine structure in absorption spectra, containing information on the vibronic levels of excited electronic states. But it is obvious that for absorption spectra other techniques are required.

One of the possibilities of revealing line structure in inhomogeneously broadened absoption spectra consists in recording excitation spectra for narrow spectral parts of the fluorescence band (Personov and Kharlamov 1973). It is well known that an excitation spectrum is obtained by scanning the wavelength of the exciting light and measuring the resulting variation in fluorescence intensity in a fixed (usually wide) spectral range. When the solution is sufficiently dilute and there is no dependence of the fluorescence quantum yield on the wavelength of the exciting light, the excitation spectrum exactly reproduces the absorption spectrum. Under conditions of inhomogeneous broadening, a sufficiently narrow spectral range can be separated out of the fluorescence spectrum so that the radiation in this range will be mainly due to centers of a single type. Then the corresponding excitation spectrum will predominantly reproduce the absorption spectrum of these centers, and it should display a fine structure. The most suitable excitation sources for such experiments are tunable lasers with a narrow line. But in a qualitative sense, this effect can even be observed when monochromators with sufficiently high resolution are employed for excitation and for recording the fluorescence. As an example we consider the fluorescence excitation spectrum for perylene in ethanol. Given in fig. 15a is the excitation spectrum with the radiation recorded in a wide spectral range ($\Delta\nu_f \approx 100\,\mathrm{cm}^{-1}$). This spectrum consists of several broad bands and reproduces the absorption spectrum of the solution being investigated. Quite a different picture is obtained if, in recording the excitation spectrum, fluorescence is recorded in a narrow spectral range ($\Delta\nu_f \approx 4\,\mathrm{cm}^{-1}$). From fig. 15b it is evident that in this case the spectrum has narrow lines accompanied by relatively wide wings on the short-wave side. A vibrational analysis of this spectrum yields the frequencies of vibration of perylene in the excited electronic state. The excitation spectrum in fig. 15b is approximately reflection symmetrical to the fluorescence line spectrum of perylene (fig. 3a). But, as indicated by measurements, the ratio of the intensities in the excitation spectrum at the maximum of the ZPL and the PW is less, and the ZPL themselves are wider, than in the fluorescence spectrum. This is an obvious consequence of the relatively large spectral width of the exciting light ($\Delta\nu \approx 4\,\mathrm{cm}^{-1}$), separated out by the monochromator, compared to the laser line width. It is also due to the insufficiently narrow range employed in the fluorescence band.

An important feature of the excitation line spectra obtained by the above-mentioned method should be pointed out. As in the emission spectra obtained upon selective excitation, the broad wings in the vicinity of the ZPL in the excitation spectra are not simply of the phonon type. Primarily,

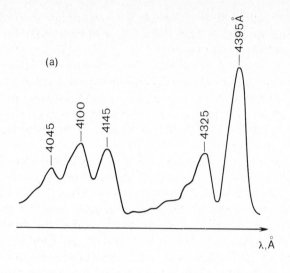

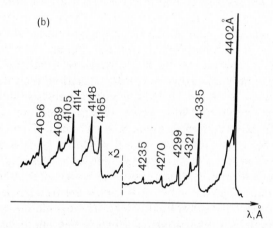

Fig. 15. Fluorescence excitation spectrum of perylene in ethanol ($T = 4.2$ K): (a) the fluorescence is recorded in a wide spectral range ($\Delta\nu_f > 100$ cm^{-1}), (b) the fluorescence is recorded in a narrow spectral range ($\Delta\nu_f \approx 4$ cm^{-1}) ($\lambda_f = 4472$ Å). The spectral width of the exciting beam was approximately 4 cm^{-1} in both cases. (From Personov and Kharlamov 1973.)

they include true short-wave PW absorption lines of the centers for which the ZPL of fluorescence are recorded. But in recording the excitation spectrum we inevitably measure the radiation of many other centers in the region of their PW. The absorption bands (ZPL + PW) of these centers are shifted towards shorter wavelengths and make a contribution to the short-wave wings of the ZPL in the excitation spectrum. Hence, the "observed

Debye–Waller factor" in the excitation spectrum will be less than the true factor. The situation here is quite similar to the one discussed in analyzing fluorescence spectra (see subsection 3.2.2.). The analysis carried out there and the relations obtained can be directly applied to excitation spectra if in the initial eqs. (18) and (19) we fix the fluorescence frequency ν and assume the excitation frequency ν_ℓ to be variable. In particular, the approximate relations (28) and (31) remain valid. The only difference is that in the excitation spectrum, the wings are located at the short-wave side of the ZPL.

It has been shown in subsection 3.2.4 that in fluorescence line spectra the ZPL intensity decreases rapidly as the temperature is raised. The lines in excitation spectra behave in a similar way. For example, in the excitation spectrum of perylene discussed above, the narrow lines practically vanish at a temperature between 40 and 50 K (Personov and Kharlamov 1973).

The method described above is applicable to the spectra of fluorescent molecules only. In the following we shall discuss another, more universal, method of revealing fine structure in absorption spectra. It is applicable, in principle, to both fluorescent and nonfluorescent molecules.

3.5. "Hole-burning" spectra

The essence of this method, developed in our laboratory, is as follows. By means of monochromatic laser radiation it is possible, in principle, to selectively modify, or "burn", the centers of predominantly a single type in an inhomogeneous system. The specific mechanism of this "burning" can vary greatly. It may be a photochemical reaction, the ionization of selectively excited molecules, reorientation of impurity molecules in the matrix by the irradiation, transfer of the molecules to a long-lived metastable state, etc. (feasible burning mechanisms are discussed in subsection 3.7). In all of these cases, changes occur in the absorption spectrum of the burnt molecules. Narrow holes appear in the broad-band absorption spectrum of the solution at the vibronic transition frequencies of the burnt impurity centers. The difference in the absorption (transmission) spectra of the specimen before and after hole burning determines the ensemble of the above-mentioned holes in their pure form. This difference spectrum is said to be the "hole-burning" spectrum. It carries direct information on the absorption line spectrum of molecules without inhomogeneous broadening.

The phenomenon of laser (nonphotochemical) "burning" of organic molecules in solid solutions was discovered by us during experimentation with spectra of perylene in ethanol (Personov et al. 1972a,b). It was established that during laser irradiation a reduction occurs in the line intensity of a fine-structure fluorescence spectrum. In other words, the centers that are absorbing at the laser frequency are "burnt". It was subsequently established that such "burning" is quite a widespread phenomenon and is observed in the spectra of many compounds in various solvents (especially in glassy ones). First registered by Kharlamov et al. (1974, 1975a,b) and subsequently investigated in detail was a narrow hole in the broad-band absorption spectra of perylene and 9-aminoacridine. Presented in fig. 16a is the long-wave portion of the absorption spectrum of perylene in ethanol; the position of the hole-burning laser line is indicated. The hole is shown in fig. 16b (on a different scale). This hole is found to be stable, and the hole-burning process to be reversible. If the specimen is held in the dark at 4.2 K, the hole is preserved for many hours. But it disappears if the specimen is heated to 20 or 30 K (with subsequent cooling to 4.2 K) or if the specimen is exposed (at 4.2 K) to the radiation of intense white light.

Evidently, narrow holes should form in the broad-band absorption spectrum, not only at the laser frequency (in the 0–0 transition region), but also at the frequencies of all the other vibronic transitions of the burnt molecules. In an allowed intense 0–0 transition, the vibronic lines may be

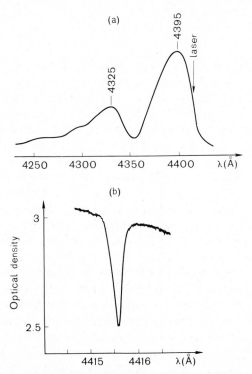

Fig. 16. Hole burning in the broad-band absorption spectrum of perylene in ethanol ($T = 4.2$ K): (a) long-wave portion of the absorption spectrum, (b) narrow hole in the absorption band after 30 min of burning with an He–Cd laser ($\lambda = 4415.6$ Å, $P = 15$ mW/cm^2). (From Kharlamov et al. 1974.)

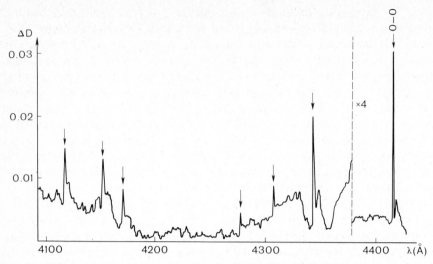

Fig. 17. "Hole-burning" spectrum of perylene in ethanol ($T = 4.2$ K) obtained by two-channel operation after irradiation of the investigated area of the specimen with a He–Cd laser for 5 min ($\lambda = 4415.6$ Å, $P = 5$ mW/cm^2). (Arrows indicate the "holes" at the vibronic transition frequencies.) (Spectrum obtained by Kharlamov.)

considerably weaker than the purely electronic line, and the corresponding holes may be less deep. Sensitive apparatus is required to register them. In this connection, a special two-channel spectral installation was developed in our laboratory for investigating hole-burning spectra. It is capable of registering changes in the optical density of frozen specimens of the order of 10^{-4}. This allowed more detailed research on hole-burning spectra (Kharlamov et al. 1977a,b; 1978).

One of the recently obtained hole-burning spectra is presented in fig. 17. Holes with $\Delta D < 10^{-2}$ are distinctly visible in this spectrum. Along with the narrow ZPL, relatively wide bands (wings) can also be observed.

Wide wings in hole-burning spectra have an essential feature that distinguishes them from the wings close to ZPL in excitation line spectra. In the case of hole-burning spectra, the presence of nonresonantly burnt centers (absorbing through the PW, not the ZPL) leads, not only to an increase of the short-wave wing (as in excitation spectra), but also to the appearance of intense wings at the long-wave side of the ZPL (see fig. 17). This becomes clear if we take into account that the ZPL frequency of the nonresonantly burnt centers is less than the laser frequency ($\nu_0 < \nu_\ell$)*.

---

*A detailed analysis of the shape of wings near ZPL in hole-burning spectra was carried out recently by Friedrich et al. (1980). Their paper deals with the use of these wings for investigation of electron–phonon coupling in glasses.

In the case discussed above hole burning was performed in the 0–0 transition region. An essentially more complicated picture emerges when laser hole burning is applied in the region of vibronic transitions of the absorption spectrum. Here complex "multiplets" appear in the spectrum instead of individual peaks. As an example are shown in fig. 18b the recordings of two multiplets in the hole-burning spectrum of Zn-tetraphenylporphin (Zn-TPP) in ethanol: the 0–0 multiplet and the vibronic multiplet in the region of the hole-burning laser line (region of vibronic transitions with excitation of vibrations at 700 to 1000 cm^{-1}). These multiplets appear because, under conditions of a considerable inhomogeneous broadening of the spectra, one and the same laser line burns several types of centers. Each of them has its vibronic transition at the laser frequency. (This is closely analogous to the formation of multiplets in fluorescence spectra, see subsection 3.2.5.) This can be elucidated by means of the simple diagram of fig. 19. Assume that in the region of the laser frequency the molecules under investigation have three vibrations in the excited electronic state with frequencies ν_1, ν_2 and ν_3 (and that the difference between these frequencies is less than the amount of inhomogeneous broadening). Then, in the solution, three types of molecules, I, II and III, can be found, which have vibronic absorption lines corresponding to the excitation of the vibrations ν_1, ν_2 and ν_3 at the laser

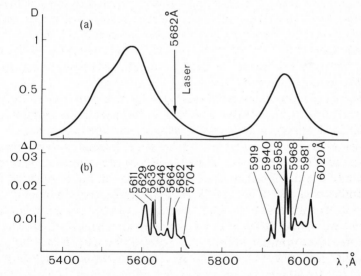

Fig. 18. Spectra of Zn-TPP in ethanol ($T = 4.2$ K): (a) absorption spectrum, (b) two multiplets (0–0 and vibronic) in a "hole-burning" spectrum after irradiation for 30 min by a Kr laser ($\lambda = 5682$ Å, $P = 200$ mW/cm^2) in the vibronic region from 700 to 1000 cm^{-1}. (From Kharlamov et al. 1977b.)

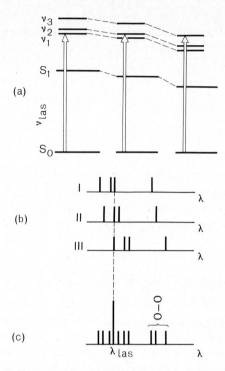

Fig. 19. Schematic diagram of "multiplet" formation in a hole-burning spectrum.

frequency (fig. 19a). Each of these types of molecules has its own hole-burning spectrum, consisting of the 0–0 line and three vibronic lines. But these spectra are shifted in wavelength with respect to one another (fig. 19b). Hence, the observed hole-burning spectrum will contain a super-position of three series of lines (fig. 19c). There should obviously be three lines in the 0–0 transition region. The distances from the laser line to each of these three lines determine the frequencies ν_1, ν_2 and ν_3. Seven lines will be observed in the shorter wavelength region of the spectrum. Three lines, located at the laser frequency, coincide. In the general case, $n^2 - n + 1$ hole-burning lines can be observed in the laser line region, where n is the number of types of centers with vibronic transitions at the laser frequency.

In the case of Zn-TPP, six lines (as a minimum) are observed in the hole-burning spectrum (fig. 18b) in the wavelength range of 5900 to 6050 Å. This means that at least six types of centers are burnt in this case. In principle, there could be 31 lines in the vibronic multiplet between 5600 and 5700 Å. But only the most intense lines appear distinctly in the spectrum. Each of these lines is referred to a definite vibronic transition in table 1.

The cited experimental results indicate that hole-burning spectra can

Table 1

Wavelengths (in Å) of the most intense lines in the hole-burning spectrum of Zn-TPP in ethanol, referred to the corresponding vibronic transitions ($\lambda_{laser} = 5682$ Å).

0–0	705 cm^{-1}	765 cm^{-1}	815 cm^{-1}	840 cm^{-1}	880 cm^{-1}	985 cm^{-1}
5919	5682		5646			
5940		5682	5664			5611
5958			5682			5629
5968				5682		5636
5981			5704		5682	
6020						5682

expediently be used to reveal hidden fine structure in the absorption spectra of complex molecules. This method can be applied with any mechanism of laser hole burning, regardless of its photophysical or photochemical nature.

3.6. Selective excitation and hole burning within the ZPL profile in quasi-line spectra of impurity molecular crystals

Selective laser excitation and hole burning can also be applied to impurity molecular crystals with quasi-line spectra, but only within the ZPL profile. The basic question of the nature of the investigated bands does not arise for such systems because ZPL are present in the spectrum from the very beginning. It is also known beforehand that at low temperatures their broadening is mainly inhomogeneous and is due to imperfections (defects) of the crystal (see, e.g., Rebane 1968, §28, Personov et al. 1971a, Al'shits et al. 1972). At liquid-helium temperatures the ZPL width is usually from 1 to 8 cm^{-1} and varies drastically depending on the method employed to prepare the specimen. The main problem in these cases is to separate out a homogeneous component of the ZPL width. This can be achieved by selective methods*.

Upon excitation by a narrow laser line within the ZPL profile, the lines in the fluorescence spectrum are narrowed (Korotaev and Personov 1974, Abram et al. 1974, Marchetti et al. 1975). But in experiments of this kind it is impossible to determine the homogeneous width of a purely electronic

---

*The present paper deals with the spectroscopy of complex organic molecules. It should be pointed out, however, that inhomogeneous ZPL broadening is a general property of all impurity crystal spectra. Note that the sharp narrowing of ZPL upon monochromatic excitation was first observed in the spectrum of impurity ions in an inorganic crystal, viz., the R-line of a chromium ion in ruby (Szabo 1970, 1971). At present, selective excitation is extensively employed in the investigation of the spectra of rare-earth ions in crystals and glasses (see, e.g., Erickson 1975, Delsart et al. 1975, Brecher and Riseberg 1976, Basiev et al. 1978).

line (due to the reason indicated in subsection 3.2.3). Hole burning was found to be extremely useful for this purpose. The first investigations of the homogeneous width of purely electronic lines of impurity molecular crystals by this method were carried out in the spectra of H_2-phthalocyanine in n-octane (Gorokhovski et al. 1974, 1976) and s-tetrazine in durene and benzene (de Vries and Wiersma 1976, 1977). A homogeneous width of 3×10^{-2} cm^{-1} was obtained for the 0–0 lines of H_2-phthalocyanine (at 5 K), and of approximately 2×10^{-3} cm^{-1} for dimethyl-s-tetrazine in durene (at 2 K). A considerable number of subsequent investigations were performed on hole burning within the ZPL profile in spectra of porphyrins in n-paraffin matrices. These compounds prove convenient because of the readily developing specific hole-burning process, associated with tautomerization (see subsection 3.7), which occurs in them. Besides the determination of the homogeneous width of purely electronic lines in the spectra of these impurity crystals, hole burning also allows detailed investigations on the temperature broadening of ZPL due to electron–phonon coupling. Temperature broadening of ZPL had been previously investigated directly from the fluorescence and absorption spectra. But in the quantitative analysis of data in the low-temperature region, it had always been necessary to approximately take the "remaining" inhomogeneous width into account. Its exact width was unknown. Measurements are substantially more reliable when hole burning is applied. Gorokhovski and Rebane (1977a) investigated the temperature broadening of the 0–0 line in the spectrum of H_2-phthalocyanine in tetradecane in the temperature range of 6 to 30 K. It was established that the homogeneous width of the ZPL increased with temperature proportionally to T^2. This broadening was interpreted by the authors as the result of interaction of the electronic transition in the impurity molecule with pseudolocalized crystal vibrations at a frequency of 36 cm^{-1}*. Especially detailed investigations of ZPL broadening by laser hole burning were performed by Voelker et al. (1977, 1978) in spectra of free-base porphin in crystals of n-octane in the temperature range of 1.5 to 4.2 K. These investigations revealed essential differences in the rate of temperature broadening of ZPL for centers of different orientations in the n-octane lattice. Extrapolation of the obtained temperature relations to the limit $T = 0$ for the width of the 0–0 lines of all the centers yields 3×10^{-4} cm^{-1}. This width is fully determined by the lifetime of the excited electronic state S_1 (17 ns).

On the one hand, these investigations made it possible to reach the theoretical limit of the ZPL width; on the other hand, they showed that even extremely complex molecules have very narrow energy levels.

---

*This interpretation was subsequently revised, and interaction was established with pseudolocalized vibrations having a frequency of 10 cm^{-1}. [Gorokhovski and Rebane (1980)].

3.7. On the "hole-burning" mechanisms

The "hole-burning" mechanisms are very diverse. Conditionally, all these processes can be divided into two groups. The first group includes all processes that do not change the structure of the impurity center, but are associated with considerable changes in the populations of the levels of the impurity molecules. If, for instance, sufficiently powerful laser radiation with $\Delta\nu_{laser} < \Delta\nu_{inhom}$ provides saturation of some transition for a definite portion of the molecules within the limits of the inhomogeneous distribution, the corresponding holes appear in the absorption spectrum. After pumping is switched off these holes vanish rapidly (during the lifetime of the excited level). Such processes are well known in laser physics. Holes similar in origin can be burned in the absorption spectrum by transferring part of the molecules to the metastable triplet state. Such holes were observed in the spectra of frozen solutions of chlorophyll "a" in ether (Avarmaa and Mauring 1976) and of Zn-porphin in *n*-octane (Shelby and Macfarlane 1979). Processes of this first group do not, however, lead to the formation of stable holes.

The second group includes hole-burning processes associated with a change in the structure of the impurity molecule or its nearest surroundings. Pertaining to such phenomena are the following photochemical changes of the impurity molecules: photodissociation observed in the case of dimethyl-*s*-tetrazine in durene and *s*-tetrazine in benzene (de Vries and Wiersma 1976, 1977), phototransfer of a proton of a compound with hydrogen bonds (Graf et al. 1978, Drissler et al. 1980), and the formation of anion radicals during irradiation from complexes of porphyrins and pyridine ions (Maslov 1977).

An exceptionally specific hole-burning mechanism is observed in the case of free-base porphyrins (without the central atom of the metal). This hole burning was discovered in the spectra of protoporphyrin (Korotaev and Personov 1972), and was subsequently found in many other metal-less porphyrins (Solov'ev et al. 1973, Gorokhovski et al. 1974, Voelker and van der Waals 1976, Voelker and Macfarlane 1979a,b,c). It was established that in these cases hole burning is due to photo-induced displacement of two protons in the center of the porphin ring to the neighbouring nitrogen atoms (Solov'ev et al. 1973, Voelker and van der Waals 1976), which is equivalent to rotation of the whole molecule in the matrix by 90°. In this case, holes are readily observed in the spectrum after irradiation by light of low intensity (<1 mW/cm^2) during several minutes or even fractions of a minute. An investigation of hole-burning kinetics shows that this process is of a one-photon nature (Gorokhovski and Kikas, 1978). Moreover, as proposed by Voelker and van der Waals (1976), tautomerization occurs in an intercombination $S_1 \rightarrow T_1$ conversion.

Of especial interest is the above-mentioned more universal hole-burning process, observed in glassy media for a great variety of compounds that are characterized by sufficient stability (polycyclic aromatic hydrocarbons, metalloporphyrins, etc.). This hole burning requires somewhat more powerful irradiation (10 to 100 mW/cm^2) and longer time (minutes or tens of minutes). Kharlamov et al. (1974, 1975a,b) found, for example, that in the case of perylene and 9-aminoacridine in ethanol (at 4.2 K), subject to hole burning by a He–Cd laser ($\lambda = 4415.6$ Å and $\Delta\nu \approx 0.05$ cm^{-1}) with an energy density of 15 mW/cm^2 on the specimen, the hole depth reached saturation (about 20% of the initial optical density) in 30 minutes. The burnt holes are found to be exceptionally stable and in the dark at 4.2 K exist for many hours (possibly days). This hole-burning process is reversible in the sense that the initial spectrum can be easily restored. In order to "erase" the holes and to restore the initial absorption spectrum, it proves sufficient to heat the specimen to 20 or 30 K or to irradiate it with intense white light (for instance, by a 100 W incandescent lamp) for several minutes. Kinetic experiments (Kharlamov et al. 1975a) are indicative of the one-photon nature of hole burning in this case as well. A number of interesting features of this "nonphotochemical" hole burning were established by Hayes and Small (1978a,b, 1979), who investigated the spectra of naphthalene, anthracene and tetracene in various glassy solvents. They found, for example, that holes burnt in the absorption spectrum of tetracene at 2 K and at 13 K not only have different widths, but different shapes as well, and that they disappear at different temperatures (at 10 K and at 35 K, respectively). The maximum of the hole burnt at 13 K was found to be displaced by 0.3 Å toward the long-wave side of the maximum of the hole burning laser line. Other features of hole burning they observed were associated with the effect of temperature, the cooling rate of the specimen, the power of the laser, etc. Hayes and Small (1978a,b, 1979, see also ch. 9 of this volume) associate hole burning in these cases with a change in the relative positions of the impurity molecules and their immediate surroundings. They qualitatively consider a model in which the adiabatic potentials of the impurity molecule in the ground and excited electronic states have two minima each. In hole burning, the molecules pass over from one minimum to the other. In the ground state this transition is accomplished in a time of the order of several hours and more, whereas in the excited state the transition time to the other minimum is comparable to the fluorescence time. In connection with the question of the nature of the hole-burning process, we point out that recently a quantitative theory was developed for transitions in impurity centers with two-well adiabatic potentials [see ch. 8 of this volume]. Numerical calculations based on this model revealed, in particular, a strong dependence of the hole lifetime on temperature. The results of these calculations explain the very long lifetime of the holes at 2 or 4 K and their disappearance at relatively little heating to 10 or 20 K. Though the general picture of the hole-burning process, developed in the cited papers, seems plausible, many important details are not quite clear and require further investigations.

In conclusion, we point out that a recent paper by de Vries and Wiersma (1980) developed a general theoretical approach to the photophysical and photochemical processes of hole burning in the spectra of impurity centers. Derived among others in this paper are equations for the profile and depth of the hole depending upon the kinetic parameters of the various hole-burning procedures. It is to be expected that further theoretical and experimental investigations in this field will lead to a better understanding of the nature of hole-burning processes.

4. *Certain applications*

The methods discussed above and the established relations offer new opportunities in the spectroscopy of complex molecules. Quite obvious is the possibility of a more accurate determination of the electronic and vibrational energy levels of the molecules (and, in the case of symmetrical molecules, the symmetry of these levels) from fine-structure spectra of the system. In the following we shall consider only certain problems for the investigation of which the usefulness of selective methods has already been demonstrated experimentally.

4.1. *Electron–vibrational interaction*

Fine-structure spectra enable one to investigate various aspects and effects of electron–vibrational interaction in detail. Above (subsection 3.6) we have discussed certain results obtained in investigations of the homogeneous width of purely electronic ZPL and its temperature dependence by the hole-burning method. These investigations yield information on the quadratic electron–phonon coupling (see subsection 2.4). Besides this, the ZPL also carry information on the intramolecular vibronic interactions. Gorokhovski and Rebane (1977b), for instance, applied hole burning to investigate the temperature broadening of the vibronic line $\nu_{0-0} + 515 \text{ cm}^{-1}$ in the absorption spectrum of phthalocyanine in *n*-nonane between 1.7 and 60 K. The broadening of this line was found to be proportional to T^n, where $n = 2.2$ to 2.7. To explain the width of this vibronic line (especially as $T \to 0$) the authors had to take into account anharmonic interaction of intramolecular modes along with the electron–phonon coupling. Voelker and Macfarlane (1979a) investigated the homogeneous width of 14 vibronic lines in the absorption spectrum of porphin in *n*-octane at 4.2 K. They have established that the widths of the vibronic lines differ drastically (by a factor of 40), but no correlation could be found between the line width and the vibrational energy. Though the reason for such differences is not yet quite clear, it can be supposed that the interaction of intramolecular vibrations plays some definite role.

It is obvious that the measurement of the intensity distribution in vibronic sequences enables one to determine the constants of the intramolecular electron–vibrational interaction. Sufficiently reliable quantitative measurements of this distribution have thus far been possible only in the quasi-line spectra of certain impurity molecular crystals (see, e.g., Strokach et al. 1973, Gradyushko et al. 1975, Nersesova and Shtrokirkh 1978). Selective methods should substantially extend the range of objects that can be investigated along these lines.

Along with the ZPL, essential information on the electron–phonon coupling is contained in PW. Thus, for instance, the intensity ratio of ZPL and PW (or the Debye–Waller factor) and its temperature behavior carry information on the linear electron–phonon coupling. From investigation of the spectral distribution in PW it is possible, in principle, to calculate the weighted density of phonon states for various systems, as has been done previously in the quasi-line spectra of impurity n-paraffin crystals (Personov et al. 1971b, Osad'ko et al. 1974). It should be noted, however, that inhomogeneous broadening is not completely eliminated from the spectra upon selective excitation. As mentioned above (subsection 3.2.2), non-resonantly excited centers make a considerable contribution to the wings observed close to the ZPL. This raises the problem of separating the true homogeneous spectrum from the one observed upon selective excitation. This problem is closely associated with the problem of determining the distribution function of the impurity centers with respect to the frequency of their 0–0 transition, and can be solved under certain conditions.

4.2. True homogeneous spectrum and distribution of centers with respect to the energy of the 0–0 transition

In order to solve the problem of extracting the true homogeneous spectrum, it is important to know whether the true spectra of various impurity centers do or do not differ in shape in an inhomogeneous system. In investigating the dependence of the shape of spectral fluorescence bands of anthracene in 2-methyltetrahydrofuran (at 4.2 K) on the wavelength of monochromatic excitation, Flatscher and Friedrich (1977) observed considerable changes in the ratio of the ZPL and PW intensities, as well as a change in the distance between the ZPL and PW maxima and a change in the PW width. The authors associated these facts with the difference in strength of electron–phonon coupling for centers with various positions of the 0–0 transition. But in analyzing the experimental data they did not take into account the contribution of the nonresonantly excited centers. When this contribution is taken into consideration, all the observed effects can be explained naturally, even if the spectra of the impurity centers have the same shape. This was clearly demonstrated by numerical calculations on simple models (Kikas 1978). Though it is impossible, in the general case, to exclude the possibility of certain differences in electron–phonon coupling for different centers, even in one and the same solvent, it may be inferred that these differences are not very large. Hence, in what follows we shall assume that the shapes of the true absorption and fluorescence spectra of different centers are the same. In this case, the shape of the emission spectra upon selective excitation is determined by relation (18). Assuming the exciting line to be a δ-function, this relation can be written in the form

$$F(\nu, \nu_\ell) = \int_0^\infty d\nu_0 \, \varphi(\nu - \nu_0)\epsilon(\nu_\ell - \nu_0)n(\nu_0). \tag{33}$$

If the emission frequency ν is fixed and the excitation frequency ν_ℓ is considered to be variable, relation (33) describes the excitation spectrum.

If we further assume that the true fluorescence spectrum φ and absorption spectrum ϵ of the impurity centers have mirror symmetry (which, in many cases, is fully justified, at least for the 0–0 transition region), there remain two unknown functions under the integral in eq. (33), φ (or ϵ) and $n(\nu_0)$, which is the distribution function of the number of centers with respect to the frequency of their 0–0 transition. If one of these functions has been established in some way, the other can be calculated (after first measuring F)*.

Tamm et al. (1976) proposed a method for the experimental determination of $n(\nu_0)$ by double scanning (i.e. simultaneously varying the excitation and recording frequencies, maintaining a constant difference between these frequencies). But the problem of re-establishing the true homogeneous spectrum was not dealt with in their paper.

The more general problem of determining two functions, φ and n, at the same time from experimental data was treated by Fünfschilling (1979) and Fünfschilling et al. (1981). In addition to eq. (33), they considered two more integral relations that express the ordinary fluorescence spectrum (with wide-band excitation) and the excitation spectrum (recording the total

---

*This problem is analogous to the classical problem in which the smearing function of the spectral instrument is excluded. If $F(\nu, \nu_\ell)$ is regarded as a known function of two variables ν and ν_ℓ, one can formulate the problem of re-establishing two functions φ and n at once on the basis of eq. (33). A number of general mathematical relations concerning the solution of eq. (33) were obtained by Kikas (1976).

fluorescence):

$$F(\nu) = \int_0^\infty F(\nu, \nu_\ell) \, d\nu_\ell = \int_0^\infty d\nu_0 \, \varphi(\nu - \nu_0) n(\nu_0), \tag{34}$$

and

$$F(\nu_\ell) = \int_0^\infty F(\nu, \nu_\ell) \, d\nu = \int_0^\infty d\nu_0 \, \epsilon(\nu_\ell - \nu_0) n(\nu_0). \tag{35}$$

Proposed in these papers (under the assumption that φ and ϵ have mirror symmetry) is a self-consistent procedure for the numerical solution of eqs. (33), (34) and (35). The initial data in this case are the emission spectrum $F(\nu, \nu_\ell)$ upon selective excitation (one of the laser lines) and the wide-band spectra $F(\nu)$ and $F(\nu_\ell)$. This procedure was applied to analyze the phosphorescence spectra of 1-iodonaphthalene in glassy 1-butyl bromide at 4.2 K. (The phosphorescence line spectrum was obtained by T_1–S_0 excitation by an intense line of an argon laser with $\lambda = 4880$ Å.) The obtained distribution function $n(\nu_0)$ is of Gaussian form with half width 200 cm^{-1}. As could be expected, the phonon wings in the homogeneous spectrum $\varphi(\nu - \nu_0)$, obtained in the calculations, were narrower and less intense than in the initial line spectrum.

The approach developed in the above-mentioned papers is, in our opinion, of vital importance for detailed investigations of electron–phonon coupling on the basis of PW, for comparative research on the properties of glassy media by means of $n(\nu_0)$, for investigation of the energy transfer processes in systems with inhomogeneously broadened spectra (Fünfschilling et al. 1978, Fünfschilling 1979), etc. But the mathematical aspects of the problem (stability of solutions, accuracy of the results, etc.) evidently require further analysis.

4.3. Spin structure of triplet states

4.3.1. Zero-field splitting of phosphorescence lines
The lowest triplet level T_1 plays a most important role in processes in which the energy of electronic excitation is transformed and transferred in complex molecular systems. Comprehensive research on the properties of triplet states of complex molecules is of particular interest.

Strictly speaking, the triple degeneracy of the triplet level of a molecule is eliminated, even in the absence of an external magnetic field, as a result of the spin–spin and spin–orbit interactions, as well as the effect of the crystalline field [see, e.g., McGlynn et al. (1969)]. In complex organic molecules, spin–spin magnetic dipole interaction plays the decisive role. It is known that the splitting of the triplet level in a zero field is characterized by the parameters D and E [zero-field splitting (ZFS) parameters, see fig. 20a].

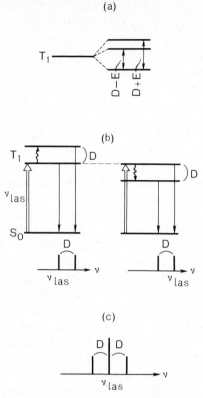

Fig. 20. Spin structure of a phosphorescence line: (a) ZFS parameters, (b) levels and spectral doublets of two types of centres, excited by the same laser line ($D \neq 0$ and $E = 0$), (c) "triplet" appearing in the spectrum.

ZFS parameters (which are the main parameters in triplet state theory) are extremely small and ordinarily range from about 0.01 to 0.1 cm^{-1}. These parameters are experimentally determined by electron spin resonance (ESR) methods. It was impossible (until recently) to observe ZFS in optical phosphorescence spectra because the width of the bands (lines) in these spectra exceeds the above quantities by several orders of magnitude. The development of selective methods of obtaining spectra with narrow lines raised the problem of direct observation of ZFS in the optical spectra of complex molecules. This problem was solved in a paper by Al'shits et al. (1977), who were the first to record and measure the zero-field splitting of lines in the phosphorescence spectrum of a complex molecule. The system investigated in this work was coronene in butyl bromide at 4.2 K. This system proved to be convenient because the intense $\lambda = 5145$ Å line of an argon laser is located precisely in the 0–0 transition region of its in-homogeneously broadened phosphorescence spectrum. The long phos-phorescence lifetime ($\tau \approx 1.3$ s) permitted easy measurement of the resonance phosphorescence line at the laser frequency.

A free coronene molecule belongs to the symmetry point group D_{6h}. For a molecule with axes of three-fold or higher symmetry (perpendicular to the plane of the molecule) $E = 0$. Hence, for coronene the triplet level should split into two components with $\Delta\nu = D$. In this case, one and the same laser line should excite two types of centers (fig. 20b). Each type of center should produce doublets in the emission spectrum*. But the com-ponents of the doublets located at the laser frequency coincide and there should be a triplet in the spectrum (fig. 20c). (In the case $E \neq 0$, a multiplet of seven lines should be present in the spectrum.) This is actually observed experimentally. Figure 21 illustrates the 0–0 transition region of the coronene phosphorescence spectrum, excited and recorded by various methods. When a sufficiently narrow exciting laser line ($\Delta\nu \approx 10^{-2}\,\mathrm{cm}^{-1}$) is used and the spectrum is recorded with high resolution (employing a Fabry–Perot interferometer), the above-mentioned triplet is distinctly observed in the spectrum. The distances of the lateral components to the central line determine the parameter $D = 0.095 \pm 0.005\,\mathrm{cm}^{-1}$. The obtained value agrees well with electron spin resonance (ESR) data for this mole-cule: $D = 0.096\,\mathrm{cm}^{-1}$ (de Groot and van der Waals 1960). Hence, the cited data demonstrate the feasibility of direct observation of ZFS in the optical spectra of organic molecules.

It is important to point out that the investigation of ZFS in an optical spectrum is not a simple duplication of ESR data, but may supplement them substantially. Thus, for instance, the ratio of the radiative lifetimes of the individual spin sublevels can be determined from the ratio of the ZFS component intensities in the optical spectrum. Such data cannot be directly obtained from ESR experiments. Moreover, optical spectra can be used, for example, to study the effects on the ZFS parameters of the crystalline field and the statistical scatter in the values of these parameters for various impurity centers in an inhomogeneous system (by investigating the broadening of the lateral components of the ZFS multiplet as compared to the central component). Finally, in the specific cases when ESR measure-ment is complicated by the difficulty of obtaining a sufficiently large population of the T_1 state (if the triplet level lifetime is short), the conditions for optical measurements are more favorable (because larger

---

*In the phosphorescence spectrum of coronene, transitions with two zero-field components from the T_1 level are forbidden by symmetry (McGlynn et al. 1969). But in butyl bromide this suppression is removed, evidently as a consequence of violation of the symmetry of the free molecule. This is confirmed by the fact that in this solvent the 0–0 line, forbidden by symmetry, becomes the most intense line in the spectrum.

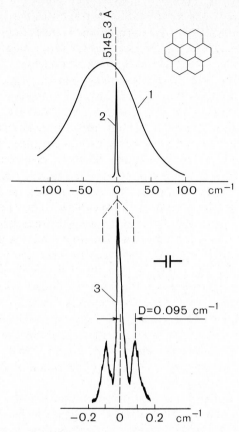

Fig. 21. Band of the 0–0 transition in the phosphorescence spectrum of coronene in butyl bromide ($T = 4.2$ K): (1) with ordinary S_1–S_0 excitation ($\lambda_{Hg} = 313$ and 365 nm), (2) with laser T_1–S_0 excitation, (3) interferometric recording of the 0–0 line. (From Al'shits et al. 1977.)

absorption coefficients at the T_1–S_0 transition correspond to shorter τ values).

4.3.2. *The Zeeman effect*
The feasibility of observing ZFS in an optical spectrum necessarily implies the possibility of measuring the Zeeman effect of complex molecules in any magnetic field (beginning with $H = 0$). The first investigations of the Zeeman effect in the spectra of complex molecules upon laser T_1–S_0 excitation were performed by Kharlamov et al. (1978b). At present these investigations are being conducted in our laboratory both in inter- mediate (in magnitude) constant magnetic fields (up to 50 or 60 kG) and in strong pulsed magnetic fields (up to 500 kG)*. Here we shall briefly con- sider only certain examples.

The Zeeman effect on purely electronic lines in the phosphorescence spectra of coronene and 5-bromacenaphthene is demonstrated in fig. 22. In our case, the Zeeman effect is characterized by the following features.

1. As is evident from fig. 22, a Zeeman multiplet has five components instead of three. This can be readily understood. In magnetic fields above 10 kG, Zeeman splitting in the external field $\Delta E = g\beta H$ (where $g = 2$ and β is the Bohr magneton) is much larger than the ZFS and the magnitude of Zeeman splitting is the same for all molecules in the solution (regardless of their orientation, since, in the above fields, spin–orbit bonds are broken). With monochromatic T_1–S_0 excitation, the laser "selects" three types of centers in the inhomogeneous system, each of which is excited to one of its Zeeman sublevels. Since kT is of the order of ΔE at 4.2 K, two other Zeeman sublevels are also populated in each type of center by radiationless transitions. Three lines are present in the emission spectrum for each type

---

*The experimental installation for low-temperature spectral investigations in strong pulsed magnetic fields is described in Kharlamov et al. (1981).

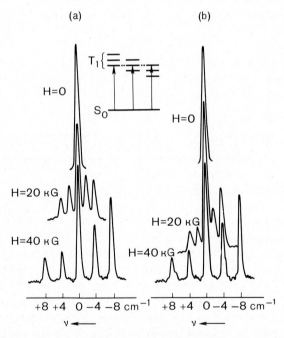

Fig. 22. The Zeeman effect on resonance 0–0 lines in phosphorescence spectra upon laser T_1–S_0 excitation ($T = 4.2$ K): (a) coronene in butyl bromide ($\lambda_{exc} = \lambda_{0-0} = 5145$ Å), (b) 5-bromacenaphthene in butyl bromide ($\lambda_{exc} = \lambda_{0-0} = 4880$ Å). (Recording obtained by Al'shits and Kharlamov.)

of center. Since part of the triplet components of the various types of centers coincide, there are five lines in the total phosphorescence spectrum.

The magnitude of the Zeeman splitting depends linearly on the magnetic field and is equal to $0.1\,cm^{-1}/kG$ (as could be expected for triplet states).

2. Of most interest in the Zeeman multiplet is the distribution of the intensity between the components. It can be readily understood that this distribution can, in general, depend upon the rate of spin–lattice relaxation. (As a matter of fact, in the absence of relaxation between the Zeeman sublevels, only a single resonance line should be visible in the spectrum.)

In the present case, the Zeeman sublevels are populated directly, rather than through the singlet state S_1. Hence, the ratio of the intensities of the Zeeman components does not depend upon the intercombination conversion constants (in contrast to ordinary cases of triplet level T_1 population upon S_1–S_0 excitation). This circumstance enables one to obtain information on the dynamics of the triplet state, even under steady-state conditions of exciting and recording spectra (without measuring the kinetics of phosphorescence decay), on the basis of the relative intensity of the Zeeman components. In particular, direct measurement of the degree of deviation from Boltzmann equilibrium in a spin system becomes possible. With such a deviation, the rate of spin–lattice relaxation can be calculated from line intensity data.

3. A theoretical analysis of the distribution of intensities in a Zeeman quintet (for isotropic distribution of molecules in the solution) indicates that the ratio I_1/I_5 and I_2/I_4 of the line intensities are given by the relations

$$I_1/I_5 = A = e^{-\beta gH/kT}, \qquad I_2/I_4 = A^2, \tag{36}$$

and are independent of the presence or absence of thermal equilibrium in the spin system.* Relations (36) are also independent of the direction of polarization of the exciting and recording radiation, and of the geometry of the experiment.

In contrast, the ratios I_3/I_4 and I_3/I_5 depend strongly on the presence or absence of Boltzmann equilibrium, on the polarization of the light and on the direction of excitation and observation of phosphorescence.

Referring to the examples shown in fig. 22, we point out that a quantitative analysis of the spectra indicates that in the case of coronene Boltzmann equilibrium exists, whereas in the case of 5-bromacenaphthene (whose phosphorescence lifetime is considerably shorter) the situation is quite different. While at $H = 40\,kG$ the spin system of 5-bromacenaphthene is close to Boltzmann equilibrium, appreciable deviations from this equilibrium are observed when the magnetic field is reduced. It proved possible

---

*The lines in fig. 22 are numbered 1 through 5 from left to right.

in this case to determine certain interesting features of the spin–lattice relaxation and its dependence on the magnetic field and on temperature.

Thus, the data cited above indicate that direct laser T_1–S_0 excitation substantially extends the range of problems accessible by Zeeman experiments, and beyond doubt offers new opportunities for the study of the structures and kinetic parameters of the triplet states of complex molecules.

4.4. Spectra of biologically active porphyrins

Porphyrins constitute an extensive class of compounds, many of which play a vital role in biological processes (chlorophyll, hemin, etc.). All porphyrins have distinctive absorption spectra in the visible region (5000 to 6800 Å), making them extremely convenient for spectral research by means of tunable dye lasers.

The most important of the porphyrins is chlorophyll. The fine-structure spectra of chlorophyll and its analogs are of exceptional interest from the viewpoint of the structure of photosynthetic pigments and the investigation of intermolecular interactions (aggregation, solvation, etc.). Hence, many papers have dealt with the generation and investigation of fine-structure spectra of chlorophyll. In the majority of them, upon exciting luminescence by the discrete lines of gas lasers, only separate groups of lines could be revealed in the spectrum and investigated. With the application of finely tunable lasers in recent years for this purpose, more comprehensive and detailed investigations become possible. Line spectra were obtained and the frequencies of normal vibrations, active in electronic spectra, were determined (both in the ground and excited electronic states) for chlorophyll "a" and "b", protochlorophylls and pheophytins (Fünfschilling and Williams 1977, Avarmaa et al. 1980, Bykovskaya et al. 1980b). Though the fundamental frequencies determined in different studies agree sufficiently well, there are some differences for the weaker transitions. These differences attest, in particular, to the high sensitivity of line spectra of chlorophyll to variations in the preparation and purification of the specimens, to the effect of the solvent, to solvation, etc. In spectra obtained by selective excitation, it was found possible to trace the effect of the central metal atom on the vibronic structure of the spectrum, and to reveal a number of fine effects of the interaction between the pigment and its molecular environment.

Recently, Romanovski et al. (1981) obtained and investigated the fine-structure spectra of a number of porphyrins of biological importance. These included etioporphyrin I, II, III·and IV; coproporphyrin I and III; mesoporphyrin IX; as well as their ionic forms. One of the aims of this research was to establish the feasibility of identifying porphyrins closely similar in structure or isomers of the same porphyrin (whose ordinary broad-band spectra are indistinguishable) from fluorescence spectra generated by selective excitation. Shown in fig. 23 are ordinary fluorescence spectra of solutions of three different porphyrins (at 4.2 K) upon excitation in the near-ultraviolet region. It is evident that the position of the maxima of the intense short-wave bands of the spectra is practically the same, and that the differences in the shapes of the vibronic bands are extremely slight. (The spectra of isomers of a single porphyrin are completely identical.) Upon monochromatic laser excitation (in the 0–0 transition region), the spectra of all three compounds acquire fine structure and consist of several dozens of narrow lines (fig. 24). Each porphyrin can be reliably identified from these spectra.

Revealing the differences in the spectra of isomers of a single porphyrin is a more complicated problem. Investigations have shown that this prob-

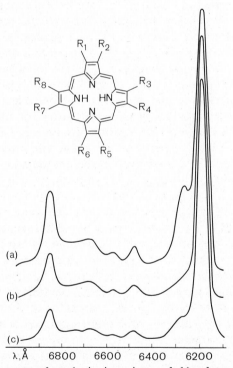

Fig. 23. Fluorescence spectra of porphyrins in a mixture of chloroform and ether (10% + 90%) upon broad-band excitation in the range 3300 to 3400 Å ($T = 4.2$ K): (a) etioporphyrin III ($R_2 = R_4 = R_6 = R_7 = C_2H_5$; $R_1 = R_3 = R_5 = R_8 = CH_3$), (b) coproporphyrin III ($R_2 = R_4 = R_6 = R_7 = CH_2CH_2COOH$; $R_1 = R_3 = R_5 = R_8 = CH_3$), (c) mesoporphyrin IX ($R_6 = R_7 = CH_2CH_2COOH$; $R_2 = R_4 = C_2H_5$; $R_1 = R_3 = R_5 = R_8 = CH_3$). (From Romanovskii et al. 1981.)

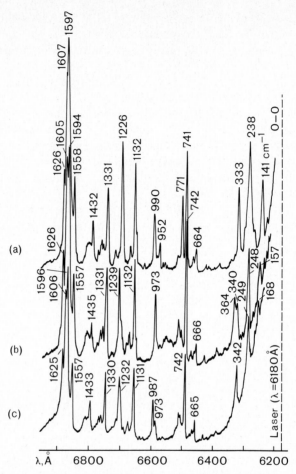

Fig. 24. Fluorescence spectra of porphyrins in a mixture of chloroform and ether upon monochromatic excitation ($T = 4.2$ K): (a) etioporphyrin III, (b) coproporphyrin III, (c) mesoporphyrin IX. (From Romanovskii et al. 1981.)

lem can also be solved. Shown as an example in fig. 25 are two portions in the spectra of four isomers of etioporphyrin, illustrating the afore said. In a detailed comparison of these spectra, features typical of each isomer can be revealed. Similar differences in the fine structure of isomer spectra are also observed for the ionic forms of etioporphyrins and coproporphyrins.

4.5. Photochemical reactions

A number of papers have been published in recent years on investigations in which selective hole burning in inhomogeneously broadened absorption

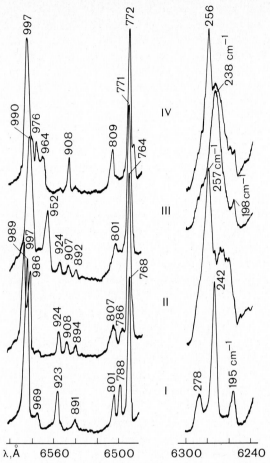

Fig. 25. Fragments of fine-structure fluorescence spectra of four isomers of etioporphyrin (I, II, III, IV) in a mixture of chloroform and ether ($T = 4.2$ K, and $\lambda_{laser} = 6180$ Å). (From Romanovskii et al. 1981.)

spectra was employed in studies of photochemical processes in the condensed phase at low temperatures. In molecular systems, photochemical processes proceed with the participation of excited electronic states. Here the photochemical process leads to the appearance of a supplementary channel for the relaxation of the excited state. If the above-mentioned process is accomplished sufficiently rapidly, the homogeneous width of the absorption line (and the hole width, appropriately related to the line width) is determined by the rate of the process. Hence, it is possible, in principle, to gain information on the rate of the photochemical reaction from the width of the burnt hole. Maslov (1978), for example, investigated the complex consisting of Cd-etiochlorine with the dimer anion–radical of pyridine (in a mixture of tetrahydrofuran with ether) at 4.2 K. A rapid intracomplex phototransfer of an electron with the formation of the anion–radical of Cd-etiochlorine is readily accomplished in this system. This process is accompanied by the appearance of a relatively wide hole in the inhomogeneously broadened absorption band of the complex. The homogeneous width of the 0–0 absorption line, found by measurements of the hole, turned out to be $5 \, cm^{-1}$, and the corresponding relaxation time was 1.1 ps. In the same investigations, it was established, by burning a hole in the vibronic absorption band, that an increase in the excess of vibrational energy of the molecule leads to an increased rate of conversion to a state with charge transfer.

Applying hole-burning techniques, Drissler et al. (1980) investigated the phototransfer of a proton in solutions of 1,4-dihydroxyanthraquinone. The photochemical reaction leads to the formation of holes 2.2 and $2.7 \, cm^{-1}$ wide (at temperatures 1.3 and 1.9 K). An assessment of the characteristic relaxation time on the basis of these measurements yields a value of 5 to 10 ps. In the opinion of Drissler et al. the photochemical reaction consists of the breaking of one of the intramolecular hydrogen bonds and the formation of an intermolecular hydrogen bond.

In an investigation of hole burning in the absorption spectrum of 1,4-dihydroxyanthraquinone in alcohol glasses, Friedrich and Haarer (1980) established differences in the rates of photochemical reactions for centers with different electron–phonon coupling.

It is of interest to note that recently hole burning was expediently applied to investigate the rate of phototransfer of an electron in the reaction centers of the photosystem I (P 700) in chlorella cells (Maslov 1979). The estimate obtained for the lower limit of the electron phototransfer time is $\tau \geqslant 13$ ps. This means, in particular, that electron phototransfer in reaction centers is slower than in certain model systems in vitro (in complexes of metalloporphyrin with pyrimidine anions).

In this way these (as yet few) examples convincingly demonstrate the new possibilities to study the rates of photochemical reactions in condensed media (without resorting to any picosecond techniques).

4.6. Spectrochemical analysis of complex organic products

In concluding the present review we shall briefly dwell on the new possibilities of practical application of spectrochemical analysis using selective methods.

In many cases the luminescence analysis of complex organic products meets with serious difficulties. The large width of the luminescence bands (hundreds of cm^{-1}) leads to overlap of the spectra of the various components. As a result, the complete spectrum has almost no specific features. Even after preliminary extraction or chromatographic separation of a complex mixture, the obtained fractions usually consist of many components and the difficulty of determining the individual compounds in the mixture cannot be completely eliminated.

The methods of selective excitation discussed above enable one to obtain luminescence spectra with very narrow lines in a great variety of media and in the presence of a large number of different compounds. The extremely characteristic nature of the line spectra and the possibility of selectively exciting one or another compound provide extraordinarily favorable conditions for developing highly selective and sensitive techniques for analyzing complex mixtures (Brown et al. 1978, Al'shits et al. 1980, Bykovskaya et al. 1981).

Using tunable lasers for exciting fluorescence it was found to be possible in many cases to directly analyze the natural product without resorting to any labor-consuming operations for preliminary separation and without using any solvents (Bykovskaya et al. 1981). Here we shall restrict ourselves to a single example: the determination of small amounts of polycyclic aromatic hydrocarbons in automobile gasoline. Shown in fig. 26a is a fluorescence spectrum of gasoline in the visible region upon ordinary

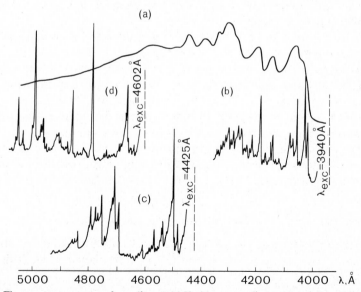

Fig. 26. Fluorescence spectra of gasoline at 4.2 K: (a) upon ordinary ultraviolet excitation, (b), (c), (d) upon monochromatic excitation by a dye laser. (Bykovskaya et al. 1981.)

ultraviolet excitation (at 4.2 K). Though there are several wide maxima in this spectrum, it is practically impossible to use the spectrum to identify the individual compounds. An entirely different picture is obtained when the fluorescence of the gasoline is excited by a tunable dye laser. In this case high-resolution spectra are observed with dozens of lines from 1 to 5 cm^{-1} wide. The structure of these spectra depends essentially on the excitation frequency. The number of different line spectra obtained in this manner is determined by the tuning step of the laser and, consequently, can be very large. Only three such spectra are shown in figs. 26b, c and d. The individual compounds in a complex mixture can be identified quite reliably from such spectra.

In conducting an analysis, the laser frequencies providing the most

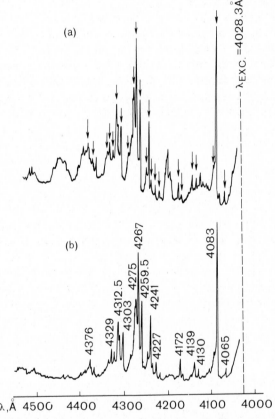

Fig. 27. Identification of 3,4-benzpyrene in gasoline from the fluorescence spectrum upon selective laser excitation ($T = 4.2$ K): (a) portion of the gasoline spectrum in which the 3,4-benzpyrene lines are indicated by arrows, (b) reference spectrum of 3,4-benzpyrene in gasoline ($C = 10^{-4}$ g/ml). (From Bykovskaya et al. 1981.)

effective excitation of fluorescence of the compounds under investigation are selected in preliminary experiments. (The excitation conditions can be selected either according to the spectra of reference solutions or directly on the basis of the spectrum of the compound under investigation. In the latter case, the investigated compound should be specially introduced into the system in a concentration high enough for its luminescence to be substantially more intense than the intrinsic luminescence of the system itself.) Spectra obtained under these conditions are used as reference spectra for comparison.

Shown in fig. 27 as an example is the reference fluorescence spectrum of the carcinogenic hydrocarbon 3,4-benzpyrene and the corresponding portion of the fluorescence spectrum of gasoline (upon excitation by the same laser line $\lambda_{exc} = 4028$ Å). It is evident that all the principal lines of 3,4-benzpyrene are present in the gasoline spectrum. Measurements indicate that over 20 lines in the two spectra coincide (within the limits of experimental error of approximately 1 Å). There can be no doubt of the presence of 3,4-benzpyrene in the gasoline.

Various techniques can be applied to obtain a quantitative analysis. Deserving special attention is the standard additions method, in which the specimen being investigated and the reference specimen are given the same total composition. This enables one to exclude (or substantially reduce) systematic errors due to the effect of other components of the mixture on the intensity of the analytical lines (as a result of luminescence quenching, reabsorption, etc.).

In order to reduce random errors, due to variations in experimental conditions in recording the spectra of the investigated specimen and the reference specimen (for example, conditions that are difficult to control in freezing the specimen, etc.), it proves expedient to carry out the determination on the basis of relative intensities of the lines. This is done by measuring the ratio of the intensities of the analytical line and a reference line (or band). Any line in the spectrum of the analyzed specimen that does not belong to a definite compound can be chosen as reference line.

We shall illustrate these techniques by the example of the quantitative determination of perylene in gasoline. The analysis was carried out with five samples of gasoline (1 ml each) and perylene was added to four of them in the concentrations: $2.5 \times 10^{-8}, 5 \times 10^{-8}, 10 \times 10^{-8}$ and 18×10^{-8} g/ml. Fluorescence (at 4.2 K) was excited by a laser with $\lambda = 4416$ Å. The intense vibronic line at 4486 Å in the fluorescence spectrum of perylene served as the analytical line, and the line at 4471 Å in the gasoline spectrum was selected as the reference line. Given in fig. 28 are the recording of the analytical pair of lines in the gasoline spectrum and the corresponding calibration graph. The concentration of perylene in gasoline, determined from the graph, was found to be $C_x = 5 \times 10^{-8}$ g/ml.

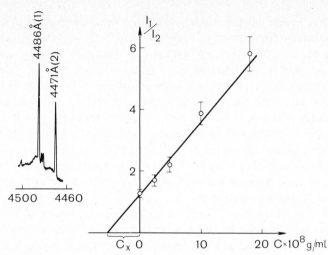

Fig. 28. (1)Analytical and (2) reference lines, and calibration graph for the determination of the concentration of perylene in gasoline ($C_x = 5 \times 10^{-8}$ g/ml). (From Bykovskaya et al. 1981.)

The most important characteristics of any method of analysis are the detection limit and the accuracy of analysis. In analyzing complex mixtures whose composition may vary in a wide range, these characteristics may also differ substantially. It is clear, for instance, that the detection limit depends upon the effective absorption cross section of the compound investigated, on the quantum yield of its luminescence, on the intensity of the continuous background in the spectrum of the specimen being investigated, etc. Hence, both the detection limit and the accuracy of analysis should be established separately in each specific case. But, as a general characteristic of the method we can cite the results obtained with model solutions of perylene in ethanol. In our installation (see Bykovskaya et al. 1981) we easily recorded the perylene line at 4486 Å down to a concentration of 10^{-11} g/ml. The average error of analysis at a concentration of 10^{-8} g/ml was less than 20%. If necessary the detection limit can be reduced easily and the accuracy of analysis increased by employing more powerful and more stabilized lasers and signal accumulation techniques.

The cited data illustrate the dramatic possibilities of selective excitation for the spectrochemical analysis of complex products. There is no doubt whatsoever that a further development of specific analysis procedures will yield useful results.

5. Conclusion

The present chapter deals with the basic principles of site selection spectroscopy of complex molecules in solutions and with the new possibilities it provides for fine spectroscopic research. From the point of view of the prospects for further development of these selective methods and extension of their fields of application, the following question is of prime importance: just how universal are the established relationships and the described methods? It is impossible as yet to give a comprehensive answer to this question. But it is already clear that at low temperatures in-homogeneous broadening of spectra is the decisive factor in a huge number of cases. These include the spectra of solutions of compounds (and their ions) of various classes: aromatic hydrocarbons, heterocyclic compounds, porphyrins and phthalocyanines and their complexes with metals, many dyes, etc. Hence, the field in which selective methods enable one to obtain line spectra is sufficiently large.

The data discussed above clarify the general mechanism of the formation of broad spectral bands for organic molecules in solutions. At sufficiently low temperatures (in the liquid-helium temperature range) and for rela-tively weak electron–phonon coupling, the absorption and emission spectra of separate molecules consist of narrow ZPL accompanied by PW. But inhomogeneous broadening blurs this picture. (In special cases in-homogeneous broadening is small for certain impurity crystals and a quasi-line structure is observed in the spectra.) The ZPL disappear when the temperature is raised. Therefore, the wide spectral bands observed at room temperature consist of phonon wings*.

It should be pointed out, however, that the above situation is evidently not universal. As a matter of fact, cases are known in which line structure in spectra cannot be revealed even at low temperatures and with the application of selective laser excitation. For example, experiments con-ducted in our laboratory (at 2 and 4.2 K) have not yet been able to reveal line structure in the spectra of solutions of a number of dyes, such as proflavine, acridine orange and certain rhodamines. One of the most probable reasons for this is the strong electron–phonon coupling. Of doubtless interest in this connection are the search for new media that interact more weakly with the dissolved molecules and for possibilities to carry out the measurements at even lower temperatures.

Certain applications of fine-structure spectra upon selective excitation were indicated in the afore said. Of course, not all possibilities of selective methods have been discovered as yet. The new possibilities may, in principle, turn out to be totally unexpected and lie outside the scope of pure spectroscopy. As an example, we cite the recent proposal (Castro et al. 1978) to make use of hole burning in inhomogeneously broadened

---

*This should be kept in mind in analyzing (by means of broad-band spectra) the general relationships in the spectra of complex molecules, such as the width and shape of the bands, the temperature dependence of the spectra, effects of the solvent on the spectra, etc.

absorption bands to expand the storage volume of optical memory devices*. Especially promising are the applications of site selection spectroscopy to the fine research of various kinds of intra- and intermolecular interaction and to the spectrochemical analysis of a great variety of organic products.

There is not the slightest doubt that further developments along the trends discussed in the present review will yield many new and interesting results in the near future.

References

Abram, I.I., R.A. Auerbach, R.R. Birge, B.E. Kohler and J.M. Stevenson, 1974, J. Chem. Phys. **61**, 3875.

Abram, I.I., R.A. Auerbach, R.R. Birge, B.E. Kohler and J.M. Stevenson, 1975, J. Chem. Phys. **63**, 2473.

Al'shits, E.I., L.A. Bykovskaya, R.I. Personov, Yu.V. Romanovskii and B.M. Kharlamov, 1980, J. Mol. Struct. **60**, 219.

Al'shits, E.I., E.D. Godyaev and R.I. Personov, 1972, Fiz. Tverd. Tela **14**, 1605.

Al'shits, E.I., R.I. Personov and B.M. Kharlamov, 1976a, Opt. Spektrosk. **41**, 803.

Al'shits, E.I., R.I. Personov and B.M. Kharlamov, 1976b, Chem. Phys. Lett. **40**, 116.

Al'shits, E.I., R.I. Personov and B.M. Kharlamov, 1977, Zh. Eksp. Teor. Fiz. Pis'ma **26**, 751.

Al'shits, E.I., R.I. Personov, A.M. Pyndyk and V.I. Stogov, 1975, Opt. Spektrosk. **39**, 274.

Avarmaa, R.A., 1974a, Izv. Akad. Nauk ESSR Fiz.-Mat. Ser. **23**, 93.

Avarmaa, R.A., 1974b, Izv. Akad. Nauk ESSR Fiz.-Mat. Ser. **23**, 238.

Avarmaa, R.A. and K.Kh. Mauring, 1976, Opt. Spektrosk. **41**, 393.

Avarmaa, R.A. and A. Suisalu, 1975, Izv. Akad. Nauk ESSR Fiz.-Mat. Ser. **24**, 444.

Avarmaa, R.A., R. Tamkivi, S. Kiisler and V. Nymm, 1980, Izv. Akad. Nauk ESSR Fiz.-Mat. Ser. **29**, 39.

Basiev, T.T., Yu.K. Voron'ko and A.M. Prokhorov, 1978, Spectroscopy of Crystals (Nauka Publishers, Leningrad), p. 83 (in Russian).

Brecher, C. and L.A. Riseberg, 1976, Phys. Rev. **B13**, 81.

Brown, J.C., M.C. Edelson and G.J. Small, 1978, Anal. Chem. **50**, 1394.

Bykovskaya, L.A., A.T. Gradyushko, R.I. Personov, Yu.V. Romanovskii, K.N. Solov'ev, A.S. Starukhin and A.M. Shul'ga, 1978, Zh. Priklad, Spektrosk. **29**, 1088.

Bykovskaya, L.A., A.T. Gradyushko, R.I. Personov, Yu.V. Romanovskii, K.N. Solov'ev, A.S. Starukhin and A.M. Shul'ga, 1980a, Izv. Akad. Nauk SSSR Fiz. Ser. **44**, 822.

Bykovskaya, L.A., F.F. Litvin, R.I. Personov and Yu.V. Romanovskii, 1980b, Biofiz. **25**, 13.

Bykovskaya, L.A., R.I. Personov and B.M. Kharlamov, 1974, Chem. Phys. Lett. **27**, 80.

Bykovskaya, L.A., R.I. Personov and Yu.V. Romanovskii, 1979, Zh. Priklad. Spektrosk. **31**, 910.

Bykovskaya, L.A., R.I. Personov and Yu.V. Romanovskii, 1981, Anal. Chim. Acta, **125**, 1.

---

*In recording information a large number of narrow stable holes are burned in an absorption band by means of a tunable laser with a narrow line. The information is retrieved by means of the same laser (with lower intensity) from the presence or absence of a hole at each frequency. The introduction of an additional frequency measurement can expand the storage volume by a factor of 10^3 to 10^4.

Castro, G., D. Haarer, R.M. Macfarlane and H.P. Trommsdorff, 1978, United States Patent, 4.101.976.

Cunningham, K., J.M. Morris, J. Fünfschilling and D.F. Williams, 1975, Chem. Phys. Lett. **32**, 581.

de Groot, M.S. and J.H. van der Waals, 1960, Mol. Phys. **3**, 190.

Delsart, C., N. Pelletier-Allard and R. Pelletier, 1975, J. Phys. B: Mol. Phys. **8**, 2771.

de Vries, H. and D.A. Wiersma, 1976, Phys. Rev. Lett. **36**, 91.

de Vries, H. and D.A. Wiersma, 1977, Chem. Phys. Lett. **51**, 565.

de Vries, H. and D.A. Wiersma, 1980, J. Chem. Phys. **72**, 1851.

Drissler, F., F. Graf and D. Haarer, 1980, J. Chem. Phys. **72**, 4996.

Eberly, J.H., W.C. McColgin, K. Kawaoka and A.P. Marchetti, 1974, Nature (London), **251**, 214.

Erickson, L.E., 1975, Phys. Rev. **B11**, 77.

Flatscher, G. and J. Friedrich, 1977, Chem. Phys. Lett. **50**, 32.

Friedrich, J. and D. Haarer, 1980, Chem. Phys. Lett. **74**, 503.

Friedrich, J., J.D. Swalen and D. Haarer, 1980, J. Chem. Phys. **73**, 705.

Fünfschilling, J., 1979, The Technique of Site-Selection Spectroscopy and Some of its Applications; Habilitationsschrift; Institut für Physik der Universität Basel.

Fünfschilling, J., E. Wasmer and I. Zschokke-Gränacher, 1978, J. Chem. Phys. **69**, 2949.

Fünfschilling, J., E. Wasmer and I. Zschokke-Gränacher, 1981, J. Chem. Phys. (to be published).

Fünfschilling, J. and D.F. Williams, 1977, Photochem. Photobiol. **26**, 109.

Gorokhovski, A.A., R.K. Kaarli and L.A. Rebane, 1974, Zh. Eksp. Teor. Fiz. Pis'ma **20**, 474.

Gorokhovski, A.A., R.K. Kaarli and L.A. Rebane, 1976, Opt. Commun. **16**, 282.

Gorokhovski, A.A. and Ya.V. Kikas, 1977, Opt. Commun. **21**, 272.

Gorokhovski, A.A. and Ya.V. Kikas, 1978, Zh. Priklad. Spektrosk. **28**, 832.

Gorokhovski, A.A. and L.A. Rebane, 1977a, Opt. Commun. **20**, 144.

Gorokhovski, A.A. and L.A. Rebane, 1977b, Fiz. Tverd. Tela **19**, 3417.

Gorokhovski, A.A. and L.A. Rebane, 1980, Izv. Akad. Nauk SSSR Fiz. Ser. **44**, 859.

Gradyushko, A.T., K.N. Solov'ev, A.S. Starukhin and A.M. Shul'ga, 1975, Izv. Akad. Nauk SSSR Fiz. Ser. **39**, 1938.

Graf, F., H.K. Hong, A. Nazzal and D. Haarer, 1978, Chem. Phys. Lett. **59**, 217.

Hayes, J.M. and G.J. Small, 1978a, Chem. Phys. **27**, 151.

Hayes, J.M. and G.J. Small, 1978b, Chem. Phys. Lett. **54**, 435.

Hayes, J.M. and G.J. Small, 1979, J. Lumin. **18/19**, 219.

Kharlamov, B.M., R.I. Personov and L.A. Bykovskaya, 1974, Opt. Commun. **12**, 191.

Kharlamov, B.M., R.I. Personov and L.A. Bykovskaya, 1975a, Opt. Spektrosk. **39**, 240.

Kharlamov, B.M., R.I. Personov and L.A. Bykovskaya, 1975b, Izv. Akad. Nauk SSSR Fiz. Ser. **39**, 1922.

Kharlamov, B.M., L.A. Bykovskaya and R.I. Personov, 1977a, Opt. Spektrosk. **42**, 755.

Kharlamov, B.M., L.A. Bykovskaya and R.I. Personov, 1977b, Chem. Phys. Lett. **50**, 407.

Kharlamov, B.M., L.A. Bykovskaya and R.I. Personov, 1978a, Zh. Priklad. Spektrosk. **28**, 840.

Kharlamov, B.M., E.I. Al'shits, R.I. Personov, V.I. Nizhankovski and V.G. Nazin, 1978b, Opt. Commun. **24**, 199.

Kharlamov, B.M., N.I. Ulitskii, A.M. Pyndyk, V.B. Podobedov and R.I. Personov, 1981, Pribory i Tekhnika Eksperimenta, No. 1, 204.

Kikas, J., 1976, Izv. Akad. Nauk ESSR Fiz.-Mat. Ser. **25**, 374.

Kikas, J., 1978, Chem. Phys. Lett. **57**, 511.

Korotaev, O.N. and R.I. Personov, 1972, Opt. Spektrosk. **32**, 900.

Korotaev, O.N. and R.I. Personov, 1974, Opt. Spektrosk. **37**, 886.

Krivoglaz, M.A., 1964, Fiz. Tverd. Tela 6, 1707.

Kukushkin, L.S., 1963, Fiz. Tverd. Tela 5, 2170.

Kukushkin, L.S., 1965, Fiz. Tverd. Tela 7, 54.

Lax, M., 1952, J. Chem. Phys. 20, 1752.

McCumber, D.E. and M.D. Sturge, 1963, J. Appl. Phys. 34, 1682.

McGlynn, S.P., T. Azumi and M. Kinoshita, 1969, Molecular Spectroscopy of the Triplet State (Prentice-Hall, Inc., Englewood Cliffs, New Jersey).

Maradudin, A.A., 1966, Theoretical and Experimental Aspects of the Effects of Point Defects and Disorder on the Vibrations of Crystals (Academic Press, New York, London).

Marchetti, A.P., W.C. McColgin and J.H. Eberly, 1975, Phys. Rev. Lett. 35, 387.

Maslov, V.G., 1977, Opt. Spektrosk. 43, 388.

Maslov, V.G., 1978, Opt. Spektrosk. 45, 824.

Maslov, V.G., 1979, Dokl. Akad. Nauk SSSR 246, 1511.

Nersesova, G.N. and O.F. Shtrokirkh, 1978, Opt. Spektrosk. 44, 102.

Osad'ko, I.S., 1972, Fiz. Tverd. Tela 14, 2927.

Osad'ko, I.S., 1975, Fiz. Tverd. Tela 17, 3180.

Osad'ko, I.S., 1977, Zh. Eksp. Teor. Fiz. 72, 1575.

Osad'ko, I.S., 1979, Usp. Fiz. Nauk 128, 31.

Osad'ko, I.S., E.I. Al'shits and R.I. Personov, 1974, Fiz. Tverd. Tela 16, 1974.

Osad'ko, I.S., R.I. Personov and E.V. Shpol'skii, 1973, J. Lumin. 6, 369.

Personov, R.I., 1978, Izv. Akad. Nauk SSSR Fiz. Ser. 42, 242.

Personov, R.I. and E.I. Al'shits, 1975, Chem. Phys. Lett. 33, 85.

Personov, R.I. and B.M. Kharlamov, 1973, Opt. Commun. 7, 417.

Personov, R.I. and V.V. Solodunov, 1968, Opt. Spektrosk. 24, 142.

Personov, R.I., E.D. Godyaev and O.N. Korotaev, 1971a, Fiz. Tverd. Tela 13, 111.

Personov, R.I., I.S. Osad'ko, E.D. Godyaev and E.I. Al'shits, 1971b, Fiz. Tverd. Tela 13, 2653.

Personov, R.I., E.I. Al'shits and L.A. Bykovskaya, 1972a, Opt. Commun. 6, 169.

Personov, R.I., E.I. Al'shits and L.A. Bykovskaya, 1972b, Zh. Eksp. Teor. Fiz. Pis'ma 15, 609.

Personov, R.I., E.I. Al'shits, L.A. Bykovskaya and B.M. Kharlamov, 1973, Zh. Eksp. Teor. Fiz. 65, 1825.

Rebane, K.K., 1968, Elementary Theory of the Vibrational Structure of Spectra of the Impurity Centers of Crystals (Nauka Publishers, Moscow) (in Russian).

Rebane, K.K., R.A. Avarmaa and A.A. Gorokhovski, 1975, Izv. Akad. Nauk SSSR Fiz. Ser. 39, 1793.

Rebane, K.K. and V.V. Khizhnyakov, 1963, Opt. Spektrosk. 14, 362, 491.

Richards, J.L. and S.A. Rice, 1971, J. Chem. Phys. 54, 2014.

Romanovskii, Yu.V., L.A. Bykovskaya and R.I. Personov, 1981, Biofiz. 26, 621.

Saari, P.M. and T.B. Tamm, 1976, Opt. Spektrosk. 40, 691.

Shelby, R.M. and R.M. Macfarlane, 1979, Chem. Phys. Lett. 64, 545.

Shpol'skii, E.V., 1960, Usp. Fiz. Nauk 71, 215.

Shpol'skii, E.V., A.A. Il'ina and L.A. Klimova, 1952, Dokl Akad. Nauk SSSR 87, 935.

Silsbee, R.H., 1962, Phys. Rev. 128, 1726.

Solov'ev, K.N., N.E. Zalesski, V.N. Kotlo and S.F. Shkirman, 1973, Zh. Eksp. Teor. Fiz. Pis'ma 17, 463.

Strokach, N.S., E.A. Gastilovich and D.N. Shigorin, 1973, Opt. Spektrosk. 35, 238.

Suter, G.W. and U.P. Wild, 1981, J. Lumin. 25/24, 497.

Szabo, A., 1970, Phys. Rev. Lett. 25, 924.

Szabo, A., 1971, Phys. Rev. Lett. 27, 323.

Tamm, T.B., Ya.V. Kikas and A.E. Sirk, 1976, Zh. Priklad. Spektrosk. 24, 315.

Tamm, T.B. and P.M. Saari, 1974, Opt. Spektrosk. 36, 328.

Tamm, T.B. and P.M. Saari, 1975a, Chem. Phys. Lett. 30, 219.

Tamm, T.B. and P.M. Saari, 1975b, Opt. Spektrosk. **38**, 1029.

Voelker, S. and R.M. Macfarlane, 1979a, Chem. Phys. Lett. **61**, 421.

Voelker, S. and R.M. Macfarlane, 1979b, Mol. Cryst. Liq. Cryst. **50**, 213.

Voelker, S. and R.M. Macfarlane, 1979c, IBM J. Res. Dev. **23**, 547.

Voelker, S., R.M. Macfarlane, A.Z. Genak, H.P. Trommsdorff and J.H. van der Waals, 1977, J. Chem. Phys. **67**, 1759.

Voelker, S., R.M. Macfarlane and J.H. van der Waals, 1978, Chem. Phys. Lett. **53**, 8.

Voelker, S. and J.H. van der Waals, 1976, Mol. Phys. **32**, 1703.

CHAPTER 11

Effects of Electric Fields on the Spectroscopic Properties of Molecular Solids*

DAVID M. HANSON**

Institute for Physical Science and Technology
University of Maryland
College Park, MD 20742
U.S.A.

J.S. PATEL, I.C. WINKLER AND A. MORROBEL-SOSA

Department of Chemistry
State University of New York
Stony Brook, NY 11794
U.S.A.

*Acknowledgement is made to the National Science Foundation and to the donors of the Petroleum Research Fund, administered by the American Chemical Society, for partial support of this research.

**1980–81 Synchrotron Ultraviolet Radiation Facility Fellow at the National Bureau of Standards, Washington, DC 20234. Permanent address: Department of Chemistry, State University of New York, Stony Brook, NY 11794, U.S.A.

Spectroscopy and Excitation Dynamics
of Condensed Molecular Systems
Edited by
V.M. Agranovich and R.M. Hochstrasser

Contents

1. Introduction

The influence of an external electric field on atomic spectra was discovered by Stark (1913) and is known as the Stark effect (Condon and Shortley 1964). Since the Stark effect arises from the addition of the following term to the system Hamiltonian,

$$H' = -\boldsymbol{\mu} \cdot \boldsymbol{F}, \tag{1}$$

where $\boldsymbol{\mu}$ is the dipole moment operator and $\boldsymbol{F}$ is the electric field, we prefer to identify all electric field effects that depend upon the presence of this term as Stark effects, independent of the spectral region and whether the system is atomic or molecular or whether the phase is vapor, liquid or solid, although the term electrochromism also has been used (Platt 1961, Liptay 1969). In this article we primarily consider Stark effects in the electronic spectra of molecular crystals.

Since the magnitude of the external field usually is small compared to the internal fields of atoms and molecules, the Stark effects can be analyzed in terms of perturbation theory while accounting for state degeneracies and near degeneracies by matrix diagonalization (Schiff 1955). Unless the electric fields are generated by high intensity lasers (which offers interesting possibilities for research that are not considered here), state energies expressed through second order and wave functions through first order in the field usually are adequate to analyze the observations. Of course, if fields are nonuniform then additional multipole moments contribute to the effect (Buckingham 1970), but this situation has not been investigated in the spectroscopy of molecular solids except as described in section 2. A particular molecular state i then is described by a wave function ψ_i and has an energy E_i in a field that is uniform over the volume of the molecule, where these quantities are given by

$$\psi_i = \psi_i^0 + \sum_k{}' \lambda_{ik} \psi_k^0, \tag{2a}$$

$$\lambda_{ik} = \boldsymbol{\mu}_{ki} \cdot \boldsymbol{F} / \Delta E_{ki}, \tag{2b}$$

$$E_i = E_i^0 - \boldsymbol{\mu}_i \cdot \boldsymbol{F} - \tfrac{1}{2} \boldsymbol{F} \cdot \boldsymbol{\alpha}_i \cdot \boldsymbol{F}, \tag{3a}$$

$$\boldsymbol{\alpha}_i = 2 \sum_k{}' \boldsymbol{\mu}_{ik} \boldsymbol{\mu}_{ki} / \Delta E_{ki}. \tag{3b}$$

The superscript 0 designates the unperturbed quantities, the prime on the sum omits state i, $\boldsymbol{\mu}_i$ is the permanent dipole moment of the state, $\boldsymbol{\mu}_{ik}$ is the transition dipole moment between the two states, $\boldsymbol{\alpha}_i$ is the polarizability of the state and ΔE_{ki} is the difference in energy of the two states.

A Stark shift, S, of a line in the electronic spectrum of a molecule fixed in space is then associated by eqs. (3) with the difference in energy of the two states involved in the transition f ← i:

$$S = -\Delta\boldsymbol{\mu}_{\text{fi}} \cdot \boldsymbol{F} - \tfrac{1}{2}\boldsymbol{F} \cdot \Delta\boldsymbol{\alpha}_{\text{fi}} \cdot \boldsymbol{F}, \tag{4}$$

where $\Delta\boldsymbol{\mu}_{\text{fi}}$ is the vector difference in the permanent dipole moments of the two states and $\Delta\boldsymbol{\alpha}_{\text{fi}}$ is the tensor difference in the polarizabilities of the two states. We term these quantities the dipole moment difference and the polarizability difference. Clearly, measurements of linear and quadratic shifts of lines as a function of field can provide values for these quantities. The actual situation is more complicated in solids because there usually is more than one space fixed orientation in crystals and an isotropic array in glasses. In addition, because of the polarization of the solid by the field, the electric field at the molecular level, termed the local field or microscopic field, is not the applied or macroscopic field. Furthermore, since the dipole of a guest molecule also polarizes the environment, and both the dipole of the guest and this polarization change as a result of an electronic transition in the guest, it is not clear that the Stark shift is attributable to the intuitive idea of a guest dipole moment difference interacting with a local field. While dealing with the multiplicity of orientations is straightforward—one simply considers each possibility in terms of eq. (4)—the polarization effects are no so straightforward. Research on this problem is described in section 2.

The change in the wave function given by eq. (2) most often is manifested as changes in the intensity distribution in a spectrum due to the accompanying changes in the transition moments. A change in the transition moment of course affects the radiative lifetime of the state and hence also the natural linewidth. Of more significance is the change in the nonradiative lifetime caused either by a mixing of states or shifts in energy of the state relative to a nonuniform background set of states. Observations of this effect, which recently have been made in the vapor phase, are discussed in section 11 along with implications for solids.

The Stark effect on molecules has been utilized extensively in rotational spectroscopy (Townes and Schawlow 1955), but little information about the molecular electronic Stark effect had been reported until the last two decades. The difficulty here was the minuteness of the effect combined with the limited resolving power of readily available spectrometers. Work prior to 1962 was reviewed by Kopelman and Klemperer (1962). The situation was changed by the availability of commercial spectrometers with resolving powers approaching one million, by the use of electric field modulation techniques, by the use of liquid and solid solutions with high dielectric strengths, thereby allowing fields of one or two million volts per centimeter to be applied and by the use of molecular crystals immersed in liquid helium, thereby providing a combination of sharp spectral lines (widths on the order of one wavenumber) and moderately high electric fields (100 kV/cm).

Some of the earliest examples of molecular electronic Stark effects in the vapor phase were reported by Freeman and Klemperer (1964, 1966), Irwin and Dalby (1965) and Phelps and Dalby (1966). Experiments on crystals immersed in liquid helium were reported by Hochstrasser and Noe (1968, 1969) and Hochstrasser and Lin (1968), Early modulation experiments on molecules in the vapor phase were done by Dows and Buckingham (1964), Buckingham and Ramsay (1965), and Bridge et al. (1966, 1968), about this same time, the results of modulation experiments on molecules in solution were reviewed (Liptay 1965, 1969, 1974, Labhart 1967), and the results of modulation experiments on molecules in solids were being reported (Kumanoto et al. 1962, Malley et al. 1968, Sauter and Albrecht 1968, Bucher et al. 1969, Hochstrasser and Noe 1970).

All of this early work was concerned with the use of electric field effects to determine molecular properties, especially with the determination of excited state dipole moments (Bakhshiev et al. 1969, Minkin et al. 1970, Buckingham 1972). The manifestation of a Stark effect in the spectrum of a molecular solid is extremely simple. A line shifts, splits or broadens (usually due to unresolved shifts or splittings) due to the interaction of the electric field with permanent dipole moments of the molecular system (first order Stark effect) or with field induced dipole moments (second order Stark effect), or the line can change intensity due to a field induced mixing of states with different transition moments. Clearly one can evaluate the permanent dipole moments, the induced moments or polarizability and locate forbidden or unobserved states from such data, but these effects offer opportunities to obtain much additional information as well.

As previously described (Hochstrasser 1973a), some of these possibilities include the use of Stark effects to modify the resonance energy transfer interaction between components of a system, to identify degenerate states, to determine molecular orientations, to influence molecular motion and change lattice equilibrium positions, to investigate environmental effects on charge distributions and polarizabilities, to count the number of molecules per primitive unit cell in crystal structure determinations and to identify polar and nonpolar isomers in solids. In addition, one can use Stark effects to measure microscopic electric fields in condensed phases stemming from both polarization and real charge distributions, to locate and characterize secondary energy trapping states, to locate and characterize ionic or charge transfer exciton states, to identify the electronic parentage of overlapping vibronic progressions and to identify vibronic coupling in these progressions. (Secondary energy trapping states arise from host molecules perturbed by a defect, the primary energy trap, in the lattice. This perturbation should induce a dipole moment in these host molecules which then are subject to a first order Stark effect. Secondary energy trapping states have been postulated to create an energy funnel and increase the effective trapping radius of the primary trap (Zoos and Powell 1972), but recent work (Argyrakis and Kopelman 1980b) has questioned the need for such a postulate.) In addition, electric fields may induce a neutral to ionic phase transition in optimum systems thereby changing the spectroscopic characteristics drastically.

While extensive work in some of these areas has been reported, others have not yet been developed. The following sections review what has been done and describe some of the prospects for future research. The topics of local fields and excited state dipole moments are so extensive and of such broad relevance that a complete presentation would require separate chapters for each. Consequently only some of the highlights are described here.

2. Local fields

The relationship between the macroscopic dielectric properties of condensed phases and the microscopic characteristics of the molecules has been a subject of concern and interest since the turn of the century. In recent years this topic has been involved with the determination of molecular polarizabilities (Vuks 1966, Dunmur 1972) and transition moments (Bridge and Gianneschi 1976), the analysis of laser light scattering experiments (Burham et al. 1975, Sullivan and Deutch 1976), the determination of orientational correlations in liquid crystals (Jen et al. 1977, Dunmur 1971, Palffy-Muhoray 1977), electro-optical phenomena in biological systems (Shiner et al. 1977, Reich and Schmidt 1972, Witt 1971), and the interpretation of Stark effects on molecular spectra, which is the topic of concern here.

The first analyses of Stark splittings in the electronic spectra of molecular crystals were based on the intuitive idea of a molecular property, the dipole moment, interacting with a local field (Hochstrasser 1973a). The local field was obtained using the isotropic Lorentz field equation (Kittel 1966) with the appropriate refractive index for each field direction, as had been done before (Sundararajan 1936, Bunn and Daubeny 1954). An alternative idea has been proposed by Vuks (1966) who considered the local field to be the same in all directions and use the mean square refractive index in the Lorentz equations. The inadequacy of these approximations was clearly demonstrated by Udagawa and Hanson (1976a) from the anisotropy of the Stark splittings observed in the electronic spectra of quinoxaline in a durene host crystal.

The problems associated with interpreting first and second order Stark effects in molecular crystal spectra in terms of a local dipole moment interacting with a local electric field have been described in detail (Chen et al. 1975). An exact analysis using the point dipole model and accounting for all guest–host interactions and polarizability differences was presented for the case of one molecule per unit cell (Fox 1976). This analysis was generalized to the case of two or more molecules per unit cell and applied to durene and naphthalene host crystals (Chen et al. 1977). Work along the same lines was reported by Dunmur, Munn and co-workers (Dunmur and Munn 1975, Cummins et al. 1975, Dunmur et al. 1977, Price et al. 1976).

For the case of the linear Stark splitting in the electronic spectrum of a polar guest in a nonpolar host crystal, for example (Chen et al. 1977), one finds from this work the exact result that within the point dipole model, the splitting depends upon the change in dipole moment of the guest and the change in local field accompanying the electronic transition between states i and f of the guest:

$$\Delta S = 2(\boldsymbol{\mu}_f \cdot \boldsymbol{F}_f - \boldsymbol{\mu}_i \cdot \boldsymbol{F}_i). \tag{5}$$

The local field depends upon the state of the guest because the polarizability of the guest changes as a result of the transition and is different from that for the host molecules. The local field at the site of the guest is given by

$$\boldsymbol{F} = \boldsymbol{Q} \cdot \boldsymbol{D} \cdot \boldsymbol{E}, \tag{6}$$

where the tensor $\boldsymbol{D}$ relates the local field $\boldsymbol{F}$ to the macroscopic field $\boldsymbol{E}$ in the perfect host, and $\boldsymbol{Q}$ accounts for the guest–host polarizability differences. While this analysis accounts for all the interactions, shown in fig. 1, that contribute to the linear Stark splitting involving the field, the permanent dipole moments of the guest, and all the induced dipole moments, with all the reaction field effects, it does not account for the finite spatial extent of the molecules and the nonuniformity of the electric field over the molecular volume. Problems also exist because the polarizabilities of molecules in the crystalline environment are generally unknown.

Without knowledge of the host polarizability, the local field cannot be determined even if it is assumed or demonstrated that the differences in guest and host polarizabilities are insignificant. It generally had been considered that the host polarizability and local field could be obtained from the crystal electric susceptibility (Kittel 1966, Cummins et al. 1973), but in the susceptibility experiment one measures the lattice polarization, the sum of induced dipoles of all molecules in the unit cell. It therefore is

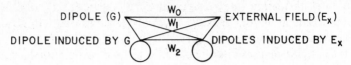

Fig. 1. The energy terms that contribute to the first order Stark effect for a polar guest in a nonpolar host crystal. W_0 is the direct interaction energy between the field and the guest, W_1 is the interaction energy of the induced dipole with the field and with the permanent dipole and W_2 is the interaction energy of the induced dipoles with each other. The unlabeled lines in the diagram represent other interactions that contribute to the crystal energy but not to the first order Stark effect.

possible to vary the molecular polarizability and local field continuously with the sum of the induced dipoles remaining constant. The crystal structure and susceptibility therefore, in general, do not determine the polarizability and local field uniquely (Chen et al. 1975). Earlier work had taken the local field (Munn 1972) or the induced moments (Dunmur 1972) for the different molecules in the unit cell to be identical even though this equivalence was not required by symmetry or other physical considerations.

In the absence of the needed experimental information, several methods have been proposed to deal with the infinite number of possible solutions to the equations relating the effective molecular polarizability, the local electric field, the crystal structure and the electric susceptibility. Since in molecular crystals the intermolecular interactions are weak, one expects the effective molecular polarizability in the crystal to be similar to the polarizability of the molecule in the vapor phase. Chen et al. (1975) therefore defined a domain of the most probable physical solution for the cases of durene and naphthalene. In this domain of solutions, the relative magnitudes of the diagonal polarizability tensor elements are the same as for the vapor phase and the off-diagonal tensor elements, no longer restricted to 0 by symmetry, are kept small. The range of possible solutions in this domain were then used to specify an uncertainty in the local field and in the effective molecular polarizability. Introducing additional, but arbitrary, constraints on the solutions has also been tried. One axis of the principal polarizability tensor has been fixed (Cummins et al. 1975) or the solution at the "center" of all the possibilities has been selected (Price et al. 1976), but for the case of naphthalene, the solutions selected in these ways lie outside the domain of the most probable physical solution (Chen et al. 1975).

The limitations of the point dipole model, arising either from the nonuniformity of the field over the molecular volume or from the difficulty of adequately accounting for the differences in guest and host polarizabilities were apparent in the analysis of the Stark effect on the 0,0 line of the first singlet–singlet transition of quinoxaline as a guest in naphthalene (Udagawa and Hanson 1976). Neither the orientation of the dipole moment difference vector nor its magnitude as obtained from the analysis were satisfactory. Additional evidence for the importance of accounting for the finite spatial extent of the molecules and the nonuniformity of the electric field has been obtained from measurements of the second order Stark effect on polycyclic aromatic hydrocarbons described in section 4.

One approach to this problem has been to treat each molecule as a set of polarizable points in order to account for the spatial extent of the molecule while preserving the formalism of the point dipole model (Chen 1975, Luty 1976, Bounds and Munn 1977). This approach, however, presents two major difficulties for general application. With more than one anisotropic polarizable point per unit cell, as described above and by Chen et al. (1975), the polarizability cannot be evaluated uniquely from available experimental information, and some ambiguity exists in evaluating the average effect of the nonuniform field on the molecule. For example, the procedure suggested by Bounds and Munn (1977) when applied to a *p*-terphenyl host crystal (Meyling et al. 1977) uses an average field rather than an average effect of the field. The local field is averaged over the centers of the aromatic rings of the host, rather than the highly polarizable regions of the guest which would be more appropriate. The host polarizability tensors and the local field tensors for the two molecules in the unit cell are constrained to be identical rather than the correct and less restrictive condition of being related by the symmetry operation that interchanges the two sublattices (Chen et al. 1975). In this procedure the total molecular polarizability is not given as the sum of the distributed submolecule polarizabilities but rather as the inverse of the sum of the inverse polarizabilities of the submolecules. A recent paper (Munn 1980) is directed at putting this approach on a more rigorous base, but the use of the concept of a distributed polarizability density in molecules also has been criticized (Sipe and Van Kranendonk 1978).

3. Dipole moments of excited states

The use of electric field effects in the spectroscopy of molecular solids to study the dipole moments of excited electronic and vibronic states has been an active field of research since the pioneering experiments of Hochstrasser and Noe (1968, 1969). Mixed crystal systems at low temperatures are free of the rotational congestion present in the electronic spectra of large molecules in the vapor phase, and generally provide sharp vibronic structure (linewidths on the order of one wavenumber), making anisotropic properties easily accessible to observation.

3.1. The determination of dipole moments by Stark spectroscopy

A typical study of an excited state dipole moment consists of measuring the splitting of a spectral line as a function of the applied electric field strength. For a mixed crystal with an inversion center, the observed splitting and applied field values are related to the excited state dipole moment by

$$\Delta S = 2B\, \Delta\mu \cdot L \cdot E, \tag{7}$$

where ΔS is the Stark splitting, $\Delta\mu$ is the vector difference between the dipole moments of the states which give rise to the spectral line being studied, E is the applied electric field, B is a unit conversion factor $(1.6792 \times 10^{-5}/\mathrm{D\ V})$ included to accommodate the units (wavenumbers, D (Debye) and V/cm) that spectroscopists use in measuring these quantities, and L is a tensor which couples $\Delta\mu$ to the applied field, accounting for all interactions and local field effects within the point dipole model but neglecting the polarizability difference between the two states of the guest. If the host local field tensor is used for L then the polarizability difference between the guest and host molecules is also neglected (Hanson and Chen 1973, Chen et al. 1975). Also see section 2. The origin of the splitting lies in the orientational degeneracy of different molecules in the crystal as discussed by Hochstrasser (1973a).

Measurements are made by placing a carefully cut crystal between two highly polished electrodes which are lowered into a liquid helium dewar. The sample is usually mounted such that the field, provided by a high voltage power supply, can be applied along one of the crystallographic axes. Data can be obtained by the techniques of absorption, emission, and excitation spectroscopy, and the spectrum can be recorded photographically or photoelectrically. When the field induced changes in the spectrum are small, field modulation can be employed in conjunction with phase sensitive detection. Equation (7) is then used to obtain the components of $\Delta\mu$ along the field directions.

The number of measurements needed to uniquely specify $\Delta\mu$ depends on whether the orientation of this vector can be deduced from the geometry of the molecule being studied (the guest), the crystal structure of the host, and the orientation of the guest in the host crystal. If the orientation of $\Delta\mu$ is known, a single measurement will yield a value, provided L is known (Hochstrasser 1973a). The sign of $\Delta\mu$, however, is not obtained.

When the orientation of $\Delta\mu$ cannot be deduced from symmetry considerations alone, the projections of $\Delta\mu$ on three orthogonal field directions must be determined. The work of Sheng and Hanson (1974b) on 2,4,5-trimethylbenzaldehyde isolated in a durene host provides an example of one such case. Measurements were made at several field strengths with the field directed along each of three orthogonal crystallographic axes. A plot of ΔS vs E was then made for each field direction. The linearity of this plot verifies that one is observing a first order Stark effect and illustrates the precision of the measurements. The slopes of these graphs were used to calculate the component of $\Delta\mu$ along each axis according to eq. (7). The ratios of the magnitudes of the components to the magnitude of the vector give the absolute direction cosines of the vector with respect to the field directions. The results therefore determine the magnitude, $|\Delta\mu|$, uniquely, but eight possible orientations remain, corresponding to different signs for the direction cosines. Four of these are related to the other four by inversion and in principle these four pairs can be distinguished from each other by additional experiments in which the field is applied parallel or perpendicular to each of the four possibilities. However, this procedure can be complicated by having more than one molecule per unit cell, as is the case for the durene host crystal.

In lieu of this procedure, a selection of the most reasonable $\Delta\mu$ can be made on the basis of the known (or assumed) molecular symmetry, substitutional orientation, and simple molecular orbital considerations. From a knowledge of the durene crystal structure and the fact that trimethylbenzaldehyde enters into the durene lattice substitutionally, the $\Delta\mu$ vectors in the crystal coordinate system can be related to the molecular frame. Considerable evidence exists for the near planarity of the state of interest, $S_1(n\pi^*)$, therefore the four vectors with large out-of-plane components are eliminated. From simple molecular orbital theory as verified by more eleborate calculations for benzaldehyde (Goodman and Koyanagi 1972, Mijoule and Yuan 1976), a decrease in dipole moment is expected for the first $^1n\pi^*$ state. This expectation eliminates two more possibilities, and allows the final selection to be made on the basis of the predicted direction of electron transfer upon excitation to an $n\pi^*$ state. Adding this $\Delta\mu$ to the ground state dipole moment, allows the magnitude and direction of the excited state dipole moment to be specified. The dipole moment of the lowest triplet state was also determined by this approach. See fig. 2. This analysis was substantiated by later experiments employing hexamethyl-benzene, with only one molecule per unit cell, as a host and also by the selective laser excitation techniques that are described in subsection 3.2 (Sheng and Hanson 1975, Udagawa and Hanson 1977).

If spectral lines are broad relative to the Stark splitting, high fields are required to resolve the Stark components. The fields are limited, however, by the dielectric strength of the sample and liquid helium. When the highest possible field does not give adequate resolution, the splitting can be obtained through a modulation experiment. This technique has been used, for example, to measure the change in dipole moment upon excitation to the phosphorescent state of 1-indanone and 1-tetralone. The splittings were

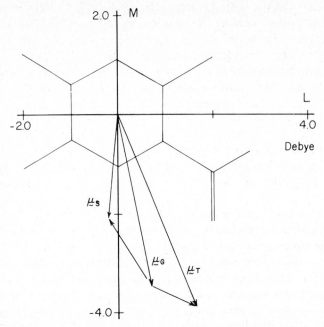

Fig. 2. The ground state and first excited singlet and triplet states of 2,4,5-trimethylbenz-aldehyde have dipole moment magnitudes of 3.53 D, 2.09 D and 4.23 D, respectively. The dipole moment difference vectors shown were uniquely determined by Stark data obtained by measuring the splitting of the absorption and phosphorescence origin lines with the field applied along three orthogonal directions in two different host crystals, and by using selective laser excitation techniques described in the text. The vectors point from positive to negative charge density.

obtained by fitting the experimental modulation data to a computer simula-tion of the experiment. Static field measurements were made on the same systems and agreed well with the modulation results. Optimization of a Stark modulation experiment was considered by Noe et al. (1976). The modulation technique has been of most use in measurements of excited state polarizabilities and second order Stark effects. Some of these results are described in section 4.

3.2. Problems associated with the Stark experiment

The interpretation of Stark experiments on molecular crystals that aim to evaluate excited state dipole moments has generally involved the use of a particularly simple model for the origin of the Stark splitting. This model assumes a uniform electric field over the volume of the molecule, treats each molecule as a point dipole and point polarizability, uses an anisotropic Lorentz field approximation to relate the microscopic field to the macroscopic field, and in mixed crystal systems assumes the guest and host polarizabilities to be identical (Hochstrasser 1973a). Environmental influences on the molecular properties are furthermore considered to be small. Although these considerations have led to plausible results, the apparent change in dipole moment for the first singlet–singlet transition in parafluoroaniline isolated in a durene host, found with this model, is approximately 100% larger (Hanson and Chen 1973) than the value reported for the molecule in the vapor phase (Huang and Lombardi 1969). This measurement was the first, and remains the only one, allowing a comparison between the results of vapor phase and condensed phase Stark effect experiments.

Stark experiments on the lowest singlet–singlet transition of quinoxaline in both durene and naphthalene host crystals also demonstrated the limitations of the simple model. For the case of the durene host, the symmetry axis of quinoxaline is known to coincide with the long axis of the durene site. However, when the Lorentz local field tensor is used to treat the data, the results are inconsistent with the anticipated orientation of $\Delta\mu$. A local field tensor for durene obtained from a point dipole approximation, which assumes the guest and host polarizabilities to be equal, gives the predicted orientation. For the case of the naphthalene host, neither approximation gave satisfactory results (Udagawa and Hanson 1976a). Similar difficulties with using point dipole approximations to local fields in crystals of polycyclic aromatic hydrocarbons were revealed in measurements of second order Stark effects described in section 4. Questions thus have been raised as to the validity of simple models for the microscopic polarization of matter, and more rigorous theories have been and are being developed. See section 2. It should be noted that the ratios of $|\Delta\mu|$ for different states of the same molecule are nearly independent of the local field and therefore are more reliable parameters than the dipole moment differences (Hochstrasser 1973a, Clark and Small 1977).

Another difficulty that arises in Stark experiments is the development of space charge. It takes only a few charged species to greatly affect the electric field within the sample and the outcome of the experiment. For example, in a crystal $10 \times 10 \times 2$ mm on a side with a dielectric constant of 2.5, only one of 3×10^{11} molecules need be ionized to produce a field of 25 kV/cm if the charges generated by photoionization in the bulk collect on the two closest parallel surfaces. This field is comparable to that of 50 to 100 kV/cm used in the experiment. Space charge formation can be detected photographically if the electric field is applied parallel to the spectrograph slit. See section 6. The problem of space charge can be avoided by using a pulsed laser as the light source to record the spectrum, provided the laser pulse width is short compared to the time required for space charge formation (Udagawa and Hanson 1976a). Other methods for preventing space charge formation also have been discussed (Chen et al. 1974).

The Stark measurements considered thus far do not provide information on the sign of the dipole moment difference (Hochstrasser 1973a). However, if the sign of the polarizability difference for the electronic transition and the sign of the crystal field are known, one can deduce the sign of the dipole moment difference by comparing measurements for a given molecule in both polar and nonpolar host crystals (Hochstrasser 1973a). This idea has been applied to azulene (Barker et al. 1973). Solvent shift data for liquid (Bayliss and McRae 1954) and solid solutions (Hochstrasser 1973a) can be used to determine the sign of the dipole moment difference as well. In the vapor phase the sign can be obtained in favourable cases by comparing the Stark splitting of rotational lines belonging to two different rotational branches (Huang and Lombardi 1970).

The relative signs of $\Delta\mu$ for absorbing and emitting states can be determined directly by a selective excitation experiment. This method consists of selectively exciting one of the two Stark components associated with an upper excited state and observing emission from a lower state. If the dipole moment changes for the absorptive and emissive transitions have the same sign, and if the same set of molecules that absorb light emit, the higher energy Stark component will be observed in emission when the higher energy Stark component is excited. If the two dipole moment changes have opposite signs, the lower energy component will emit when the higher component is excited. These ideas are best illustrated with an energy level diagram, fig. 3. Overlap of the two Stark components in the absorbing state will produce unequal intensities in the luminescent doublet. Once the relative sign of the dipole moment parameter is known for the two states, the orbital configuration (e.g. $n\pi^*$ vs $\pi\pi^*$ for aromatic carbonyl compounds or aromatic nitrogen heterocycles) of one of the states can be deduced from a knowledge of the orbital character of the other state (Udagawa and Hanson 1977).

This technique has recently been expanded into a method for obtaining the first order Stark splitting of spectral lines too broad to show an increase in linewidth, even at fields greater than 100 kV/cm (Winkler and Hanson 1981). With the field on, a laser is tuned to various frequencies along the broad line. At each excitation frequency the Stark emission spectrum is recorded. See fig. 4 for an example at one of the excitation frequencies. In the absence of energy transfer, the relative intensities of the emitting Stark components should be equal to the relative absorption coefficients of the Stark components of the absorbing state at the laser wavelength. The experimentally determined emission ratios are then used to fit ratios of absorption coefficients obtained in a computer model of the splitting with the splitting as the adjustable parameter. This approach requires that the second order Stark shift be negligible. Preliminary results for the spectral band assigned to the origin of the second triplet state of 2,4,5-trimethyl-benzaldehyde have been obtained by this method and show that, if the assignment is correct, then the signs of $\Delta\mu(T_1)$ and $\Delta\mu(T_2)$ are the same and their magnitudes are similar. These findings are at variance with assumptions made in previous work (Sheng and El-Sayed 1977) and are somewhat surprising since the first triplet state is known to be $\pi\pi^*$ in character while the second is expected to be $n\pi^*$.

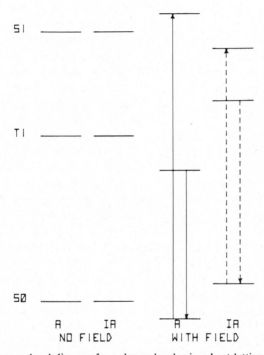

Fig. 3. A model energy level diagram for polar molecules in a host lattice with inversion site symmetry. Half the guest molecules (labeled A) are related to the other half (labeled IA) by the inversion symmetry element. The external electric field removes this symmetry and concomitant energy level degeneracy producing two Stark components split by $2\Delta\mu \cdot F$ in the absorption and phosphorescence spectra. For the example illustrated here, if excitation selectively occurs at the higher energy transition to S_1 (corresponding to type A molecules), then the lower energy component appears in phosphorescence as shown by the solid arrows in the figure. Similarly, if type IA molecules are excited (lower energy transition to S_1), then the higher energy component appears in phosphorescence as shown by the dotted arrows in this figure. This upper–lower and lower–upper relationship arises because the projection of the dipole moment difference vector onto the field for the singlet state has a sign opposite to the projection for the triplet state.

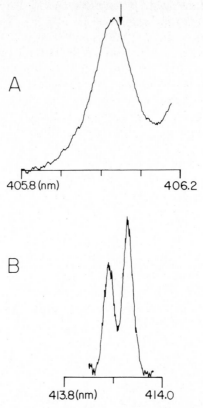

Fig. 4. (A) The most intense peak in the excitation spectrum of 2,4,5-trimethylbenzaldehyde in a durene host in the spectral region thought to contain the second triplet state. (B) The Stark split origin of the phosphorescence from the first triplet state that is obtained when the laser is tuned to the frequency indicated by the arrow in A. A field of 52 kV/cm has been applied along the *b*-axis of the crystal. Note that the lower energy phosphorescence component is more intense when the laser is tuned to the low energy side of the absorption band. For the reasons discussed and illustrated in fig. 3, this result indicates that the projections of the dipole moment difference vectors, for both transitions, onto the applied field direction have the same sign. This result is surprising because the first triplet state is thought to be a $\pi\pi^*$ and the second triplet state is thought to be a nπ^* configuration.

3.3. Applications and uses of the dipole moment parameter

In addition to providing information on excited state charge distributions, the dipole moment parameter has been used to probe the mixing of states that affect these distributions and to test specific theories of molecular electronic structure. Ideas have been advanced that attempt to explain the relative magnitudes of $\Delta\mu$ for singlet and triplet states arising from the same orbital configuration, e.g. the $n\pi^*$ configuration in carbonyl compounds. Experimentally the examples of formaldehyde (Buckingham et al. 1970), benzophenone (Hochstrasser and Noe 1971), 4,4'-dichlorobenzophenone (Hochstrasser and Michaluk 1972), and p-chloro- and p-methylbenzaldehyde (Hossain and Hanson 1978) have been studied. Qualitative electron correlation arguments have been used to explain the results for formaldehyde and the benzophenones. Although each idea correctly predicts the observed effect for the molecule in question, the explanation given for formaldehyde does not apply to the benzophenones and the converse. Ideas proposed by Coulson (1972) are in agreement with the findings for formaldehyde but not for the benzaldehydes or the benzophenones. It therefore appears that simple electron correlation arguments have little predictive value for large polyatomic molecules, and the details of electron correlation must be considered for each molecule. Hossain and Hanson (1978) explain their results on the substituted benzaldehydes in terms of an electrostatically induced mixing of the $^3n\pi^*$ state with the $^3\pi\pi^*$ state. This idea can be successfully applied to the benzophenones and can be tested by measuring the dipole moment change for the vibronic bands of the $^3n\pi^*$ state or by locating the $^3\pi\pi^*$ state. Goodman and Ozkan (1979) similarly describe such configurational mixing by Herzberg–Teller vibronic coupling involving non-totally symmetric vibrational modes and suggest Stark experiments using deuterated aldehydes to distinguish the vibronic from other possible coupling mechanisms.

In electric field experiments on the $^1A_1 \leftarrow {}^1A_1$ absorptive system in azulene and the $^1B_1 \leftarrow {}^1A_1$ system in 1,3-diaza-azulene, Clark and Small (1977) observed a vibronic state dependence of the dipole moment. The states that exhibited a significantly different dipole moment were assigned to totally symmetric vibrations, and from studies of the medium dependence of the absorption spectra, these states are thought to be vibronically induced in both molecules (Lacy et al. 1973, Small and Burke 1977b). Two mechanisms can be delineated by which vibronic coupling can lead to changes in the dipole moment. One is the borrowing of a permanent dipole from the electronic state being coupled to the observed state, and the other is borrowing from the transition dipole between the two states. Calculations suggested that only the second mechanism is operative, and since this mechanism is open only to totally symmetric vibrations, the calculations may explain why the non-totally symmetric vibronic components in the 3500 Å system of azulene have the same dipole moment as the origin.

It also has been shown that the dipole moment of a triplet state depends on the spin level (Sheng et al. 1975). This dependence was attributed to differences in the spin–orbit coupling of the sublevels to higher triplet states. In favorable cases, it was possible to extract the dipole moment of the higher state from the data (Sheng and El-Sayed 1977b). The results for

2,4,5-trimethylbenzaldehyde, however, do not agree with the more direct and recent measurements described in subsection 3.2 (Winkler and Hanson 1981).

Stark measurements have been used to help explain the dependence of the 2,4,5-trimethylbenzyl radical fluorescence on the photolysis temperature. When this radical is produced by the photolysis of durene at 80 K, the fluorescence recorded at 4.2 K is relatively simple. Photolysis at 4.2 K, however, results in a far more complicated spectrum with an additional origin line appearing at higher energy than the original one. Stark experiments showed that the dipole moment changes associated with these two origins have different magnitudes and directions. This difference implies that the spectral structure produced by photolysis at the low temperature arises from different orientations of the radical in the host crystal which may result from recoil in the photolysis. When the temperature of the crystal is raised, these sites anneal out, and the spectrum is identical to that obtained by photolysis at the higher temperature (Udagawa and Hanson 1976b).

Stark measurements have also been used to probe the structure of molecules in crystalline environments. While a distortion in parafluoroaniline might be anticipated (Hanson and Chen 1973), the cases of parabenzoquinone (Veenvliet and Wiersma 1973a,b, 1975, Hochstrasser et al. 1973, Trommsdorff and Galaup 1976, Sheng and El-Sayed 1975, Sheng et al. 1977) and 1,3,5-triazine (Hochstrasser and Zewail 1971, Wiersma 1972, Aartsma and Wiersma 1973, Bernstein and Smalley 1973, Hochstrasser and Zewail 1973) have been more controversial and have been studied in more depth.

Recently the Stark effect has been used to help understand the origin of inhomogeneous broadening of spectral lines in solids (Clark and Tinti 1980). A simple model was used attributing the spectral linewidths to the effects of random electric field fluctuations in the solid. Experimental data were presented and interpreted with the model for the crystalline system of sodium nitrite doped with silver nitrite. External Stark perturbations were used to measure shifts in known fields to show consistency with the model.

New experimental techniques for excited state dipole moment measurements are being developed for use in the vapor phase. Also see section 5. Hopefully these techniques will stimulate increased work in this area. A greater overlap of measurements made on molecules in both vapor and condensed phases will provide information about environmental effects on excited state charge distributions and will allow the premises on which the analysis of condensed phase Stark experiments is based to be tested. Along these lines, electric field level crossing experiments have been reported for LiH (Dagdigian 1980) and BaO (Dohnt et al. 1979). Similarly the use of supersonic molecular beams to eliminate the spectral congestion common for polyatomic molecules in the vapor phase (Smalley et al. 1975) should prove of great benefit. In previous measurements of Stark effects on such systems (Lombardi 1968), the splitting or shift of a line appears in the spectrum as a shoulder or part of an overlapping doublet. While these measurements have been of great value, new opportunities are provided by the use of molecular beam and high resolution laser techniques.

Spectrally forbidden states of molecules have been located in the vapor phase by means of field induced mixing (Huang and Lombardi 1971). Through the observation of nonlinear field induced changes in rotational structure, it was possible to extract the zero field separation, the magnitude of the transition moment, and the dipole moment of the perturbing state. One therefore expects to see similar results for these and other molecules in glasses and mixed crystals.

4. Polarizabilities of excited states

Numerous studies have utilized the second order Stark effect to obtain a measure of the polarizabilities of excited electronic states of molecules, primarily aromatic hydrocarbons (Seibold et al. 1969, Liptay et al. 1971, Hill and Malley 1971, Barnett et al. 1973, Kurzmack and Malley 1973, Mathies and Albrecht 1972, 1974, Meyling and Wiersma 1973). Information about the excited state polarizability also has been obtained from the magnitude of the dipole moment induced by the crystal field in polar lattices (Barker and Noe 1972, Barker et al. 1973, Marchetti and Scozzafava 1975). In this work measurements were made of the first order Stark effect due to the permanent moment, which may be zero, plus the induced moment. The value of the induced moment and the crystal field, and hence the polarizability, were extracted from the data by using well-defined approximations.

Among the highlights of the work on excited state polarizabilities is the elegant multichannel detection system assembled by Wiersma and coworkers (Meyling et al. 1976) and used with modulation techniques to obtain the full anisotropy of the polarizability difference tensor for pentacene and tetracene isolated in a p-terphenyl host. Since the observed effect is quadratic in the field, these measurements also served as a sensitive probe of the local field and apparently of the nonuniformity of the field over the molecular volume (Meyling et al. 1977). Similar measurements and conclusions regarding the local field were obtained for anthracene in a naphthalene crystal by Hochstrasser and Klimcak (1980).

The second order Stark shifts for tetracene and pentacene in the p-terphenyl host crystal were first analyzed by using the modified Lorentz formula for the local field (Meyling et al. 1976). The principal values for the polarizability difference tensor for the first singlet–singlet transition were found to be 29 Å³, 25 Å³ and 5 Å³ for tetracene along the long, short, and normal molecular axes, respectively, and 53 Å³, 90 Å³ and −11 Å³ for pentacene, in the same respective order. These values disagree with the expectation that the long axis component should be much larger than the short axis component. It therefore was suggested that the approximation used for the local field was responsible for this discrepancy. Later, the description of the local field proposed by Bounds and Munn (1977) was used in a new analysis of the data (Meyling et al. 1977) in order to average in some way the local field over the molecular volume. The revised polarizability difference values for tetracene are 65 Å, 23 Å³ and −8 Å³ and for pentacene 131 Å³, 76 Å³ and −32 Å³, again in the order long axis, short axis and normal axis. These values now agree with the expectation that the long axis value should exceed the short axis value by a considerable amount, and now lead to the conclusion, with somewhat more certainty, that the polarizability component normal to the molecular plane decreases on excitation. These results are, however, subject to the limitations of this description that were discussed in section 2. The fact that the results appear to be reasonable does not, of course, demonstrate the validity of the method.

For the case of anthracene in naphthalene (Hochstrasser and Klimcak 1980), the use of the modified Lorentz formula to analyze the Stark shift data produces polarizability values of 48 Å³, 24 Å³ and 5 Å³ for the first singlet–singlet transition. While these are reasonable results, it must be remembered that this formula did not provide a satisfactory result in analyzing the first order Stark effect of quinoxaline in the naphthalene host. See section 2. This approach also does not have a rigorous theoretical derivation. The analysis of the data also was done using five other descriptions of the local field in the naphthalene crystal. In comparison with the results of theoretical calculations of the polarizability difference as summarized by Hochstrasser and Klimcak (1980), none of the different descriptions of the local field appears to give unreasonable results for the polarizability difference tensor. The authors, however, postulate that the trace of the polarizability difference tensor for anthracene is 40 to 60 Å³, based on experiments in isotropic media where they consider the Lorentz field analysis to be justified. This range for the trace is then used as the standard for identifying the preferred description of the local field. They consequently conclude: "The best agreement is obtained with the local field calculation of Munn and Williams (1973). In their analysis the local field is computed from the relations between the macroscopic and microscopic polarizations induced by an applied field. The polarizations are obtained from the measured dielectric permittivities, molecular polarizabilities, and unit cell volume. This technique accomplishes the spatial averaging (that was shown to be essential) without having to evaluate lattice dipole sums. In view of the fact that use of this tensor unifies experimental studies from isotropic and anisotropic media, we propose it as the preferred internal electric field in the mixed crystal system." (Hochstrasser and Klimcak 1980).

Before accepting this conclusion, one should be aware of several points. First, while the Lorentz field may be the average field obtained by averaging over all the local configurations in a glass or solution, it has not been shown that the measured observable in the Stark experiment is a result of just this averaging. Certainly for a specific guest molecule in a glass, the local field depends upon the local configuration of the environment and not the average configuration. Furthermore the nonuniformity of the field over the molecular volume probably is not greatly different in a glassy host or in solution compared to a crystalline host, and one still needs to account for the differences in guest and host polarizabilities in both the glass and in solution. Consequently, the results of Stark experiments on molecules in isotropic media, such as rigid glasses and fluid solutions, are subject to the same uncertainties in the analysis as are the results for experiments on crystals, and hence are not well suited to be used as a standard for comparison.

Second, the method used by Munn and Williams (1973) originally was proposed by Munn (1972) and is based on a point dipole formalism, since the microscopic polarization is taken to be given by a point polarizability times a local field. The spatial averaging, which is obtained by this method and which was considered to be essential for the anthracene in naphthalene analysis, is an average of the point dipole local field over the two molecules in the primitive unit cell. Such an average was considered by Munn and Williams (1973) to be appropriate for describing electron mobilities in conductivity measurements where the electron visits both sites in the unit cell. It does not seem appropriate, however, for describing the effect of an electric field on a guest molecule substituted into only one site in the unit cell.

Finally, both the method proposed by Bounds and Munn (1977) and the method preferred by Hochstrasser and Klimcak (1980) require tensor quantities (e.g. the polarizability tensor in the first case, and the local field tensor in the second) for each of the two molecules in the unit cell to be identical whereas all that is required by the physical situation is that the tensor quantities for the two molecules be related by the symmetry operation that interchanges the two sublattices (Chen et al. 1975). The requirement that the tensors be identical has the effect of making certain elements in the local field tensor identically zero. For the case of a durene host, it was shown by the anisotropy of the Stark splitting of a quinoxaline guest (Udagawa and Hanson 1976a) that these elements were necessarily different from zero.

In conclusion, ample evidence exists to document the shortcomings of the point dipole description of local fields in anisotropic molecular crystals. These shortcomings are not surprising in view of the long recognized simplicity of the model. The model was developed and applied to this problem, not to provide the ultimate answer, but to put the analysis of Stark effect data on a rigorous theoretical base so that the nature of the approximations and assumptions inherent in the analysis would be clear. As the discussions here and in section 2 show, the model has been remarkably successful in this regard. In fact the model has revealed that a sufficiently complete data set does not exist to allow for a unique analysis of Stark effect experiments on most crystals unless arbitrary constraints are imposed. While these constraints tend to be based on "physical intuition" and are justified in terms of producing an "expected" result, they cannot reliably be transferred from system to system, e.g. the "preferred" analysis for the case of anthracene in naphthalene clearly is deficient when applied to quinoxaline in durene. Consequently it is clear that both novel experiments probing directly the microscopic polarization field in condensed phases and the application of rigorous theoretical developments that clearly overcome the shortcomings of the point dipole model would be of great significance.

5. Coherent transient and hole burning techniques

The Stark switching techniques developed by Brewer and co-workers with infrared lasers (Brewer and Shoemaker 1971, 1972, Brewer 1977) can be applied to the study of the electronic states of molecules and molecular solids provided one has visible and ultraviolet lasers with the requisite frequency stability. In addition to extending the precision of Stark effect measurements by a few orders of magnitude below the limits set by inhomogeneous linewidths and conventional spectrometers, these techniques allow dephasing times (the transverse relaxation time T_2) to be measured. The increase in precision can be used to measure smaller dipole moment and polarizability changes and to make measurements at lower field strengths, thereby largely eliminating the problems of dielectric breakdown and space charge formation. The pulsed nature of the measurement, allowing it to be made on a microsecond or faster time scale, also contributes to eliminating these problems (Udagawa and Hanson 1976a).

In many systems where the homogeneous linewidth is much smaller than the observed inhomogeneous linewidth, a narrow band laser can be used to burn holes in the inhomogeneous line. These holes are produced by depleting, through any one of several mechanisms, the population of molecules that absorb radiation at a given frequency. The holes can then be split, shifted or broadened by an electric field. In favorable cases the dephasing time can be obtained from the width of the hole, and the precision of the Stark measurement will be limited only by the homogeneous linewidth and the laser bandwidth. In early hole burning experiments, the population was depleted by pumping some metastable state that reequilibrated slowly or by inducing some irreversible photochemical transformation. Consequently the technique did not seem of very general applicability until the discovery of nonphotochemical hole burning in glasses (Hayes and Small 1978a,b).

Such hole burning should be common to many diverse systems, thereby presenting many advantages in Stark experiments. Many molecules for which suitable host crystals are difficult or impossible to find for mixed crystal experiments are readily soluble in many glasses. Stark cells for use with glasses can be designed so the glass entirely surrounds the electrodes providing dielectric strengths of one or two million volts per centimeter, characteristic of the solid environment. The Stark experiment also offers the opportunity to probe some of the microscopic properties of the glass itself.

The first Stark effect observed on a burnt hole was reported by Marchetti et al. (1977). Thin films of the dye resorufin dispersed in polymethylmethacrylate were coated over a semitransparent nickel coated Estar support. The counter electrode of NESA coated quartz was placed in contact with the polymethylmethacrylate. The sample was immersed in liquid helium at about 1.8 K, and holes burnt with a narrow bandwidth (0.01 to $0.1\,cm^{-1}$) tunable dye laser. The spectra were recorded using a 1 m spectrometer with a resolution of about $0.5\,cm^{-1}$. The spectral width of the burnt hole was about $0.2\,cm^{-1}$, and from the broadening observed in electric fields up to about 200 kV/cm, a dipole moment change upon excitation of 0.2 D was deduced. Such a measurement would have been impossible, because of 150 Å widths of the features in the absorption spectrum, without the advantages of the narrow spectral width of the hole and the high electric fields attained by coating the entire electrode with the insulating film.

The results of Stark effect measurements using both population hole burning and coherent transient techniques on praseodymium ions doped into lanthanum fluoride crystals clearly demonstrate the advantages offered by these techniques (Shelby and Macfarlane 1978). Similar experiments on ruby have also been reported (Szabo and Kroll 1978, Muramoto et al. 1978). In the case of praseodymium, the holes arise from an optical pumping cycle that transfers population between nuclear hyperfine levels.

Thus, the holes last for several minutes, decaying with the nuclear longitudinal relaxation time. The resolution, limited by the laser linewidth of about 2 MHz, was adequate to resolve Stark splitting arising from a dipole moment change of 0.056 D using fields of only a few kV/cm. In comparison, conventional techniques often require fields up to 100 kV/cm to clearly resolve Stark splittings arising from a dipole moment change of about 1 D.

The Stark shifts of the praseodymium levels were observed also by measuring the optical free induction decay using Stark switching. The free induction decay was induced by exciting the ions for a time long compared to the dephasing time, and then quickly switching them out of resonance with the laser by applying an electric field pulse with a rise time of less than 10 ns. A coherent component of the luminescence from these ions beats with the laser. The frequency of these beats from the detector is just the Stark shift. The decay of the amplitude of these beats is determined by the dephasing time of the ions, the frequency jitter of the laser, and the inhomogeneity of the Stark field. The results for the Stark shift were consistent with those obtained from the hole splittings, and the free induction decay time was shown to be determined by the laser frequency jitter.

The first molecular crystal system to be studied by the Stark switching technique was the 2,4,5-trimethylbenzyl radical in a durene host crystal (Burland et al. 1979). The use of this technique produced an optical nutation signal as new radicals were brought into resonance with the laser at the beginning and end of the electric field pulse.

6. Charge distributions in condensed phases and at phase boundaries

The magnitude and direction of electric fields in condensed phases and at phase boundaries are important in many areas of physics, chemistry and biology. Clearly Stark effects, the shifts and splittings of spectral lines or bands in an electic field, can provide information about these fields. The microscopic field, called the local field, arising from an external field and the concomitant polarization of the solid is discussed in another section. Here the effects of fields due to charges are considered for molecular solids. The techniques, however, are also applicable to biological systems (Witt 1975).

Variations in fields, and hence the charge density, through a sample can be observed by the spatial resolution of the spectroscopic signal in the detection system. The applicability of the technique in specific instances depends upon the detection of a spectral Stark splitting or shift and upon the spatial resolution of the signal to the extent required by the problem at hand. The availability of extremely monochromatic sources or high resolution techniques through most of the electromagnetic spectrum, the use of selected atomic, ionic, or molecular dopants as spectral probes, and the use of acoustic and electronic modulation and imaging techniques as well as optical imaging speak for the potential wide-spread applicability of the method.

Sheng and Hanson (1974a) used the orientational Stark splitting of the first spin allowed electronic transition of parafluoroaniline, introduced as a dopant probe, to measure the electric field distribution and space charge in an organic insulator, durene. The spatial resolution of the Stark splitting through the crystal was provided by imaging the crystal on the slit of a stigmatic spectrograph. Points along the spectral lines, recorded on photographic plates, parallel to the slit correspond to different points along the crystal. The observed variation of Stark splitting was attributed to variations in electric field caused by a space charge distribution produced by photoionization and electrode injection processes. The space charge distribution was fit to a model to provide quantitative information about the processes of charge generation, migration, and trapping.

If an external electric field is modulated, then an existing internal field may be modulated, and an observed modulation of the transmission or reflection of light can be used to provide information about the internal field or electrical potential. This technique has found application in the study of interfacial potentials of inorganic semiconductors (Frova and Handler 1965, Aspnes 1972, Holtz 1974, Yokomoto et al. 1976, Wolf et al. 1977, Bottka and Hills 1978).

The effect of the external field, as well as the effect of light intensity, on an internal potential in a molecular photo-electrical cell was reported by Killesreiter and Schneider (1977). The spectral shifts were so large (6 or 7 nm) that it was not necessary to use modulation techniques to observe them. The cell was composed of layers of arachidic acid on a merocyanine dye on a p-chloranil crystal between aluminum electrodes. A red shift of the photocurrent action spectrum was observed with increasing electric field and light intensity. This shift was attributed to the shift in the absorption spectrum of the dye molecules caused by the electric field from a layer of positively charged dye molecules at this interface. These observations allowed the density of ionized molecules in the layer to be evaluated and provided a way to investigate rate constants for charge carrier injection processes.

Since a modulated external field adds to an existing internal field, a Stark effect, such as the spectral shift due to the polarizability–field interaction, that depends upon the square of the total electric field involves a cross term of the internal and external fields. Electric fields in an organic photovoltaic cell made from aluminum and tin oxide electrodes and a fuchsin dye layer have been studied by using this effect (Dubinin et al. 1976). For a cell with a tetracene layer between aluminum and nesatron (indium oxide coated glass) electrodes, Scott and Albrecht (1979) found from the modulation of the absorption and reflection spectra that the difference in electrical potential between the electrode interfaces was 2.0 V with the aluminum interface at the higher potential. The contact field at each interface also was determined to be 200 000 V/cm at nesatron and 300 000 V/cm at aluminum, both with the tetracene being positive.

7. *Excitation energy transfer*

While excitation energy transfer in molecular systems has been of interest for some time (Frenkel 1931, Franck and Teller 1938), the influence of electric fields has not been widely considered. The first detailed theoretical explorations of some specific mechanisms for electric field effects on exciton motion were reported by Bierman (1963a,b). He considered a linear chain of atoms in the presence of a nonuniform electric field applied along the chain axis. Nearest neighbor excitation transfer interactions were assumed, and the field was represented by a power series expansion of the electric potential. Since an atom in a parity symmetric environment has no permanent dipole moment, Bierman found that the exciton wave packet was accelerated along the chain, in the gradient of the field, as a result of the dipole moments induced by the field. He also found that the width of the wave packet was not affected by the field unless the environment induces a dipole moment in the atom. Since the magnitudes of induced moments usually are small, as is a field gradient over atomic dimensions when produced by a nonuniform macroscopic field, the effects treated probably will not be observed in most molecular solids. However, as Bierman pointed out, such effects may be produced by the intense fields that sometimes arise from microscopic charge distributions.

In contrast one can expect significant effects of electric fields on energy transfer in systems composed of molecules with large dipole moments. In these cases an electric field (Hochstrasser 1973b) or field gradient can be used to make molecules inequivalent and eliminate or reduce the possibility for resonance energy transfer between them, provided the energy mismatch is comparable to or larger than the excitation transfer interaction. The energy mismatch, or difference in excitation energies of the two molecules, is given by

$$\Delta E = \Delta \boldsymbol{\mu}_i \cdot \boldsymbol{F}_i - \Delta \boldsymbol{\mu}_j \cdot \boldsymbol{F}_j, \tag{8}$$

where $\boldsymbol{F}_i$ is the field at molecule i, and $\Delta \boldsymbol{\mu}_i$ is the vector difference in dipole moments between the ground and excited electronic states. Of course, if the dipole moments of the two molecules are parallel and the fields are equal, there will be no mismatch.

This picture of an energy mismatch between pairs of molecules is directly relevant to the hopping model for energy transfer where one considers the excitation energy to be localized and hopping from molecule to molecule (Wolf 1967). In recent years research has been directed at describing the case of coherent energy transfer and explicitly incorporating the scattering by lattice defects and phonons that lead to the loss of coherence (Silbey 1976, Harris and Zwemer 1978). In what follows we review the effects of these energy mismatches on coherent energy transfer and then consider the hopping model and the intermediate regime of quasi-coherence. While these considerations are based on linear or one-dimensional models, the extension to two or three dimensions is not difficult. Furthermore, the description is purposely simple in order to bare the essential features of the phenomenon. Incorporating the effect of an electric field into more rigorous and mathematically complex descriptions of energy transfer clearly is possible. While other mechanisms for electric field effects on excitation energy transfer exist (Hanson et al. 1977), the discussion here is limited to the case of energy mismatches induced by a uniform macroscopic field. The field thereby can influence the energy transfer topology, spin–lattice relaxation and luminescence properties.

7.1. The limit of coherence

In the limit of coherent energy transfer, the migration of energy can be described in terms of a wave packet formed by a combination of the exciton eigenfunctions of a perfect crystal. The wave packet travels through the crystal at its characteristic group velocity, which is determined by the exciton band dispersion. If scattering events are independent, then the group velocity times the mean time between scattering events gives the exciton mean free path. In the incoherent limit, where the site-to-site hopping model is expected to apply, the mean free path is only one lattice site (Trlifaj 1958, Harris and Zwemer 1978).

The effect of an external electric field on the exciton group velocity and hence on the rate of energy transfer has been described for a linear antiferroelectric array of dipolar molecules, shown in fig. 5, with only nearest neighbor excitation transfer interactions (Hanson 1980). This model is a simple but widely used one (e.g. Shelby et al. 1976), that contains the essential features of the problem. The external field, applied in the direction of the dipole moments causes the electronic excitation energy of neighboring molecules to differ $2\Delta\mu \cdot F$ which leads to a narrowing of the exciton band and a decrease in the exciton group velocity. The decrease in the

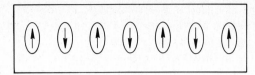

Fig. 5. The arrangement of polar molecules in a model linear crystal with two molecules per unit cell.

energy transfer rate, shown by curve E in fig. 6 is proportional to the decrease in the thermal average exciton group velocity, which for the model is determined entirely by the Davydov splitting and the Stark splitting, both of which can be determined spectroscopically. One then has a unique correspondence between exciton band structure and exciton dynamics with no adjustable parameters. Observations of changes in the energy transfer rate caused by an external electric field will be of great interest, not only because they test the applicability of the simple model, but also because they offer the opportunity to test the basic idea that an exciton travels through the lattice with a characteristic group velocity. Since the effect of the electric field is to decrease the exciton bandwidth, in

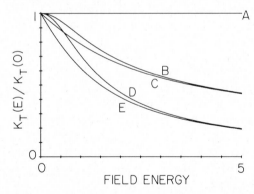

Fig. 6. The energy transfer rate function normalized to the zero field function vs the field energy $\Delta \boldsymbol{\mu} \cdot \boldsymbol{F}$ in units of V_{12} for different models. In each case only nearest neighbor interactions are included. (A) Random walk model in which coherence is destroyed in a time much less than the coherent pairwise transfer time. (B) Model in which the coherence is limited by exciton–defect scattering and the rate of coherent energy transfer is determined by the coherent pairwise transfer time. (C) Model in which the coherence is limited by exciton–defect scattering and the rate of coherent energy transfer is determined by the thermal average group velocity. (D) Model in which the coherence is limited by exciton–phonon scattering and the rate of coherent energy transfer is determined by the coherent pairwise transfer time. (E) Model in which the coherence is limited by exciton–phonon scattering and the rate of coherent energy transfer is determined by the thermal average group velocity. If the excitation hops from molecule to molecule with a jump time taken to be the coherent pairwise transfer time then the effect of the electric field follows curve B.

some systems the electric field may induce an Anderson transition from extended to localized states (Hanson 1980) if the bandwidth can be made smaller than the inhomogeneous broadening (Klafter and Jortner 1977).

If in addition to nearest neighbor interactions, next nearest neighbor interactions are included as well, then the electric field affects the thermal average group velocity in the two branches of the exciton band differently. The two branches arise from having two molecules per unit cell. This difference can then lead to differences in the energy transfer and trapping rates for the two sublattices, provided these rates are fast compared to the rate of energy transfer between the two sublattices. The dispersion relation for this case is

$$E_\pm = E_0 + 2V_{11}\cos(\boldsymbol{k}\cdot\boldsymbol{a}) \pm [(\Delta\boldsymbol{\mu}\cdot\boldsymbol{F})^2 + 4V_{12}^2\cos^2(\tfrac{1}{2}\boldsymbol{k}\cdot\boldsymbol{a})]^{1/2}, \qquad (9)$$

where V_{11} and V_{12} are the nearest translationally equivalent and translationally inequivalent excitation transfer interactions, respectively (Robinson 1970). This expression reduces to the one given by Hanson (1980) when $V_{11} = 0$ and to the one given by Hochstrasser (1973b) when $V_{12} = 0$. The equation can be rewritten as

$$E_\pm = E_0 + 2V_{11}\cos(\boldsymbol{k}\cdot\boldsymbol{a}) \pm |\Delta\boldsymbol{\mu}\cdot\boldsymbol{F}|\left[1 + \left(\frac{2V_{12}}{\Delta\boldsymbol{\mu}\cdot\boldsymbol{F}}\right)^2\cos^2(\tfrac{1}{2}\boldsymbol{k}\cdot\boldsymbol{a})\right]^{1/2} \qquad (10)$$

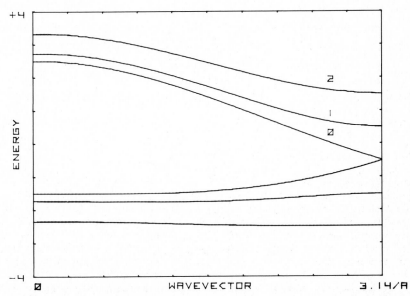

Fig. 7. The exciton energy measured in units of V_{12} as a function of wave vector for three values of the field energy: $\Delta\boldsymbol{\mu}\cdot\boldsymbol{F} = 0$, 1 and 2 V_{12}. Nearest neighbor interactions V_{12} and next nearest neighbor interactions V_{11} are included with $V_{11} = 0.25\ V_{12}$. The ideal mixed crystal level is taken as the zero of energy.

to utilize a series expansion in the limit $2V_{12}/|\Delta\boldsymbol{\mu}\cdot\boldsymbol{F}| \ll 1$. One then has

$$E_{\pm} \approx E_0 + 2V_{11}\cos(\boldsymbol{k}\cdot\boldsymbol{a}) \pm \left(|\Delta\boldsymbol{\mu}\cdot\boldsymbol{F}| + \frac{V_{12}^2}{|\Delta\boldsymbol{\mu}\cdot\boldsymbol{F}|} + \frac{V_{12}^2}{|\Delta\boldsymbol{\mu}\cdot\boldsymbol{F}|}\cos(\boldsymbol{k}\cdot\boldsymbol{a})\right). \qquad (11)$$

The electric field makes the two sublattices inequivalent, thereby splitting the two branches of the exciton band into two separated bands. The dispersion in these bands arises from two different interactions. A direct one that depends on V_{11}, and an indirect one that accounts for the coupling of two molecules on one sublattice with a molecule on the other. This interaction scales as $V_{12}^2/|\Delta\boldsymbol{\mu}\cdot\boldsymbol{F}|$ and has been called a superexchange interaction (Hong and Kopelman 1970, 1971, Hanson 1980). The dispersion relation for $V_{11}=\frac{1}{4}V_{12}$ is shown in fig. 7.

By differentiating eq. (9) one obtains the group velocities for the two bands,

$$v_g(k) = \frac{-2a}{\hbar}\left(V_{11}\sin(\boldsymbol{k}\cdot\boldsymbol{a}) \pm \frac{V_{12}^2\sin(\frac{1}{2}\boldsymbol{k}\cdot\boldsymbol{a})\cos(\frac{1}{2}\boldsymbol{k}\cdot\boldsymbol{a})}{[4V_{12}^2\cos^2(\frac{1}{2}\boldsymbol{k}\cdot\boldsymbol{a})+(\Delta\boldsymbol{\mu}\cdot\boldsymbol{F})^2]^{1/2}}\right), \qquad (12)$$

which is shown in fig. 8 as a function of wave vector and electric field. Since the dispersion in the two bands is different, so are the group velocities and the energy transfer rate, since this quantity is proportional to

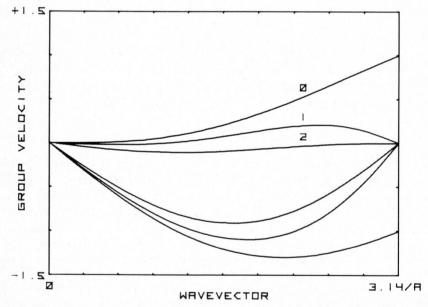

Fig. 8. The exciton group velocity as a function of wave vector for three values of the field energy: $\Delta\boldsymbol{\mu}\cdot\boldsymbol{F}=0$, 1 and $2\,V_{12}$ with $V_{11}=0.25\,V_{12}$. The velocity is in units of $aV_{12}/\hbar$.

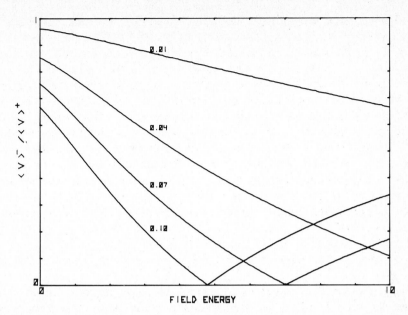

Fig. 9. The ratio of the thermal average absolute group velocity in the two branches of the exciton band as a function of field energy in units of V_{12} for four different values of the translationally equivalent excitation transfer interactions: $V_{11} = 0.01$, 0.04, 0.07 and 0.10 V_{12}. The ratio tends to zero for certain values of the field energy because the width of the branch tends to zero.

the group velocity in the coherent model. To better characterize this difference, the ratio of the average group velocity in one band to that in the other is plotted in fig. 9 as a function of electric field and as a function of the translationally equivalent interactions.

7.2. The regime of incoherence

In the limit of incoherent energy transfer, the excitation is considered to hop randomly from molecule to molecule. This hopping process usually is considered to be Markovian, i.e., there are no memory effects, each hop is independent of what has gone before. Aside from the topology of the lattice, the fundamental physical parameter necessary to describe incoherent energy transfer by using the extensive theoretical development of random walks on lattices (Montroll 1964, 1969, Montroll and Weiss 1965) is the average number of hops the excitation makes. This number is given by the exciton lifetime, t_L, divided by the time between hops, t_H.

Memory effects can be incorporated into the hopping model by postulating that the excitation preserves the direction of its motion over a mean free path of l lattice sites, which can be called the coherence length. A distribution of free paths characterized by a standard deviation can also be included in the model. At the end of a free path, the exciton is strongly scattered by a lattice defect or phonon and suffers complete loss of directional memory. The correlation time or memory time is the mean time between these scattering events $l \times t_H$. The consequences of this modification to the hopping model are described by theories of random walks on lattices with variable step size (Reif 1965, Lakatos-Lindenberg and Schuler 1971). In this excitation transfer process, each Markovian step consists of several non-Markovian hops (Kopelman 1976).

It is also possible to use this idea of "quasi-coherence" with the coherent model by postulating that between the Markovian steps, the exciton moves with a characteristic group velocity (Shelby et al. 1976). It is shown in subsection 7.3 that the nature of the electric field effect on energy transfer for these two ideas of "quasi-coherence" is qualitatively different, i.e., the nature of the effect depends on whether the excitation hops with memory or moves with a characteristic group velocity between the memory destroying scattering events. The electric field effect therefore probes the degree of localization of the excitation during the energy transfer process. We now describe the effect of the electric field in the site-to-site hopping model.

If an excitation is localized on one molecule of a pair, one has a nonstationary state of the system. The excitation oscillates from molecule to molecule with the probability of having the second molecule excited being given by

$$\Pr(2) = (V_{12}/\Gamma)^2 \sin^2(\Gamma t/\hbar), \tag{13}$$

where $\Gamma^2 = V_{12}^2 + \frac{1}{4}(\Delta E)^2$ (Silbey 1976). This phenomenon is just an example of quantum oscillations in two-state systems (Feynman 1965). The time required to transfer energy from one molecule to the other when $\Delta E = 0$ is just the time required for the $\Pr(2)$ to go from 0 to 1 (Hanson 1975, Kenkre and Knox 1974, Kopelman 1976). We call this interval the coherent pairwise transfer time,

$$t_{CP} = \frac{1}{4} h V_{12}^{-1} = \frac{1}{4} M^{-1}, \tag{14}$$

where M is the excitation transfer interaction in frequency units.

In the presence of the electric field, ΔE is not zero but equals $2\Delta\boldsymbol{\mu} \cdot \boldsymbol{F}$ for the linear antiferroelectric array. As the field increases, Γ thereby increases, and from eq. (13) one sees that the oscillation frequency for the probability increases but the probability amplitude decreases, as shown in fig. 10. The amplitude factor dominates, and the net result is that it takes longer for energy transfer to occur. A coherent pairwise transfer time can be defined for this situation by multiplying the oscillation period by the

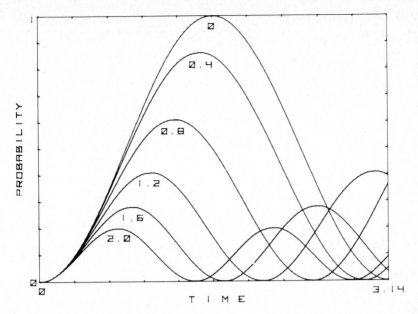

Fig. 10. Probability of excitation energy transfer from one molecule to the second of a pair, given that the first is excited at time $t = 0$. Time is in units of $\hbar/V_{12}$, where V_{12} is the excitation transfer matrix element. The six curves are for six different values of field energy: $\Delta\boldsymbol{\mu} \cdot \boldsymbol{F} = 0$, 0.4, 0.8, 1.2, 1.6 and 2.0 V.

number of periods necessary before one can expect to find the second molecule excited. Here one is simply saying that if the Pr(2) goes from 0 to 0.01 during each oscillation period, then on the average in an ensemble, the excitation will be transferred only once in every 100 periods. We therefore have

$$t_{CP} = \tfrac{1}{4} h \Gamma V_{12}^{-2}. \tag{15}$$

We call the reciprocal of this time the coherent pairwise transfer rate. The fractional decrease in this transfer rate is plotted in fig. 6 as a function of electric field. The use of the coherent pairwise transfer time as the hopping time in random walk models for excitation energy transfer in molecular crystal systems appears to have been quite successful in explaining numerous observations (Kopelman 1976, Argyrakis and Kopelman 1977, 1978, 1980a,b), and it will be interesting to see if this concept can also account for the influence of an electric field, although it is not rigorously applicable to a many-body system (Kenkre and Knox 1974, Stepanov and Gribkovskii 1968, Magee and Funabashi 1961).

The connection of the hopping time with a measurable energy transfer rate has been made most simply by Wieting et al. (1978). Under the appropriate conditions the exciton population N_E is described by the rate equation

$$\dot{N}_E(t) = -[K_E + K_T(t)] N_E(t), \tag{16}$$

where K_E describes the decay in the absence of energy transfer and trapping and $K_T(t)$ is the time dependent rate function that accounts for the transfer of energy through the lattice to a trap. Since the rate of relaxation into the trap is supposed to be fast compared to the rate of energy transfer to the trap, we call this parameter the energy transfer rate function. The decay rate parameters K_E and $K_T(t)$ can be evaluated experimentally by measuring the decay of exciton luminescence, for example. If c is the fraction of molecules that are traps in a crystal, then the probability that the excitation is not trapped if another new site is sampled on an excitation hop is given by $(1 - c)$. In time t, the number of new sites sampled is given by $S(t)$, the sampling function, so the probability that tht excitation has not decayed or been trapped is given by

$$P_E = (1 - c)^{S(t)} e^{-K_E t}. \tag{17}$$

Defining K, differentiating eq. (17) to get the rate of change of the probability, and comparing with eq. (16), identifies the time derivative of K with the energy transfer rate function:

$$K \equiv -S(t) \ln(1 - c), \tag{18}$$

$$\dot{P}_E = -(K_E + \dot{K}) P_E, \tag{19}$$

$$K_T(t) = \dot{K} = -\ln(1 - c)\, \dot{S}(t) \approx c\dot{S}(t). \tag{20}$$

The leading term of the sampling function has been calculated by Montroll and Weiss (1965) to be

$$S(t) = [8n(t)/\pi]^{1/2}, \tag{21}$$

where $n(t)$ is the number of hops made up to time t which is given by the time elapsed since the excitation was created divided by the hopping time. Using the coherent pairwise transfer time for the hopping time gives the following energy transfer rate function:

$$K_T(t) = c \left(\frac{8 V_{12}^2}{\pi h [V_{12}^2 + (\Delta \boldsymbol{\mu} \cdot \boldsymbol{F})^2]^{1/2}} \right)^{1/2} t^{-1/2}. \tag{22}$$

The fractional decrease in this rate as a function of the electric field is shown by curve B in fig. 6. Note the qualitative difference between this result and the result obtained by taking the energy transfer rate proportional to the group velocity.

7.3. *Quasi-coherence*

The above discussion has described electric field effects on energy transfer in the two limits of coherent wave packet motion and incoherent site-to-site hopping motion. One can expect the coherent motion to be limited by exciton–phonon and exciton–defect scattering processes. On the other hand, the hopping motion may preserve directional memory over several lattice sites with loss of directional memory being caused by an exciton–phonon or exciton–defect scattering event. In what follows we describe this intermediate regime of quasi-coherence in a simple way. Similar concepts have been used before (Zwemer and Harris 1978, Dlott et al. 1977, 1978, Kopelman 1976, Shelby et al. 1976), and we expect that more sophisticated and rigorous descriptions will be developed. We have seen that the electric field effect depends on whether the motion is coherent or hopping, and it now will be demonstrated that the electric field effect is sensitive to whether the coherence is limited by exciton–phonon or by exciton–defect scattering.

We suppose that exciton–phonon scattering can be characterized by a mean time between scattering events, t_S, that is independent of the electric field. The coherence length l is then field dependent because of the field dependence of the exciton velocity for either wave packet of site-to-site hopping motion. As a consequence of the scattering we imagine the exciton to move through the lattice in a random walk characterized by a mean step size l. Now we use the site sampling function

$$S(t) = l[8n(t)/\pi]^{1/2},$$ (23)

where $n(t)$ is the number of steps taken, and the square root gives the number of steps that cover new ground. This quantity is multiplied by l because in the random walk with variable step size, the mean square displacement is $l^2 n(t)$, where l^2 is the mean square step size (Reif 1965). Since the interval between scattering events determines a step, the number of steps taken in time t is just t/t_S. The energy transfer rate function is then

$$K_T(t) = \tfrac{1}{2}cl(8/\pi)^{1/2}(t_S t)^{-1/2}.$$ (24)

In the hopping model $l = t_S/t_{CP}$ to give the result

$$K_T(t) = \tfrac{1}{2}c(8/\pi)^{1/2}(t_{CP}^2 t/t_S)^{-1/2}.$$ (25)

In the wave packet model $l = \bar{v}_g t_S$ to give the result

$$K_T(t) = \tfrac{1}{2}c(8/\pi)^{1/2}\bar{v}_g(t/t_S)^{-1/2}.$$ (26)

The fractional decrease of these rates as a function of electric field is shown by curves D and E in Fig. 6, respectively. The decrease of the energy transfer rate in the electric field for wave packet motion has the same field dependence as the coherent limit, while the field dependence for the quasi-coherent hopping model differs from that for the incoherent limit where $n = t/t_{CP}$.

We suppose that the exciton–defect scattering can be characterized by a mean distance between scattering events, and that this coherence length l is independent of the electric field. Now the time between scattering events, t_S, is field dependent through the field dependence of the exciton velocity. The exciton velocity now affects the new site sampling function by affecting the number of steps and not the step length, where each step consists of several hops in the same direction:

$$K_T(t) = \tfrac{1}{2}c(8lv/\pi)^{1/2}t^{-1/2}. \tag{27}$$

In the hopping model $v = t_{CP}^{-1}$ and

$$K_T(t) = \tfrac{1}{2}c(8l/\pi)^{1/2}(t_{CP}t)^{-1/2}, \tag{28}$$

giving the same field dependence as for the incoherent limit. In the wave packet group velocity model $v = \bar{v}_g$ and

$$K_T(t) = \tfrac{1}{2}c(8l/\pi)^{1/2}(t/\bar{v}_g)^{-1/2}, \tag{29}$$

giving a different result. The field dependence of these rates is shown by curves B and C in fig. 6, respectively. Observations of the field dependence

Table 1
Electric field effects on energy transfer.

Exciton–phonon scattering
Constant scattering time
Field dependent scattering length

Hopping model	eq. (25)	fig. 6(D)
Group velocity model	eq. (26)	fig. 6(E)

Exciton–defect scattering
Constant scattering length
Field dependent scattering time

Hopping model	eq. (28)	fig. 6(B)
Group velocity model	eq. (29)	fig. 6(C)

For coherent energy transfer, the scattering time is a constant and is longer than the exciton lifetime. The electric field effect then follows the form for the case of coherence limited by exciton–phonon scattering.

For a constant step length of one lattice spacing in the hopping model, the electric field effect follows eq. (28) and curve B.

For rapid scattering, little or no electric field effect is predicted.

of the transfer rate therefore may offer the opportunity to discover whether exciton motion is described by a coherent hopping time or a group velocity and whether the coherence in the motion is limited by exciton–phonon or exciton–defect scattering.

In the incoherent limit where the scattering time is much shorter than the coherent pairwise transfer time ($t_S \ll t_{CP}$), little if any electric field effect on the transfer rate is expected because, as can be seen from fig. 10, for times short compared to the coherent pairwise transfer time, the probability for energy transfer is small and is little affected by the field. The field dependence vanishes entirely, as shown by curve A in fig. 6, if scattering times are so short that the sine function in eq. (13) can be represented by the first term in the series expansion. A summary of the effects of an external electric field on the various mechanisms of energy transfer is given in table 1.

7.4. *The role of phonons*

In the above discussion, the scattering processes that destroy the coherence of the energy transfer can be due to phonons or lattice defects. The energy transfer rate was determined by the excitation transfer interaction and the time between scattering events. Account was not taken of the ability of the phonon system to exchange energy with the electronic system to compensate energy mismatches. This neglect of inelastic scattering processes may be valid under conditions of low temperatures where exciton–phonon interactions produce phase or coherence destroying collisions without net energy transfer between the exciton and phonon systems. In this temperature regime the exciton–phonon scattering rate may be little dependent on the electric field.

At higher temperatures phonon assisted energy transfer can take place. In the simplest model one can consider a one-phonon process. The rate for one-phonon absorption and emission to compensate an energy mismatch of $2\Delta\boldsymbol{\mu} \cdot \boldsymbol{F}$ is proportional to the square of the exciton–phonon coupling matrix element, to the density of states, and to their occupation number n, or $n + 1$ for emission (DiBartolo 1968, Orbach 1967). Since the energy mismatches are expected to be in the range of 0 to 10 wavenumbers, the phonons involved are the acoustic modes, which in the Debye model have a density of states that increases as $(2\Delta\boldsymbol{\mu} \cdot \boldsymbol{F})^{d-1}$, where d is the dimensionality of the crystal. The occupation number varies as

$$[\exp(2\Delta\boldsymbol{\mu} \cdot \boldsymbol{F}/k_B T) - 1]^{-1}.$$

If the effect of exciton–phonon coupling is taken to be proportional to the phonon frequency then the one-phonon assisted rate is expected to be proportional to

$$(2\Delta\boldsymbol{\mu} \cdot \boldsymbol{F})^d [\exp(2\Delta\boldsymbol{\mu} \cdot \boldsymbol{F}/k_B T) - 1]^{-1}. \tag{30}$$

The rate then is obviously small for a small energy mismatch, and in fact, as pointed out by Orbach (1967), the transfer rate increases with increasing mismatch. The other phonon process that may contribute to the energy transfer rate is the Raman process in which two phonons are involved, and the difference in the phonon energy equals the energy mismatch (Holstein et al. 1976, 1977, 1978). In this case the rate is inversely proportional to the square of the energy mismatch and has a T^7 dependence. The Raman process allows a small energy mismatch to be compensated by much higher frequency phonons that may couple strongly to the exciton system because of the increased relative motion of neighboring molecules at the higher phonon frequencies. Thus one can see that the effect of the electric field on phonon assisted energy transfer can provide information about the nature of the phonon process involved, the density of phonon states, and the variation in exciton–phonon coupling matrix elements.

7.5. Experimental observations

For the models of energy transfer considered above, the application of an external electric field results in a decrease in the extent of the exciton motion. This decrease should give rise to an increase in exciton luminescence and a decrease in trap luminescence. Such increases in the exciton intensity have been observed in 4,4'-dichlorobenzophenone (Hanson et al. 1977) and in 4,4'-dimethylbenzophenone (Patel and Hanson 1981a). The trap intensities, however, exhibit a much more complex behavior. Sheng and El-Sayed (1977a) have observed that at 4.2 K the intensities of the shallow traps increase, which they explain results from an increase in the radiative properties of the x and y spin levels. Hochstrasser et al. (1980) have also observed such changes in intensity in 1,4-dibromonaphthalene, which they attribute either to energy transfer or to changes in the absorption cross section that may result from a field induced structural modification.

Michaluk (1972) was first to observe variations in the intensities of the different Stark components of the deep trap in dichlorobenzophenone as a function of the external field. This observation may be explained by the model presented here by noting that the two branches of the exciton band can have different widths. This results in different thermal average group velocities for the two branches. If one Stark component of the trap is coupled strongly to one of the branches and the other to the other branch, and if the scattering between the two branches is small, then the rate of energy transfer to one component would differ from the rate to the other and unequal luminescence intensities would result. (The lack of thermal equilibrium between two branches of an exciton band has been observed (Hanson et al. 1977) in emission excitation experiments)

8. Pyroelectric luminescence

A pyroelectric crystal is characterized by a unit cell with a net dipole moment, and unlike a ferroelectric the direction of the resulting lattice polarization cannot be changed by an external field. The spontaneous polarization of such crystals is continuously compensated by acquisition of free electric charges from the surrounding medium, by internal conduction of free charges, and by migration of surface charges (Lang 1974, Nye 1960). Heating or cooling such a crystal returns it to a polar state until the charge distribution again reaches equilibrium. During this time, the magnitude of the resulting polarization fields can be large enough to produce dielectric breakdown in the ambient atmosphere and induce luminescence not only from the breakdown but also from the solid itself.

This phenomenon, which we call pyroelectric luminescence, was discovered when pyroelectric crystals were dropped into liquid helium in attempts to deionize it. The deionization was sought to increase the dielectric strength of the helium for Stark effect experiments. These attempts were successful but no quantitative measurements of the ion concentration were made.

Experiments to characterize the phenomenon of pyroelectric luminescence were done with a cryorefrigerator in which the sample was mounted on a copper cold finger surrounded by an evacuated shroud. The luminescence was monitored using a photomultiplier tube. The apparatus allowed the temperature of the sample to be changed slowly and continuously and the pressure and composition of the ambient atmosphere to be varied.

We have studied a number of organic pyroelectrics, all of which show light emission when the compound is either heated or cooled. Luminescence induced by cooling is highly unusual and distinguishes this phenomenon from the well-studied phenomenon of thermoluminescence. In thermoluminescence raising the temperature detraps charge carriers and results in light emission when they recombine. The intensity and dependence on time of the pyroelectric luminescence is a function of several variables including the rate of temperature change, the size of the crystal, and the pressure of the atmosphere surrounding the crystal. We have observed that the pressure of air in the vacuum shroud dramatically alters the luminescence properties and can be used to classify the luminescence. Three broad catagories are characterized by the pressure ranging from (a) atmosphere to 1 Torr, (b) 1 Torr to 1 mTorr, and (c) less than 1 mTorr.

The luminescence intensity gradually decreases from class (a) to class (c). In class (a) the luminescence is in the form of bursts of light, probably resulting from dielectric breakdown. Surface charge measurements indicate that the charge changes synchronously with light emission. In class (b) the light is still emitted in the form of bursts, but the bursts are preceded by short, rapid, less intense pulses. This sequence repeats itself as the temperature gradually changes with time. See fig. 11. Although this light emission can be attributed to dielectric breakdown, other possibilities exist. In class (c) the luminescence intensity varies more gradually with temperature and time and is observed only over a small temperature range. The intensity peaks at particular temperatures that are characteristic of the compound. Such luminescence is shown in fig. 12 for a crystal of coumarin. We have examined the Raman spectrum of coumarin in the region of the phonons for evidence of a phase transition over this temperature range but have failed to find any.

The above classification has been made to bring out different features of this luminescence, but it should be noted that different classes of the luminescence can appear simultaneously. Details of our investigations of

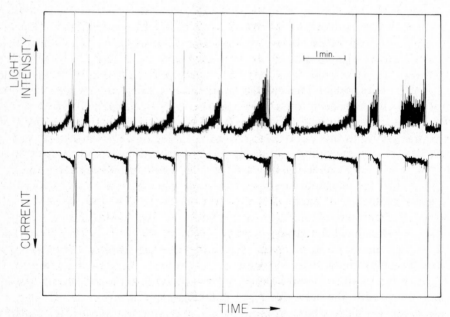

Fig. 11. Top: Intensity of light emitted by a crystal of coumarin when heated at the rate of 5°/min starting at 270 K. The pressure of the air surrounding the crystal was about 1 mTorr. Bottom: Simultaneous measurement of the current flow in an external circuit connecting opposite faces of the crystal.

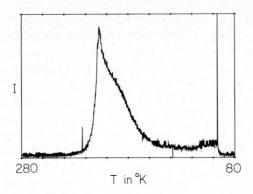

Fig. 12. Intensity of light emitted by a crystal of coumarin when cooled at the rate of 7°/min. The pressure of the air surrounding the crystal was about 0.05 mTorr.

pyroelectric luminescence are described elsewhere (Patel and Hanson, 1981b).

9. Exciton band structures

In cases where molecules in a crystal are physically equivalent by virtue of a symmetry operation transposing them, the resulting degeneracy of excited states is split into a band by the resonance excitation transfer interactions. This band of states is called an exciton band and is characterized by a dispersion relation and a density of states function. In the past, much effort has been directed at evaluating the parameters in these functions experimentally (Robinson 1970). Stark effects on exciton states can contribute to this effort by inducing forbidden transitions to appear in spectra and by providing values for the dipole moment and polarizability difference of the collective exciton state to compare with corresponding values for the isolated molecule. This comparison serves to test the underlying theory of excitons used in the analysis of the Stark effects. A general discussion of the effect of an electric field on the structure of exciton bands has been given by Knox (1963). Changes in the exciton band structure can also lead to changes in energy transfer properties, some of which are described in section 7.

Since intermolecular interactions generally are weak in molecular crystals compared to the intramolecular interactions, the results of experiments on mixed crystals, where the resonance interactions between molecules have been eliminated, are considered to provide information about molecular properties as described in sections 3 and 4. In neat crystals, however, the resonance excitation transfer interactions are of prime importance in accounting for the properties of the excited electronic states. If these interactions are sufficiently large, the first order Stark effect characteristic of polar molecules oriented in space can be eliminated completely. For example, the first order Stark splitting of the 0,0 phosphorescence line of 1-indanone in a durene host crystal is easily observed to be 4.9 wavenumbers in a field of 80 kV/cm when the field is applied along the *b* crystal axis (Sheng and Hanson 1975); whereas, in the neat 1-indanone crystal no splitting or shift of the corresponding line is observed in the absorption spectrum in fields up to 80 kV/cm applied along any of three orthogonal directions (Patel and Hanson 1982). This difference cannot be explained by differences in the geometrical or local field factors but is a manifestation of the nonpolar space group of the crystal and the magnitude of the resonance excitation transfer interactions. If these interactions are not too large, then the spectral properties of the neat crystal can be described in terms of the molecular dipole moments and the excitation transfer interactions (Hochstrasser and Michaluk 1972). Also see section 7, Dubinin (1977) and Lee (1975) for the algebra of this description. In some cases, where these interactions are large, the observed Stark effects can be expected to resemble those arising from the polarizability, and the identification of the nonpolar crystal as being composed of polar molecules will be lost. In such cases, an analysis of Stark effect data by using the simple theory valid for the regime of small interactions will result in a value for the dipole moment difference vector that does not agree with the value obtained in a mixed crystal experiment. Of course, crystal fields can distort the molecule and change the dipole moment values, but this effect can be expected in going from one mixed crystal to another and is not related to the absence and presence of resonance interactions in mixed crystals and neat crystals, respectively. In noncrystalline aggregates, where molecules with different orientations and possibly different dipole moments contribute unequally to the various excited states of the aggregate, the dipole moment difference will depend upon the excited state of the aggregate and will differ from that of the monomer (Dubinin 1977).

Such changes between the dipole moment difference vector of a collective excited state and that of the corresponding monomer state have been reported for the case of aggregates of the dye rhodamine 6G (Dubinin et al. 1978), but not for the case of molecular crystals, although one must be wary of statements in the literature like "we have discovered the Stark splitting for the quasilinear exciton to be very slightly larger than for the perdeutero isotropic trap transition even though the experimental uncertainties are such that the two values could be the same" (Hochstrasser et al. 1980).

Electric fields have also been used to induce transitions to spectroscopically forbidden factor group components and to aid in the analysis of exciton band structures. The first such experiments were performed on benzophenone (Hochstrasser and Lin 1968). This technique allowed pairwise excitation transfer interactions and the molecular dipole moment difference vector to be evaluated. In addition to the work on triazine and benzoquinone referenced in section 3, Stark effects on neat crystals also have been reported for 1,4-dibromonaphthalene (Hochstrasser et al. 1980), 1,4-naphthaquinone (Galaup et al. 1978), *p*-chloroaniline (Marchetti 1973), sulfur dioxide (Hochstrasser and Marchetti 1970), sodium nitrate (Hochstrasser and Klimcak 1978), 4,4′-dichlorobenzophenone (Hochstrasser and Michaluk 1972), 4,4′-dibromobenzophenone (Noe et al. 1976) and a merocyanine dye (Marchetti 1974).

10. Charge transfer excitons

The separation of positive and negative charges involving the transfer of an electron from one molecule to another and the reverse process of electron–hole recombination have been difficult to characterize in a quantitative and detailed microscopic way by experiments on organic materials. The importance of these processes in chemistry, physics, and biology is well recognized. When the charge transfer state is an excited state of the system, the charge transfer process can be induced by the absorption of light. This situation makes it possible to study the charge transfer and recombination by a variety of powerful spectroscopic techniques. While exciting work along these lines is being done in the areas of thermal and photo-assisted electron transfer in biological systems (Redi and Hopfield 1980, and references cited therein), in the areas of charge transfer at surfaces and interfaces and in organic metals and semiconductors, and on electric field induced quenching of fluorescence (Popovic and Menzel 1979, Petelenz 1979, Bullot et al. 1978, and Yokoyama et al. 1975), the discussion in this section is limited to Stark effects on charge transfer states that appear in the absorption and reflection spectra of molecular crystals and thin amorphous or crystalline films.

Charge transfer exciton states in crystals arise when an electron–hole pair can be bound by their Coulomb field and the polarization field of the lattice with the electronic wave function extending over more than one lattice site. The importance of these states in molecular crystal research stems from their use in configuration interaction schemes to represent the delocalization of electrons and from their participation in many energy conversion processes as documented in a review article (Hanson 1973). For example, using calculated or parameterized off-diagonal matrix elements and energy denominators, mixing of charge transfer and neutral exciton states has been used to account for factor group splittings (Choi et al. 1964, Tiberghien et al. 1973), energy transfer rates (Jortner et al. 1965), excimer formation (Hochstrasser 1962), and pressure effects (Beardslee and Offen 1970). Charge transfer exciton states have been proposed to play a role in the intrinsic generation of charge carriers, the photoinjection of charge carriers from electrodes, exciton ionization at impurities, charge carrier trapping, electron–hole recombination, electroluminescence, crystal photoionization, exciton–exciton annihilation and exciton fission.

It generally has been impossible, however, to obtain much experimental information about the charge transfer exciton states. In most experiments the observable properties were the end results of a complicated series of processes and could not be unambiguously related to the characteristics of charge transfer excitons (Hanson 1973). More direct techniques, like absorption spectroscopy, have not been successful because the spectral absorption is dominated by transitions to neutral Frenkel exciton states (Berry et al. 1965), states that are also present in the free molecule, even when optical pumping or multiphoton phenomena were utilized (Webman and Jortner 1969, Bergmen and Jortner 1972).

Some time ago it was demonstrated that electric field modulated absorption or reflectance spectroscopy is sensitive to the charge transfer transitions and insensitive to the Frenkel transitions (Abbi and Hanson 1974). This differentiation is possible because the charge transfer states have large dipole moments, or, in crystals with a center of inversion, large pseudo-dipole moments somewhat analogous to the first order Stark effect in the hydrogen atom. In this work on 9,10-dichloroanthracene, it was reported that the transition energy and polarization observed in the spectra, the second derivative line shape of the Stark modulation spectrum, the modulation signal occurring at twice the field modulation frequency, the dependence of the signal on the magnitude of the electric field, and the magnitude of the dipole moment difference derived from the data, all were explained by a transition to a charge transfer state equivalent to an electron transfer of 5 Å along the a-axis of the crystal. 9,10-dichloroanthracene was chosen for this first study because of the anomalously polarized absorption bands that were observed by Tanaka and Shibata (1968) and assigned to a charge transfer transition.

In crystals of electron donor–acceptor complexes, the charge transfer states are spectroscopically accessible because of the favorable balance between the ionization potential of the donor and the electron affinity of the acceptor. The optical spectra, however, are characteristically broad and structureless because the Coulomb attraction of positive and negative charges on neighboring molecules results in strong electron–phonon coupling. Transitions to intramolecular vibrations are thereby broadened into structureless envelopes by a series of unresolved phonon transitions. As a consequence little detailed information about the charge transfer state has been extracted from these spectra. The one exception is the first singlet charge transfer transition of the complex anthracene–pyromellitic acid dianhydride (PMDA) where the absorption and fluorescence spectra contain sharp lines on top of broad continua (Haarer 1974).

In this system, the sharp zero phonon line made it possible to characterize the charge transfer state by using a static electric field perturbation (Haarer 1975) where the modulation technique had been required for 9,10-dichloroanthracene. The observations of the Stark effect and the structure of the phonon sideband combined with theoretical models allowed the exciton bandwidth and the electron–phonon coupling parameters for anthracene–PMDA to be deduced (Haarer et al. 1975, Haarer 1977).

In spite of this promising beginning, the study of charge transfer states in molecular solids by the techniques of Stark spectroscopy has been hindered by numerous experimental problems. In addition to the difficulty of detecting relatively small shifts and splittings of broad spectral bands which are overlapped by more intense transitions in crystals of a single component, crystals where a charge transfer interaction is important generally grow in the shape of small needles. For the Stark experiment these crystals must be mounted rigidly between electrodes in a way that prevents the formation of a space charge in the crystal or at the electrode–crystal interface. Any small motion of the crystal in the field obviously will modulate the light and obscure the desired field modulation of the spectrum.

Recent results, however, on both 9,10-dichloroanthracene and anthracene–PMDA have provided additional motivation for continued work on these electric field effects. Polarized reflection spectra of single crystals of anthracene–PMDA have been measured at 2 K, and the absorption spectrum derived by the Kramers–Kronig transformation. The spectrum consists of the electronic origin and a progression of vibronic states involving vibrational excitations of the molecules. Each of these serves as an origin for lattice phonon modes. The absorption spectrum was interpreted in terms of an appropriate theory of exciton–phonon coupling, and values for an exciton–phonon coupling parameter for the electronic and vibronic origins were obtained (Brillante et al. 1978, Brillante and Philpott 1980). The electric field effects on all these spectral lines are of obvious interest, especially for comparison with the variation of exciton–phonon coupling with vibronic origin.

Absorption spectra obtained using light polarized along the *a*-axis of extremely thin single crystals of 9,10-dichloroanthracene grown on a variety of substrates also exhibit much structure (Cipollini 1975) not observed previously either in absorption or reflection spectra. This structure appears in the same wavelength region as the charge transfer transition reported by Abbi and Hanson (1974) and as the broad features in the low temperature, *a*-polarized reflection spectrum reported by

Syassen and Philpott (1977). A possible explanation of the structure involves mixing of the charge transfer state with the neutral exciton state. The structure comes from the neutral state and the a-polarized transition moment comes from the charge transfer state, since the transition moment to the neutral state lies in the molecular plane which is normal to the a-axis. Again, electric field effects on this structure and on the structure associated with the neutral exciton transitions observed in other polarizations would be informative.

Photoemission at energies below the ionization threshold of 9,10-dichloroanthracene has been investigated (Rozenshtein et al. 1980) by the Millikan–Pope–Arnold method (Arnold et al. 1979). These studies have provided additional evidence for the existence of a metastable state at the energies indicated by the spectroscopic experiments described in the previous paragraph. The characteristics of the photoemission were reported to be consistent with an Auger mechanism involving a free electron and a trapped charge transfer state.

The only other studies of charge transfer states in single crystals using the techniques of Stark spectroscopy have been on polydiacetylene-bis-(p-toluene sulfonate) or PTS. Electroreflectance and electroabsorption spectra were obtained by electric field modulation techniques (Sebastian and Weiser 1979). The resulting spectra were not analyzed quantitatively, and unlike the cases of 9,10-dichloroanthracene and anthracene–PMDA, a clear picture of the nature of the states involved in the electronic transitions giving rise to the Stark effects was not proposed. Stark effects on the low energy region of the spectrum are attributed to an excitation at a defect of the polymer chain. One feature remains unassigned. The middle region of the spectrum involves π electron excitations of the chain and associated vibronic structure, and the Stark effect in this region resembles that obtained from the polarizability differences of the excited state and the ground state. The most interesting feature in the spectra is a doublet at the high energy end, near 2.4 eV. This doublet appears in the electroreflectance spectrum, observed using modulation techniques, but there are no obviously corresponding features in the reflectance spectrum. Since the position of the doublet agrees with the threshold for intrinsic photoconductivity, the doublet is assigned to a weak transition between the valence band and the conduction band of the polymer.

Future studies of the Stark effect on charge transfer transitions appearing in the spectra of crystals consisting both of one component and of electron donor–acceptor complexes can be expected to help answer numerous questions:

What are the energies of charge transfer exciton states in compounds other than 9,10-dichloroanthracene? Observations of resonances with characteristics similar to the ones reported for 9,10-dichloroanthracene will be of interest in this regard.

What is the nature of the electronic charge distribution in these states? Measurements of the anisotropy of the Stark effect and the dependence on field strength will provide relevant information. Mixing of the charge transfer state with neutral exciton states (Pollans and Choi 1970) and field induced mixing of these states (Petelenz and Zgierski 1975, Petelenz and Petelenz 1977, Petelenz 1977a,b) may be revealed by such studies. With more than one molecule or donor–acceptor pair per unit cell, the mixing of different charge transfer configurations between these molecules should also be revealed.

What are the parameters of electron–phonon coupling and how do they correlate with the degree of charge transfer? Measurements of the temperature dependence of the Stark effect, the dependence of vibronic state, and comparison with Debye–Waller factors or other parameters derived from theories of exciton–phonon coupling are of interest here.

What is the structure (dispersion relation and density of states) of the charge transfer exciton band? The electric field perturbation can induce forbidden components of the exciton band to appear in spectra, and the observed splittings can be interpreted in terms of electron and excitation transfer matrix elements that determine the band structure.

In cases where there is a large gap between the absorption spectrum and the fluorescence spectrum, what is the difference between the charge transfer state created initially in the absorption process and the relaxed state that luminesces? A comparison of electric field effects on the absorption and fluorescence spectra will reveal differences in polarizabilities, dipole moments, and the mixing of different charge transfer configurations.

How and why do the answers to the above questions vary from compound to compound? Clearly studies on a wider variety of systems are needed.

In view of the above mentioned difficulties in doing these experiments on single crystals, additional effort should be devoted to studies of thin crystalline films. The spectra obtained by Cipollini (1975) for thin substrate supported crystals of 9,10-dichloroanthracene are of extremely high quality compared to what is generally seen with this technique. Perhaps more attention should be given to optimizing the conditions under which such films can be prepared. The use of crystalline films in Stark experiments aimed at detecting charge transfer excitons seems to have been initiated by Blinov and Kirichenko (1973). Like Abbi and Hanson (1974), they reasoned that the charge transfer transitions should be much more affected by an electric field than overlying neutral exciton transitions. Electric field modulation techniques could then bring out the small contribution to the absorption or reflection from the charge transfer transition. Several molecular compounds were studied, specifically films of phthalocyanine and its metallic complexes, aromatic hydrocarbons like violanthrone and violanthrene, and some dyes. The electroabsorption spectrum in almost all the films reportedly had the same form with a sharp peak at the long wavelength edge of the absorption band and oscillations within the band. Although the spectra were not analyzed quantitatively, the sharp peak was ascribed to a charge transfer exciton. Spectra for copper phthalocyanine were shown as an example of the results.

The better characterized systems of tetracene and perylene were investigated next by this technique (Blinov and Kirichenko 1974). The Stark effects for tetracene were identified as being due to the polarizability difference between the ground and excited states, and no clear evidence for a Stark effect arising from transitions to a charge transfer state could be identified. Later studies of the excitation spectrum of excimer luminescence in the tetracene crystal (Kolendritskii et al. 1977, 1978) revealed a maximum at 378 nm (3.28 eV) that did not correspond to any structure in the absorption spectrum. This wavelength is just beyond the spectral region studied by Blinov and Kirichenko (1974). It was proposed that this maximum came from the excitation of an ion pair or charge transfer state that relaxed to the emitting excimer state with high efficiency. The onset of an intrinsic ionization process at above 3.2 eV had previously been proposed to explain the drop in tetracene fluorescence efficiency (Geacintov et al. 1966). Measurements of the double quantum external photoelectric effect in tetracene (Arnold et al. 1979) have provided evidence for a trapped charge transfer state at 2.85 eV. The excimer excitation band extends through this region so all these results seem consistent with locating a charge transfer state there.

The Stark effects on the perylene films (Blinov and Kirichenko 1974) in the long wavelength region were characteristic of the polarizability difference effect with the absorption bands shifting to the red in the electric field, giving a first derivative lineshape to the modulation spectrum at twice the modulation frequency. Between 360 and 420 nm the modulation spectrum deviated greatly from the derivative of the absorption spectrum implicating a different effect. It is in this region that the existence of a charge transfer transition was proposed (Tanaka 1963, Tanaka et al. 1974) to explain the polarized absorption spectra of perylene crystals. No detailed analysis of the spectra was done, however, to extract properties of the charge transfer state in these films.

Similar studies on amorphous films of the charge transfer complex polyvinylcarbazole and trinitrofluorenone (PVK–TNF) were reported by Weiser (1973). Two charge transfer bands were identified in the absorption spectrum, and the data were analyzed in detail. The observed Stark effects could be explained by a broadening of the charge transfer bands due to the different energy shifts that arise from the interaction of the permanent dipole moments in this amorphous system with the electric field. A shift of the bands due to the polarizability difference of the ground and excited states was included as well. Values for the dipole moment difference and the polarizability difference for the two transitions were obtained from the analysis. Since the Stark spectra of this complex contained features similar to those observed in the spectra of amorphous selenium (Weiser and Stuke 1969), it was suggested that the model of a localized charge transfer state used to explain the PVK–TNF results might apply to some transitions in inorganic materials, especially amorphous selenium, which consists partly of ring molecules.

[Note added in proof: Recent Stark modulation experiments on substrate-supported crystalline films of anthracene, tetracene, and pentacene seem to have successfully located charge transfer transitions in these systems (Sebastian et al. 1981, 1983).]

11. Radiationless transitions

The basic mechanism for the nonradiative decay of excited electronic states of molecules was proposed by Robinson and Frosch (1962, 1963), although many scientists have contributed to an improved and more detailed understanding of the concepts involved. The idea of a radiationless transition follows from an approximate description of the excited states where an electron state s overlaps a manifold of other states ⟨m⟩. This manifold generally consists of the vibronic levels of lower electronic states. An optical transition from the ground state to s is allowed while those to the states ⟨m⟩ are forbidden due to symmetry or spin selection rules, or very small Franck–Condon factors. The state s and the states ⟨m⟩ can be coupled by terms in the complete molecular Hamiltonian so the molecular eigenstates are mixtures of these zero order states. If a molecule is prepared in state s initially, then this coupling causes a radiationless transition to the manifold ⟨m⟩. The description of the time evolution of the excited state, and with it the understanding of radiative and nonradiative phenomena, is dependent upon the nature of the initially prepared state, the density of states in the manifold, the nature and strength of the coupling, and the energy spacing between the states (Bixon and Jortner 1968, Englman and Jortner 1970, Freed 1972, 1976c, Lim 1977).

Two characteristic limits describe the behavior of many systems (Tramer and Voltz 1979). In the small molecule limit, the system is characterized by a manifold with a low density. The states are discrete, and when light is absorbed the molecule makes a transition from one eigenstate to another. In the absence of collisions or an environment, the excited state decays only by emitting radiation. In the large molecular or statistical limit, the system is characterized by a very dense manifold. An excited state produced by the absorption of light from most radiation sources is thought to be nonstationary. The initial state is described by a linear combination of the molecular eigenstates which are so closely spaced that they cannot be excited separately. If this prepared state is the state s, then coupling to the set of states ⟨m⟩ allows s to decay nonradiatively and irreversibly with the nonradiative decay rate given (Freed 1976c) in terms of the coupling matrix elements and the density of states:

$$k_{NR} = \frac{2\pi}{\hbar} \sum_m |\langle s|V|m\rangle|^2 \rho_m. \tag{31}$$

One thus expects rapid radiationless relaxation characterized by a single exponential decay.

The intermediate cases between these limits are of interest because there may be a high enough density of states to allow for radiationless decay but low enough to allow small perturbations such as an external magnetic or electric field to have a significant effect on the relaxation rate. Finding molecules in the intermediate regime is not straightforward because the effective density of states may well be lower than the total density because some states may be only weakly coupled to the initial state due to symmetry or other restrictions.

External fields can be expected to affect the relaxation of an electronically excited state by shifting states into or out of resonance with it, by mixing vibronic states, and by mixing rovibronic states. The combinations of studying magnetic field effects and electric field effects on the absorption spectra and relaxation processes in vapor and solid phases should contribute to determining the details of the relaxation mechanisms and the origin of the field effects. Since the selection rules on matrix elements for magnetic and electric perturbations differ, a similarity or dissimilarity in magnetic and electric field effects will serve to identify properties of the states that are involved. Furthermore, a contrast in vapor and solid phase results will implicate the necessary involvement of rotational levels in the mechanism. The combination of electric and magnetic field perturbations with single rovibronic level excitation by monochromatic laser sources should make the most significant contributions to our knowledge of these relaxation processes and associated field effects. In this experiment, the character and energy of a single rovibronic level can be continuously tuned by the field. This tuning can be monitored by high resolution absorption spectroscopy, and the relaxation rates can be measured and correlated with the character of the mixed state. The anticipated effects need to be considered in terms of the small molecule limit, the large molecule limit and the intermediate case.

In the small molecule limit, the energy states are well separated, and the comparatively small field induced mixing, shifts and splittings are expected to have a negligible effect on the radiative and nonradiative decay of the excited states. Field induced orientational effects on the collisions that remove energy from the excited molecule, however, may be significant.

In the large molecule limit, nonradiative transitions occur even in the absence of collisions due to the high density of nonemitting states that are coupled to the initially prepared state. This high density of already coupled states may overshadow an added external field effect. On the other hand, the density distribution may not be uniform so energy level shifts will have an observable effect, and there may be a corresponding high density of uncoupled states that can be coupled by the field.

In the intermediate case, the quenching effect of collisions is very sensitive to the character of the initial state (Freed 1976a,b, Freed and Tric 1978, Gelbart and Freed 1973). In addition, collisions may serve to perturbatively couple states, to broaden or shift them into resonance, or to rearrange the energy distribution over the states. An external field may then affect nonradiative decay not only directly, as described above, but also by influencing the effect of collisions by changing the character of the initial state.

One thus expects that an external field may affect radiationless decay by shifting the manifold relative to the prepared state and by changing the coupling between the manifold and the state either directly or indirectly by mixing with a third state. In addition, there may be effects of fields on collision induced quenching.

11.1. Magnetic field effects

The effect of a magnetic field on the coupling between the initial state and the manifold has been described for both the direct and indirect mechanisms (Atkins and Stannard 1977, Stannard 1978). In both cases, changes in the density of states were not considered. For the direct mechanism, the Zeeman interaction was added to the zero field interaction between the initial state and the manifold. The nonradiative rate, depending upon the absolute square of matrix elements involving this sum, was found to increase quadratically with the field. The term linear in the applied field vanished after summing over the rotational levels. The indirect mechanism accounts for field induced changes in the zero field coupling by mixing with a third state. It was found that this mechanism increases the nonradiative rate quadratically at low fields, reaching a high field limiting value. The quadratic dependence arises when the Zeeman interaction is small compared to the energy differences between the states that are being mixed by the field. The limiting value is reached at high fields when the mixing is complete. Stannard (1978) discusses the magnetic field quenching of fluorescence from iodine, glyoxal, carbon disulfide and nitrogen dioxide in terms of these mechanisms.

Mechanisms for the magnetic field quenching of fluorescence have also been considered by Matsuzaki and Nagakura (1978). Their mechanism I accounts for the direct coupling of the initial state to the manifold by the Zeeman interaction. Mechanism II accounts for the relative shift of the rovibronic levels of the manifold by the Zeeman effect. The latter mechanism was suggested to be particularly effective for a low density of states by acting to decrease the energy spacing between states.

Observations of the simultaneous effect of a magnetic field and collisions on radiationless transitions have been discussed recently (Selzle et al. 1979, Michel and Tric 1980).

11.2. *Electric field effects*

Quenching of fluorescence by an electric field has been reported only for two molecular systems. Sullivan and Dows (1980) observed electric field modulated fluorescence from iodine excited by a cw dye laser in the range 580 to 640 nm. The field coupled the B state of iodine to unbound states thereby inducing predissociation and quenching fluorescence. The magnitude of the quenching rate constant and its variation with vibronic level provided a value of 0.03 to 0.04 D for the electric dipole coupling matrix element and located the crossing of the repulsive potential energy surface and the B state surface between vibronic levels 3 and 4.

Weisshaar and Moore (1980a,b) observed that fluorescence lifetimes of single rotational levels of formaldehyde depended on an electric field. Since the mixing of states by the electric field was considered to be negligible, the changes in fluorescence decay times were attributed to shifts in energy level spacings between the initial state and the manifold consisting of the rovibronic levels of the ground electronic state. The lifetime associated with one line in the spectrum was observed to decrease by about 50% and then increase again over a 500 V/cm change in the field. This behavior is consistent with a particular level or levels of the manifold being shifted into and then out of resonance with the initial state. These results clearly demonstrate for this case that the manifold is characterized by levels well separated compared to their widths. A smooth continuum model does not seem appropriate. Values for the coupling matrix elements and the energy level widths also can be obtained from such experiments.

Coupling between the rovibronic levels of two electronic states by an electric field to induce radiationless transitions in isolated molecules has been considered by Strek (1978). By a lengthy analysis, he came to the expected conclusion that the enhancement of the nonradiative decay caused by a direct mixing of states by the electric field depends upon the square of the product of the electric dipole transition moment, the electric field, and the inverse of the energy difference between the electronic states.

11.3. Relaxation in solids

Studies of the effects of electric fields on the F center luminescence of alkali halide crystals (Bogan and Fitchen 1970, Stiles et al. 1970) revealed a shortening of the lifetime upon application of an electric field. This decrease was attributed to a decrease in the radiative lifetime due to mixing of the emitting 2S-like state with the 2P state. A decrease in the quantum yield was also observed in some situations, but the reasons for this decrease were not fully understood. Ionization of the F center by the electric field was considered to be one possibility. The contribution of an electric field effect to radiationless transitions in the F centers was considered by Lin (1975) as an application of his theoretical development describing electric field induced absorption, electrical dichroism and the Kerr effect, and nonradiative processes.

The use of solid state systems to study the effects of electric fields on relaxation processes in molecules offers the advantages of isolating and orienting the molecule in a well-defined environment, providing sharp spectral lines, eliminating the rotational degrees of freedom, and allowing the application of large electric fields. One might expect that the large number of degrees of freedom of the solid would put all molecules in the statistical limit, but this is not necessarily the case since not all these degrees of freedom necessarily couple to the molecular electronic excitation.

Studies of deuteroformaldehyde in the condensed phase demonstrate that this may indeed be the case (Goodman and Brus 1978). Fluorescence spectra, lifetimes and fluorescence excitation spectra of deuteroformaldehyde in solid neon, argon and nitrogen were obtained. The lifetime is only a factor of ten shorter in neon than in the vapor phase under collision free conditions and varies from $0.42 \, \mu s$ in neon to $0.66 \, \mu s$ in xenon and $0.19 \, \mu s$ in nitrogen. The decay rate in the solid is not much faster than in the vapor, and the final density of effective states is too low for the molecule to be in the statistical limit where the environment is expected to have no effect on the lifetime. Furthermore, the fluorescence quantum yield decreases as the level of vibronic excitation increases, implying that a higher density of states at the higher excitation energies increases the nonradiative decay rate. Thus, the electric field effects on fluorescence lifetimes observed in the vapor phase might well be observable in the solid state as well.

The decrease of spectral linewidths due to an external electric field has been observed in emission excitation spectra of p-chlorobenzaldehyde and p-methylbenzaldehyde in a p-dimethoxybenzene host. The origin linewidth of p-MBA decreases from 4.9 to $3.5 \, cm^{-1}$ when a field of $80 \, kV/cm$ is applied along the c-axis of the crystal. If the linewidth is attributed to homogeneous broadening, then the electric field is affecting the coupling with lower electronic states and decreasing the rate of relaxation (Hossain and Hanson 1978).

References

Aartsma, T. and D.A. Wiersma, 1973, Chem. Phys. **1**, 211.
Abbi, S.C. and D.M. Hanson, 1974, J. Chem. Phys. **60**, 319.
Argyrakis, P. and R. Kopelman, 1977, J. Chem. Phys. **66**, 3301.
Argyrakis, P. and R. Kopelman, 1978, J. Theor. Biol. **73**, 205.
Argyrakis, P. and R. Kopelman, 1980a, Phys. Rev. **B22**, 1830.
Argyrakis, P. and R. Kopelman, 1980b, Chem. Phys. **51**, 9.
Arnold, S., M. Pope and T.K.T. Hsieh, 1979, Phys. Stat. Sol. (b) **94**, 263.
Aspnes, D.E. 1972, Phys. Rev. Lett. **28**, 913.
Atkins, P.W. and P.R. Stannard, 1977, Chem. Phys. Lett. **47**, 113.
Bakhshiev, N.G., 1969, Russ. Chem. Rev. **38**, 740.
Barker, J.W. and L.J. Noe, 1972, J. Chem. Phys. **57**, 3035.
Barker, J.W., L.J. Noe and A.P. Marchetti, 1973, J. Chem. Phys. **59**, 1304.
Barnett, G.P., M.A. Kurzmack and M.M. Malley, 1973, Chem. Phys. Lett. **23**, 237.
Bayliss, N.S. and E.G. McRae, 1954, J. Phys. Chem. **58**, 1002.
Beardslee, R.A. and H.W. Offen, 1970, J. Chem. Phys. **52**, 6016.
Bergman, A. and J. Jortner, 1972, Chem. Phys. Lett. **15**, 309.
Bernstein, E.R. and R.E. Smalley, 1973, Chem. Phys. **2**, 321.
Berry, R.S., J. Jortner, J.C. Mackie, E.S. Pysh and S.A. Rice, 1965, J. Chem. Phys. **42**, 1535.
Bierman, A., 1963a, Phys. Rev. **130**, 2266.
Bierman, A., 1963b, Phys. Rev. **132**, 529.
Bixon, M. and J. Jortner, 1968, J. Chem. Phys. **48**, 715.
Blinov, L.M. and N.A. Kirichenko, 1973, Sov. Phys. Solid State **14**, 2163.
Blinov, L.M. and N.A. Kirichenko, 1974, Opt. Spectrosc. **37**, 513.
Bogen, L.D. and D.B. Fitchen, 1970, Phys. Rev. **B1**, 4122.
Bottka, N. and M.E. Hills, 1978, Appl. Phys. Lett. **33**, 765.
Bounds, P.J. and R.W. Munn, 1977, Chem. Phys. **24**, 343.
Brewer, R.G., 1977, in: Frontiers in Laser Spectroscopy, vol. I, Les Houches 1975, vol. XXVII, eds. R. Balian, S. Haroche and S. Liberman (North-Holland, Amsterdam).
Brewer, R.G. and R.L. Shoemaker, 1971, Phys. Rev. Lett. **27**, 631.
Brewer, R.G. and R.L. Shoemaker, 1972, Phys. Rev. **A6**, 2001.
Bridge, N.J. and L.R. Gianneschi, 1976, J. Chem. Soc. London Faraday Trans. II **7**, 1622.
Bridge, N.J., D.A. Harner and D.A. Dows, 1966, J. Chem. Phys. **44**, 3128.
Bridge, N.J., D.A. Harner and D.A. Dows, 1968, J. Chem. Phys. **48**, 4196.
Brillante, A. and M.R. Philpott, 1980, J. Chem. Phys. **72**, 4019.
Brillante, A., M.R. Philpott and D. Haarer, 1978, Chem. Phys. Lett. **56**, 218.
Bucher, H., J. Wiegand, B.B. Snavely, K.H. Beck and H. Kuhn, 1969, Chem. Phys. Lett. **3**, 508.
Buckingham, A.D., 1970, in: Physical Chemistry: An Advanced Treatise, vol. IV, eds. H. Eyring, D. Henderson and W. Jost (Academic Press, New York) ch. 8.
Buckingham, A.D., 1972, in: MTP International Review of Science, Physical Chemistry Series 1, vol. 3: Spectroscopy, eds. D.A. Ramsay and A.D. Buckingham (Butterworths, London) pp. 73–117.
Buckingham, A.D. and D.A. Ramsay, 1965, J. Chem. Phys. **42**, 3721.
Buckingham, A.D., D.A. Ramsay and J. Tyrrel, 1970, Can. J. Phys. **48**, 1242.

Bullot, J., P. Cordier and M. Gauthier, 1978, Chem. Phys. Lett. **54**, 77.

Bunn, C. and R. Daubeny, 1954, Trans. Faraday Soc. **50**, 1173.

Burland, D.M., F. Carmona and E. Cuellar, 1979, Chem. Phys. Lett. **64**, 5.

Burnham, A.K., G.R. Alms and W.H. Flygare, 1975, J. Chem. Phys. **62**, 3289.

Chen, F.P., 1975, Ph.D. Dissertation, State University of New York at Stony Brook.

Chen, F.P., S.J. Sheng and D.M. Hanson, 1974, Chem. Phys. **5**, 60.

Chen, F.P., D.M. Hanson and D. Fox, 1975, J. Chem. Phys. **63**, 3878; Chem. Phys. Lett. **30**, 337.

Chen, F.P., D.M. Hanson and D. Fox, 1977, J. Chem. Phys. **66**, 4954.

Choi, S., J. Jortner, S.A. Rice and R. Silbey, 1964, J. Chem. Phys. **41**, 3294.

Cipollini, N., 1975, Ph.D. Dissertation, Dartmouth College.

Clark, R. and G.J. Small, 1977, J. Chem. Phys. **66**, 1779.

Clark, S.E. and D.S. Tinti, 1980, Chem. Phys. **51**, 17.

Condon, E.U. and G.H. Shortley, 1964, The Theory of Atomic Spectra (Cambridge Univ. Press, London) ch. 17.

Coulson, C.A., 1972, Isr. J. Chem. **10**, 229.

Cummins, P.G., D.A. Dunmur and R.W. Munn, 1973, Chem. Phys. Lett. **22**, 519.

Cummins, P.G., D.A. Dunmur and R.W. Munn, 1975, Chem. Phys. Lett. **36**, 199.

Dagdigian, P.J., 1980, J. Chem. Phys. **73**, 2049.

Di Bartolo, B., 1968, Optical Interactions in Solids (Wiley, New York) pp. 456–458.

Dlott, D.D., M.D. Fayer and R.D. Wieting, 1977, J. Chem. Phys. **67**, 3808.

Dlott, D.D., M.D. Fayer and R.D. Wieting, 1978, J. Chem. Phys. **69**, 2752.

Dohnt, G., A. Hese, A. Renn and H.S. Schweda, 1979, Chem. Phys. **42**, 183.

Dows, D.A. and A.D. Buckingham, 1964, J. Mol. Spectrosc. **12**, 189.

Dubinin, N.V., 1977, Opt. Spectrosc. **43**, 49.

Dubinin, N.V., L.M. Blinov, E.L. Lutsenko and L.D. Rozenshtein, 1976, Sov. Phys. Solid State **18**, 1392.

Dubinin, N.V., L.M. Blinov and S.V. Yablonskii, 1978, Opt. Spectrosc. **44**, 473.

Dunmur, D.A., 1971, Chem. Phys. Lett. **10**, 49.

Dunmur, D.A., 1972, Mol. Phys. **23**, 109.

Dunmur, D.A. and R.W. Munn, 1975, Chem. Phys. **11**, 297.

Dunmur, D.A., W.H. Miller and R.W. Munn, 1977, Chem. Phys. Lett. **47**, 592.

Englman, R. and J. Jortner, 1970, Mol. Phys. **18**, 145.

Feynman, R.P., R.B. Leighton and M. Sands, 1965, Lectures in Physics (Addison-Wesley, Reading, MA) ch. 8.

Fox, D., 1976, Chem. Phys. **17**, 273.

Franck, J. and E. Teller, 1938, J. Chem. Phys. **6**, 861.

Freed, K., 1972, Topics in Current Chemistry, vol. 31 (Springer, Berlin) p. 105.

Freed, K., 1976a, J. Chem. Phys. **64**, 1604.

Freed, K., 1976b, Chem. Phys. Lett. **37**, 47.

Freed, K., 1976c, Topics in Applied Physics, vol. 15 (Springer, Berlin) p. 23.

Freed, K. and C. Tric, 1978, Chem. Phys. **33**, 249.

Freeman, D.E. and W. Klemperer, 1964, J. Chem. Phys. **40**, 604.

Freeman, D.E. and W. Klemperer, 1966, J. Chem. Phys. **45**, 52, 58.

Frenkel, 1931, Phys. Rev. **37**, 13, 1276.

Frova, A. and P. Handler, 1965, Phys. Rev. **137**, A1857.

Galaup, J.P., J. Megal and H.P. Trommsdorff, 1978, J. Chem. Phys. **69**, 1030.

Geacintov, N.E., M. Pope and H. Kallmann, 1966, J. Chem. Phys. **45**, 2369.

Gelbart, W.M. and K. Freed, 1973, Chem. Phys. Lett. **18**, 470.

Goodman, J. and L.E. Brus, 1978, Chem. Phys. Lett. **58**, 399.

Goodman, L. and M. Koyanagi, 1972, Mol. Photochem. **4**, 369.

Goodman, L. and I. Ozkan, 1979, Chem. Phys. Lett **61**, 216.

Haarer, D., 1974, Chem. Phys. Lett. **27**, 91.

Haarer, D., 1975, Chem. Phys. Lett. **31**, 192.

Haarer, D., 1977, J. Chem. Phys. **67**, 4076.

Haarer, D., M.R. Philpott and H. Morawitz, 1975, J. Chem. Phys. **63**, 5238.

Hanson, D.M., 1973, Crit. Rev. Solid State Sci. **3**, 243.

Hanson, D.M., 1975, Energy States and Energy Transfer in Molecular Crystals: A Primer, unpublished.

Hanson, D.M., 1980, Mol. Cryst. Liq. Cryst. **57**, 243.

Hanson, D.M. and F.P. Chen, 1973, Chem. Phys. Lett. **21**, 17.

Hanson, D.M., A. Kakuta, K. Kato and S.J. Sheng, 1977, Proc. Eighth Molecular Crystals Symp., Santa Barbara, p. 127.

Harris, C.B. and D.A. Zwemer, 1978, Ann. Rev. Phys. Chem. **29**, 473.

Hayes, J.M. and G.J. Small, 1978a, Chem. Phys. Lett. **54**, 435.

Hayes, J.M. and G.J. Small, 1978b, Chem. Phys. **27**, 151.

Hill, A.R. and M.M. Malley, 1971, J. Mol. Spectrosc. **40**, 428.

Hinze, J. and K.K. Dochen, 1972, J. Chem. Phys. **57**, 4928, 4936.

Hochstrasser, R.M., 1962, J. Chem. Phys. **36**, 1099.

Hochstrasser, R.M., 1973a, Accounts Chem. Res. **6**, 263.

Hochstrasser, R.M. 1973b, in: Molecular Energy Transfer, eds. R.D. Levine and J. Jortner (Wiley, New York) pp. 292–309.

Hochstrasser, R.M. and C.M. Klimcak, 1978, J. Chem. Phys. **69**, 2580.

Hochstrasser, R.M. and C.M. Klimcak, 1980, Mol. Cryst. Liq. Cryst. **58**, 109.

Hochstrasser, R.M. and T.S. Lin, 1968, J. Chem. Phys. **49**, 4929.

Hochstrasser, R.M. and A.P. Marchetti, 1970, J. Mol. Spectrosc. **35**, 335.

Hochstrasser, R.M. and J.W. Michaluk, 1972, J. Mol. Spectrosc. **42**, 197.

Hochstrasser, R.M. and L.J. Noe, 1968, J. Chem. Phys. **48**, 514.

Hochstrasser, R.M. and L.J. Noe, 1969, J. Chem. Phys. **50**, 1684.

Hochstrasser, R.M. and L.J. Noe, 1970, J. Chem. Phys. Lett. **5**, 489.

Hochstrasser, R.M. and L.J. Noe, 1971, J. Mol. Spectrosc. **38**, 175.

Hochstrasser, R.M. and A.H. Zewail, 1971, Chem. Phys. Lett. **11**, 157.

Hochstrasser, R.M. and A.H. Zewail, 1973, Chem. Phys. Lett. **21**, 15.

Hochstrasser, R.M., L.W. Johnson and H.P. Trommsdorff, 1973, Chem. Phys. Lett. **21**, 251.

Hochstrasser, R.M., L.W. Johnson and C.M. Klimcak, 1980, J. Chem. Phys. **73**, 156.

Holstein, T., S.K. Lyo and R. Orbach, 1976, Phys. Rev. Lett. **36**, 891.

Holstein, T., S.K. Lyo and R. Orbach, 1977, Phys. Rev. **B15**, 4693.

Holstein, T., S.K. Lyo and R. Orbach, 1978, Commun. Solid State Phys. **8**, 119.

Holts, F., 1974, Phys. Stat. Sol. **A21**, 469.

Hong, H.K. and R. Kopelman, 1970, Phys. Rev. Lett. **25**, 1030.

Hong, H.K. and R. Kopelman, 1971, J. Chem. Phys. **55**, 724.

Hossain, M. and D.M. Hanson, 1978, Chem. Phys. **30**, 155.

Huang, K.T. and J.R. Lombardi, 1969, J. Chem. Phys. **51**, 1228.

Huang, K.T. and J.R. Lombardi, 1970, J. Chem. Phys. **52**, 5613.

Huang, K.T. and J.R. Lombardi, 1971, J. Chem. Phys. **55**, 4072.

Irwin, T.A.R. and F.W. Dalby, 1965, Can. J. Phys. **43**, 1766.

Jen, S., N.A. Clark, P.S. Pershan and E.B. Priestly, 1977, J. Chem. Phys. **66**, 4635.

Jortner, J., S.A. Rice, J.L. Katz and S. Choi, 1965, J. Chem. Phys. **42**, 309.

Kenkre, V.M. and R.S. Knox, 1974, Phys. Rev. Lett. **33**, 803.

Killesreiter, H. and S. Schneider, 1977, Chem. Phys. Lett. **52**, 191.

Kittel, C., 1966, Introduction to Solid State Physics, 3rd Ed. (Wiley, New York) pp. 374–382.

Klafter, J. and J. Jortner, 1977, Chem. Phys. Lett. **49**, 410.

Knox, R.S., 1963, Theory of Excitons (Academic Press, New York) pp. 74–79.

Kolendritskii, D.D., M.V. Kurik and Yu. P. Piryatinskii, 1977, Sov. Phys. Solid State **19**, 1296.

Kolendritskii, D.D., M.V. Kurik and Yu.P. Piryatinskii, 1978, Opt. Spectrosc. **44**, 162.

Kopelman, R., 1976, in: Topics in Applied Physics, Vol. 15: Radiationless Processes in Molecules and Condensed Phases, ed. F.K. Fong (Springer, Berlin) ch. 5.

Kopelman, R. and W. Klemperer, 1962, J. Chem. Phys. **36**, 1693.

Kumanoto, J., J.C. Powers, Jr. and W.R. Heller, 1962, J. Chem. Phys. **36**, 2893.

Kurzmack, M.A. and M.M. Malley, 1973, Chem. Phys. Lett. **21**, 385.

Labhart, H., 1967, in: Advances in Chemical Physics, vol. 13, ed. I. Prigogine (Interscience, New York) p. 179.

Lacey, A.R., E.F. McCoy and I.G. Ross, 1973, Chem. Phys. Lett. **21**, 233.

Lakatos-Lindenberg, K. and K.E. Shuler, 1971, J. Math. Phys. **12**, 633.

Lang, S.B., 1974, Sourcebook of Pyroelectricity (Gordon and Breach, New York) ch. 1.

Lee, J.W., 1975, J. Korean Chem. Soc. **19**, 304.

Lim, E.C., 1977, in: Excited States, vol. 3, ed. E.C. Lim (Academic Press, New York) pp. 305–337.

Lin, S.H., 1975, J. Chem. Phys. **62**, 4500.

Liptay, W., 1965, in: Modern Quantum Chemistry III, ed. O. Sinanoglu (Academic Press, New York) ch. 3.

Liptay, W., 1969, Angew, Chem., Int. Ed. **8**, 177.

Liptay, W., 1974, in: Excited States, ed. E.C. Lim (Academic Press, New York) pp. 129–231.

Liptay, W., G. Walz, W. Baumann, H.J. Schlosser, H. Deckers and N. Detzer, 1971, Z. Naturforsch. **26A**, 2020.

Lombardi, J.R., 1968, J. Chem. Phys. **50**, 3780.

Luty, T., 1976, Chem. Phys. Lett. **44**, 335.

Magee, J.L. and K. Funabashi, 1961, J. Chem. Phys. **34**, 1715.

Malley, M., G. Feher and D. Mauzerall, 1968, J. Mol. Spectrosc. **25**, 544.

Marchetti, A.P., 1973, Chem. Phys. Lett. **23**, 213.

Marchetti, A.P., 1974, Mol. Cryst. Liq. Cryst. **29**, 147.

Marchetti, A.P. and M. Scozzafava, 1975, Mol. Cryst. Liq. Cryst. **31**, 115.

Marchetti, A.P., M. Scozzafava and R.H. Young, 1977, Chem. Phys. Lett. **51**, 424.

Mathies, R. and A.C. Albrecht, 1972, Chem. Phys. Lett. **16**, 231.

Mathies, R. and A.C. Albrecht, 1974, J. Chem. Phys. **60**, 2500.

Matsuzaki, A. and S. Nagakura, 1978, Helv. Chim. Acta **61**, 675.

Meyling, J.H. and D.A. Wiersma, 1973, Chem. Phys. Lett. **20**, 383.

Meyling, J.H., W.H. Hesselink and D.A. Wiersma, 1976, Chem. Phys. **17**, 353.

Meyling, J.H., P.J. Bounds and R.W. Munn, 1977, Chem. Phys. Lett. **51**, 234.

Michaluk, J.W., 1972, Ph.D. Dissertation, Univ. of Pennsylvania.

Michel, C. and C. Tric, 1980, Chem. Phys. **50**, 341.

Mijoule, C. and P. Yuan, 1976, Chem. Phys. Lett. **43**, 524.

Minkin, V.I., O.A. Osipov and Y.A. Zhdanov, 1970, Dipole Moments in Organic Chemistry (Plenum Press, New York) transl. B.J. Hazzard.

Montroll, E.W., 1964, Proc. Symp. Appl. Math **XVI**, 193.

Montroll, E.W., 1969, J. Math. Phys. **10**, 753.

Montroll, E.W. and G.H. Weiss, 1965, J. Math. Phys. **6**, 167.

Munn, R.W., 1972, Chem. Phys. Lett. **16**, 429.

Munn, R.W., 1980, Chem. Phys. **50**, 119.

Munn, R.W. and D.F. Williams, 1973, J. Chem. Phys. **59**, 1742.

Muramoto, T., S. Nakanishi, and T. Hashi, 1978, Opt. Commun. **24**, 316.

Noe, L.J., T.F. Turner and L. Wojdac, 1976, Chem. Phys. **17**, 285.

Nye, J.F., 1960, Physical Properties of Crystals and Their Representation by Tensors and Matrices (Clarendon, Oxford).

Orbach, R., 1966, in: Optical Properties of Ions in Crystals, eds. H.M. Crosswhite and H.W.
  Moos (Interscience, New York) pp. 445–455.
Palffy-Muhoray, P., 1977, Chem. Phys. Lett. **48**, 315.
Patel. J.S. and D.M. Hanson, 1981a, J. Chem. Phys. **75**, 5203.
Patel, J.S. and D.M. Hanson, 1981b Nature **293**, 445.
Patel, J.S. and D.M. Hanson, 1982, Chem. Phys. **69**, 249.
Petelenz, P., 1977a, Phys. Stat. Sol. (b) **83**, 169.
Petelenz, P., 1977b, Chem. Phys. Lett. **47**, 603.
Petelenz, P., 1979, Chem. Phys. Lett. **65**, 579.
Petelenz, P. and B. Petelenz, 1977, Phys. Stat. Sol. (b) **83**, 49.
Petelenz, P. and M.Z. Zgierski, 1975, Phys. Stat. Sol. (b) **81**, 133.
Phelps, D.A. and F.W. Dalby, 1966, Phys. Rev. Lett. **16**, 3.
Platt, J.R., 1961, J. Chem. Phys. **34**, 862.
Pollans, W.L. and S.I. Choi, 1970, J. Chem. Phys. **52**, 3691.
Popovic, Z.D. and E.R. Menzel, 1979, J. Chem. Phys. **71**, 5090.
Price, A.H., J.O. Willams and R.W. Munn, 1976, Chem. Phys. **14**, 413.
Redi, M. and J.J. Hopfield, 1980, J. Chem. Phys. **72**, 6651.
Reich, R. and S. Schmidt, 1972, Ber. Bunsenges. Phys. Chem. **76**, 589, 1202.
Reif, F., 1965, Fundamentals of Statistical and Thermal Physics (McGraw-Hill, New York).
Robinson, G.W., 1970, Ann. Rev. Phys. Chem. **21**, 429.
Robinson, G.W. and R.P. Frosch, 1962, J. Chem. Phys. **37**, 1962.
Robinson, G.W. and R.P. Frosch, 1963, J. Chem. Phys. **38**, 1187.
Rozenshtein, L., M. Pope and S. Arnold, 1980, Bull. Am. Phys. Soc. **25**(3) 373.
Sauter, H. and A.C. Albrecht, 1968, Chem. Phys. Lett. **2**, 8.
Schiff, L.I., 1955, Quantum Mechanics (McGraw-Hill, New York) p. 156.
Scott, T.W. and A.C. Albrecht, 1979, J. Chem. Phys. **70**, 3657.
Sebastian, L. and G. Weiser, 1979, Chem. Phys. Lett. **64**, 396.
Sebastian, L., G. Weiser and H. Bässler, 1981, Chem. Phys. **61**, 125.
Sebastian, L., G. Weiser, G. Peter and H. Bässler, 1983, Chem. Phys., in press.
Seibold, K., H. Navangul and H. Labhart, 1969, Chem. Phys. Lett. **3**, 275.
Selzle, H.L., S.H. Lin and E.W. Schlag, 1979, Chem. Phys. Lett. **62**, 230.
Shelby, R.M. and R.M. MacFarlane, 1978, Opt. Commun. **27**, 399.
Shelby, R.M., A.H. Zewail, and C.B. Harris, 1976, J. Chem. Phys. **64**, 3192.
Sheng, S.H., M.A. El-Sayed and H.P. Trommsdorff, 1977, Chem. Phys. Lett. **45**, 404.
Sheng, S.J. and M.A. El-Sayed, 1975, Chem. Phys. Lett. **34**, 216.
Sheng, S.J. and M.A. El-Sayed, 1977a, Chem. Phys. Lett. **45**, 6.
Sheng, S.J. and M.A. El-Sayed, 1977b, Chem. Phys. **20**, 61.
Sheng, S.J. and D.M. Hanson, 1974a, J. Appl. Phys. **45**, 4954.
Sheng, S.J. and D.M. Hanson, 1974b, J. Chem. Phys. **60**, 368.
Sheng, S.J. and D.M. Hanson, 1975, Chem. Phys. **10**, 51.
Sheng, S.J., M.A. El-Sayed and M. Leung, 1975, J. Chem. Phys. **62**, 1988.
Shiner, R., S. Druckmann, M. Ottolenghi and R. Korenstein, 1977, Biophys. J. **19**, 1.
Silbey, R., 1976, Ann. Rev. Phys. Chem. **27**, 203.
Sipe, J.E. and J. Van. Kranendonk, 1978, Mol. Phys. **35**, 1579.
Small, G.J. and P. Burke, 1977, J. Chem. Phys. **66**, 1767.
Smalley, R., L. Wharton and D.H. Levy, 1975, J. Chem. Phys. **63**, 4977.
Stannard, P.R., 1978, J. Chem. Phys. **68**, 3932.
Stark, J., 1913, Sitzungsber Akad. Wiss. Berlin **47**, 932.
Stepanov, B.I. and V.P. Gribkovskii, 1968, Theory of Luminescence (Iliffe Books, London)
  pp. 147–152.
Stiles, Jr., L.F., M.P. Fontana and D.B. Fitchen, 1970, Phys. Rev. **B2**, 2077.
Strek, W., 1978, Chem. Phys. Lett. **57**, 121.

Sullivan, B.J. and D.A. Dows, 1980, Chem. Phys. **46**, 231.

Sullivan, D.E. and J.M. Deutch, 1976, J. Chem. Phys. **64**, 3878.

Sundararajan, K.S., 1936, Z. Krist. **93**, 238.

Syassen, K. and M.R. Philpott, 1977, Chem. Phys. Lett. **50**, 14.

Szabo, A. and M. Kroll, 1978, Opt. Lett. **2**, 10.

Tanaka, J., 1963, Bull. Chem. Soc. Jpn. **36**, 1237.

Tanaka, J. and M. Shibata, 1968, Bull. Chem. Soc. Jpn. **41**, 34.

Tanaka, J., T. Kishi and M. Tanaka, 1974, Bull. Chem. Soc. Jpn. **47**, 2376.

Tiberghien, A., G. Delacote and M. Schott, 1973, J. Chem. Phys. **59**, 3762.

Townes, C.H. and A.L. Schawlow, 1955, Microwave Spectroscopy (McGraw-Hill, New York).

Tramer, A. and R. Voltz, 1979, in: Excited States, vol. 4, ed. E.C. Lim (Academic Press, New York) pp. 281–394.

Trlifaj, M., 1958, Czech. J. Phys. **8**, 511.

Trommsdorff, H.P. and J.P. Galaup, 1976, Chem. Phys. **12**, 463.

Udagawa, Y. and D.M. Hanson, 1976a, J. Chem. Phys. **65**, 5367.

Udagawa, Y. and D.M. Hanson, 1976b, J. Chem. Phys. **64**, 3753.

Udagawa, Y. and D.M. Hanson, 1977, Chem. Phys. Lett. **45**, 228.

Veenvliet, H. and D.A. Wiersma, 1973a, Chem. Phys. **2**, 69.

Veenvliet, H. and D.A. Wiersma, 1973b, Chem. Phys. Lett. **22**, 87.

Veenvliet, H. and D.A. Wiersma, 1975, Chem. Phys. Lett. **33**, 305.

Vuks, M.F. 1966, Opt. Spectrosc. **20**, 361.

Webman, J. and J. Jortner, 1969, J. Chem. Phys. **50**, 2706.

Weiser, G. and J. Stuke, 1969, Phys. Stat. Sol. **35**, 747.

Weiser, G., 1973, Phys. Stat. Sol. (a) **18**, 347.

Weisshaar, J.C. and C.B. Moore, 1980a, J. Chem. Phys. **72**, 2875.

Weisshaar, J.C. and C.B. Moore, 1980b, J. Chem. Phys. **72**, 5415.

Wiersma, D.A., 1972, Chem. Phys. Lett. **16**, 517.

Wieting, R.D., M.D. Fayer, and D.D. Dlott, 1978, J. Chem. Phys. **69**, 1996.

Winkler, I.C. and D.M. Hanson, 1981, J. Am. Chem. Soc. **103**, 6264.

Witt, H.T., 1971, Quart. Rev. Biophys. **4**, 365.

Witt, H.T., 1975, in: Bioenergetics of Photosynthesis, ed. Govindjee (Academic Press, New York) ch. 10.

Wolf, H.C., 1967, in: Advances in Atomic and Molecular Physics, vol. 3, ed. D.R. Bates (Academic Press, New York) p. 319.

Wolf, R., W. Kuhn, R. Enderlein and H. Lange, 1977, Phys. Stat. Sol. **A42**, 629.

Yokomoto, H., K. Kondo and J. Nakai, 1976, J. Appl. Phys. **15**, 2137.

Yokoyama, M., Y. Endo and H. Mikawa, 1975, Chem. Phys. Lett. **34**, 597.

Zoos, Z.G. and R.C. Powell, 1972, Phys. Rev. **B6**, 4035.

Zwemer, D.A. and C.B. Harris, 1978, J. Chem. Phys. **68**, 2184.

SUBJECT INDEX